FOUNDATIONS OF

COLLEGE CHEMISTRY

15E

FOUNDATIONS OF COLLEGE CHEMISTRY

Morris Hein
Mount San Antonio College

Susan Arena
University of Illinois, Urbana-Champaign

Cary Willard
Grossmont College

VP & Director: Petra Recter
Senior Acquisition Editor: Nicholas Ferrari
Product Designer: Sean Hickey
Senior Product Designer: Geraldine Osnato
Senior Marketing Manager: Kristine Ruff
Associate Director, Product Delivery: Kevin Holm
Senior Production Editor: Elizabeth Swain
Design Director: Harry Nolan
Senior Designer: Maureen Eide
Senior Market Solutions Assistant: Mallory Fryc
Senior Photo Editor: Mary Ann Price
Production Management Services: codeMantra, CSR Sofia Buono and Gandhimathi Ganesan
Cover: Paul Nicklen / Getty Images

This book was typeset in 10/12 Times New Roman MT Std at codeMantra and printed and bound by Quad Graphics, Versailles. The cover was printed by Quad Graphics, Versailles.

The paper in this book was manufactured by a mill whose forest management programs include sustained yield—harvesting of its timberlands. Sustained yield harvesting principles ensure that the number of trees cut each year does not exceed the amount of new growth.

This book is printed on acid free paper. ∞

Library of Congress Cataloging-in-Publication Data
Names: Hein, Morris. | Arena, Susan. | Willard, Cary.
Title: Foundations of college chemistry.
Description: 15th edition / Morris Hein, Susan Arena, Cary Willard. |
Hoboken, NJ : John Wiley & Sons, Inc., [2016] | Includes bibliographical references and index.
Identifiers: LCCN 2015041455 | ISBN 9781119083900 (looseleaf: alk. paper)
Subjects: LCSH: Chemistry—Textbooks.
Classification: LCC QD33.2 .H45 2016 | DDC 540—dc23 LC record available at http://lccn.loc.gov/2015041455

Binder-ready version	ISBN 978-1-119-08390-0
Evaluation/desk copy	ISBN 978-1-119-24673-2
High School binding	ISBN 978-1-119-23455-5

The inside back cover will contain printing identification and country of origin if omitted from this page. In addition, if the ISBN on the back cover differs from the ISBN on this page, the one on the back cover is correct.

Printed in the United States of America
V10009844_051519

ABOUT THE AUTHOR

MORRIS HEIN earned his PhD at the University of Colorado, Boulder. He was Professor Emeritus of Chemistry at Mt. San Antonio College, where he regularly taught the preparatory chemistry course and organic chemistry. He was the original author of *Foundations of College Chemistry*, and his name has become synonymous with clarity, meticulous accuracy, and a step-by-step approach that students can follow. Morris passed away in late 2014 and was actively working on our texts till the very end. His sharp eyes and attention to detail will be sorely missed.

SUSAN ARENA earned a BS and MA in Chemistry at California State University-Fullerton. She has taught science and mathematics at all levels, including middle school, high school, community college, and university. At the University of Illinois she developed a program for increasing the retention of minorities and women in science and engineering. This program focused on using active learning and peer teaching to encourage students to excel in the sciences. She has coordinated and led workshops and programs for science teachers from elementary through college levels that encourage and support active learning and creative science teaching techniques. For several years she was director of an Institute for Chemical Education (ICE) field center in Southern California. In addition to *Foundations of College Chemistry*, 15th edition, she is co-author of *Introduction to General, Organic and Biochemistry*, 10th edition. Susan enjoys reading, knitting, traveling, classic cars, and gardening in her spare time when she is not playing with her grandchildren.

CARY WILLARD received her BS in chemistry from California State Polytechnic Institute, Pomona, and her PhD from the University of California, Davis. She has been teaching chemistry at Grossmont College in El Cajon, California for over 20 years. She teaches the chemistry courses for both science and nonscience majors. Her biggest successes are the students who come into chemistry afraid of the subject and leave having discovered that chemistry really is fun. Her interest in sharing the excitement of science to the community resulted in Grossmont College hosting a Science Festival, multiple Science Decathlons, model airplane flying, and robotics competitions. Cary is also a member of the BE WiSE (Better Education for Women in Science and Engineering) steering committee and works with the group to share the excitement of scientific research with a new generation of young women. In her spare time, Cary enjoys exploring the San Diego region both on foot and in her kayak.

BRIEF CONTENTS

CONTENTS

APPENDICES A1

GLOSSARY G1

INDEX I1

PREFACE

This new Fifteenth Edition of *Foundations of College Chemistry* presents chemistry as a modern, vital subject and is designed to make introductory chemistry accessible to all beginning students. The central focus is the same as it has been from the first edition: to make chemistry interesting and understandable to students and teach them the problem-solving skills they will need. In preparing this new edition, we considered the comments and suggestions of students and instructors to design a revision that builds on the strengths of previous editions including clear explanations and step-by-step problem solving. We have especially tried to relate chemistry to the real lives of our students as we develop the principles that form the foundation for the further study of chemistry, and to provide them with problem-solving skills and practice needed in their future studies. We have focused on using electronic materials to enhance the student experience. To that end we have authored and added new interactive features ourselves and enhanced our existing electronic materials to draw the students into direct involvement in an active learning experience.

Foundations of College Chemistry, 15th Edition, is intended for students who have never taken a chemistry course or those who have had a significant interruption in their studies but plan to continue with the general chemistry sequence. Since its inception, this book has helped define the preparatory chemistry course and has developed a much wider audience. In addition to preparatory chemistry, our text is used extensively in one-semester general-purpose courses (such as those for applied health fields) and in courses for nonscience majors.

Development of Problem-Solving Skills

We all want our students to develop real skills in solving problems. We believe that a key to the success of this text is the fact that our problem-solving approach works for students. It is a step-by-step process that teaches the use of units and shows the change from one unit to the next. We have used this problem-solving approach in our examples throughout the text to encourage students to think their way through each problem. In this edition we continue to use examples to incorporate fundamental mathematical skills, scientific notation, and significant figures. We have added Problem-Solving Strategy boxes in the text to highlight the steps needed to solve chemistry problems. Painstaking care has been taken to show each step in the problem-solving process and to use these steps in solving example problems. We continue to use four significant figures for atomic and molar masses for consistency and for rounding off answers appropriately. We have been meticulous in providing answers, correctly rounded, for students who have difficulty with mathematics.

FOSTERING STUDENT SKILLS *Attitude* plays a critical role in problem solving. We encourage students to learn that a systematic approach to solving problems is better than simple memorization. Throughout the book we emphasize the use of our approach to problem solving to encourage students to think through each problem. Once we have laid the foundations of concepts, we highlight the steps so students can locate them easily. Important rules and equations are highlighted for emphasis and ready reference.

STUDENT PRACTICE Practice problems follow the examples in the text, with answers provided at the end of the chapter. The end of each chapter begins with a *Chapter Review* and *Review Questions* section, which help students review key terms and concepts, as well as material presented in tables and figures. This is followed by *Paired Exercises*, covering concepts and numerical exercises, where two similar exercises are presented side by side. The section titled *Additional Exercises* includes further practice problems presented in a more random order. The final section of exercises is titled Challenge Exercises and contains problems designed to stretch the student's understanding of concepts and to integrate concepts from other chapters. In our new edition we have changed a number of exercises per chapter, In addition we have expanded the electronic Enhanced Examples to a total of 178 throughout the chapters and developed two new electronic practice methods.

PRACTICE 7.7

Calculate the percent composition of $Ca(NO_3)_2$.

PRACTICE 7.8

Calculate the percent composition of K_2CrO_4.

Practice Problems

PROBLEM-SOLVING STRATEGY

For Calculating Percent Composition from Formula

1. Calculate the molar mass (Section 7.2).
2. Divide the total mass of each element in the formula by the molar mass and multiply by 100. This gives the percent composition:

$$\frac{\text{total mass of the element}}{\text{molar mass}} \times 100 = \text{percent of the element}$$

Problem-solving Strategy

Organization

We continue to emphasize the less theoretical aspects of chemistry early in the book, leaving the more abstract theory for later. This sequence seems especially appropriate in a course where students are encountering chemistry for the very first time. Atoms, molecules, and reactions are all an integral part of the chemical nature of matter. A sound understanding of these topics allows the student to develop a basic understanding of chemical properties and vocabulary.

Chapters 1 through 3 present the basic mathematics and the language of chemistry, including an explanation of the metric system and significant figures. We added a new section in Chapter 2 (2.7) on Percents to assist students in understanding how the concept is applied in chemistry. In Chapter 4 we present chemical properties—the ability of a substance to form new substances. Then, in Chapter 5, students encounter the history and language of basic atomic theory. In Chapter 8 we added a new section 8.3 "Why Reactions Occur" and revised the section discussing types of reactions extensively to better reflect the level and needs of our students.

We continue to present new material at a level appropriate for the beginning student by emphasizing nomenclature, composition of compounds, and reactions in Chapters 6 through 9 before moving into the details of modern atomic theory. Some applications of the Periodic Table are shown in early chapters and discussed in detail in Chapters 10 and 11. Students gain confidence in their own ability to identify and work with chemicals in the laboratory before tackling the molecular models of matter. As practicing chemists we have little difficulty connecting molecular models and chemical properties. Students, especially those with no prior chemistry background, may not share this ability to connect the molecular models and the macroscopic properties of matter. Those instructors who feel it is essential to teach atomic theory and bonding early in the course can cover Chapters 10 and 11 immediately following Chapter 5.

New to This Edition

In the Fifteenth Edition we have tried to build on the strengths of the previous editions. We have added a new author, **Cary Willard**, from Grossmont College in California. Cary revised the end-of-chapter materials and added exercises, including many applications in fields of interest for our students. She has focused on expanding and developing important and meaningful media assets for the *WileyPLUS* course. Her focus has created deeper synergies between the content in print format and the electronic resources in our *WileyPLUS* course.

We continually strive to keep the material at the same level so that students can easily read and use the text and supplemental material to learn chemistry. With a focus on problem solving, student engagement, and clarity, some of the specific changes are highlighted below:

- Chemistry In Action boxes have been updated, and new boxes have been added to include different applications of the concepts in the text.
- Some of the older industrial chemistry applications have been removed and newer applications added as appropriate throughout the text.
- 229 Check Your Understanding Questions appear throughout the text to provide an opportunity for the students to determine their level of understanding of concepts. These are highlighted in the margins and refer the student to *WileyPlus* for completion.

CHECK YOUR UNDERSTANDING
7.6 Molecules and Formula Units
WileyPLUS

- 178 Enhanced Examples are also located in *WileyPlus* and provide a more interactive experience and practice. Students receive immediate feedback on their answers and proceed stepwise through the example to its conclusion.

ENHANCED EXAMPLE 7.6 WileyPLUS

How many oxygen atoms are present in 1.00 mol of oxygen molecules?

SOLUTION

Plan Oxygen is a diatomic molecule with the formula O_2. Therefore, a molecule of oxygen contains 2 oxygen atoms: $\frac{2 \text{ atoms O}}{1 \text{ molecule } O_2}$.

Solution map: moles O_2 → molecules O_2 → atoms O

The conversion factors needed are

$$\frac{6.022 \times 10^{23} \text{ molecules } O_2}{1 \text{ mole } O_2} \quad \text{and} \quad \frac{2 \text{ atoms O}}{1 \text{ molecule } O_2}$$

Calculate

$$(1.00 \text{ mol } O_2)\left(\frac{6.022 \times 10^{23} \text{ molecules } O_2}{1 \text{ mole } O_2}\right)\left(\frac{2 \text{ atoms O}}{1 \text{ molecule } O_2}\right) = 1.20 \times 10^{24} \text{ atoms O}$$

PRACTICE 7.2

What is the mass of 2.50 mol of helium (He)?

PRACTICE 7.3

How many atoms are present in 0.025 mol of iron?

- 131 new Online Learning Modules movies provide a one-on-one learning experience that mimics the assistance an instructor provides in a help session. The modules are paired with in-text examples to provide an alternative way to learn problem solving.
- Each chapter is complemented by ORION, an adaptive learning module, available within *WileyPLUS*. These can be used individually for student review or can be assigned by instructors as a specific review assignment tailored to the chapters and sections on an exam.
- Chapters 19 and 20 (Introduction to Organic Chemistry and Introduction to Biochemistry) are now available online only. This provides an option for those desiring a shorter version of the text and for those wishing to cover this material in their course.

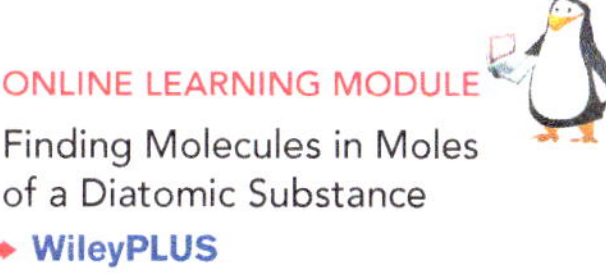

- The Putting It Together sections have been removed in order to provide a more flexible review source for students and instructors. This new resource is called Orion and is found on *WileyPlus*.
- New, modern design. The entire text has been redesigned to foster greater accessibility and increase student engagement. New icons and notes direct students to *WileyPlus* for more individual direct interaction with the material.
- Appendix II: Using a Scientific Calculator has been removed because most students begin using calculators in elementary school and there are so many kinds of calculators that specific instruction on their use is no longer possible.

Learning Aids

To help the beginning student gain the confidence necessary to master technical material, we have refined and enhanced a series of learning aids:

- **Learning Objectives** highlight the concept being taught in each section. These objectives are tied to **Example**, **Practice Problems**, **Review Exercises**, and **Exercises** to assist the student in mastering each concept module and objective.
- Important **terms** are set off in bold type where they are defined and are listed in gray at the beginning of each section. All **Key Terms** listed in the **Chapter Review** are also defined in the **Glossary**.
- Worked **examples** show students the how of problem solving using **Problem-Solving Strategies** and **Solution Maps** before they are asked to tackle problems on their own.
- **Practice problems** permit immediate reinforcement of a skill shown in the example problems. Answers are provided at the end of the chapter to encourage students to check their problem solving immediately.

- **Marginal notations** help students understand basic concepts and problem-solving techniques. These are printed in blue to clearly distinguish them from text and vocabulary terms.

LEARNING AIDS: MATH SKILLS For students who may need help with the mathematical aspects of chemistry, the following learning aids are available:

- A **Review of Mathematics**, covering the basic functions, is provided in Appendix I.
- **Math Survival Guide: Tips and Tricks for Science Students**, 2nd Edition, by Jeffrey R. Appling and Jean C. Richardson, a brief paperback summary of basic skills that can be packaged with the text, provides an excellent resource for students who need help with the mathematical aspects of chemistry.

Supplements Package

FOR THE STUDENT **Study Guide** by Rachael Henriques Porter is a self-study guide for students. For each chapter, the **Study Guide** includes a self-evaluation section with student exercises, a summary of chapter concepts, one or more "challenge problems," and answers and solutions to all **Study Guide** exercises.

Math Survival Guide: Tips and Tricks for Science Students, 2nd Edition, by Jeffrey Appling and Jean Richardson, is a paperback summary of basic skills, with practice exercises in every chapter.

Foundations of Chemistry in the Laboratory, 14th Edition, by Morris Hein, Judith N. Peisen, and Robert L. Miner includes 28 experiments for a laboratory program that may accompany the lecture course. Featuring updated information on waste disposal and emphasizing safe laboratory procedures, the lab manual also includes study aids and exercises.

FOR THE INSTRUCTOR **Test Bank**, by Harpreet Malhotra, includes chapter tests with additional test questions and answers to all test questions.

Computerized Test Bank. The test bank contains true-false, multiple-choice, and open-ended questions and is available in two formats.

Digital Image Library: Images from the text are available online in JPEG format. Instructors may use these to customize their presentations and to provide additional visual support for quizzes and exams.

Power Point Lecture Slides: Updated for this version by William Douglas Urban, these slides contain lecture outlines and key topics from each chapter of the text, along with supporting artwork and figures from the text.

WileyPLUS

WileyPLUS is an innovative, research-based online environment for effective teaching and learning.

WileyPLUS builds students' confidence because it takes the guesswork out of studying by providing students with a clear roadmap: **what to do, how to do it, if they did it right**. This interactive approach focuses on:

CONFIDENCE: Research shows that students experience a great deal of anxiety studying. That's why we provide a structured learning environment that helps students focus on **what to do**, along with the support of immediate resources.

MOTIVATION: To increase and sustain motivation throughout the semester, *WileyPLUS* helps students learn **how to do it** at a pace that's right for them. Our integrated resources—available 24/7—function like a personal tutor, directly addressing each student's demonstrated needs with specific problem-solving techniques.

WileyPLUS is now equipped with an adaptive learning module called ORION.

WileyPLUS with ORION is:

BEGIN

A **personalized study guide** that helps students understand both strengths and areas where they need to invest more time, especially in preparation for quizzes and exams.

PRACTICE

For each topic, students can either Study or Practice. **Study** directs the student to the specific topic they choose in *WileyPLUS*, where they can read from the e-textbook, or use the variety of relevant resources available there. Students can also **practice**, using questions and feedback powered by ORION's adaptive learning engine.

MAINTAIN

ORION includes a number of reports and ongoing recommendations for students to help them maintain their proficiency over time for each topic.

Students can easily access ORION from within *WileyPLUS*. It does not require any additional registration, and there is no additional charge for students using this adaptive learning system.

What Do Students Receive with *WileyPLUS*?

- The complete digital textbook, saving students up to 50% off the cost of a printed text.
- Math Skills module provides a comprehensive review of topics including: ratios, scientific notation, significant figures, and graphing.
- The Check Your Understanding Questions, Enhanced Examples, and Online Learning Modules outlined above are available by default to students for self-study, no instructor set-up is required.
- Visualizations of key concepts with audio narration.

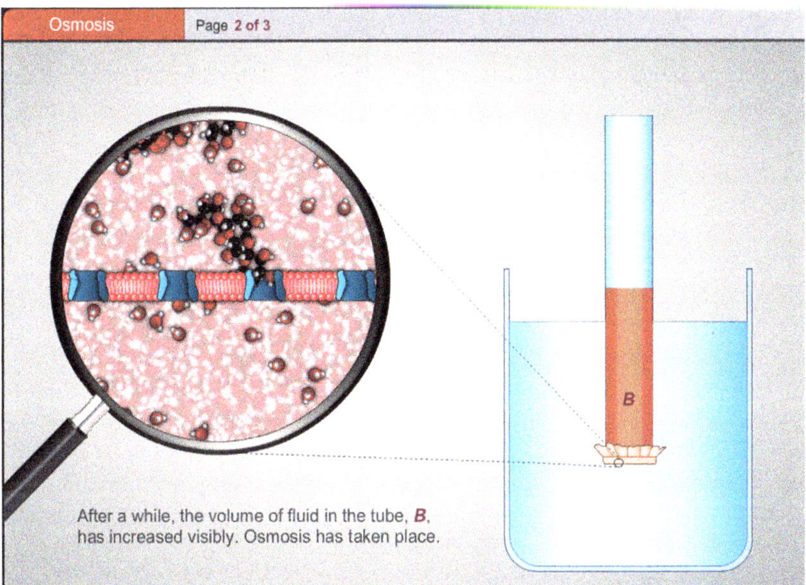

What Do Instructors Receive with *WileyPLUS*?

- All end-of-chapter questions are available for assignment and automatic grading.
- Every assignable question has been enhanced with additional content such as hints, solutions, and guided tutorials which can be made available to students at the instructor's discretion.

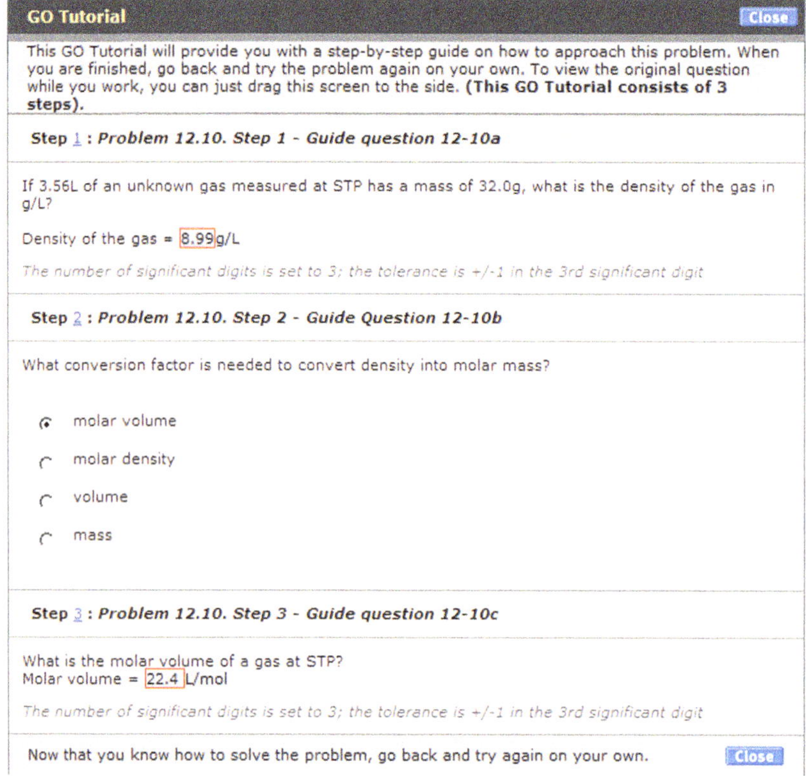

- Pre-built assignments with two options per chapter that can be assigned with a single click.
- Prelecture checkpoint questions
- The complete instructor ancillary package can be accessed directly from *WileyPLUS* including: Lecture PowerPoints, Solutions Manual, Classroom Response System (Clickers) questions and Test Bank.
- *WileyPLUS* can be integrated with many popular Learning Management Systems such as BlackBoard and Canvas. Integration allows for single-sign-on and gradebook synchronization.

Customization and Flexible Options to Meet Your Needs

Wiley Custom Select allows you to create a textbook with precisely the content you want, in a simple, three-step online process that brings your students a cost-efficient alternative to a traditional textbook. Select from an extensive collection of content at **http://customselect.wiley.com**, upload your own materials as well, and select from multiple delivery formats—full color or black and white print with a variety of binding options, or eBook. Preview the full text online, get an instant price quote, and submit your order; we'll take it from there. *WileyFlex* offers content in flexible and cost-saving options to students. Our goal is to deliver our learning materials to our customers in the formats that work best for them, whether it's traditional text, eTextbook, *WileyPLUS*, loose-leaf binder editions, or customized content through Wiley Custom Select.

Acknowledgments

Books are the result of a collaborative effort of many talented and dedicated people. We particularly want to thank our editor, Nick Ferrari, who guided the project through revision with the challenge of creating several new electronic features. We are grateful to Mary Ann

Price, Senior Photo Editor, for finding new and interesting photos that add so much to the pages of our text. We also want to thank Elizabeth Swain, Senior Production Editor, who kept us on track and production running smoothly. Sofia Buono, Project Manager/Copy Editor, in Italy was amazing and turned our revisions into the beautiful text in our 15th edition. We also especially appreciate the work of Sean Hickey, Product Designer, who helped us with the Online Learning Modules and incorporating new Check Your Understanding questions into *WileyPlus*. Thanks to others who provided valuable assistance on this revision, including Senior Marketing Manager, Kristine Ruff, Senior Designer, Maureen Eide, and Mallory Fryc, Senior Market Solutions Assistant, who was always ready to find an answer to anything we needed. We are grateful for the many helpful comments from colleagues and students who, over the years, have made this book possible. We hope they will continue to share their ideas for change with us, either directly or through our publisher.

We are especially thankful for the help and support of Tom Martin, our developmental editor. His positive attitude, attention to detail, efficiency, good humor, and willingness to help in any way were indispensable in this revision.

Our sincere appreciation goes to the following reviewers who were kind enough to read and give their professional comments.

Reviewers

FOR THE15TH EDITION

Jing Yi Chin
Suffolk County Community College

Kara Kuvakas
Brandman University

Ralph Morasch
Pierce College

Rajiv Narula
SUNY Canton

Alexander Nazarenko
SUNY Buffalo State

Ethel Owusu-Nukunya
Santa Fe College

Sharadha Sambasivan
Suffolk County Community College

Anne Marie Sokol
SUNY Buffalo State

FOR THE14TH EDITION

Jeffrey Allison
Austin Community College

Jeanne Arquette
Phoenix College

Rebecca Broyer
University of Southern California

Michael Byler
Community College of Philadelphia

Kevin Cannon
Penn State Abington

Rong Cao
Community College of Allegheny City

Ken Capps
College of Central Florida

Charles Carraher
Florida Atlantic University

Loretta Dorn
Fort Hays State University

Robert Eves
Southern Utah University

Mitchel Fedak
Duquesne University

Paul Fox
Danville Community College

Erick Fuoco
Richard J. Daley College

Amy Grant
El Camino College

Tamara Hanna
Texas Tech Lubbock

Chris Hamaker
Illinois State University

Claudia Hein
Diablo Valley College

Donna Iannotti
Brevard Community College

Crystal Jenkins
Santa Ana College

Jodi Kreiling
University of Nebraska at Omaha

Julie Larson
Bemidji State University

Anne Lerner
Santa Fe College

Lauren McMills
Ohio University—Main Campus

Mitchel Millan
Casper College

Timothy Minger
Mesa Community College

Franklin Ow
East LA City College

Ethel (April) Owusu
Santa Fe College

Fumin Pan
Mohawk Valley Community College

David Peitz
Wayne State College

Sharadha Sambasivan
Suffolk Community College

Hussein Samha
Southern Utah University

Mary Shoemaker
Pennsylvania State University—University Park

Lee Silverberg
Penn State—Schuylkill

Gabriela Smeureanu
Hunter College

Sunanda Sukumar
Albany College of Pharmacy

Paris Svoronos
QCC of Cuny

Susan Thomas
University of Texas—San Antonio

Sergey Trusov
Oxnard College

Elaine Vickers
Southern Utah University

Loretta Vogel
Ocean County College

Liwen Yu
Inver Hills Community College

Karl Wallace
University of Southern Mississippi

1

An Introduction to Chemistry

Jon Hilmarsson/Getty Images

CHAPTER OUTLINE

Do you know how the beautiful, intricate fireworks displays are created? Have you ever wondered how a tiny seedling can grow into a cornstalk taller than you in just one season? Perhaps you have been mesmerized by the flames in your fireplace on a romantic evening as they change color and form. The spectacular colors of the aurora borealis shown above are the result of chemistry in our atmosphere. And think of your relief when you dropped a container and found that it was plastic, not glass. These phenomena are the result of chemistry that occurs all around us, all the time. Chemical changes bring us beautiful colors, warmth, light, and products to make our lives function more smoothly. Understanding, explaining, and using the diversity of materials we find around us is what chemistry is all about.

1.1 The Nature of Chemistry

LEARNING OBJECTIVE State the definition of chemistry and why the study of chemistry is important.

KEY TERM
Chemistry

A knowledge of chemistry is useful to virtually everyone—we see chemistry occurring around us every day. An understanding of chemistry is useful to engineers, teachers, health care professionals, attorneys, homemakers, businesspeople, firefighters, and environmentalists, just to name a few. Even if you're not planning to work in any of these fields, chemistry is important and is used by people every day. Learning about the benefits and risks associated with chemicals will help you to be an informed citizen, able to make intelligent choices concerning the world around you. Studying chemistry teaches you to solve problems and communicate with others in an organized and logical manner. These skills will be helpful in college and throughout your career.

Key terms are **highlighted in bold** to alert you to new terms defined in the text.

What is chemistry? One dictionary gives this definition: "**Chemistry** is the science of the composition, structure, properties, and reactions of matter, especially of atomic and molecular systems." A somewhat simpler definition is "Chemistry is the science dealing with the composition of *matter* and the changes in composition that matter undergoes." Neither of these definitions is entirely adequate. Chemistry and physics form a fundamental branch of knowledge. Chemistry is also closely related to biology, not only because living organisms are made of material substances but also because life itself is essentially a complicated system of interrelated chemical processes.

The scope of chemistry is extremely broad. It includes the whole universe and everything, animate and inanimate, in it. Chemistry is concerned with the composition and changes in the composition of matter and also with the energy and energy changes associated with matter. Through chemistry we seek to learn and to understand the general principles that govern the behavior of all matter.

The chemist, like other scientists, observes nature and attempts to understand its secrets: What makes a tulip red? Why is sugar sweet? What is occurring when iron rusts? Why is carbon monoxide poisonous? Problems such as these—some of which have been solved, some of which are still to be solved—are all part of what we call chemistry.

A chemist may interpret natural phenomena, devise experiments that reveal the composition and structure of complex substances, study methods for improving natural processes, or synthesize substances. Ultimately, the efforts of successful chemists advance the frontiers of knowledge and at the same time contribute to the well-being of humanity.

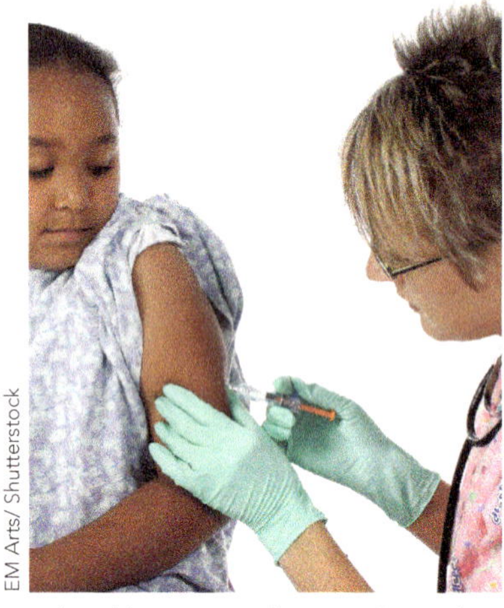

EM Arts/ Shutterstock

A health care professional needs to understand chemistry in order to administer the correct dose of medication.

Thinking Like a Chemist

Chemists take a special view of things in order to understand the nature of the chemical changes taking place. Chemists "look inside" everyday objects to see how the basic components are behaving. To understand this approach, let's consider a lake. When we view the lake from a distance, we get an overall picture of the water and shoreline. This overall view is called the *macroscopic* picture.

As we approach the lake we begin to see more details—rocks, sandy beach, plants submerged in the water, and aquatic life. We get more and more curious. What makes the rocks and sand? What kind of organisms live in the water? How do plants survive underwater? What lies hidden in the water? We can use a microscope to learn the answers to some of these questions. Within the water and the plants, we can see single cells and inside them organelles working to keep the organisms alive. For answers to other questions, we need to go even further inside the lake. A drop of lake water can itself become a mysterious and fascinating *microscopic* picture full of molecules and motion. (FIGURE 1.1) A chemist

Don Mason/Getty Images

FIGURE 1.1 Inside a drop of lake water we find water molecules and some dissolved substances.

looks into the world of atoms and molecules and their motions. Chemistry makes the connection between the *microscopic* world of molecules and the *macroscopic* world of everyday objects.

Think about the water in the lake. On the surface it has beauty and colors, and it gently laps the shore of the lake. What is the microscopic nature of water? It is composed of tiny molecules represented as

In this case H represents a hydrogen atom and O an oxygen atom. The water molecule is represented by H_2O since it is made up of two hydrogen atoms and one oxygen atom.

ENHANCED EXAMPLE 1.1 **WileyPLUS**

You are given eight oxygen atoms and fifteen hydrogen atoms. How many water molecules can you make from them?

SOLUTION

From the model shown above for a water molecule you can see that one molecule of water contains one atom of oxygen and two atoms of hydrogen. Using this model as reference, you can make eight water molecules from eight oxygen atoms. But you can make only seven water molecules from fifteen hydrogen atoms with one H atom and one O atom left over. The answer is seven water molecules.

PRACTICE 1.1

You are given ten hydrogen atoms and eight oxygen atoms. How many water molecules can you make from them?

CHECK YOUR UNDERSTANDING

1.1 Making Methane Molecules 1

1.2 Making Methane Molecules 2

➡ **WileyPLUS**

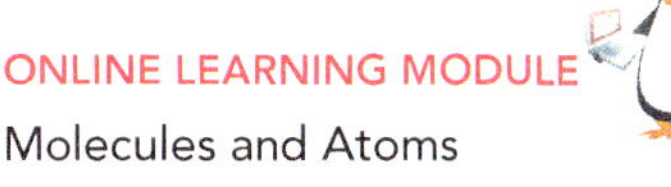

Molecules and Atoms

➡ **WileyPLUS**

1.2 A Scientific Approach to Problem Solving

Describe the steps involved in the scientific method.

LEARNING OBJECTIVE

KEY TERMS

scientific method
hypothesis
theory
scientific laws

One of the most common and important things we do every day is to solve problems. For example,

- You have two exams and a laboratory report due on Monday. How should you divide your time?
- You leave for school and learn from the radio that there is a big accident on the freeway. What is your fastest alternate route to avoid the traffic problem?
- You need to buy groceries, mail some packages, attend your child's soccer game, and pick up the dry cleaning. What is the most efficient sequence of events?

We all face these kinds of problems and decisions. A logical approach can be useful for solving daily problems:

1. Define the problem. We first need to recognize we have a problem and state it clearly, including all the known information. When we do this in science, we call it *making an observation*.
2. Propose possible solutions to the problem. In science this is called making a *hypothesis*.
3. Decide which is the best way to proceed or solve the problem. In daily life we use our memory of past experiences to help us. In the world of science we *perform an experiment*.

Using a scientific approach to problem solving is worthwhile. It helps in all parts of your life whether you plan to be a scientist, doctor, businessperson, or writer.

CHEMISTRY *IN ACTION* Egyptians, the First Medicinal Chemists

Terra-cotta sculpture of Nefertiti.

Look at any images of the ancient Egyptians and notice the black eyeliner commonly worn at that time. As chemists analyzed the composition of a sample of this eyeliner in the antiquities collection at the Louvre Museum in Paris, they were appalled to discover the high concentration of lead in the samples. Today lead is routinely removed from most consumer products because it is very toxic even in low concentrations. It is toxic to many organs and can cause symptoms such as abdominal pain, dementia, anemia, seizures, and even death. It turns out the lead compounds found in the Egyptian eyeliner are not found in nature but must be synthesized. The synthesis of these lead salts is complicated and the products are not lustrous. This led chemists to question why the Egyptians would add these compounds to their eyeliner. The answer was revealed after reading some of the ancient manuscripts from that time. Lead salts were synthesized for use in treating eye ailments, scars, and discolorations. So even if the lead salts were not the best ingredients for beauty, they were added for the perceived health benefits.

Since we now know that lead compounds are very toxic, Christian Amatore, an analytical chemist at the Ecole Normale Supérieure in Paris, wondered if the lead compounds in Egyptian eyeliner could have actually conferred any health benefits. He introduced lead salts into samples of human tissue growing in the laboratory and observed that the cells began forming compounds that trigger an immune response. Perhaps the ancient Egyptians did know something about medicinal chemistry after all. So should we follow the Egyptians' example and add lead to our cosmetics? This is probably not a good idea because the risks associated with prolonged lead exposure outweigh the benefits.

Scientists employ the scientific method every day in their laboratory work.

The Scientific Method

Chemists work together and also with other scientists to solve problems. As scientists conduct studies they ask many questions, and their questions often lead in directions that are not part of the original problem. The amazing developments from chemistry and technology usually involve what we call the **scientific method**, which can generally be described as follows:

1. **Collect the facts or data** that are relevant to the problem or question at hand. This is usually done by planned experimentation. The data are then analyzed to find trends or regularities that are pertinent to the problem.
2. **Formulate a hypothesis** that will account for the data and that can be tested by further experimentation.
3. **Plan and do additional experiments to test the hypothesis.**
4. **Modify the hypothesis** as necessary so that it is compatible with all the pertinent data.

Confusion sometimes arises regarding the exact meanings of the words *hypothesis*, *theory*, and *law*. A **hypothesis** is a tentative explanation of certain facts that provides a basis for further experimentation. A well-established hypothesis is often called a **theory** or model. Thus, a theory is an explanation of the general principles of certain phenomena with considerable evidence or facts to support it. Hypotheses and theories explain natural phenomena, whereas **scientific laws** are simple statements of natural phenomena to which no exceptions are known under the given conditions.

These four steps are a broad outline of the general procedure that is followed in most scientific work, but they are not a "recipe" for doing chemistry or any other science (FIGURE 1.2). Chemistry is an experimental science, however, and much of its progress has been due to application of the scientific method through systematic research.

We study many theories and laws in chemistry; this makes our task as students easier because theories and laws summarize important aspects of the sciences. Certain theories and models advanced by great scientists in the past have since been substantially altered and

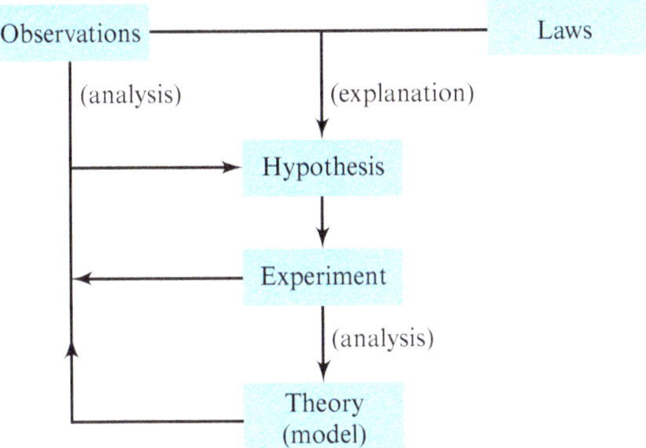

FIGURE 1.2 The scientific method.

modified. Such changes do not mean that the discoveries of the past are any less significant. Modification of existing theories and models in the light of new experimental evidence is essential to the growth and evolution of scientific knowledge. Science is dynamic.

1.3 The Particulate Nature of Matter

LEARNING OBJECTIVE

Describe the characteristics of matter, including the states of matter.

KEY TERMS

matter
solid
amorphous
liquid
gas

The entire universe consists of matter and energy. Every day we come into contact with countless kinds of matter. Air, food, water, rocks, soil, glass, and this book are all different types of matter. Broadly defined, **matter** is *anything* that has mass and occupies space.

Matter may be quite invisible. For example, if an apparently empty test tube is submerged mouth downward in a beaker of water, the water rises only slightly into the tube. The water cannot rise further because the tube is filled with invisible matter: air (FIGURE 1.3).

To the macroscopic eye, matter appears to be continuous and unbroken. We are impressed by the great diversity of matter. Given its many forms, it is difficult to believe that on a microscopic level all of matter is composed of discrete, tiny, fundamental particles called *atoms* (FIGURE 1.4). It is truly amazing to understand that the fundamental particles in ice cream are very similar to the particles in air that we breathe. Matter is actually discontinuous and is composed of discrete, tiny particles called *atoms*.

FIGURE 1.3 An apparently empty test tube is submerged, mouth downward, in water. Only a small volume of water rises into the tube, which is actually filled with invisible matter: air.

FIGURE 1.4 Silicon atoms on a silicon chip produced this image using a scanning tunneling microscope.

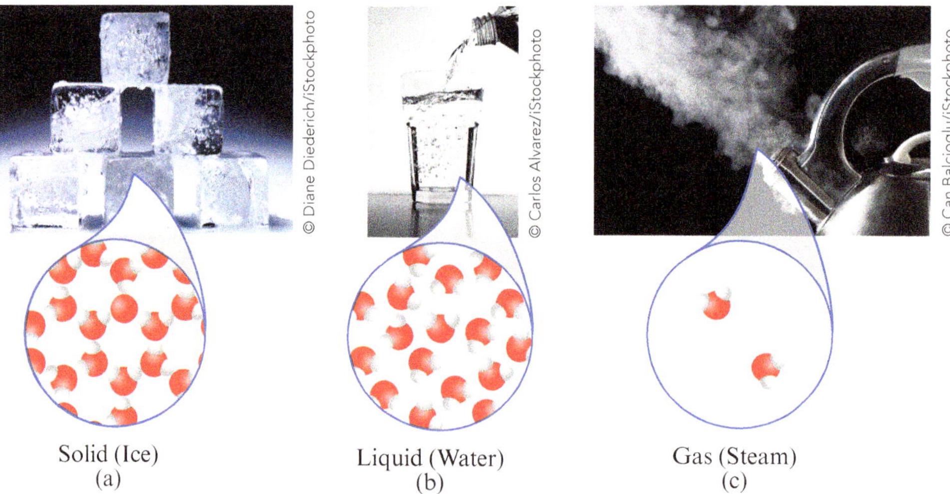

FIGURE 1.5 The three states of matter. (a) Solid—water molecules are held together rigidly and are very close to each other. (b) Liquid—water molecules are close together but are free to move around and slide over each other. (c) Gas—water molecules are far apart and move freely and randomly.

Physical States of Matter

Matter exists in three physical states: solid, liquid, and gas (FIGURE 1.5). A **solid** has a definite shape and volume, with particles that cling rigidly to one another. The shape of a solid can be independent of its container. In Figure 1.5a we see water in its solid form. Another example, a crystal of sulfur, has the same shape and volume whether it is placed in a beaker or simply laid on a glass plate.

Crystalline solid

Amorphous solid

Most commonly occurring solids, such as salt, sugar, quartz, and metals, are *crystalline*. The particles that form crystalline materials exist in regular, repeating, three-dimensional, geometric patterns (FIGURE 1.6). Some solids such as plastics, glass, and gels do not have any regular, internal geometric pattern. Such solids are called **amorphous** solids. (*Amorphous* means "without shape or form.")

A **liquid** has a definite volume but not a definite shape, with particles that stick firmly but not rigidly. Although the particles are held together by strong attractive forces and are in close contact with one another, they are able to move freely. Particle mobility gives a liquid fluidity and causes it to take the shape of the container in which it is stored. Note how water looks as a liquid in Figure 1.5b.

A **gas** has indefinite volume and no fixed shape, with particles that move independently of one another. Particles in the gaseous state have gained enough energy to overcome the attractive forces that held them together as liquids or solids. A gas presses continuously in all directions on the walls of any container. Because of this quality, a gas completely fills a container. The particles of a gas are relatively far apart compared with those of solids and liquids. The actual volume of the gas particles is very small compared with the volume of the space occupied by the gas. Observe the large space between the water molecules in Figure 1.5c compared to ice and liquid water. A gas therefore may be compressed into a very small volume or expanded almost indefinitely. Liquids cannot be compressed to any great extent, and solids are even less compressible than liquids.

If a bottle of ammonia solution is opened in one corner of the laboratory, we can soon smell its familiar odor in all parts of the room. The ammonia gas escaping from the solution demonstrates that gaseous particles move freely and rapidly and tend to permeate the entire area into which they are released.

FIGURE 1.6 A large crystal of table salt. A salt crystal is composed of a three-dimensional array of particles.

Although matter is discontinuous, attractive forces exist that hold the particles together and give matter its appearance of continuity. These attractive forces are strongest in solids, giving them rigidity; they are weaker in liquids but still strong enough to hold liquids to definite volumes. In gases, the attractive forces are so weak that the particles of a gas are practically independent of one another. TABLE 1.1 lists common materials that exist as solids, liquids, and gases. TABLE 1.2 compares the properties of solids, liquids, and gases.

TABLE 1.1 Common Materials in the Solid, Liquid, and Gaseous States of Matter

Solids	Liquids	Gases
Aluminum	Alcohol	Acetylene
Copper	Blood	Air
Gold	Gasoline	Butane
Polyethylene	Honey	Carbon dioxide
Salt	Mercury	Chlorine
Sand	Oil	Helium
Steel	Syrup	Methane
Sugar	Vinegar	Nitrogen
Sulfur	Water	Oxygen

TABLE 1.2 Physical Properties of Solids, Liquids, and Gases

State	Shape	Volume	Particles	Compressibility
Solid	Definite	Definite	Rigidly clinging; tightly packed	Very slight
Liquid	Indefinite	Definite	Mobile; adhering	Slight
Gas	Indefinite	Indefinite	Independent of each other and relatively far apart	High

1.4 Classifying Matter

LEARNING OBJECTIVE

Distinguish among a pure substance, a homogeneous mixture, and a heterogeneous mixture.

KEY TERMS

substance
homogeneous
heterogeneous
phase
system
mixture

The term *matter* refers to all materials that make up the universe. Many thousands of distinct kinds of matter exist. A **substance** is a particular kind of matter with a definite, fixed composition. Sometimes known as *pure substances*, substances are either elements or compounds. Familiar examples of elements are copper, gold, and oxygen. Familiar compounds are salt, sugar, and water. We'll discuss elements and compounds in more detail in Chapter 3.

We classify a sample of matter as either *homogeneous or heterogeneous* by examining it. **Homogeneous** matter is uniform in appearance and has the same properties throughout. Matter consisting of two or more physically distinct phases is **heterogeneous**. A **phase** is a homogeneous part of a system separated from other parts by physical boundaries. A **system** is simply the body of matter under consideration. Whenever we have a system in which visible boundaries exist between the parts or components, that system has more than one phase and is heterogeneous. It does not matter whether these components are in the solid, liquid, or gaseous states.

A pure substance may exist as different phases in a heterogeneous system. Ice floating in water, for example, is a two-phase system made up of solid water and liquid water. The water in each phase is homogeneous in composition, but because two phases are present, the system is heterogeneous.

A **mixture** is a material containing two or more substances and can be either heterogeneous or homogeneous. Mixtures are variable in composition. If we add a spoonful of sugar to a glass of water, a heterogeneous mixture is formed immediately. The two phases are a solid (sugar) and a liquid (water). But upon stirring, the sugar dissolves to form a homogeneous mixture or solution. Both substances are still present: All parts of the solution are sweet and wet. The proportions of sugar and water can be varied simply by adding more sugar and stirring to dissolve. Solutions do not have to be liquid. For example, air is a homogeneous mixture of gases. Solid solutions also exist. Brass is a homogeneous solution of copper and zinc.

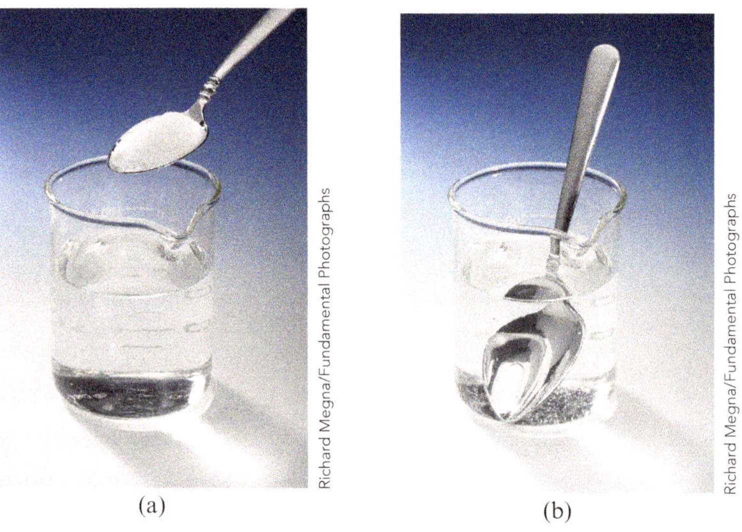

(a) Water is the liquid in the beaker, and the white solid in the spoon is sugar. (b) Sugar can be dissolved in the water to produce a solution.

Many substances do not form homogeneous mixtures. If we mix sugar and fine white sand, a heterogeneous mixture is formed. Careful examination may be needed to decide that the mixture is heterogeneous because the two phases (sugar and sand) are both white solids. Ordinary matter exists mostly as mixtures. If we examine soil, granite, iron ore, or other naturally occurring mineral deposits, we find them to be heterogeneous mixtures. FIGURE 1.7 illustrates the relationships of substances and mixtures.

Flowcharts can help you to visualize the connections between concepts.

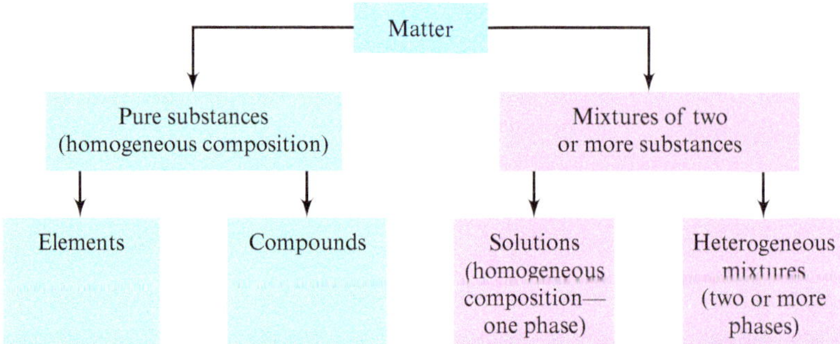

FIGURE 1.7 Classification of matter. A pure substance is always homogeneous in composition, whereas a mixture always contains two or more substances and may be either homogeneous or heterogeneous.

Distinguishing Mixtures from Pure Substances

Single substances—elements or compounds—seldom occur naturally in a pure state. Air is a mixture of gases; seawater is a mixture of a variety of dissolved minerals; ordinary soil is a complex mixture of minerals and various organic materials.

Mixture	**Pure Substance**
1. A mixture always contains two or more substances that can be present in varying amounts.	**1.** A pure substance (element or compound) always has a definite composition by mass.
2. The components of a mixture do not lose their identities and may be separated by physical means.	**2.** The elements in a compound lose their identities and may be separated only by chemical means.

PRACTICE 1.2

Which of the following is a mixture and which is a pure substance? Explain your answer.
(a) vinegar (4% acetic acid and 96% water)
(b) sodium chloride (salt) solution
(c) gold
(d) milk

How is a mixture distinguished from a pure substance? A mixture always contains two or more substances that can be present in varying concentrations. Let's consider two examples.

Homogeneous Mixture Homogeneous mixtures (solutions) containing either 5% or 10% salt in water can be prepared simply by mixing the correct amounts of salt and water. These mixtures can be separated by boiling away the water, leaving the salt as a residue.

Heterogeneous Mixture The composition of a heterogeneous mixture of sulfur crystals and iron filings can be varied by merely blending in either more sulfur or more iron filings. This mixture can be separated physically by using a magnet to attract the iron.

Ken Karp

(a)

Ken Karp

(b)

(a) When iron and sulfur exist as pure substances, only the iron is attracted to a magnet. (b) A mixture of iron and sulfur can be separated by using the difference in magnetic attraction.

CHAPTER 1 REVIEW

1.1 The Nature of Chemistry

KEY TERM

chemistry

- Chemistry is important to everyone because chemistry occurs all around us in our daily lives.
- Chemistry is the science dealing with matter and the changes in composition that matter undergoes.
- Chemists seek to understand the general principles governing the behavior of all matter.
- Chemistry "looks inside" ordinary objects to study how their components behave.
- Chemistry connects the macroscopic and microscopic worlds.

1.2 A Scientific Approach to Problem Solving

KEY TERMS

scientific method
hypothesis
theory
scientific laws

- Scientific thinking helps us solve problems in our daily lives.
- General steps for solving problems include:
  - Defining the problem
  - Proposing possible solutions
  - Solving the problem
- The scientific method is a procedure for processing information in which we:
  - Collect the facts
  - Formulate a hypothesis
  - Plan and do experiments
  - Modify the hypothesis if necessary

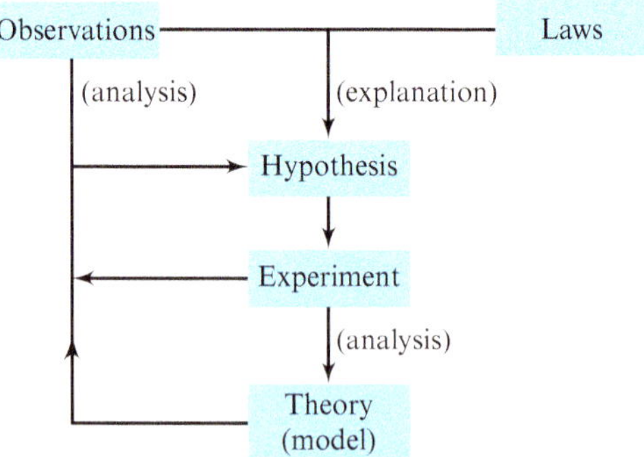

1.3 The Particulate Nature of Matter

KEY TERMS

matter
solid
amorphous
liquid
gas

- Matter is anything with the following two characteristics:
  - Has mass
  - Occupies space
- On the macroscopic level matter appears continuous.
- On the microscopic level matter is discontinuous and composed of atoms.
- Solid—rigid substance with a definite shape
- Liquid—fluid substance with a definite volume that takes the shape of its container
- Gas—takes the shape and volume of its container

1.4 Classifying Matter

KEY TERMS

substance
homogeneous
heterogeneous
phase
system
mixture

- Matter can be classified as a pure substance or a mixture.
- A mixture has variable composition:
  - Homogeneous mixtures have the same properties throughout.
  - Heterogeneous mixtures have different properties in different parts of the system.
- A pure substance always has the same composition. There are two types of pure substances:
  - Elements
  - Compounds

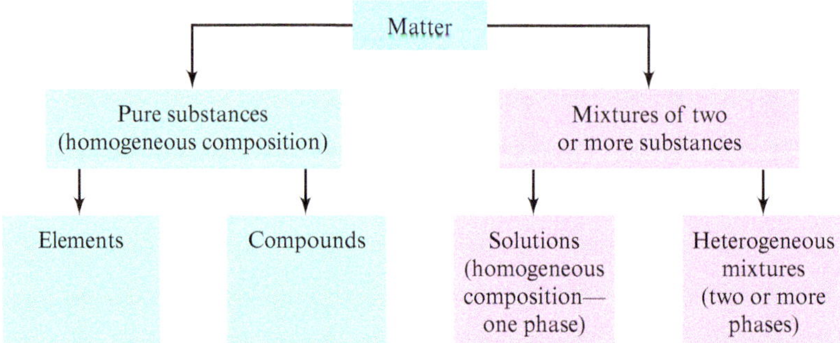

REVIEW QUESTIONS

1. Explain the difference between
   (a) a hypothesis and a theory
   (b) a theory and a scientific law
2. Consider each of the following statements and determine whether it represents an observation, a hypothesis, a theory, or a scientific law:
   (a) The battery in my watch must be dead since it is no longer keeping time.
   (b) My computer must have a virus since it is not working properly.
   (c) The air feels cool.
   (d) The candle burns more brightly in pure oxygen than in air because oxygen supports combustion.
   (e) My sister wears red quite often.
   (f) A pure substance has a definite, fixed composition.
3. Determine whether each of the following statements refers to a solid, a liquid, or a gas:
   (a) It has a definite volume but not a definite shape.
   (b) It has an indefinite volume and high compressibility.
   (c) It has a definite shape.
   (d) It has an indefinite shape and slight compressibility.
4. Some solids have a crystalline structure, while others have an amorphous structure. For each of the following five descriptions, determine whether it refers to a crystalline solid or an amorphous solid:
   (a) has a regular repeating pattern
   (b) plastic
   (c) has no regular repeating pattern
   (d) glass
   (e) gold
5. Define a phase.
6. How many phases are present in the graduated cylinder shown here?

7. What is another name for a homogeneous mixture?
8. Which liquids listed in Table 1.1 are not mixtures?
9. Which of the gases listed in Table 1.1 are not pure substances?
10. When the stopper is removed from a partly filled bottle containing solid and liquid acetic acid at 16.7°C, a strong vinegar-like odor is noticeable immediately. How many acetic acid phases must be present in the bottle? Explain.
11. Is the system enclosed in the bottle in Question 10 homogeneous or heterogeneous? Explain.
12. Is a system that contains only one substance necessarily homogeneous? Explain.
13. Is a system that contains two or more substances necessarily heterogeneous? Explain.
14. Distinguish between homogeneous and heterogeneous mixtures.
15. Which of the following are pure substances?
   (a) sugar
   (b) sand
   (c) gold
   (d) maple syrup
   (e) eggs
16. Use the steps of the scientific method to help determine the reason that your cell phone has suddenly stopped working:
   (a) observation
   (b) hypothesis
   (c) experiment
   (d) theory

PAIRED EXERCISES

Most of the exercises in this chapter are available for assignment via the online homework management program, WileyPLUS (www.wileyplus.com) All exercises with blue numbers have answers in Appendix 5.

1. Refer to the illustration and determine which state(s) of matter are present.

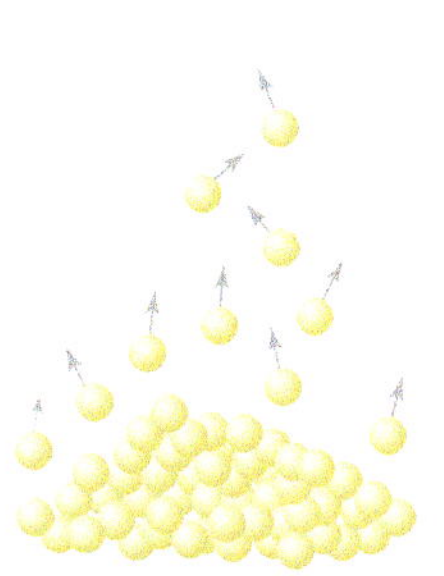

2. Refer to the illustration and determine which states(s) of matter are present.

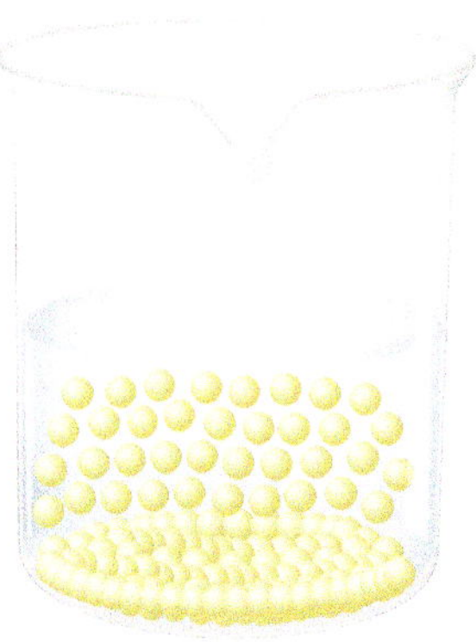

3. Look at the photo and determine whether it represents a homogeneous or heterogeneous mixture.

Richard Megna/Fundamental Photographs

4. Look at the maple leaf below and determine whether it represents a homogeneous or heterogeneous mixture.

© Justin Horrocks/iStockphoto

5. For each of the following mixtures, state whether it is homogeneous or heterogeneous:
(a) tap water
(b) carbonated beverage
(c) oil and vinegar salad dressing
(d) people in a football stadium

6. For each of the following mixtures, state whether it is homogeneous or heterogeneous:
(a) stainless steel
(b) motor oil
(c) soil
(d) a tree

ADDITIONAL EXERCISES

7. At home, check your kitchen and bathroom cabinets for five different substances; then read the labels and list the first ingredient of each.

8. During the first week of a new semester, consider that you have enrolled in five different classes, each of which meets for 3 hours per week. For every 1 hour that is spent in class, a minimum of 1 hour is required outside of class to complete assignments and study for exams. You also work 20 hours per week, and it takes you 1 hour to drive to the job site and back home. On Friday nights, you socialize with your friends. You are fairly certain that you will be able to successfully complete the semester with good grades. Show how the steps in the scientific method can help you predict the outcome of the semester.

Use the following food label to answer Exercises 9–10.

Nutrition Facts	Amount/serving	%DV*	Amount/serving	%DV*
Serv. Size 1 cup (249g)	**Total Fat 12g**	**18%**	**Sodium 940mg**	**39%**
Servings About 3	Sat. Fat 6g	**30%**	**Total Carb. 24g**	**8%**
Calories 250	Polyunsat. Fat 1.5g		Dietary Fiber 1g	**4%**
Fat Cal. 110	Monounsat. Fat 2.5g		Sugars 1g	
*Percent Daily Values (DV) are based on a 2,000 calorie diet.	**Cholest. 60mg**	**20%**	**Protein 10g**	**20%**
	Vitamin A 0% • Vitamin C 0% • Calcium 6% • Iron 8%			

INGREDIENTS: WATER, CHICKEN STOCK, ENRICHED PASTA (SEMOLINA WHEAT FLOUR, EGG WHITE SOLIDS, NIACIN, IRON, THIAMINE MONONITRATE [VITAMIN B1], RIBOFLAVIN [VITAMIN B2] AND FOLIC ACID), CREAM (DERIVED FROM MILK), CHICKEN, CONTAINS LESS THAN 2% OF: CHEESES (GRANULAR, PARMESAN AND ROMANO PASTE [PASTERURIZED COW'S MILK, CULTURES, SALT, ENZYMES], WATER, SALT, LACTIC ACID, CITRIC ACID AND DISODIUM PHOSPHATE), BUTTER (PASTERURIZED SWEET CREAM [DERIVED FROM MILK] AND SALT), MODIFIED CORN STARCH, SALT, WHOLE EGG SOLIDS, SUGAR, RICE STARCH, GARLIC, SPICE, XANTHAM GUM, MUSTARD FLOUR, ISOLATED SOY PROTEIN AND SODIUM PHOSPHATE.

9. Identify the following ingredients as either a pure substance or a mixture.
(a) water
(b) chicken stock (the liquid that remains in the pot after cooking the chicken)
(c) salt
(d) mustard flour

10. Using the food label, make a hypothesis regarding the nutritional quality of this food. Propose a way to determine whether your hypothesis is valid.

11. Read the following passage (from *Science News*) and identify the observation and hypothesis.

Could an invisibility cloak, like the one used by Harry Potter, really exist? For scientists in Cambridge, Massachusetts, it may be a reality. Researchers at MIT have developed an invisibility cloak for small objects such as an ant or a grain of sand. Calcite crystals have the ability to reflect light around an object, rendering it invisible, or at least "unseeable." If a larger crystal were to be used, it should be able to hide larger objects. The invisibility cloak works only with laser light aimed directly at the crystal. Future work will improve the effectiveness of these invisibility cloaks.

Use these pictures to answer Question 12.

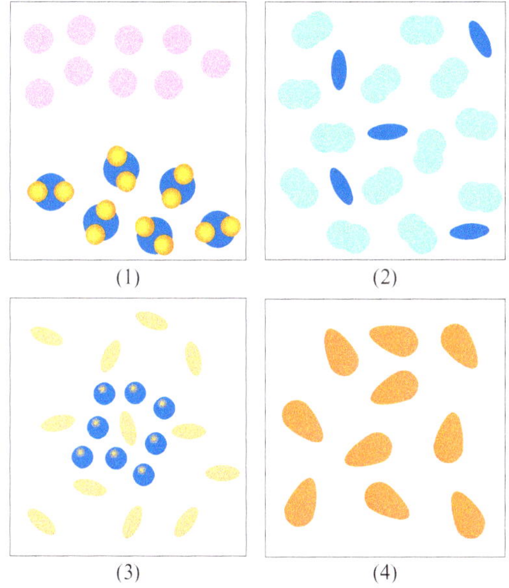

12. (a) Which picture best describes a homogeneous mixture?
(b) How would you classify the contents of the other containers?
(c) Which picture contains a compound? Explain how you made your choice.

13. Various chemical elements are pictured. For each one a microscopic view is also shown. Determine the number of phases for each substance below and identify them.
(a) iodine

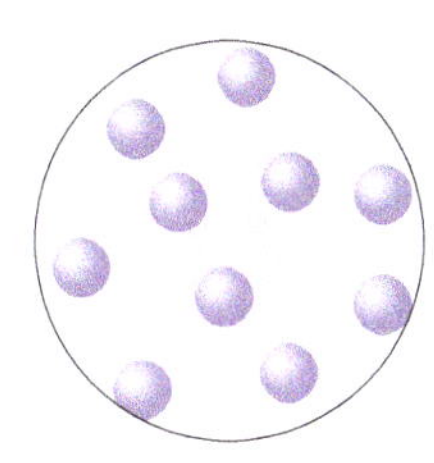

© 2001 Richard Megna/ Fundamental Photographs

(b) bromine

© 1995 Richard Megna/ Fundamental Photographs

(c) sulfur

Stocktrek Images/Richard Roscoe/Getty Images, Inc.

ANSWERS TO PRACTICE EXERCISES

1.1 five water molecules: H_2O, H_2O, H_2O, H_2O, H_2O

1.2 (a) mixture; concentration can be changed by adding more acetic acid or more water. (b) mixture; concentration can be changed by adding more or less salt. (c) pure substance; gold is 100% gold. (d) mixture; milk contains several substances.

2

Standards for Measurement

Tim Hall/Getty Images

CHAPTER OUTLINE

Doing an experiment in chemistry is very much like cooking a meal in the kitchen. It's important to know the ingredients and the amounts of each in order to have a tasty product. Working on your car requires specific tools in exact sizes. Buying new carpeting or draperies is an exercise in precise and accurate measurement for a good fit. A small difference in the concentration or amount of medication a pharmacist gives you may have significant effects on your well-being. As we saw in Chapter 1, observation is an important part of the scientific process. Observations can be qualitative (the substance is a blue solid) or quantitative (the mass of the substance is 4.7 grams). A quantitative observation is called a measurement. Both a number and a unit are required for a measurement. For example, at home in your kitchen measuring 3 flour is not possible. We need to know the unit on the 3. Is it cups, grams, or tablespoons? A measurement of 3 cups tells us both the amount and the size of the measurement. In this chapter we will discuss measurements and the rules for calculations using these measurements.

2.1 Scientific Notation

Write decimal numbers in scientific notation.

LEARNING OBJECTIVE

KEY TERMS

measurement
scientific notation

Scientists often use numbers that are very large or very small in **measurements**. For example, the Earth's age is estimated to be about 4,500,000 (4.5 billion) years. Numbers like these are bulky to write, so to make them more compact scientists use powers of 10. Writing a number as the product of a number between 1 and 10 multiplied by 10 raised to some power is called **scientific notation**.

To learn how to write a number in scientific notation, let's consider the number 2468. To write this number in scientific notation:

1. Move the decimal point in the original number so that it is located after the first nonzero digit.

   2468 → 2.468 (decimal moves three places to the left)

2. Multiply this new number by 10 raised to the proper exponent (power). The proper exponent is equal to the number of places that the decimal point was moved.

   $$2.468 \times 10^3$$

3. The sign on the exponent indicates the direction the decimal was moved.

   moved right → negative exponent
   moved left → positive exponent

CHECK YOUR UNDERSTANDING

2.1 Scientific Notation for Numbers Larger Than 10

2.2 Scientific Notation for Numbers Between 0 and 1

➡ WileyPLUS

Examples show you problem-solving techniques in a step-by-step form. Study each one and then try the Practice Exercises.

ONLINE LEARNING MODULE

Writing Numbers in Scientific Notation

➡ WileyPLUS

ENHANCED EXAMPLE 2.1 WileyPLUS

Write 5283 in scientific notation.

SOLUTION 5283. (3) Place the decimal between the 5 and the 2. Since the decimal was moved three places to the left, the power of 10 will be 3 and the number 5.283 is multiplied by 10^3.

5.283×10^3

ENHANCED EXAMPLE 2.2 WileyPLUS

Write 4,500,000,000 in scientific notation (two digits).

SOLUTION 4 500 000 000. (9) Place the decimal between the 4 and the 5. Since the decimal was moved nine places to the left, the power of 10 will be 9 and the number 4.5 is multiplied by 10^9.

4.5×10^9

NASA

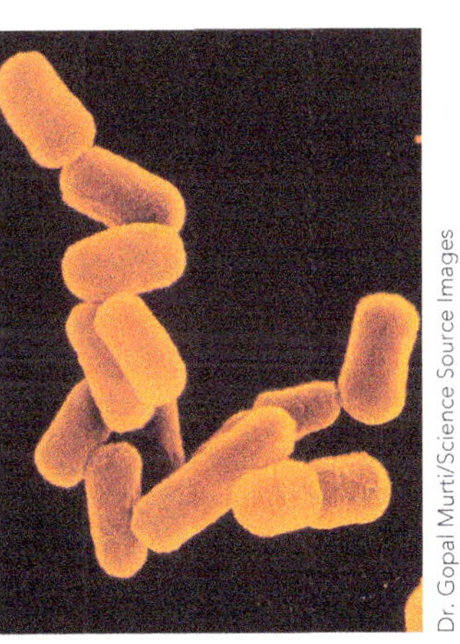

Dr. Gopal Murti/Science Source Images

Scientific notation is a useful way to write very large numbers, such as the distance between the Earth and the moon, or very small numbers, such as the length of these *E. coli* bacteria (shown here as a colored scanning electron micrograph × 14,000).

ONLINE LEARNING MODULE
Writing Small Numbers in Scientific Notation
WileyPLUS

ENHANCED EXAMPLE 2.3 WileyPLUS

Write 0.000123 in scientific notation.

SOLUTION 0.000123 (4) Place the decimal between the 1 and the 2. Since the decimal was moved four places to the right, the power of 10 will be −4 and the number 1.23 is multiplied by 10^{-4}.

1.23×10^{-4}

PRACTICE 2.1

Write the following numbers in scientific notation:

(a) 1200 (four digits)
(b) 6,600,000 (two digits)
(c) 0.0468
(d) 0.00003

2.2 Measurement and Uncertainty

LEARNING OBJECTIVE Explain the significance of uncertainty in measurement in chemistry and how significant figures are used to indicate a measurement's certainty.

KEY TERM
significant figures

To understand chemistry, it is necessary to set up and solve problems. Problem solving requires an understanding of the mathematical operations used to manipulate numbers. Measurements are made in an experiment, and chemists use these data to calculate the extent of the physical and chemical changes occurring in the substances that are being studied. By appropriate calculations, an experiment's results can be compared with those of other experiments and summarized in ways that are meaningful.

A measurement always requires a unit.

A measurement is expressed by a numerical value together with a unit of that measurement. For example,

numerical value
70.0 kilograms = 154 pounds
unit

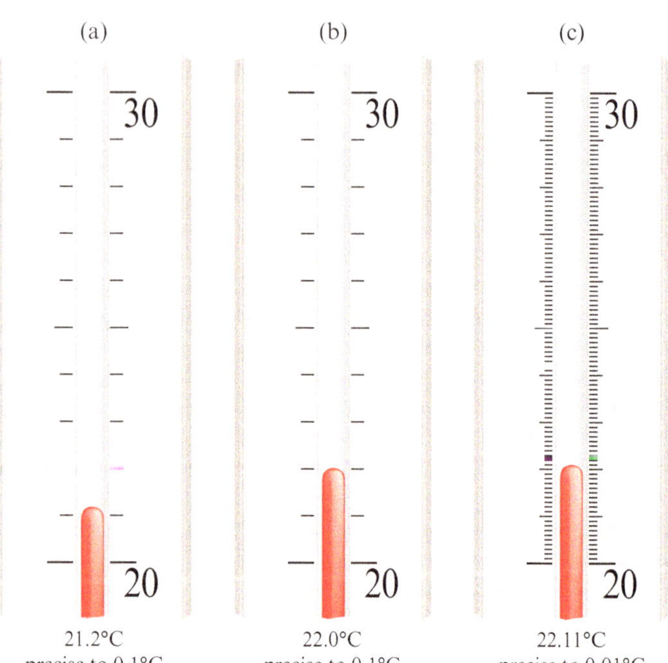

FIGURE 2.1 Measuring temperature (°C) with various degrees of precision.

Whenever a measurement is made with an instrument such as a thermometer or ruler, an estimate is required. We can illustrate this by measuring temperature. Suppose we measure temperature on a thermometer calibrated in degrees and observe that the mercury stops between 21 and 22 (FIGURE 2.1A). We then know that the temperature is at least 21 degrees and is less than 22 degrees. To express the temperature with one more digit, we estimate that the mercury is about two-tenths the distance between 21 and 22. The temperature is therefore 21.2 degrees. The last digit (2) has some uncertainty because it is an estimated value. Because the last digit has some uncertainty (we made a visual estimate), it may be different when another person makes the same measurement. If three more people make this same reading, the results might be

Person	Measurement
1	21.2
2	21.3
3	21.1
4	21.2

Notice that the first two digits of the measurements did not change (they are *certain*). The last digit in these measurements is *uncertain*. The custom in science is to record all of the certain digits and the first uncertain digit.

Numbers obtained from a measurement are never exact values. They always have some degree of uncertainty due to the limitations of the measuring instrument and the skill of the individual making the measurement. It is customary when recording a measurement to include all the digits that are known plus one digit that is estimated. This last estimated digit introduces some uncertainty. Because of this uncertainty, every number that expresses a measurement can have only a limited number of digits. These digits, used to express a measured quantity, are known as **significant figures**.

CHECK YOUR UNDERSTANDING

2.3 Significant Figures

➡ WileyPLUS

Now let's return to our temperature measurements. In Figure 2.1a the temperature is recorded as 21.2 degrees and is said to have three significant figures. If the mercury stopped exactly on the 22 (Figure 2.1b), the temperature would be recorded as 22.0 degrees. The zero is used to indicate that the temperature was estimated to a precision of one-tenth degree. Finally, look at Figure 2.1c. On this thermometer, the temperature is recorded as 22.11°C (four significant figures). Since the thermometer is calibrated to tenths of a degree, the first estimated digit is the hundredths.

ENHANCED EXAMPLE 2.4 **WileyPLUS**

The length of a ballpoint pen was recorded by three students as 14.30 cm, 14.33 cm, and 14.34 cm.

(a) What are the estimated digits in these measurements?
(b) How many figures are certain?

SOLUTION

(a) The estimated digits are 0, 3, and 4, respectively.
(b) Three figures are certain in each measurement.

Ingvar Bjork/Shutterstock

PRACTICE 2.2

Three measurements for the boiling point of water are 100.4°C, 100.1°C, and 100.0°C. The accepted boiling point is 100.0°C.

(a) What are the estimated digits?
(b) How many figures are certain?

2.3 Significant Figures

LEARNING OBJECTIVE

Determine the number of significant figures in a given measurement and round measurements to a specific number of significant figures.

KEY TERM

rounding off numbers

Because all measurements involve uncertainty, we must use the proper number of significant figures in each measurement. In chemistry we frequently do calculations involving measurements, so we must understand what happens when we do arithmetic on numbers containing uncertainties. We'll learn several rules for doing these calculations and figuring out how many significant figures to have in the result. You will need to follow these rules throughout the calculations in this text.

The first thing we need to learn is how to determine how many significant figures are in a number.

RULES FOR COUNTING SIGNIFICANT FIGURES

1. *Nonzero digits.* All nonzero digits are significant.
2. *Exact numbers.* Some numbers are exact and have an infinite number of significant figures. Exact numbers occur in simple counting operations; when you count 25 dollars, you have exactly 25 dollars. Defined numbers, such as 12 inches in 1 foot, 60

minutes in 1 hour, and 100 centimeters in 1 meter, are also considered to be exact numbers. Exact numbers have no uncertainty.

Rules for significant figures should be memorized for use throughout the text.

3. *Zeros.* A zero is *significant* when it is
   - between nonzero digits:
     205 has three significant figures (2, 0, 5)
     2.05 has three significant figures (2, 0, 5)
     61.09 has four significant figures (6, 1, 0, 9)
   - at the end of a number that includes a decimal point:
     0.500 has three significant figures (5, 0, 0)
     25.160 has five significant figures (2, 5, 1, 6, 0)
     3.00 has three significant figures (3, 0, 0)
     20. has two significant figures (2, 0)

   A zero is *not significant* when it is
   - before the first nonzero digit. These zeros are used to locate a decimal point:
     0.0025 has two significant figures (2, 5)
     0.0108 has three significant figures (1, 0, 8)
   - at the end of a number without a decimal point:
     1000 has one significant figure (1)
     590 has two significant figures (5, 9)

CHECK YOUR UNDERSTANDING

2.4 Non-Zero Significant Figures

2.5 Exact Numbers

2.6 Zeros

➡ WileyPLUS

One way of indicating that these zeros are significant is to write the number using scientific notation. Thus if the value 1000 has been determined to four significant figures, it is written as 1.000×10^3. If 590 has only two significant figures, it is written as 5.9×10^2.

ONLINE LEARNING MODULE

Determining Significant Figures

➡ WileyPLUS

ENHANCED EXAMPLE 2.5 WileyPLUS

How many significant figures are in each of the following measurements?

(a) 45 apples
(b) 0.02050 cm
(c) 3500 ft

SOLUTION

(a) The number is exact and has no uncertainty associated with it (Rule 2).
(b) Four significant figures. The zeroes before the 2 are not significant, the 2 is significant, the zero between the 2 and the 5 is significant, the 5 is significant, and the final zero is significant.
(c) Two significant figures. The zeroes at the end of a number without a decimal point are not significant.

PRACTICE 2.3

How many significant figures are in each of these measurements?

(a) 4.5 inches
(b) 3.025 feet
(c) 125.0 meters
(d) 0.001 mile
(e) 25.0 grams
(f) 12.20 liters
(g) 100,000 people
(h) 205 birds

Rounding Off Numbers

When we do calculations on a calculator, we often obtain answers that have more digits than are justified. It is therefore necessary to drop the excess digits in order to express the answer with the proper number of significant figures. When digits are dropped from a number, the value of the last digit retained is determined by a process known as **rounding off numbers**. Two rules will be used in this book for rounding off numbers.

RULES FOR ROUNDING OFF

1. When the first digit after those you want to retain is 4 or less, that digit and all others to its right are dropped. The last digit retained is not changed. The following examples are rounded off to four digits:

   74.693 = 74.69 — This digit is dropped.

   1.00629 = 1.006 — These two digits are dropped.

2. When the first digit after those you want to retain is 5 or greater, that digit and all others to the right are dropped and the last digit retained is increased by one. These examples are rounded off to four digits:

   1.026868 = 1.027 — These three digits are dropped. This digit is changed to 7.

   18.02500 = 18.03 — These three digits are dropped. This digit is changed to 3.

   12.899 = 12.90 — This digit is dropped. These two digits are changed to 90.

Not all schools use the same rules for rounding. Check with your instructor for variations in these rules.

PRACTICE 2.4

Round off these numbers to the number of significant figures indicated:

(a) 42.246 (four)
(b) 88.015 (three)
(c) 0.08965 (three)
(d) 0.08965 (two)
(e) 225.3 (three)
(f) 14.150 (three)

2.4 Significant Figures in Calculations

LEARNING OBJECTIVE

Apply the rules for significant figures in calculations involving addition, subtraction, multiplication, and division.

The results of a calculation based on measurements cannot be more precise than the least precise measurement.

Multiplication or Division

In calculations involving multiplication or division, the answer must contain the same number of significant figures as in the measurement that has the least number of significant figures. Consider the following examples:

CHECK YOUR UNDERSTANDING
2.7 Significant Figures in Multiplication
WileyPLUS

ENHANCED EXAMPLE 2.6 WileyPLUS

(190.6)(2.3) = 438.38

SOLUTION

The value 438.38 was obtained with a calculator. The answer should have two significant figures because 2.3, the number with the fewest significant figures, has only two significant figures.

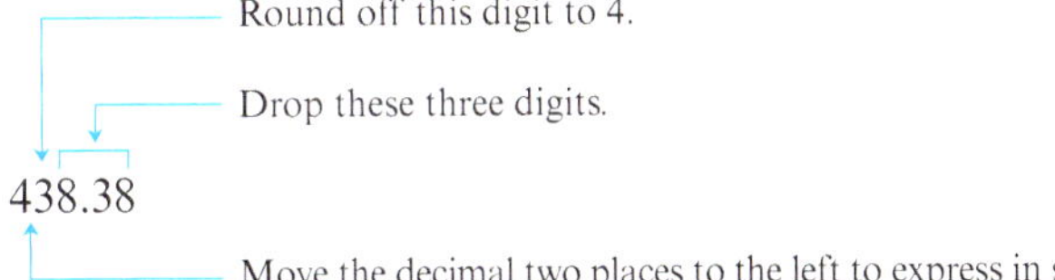

The correct answer is 440 or 4.4×10^2.

ONLINE LEARNING MODULE
Significant Figures in Multiplication
WileyPLUS

Use your calculator to check your work in the examples. Compare your results to be sure you understand the mathematics.

ENHANCED EXAMPLE 2.7 WileyPLUS

$$\frac{(13.59)(6.3)}{12} = 7.13475$$

SOLUTION

The value 7.13475 was obtained with a calculator. The answer should contain two significant figures because 6.3 and 12 each have only two significant figures.

7.13475 — Drop these four digits (3475). This digit remains the same (1).

The correct answer is 7.1.

PRACTICE 2.5

(a) (134 in.)(25 in.) = ?

(b) $\frac{213 \text{ miles}}{4.20 \text{ hours}} = ?$

(c) $\frac{(2.2)(273)}{760} = ?$

(d) $0.0321 \times 42 = ?$

(e) $\frac{0.0450}{0.00220} = ?$

(f) $\frac{1.280}{0.345} = ?$

Addition or Subtraction

CHECK YOUR UNDERSTANDING
2.8 Significant Figures in Addition
WileyPLUS

The results of an addition or a subtraction must be expressed to the same precision as the least precise measurement. This means the result must be rounded to the same number of decimal places as the value with the fewest decimal places (blue line in examples).

ONLINE LEARNING MODULE
Significant Figures in Addition
WileyPLUS

ENHANCED EXAMPLE 2.8 WileyPLUS

Add 125.17, 129, and 52.2.

SOLUTION

```
125.|17
129 |
 52.|2
306.|37
```

The number with the least precision is 129. Therefore the answer is rounded off to the nearest unit: 306.

ENHANCED EXAMPLE 2.9 WileyPLUS

Subtract 14.1 from 132.56.

SOLUTION

```
  132.5|6
−  14.1|
  118.4|6
```

The number with the least precision is 14.1. Therefore the answer is rounded off to the nearest tenth: 118.5.

ENHANCED EXAMPLE 2.10 WileyPLUS

Subtract 120 from 1587.

SOLUTION

$$\begin{array}{r} 158|7 \\ -\ 12|0 \\ \hline 146|7 \end{array}$$

The number with the least precision is 120. The zero is not considered significant; therefore the answer must be rounded to the nearest ten: 1470 or 1.47×10^3.

ENHANCED EXAMPLE 2.11 WileyPLUS

Add 5672 and 0.00063.

SOLUTION

$$\begin{array}{r} 5672\ \ \ \ \ \ \ \ \\ +\ \ 0.|00063 \\ \hline 5672.|00063 \end{array}$$

The number with the least precision is 5672. So the answer is rounded off to the nearest unit: 5672.

Note: When a very small number is added to a large number, the result is simply the original number.

ENHANCED EXAMPLE 2.12 WileyPLUS

$$\frac{1.039 - 1.020}{1.039} = ?$$

SOLUTION

$$\frac{1.039 - 1.020}{1.039} = 0.018286814$$

The value 0.018286814 was obtained with a calculator. When the subtraction in the numerator is done,

$$1.039 - 1.020 = 0.019$$

the number of significant figures changes from four to two. Therefore the answer should contain two significant figures after the division is carried out:

Drop these six digits.

0.018286814

This digit remains the same.

The correct answer is 0.018, or 1.8×10^{-2}.

ONLINE LEARNING MODULE

Significant Figures in Complex Calculations

➡ WileyPLUS

PRACTICE 2.6

How many significant figures should the answer in each of these calculations contain?

(a) (14.0)(5.2)

(b) (0.1682)(8.2)

(c) $\frac{(160)(33)}{4}$

(d) 8.2 + 0.125

(e) 119.1 − 3.44

(f) $\frac{94.5}{1.2}$

(g) 1200 + 6.34

(h) 1.6 + 23 − 0.005

If you need to brush up on your math skills, refer to the "Mathematical Review" in Appendix I.

RULES FOR SIGNIFICANT FIGURES IN CALCULATIONS

Multiplication or Division

The answer contains the same number of significant figures as the measurement with the *least number* of significant figures.

Addition or Subtraction

The answer contains the same number of significant figures as the *least precise* measurement.

Additional material on mathematical operations is given in Appendix I, "Mathematical Review." Study any portions that are not familiar to you. You may need to do this at various times during the course when additional knowledge of mathematical operations is required.

2.5 The Metric System

LEARNING OBJECTIVE Name the units for mass, length, and volume in the metric system and convert from one unit to another.

KEY TERMS
metric system or International System (SI)
meter (m)
conversion factor
solution map
mass
weight
kilogram (kg)
volume
liter (L)

The **metric system**, or **International System** (**SI**, from *Système International*), is a decimal system of units for measurements of mass, length, time, and other physical quantities. Built around a set of standard units, the metric system uses factors of 10 to express larger or smaller numbers of these units. To express quantities that are larger or smaller than the standard units, prefixes are added to the names of the units. These prefixes represent multiples of 10, making the metric system a decimal system of measurements. TABLE 2.1 shows the names, symbols, and numerical values of the common prefixes. Some examples of the more commonly used prefixes are

$$1\ \textit{kilo}\text{meter} = 1000\ \text{meters}$$

$$1\ \textit{kilo}\text{gram} = 1000\ \text{grams}$$

$$1\ \textit{milli}\text{meter} = 0.001\ \text{meter}$$

$$1\ \textit{micro}\text{second} = 0.000001\ \text{second}$$

The prefixes most commonly used in chemistry are shown in bold in Table 2.1.

The common standard units in the International System, their abbreviations, and the quantities they measure are given in TABLE 2.2. Other units are derived from these units. The metric system, or International System, is currently used by most of the countries in the world, not only in scientific and technical work but also in commerce and industry.

TABLE 2.1 Common Prefixes and Numerical Values for SI Units

Prefix	Symbol	Numerical value	Power of 10 equivalent
giga	G	1,000,000,000	10^9
mega	M	1,000,000	10^6
kilo	k	1,000	10^3
hecto	h	100	10^2
deka	da	10	10^1
—	—	1	10^0
deci	d	0.1	10^{-1}
centi	c	0.01	10^{-2}
milli	m	0.001	10^{-3}
micro	μ	0.000001	10^{-6}
nano	n	0.000000001	10^{-9}
pico	p	0.000000000001	10^{-12}
femto	f	0.000000000000001	10^{-15}

TABLE 2.2 International System's Standard Units of Measurement

Quantity	Name of unit	Abbreviation
Length	meter	m
Mass	kilogram	kg
Temperature	kelvin	K
Time	second	s
Amount of substance	mole	mol
Electric current	ampere	A
Luminous intensity	candela	cd

Sheila Fitzgerald / Shutterstock

Most products today list both systems of measurement on their labels.

Measurement of Length

The standard unit of length in the metric system is the **meter (m)**. When the metric system was first introduced in the 1790s, the meter was defined as one ten-millionth of the distance from the equator to the North Pole, measured along the meridian passing through Dunkirk, France. The latest definition describes a meter as the distance that light travels in a vacuum during 1/299,792,458 of a second.

A meter is 39.37 inches, a little longer than 1 yard. One meter equals 10 decimeters, 100 centimeters, or 1000 millimeters (FIGURE 2.2). A kilometer contains 1000 meters. TABLE 2.3 shows the relationships of these units.

The nanometer (10^{-9} m) is used extensively in expressing the wavelength of light, as well as in atomic dimensions. See the inside back cover for a complete table of common conversions.

CHECK YOUR UNDERSTANDING

2.9 Using the Prefix kilo-
2.10 Using the Prefix mega-
2.11 Using the Prefix milli-
2.12 Using the Prefix micro-
2.13 Using the Prefix centi-
➡ WileyPLUS

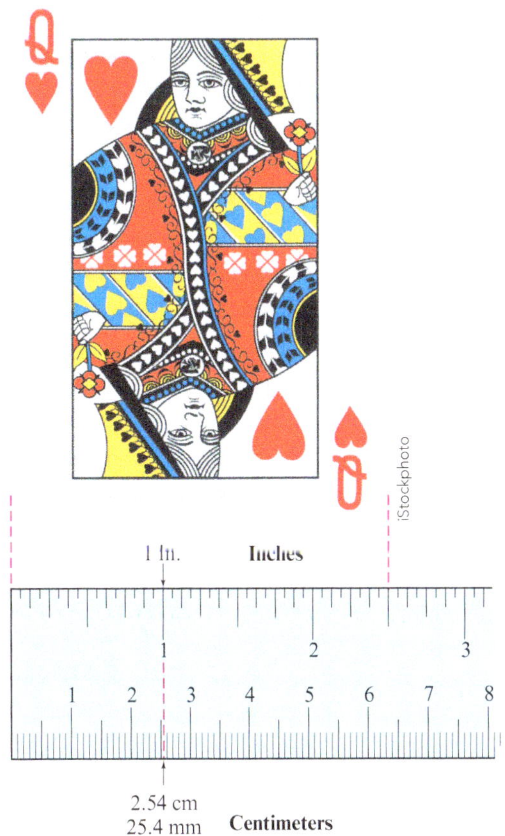

iStockphoto

FIGURE 2.2 Comparison of the metric and U.S. systems of length measurement: 2.54 cm = 1 in. A playing card measures $2\frac{1}{2}$ in. or 6.3 cm.

TABLE 2.3 Metric Units of Length

Unit	Abbreviation	Meter equivalent	Exponential equivalent
kilometer	km	1000 m	10^3 m
meter	m	1 m	10^0 m
decimeter	dm	0.1 m	10^{-1} m
centimeter	cm	0.01 m	10^{-2} m
millimeter	mm	0.001 m	10^{-3} m
micrometer	μm	0.000001 m	10^{-6} m
nanometer	nm	0.000000001 m	10^{-9} m
angstrom	Å	0.0000000001 m	10^{-10} m

Common length relationships:

1 m = 10^6 μm = 10^{10} Å
= 100 cm = 1000 mm
1 cm = 10 mm = 0.01 m
1 in. = 2.54 cm
1 mile = 1.609 km

See inside back cover for a table of conversions.

TABLE 2.4	Length Equivalent Units
1 m = 10 dm	
1 m = 100 cm	
1 m = 1000 mm	
1 kg = 1000 g	
1 cm = 10 mm	

Unit Conversions

One of the benefits of using the metric system is the ease with which we can convert from one unit to another. We do this by using the relationships between units—math equations called equivalent units. TABLE 2.4 shows common equivalent units in the metric system.

To convert from one unit to another we must use a **conversion factor**. A conversion factor is a ratio of equivalent quantities. Where do we find these conversion factors? They are formed from equivalent units. For example, since 1 m = 100 cm (Table 2.4), we could write two ratios: $\frac{1\text{ m}}{100\text{ cm}}$ or $\frac{100\text{ cm}}{1\text{ m}}$. In both cases the ratio is 1 since the units are equal.

We can convert one metric unit to another by using conversion factors.

Important equations and statements are boxed or highlighted in color.

$$\textbf{unit}_1 \times \textbf{conversion factor} = \textbf{unit}_2$$

If you want to know how many millimeters are in 2.5 meters, you need to convert meters (m) to millimeters (mm). Start by writing

$$\text{m} \times \text{conversion factor} = \text{mm}$$

This conversion factor must accomplish two things: It must cancel (or eliminate) meters, *and* it must introduce millimeters—the unit wanted in the answer. Such a conversion factor will be in fractional form and have meters in the denominator and millimeters in the numerator:

$$\cancel{\text{m}} \times \frac{\text{mm}}{\cancel{\text{m}}} = \text{mm}$$

We know that 1 m = 1000 mm. From this relationship we can write two conversion factors.

We know that multiplying a measurement by 1 does not change its value. Since our conversion factors both equal 1, we can multiply the measurement by the appropriate one to convert units.

$$\frac{1\text{ m}}{1000\text{ mm}} \quad \text{and} \quad \frac{1000\text{ mm}}{1\text{ m}}$$

Choosing the conversion factor $\frac{1000\text{ mm}}{1\text{ m}}$, we can set up the calculation for the conversion of 2.5 m to millimeters:

$$(2.5\ \cancel{\text{m}})\left(\frac{1000\text{ mm}}{1\ \cancel{\text{m}}}\right) = 2500\text{ mm} \quad \text{or} \quad 2.5 \times 10^3\text{ mm} \quad \text{(two significant figures)}$$

Note that, in making this calculation, units are treated as numbers; meters in the numerator are canceled by meters in the denominator.

ONLINE LEARNING MODULE 
Metric Conversions for Length
➡ WileyPLUS

ENHANCED EXAMPLE 2.13 WileyPLUS

Change 215 centimeters to meters.

SOLUTION

We will use a **solution map** to outline our path for the conversion for this example:

Solution map: cm → m

We start with

$$\text{cm} \times \text{conversion factor} = \text{m}$$

The conversion factor must have centimeters in the denominator and meters in the numerator:

$$\cancel{\text{cm}} \times \frac{\text{m}}{\cancel{\text{cm}}} = \text{m}$$

From the relationship 100 cm = 1 m, we can write a factor that will accomplish this conversion:

$$\frac{1\text{ m}}{100\text{ cm}}$$

Now set up the calculation using all the data given:

$$(215\ \cancel{\text{cm}})\left(\frac{1\ \text{m}}{100\ \cancel{\text{cm}}}\right) = 2.15\ \text{m}$$

PRACTICE 2.7

Convert the following measurements:
(a) 567 mm to m
(b) 68 cm to mm
(c) 125 m to km

Measurement of Mass

Although we often use mass and weight interchangeably in our everyday lives, they have quite different meanings in chemistry. In science we define the **mass** of an object as the amount of matter in the object. Mass is measured on an instrument called a balance. The **weight** of an object is a measure of the effect of gravity on the object. Weight is determined by using an instrument called a scale, which measures force against a spring. This means mass is independent of the location of an object, but weight is not. In this text we will use the term *mass* for all of our metric mass measurements. The gram is a unit of mass measurement, but it is a tiny amount of mass; for instance, a U.S. nickel has a mass of about 5 grams. Therefore the *standard unit* of mass in the SI system is the **kilogram (kg)** (equal to 1000 g). The amount of mass in a kilogram is defined by international agreement as exactly equal to the mass of a platinum-iridium cylinder (international prototype kilogram) kept in a vault at Sèvres, France. Comparing this unit of mass to 1 lb (16 oz), we find that 1 kg is equal to 2.205 lb. A pound is equal to 453.6 g (0.4536 kg). The same prefixes used in length measurement are used to indicate larger and smaller gram units (TABLE 2.5). A balance is used to measure mass. Two examples of balances are shown in FIGURE 2.3.

TABLE 2.5 | Metric Units of Mass

Unit	Abbreviation	Gram equivalent	Exponential equivalent
kilogram	kg	1000 g	10^3 g
gram	g	1 g	10^0 g
decigram	dg	0.1 g	10^{-1} g
centigram	cg	0.01 g	10^{-2} g
milligram	mg	0.001 g	10^{-3} g
microgram	μg	0.000001 g	10^{-6} g

Martyn F. Chillmaid/Science Source

(a)

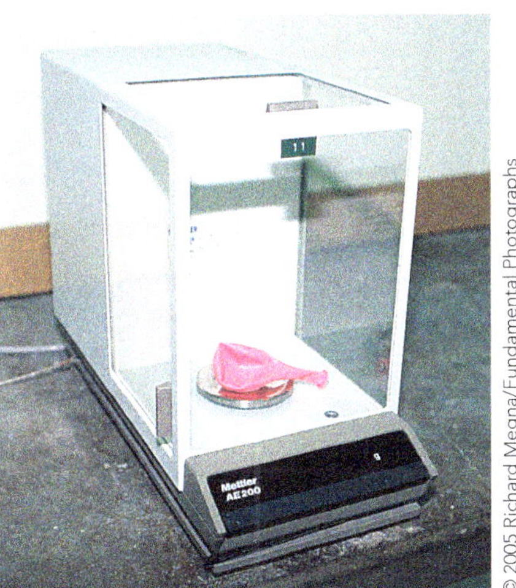

© 2005 Richard Megna/Fundamental Photographs

(b)

FIGURE 2.3 (a) Digital electronic, top-loading balance with a precision of 0.001 g (1 mg); and (b) a digital electronic analytical balance with a precision of 0.0001 g (0.1 mg).

CHEMISTRY IN ACTION Keeping Track of Units

Mars Climate Orbiter.

Why do scientists worry about units? The National Aeronautics and Space Administration (NASA) recently was reminded of just why keeping track of units is so important. In 1999 a $125 million satellite was lost in the atmosphere of Mars because scientists made some improper assumptions about units. NASA's scientists at the Jet Propulsion Lab (JPL) in Pasadena, California, received thrust data from the satellite's manufacturer, Lockheed Martin Aeronautics in Denver, Colorado. Unfortunately, the Denver scientists used U.S. units in their measurements and the JPL scientists assumed the units were metric. This mistake caused the satellite to fall 100 km lower into the Mars atmosphere than planned. The spacecraft burned up from the friction with the atmosphere.

Measuring and using units correctly is very important. In fact, it can be critical, as we have just seen. For example, a Canadian jet almost crashed when the tanks were filled with 22,300 pounds (instead of kilograms) of fuel. Calculations for distance were based on kilograms, and the jet almost ran out of fuel before landing at the destination. Correct units are also important to ensure perfect fits for household purchases such as drapes, carpets, or appliances. Be sure to pay attention to units in both your chemistry problems and in everyday life!

Common mass relationships:
1 g = 1000 mg
1 kg = 1000 g
1 kg = 2.205 lb
1 lb = 453.6 g

ENHANCED EXAMPLE 2.14 WileyPLUS

Change 25 grams to milligrams.

SOLUTION

We use the conversion factor 1000 mg/g.
Solution map: g → mg

$$(25\ \cancel{g})\left(\frac{1000\ \text{mg}}{1\ \cancel{g}}\right) = 25{,}000\ \text{mg} \qquad (2.5 \times 10^4\ \text{mg})$$

Note that multiplying a number by 1000 is the same as multiplying the number by 10^3 and can be done simply by moving the decimal point three places to the right:

$$(6.428)(1000) = 6428 \qquad (6.\underbrace{428}_{3})$$

PRACTICE 2.8

Convert the following measurements:
(a) 560 mg to g
(b) 525 g to kg
(c) 175 g to mg

To change milligrams to grams, we use the conversion factor 1 g/1000 mg. For example, we convert 155 mg to grams as follows:

$$(155\ \cancel{\text{mg}})\left(\frac{1\ \text{g}}{1000\ \cancel{\text{mg}}}\right) = 0.155\ \text{g}$$

Measurement of Volume

Volume, as used here, is the amount of space occupied by matter. The SI unit of volume is the *cubic meter* (m^3). However, the liter (pronounced *lee-ter* and abbreviated L) and the milliliter (abbreviated mL) are the standard units of volume used in most chemical laboratories. A **liter (L)** is usually defined as 1 cubic decimeter (1 kg) of water at 4°C.

The most common instruments or equipment for measuring liquids are the graduated cylinder, volumetric flask, buret, pipet, and syringe, which are illustrated in FIGURE 2.4. These pieces are usually made of glass and are available in various sizes.

Common volume relationships:
1 L = 1000 mL = 1000 cm^3
1 mL = 1 cm^3
1 L = 1.057 qt
946.1 mL = 1 qt

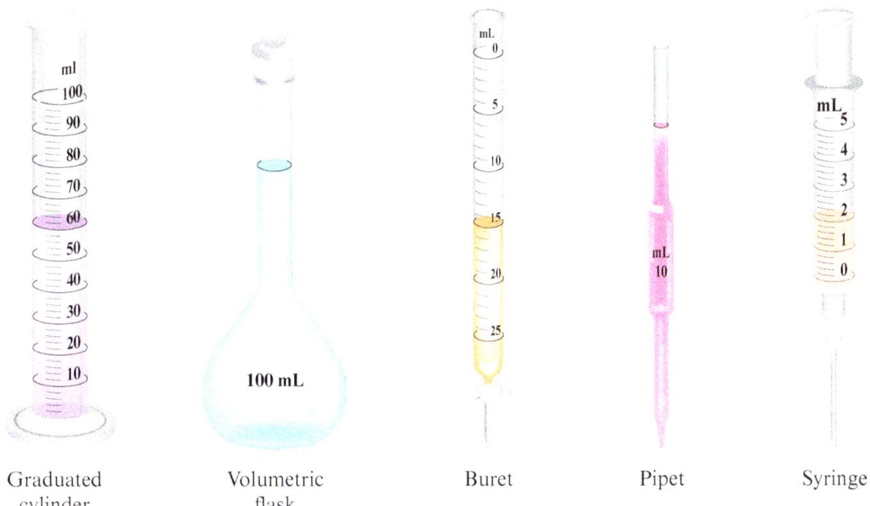

FIGURE 2.4 Calibrated glassware for measuring the volume of liquids.

ENHANCED EXAMPLE 2.15 WileyPLUS

How many milliliters are contained in 3.5 liters?

SOLUTION

The conversion factor to change liters to milliliters is 1000 mL/L:

Solution map: L → mL

$$(3.5\ \cancel{L})\left(\frac{1000\ mL}{\cancel{L}}\right) = 3500\ mL \qquad (3.5 \times 10^3\ mL)$$

Liters may be changed to milliliters by moving the decimal point three places to the right and changing the units to milliliters:

$$\underbrace{1.500}_{3}\ L = 1500.\ mL$$

PRACTICE 2.9

Convert the following measurements:
(a) 25 mL to L
(b) 1.55 L to mL
(c) 4.5 L to dL

2.6 Dimensional Analysis: A Problem-Solving Method

Use dimensional analysis to solve problems involving unit conversions. LEARNING OBJECTIVE

Many chemical principles can be illustrated mathematically. Learning how to set up and solve numerical problems in a systematic fashion is *essential* in the study of chemistry.

Usually a problem can be solved by several methods. But in all methods it is best to use a systematic, orderly approach. The *dimensional analysis method* is emphasized in this book because it

- provides a systematic, straightforward way to set up problems.
- gives a clear understanding of the principles involved.
- trains you to organize and evaluate data.
- helps to identify errors, since unwanted units are not eliminated if the setup of the problem is incorrect.

Dimensional analysis converts one unit to another unit by using conversion factors.

Label all factors with the proper units.

The following examples show the conversion between U.S. units and metric units.

ONLINE LEARNING MODULE

English to Metric Mass Conversions

➥ WileyPLUS

ENHANCED EXAMPLE 2.16 WileyPLUS

A 1.50-lb package contains how many grams of baking soda?

SOLUTION

We are solving for the number of grams equivalent to 1.50 lb. Since 1 lb = 453.6 g, the conversion factor is 453.6 g/lb:

Solution map: lb → g

$$(1.50\ \cancel{\text{lb}})\left(\frac{453.6\ \text{g}}{1\ \cancel{\text{lb}}}\right) = 680.\ \text{g}$$

PRACTICE 2.10

A tennis ball has a mass of 65 g. Determine the U.S. equivalent in pounds.

Units are emphasized in problems by using color and solution maps to help visualize the steps in the process.

Some problems require a series of conversions to reach the correct units in the answer. For example, suppose we want to know the number of seconds in 1 day. We need to convert from the unit of days to seconds using this solution map:

Solution map: day → hours → minutes → seconds

Each arrow in the solution map represents a conversion factor. This sequence requires three conversion factors, one for each step. We convert days to hours (hr), hours to minutes (min), and minutes to seconds (s). The conversions can be done individually or in a continuous sequence:

$$\cancel{\text{day}} \times \frac{\text{hr}}{\cancel{\text{day}}} \longrightarrow \cancel{\text{hr}} \times \frac{\text{min}}{\cancel{\text{hr}}} \longrightarrow \cancel{\text{min}} \times \frac{\text{s}}{\cancel{\text{min}}} = \text{s}$$

$$\cancel{\text{day}} \times \frac{\cancel{\text{hr}}}{\cancel{\text{day}}} \times \frac{\cancel{\text{min}}}{\cancel{\text{hr}}} \times \frac{\text{s}}{\cancel{\text{min}}} = \text{s}$$

Inserting the proper factors, we calculate the number of seconds in 1 day to be

$$(1\ \cancel{\text{day}})\left(\frac{24\ \cancel{\text{hr}}}{1\ \cancel{\text{day}}}\right)\left(\frac{60\ \cancel{\text{min}}}{1\ \cancel{\text{hr}}}\right)\left(\frac{60\ \text{s}}{1\ \cancel{\text{min}}}\right) = 86{,}400.\ \text{s}$$

All five digits in 86,400. are significant, since all the factors in the calculation are exact numbers.

ENHANCED EXAMPLE 2.17 WileyPLUS

How many centimeters are in 2.00 ft?

SOLUTION

The stepwise conversion of units from feet to centimeters may be done in this sequence. Convert feet to inches; then convert inches to centimeters:

Solution map: ft → in. → cm

The conversion factors needed are

$$\frac{12 \text{ in.}}{1 \text{ ft}} \quad \text{and} \quad \frac{2.54 \text{ cm}}{1 \text{ in.}}$$

$$(2.00 \cancel{\text{ft}})\left(\frac{12 \text{ in.}}{1 \cancel{\text{ft}}}\right) = 24.0 \text{ in.}$$

$$(24.0 \cancel{\text{in.}})\left(\frac{2.54 \text{ cm}}{1 \cancel{\text{in.}}}\right) = 61.0 \text{ cm}$$

NOTE Since 1 ft and 12 in. are exact numbers, the number of significant figures allowed in the answer is three, based on the number 2.00.

CHECK YOUR UNDERSTANDING

2.14 Converting Inches to Centimeters

2.15 Converting Pounds to Grams

2.16 Converting Quarts to Milliliters

2.17 Writing Conversion Factors

➥ WileyPLUS

ENHANCED EXAMPLE 2.18 WileyPLUS

How many meters are in a 100.-yd football field?

SOLUTION

The stepwise conversion of units from yards to meters may be done by this sequence, using the proper conversion factors:

Solution map: yd → ft → in. → cm → m

$$(100. \cancel{\text{yd}})\left(\frac{3 \text{ ft}}{1 \cancel{\text{yd}}}\right) = 300. \text{ ft}$$

$$(300. \cancel{\text{ft}})\left(\frac{12 \text{ in.}}{1 \cancel{\text{ft}}}\right) = 3600 \text{ in.}$$

$$(3600 \cancel{\text{in.}})\left(\frac{2.54 \text{ cm}}{1 \cancel{\text{in.}}}\right) = 9144 \text{ cm}$$

$$(9144 \cancel{\text{cm}})\left(\frac{1 \text{ m}}{100 \cancel{\text{cm}}}\right) = 91.4 \text{ m} \quad \text{(three significant figures)}$$

ONLINE LEARNING MODULE

English to Metric Length Conversions

➥ WileyPLUS

PRACTICE 2.11

(a) How many meters are in 10.5 miles?
(b) What is the area of a 6.0-in. × 9.0-in. rectangle in square meters?

Examples 2.17 and 2.18 may be solved using a linear expression and writing down conversion factors in succession. This method often saves one or two calculation steps and leads to simpler calculations. The single linear expressions for Examples 2.17 and 2.18 are

$$(2.00 \cancel{\text{ft}})\left(\frac{12 \cancel{\text{in.}}}{1 \cancel{\text{ft}}}\right)\left(\frac{2.54 \text{ cm}}{1 \cancel{\text{in.}}}\right) = 61.0 \text{ cm}$$

$$(100. \cancel{\text{yd}})\left(\frac{3 \cancel{\text{ft}}}{1 \cancel{\text{yd}}}\right)\left(\frac{12 \cancel{\text{in.}}}{1 \cancel{\text{ft}}}\right)\left(\frac{2.54 \cancel{\text{cm}}}{1 \cancel{\text{in.}}}\right)\left(\frac{1 \text{ m}}{100 \cancel{\text{cm}}}\right) = 91.4 \text{ m}$$

Using the units alone (Example 2.18), we see that the stepwise cancellation proceeds in succession until the desired unit is reached:

$$\cancel{\text{yd}} \times \frac{\cancel{\text{ft}}}{\cancel{\text{yd}}} \times \frac{\cancel{\text{in.}}}{\cancel{\text{ft}}} \times \frac{\cancel{\text{cm}}}{\cancel{\text{in.}}} \times \frac{\text{m}}{\cancel{\text{cm}}} = \text{m}$$

ONLINE LEARNING MODULE
Complex English to Metric Unit Conversions
➡ WileyPLUS

ENHANCED EXAMPLE 2.19 WileyPLUS

A driver of a car is obeying the speed limit of 55 miles per hour. How fast is the car traveling in kilometers per second?

SOLUTION

Several conversions are needed to solve this problem:

Solution map: mi → km

hr → min → s

To convert mi → km,

$$\left(\frac{55\ \cancel{\text{mi}}}{\text{hr}}\right)\left(\frac{1.609\ \text{km}}{1\ \cancel{\text{mi}}}\right) = 88\frac{\text{km}}{\text{hr}}$$

Next we must convert hr → min → s. Notice that hours is in the denominator of our quantity, so the conversion factor must have hours in the numerator:

$$\left(\frac{88\ \text{km}}{\cancel{\text{hr}}}\right)\left(\frac{1\ \cancel{\text{hr}}}{60\ \cancel{\text{min}}}\right)\left(\frac{1\ \cancel{\text{min}}}{60\ \text{s}}\right) = 0.024\frac{\text{km}}{\text{s}}$$

PRACTICE 2.12

How many cubic meters of air are in a room measuring 8 ft × 10 ft × 12 ft?

The volume of a cubic or rectangular container can be determined by multiplying its length × width × height. Thus a box 10 cm on each side has a volume of (10 cm)(10 cm) (10 cm) = 1000 cm^3.

ONLINE LEARNING MODULE
English to Metric Volume Conversions
➡ WileyPLUS

ENHANCED EXAMPLE 2.20 WileyPLUS

How many cubic centimeters are in a cube that is 11.1 inches on a side?

SOLUTION

First we change inches to centimeters; our conversion factor is 2.54 cm/in.:

Solution map: in. → cm

$$(11.1\ \cancel{\text{in.}})\left(\frac{2.54\ \text{cm}}{1\ \cancel{\text{in.}}}\right) = 28.2\ \text{cm on a side}$$

Then we determine volume (length × width × height):

$$(28.2\ \text{cm})(28.2\ \text{cm})(28.2\ \text{cm}) = 22{,}426\ \text{cm}^3 \qquad (2.24 \times 10^4\ \text{cm}^3)$$

When doing problems with multiple steps, you should round only at the end of the problem. We are rounding at the end of each step in example problems to illustrate the proper significant figures.

ENHANCED EXAMPLE 2.21 WileyPLUS

How many cubic centimeters (cm^3) are in a box that measures 2.20 in. by 4.00 in. by 6.00 in.?

SOLUTION

First we need to determine the volume of the box in cubic inches ($in.^3$) by multiplying the length × width × height:

$$(2.20\ \text{in.})(4.00\ \text{in.})(6.00\ \text{in.}) = 52.8\ \text{in.}^3$$

Now we convert in.^{3} to cm^3 by using the inches and centimeters relationship (1 in. = 2.54 cm) three times:

Solution map: $(\text{in.} \rightarrow \text{cm})^3$

$$\cancel{\text{in.}^3} \times \frac{\text{cm}}{\cancel{\text{in.}}} \times \frac{\text{cm}}{\cancel{\text{in.}}} \times \frac{\text{cm}}{\cancel{\text{in.}}} = \text{cm}^3$$

$$(52.8\ \cancel{\text{in.}^3})\left(\frac{2.54\ \text{cm}}{1\ \cancel{\text{in.}}}\right)\left(\frac{2.54\ \text{cm}}{1\ \cancel{\text{in.}}}\right)\left(\frac{2.54\ \text{cm}}{1\ \cancel{\text{in.}}}\right) = 865\ \text{cm}^3$$

ENHANCED EXAMPLE 2.22 **WileyPLUS**

Suppose four ostrich feathers weigh 1.00 lb. Assuming that each feather is equal in mass, how many milligrams does a single feather weigh?

Burke Triolo Productions/Getty Images

SOLUTION

The unit conversion in this problem is from 1.00 lb/4 feathers to milligrams per feather. Since the unit *feathers* occurs in the denominator of both the starting unit and the desired unit, the unit conversions are

Solution map: $\text{lb} \rightarrow \text{g} \rightarrow \text{mg}$

$$\left(\frac{1.00\ \cancel{\text{lb}}}{4\ \text{feathers}}\right)\left(\frac{453.6\ \cancel{\text{g}}}{1\ \cancel{\text{lb}}}\right)\left(\frac{1000\ \text{mg}}{1\ \cancel{\text{g}}}\right) = \frac{113{,}400\ \text{mg}}{\text{feather}} (1.13 \times 10^5\ \text{mg/feather})$$

PRACTICE 2.13

You are traveling in Europe and wake up one morning to find your mass is 75.0 kg. Determine the U.S. equivalent (in pounds) to see whether you need to go on a diet before you return home.

PRACTICE 2.14

A bottle of excellent Chianti holds 750 mL. What is its volume in quarts?

PRACTICE 2.15

Milk is often purchased by the half gallon. Determine the number of liters equal to this amount.

2.7 Percent

Solve problems involving percent.

LEARNING OBJECTIVE

KEY TERM

percent

The composition of many mixtures is given in percent. **Percent** is defined as parts per hundred. For example, a bag of jelly beans that is 35% mango flavored means that there are 35 mango-flavored jelly beans in every 100 jelly beans. When we don't have 100 parts of a sample, to find percent we need to convert to parts per 100 parts. We can do this using the following expression: $\frac{\text{parts}}{\text{total parts}} = \frac{x}{100}$ where x = percent. Use this relationship to solve for percent to give the following:

$$\text{percent} = \left(\frac{\text{parts}}{\text{total parts}}\right) \times 100\%$$

In chemistry, we often use mass percent where

$$\text{mass percent} = \left(\frac{\text{mass part}}{\text{mass whole}}\right) \times 100\%$$

This is a convenient way to specify the composition of a substance because it can be used with a wide variety of mass units.

@Alexey Solodov/Alamy

ENHANCED EXAMPLE 2.23 WileyPLUS

In a genetics experiment there are 28 red flowers, 41 white flowers, and 12 yellow flowers. What is the percent of red flowers produced in this experiment?

SOLUTION

We need to solve for percent, so we start by writing the formula for calculating percent:

$$\text{percent} = \left(\frac{\text{amount part}}{\text{amount whole}}\right) \times 100\%$$

We then determine the number of red flowers (the part) and the number of total flowers (the whole).

The number of red flowers is 28.

The total number of flowers is $28 + 41 + 12 = 81$.

Substituting these values into the equation we get

$$\text{percent red flowers} = \left(\frac{28 \text{ red flowers}}{81 \text{ flowers}}\right) \times 100\% = 35\% \text{ red flowers}$$

ONLINE LEARNING MODULE

Calculating Percent from Mass

➡ WileyPLUS

ENHANCED EXAMPLE 2.24 WileyPLUS

A sample of nickel oxide is composed of 7.00 g of nickel and 3.82 g of oxygen. Calculate the percent nickel and oxygen in the oxide.

SOLUTION

We are solving for mass percent, so we will use this equation:

$$\text{mass percent} = \left(\frac{\text{mass part}}{\text{mass whole}}\right) \times 100\%$$

Data given in the problem:

Mass Ni = 7.00 g   mass O = 3.82 g

The total mass is 7.00 g + 3.82 g = 10.82 g.

Substituting into the equation we get

$$\% \text{ Ni} = \left(\frac{7.00 \text{ g Ni}}{10.82 \text{ g}}\right) \times 100\% = 64.7\% \text{ Ni}$$

$$\% \text{ O} = \left(\frac{3.82 \text{ g O}}{10.82 \text{ g}}\right) \times 100\% = 35.3\% \text{ O}$$

PRACTICE 2.16

A sulfide of gold is composed of 6.31 g of sulfur and 25.83 g of gold. Calculate the percent of both elements in the compound.

PRACTICE 2.17

An alloy of bronze is manufactured by melting 50.0 g of copper with 6.82 g of tin. What is the percent copper in the bronze?

Percent is also useful as a conversion factor and may be written in dimensional analysis form as

$$\frac{\text{units part}}{\text{100 units total}}$$

For example, an alloy of steel composed of 8.18% chromium would have 8.18 grams of chromium in 100 grams of steel giving rise to the two conversion factors

$$\frac{8.18\text{ g Cr}}{100\text{ g steel}} \text{ and } \frac{100\text{ g steel}}{8.18\text{ g Cr}}$$

Each of these conversion factors would have three significant figures because the 100 g is an exact number and the 8.18 g of chromium is known to three significant figures for an exact 100-g sample of steel.

ENHANCED EXAMPLE 2.25 WileyPLUS

Peaches are 8.6% sugar by mass. If a serving of peaches is 170 grams, how many grams of sugar are in one serving?

SOLUTION

Write out the possible dimensional analysis factors for the percentage given.

$$\frac{8.6\text{ g sugar}}{100\text{ g peach}} \text{ and } \frac{100\text{ g peach}}{8.6\text{ g sugar}}$$

Now use this to convert the mass of a serving of peaches to the mass of sugar.

$$(170\text{ }\cancel{\text{g peaches}})\left(\frac{8.6\text{ g sugar}}{100\text{ }\cancel{\text{g peach}}}\right) = 15\text{ g sugar}$$

Alternatively, the problem could be solved by using the percent formula:

$$\text{mass percent} = \left(\frac{\text{mass part}}{\text{mass total}}\right) \times 100\%$$

$$8.6\%\text{ sugar} = \left(\frac{x}{170\text{ g peaches}}\right) \times 100 \rightarrow x = 15\text{ g sugar}$$

ENHANCED EXAMPLE 2.26 WileyPLUS

ONLINE LEARNING MODULE
Using Percent
➡ WileyPLUS

A pharmaceutical chemist is using powdered orange juice to make vitamin C tablets. How much powdered orange juice needs to be used to produce a mixture that contains 52.3 kg of vitamin C if powdered orange juice is 39.5% by mass vitamin C?

SOLUTION

Write out the possible dimensional analysis factors for the percentage given.

$$\frac{39.5\text{ kg vitamin C}}{100\text{ kg powdered orange juice}} \text{ and } \frac{100\text{ kg powdered orange juice}}{39.5\text{ kg vitamin C}}$$

Now use this to convert the mass of vitamin C required to the amount of powdered orange juice to add to the batch of vitamins.

$$(52.3\text{ }\cancel{\text{kg vitamin C}})\left(\frac{100\text{ kg powdered orange juice}}{39.5\text{ }\cancel{\text{kg vitamin C}}}\right) = 132\text{ kg powdered orange juice}$$

Alternatively, the problem could be solved by using the percent formula:

$$\text{mass percent} = \left(\frac{\text{mass part}}{\text{mass total}}\right) \times 100\%$$

$$39.5\% \text{ vitamin C} = \left(\frac{52.3 \text{ kg kg vitamin C}}{x}\right) \times 100\%$$

$$x = 132 \text{ kg powdered orange juice}$$

PRACTICE 2.18

Soldering guns often use a soft solder alloy composed of 35.0% lead. If you are soldering parts onto a wagon and are only allowed to have a maximum of 2.50 g of lead in the finished product, how much solder can you use to attach all of the parts?

PRACTICE 2.19

A chocolate bar contains 0.047% caffeine. How many mg of caffeine will you consume if you eat a 1.5-oz chocolate bar?

2.8 Measurement of Temperature

LEARNING OBJECTIVE Convert measurements among the Fahrenheit, Celsius, and Kelvin temperature scales.

KEY TERMS
thermal energy
temperature
heat

Thermal energy is a form of energy associated with the motion of small particles of matter. Depending on the amount of thermal energy present, a given system is said to be hot or cold. **Temperature** is a measure of the intensity of thermal energy, or how hot a system is, regardless of its size. The term **heat** refers to the flow of energy due to a temperature difference. Heat always flows from a region of higher temperature to one of lower temperature. The SI unit of temperature is the kelvin. The common laboratory instrument for measuring temperature is a thermometer (FIGURE 2.5).

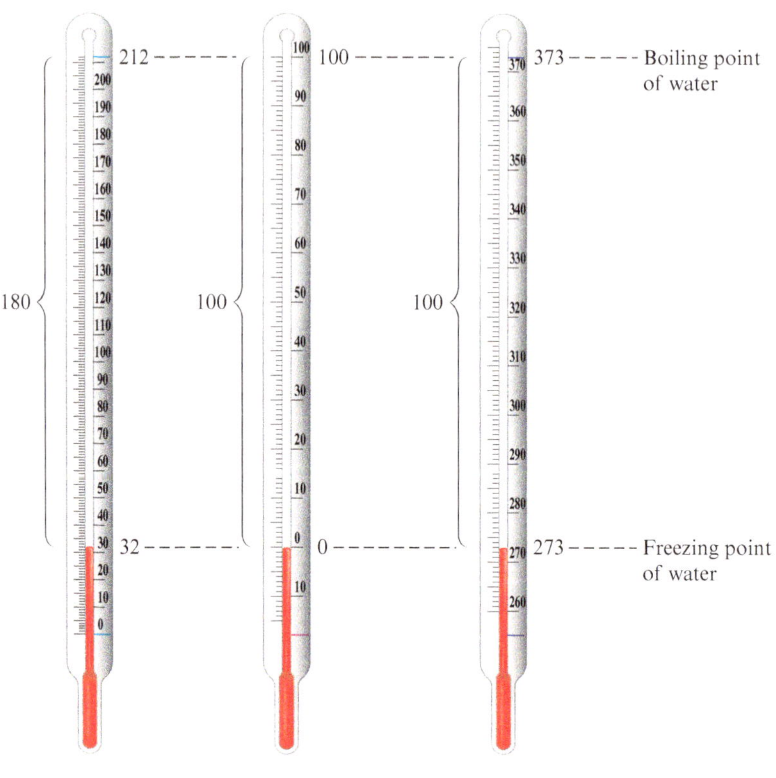

FIGURE 2.5 Comparison of Celsius, Kelvin, and Fahrenheit temperature scales.

The temperature of a system can be expressed by several different scales. Three commonly used temperature scales are Celsius (pronounced *sell-see-us*), Kelvin (also called absolute), and Fahrenheit. The unit of temperature on the Celsius and Fahrenheit scales is called a *degree*, but the size of the Celsius and the Fahrenheit degree is not the same. The symbol for the Celsius and Fahrenheit degrees is °, and it is placed as a superscript after the number and before the symbol for the scale. Thus, 10.0°C means 10.0 *degrees Celsius*. The degree sign is not used with Kelvin temperatures:

$$\text{degrees Celsius} = {}^\circ\text{C}$$
$$\text{Kelvin (absolute)} = \text{K}$$
$$\text{degrees Fahrenheit} = {}^\circ\text{F}$$

On the Celsius scale the interval between the freezing and boiling temperatures of water is divided into 100 equal parts, or degrees. The freezing point of water is assigned a temperature of 0°C and the boiling point of water a temperature of 100°C. The Kelvin temperature scale is known as the absolute temperature scale because 0 K is the lowest temperature theoretically attainable. The Kelvin zero is 273.15 kelvins below the Celsius zero. A kelvin is equal in size to a Celsius degree. The freezing point of water on the Kelvin scale is 273.15 K. The Fahrenheit scale has 180 degrees between the freezing and boiling temperatures of water. On this scale the freezing point of water is 32°F and the boiling point is 212°F:

$$0^\circ\text{C} \cong 273\ \text{K} \cong 32^\circ\text{F} \qquad 100^\circ\text{C} \cong 373\ \text{K} \cong 212^\circ\text{F}$$

CHECK YOUR UNDERSTANDING

2.18 Temperature Scales

2.19 Comparing Celsius and Fahrenheit

➥ **WileyPLUS**

The three scales are compared in Figure 2.5. Although absolute zero (0 K) is the lower limit of temperature on these scales, temperature has no upper limit. Temperatures of several million degrees are known to exist in the sun and in other stars.

By examining Figure 2.5, we can see that there are 100 Celsius degrees and 100 kelvins between the freezing and boiling points of water, but there are 180 Fahrenheit degrees between these two temperatures. Hence, the size of the Celsius degree and the kelvin are the same, but 1 Celsius degree is equal to 1.8 Fahrenheit degrees.

$$\frac{180}{100} = 1.8$$

CHEMISTRY *IN ACTION* Setting Standards

The kilogram is the standard base unit for mass in the SI system. But who decides just what a kilogram is? Before 1880, a kilogram was defined as the mass of a cubic decimeter of water. But this is difficult to reproduce very accurately since impurities and air are dissolved in water. The density of water also changes with temperature, leading to inaccuracy. So in 1885 a Pt-Ir cylinder was made with a mass of exactly 1 kilogram. This cylinder is kept in a vault at Sèvres, France, outside Paris, along with six copies. In 1889 the kilogram was determined to be the base unit of mass for the metric system and has been weighed against its copies three times in the past 100 years (1890, 1948, 1992) to make sure that it is accurate. The trouble with all this is that the cylinder itself can vary. Experts in the science of measurement want to change the definition of the kilogram to link it to a property of matter that does not change.

The kilogram is the only one of seven SI base units defined in terms of a physical object instead of a property of matter. The meter (redefined in 1983) is defined in terms of the speed of light (1 m = distance light travels in $(\frac{1}{299,792,458})$ s). The second is defined in terms of the natural vibration of the cesium atom. Unfortunately, three of the SI base units (mole, ampere, and candela) depend on the definition of the kilogram. If the mass of the kilogram is uncertain, then all of these units are uncertain.

How else could we define a kilogram? Scientists have several ideas. They propose to fix the kilogram to a set value of a physical constant (which does not change). These constants connect experiment to theory. Right now they are continuing to struggle to find a new way to determine the mass of the kilogram.

Kilogram artifact in Sèvres, France.

CHECK YOUR UNDERSTANDING

2.20 Comparing Celsius and Kelvins

WileyPLUS

From these data, mathematical formulas have been derived to convert a temperature on one scale to the corresponding temperature on another scale:

$$K = {}^\circ C + 273.15$$

$$^\circ F = (1.8 \times {}^\circ C) + 32$$

$$^\circ C = \frac{^\circ F - 32}{1.8}$$

ENHANCED EXAMPLE 2.27 WileyPLUS

The temperature at which table salt (sodium chloride) melts is 800.°C. What is this temperature on the Kelvin and Fahrenheit scales?

SOLUTION

To calculate K from °C, we use the formula

$$K = {}^\circ C + 273.15$$

$$K = 800.{}^\circ C + 273.15 = 1073\ K$$

To calculate °F from °C, we use the formula

$$^\circ F = (1.8 \times {}^\circ C) + 32$$

$$^\circ F = (1.8)(800.{}^\circ C) + 32$$

$$^\circ F = 1440 + 32 = 1472^\circ F$$

Summarizing our calculations, we see that

$$800.{}^\circ C = 1073\ K = 1472^\circ F$$

Remember that the original measurement of 800.°C was to the units place, so the converted temperature is also to the units place.

ONLINE LEARNING MODULE

Temperature Conversions

WileyPLUS

ENHANCED EXAMPLE 2.28 WileyPLUS

The temperature on December 1 in Honolulu, Hawaii, was 110.°F, a new record. Convert this temperature to °C.

SOLUTION

We use the formula

$$^\circ C = \frac{^\circ F - 32}{1.8}$$

$$^\circ C = \frac{110. - 32}{1.8} = \frac{78}{1.8} = 43^\circ C$$

ENHANCED EXAMPLE 2.29 WileyPLUS

What temperature on the Fahrenheit scale corresponds to −8.0°C? (Notice the negative sign in this problem.)

SOLUTION

$$^\circ F = (1.8 \times {}^\circ C) + 32$$

$$^\circ F = (1.8)(-8.0) + 32 = -14 + 32$$

$$^\circ F = 18^\circ F$$

PRACTICE 2.20

Helium boils at 4 K. Convert this temperature to °C and then to °F.

PRACTICE 2.21

"Normal" human body temperature is 98.6°F. Convert this to °C and K.

Temperatures used throughout this book are in degrees Celsius (°C) unless specified otherwise. The temperature after conversion should be expressed to the same precision as the original measurement.

2.9 Density

Solve problems involving density.

LEARNING OBJECTIVE

KEY TERM

density

Density (d) is the ratio of the mass of a substance to the volume occupied by that mass; it is the mass per unit of volume and is given by the equation

$$d = \frac{\text{mass}}{\text{volume}}$$

Density is a physical characteristic of a substance and may be used as an aid to its identification. When the density of a solid or a liquid is given, the mass is usually expressed in grams and the volume in milliliters or cubic centimeters.

$$d = \frac{\text{mass}}{\text{volume}} = \frac{\text{g}}{\text{mL}} \quad or \quad d = \frac{\text{g}}{\text{cm}^3}$$

Since the volume of a substance (especially liquids and gases) varies with temperature, it is important to state the temperature along with the density. For example, the volume of 1.0000 g of water at 4°C is 1.0000 mL; at 20°C, it is 1.0018 mL; and at 80°C, it is 1.0290 mL. Density therefore also varies with temperature.

The density of water at 4°C is 1.0000 g/mL, but at 80°C the density of water is 0.9718 g/mL:

$$d^{4°\text{C}} = \frac{1.0000 \text{ g}}{1.0000 \text{ mL}} = 1.0000 \text{ g/mL}$$

$$d^{80°\text{C}} = \frac{1.0000 \text{ g}}{1.0290 \text{ mL}} = 0.97182 \text{ g/mL}$$

The density of iron at 20°C is 7.86 g/mL.

$$d^{20°\text{C}} = \frac{7.86 \text{ g}}{1.00 \text{ mL}} = 7.86 \text{ g/mL}$$

The densities of a variety of materials are compared in FIGURE 2.6.

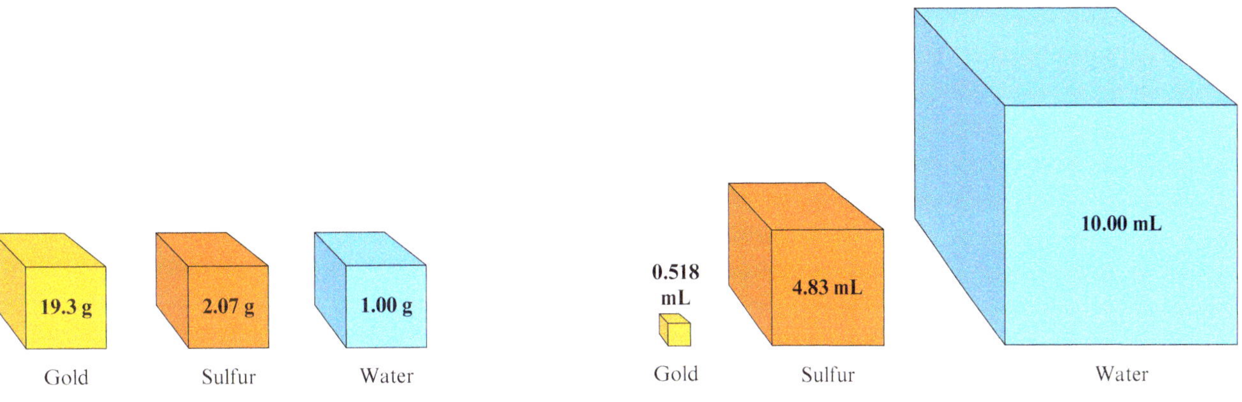

FIGURE 2.6 Comparison of mass and volume.

Densities for liquids and solids are usually represented in terms of grams per milliliter (g/mL) or grams per cubic centimeter (g/cm^3). The density of gases, however, is expressed in terms of grams per liter (g/L). Unless otherwise stated, gas densities are given for 0°C and 1 atmosphere pressure (discussed further in Chapter 13). **TABLE 2.6** lists the densities of some common materials.

When an insoluble solid object is dropped into water, it will sink or float, depending on its density. If the object is less dense than water, it will float, displacing a *mass* of water equal to the mass of the object. If the object is more dense than water, it will sink, displacing a *volume* of water equal to the volume of the object. This information can be used to determine the volume (and density) of irregularly shaped objects.

TABLE 2.6 Densities of Some Selected Materials

Liquids and solids		Gases	
Substance	Density (g/mL at 20°C)	Substance	Density (g/L at 0°C)
Wood (Douglas fir)	0.512	Hydrogen	0.090
Ethyl alcohol	0.789	Helium	0.178
Vegetable oil	0.91	Methane	0.714
Water (4°C)	1.000*	Ammonia	0.771
Sugar	1.59	Neon	0.90
Glycerin	1.26	Carbon monoxide	1.25
Karo syrup	1.37	Nitrogen	1.251
Magnesium	1.74	Air	1.293*
Sulfuric acid	1.84	Oxygen	1.429
Sulfur	2.07	Hydrogen chloride	1.63
Salt	2.16	Argon	1.78
Aluminum	2.70	Carbon dioxide	1.963
Silver	10.5	Chlorine	3.17
Lead	11.34		
Mercury	13.55		
Gold	19.3		

*For comparing densities, the density of water is the reference for solids and liquids; air is the reference for gases.

ENHANCED EXAMPLE 2.30 **WileyPLUS**

What is the density of a mineral if 427 g of the mineral occupy a volume of 35.0 mL?

SOLUTION

We need to solve for density, so we start by writing the formula for calculating density:

$$d = \frac{\text{mass}}{\text{volume}}$$

Then we substitute the data given in the problem into the equation and solve:

$$\text{mass} = 427\text{ g} \qquad \text{volume} = 35.0\text{ mL}$$

$$d = \frac{\text{mass}}{\text{volume}} = \frac{427\text{ g}}{35.0\text{ mL}} = 12.2\text{ g/mL}$$

ENHANCED EXAMPLE 2.31 WileyPLUS

The density of gold is 19.3 g/mL. What is the mass of 25.0 mL of gold?

SOLUTION

Use density as a conversion factor, converting

Solution map: mL → g

The conversion of units is

$$\text{mL} \times \frac{\text{g}}{\text{mL}} = \text{g}$$

$$(25.0\ \cancel{\text{mL}})\left(\frac{19.3\ \text{g}}{\cancel{\text{mL}}}\right) = 483\ \text{g}$$

ONLINE LEARNING MODULE

Using Density to Determine Mass

➡ WileyPLUS

ENHANCED EXAMPLE 2.32 WileyPLUS

Calculate the volume (in mL) of 100. g of ethyl alcohol.

SOLUTION

From Table 2.6 we see that the density of ethyl alcohol is 0.789 g/mL. This density also means that 1 mL of the alcohol has a mass of 0.789 g (1 mL/0.789 g).
For a conversion factor, we can use either

$$\frac{\text{g}}{\text{mL}} \quad \text{or} \quad \frac{\text{mL}}{\text{g}}$$

Solution map: g → mL,

$$(100.\ \cancel{\text{g}})\left(\frac{1\ \text{mL}}{0.789\ \cancel{\text{g}}}\right) = 127\ \text{mL of ethyl alcohol}$$

ENHANCED EXAMPLE 2.33 WileyPLUS

ONLINE LEARNING MODULE

Finding Density from Data

➡ WileyPLUS

The water level in a graduated cylinder stands at 20.0 mL before and at 26.2 mL after a 16.74-g metal bolt is submerged in the water. (a) What is the volume of the bolt? (b) What is the density of the bolt?

SOLUTION

(a) The bolt will displace a volume of water equal to the volume of the bolt. Thus, the increase in volume is the volume of the bolt:

$$\begin{array}{rl} 26.2\ \text{mL} & = \text{volume of water plus bolt} \\ -20.0\ \text{mL} & = \text{volume of water} \\ \hline 6.2\ \text{mL} & = \text{volume of bolt} \end{array}$$

(b) $d = \dfrac{\text{mass of bolt}}{\text{volume of bolt}} = \dfrac{16.74\ \text{g}}{6.2\ \text{mL}} = 2.7\ \text{g/mL}$

PRACTICE 2.22

Pure silver has a density of 10.5 g/mL. A ring sold as pure silver has a mass of 18.7 g. When it is placed in a graduated cylinder, the water level rises 2.0 mL. Determine whether the ring is actually pure silver or whether the customer should contact the Better Business Bureau.

PRACTICE 2.23

The water level in a metric measuring cup is 0.75 L before the addition of 150. g of shortening. The water level after submerging the shortening is 0.92 L. Determine the density of the shortening.

CHAPTER 2 REVIEW

2.1 Scientific Notation

KEY TERMS

measurement
scientific notation

- Quantitative observations consist of a number and a unit and are called measurements.
- Very large and very small numbers can be represented compactly by using scientific notation:
  - The number is represented as a decimal between 1 and 10 and is multiplied by 10 raised to the appropriate exponent.
  - The sign on the exponent is determined by the direction the decimal point is moved in the original number.

2.2 Measurement and Uncertainty

KEY TERM

significant figures

- All measurements reflect some amount of uncertainty, which is indicated by the number of significant figures in the measurement.
- The significant figures include all those known with certainty plus one estimated digit.

2.3 Significant Figures

KEY TERM

rounding off numbers

- Rules exist for counting significant figures in a measurement:
  - Nonzero numbers are always significant.
  - Exact numbers have an infinite number of significant figures.
  - The significance of a zero in a measurement is determined by its position within the number.
- Rules exist for rounding off the result of a calculation to the correct number of significant figures.
  - If the first number after the one you want to retain is 4 or less, that digit and all those after it are dropped.
  - If the first number after the one you want to retain is 5 or greater, that digit and all those after it are dropped and the last digit retained is increased by one.

2.4 Significant Figures in Calculations

- The results of a calculation cannot be more precise than the least precise measurement.
- Rules exist for determining the correct number of significant figures in the result of a calculation.

2.5 The Metric System

KEY TERMS

metric system or International System (SI)
meter (m)
conversion factor
solution map
mass
weight
kilogram (kg)
volume
liter (L)

- The metric system uses factors of 10 and a set of standard units for measurements.
- Length in the metric system is measured by the standard unit of the meter.
- The standard unit for mass in the metric system is the kilogram. In chemistry we often use the gram instead, as we tend to work in smaller quantities.
- Volume is the amount of space occupied by matter.
- The standard unit for volume is the cubic meter. In chemistry we usually use the volume unit of the liter or the milliliter.

2.6 Dimensional Analysis: A problem-solving method

- Dimensional analysis is a common method used to convert one unit to another.
  - Conversion factors are used to convert one unit into another:

$$\text{unit}_1 \times \text{conversion factor} = \text{unit}_2$$

  - A solution map is used to outline the steps in the unit conversion.

2.7 Percent

KEY TERM

percent

- The general formula for percent is $\text{percent} = \left(\frac{\text{amount part}}{\text{amount total}}\right) \times 100$

2.8 Measurement of Temperature

KEY TERMS

thermal energy
temperature
heat

- There are three commonly used temperature scales: Fahrenheit, Celsius, and Kelvin.
- We can convert among the temperature scales by using mathematical formulas:
  - $K = °C + 273.15$
  - $°F = (1.8 \times °C) + 32$
  - $°C = \frac{°F - 32}{1.8}$

2.9 Density

KEY TERM

density

- The density of a substance is the amount of matter (mass) in a given volume of the substance:

$$d = \frac{\text{mass}}{\text{volume}}$$

REVIEW QUESTIONS

1. When a very large number is written in scientific notation, should the exponent be a positive or a negative number?
2. To write 635.2×10^{-8} in the proper scientific notation form, should the exponent increase or decrease?
3. Explain why the last digit in a measurement is uncertain.
4. How can the number 642,000 g be written to indicate that there are four significant figures?
5. How are significant zeroes identified?
6. State the rules used in this text for rounding off numbers.
7. Is it possible for the calculated answer from a multiplication or division problem to contain more significant figures than the measurements used in the calculation?
8. Is it possible for the calculated answer from an addition or subtraction problem to contain more significant figures than the measurements used in the calculation?
9. How many nanometers are in 1 cm? (Table 2.3)
10. How many milligrams are in 1 kg?
11. Why does an astronaut weigh more on Earth than in space when his or her mass remains the same in both places?
12. What is the relationship between milliliters and cubic centimeters?
13. What is the metric equivalent of 3.5 in.?
14. Distinguish between heat and temperature.
15. Compare the number of degrees between the freezing point of water and its boiling point on the Fahrenheit, Kelvin, and Celsius temperature scales. (Figure 2.5)
16. Describe the order of the following substances (top to bottom) if these three substances were placed in a 100-mL graduated cylinder: 25 mL glycerin, 25 mL mercury, and a cube of magnesium 2.0 cm on an edge. (Table 2.6)
17. Arrange these materials in order of increasing density: salt, vegetable oil, lead, and ethyl alcohol. (Table 2.6)
18. Ice floats in vegetable oil and sinks in ethyl alcohol. The density of ice must lie between what numerical values? (Table 2.6)
19. Ice floats in water, yet ice is simply frozen water. If the density of water is 1.0 g/mL, how is this possible?
20. If you collect a container of oxygen gas, should you store it with the mouth up or down? Explain your answer. (Table 2.6)
21. Which substance has the greater volume; 25 g gold or 25 g silver? (Table 2.6)

PAIRED EXERCISES

Most of the exercises in this chapter are available for assignment via the online homework management program, WileyPLUS (www.wileyplus.com). All exercises with blue numbers have answers in Appendix 5.

1. What is the numerical meaning of each of the following?
   (a) kilogram
   (b) centimeter
   (c) microliter
   (d) millimeter
   (e) deciliter
2. What is the correct unit of measurement for each of the following?
   (a) 1000 m
   (b) 0.1 g
   (c) 0.000001 L
   (d) 0.01 m
   (e) 0.001 L
3. State the abbreviation for each of the following units:
   (a) gram
   (b) microgram
   (c) centimeter
   (d) micrometer
   (e) milliliter
   (f) deciliter
4. State the abbreviation for each of the following units:
   (a) milligram
   (b) kilogram
   (c) meter
   (d) nanometer
   (e) angstrom
   (f) microliter

5. Determine whether the zeros in each number are significant:
(a) 2050 (d) 0.075
(b) 9.00×10^2 (e) 300.
(c) 0.0530 (f) 285.00

6. Determine whether the zeros in each number are significant:
(a) 0.005 (d) 10.000
(b) 1500 (e) 6.070×10^4
(c) 250. (f) 0.2300

7. How many significant figures are in each of the following numbers?
(a) 0.025 (c) 0.0404
(b) 22.4 (d) 5.50×10^3

8. State the number of significant figures in each of the following numbers:
(a) 40.0 (c) 129,042
(b) 0.081 (d) 4.090×10^{-3}

9. Round each of the following numbers to three significant figures:
(a) 93.246 (c) 4.644
(b) 0.02857 (d) 34.250

10. Round each of the following numbers to three significant figures:
(a) 8.8726 (c) 129.509
(b) 21.25 (d) 1.995×10^6

11. Express each of the following numbers in exponential notation:
(a) 2,900,000 (c) 0.00840
(b) 0.587 (d) 0.0000055

12. Write each of the following numbers in exponential notation:
(a) 0.0456 (c) 40.30
(b) 4082.2 (d) 12,000,000

13. Solve the following problems, stating answers to the proper number of significant figures:
(a) $12.62 + 1.5 + 0.25 = ?$
(b) $(2.25 \times 10^3)(4.80 \times 10^4) = ?$
(c) $\frac{(452)(6.2)}{14.3} = ?$
(d) $(0.0394)(12.8) = ?$
(e) $\frac{0.4278}{59.6} = ?$
(f) $10.4 + 3.75(1.5 \times 10^4) = ?$

14. Evaluate each of the following expressions. State the answer to the proper number of significant figures:
(a) $15.2 - 2.75 + 15.67$
(b) $(4.68)(12.5)$
(c) $\frac{182.6}{4.6}$
(d) $1986 + 23.84 + 0.012$
(e) $\frac{29.3}{(284)(415)}$
(f) $(2.92 \times 10^{-3})(6.14 \times 10^5)$

15. Change these fractions into decimals. Express each answer to three significant figures:
(a) $\frac{5}{6}$ (b) $\frac{3}{7}$ (c) $\frac{12}{16}$ (d) $\frac{9}{18}$

16. Change each of the following decimals to fractions in lowest terms:
(a) 0.25 (b) 0.625 (c) 1.67 (d) 0.8888

17. Solve each of these equations for x:
(a) $3.42x = 6.5$ (c) $\frac{0.525}{x} = 0.25$
(b) $\frac{x}{12.3} = 7.05$

18. Solve each equation for the variable:
(a) $x = \frac{212 - 32}{1.8}$ (c) $72°F = 1.8x + 32$
(b) $8.9\frac{g}{mL} = \frac{40.90 \text{ g}}{x}$

Unit Conversions

19. Complete the following metric conversions using the correct number of significant figures:
(a) 28.0 cm to m (e) 6.8×10^4 mg to kg
(b) 1000. m to km (f) 8.54 g to kg
(c) 9.28 cm to mm (g) 25.0 mL to L
(d) 10.68 g to mg (h) 22.4 L to μL

20. Complete the following metric conversions using the correct number of significant figures:
(a) 4.5 cm to Å (e) 0.65 kg to mg
(b) 12 nm to cm (f) 5.5 kg to g
(c) 8.0 km to mm (g) 0.468 L to mL
(d) 164 mg to g (h) 9.0 μL to mL

21. Complete the following U.S./metric conversions using the correct number of significant figures:
(a) 42.2 in. to cm (d) 42.8 kg to lb
(b) 0.64 m to in. (e) 3.5 qt to mL
(c) 2.00 in.^{2} to cm^2 (f) 20.0 L to gal

22. Make the following conversions using the correct number of significant figures:
(a) 35.6 m to ft (d) 95 lb to g
(b) 16.5 km to mi (e) 20.0 gal to L
(c) 4.5 in.^{3} to mm^3 (f) 4.5×10^4 ft^3 to m^3

23. After you have worked out at the gym on a stationary bike for 45 min, the distance gauge indicates that you have traveled 15.2 mi. What was your rate in km/hr?

24. A competitive college runner ran a 5-K (5.0-km) race in 15 min, 23 s. What was her pace in miles per hour?

25. A pharmacy technician is asked to prepare an antibiotic IV solution that will contain 500. mg of cephalosporin for every 100. mL of normal saline solution. The total volume of saline solution will be 1 L. How many grams of the cephalosporin will be needed for this IV solution?

26. An extra-strength aspirin tablet contains 0.500 g of the active ingredient, acetylsalicylic acid. Aspirin strength used to be measured in grains. If 1 grain = 60 mg, how many grains of the active ingredient are in 1 tablet? (Report your answer to three significant figures.)

27. The maximum speed recorded for a giant tortoise is 0.11 m/sec. How many miles could a gaint tortoise travel in 5.0 hr?

28. How many days would it take for a crepe myrtle tree to grow 1 cm in height if it grows 3.38 feet per year? (Report your answer to three significant figures.)

29. A giant anteater is able to eat 6.7 lb of ants in a day. If an ant has a mass of 85 mg, how many ants does the giant anteater eat in a day?

30. A large egg has an average mass of 58.5 g. If a recipe calls for 0.75 lb of eggs, how many large eggs should be added to the mix?

31. A personal trainer uses calipers on a client to determine his percent body fat. After taking the necessary measurements, the personal trainer determines that the client's body contains 11.2% fat by mass (11.2 lb of fat per 100 lb of body mass). If the client weighs 225 lb, how many kg of fat does he have?

32. The weight of a diamond is measured in carats. How many pounds does a 5.75-carat diamond weigh? (1 carat = 200. mg)

33. A competitive high school swimmer takes 52 s to swim 100. yards. What is his rate in m/min?

34. In 2005, Jarno Trulli was the pole winner of the U.S. Grand Prix Race with a speed of 133 mi per hr. What was his speed in cm/s?

35. In 2006, Christian Stengl climbed to the top of Mount Everest, elevation 29,035 ft, from a starting point of 21,002 ft in a record time of 16 hr, 42 min. Determine his average rate of climb in
(a) miles per minute (b) meters per second

36. The *Alvin*, a submersible research vessel, can descend into the ocean to a depth of approximately 4500 m in just over 5 hr. Determine its average rate of submersion in
(a) feet per minute (b) kilometers per second

37. The 2013 Rolls Royce Ghost Alpine has a top speed of 155 miles per hour. Calculate the speed in centimeters per second.

38. The 2014 Tesla Roadster, an all-electric car, has a top speed of 125 miles per hour. Calculate the speed in centimeters per second.

@Car Culture/Corbis

Teddy Leung/Shutterstock

39. The world's record for the largest cup of coffee was broken on October 15, 2010, with a 2010-gal cup of coffee in Las Vegas, Nevada. If a cup of coffee contains 473 mL of coffee, how many cups of coffee would be required to fill this coffee cup?

40. Tilapia is rapidly becoming an important source of fish around the world. This is because it can be farmed easily and sustainably. In 2011, 475 million lb of tilapia were consumed by Americans. If the average tilapia has a mass of 535 g, how many tilapia were consumed in 2011?

courtesy GourmetGiftBaskets.com

Greg Vaughn / Getty Images, Inc.

41. Assuming that there are 20. drops in 1.0 mL, how many drops are in 1.0 gallon?

42. How many liters of oil are in a 42-gal barrel of oil?

43. Calculate the number of milliliters of water in a cubic foot of water.

44. Oil spreads in a thin layer on water called an "oil slick." How much area in m^2 will 200 cm^3 of oil cover if it forms a layer 0.5 nm thick?

45. A textbook is 27 cm long, 21 cm wide, and 4.4 cm thick. What is the volume in:
(a) cubic centimeters? (c) cubic inches?
(b) liters?

46. An aquarium measures 16 in. × 8 in. × 10 in. How many liters of water does it hold? How many gallons?

47. Raspberry ketone, used to give food a raspberry or blueberry flavor, is composed of 73.14% by mass carbon. How many grams of carbon are in a 250.-g sample of this flavoring?

48. Nectaryl is a compound with a peachy, apricot-like odor. It is composed of 10.98% hydrogen. How many grams of hydrogen are in a 475-g sample of nectaryl?

49. A solution is prepared by dissolving 283.4 grams of potassium dichromate in 650.0 grams of water.
(a) What is the percent potassium dichromate in the solution?
(b) If 275. grams of the solution are poured into a beaker, how many grams of potassium dichromate are introduced into the beaker?
(c) How many grams of the solution are required to obtain 15.0 grams of potassium dichromate?

50. A solution is prepared by dissolving 95.4 grams of sodium nitrate in 753 grams of water.
(a) What is the percent sodium nitrate in the solution?
(b) If 350. grams of the solution are poured into a beaker, how many grams of sodium nitrate are introduced into the beaker?
(c) How many grams of the solution are required to obtain 50.0 grams of sodium nitrate?

51. A toddler in Italy visits the family doctor. The nurse takes the child's temperature, which reads 38.8°C.
(a) Convert this temperature to °F.
(b) If 98.6°F is considered normal, does the child have a fever?

52. Driving to the grocery store, you notice the temperature is 45°C. Determine what this temperature is on the Fahrenheit scale and what season of the year it might be.

53. Make the following conversions and include an equation for each one:
(a) 162°F to °C
(b) 0.0°F to K
(c) −18°C to °F
(d) 212 K to °C

54. Make the following conversions and include an equation for each one:
(a) 32°C to °F
(b) −8.6°F to °C
(c) 273°C to K
(d) 100 K to °F

55. A recipe calls for heating the oven to 425°F. If you are vacationing in Europe and the oven only has temperature marked in °C, to what temperature do you preheat the oven? What is this temperature in K?

56. You borrowed a thermometer from the chemistry lab to use in the kitchen. If you want to heat a mixture of sugar and cream to 247°F, what temperature do you need in °C? What is this temperature in K?

57. At what temperature are the Fahrenheit and Celsius temperatures exactly equal?

58. At what temperature are Fahrenheit and Celsius temperatures the same in value but opposite in sign?

59. The average temperature on Venus is 460°C. What is this temperature in °F?

60. The average temperature at the top of Jupiter's clouds is −244°F. What is this temperature in °C?

61. What is the density of a sample of 65.0 mL of automobile oil having a mass of 59.82 g?

62. A 25.2-mL sample of kerosene was determined to have a mass of 20.41 g. What is the density of kerosene?

63. A student weighed an empty graduated cylinder and found that it had a mass of 25.23 g. When filled with 25.0 mL of an unknown liquid, the total mass was 50.92 g. What is the density of the liquid?

64. A total of 32.95 g of mossy zinc was placed into a graduated cylinder containing 50.0 mL of water. The water level rose to 54.6 mL. Determine the density of the zinc.

65. Linseed oil has a density of 0.929 g/mL. How many mL are in 15 g of the oil?

66. Glycerol has a density of 1.20 g/mL. How many mL are in 75 g of glycerol?

67. A Mexican restaurant has the reputation of serving the hottest salsa in town. In order to maintain this standard, the restaurant must use a minimum of 8.41% by mass of cayenne pepper. If the restaurant just received a 3.26-kg shipment of cayenne pepper, how many pounds of salsa can it make?

68. In a candy factory, the nutty chocolate bars contain 22.0% by mass of pecans. If 7.0 kg of pecans were used for candy last Thursday, how many pounds of nutty chocolate bars were made?

EM Arts/Shutterstock

69. U.S. quarters are composed of 91.67% copper and 8.33% nickel. The mass of a quarter is 5.670 g.
(a) Write out the dimensional analysis factors that can be derived from the two percentages given above. Be sure to include units.
(b) If the U.S. mint buys 3.0 tons of copper for making quarters, how many quarters can it make?

70. A jewelry maker sells lots of sterling silver hearts with a mass of 35.0 g. Sterling silver is composed of 92.5% silver and 7.50% copper.
(a) Write out the dimensional analysis factors that can be derived from the two percentages given above. Be sure to include units.
(b) If the jeweler buys 50.0 lb of silver to make the hearts, how many sterling silver hearts can the jeweler make?

ADDITIONAL EXERCISES

71. You calculate that you need 10.0123576 g of NaCl for an experiment. What amount should you measure out if the precision of the balance is
(a) + or − 0.01 g?
(b) + or − 0.001 g?
(c) + or − 0.0001 g?

72. Often small objects are measured by mass in order to count them.
(a) If the mass of 1 Skittle is 1.134 g, what mass of Skittles should be packaged in a bag containing 175 Skittles?
(b) The volume of exactly 6 Skittles is 5.3 mL as measured in a 10.0-mL graduated cylinder. What volume (in liters) of Skittles should be packaged in a bag containing 175 Skittles?

(c) Determine the number of Skittles expected in a 325.0-g bag of Skittles.
(d) Determine the number of Skittles expected in a beaker containing 0.550 L of Skittles.
(e) Five bags were filled with 350.0 g of Skittles and 5 more bags were filled with 0.325 L of Skittles. The Skittles in each bag were then counted, and the data are tabulated below:

Mass Skittles	Number of Skittles	Volume Skittles	Number of Skittles
350.0 g	310	0.325 L	392
350.0 g	313	0.325 L	378
350.0 g	308	0.325 L	401
350.0 g	309	0.325 L	369
350.0 g	312	0.325 L	382

Which measurement method is more accurate? More precise? Explain your reasoning.

73. Suppose you want to add 100 mL of solvent to a reaction flask. Which piece of glassware shown in Figure 2.4 would be the **best** choice for accomplishing this task and why?

74. A reaction requires 21.5 g of $CHCl_3$. No balance is available, so it will have to be measured by volume. How many mL of $CHCl_3$ need to be taken? (Density of $CHCl_3$ is 1.484 g/mL.)

75. A 25.27-g sample of pure sodium was prepared for an experiment. How many mL of sodium is this? (Density of sodium is 0.97 g/mL.)

76. In the United States, coffee consumption averages 4.00×10^8 cups per day. If each cup of coffee contains 160 mg caffeine, how many pounds of caffeine are being consumed each day?

77. In a role-playing video game (RPG) your character is a Human Paladin that can carry 115 lb of gear. Your character is carrying 92 lb of gear and a vial of strength potion (which allows you to carry an additional 50.0 lb of gear). If you find a cave filled with mass potions (used for resisting strong winds), after using the strength potion, how many vials can you collect if the vials each contain 50.0 mL of mass potion with a density of 193 g/mL? The vials have negligible mass.

78. A cape is designed for Lady Gaga's concert with 4560 sequins. If a sequin has a volume of 0.0241 cm^3 and the sequins have a density of 41.6 g/cm^3, what is the mass in lb and kg of sequins on Lady Gaga's cape?

79. Will a hollow cube with sides of length 0.50 m hold 8.5 L of solution? Depending on your answer, how much additional solution would be required to fill the container or how many times would the container need to be filled to measure the 8.5 L?

80. The accepted toxic dose of mercury is 300 μg/day. Dental offices sometimes contain as much as 180 μg of mercury per cubic meter of air. If a nurse working in the office ingests 2×10^4 L of air per day, is he or she at risk for mercury poisoning?

81. The city of San Diego used an average of 1380 acre-feet of water per day during January of this year. Given that 1 acre-foot = 1.233×10^6 L,
(a) Calculate the kL of water used every day.
(b) Calculate the number of gallons of water used by San Diego last January.

82. Hydrogen becomes a liquid at 20.27 K. What is this temperature in
(a) °C? (b) °F?

83. Scientists led by Rob Eagle at CalTech have examined 150-million-year-old fossilized teeth of sauropods, huge four-legged dinosaurs, to determine their average body temperature. They analyzed the type and number of carbon-oxygen bonds, leading them to conclude that the sauropods' internal temperature was between 36 and 38°C. Determine this temperatue range in °F and compare the body temperature to other modern animals. Which of the following animals do these dinosaurs most closely resemble?

Animal	Temperature
dogs	100.5–102.5°F
tropical fish	78°F
bottlenose dolphin	97–99°F
tortoises	78–82°F
humans	99°F
birds	105°F
cows	102°F

84. According to the National Heart, Lung, and Blood Institute, LDL-cholesterol levels of less than 130 mg of LDL-cholesterol per deciliter of blood are desirable for heart health in humans. On the average, a human has 4.7 L of whole blood. What is the maximum number of grams of LDL-cholesterol that a human should have?

85. You have been sent to buy gold from prospectors in the West. You have no balance but you do have a flask full of mercury. You have been told that many prospectors will bring you fool's gold or iron pyrite in the hopes that you cannot tell the difference. Given the following densities for gold, mercury, and iron pyrite, how could you determine whether the samples brought to you are really gold?

gold	18.3 g/mL
mercury	13.6 g/mL
iron pyrite	5.00 g/mL

86. Paleontologists will sometimes use the unit elephant (the mass of an African bull elephant) when describing the mass of large dinosaurs. One elephant (El) is defined as 6.00 tons (1 El = 6.00 tons). If a dinosaur is estimated to have a mass of 5.3 elephants, what is its mass in pounds? (1 ton = 2000 lb)

87. The height of a horse is measured in hands (1 hand = exactly 4 in.). How many meters is a horse that measures 14.2 hands?

88. Camels have been reported to drink as much as 22.5 gal of water in 12 hr. How many liters can they drink in 30. days?

89. You are given three cubes, A, B, and C; one is magnesium, one is aluminum, and the third is silver. All three cubes have the same mass, but cube A has a volume of 25.9 mL, cube B has a volume of 16.7 mL, and cube C has a volume of 4.29 mL. Identify cubes A, B, and C.

90. When a chunk of wood burns, much more than just smoke is produced. In addition, nanotubes made of pure carbon are formed that have a structure resembling a roll of chicken wire. These nanotubes are stronger than steel, resistant to fire, and make good heat conductors. If a typical carbon nanotube has a diameter of 1.3 nm, how many nanotubes would need to be laid side by side to construct a bridge 40.0 feet wide?

91. The recommended daily allowance of magnesium for a college student is 380 mg. If ¼ cup of peanuts contains 67 mg of magnesium, how many cups of peanuts should a college student eat to get his or her daily intake of magnesium?

92. In the United States, land is measured in acres. There are 43,560 ft^2 in each acre. How many km^2 are in 125 acres?

93. A cube of aluminum has a mass of 500. g. What will be the mass of a cube of gold of the same dimensions?

94. A 25.0-mL sample of water at 90°C has a mass of 24.12 g. Calculate the density of water at this temperature.

95. The mass of an empty container is 88.25 g. The mass of the container when filled with a liquid (d = 1.25 g/mL) is 150.50 g. What is the volume of the container?

96. Which liquid will occupy the greater volume, 50 g of water or 50 g of ethyl alcohol? Explain.

97. The Sacagawea gold-colored dollar coin has a mass of 8.1 g and contains 3.5% manganese. What is its mass in ounces (1 lb = 16 oz), and how many ounces of Mn are in this coin?

98. The density of sulfuric acid is 1.84 g/mL. What volume of this acid will weigh 100. g?

99. Dark-roasted coffee contains increased amonts of N-methylpyridinium, or NMP, a ringed compound that is not present in green unroasted coffee beans. This compound seems to decrease the stomach acid production normally associated with drinking coffee. In coffee made from dark-roasted coffee, the concentration of NMP is 31.4 mg/L. How many milligrams of NMP would you consume if you drank 2.00 large cups of dark-roasted coffee? (One coffee cup contains 10.0 fluid ounces of liquid.)

100. The density of palladium at 20°C is 12.0 g/mL, and at 1550°C the density is 11.0 g/mL. What is the change in volume (in mL) of 1.00 kg Pd in going from 20°C to 1550°C?

101. As a solid substance is heated, its volume increases, but its mass remains the same. Sketch a graph of density versus temperature showing the trend you expect. Briefly explain.

102. The first Apple computer had 5.0 Mbytes of storage space on its hard drive, and the cost of this computer was $9995. An Apple iPad Air 2 has 128 Gbytes of storage for a cost of $699. Calculate the cost per byte for each of these two Apple products. Which is a better buy?

Volker Steger/Science Source

(a) Apple II computer

Justin Sullivan/Getty Images

(b) iPad Air 2

103. A gold bullion dealer advertised a bar of pure gold for sale. The gold bar had a mass of 3300 g and measured 2.00 cm × 15.0 cm × 6.00 cm. Was the bar pure gold? Show evidence for your answer.

104. A 35.0-mL sample of ethyl alcohol (density 0.789 g/mL) is added to a graduated cylinder that has a mass of 49.28 g. What will be the mass of the cylinder plus the alcohol?

105. Several years ago, pharmacists used the Apothecary System of Measurement. In this system, 1 scruple is equal to 20 grains (gr). There are 480 gr in 1 oz, and in this system, there are 373 g in 12 oz. How many scruples would be in 695 g?

CHALLENGE EXERCISES

106. You have just purchased a 500-mL bottle of a decongestant medication. The doctor prescribed 2 teaspoons 4 times a day for 10 days. Have you purchased enough medication? (1 teaspoon (tsp) = 5 mL)

107. A sample of brine collected from the ocean has a density of 1.28 g/mL and contains 7.55% sodium chloride.
(a) What is the volume in L of an 8.14-kg sample of this brine?
(b) How many grams of brine are required to obtain 150.0 grams of sodium chloride?

108. Your boss found a piece of metal in the lab and wants you to determine what the metal is. She is pretty sure that the metal is either lead, aluminum, or silver. The lab bench has a balance and a 100-mL graduated cylinder with 50 mL of water in it. You decide to weigh the metal and find that it has a mass of 20.25 g. After dropping the piece of metal into the graduated cylinder containing water, you observe that the volume increased to 57.5 mL. Identify the metal.

109. Forgetful Freddie placed 25.0 mL of a liquid in a graduated cylinder with a mass of 89.450 g when empty. When Freddie placed a metal slug with a mass of 15.454 g into the cylinder, the volume rose to 30.7 mL. Freddie was asked to calculate the density of the liquid and of the metal slug from his data, but he forgot to obtain the mass of the liquid. He was told that if he found the mass of the cylinder containing the liquid and the slug, he would have enough data for the calculations. He did so and found its mass to be 125.934 g. Calculate the density of the liquid and of the metal slug.

110. Jodi, a sculptor, has prepared a mold for casting a bronze figure. The figure has a volume of 225 mL. If bronze has a density of 7.8 g/mL and Jodi likes to use 90.0% of the bronze melted (she sometimes spills), how many ounces of bronze does Jodi need to melt in the preparation of the bronze figure?

111. Neutrinos are subatomic particles with a very low mass. Recent work at CERN, Europe's particle-physics lab near Geneva, Switzerland, suggests that neutrinos may have the ability to travel faster than the speed of light. If the speed of light is 1.86×10^8 mi/hr, how many nanoseconds should it take for light to travel from CERN to the Gran Sasso National Lab in Italy, a 730.0 km journey? If a neutrino can travel the same distance 60 nsec faster, how many significant figures would you need to detect the difference in speed?

112. You decide to go sailing in the tropics with some friends. Once there, you listen to the marine forecast, which predicts in-shore wave heights of 1.5 m, offshore wave heights of 4 m, and temperature of 27°C. Your friend the captain is unfamiliar with the metric system, and he needs to know whether it is safe for your small boat and if it will be warm. He asks you to convert the measurements to feet and degrees Fahrenheit, respectively.

ANSWERS TO PRACTICE EXERCISES

2.1 (a) $1200 = 1.200 \times 10^3$ (left means positive exponent)
(b) $6{,}600{,}000 = 6.6 \times 10^6$ (left means positive exponent)
(c) $0.0468 = 4.68 \times 10^{-2}$ (right means negative exponent)
(d) $0.00003 = 3 \times 10^{-5}$ (right means negative exponent)

2.2 (a) the estimated digits are 4, 1, 0; (b) 3

2.3 (a) 2; (b) 4; (c) 4; (d) 1; (e) 3; (f) 4; (g) 1; (h) 3

2.4 (a) 42.25 (Rule 2); (b) 88.0 (Rule 1); (c) 0.0897 (Rule 2); (d) 0.090 (Rule 2); (e) 225 (Rule 1); (f) 14.2 (Rule 2)

2.5 (a) 3350 in.^{2} = 3.4×10^3 in.^{2}; (b) 50.7 mi/hr; (c) 0.79; (d) 1.3; (e) 20.5; (f) 3.71

2.6 (a) 2; (b) 2; (c) 1; (d) 2; (e) 4; (f) 2; (g) 2; (h) 2

2.7 (a) 0.567 m; (b) 680; (c) 0.125

2.8 (a) 0.560 g; (b) 0.525 kg; (c) 1.75×10^5 mg

2.9 (a) 0.025 L; (b) 1550 mL; (c) 45 dL

2.10 0.14 lb

2.11 (a) 1.69×10^4 m; (b) 3.5×10^{-2} m^2

2.12 30 m^3 or 3×10^1 m^3

2.13 165 lb

2.14 0.793 qt

2.15 1.89 L (the number of significant figures is arbitrary)

2.16 19.6% S, 80.37% Au

2.17 88.0% Cu

2.18 7.14 g solder

2.19 20. mg caffeine

2.20 −269°C, −452°F

2.21 37.0°C, 310. K

2.22 The density is 9.35 g/mL; therefore the ring is not pure silver. The density of silver is 10.5 g/mL.

2.23 0.88 g/mL

3

Elements and Compounds

CHAPTER OUTLINE

In Chapter 1 we learned that matter can be divided into the broad categories of pure substances and mixtures. We further learned that pure substances are either elements or compounds. The chemical elements are very important to us in our daily lives. In tiny amounts they play a large role in our health and metabolism. Metallic elements are used for the skin of airplanes, buildings, and sculpture. The reclining Buddha, above, in the Grand Palace in Bangkok, Thailand, is covered in the element gold. In this chapter we explore the nature of the chemical elements and begin to learn how chemists classify them.

Chemists use a shorthand language to represent the chemical elements. Learning chemistry involves learning the symbols chemists use just as learning mathematics requires the use of symbols. In this chapter we begin learning the symbols for the elements and how to read formulas for compounds.

3.1 Elements

Define an element and write the chemical symbol for an element when given its name.

LEARNING OBJECTIVE

KEY TERMS

element
atom
symbols

All words in English are formed from an alphabet consisting of only 26 letters. All known substances on Earth—and most probably in the universe, too—are formed from a sort of "chemical alphabet" consisting of over 100 known elements. An **element** is a fundamental or elementary substance that cannot be broken down by chemical means to simpler substances. Elements are the building blocks of all substances. The elements are numbered in order of increasing complexity beginning with hydrogen, number 1. Of the first 92 elements, 88 are known to occur in nature. The other four—technetium (43), promethium (61), astatine (85), and francium (87)—either do not occur in nature or have only transitory existences during radioactive decay. With the exception of number 94, plutonium, elements above number 92 are not known to occur naturally but have been synthesized, usually in very small quantities, in laboratories. The discovery of trace amounts of element 94 (plutonium) in nature has been reported. No elements other than those on Earth have been detected on other bodies in the universe.

See the periodic table on the inside front cover.

Most substances can be decomposed into two or more simpler substances. Water can be decomposed into hydrogen and oxygen. Sugar can be decomposed into carbon, hydrogen, and oxygen. Table salt is easily decomposed into sodium and chlorine. An element, however, cannot be decomposed into simpler substances by ordinary chemical changes.

If we could take a small piece of an element, say copper, and divide it and subdivide it into smaller and smaller particles, we would finally come to a single unit of copper that we could no longer divide and still have copper (FIGURE 3.1). This smallest particle of an element that can exist is called an **atom**, which is also the smallest unit of an element that can enter into a chemical reaction. Atoms are made up of still smaller subatomic particles. However, these subatomic particles (described in Chapter 5) do not have the properties of elements.

@John Brueske/Shutterstock

FIGURE 3.1 The surface of a penny is made up of tiny identical copper atoms packed tightly together.

Natural States of the Elements

Most substances around us are mixtures or compounds. Elements tend to be reactive, and they combine with other elements to form compounds. It is rare to find elements in nature in pure form. There are some exceptions, however. Gold, for example, can be found as nuggets. Gold is one of the few elements found in nature in an uncombined state. Silver and platinum can also be found in nature in pure form. In fact, these metals are sometimes called the *noble metals* since they have a low reactivity. The noble gases are also not reactive and can be found in nature in uncombined form. Helium gas, for example, consists of tiny helium atoms moving independently.

Air can also be divided into its component gases. It is mainly composed of nitrogen and oxygen gases. But when we "look inside" these gases, we find tiny molecules (N_2 and O_2) instead of independent atoms like we see in the noble gases.

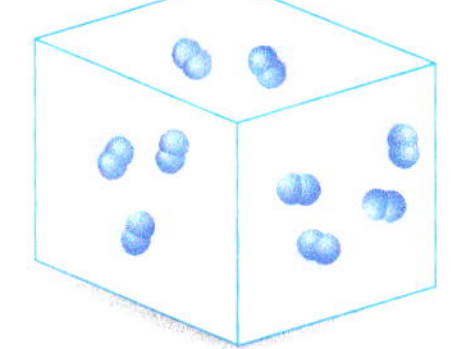

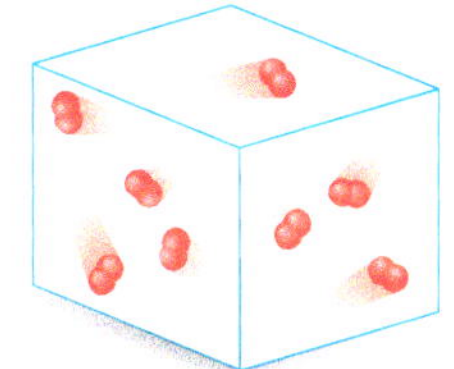

Nitrogen and oxygen gases are composed of molecules (N_2,) and (O_2,).

Distribution of Elements

At the present time, 118 elements are known and only 88 of these occur naturally. At normal room temperature only two of the elements, bromine and mercury, are liquids (FIGURE 3.2A). Eleven elements—hydrogen, nitrogen, oxygen, fluorine, chlorine, helium, neon, argon, krypton, xenon, and radon—are gases (FIGURE 3.2B). The elements are distributed unequally in nature.

Ten elements make up about 99% of the mass of the Earth's crust, seawater, and atmosphere. The distribution of the elements is listed in order of their abundance in TABLE 3.1.

Oxygen, the most abundant of these elements, accounts for about 20% of the atmosphere and is found in virtually all rocks, sand, and soil. In these places, oxygen is not present as O_2 molecules but as part of compounds usually containing silicon and aluminum atoms. The mass percents given in Table 3.1 include the Earth's crust to a depth of

iStockphoto

Helium causes these balloons to float in air.

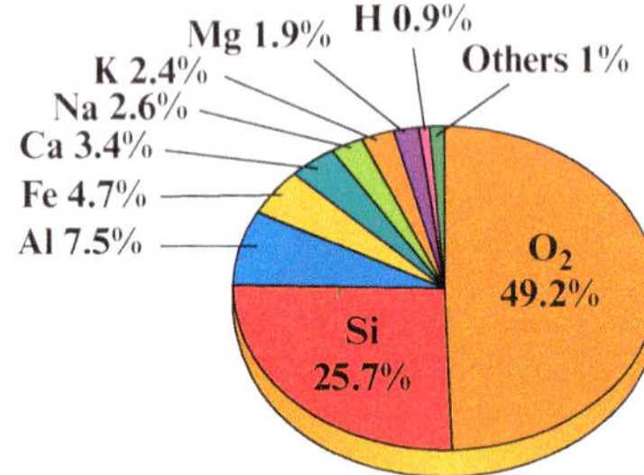

TABLE 3.1 Mass Percent of the Most Abundant Elements in the Earth's Crust, Oceans, and Atmosphere

Element	Mass percent	Element	Mass percent
Oxygen	49.2	Titanium	0.6
Silicon	25.7	Chlorine	0.19
Aluminum	7.5	Phosphorus	0.11
Iron	4.7	Manganese	0.09
Calcium	3.4	Carbon	0.08
Sodium	2.6	Sulfur	0.06
Potassium	2.4	Barium	0.04
Magnesium	1.9	Nitrogen	0.03
Hydrogen	0.9	Fluorine	0.03
		All others	0.49

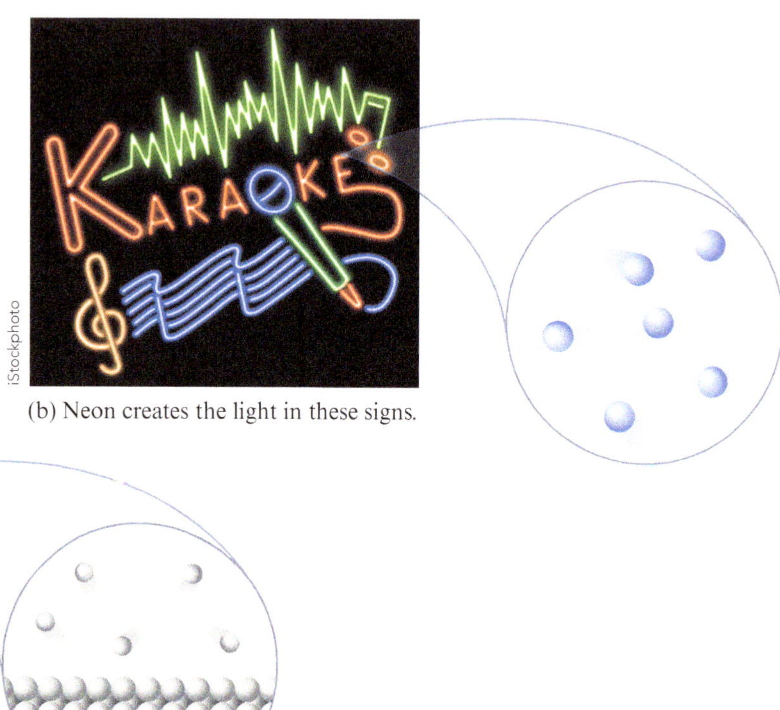

iStockphoto

(b) Neon creates the light in these signs.

Mark Evans/iStockphoto

(a) Mercury is sometimes used in thermometers.

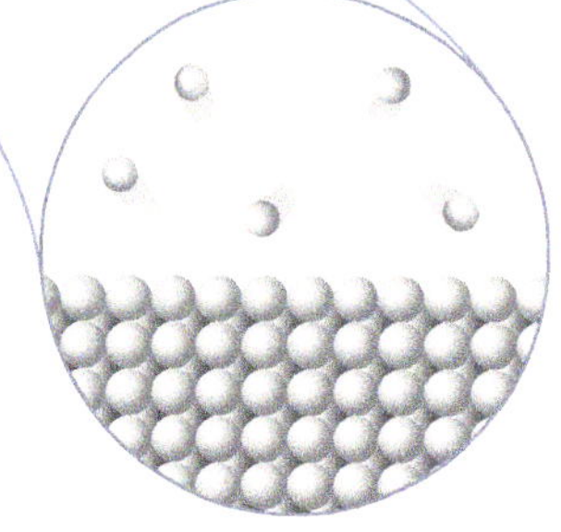

FIGURE 3.2 Examples of a liquid and a gaseous element.

10 miles, the oceans, fresh water, and the atmosphere. It does not include the mantle and core of the Earth, which consist primarily of nickel and iron.

The list of elements found in living matter is very different from those of the Earth's crust, oceans, and atmosphere. The major elements in living organisms are shown in TABLE 3.2.

The major biologically important molecules are formed primarily from oxygen, carbon, hydrogen, and nitrogen. Some elements found in the body are crucial for life, although they are present in very tiny amounts (trace elements). Some of these trace elements include chromium, copper, fluorine, iodine, and selenium.

TABLE 3.2 Mass Percent of the Elements in the Human Body

Element	Mass percent
Oxygen	65
Carbon	18
Hydrogen	10
Nitrogen	3
Calcium	1.4
Phosphorus	1
Magnesium	0.5
Potassium	0.34
Sulfur	0.26
Sodium	0.14

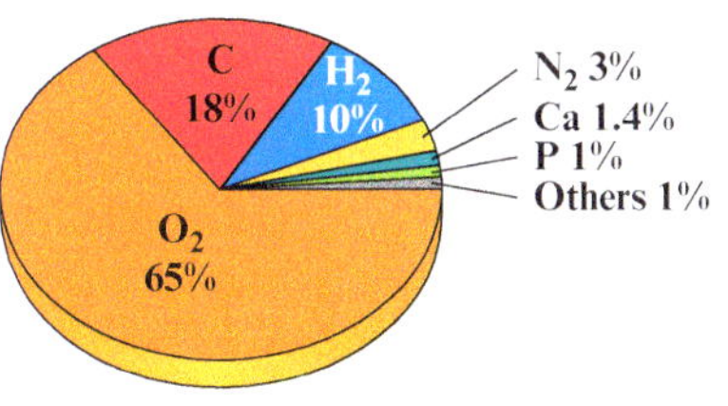

Names of the Elements

The names of the elements come to us from various sources. Many are derived from early Greek, Latin, or German words that describe some property of the element. For example, iodine is taken from the Greek word *iodes*, meaning "violetlike," and iodine is certainly violet in the vapor state. The name of the metal bismuth originates from the German words *weisse masse*, which means "white mass." Miners called it *wismat*; it was later changed to *bismat*, and finally to bismuth. Some elements are named for the location of their discovery—for example, germanium, discovered in 1886 by a German chemist. Others are named in commemoration of famous scientists, such as einsteinium and curium, named for Albert Einstein and Marie Curie, respectively.

Symbols of the Elements

We all recognize Mr., N.Y., and Ave. as abbreviations for mister, New York, and avenue, respectively. In a like manner, each element also has an abbreviation; these are called **symbols** of the elements. Fourteen elements have a single letter as their symbol, and the rest have two letters. A symbol stands for the element itself, for one atom of the element, and (as we shall see later) for a particular quantity of the element.

RULES FOR SYMBOLS OF ELEMENTS

1. Symbols have either one or two letters.
2. If one letter is used, it is capitalized.
3. If two letters are used, only the first is capitalized.

Examples: Iodine I Barium Ba

The symbols and names of all the elements are given in the table on the inside front cover of this book. TABLE 3.3 lists the more commonly used elements and their symbols.

TABLE 3.3 Symbols of the Most Common Elements

Element	Symbol	Element	Symbol	Element	Symbol
Aluminum	Al	Gold	Au	Platinum	Pt
Antimony	Sb	Helium	He	Plutonium	Pu
Argon	Ar	Hydrogen	H	Potassium	K
Arsenic	As	Iodine	I	Radium	Ra
Barium	Ba	Iron	Fe	Silicon	Si
Bismuth	Bi	Lead	Pb	Silver	Ag
Boron	B	Lithium	Li	Sodium	Na
Bromine	Br	Magnesium	Mg	Strontium	Sr
Cadmium	Cd	Manganese	Mn	Sulfur	S
Calcium	Ca	Mercury	Hg	Tin	Sn
Carbon	C	Neon	Ne	Titanium	Ti
Chlorine	Cl	Nickel	Ni	Tungsten	W
Chromium	Cr	Nitrogen	N	Uranium	U
Cobalt	Co	Oxygen	O	Xenon	Xe
Copper	Cu	Palladium	Pd	Zinc	Zn
Fluorine	F	Phosphorus	P		

Aluminum is used for recyclable cans.

Chunks of sulfur are transported in baskets from a mine in East Java.

TABLE 3.4 Symbols of the Elements Derived from Early Names*

Present name	Symbol	Former name
Antimony	Sb	Stibium
Copper	Cu	Cuprum
Gold	Au	Aurum
Iron	Fe	Ferrum
Lead	Pb	Plumbum
Mercury	Hg	Hydrargyrum
Potassium	K	Kalium
Silver	Ag	Argentum
Sodium	Na	Natrium
Tin	Sn	Stannum
Tungsten	W	Wolfram

*These symbols are in use today even though they do not correspond to the current name of the element.

Examine Table 3.3 carefully and you will note that most of the symbols start with the same letter as the name of the element that is represented. A number of symbols, however, appear to have no connection with the names of the elements they represent (TABLE 3.4). These symbols have been carried over from earlier names (usually in Latin) of the elements and are so firmly implanted in the literature that their use is continued today.

Special care must be taken in writing symbols. Capitalize only the first letter, and use a lowercase second letter if needed. This is important. For example, consider Co, the symbol for the element cobalt. If you write CO (capital C and capital O), you will have written the two elements carbon and oxygen (the *formula* for carbon monoxide), *not* the single element cobalt. Also, make sure that you write the letters distinctly; otherwise, Co (for cobalt) may be misread as Ca (for calcium).

CHECK YOUR UNDERSTANDING

3.1 Writing Symbols for the Elements

➡ WileyPLUS

Knowledge of symbols is essential for writing chemical formulas and equations and will be needed in the remainder of this book and in any future chemistry courses you

CHEMISTRY *IN ACTION* Naming Elements

Have you ever wondered where the names of the elements came from? Some of the elements were named for famous scientists or where they were first discovered and others for their characteristics. Some of the elemental symbols refer to old and rarely used names for the elements.

Elements named for places

- Americium (Am)—first created in America.
- Berkelium (Bk)—first created in Berkeley, Califorina.
- Francium (Fr)—discovered at Curie Institute in France.
- Scandium (Sc)—discovered and mined in Scandinavia.

Elements named for planets and the sun

- Helium (He)—named for the sun (helios) because this was the first place helium was detected.
- Uranium (U)—named for Uranus, which was discovered just before uranium.
- Plutonium (Pu)—named for the former planet Pluto.

Elements named for mythological characters

- Thorium (Th)—named after Thor, the Scandinavian god of war. Thorium is used as a fuel for nuclear weapons.
- Titanium (Ti)—named after Titans, supermen of Greek mythology. Titanium is a super-element because it is very resistant to acid.
- Vanadium (V)—named after the Scandinavian goddess of beauty, Vandis, because vanadium compounds form such beautiful colors.
- Tantalum (Ta) and Niobium (Nb)—named after the mythological Greek king, Tantalus, and his daughter, Princess Niobe. Like the king and his daughter, tantalum and niobium are often found together.

Elements named for their properties

- Chlorine (Cl)—from Greek *chloros* meaning "green." Chlorine is a greenish yellow gas.
- Iodine (I)—from Greek *iodos* meaning "violet."
- Argon (Ar)—from Greek *argos* meaning "lazy." Argon is an extremely unreactive gas, so it was thought too lazy to react.
- Cobalt (Co)—from Greek *kobold* meaning "goblin" or "evil spirit." Cobalt miners often died suddenly and unexpectedly, which was probably due to ingesting arsenic found with the cobalt ore in the German mines.

may take. One way to learn the symbols is to practice a few minutes a day by making flash cards of names and symbols and then practicing daily. Initially, it is a good plan to learn the symbols of the most common elements shown in Table 3.3.

ENHANCED EXAMPLE 3.1 **WileyPLUS**

Write the names and symbols for the elements that have only one letter as their symbol. (Use the periodic table on the inside front cover of your text.)

SOLUTION

Boron, B; Carbon, C; Fluorine, F; Hydrogen, H; Iodine, I; Nitrogen, N; Oxygen, O; Phosphorus, P; Potassium, K; Sulfur, S; Uranium, U; Vanadium, V; Tungsten, W; Yttrium, Y.

PRACTICE 3.1

Write the name, symbol, and vertical column location for the elements whose symbols begin with C. [Use the periodic table on the inside front cover of your text. (1A, 2A, etc.)]

ONLINE LEARNING MODULE

Identifying Elements

➡ **WileyPLUS**

3.2 Introduction to the Periodic Table

LEARNING OBJECTIVE

Explain the arrangement of the elements on the periodic table and classify elements as metal, nonmetal, or metalloid.

KEY TERMS

groups
noble gases
alkali metals
alkaline earth metals
halogens
representative elements
transition elements
metals
nonmetals
metalloids
diatomic molecules

Almost all chemistry classrooms have a chart called the *periodic table* hanging on the wall. It shows all the chemical elements and contains a great deal of useful information about them. As we continue our study of chemistry, we will learn much more about the periodic table. For now let's begin with the basics.

A simple version of the periodic table is shown in **TABLE 3.5**. Notice that in each box there is the symbol for the element and, above it, a number called the *atomic number*. For example, nitrogen is [7 N] and gold is [79 Au].

TABLE 3.5 The Periodic Table

Legend: Metals; Metalloids; Nonmetals

1 1A	2 2A	3	4	5	6	7	8	9	10	11	12	13 3A	14 4A	15 5A	16 6A	17 7A	18 8A
1 H																	2 He
3 Li	4 Be											5 B	6 C	7 N	8 O	9 F	10 Ne
11 Na	12 Mg											13 Al	14 Si	15 P	16 S	17 Cl	18 Ar
19 K	20 Ca	21 Sc	22 Ti	23 V	24 Cr	25 Mn	26 Fe	27 Co	28 Ni	29 Cu	30 Zn	31 Ga	32 Ge	33 As	34 Se	35 Br	36 Kr
37 Rb	38 Sr	39 Y	40 Zr	41 Nb	42 Mo	43 Tc	44 Ru	45 Rh	46 Pd	47 Ag	48 Cd	49 In	50 Sn	51 Sb	52 Te	53 I	54 Xe
55 Cs	56 Ba	57 La*	72 Hf	73 Ta	74 W	75 Re	76 Os	77 Ir	78 Pt	79 Au	80 Hg	81 Tl	82 Pb	83 Bi	84 Po	85 At	86 Rn
87 Fr	88 Ra	89 Ac†	104 Rf	105 Db	106 Sg	107 Bh	108 Hs	109 Mt	110 Ds	111 Rg	112 Cn	113 Uut	114 Fl	115 Uup	116 Lv	117 Uus	118 Uuo

*	58 Ce	59 Pr	60 Nd	61 Pm	62 Sm	63 Eu	64 Gd	65 Tb	66 Dy	67 Ho	68 Er	69 Tm	70 Yb	71 Lu
†	90 Th	91 Pa	92 U	93 Np	94 Pu	95 Am	96 Cm	97 Bk	98 Cf	99 Es	100 Fm	101 Md	102 No	103 Lr

1A | 2A | 3 | 4 | 5 | 6 | 7 | 8 | 9 | 10 | 11 | 12 | 13 3A | 14 4A | 15 5A | 16 6A | 17 7A | 18 8A

H

Alkali metals · Alkaline Earth Metals · Transition metals · Halogens · Noble gases

FIGURE 3.3 Some groups of elements have special names.

Noble Gases

2 **He**
10 **Ne**
18 **Ar**
36 **Kr**
54 **Xe**
86 **Rn**

The elements are placed in the table in order of increasing atomic number in a particular arrangement designed by Dimitri Mendeleev in 1869. His arrangement organizes the elements with similar chemical properties in columns called families or **groups**. An example of this is the column in the margin.

These elements are all gases and nonreactive. The group is called the **noble gases**. Other groups with special names are the **alkali metals** (under 1A on the table), **alkaline earth metals** (Group 2A), and **halogens** (Group 7A).

The tall columns of the periodic table (1A–7A and the noble gases) are known as the **representative elements**. Those elements in the center section of the periodic table are called **transition elements**. A more recent method for labeling the groups on the periodic table uses the numbers 1-18. We will show both systems on the periodic tables in this book. FIGURE 3.3 shows these groups on a periodic table.

CHECK YOUR UNDERSTANDING

3.2 Periodic Table Family Identification - 1A (1)

3.3 Periodic Table Family Identification - 2A (2)

3.4 Periodic Table Family Identification - 7A (17)

3.5 Periodic Table Family Identification - 8A (18)

3.6 Periodic Table Family Identification - 3-12

➡ WileyPLUS

Courtesy IBM Research Division

These tiny chips containing silicon, a metalloid, are being glued together to form a microprocessor many times more powerful than current chips.

Metals, Nonmetals, and Metalloids

The elements can be classified as metals, nonmetals, and metalloids. Most of the elements are metals. We are familiar with them because of their widespread use in tools, construction materials, automobiles, and so on. But nonmetals are equally useful in our everyday life as major components of clothing, food, fuel, glass, plastics, and wood. Metalloids are often used in the electronics industry.

The **metals** are solids at room temperature (mercury is an exception). They have high luster, are good conductors of heat and electricity, are *malleable* (can be rolled or hammered into sheets), and are *ductile* (can be drawn into wires). Most metals have a high melting point and a high density. Familiar metals are aluminum, chromium, copper, gold, iron, lead, magnesium, mercury, nickel, platinum, silver, tin, and zinc. Less familiar but still important metals are calcium, cobalt, potassium, sodium, uranium, and titanium.

Metals have little tendency to combine with each other to form compounds. But many metals readily combine with nonmetals such as chlorine, oxygen, and sulfur to form compounds such as metallic chlorides, oxides, and sulfides. In nature, minerals are composed of the more reactive metals combined with other elements. A few of the less reactive metals such as copper, gold, and silver are sometimes found in a native, or free, state. Metals are often mixed with one another to form homogeneous mixtures of solids called alloys. Some examples are brass, bronze, steel, and coinage metals.

Nonmetals, unlike metals, are not lustrous, have relatively low melting points and densities, and are generally poor conductors of heat and electricity. Carbon, phosphorus, sulfur, selenium, and iodine are solids; bromine is a liquid; and the rest of the nonmetals are gases. Common nonmetals found uncombined in nature are carbon (graphite and diamond), nitrogen, oxygen, sulfur, and the noble gases (helium, neon, argon, krypton, xenon, and radon).

Nonmetals combine with one another to form molecular compounds such as carbon dioxide (CO_2), methane (CH_4), butane (C_4H_{10}), and sulfur dioxide (SO_2). Fluorine, the most reactive nonmetal, combines readily with almost all other elements.

Several elements (boron, silicon, germanium, arsenic, antimony, tellurium, and polonium) are classified as **metalloids** and have properties that are intermediate between those of metals and those of nonmetals. The intermediate position of these elements is shown in Table 3.5. Certain metalloids, such as boron, silicon, and germanium, are the raw materials for the semiconductor devices that make the electronics industry possible.

EXAMPLE 3.2

Which of the following elements are not metals?

Na, Mo, Cl, S, Mg, Pt, Kr, I, C, Cu

SOLUTION

Nonmetals are generally found on the right-hand side of the periodic table: Cl, S, Kr, I, and C.

PRACTICE 3.2

Write the chemical symbols for
(a) five elements that are metals.
(b) two elements that are liquids at normal room temperature.
(c) five elements that are gases (not noble gases) at normal room temperature.

ONLINE LEARNING MODULE
Identifying Nonmetals
➡ WileyPLUS

Diatomic Elements

Diatomic molecules each contain exactly two atoms (alike or different). Seven elements in their uncombined state are **diatomic molecules**. Their symbols, formulas, and brief descriptions are listed in TABLE 3.6. Whether found free in nature or prepared in the laboratory, the molecules of these elements always contain two atoms. The formulas of the free elements are therefore always written to show this molecular composition: H_2, N_2, O_2, F_2, Cl_2, Br_2, and I_2.

Container of hydrogen molecules.

TABLE 3.6 Elements That Exist as Diatomic Molecules

Element	Symbol	Molecular formula	Normal state
Hydrogen	H	H_2	Colorless gas
Nitrogen	N	N_2	Colorless gas
Oxygen	O	O_2	Colorless gas
Fluorine	F	F_2	Pale yellow gas
Chlorine	Cl	Cl_2	Greenish-yellow gas
Bromine	Br	Br_2	Reddish-brown liquid
Iodine	I	I_2	Bluish-black solid

H_2 (gray) and O_2 (red) molecules.

It is important to see that symbols can designate either an atom or a molecule of an element. Consider hydrogen and oxygen. Hydrogen gas is present in volcanic gases and can be prepared by many chemical reactions. Regardless of their source, all samples of free hydrogen gas consist of diatomic molecules.

Free hydrogen is designated by the formula H_2, which also expresses its composition. Oxygen makes up about 20% by volume of the air that we breathe. This free oxygen is constantly being replenished by photosynthesis; it can also be prepared in the laboratory by several reactions. The majority of free oxygen is diatomic and is designated by the formula O_2. Now consider water, a compound designated by the formula H_2O (sometimes HOH). Water contains neither free hydrogen (H_2) nor free oxygen (O_2). The H_2 part of the formula H_2O simply indicates that two atoms of hydrogen are combined with one atom of oxygen to form water. FIGURE 3.4 summarizes the classification of elements.

Symbols are used to designate elements, show the composition of molecules of elements, and give the elemental composition of compounds.

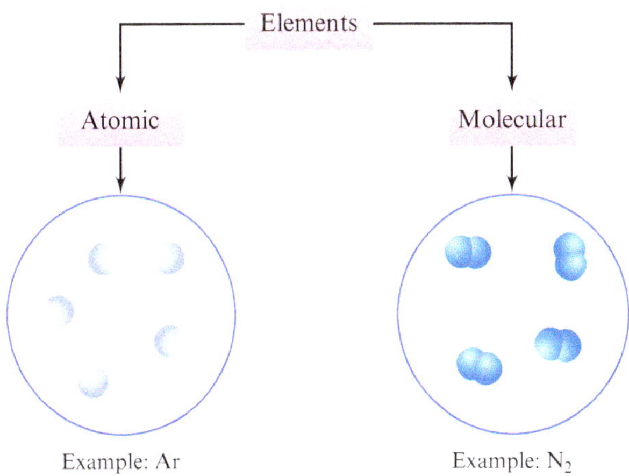

FIGURE 3.4 Elements fall into two broad categories.

CHEMISTRY IN ACTION Smartphone Elements

What does chemistry have to do with smartphone technology? A surprising number of elements from the periodic table are used in producing the essentials of your smartphone.

Let's start with the microelectronic brain of the smartphone. The microchips in your phone are made of silicon and several other elements, including arsenic, antimony, phosphorus, and gallium, which allow the conduction of electricity. These chips are connected to other microelectronic parts by copper wiring. The microelectronic parts are formed from gold, silver, and copper, pushing the cost to build the brains of our smartphones each year to more than $2.5 billion.

The microphone and speakers necessary to communicate with our phone contain magnets made of alloys (substances containing a mixture of elements having metallic properties), using praseodymium, gadolinium, and neodymium. To produce the vibrations you feel when your ringtone is off, neodymium, terbium, and dysprosium are required.

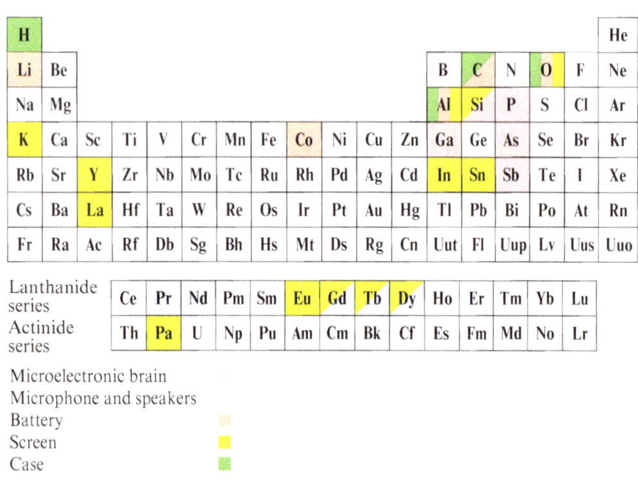

Smartphone technology uses small, long-lived batteries, so constant charging is not necessary. Most phones use lithium-ion batteries containing the elements lithium, cobalt, oxygen, and carbon. The case of the battery is aluminum.

@mbbirdy/iStockphoto

How does the screen on your smartphone work? Composed of a strong glass containing aluminum, silicon, oxygen, and potassium that strengthens it, the screen doesn't shatter when dropped. The touch screen works because of a thin film over the screen containing indium, oxygen, and tin that conducts electricity. The beautiful colors we see on the screen are the result of small amounts of the elements yttrium, lanthanum, terbium, protactinium, europium, dysprosium, and gadolinium.

If the case of the smartphone is metal, then often magnesium has been alloyed with other elements to make it strong and light. Some cases are plastic containing carbon, hydrogen, oxygen, and a variety of other elements to make the phone flame resistant.

Although exactly how a smartphone is manufactured is protected information (by patent law and trade secrets), we can be sure that the chemistry of the elements is critical in the process.

PRACTICE 3.3

Identify the physical state of each of the following elements at room temperature (20°C):

H, Na, Ca, N, S, Fe, Cl, Br, Ne, Hg

Hint: You may need to use a resource (such as the Internet or a chemical handbook) to assist you.

PRACTICE 3.4

Identify each of the following elements as a nonmetal, metal, or metalloid:

Na, F, Cr, Mo, Kr, Si, Cu, Sb, I, S

3.3 Compounds and Formulas

LEARNING OBJECTIVE Distinguish between molecular and ionic compounds and write chemical formulas for compounds.

A **compound** is a distinct substance that contains two or more elements chemically combined in a definite proportion by mass. Compounds, unlike elements, can be decomposed chemically into simpler substances—that is, into simpler compounds and/or elements.

Atoms of the elements in a compound are combined in whole-number ratios, never as fractional parts.

KEY TERMS

compound
molecule
ion
cation
anion
chemical formula
subscripts
natural law
law of definite composition
law of multiple proportions

Molecular and Ionic Compounds

Compounds fall into two general types, *molecular* and *ionic*. FIGURE 3.5 illustrates the classification of compounds.

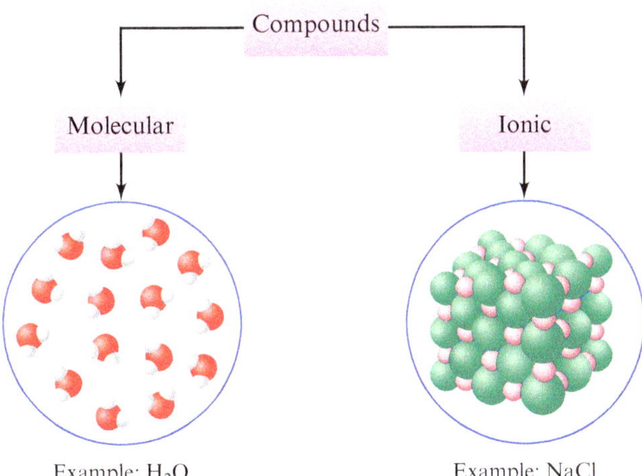

FIGURE 3.5 Compounds can be classified as molecular and ionic. Ionic compounds are held together by attractive forces between their positive and negative charges. Molecular compounds are held together by covalent bonds.

A **molecule** is the smallest uncharged individual unit of a compound formed by the union of two or more atoms. Water is a typical molecular compound. If we divide a drop of water into smaller and smaller particles, we finally obtain a single molecule of water consisting of two hydrogen atoms bonded to one oxygen atom, as shown in FIGURE 3.6. This molecule is the ultimate particle of water; it cannot be further subdivided without destroying the water molecule and forming hydrogen and oxygen.

An **ion** is a positively or negatively charged atom or group of atoms. An ionic compound is held together by attractive forces that exist between positively and negatively charged ions. A positively charged ion is called a **cation** (pronounced *cat-eye-on*); a negatively charged ion is called an **anion** (pronounced *an-eye-on*).

Sodium chloride is a typical ionic compound. The ultimate particles of sodium chloride are positively charged sodium ions and negatively charged chloride ions, shown in Figure 3.6. Sodium chloride is held together in a crystalline structure by the attractive forces existing between these oppositely charged ions. Although ionic compounds consist of large aggregates of cations and anions, their formulas are normally represented by the simplest possible ratio of the atoms in the compound. For example, in sodium chloride the ratio is one sodium ion to one chloride ion, so the formula is NaCl.

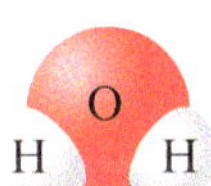

Two hydrogen atoms combined with an oxygen atom to form a molecule of water (H_2O).

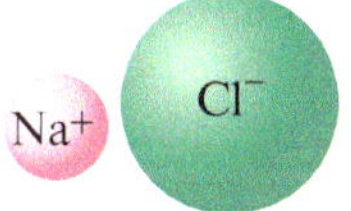

A positively charged sodium ion and a negatively charged chloride ion form the compound sodium chloride (NaCl).

FIGURE 3.6 Representation of molecular and ionic (nonmolecular) compounds.

FIGURE 3.7 A representation of the decomposition of water into oxygen and hydrogen molecules.

There are more than 50 million known registered compounds, with no end in sight as to the number that will be prepared in the future. Each compound is unique and has characteristic properties. Let's consider two compounds, water and sodium chloride, in some detail. Water is a colorless, odorless, tasteless liquid that can be changed to a solid (ice) at 0°C and to a gas (steam) at 100°C. Composed of two atoms of hydrogen and one atom of oxygen per molecule, water is 11.2% hydrogen and 88.8% oxygen by mass. Water reacts chemically with sodium to produce hydrogen gas and sodium hydroxide, with lime to produce calcium hydroxide, and with sulfur trioxide to produce sulfuric acid. When water is decomposed, it forms hydrogen and oxygen molecules (FIGURE 3.7). No other compound has all these exact physical and chemical properties; they are characteristic of water alone.

Sodium chloride is a colorless crystalline substance with a ratio of one atom of sodium to one atom of chlorine. Its composition by mass is 39.3% sodium and 60.7% chlorine. It does not conduct electricity in its solid state; it dissolves in water to produce a solution that conducts electricity. When a current is passed through molten sodium chloride, solid sodium and gaseous chlorine are produced (FIGURE 3.8). These specific properties belong to sodium chloride and to no other substance. Thus, a compound may be identified and distinguished from all other compounds by its characteristic properties. We consider these chemical properties further in Chapter 4.

FIGURE 3.8 When sodium chloride is decomposed, it forms sodium metal and chlorine gas.

Writing Formulas of Compounds

Chemical formulas are used as abbreviations for compounds. A **chemical formula** shows the symbols and the ratio of the atoms of the elements in a compound. Sodium chloride contains one atom of sodium per atom of chlorine; its formula is NaCl. The formula for water is H_2O; it shows that a molecule of water contains two atoms of hydrogen and one atom of oxygen.

The formula of a compound tells us which elements it is composed of and how many atoms of each element are present in a formula unit. For example, a unit of sulfuric acid is composed of two atoms of hydrogen, one atom of sulfur, and four atoms of oxygen. We could express this compound as HHSOOOO, but this is cumbersome, so we write H_2SO_4 instead. The formula may be expressed verbally as "H-two-S-O-four." Numbers that appear partially below the line and to the right of a symbol of an element are called **subscripts**. Thus, the 2 and the 4 in H_2SO_4 are subscripts (FIGURE 3.9).

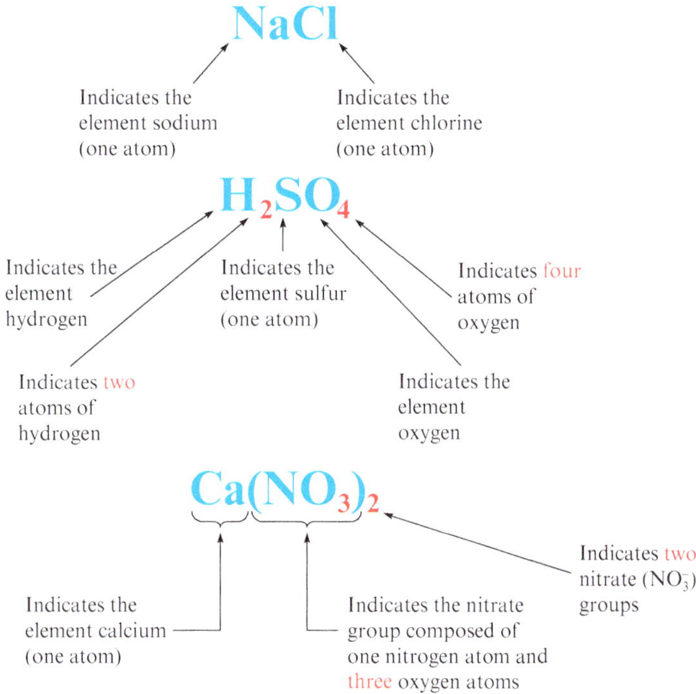

FIGURE 3.9 Explanation of the formulas NaCl, H_2SO_4, and $Ca(NO_3)_2$.

Characteristics of chemical formulas are as follows:

1. The formula of a compound contains the symbols of all the elements in the compound.
2. When the formula contains one atom of an element, the symbol of that element represents that one atom. The number 1 is not used as a subscript to indicate one atom of an element.
3. When the formula contains more than one atom of an element, the number of atoms is indicated by a subscript written to the right of the symbol of that atom. For example, the 2 in H_2O indicates two atoms of H in the formula.
4. When the formula contains more than one of a group of atoms that occurs as a unit, parentheses are placed around the group, and the number of units of the group is indicated by a subscript placed to the right of the parentheses. Consider the nitrate group, NO_3^-. The formula for sodium nitrate, $NaNO_3$, has only one nitrate group, so no parentheses are needed. Calcium nitrate, $Ca(NO_3)_2$, has two nitrate groups, as indicated by the use of parentheses and the subscript 2. $Ca(NO_3)_2$ has a total of nine atoms: one Ca, two N, and six O atoms. The formula $Ca(NO_3)_2$ is read as "C-A [pause] N-O-three taken twice."
5. Formulas written as H_2O, H_2SO_4, $Ca(NO_3)_2$, and $C_{12}H_{22}O_{11}$ show only the number and kind of each atom contained in the compound; they do not show the arrangement of the atoms in the compound or how they are chemically bonded to one another.

PRACTICE 3.5

How would you read these formulas aloud?

(a) KBr (b) $PbCl_2$ (c) $CaCO_3$ (d) $Mg(OH)_2$

ONLINE LEARNING MODULE
Chemical Formulas from Atomic Composition
WileyPLUS

ENHANCED EXAMPLE 3.3 WileyPLUS

Write formulas for the following compounds; the atomic composition is given.

(a) hydrogen chloride: 1 atom hydrogen + 1 atom chlorine;
(b) methane: 1 atom carbon + 4 atoms hydrogen;
(c) glucose: 6 atoms carbon + 12 atoms hydrogen + 6 atoms oxygen.

SOLUTION

(a) First write the symbols of the atoms in the formula: H Cl.
Since the ratio of atoms is one to one, we bring the symbols together to give the formula for hydrogen chloride as HCl.
(b) Write the symbols of the atoms: C H.
Now bring the symbols together and place a subscript 4 after the hydrogen atom. The formula is CH_4.
(c) Write the symbols of the atoms: C H O.
Now write the formula, bringing together the symbols followed by the correct subscripts according to the data given (six C, twelve H, six O). The formula is $C_6H_{12}O_6$.

PRACTICE 3.6

Write formulas for the following compounds from the compositions given.

(a) potassium bromide: 1 atom potassium + 1 atom bromine
(b) sodium chlorate: 1 atom sodium + 1 atom chlorine + 3 atoms oxygen
(c) sulfurous acid: 2 atoms hydrogen + 1 atom sulfur + 3 atoms oxygen
(d) aluminum carbonate: 2 aluminum atoms + 3 carbonate ions

Composition of Compounds

A large number of experiments extending over a long period have established the fact that a particular compound always contains the same elements in the same proportions by mass. For example, water always contains 11.2% hydrogen and 88.8% oxygen by mass (FIGURE 3.10). The fact that water contains hydrogen and oxygen in this particular ratio does not mean that hydrogen and oxygen cannot combine in some other ratio but rather that a compound with a different ratio would not be water. In fact, hydrogen peroxide is made up of two atoms of hydrogen and two atoms of oxygen per molecule and contains 5.9% hydrogen and 94.1% oxygen by mass; its properties are markedly different from those of water (Figure 3.10).

	Water	**Hydrogen peroxide**
	11.2% H	5.9% H
	88.8% O	94.1% O
Atomic composition	2 H + 1 O	2 H + 2 O

We often summarize our general observations regarding nature into a statement called a **natural law**. In the case of the composition of a compound, we use the **law of definite composition**, which states that a compound always contains two or more elements chemically combined in a definite proportion by mass.

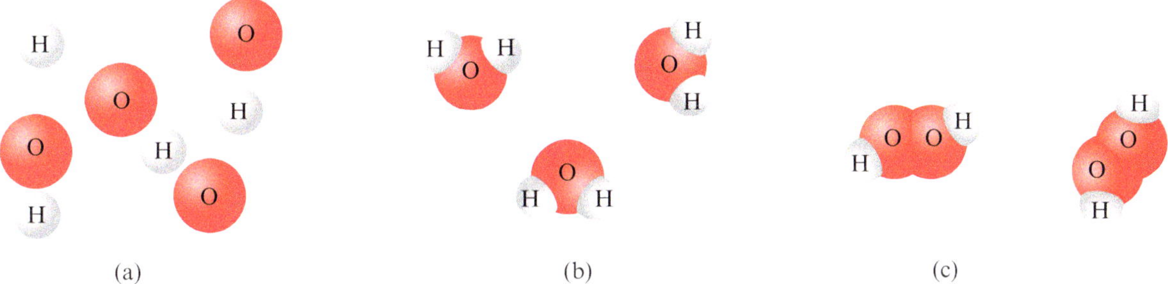

FIGURE 3.10 (a) Dalton's atoms were individual particles, the atoms of each element being alike in mass and size but different in mass and size from other elements. (b) and (c) Dalton's atoms combine in specific ratios to form compounds.

Let's consider two elements, oxygen and hydrogen, that form more than one compound. In water, 8.0 g of oxygen are present for each gram of hydrogen. In hydrogen peroxide, 16.0 g of oxygen are present for each gram of hydrogen. The masses of oxygen are in the ratio of small whole numbers, 16 : 8 or 2 : 1. Hydrogen peroxide has twice as much oxygen (by mass) as does water. Using Dalton's atomic model, we deduce that hydrogen peroxide has twice as many oxygen atoms per hydrogen atom as water. In fact, we now write the formulas for water as H_2O and for hydrogen peroxide as H_2O_2 (Figure 3.10b and c).

The **law of multiple proportions** states that atoms of two or more elements may combine in different ratios to produce more than one compound.

Some examples of the law of multiple proportions are given in TABLE 3.7. The reliability of this law and the law of definite composition is the cornerstone of the science of chemistry. In essence, these laws state that (1) the composition of a particular substance will always be the same no matter what its origin or how it is formed, and (2) the composition of different compounds formed from the same elements will always be unique.

You need to recognize the difference between a *law* and a *model* (*theory*). A law is a summary of observed behavior. A model (theory) is an attempt to explain the observed behavior. This means that laws remain constant—that is, they do not undergo modification—while theories (models) sometimes fail and are modified or discarded over time.

TABLE 3.7 Selected Compounds Showing Elements That Combine to Give More Than One Compound

Compound	Formula	Percent composition
Copper(I) chloride	$CuCl$	64.2% Cu, 35.8% Cl
Copper(II) chloride	$CuCl_2$	47.3% Cu, 52.7% Cl
Methane	CH_4	74.9% C, 25.1% H
Octane	C_8H_{18}	85.6% C, 14.4% H
Methyl alcohol	CH_4O	37.5% C, 12.6% H, 49.9% O
Ethyl alcohol	C_2H_6O	52.1% C, 13.1% H, 34.7% O
Glucose	$C_6H_{12}O_6$	40.0% C, 6.7% H, 53.3% O

CHAPTER 3 REVIEW

3.1 Elements

KEY TERMS
element
atom
symbols

- All matter consists of about 100 elements.
- An element is a fundamental chemical substance.
- The smallest particle of an element is an atom.
- Elements cannot be broken down by chemical means to a simpler substance.
- Most elements are found in nature combined with other elements.
- Elements that are found in uncombined form in nature include gold, silver, copper, and platinum as well as the noble gases (He, Ne, Ar, Kr, Xe, Rn).
- Chemical elements are not distributed equally in nature.
- Hydrogen is the most abundant element in the universe.
- Oxygen is the most abundant element on the Earth and in the human body.
- Names for the chemical elements come from a variety of sources, including Latin, location of discovery, and famous scientists.
- Rules for writing symbols for the elements are:
  - One or two letters
  - If one letter, use a capital
  - If two letters, only the first is a capital

3.2 Introduction to the Periodic Table

KEY TERMS
groups
noble gases
alkali metals
alkaline earth metals
halogens
representative elements
transition elements
metals
nonmetals
metalloids
diatomic molecules

- The periodic table was designed by Dimitri Mendeleev and arranges the elements according to their atomic numbers and in groups by their chemical properties.
- Elements can be classified as representative or as transition elements.
- Elements can also be classified by special groups with similar chemical properties. Such groups include the noble gases, alkali metals, alkaline earth metals, and halogens.
- Elements can be classified as metals, nonmetals, or metalloids.
  - Most elements are metals.
  - Metals have the following properties:
    - High luster
    - Good conductors of heat and electricity
    - Malleable
  - Nonmetals have the following properties:
    - Not lustrous
    - Poor conductors of heat and electricity
- Diatomic molecules contain exactly two atoms (alike or different).
- Seven elements exist as diatomic molecules—H_2, N_2, O_2, F_2, Cl_2, Br_2, and I_2.

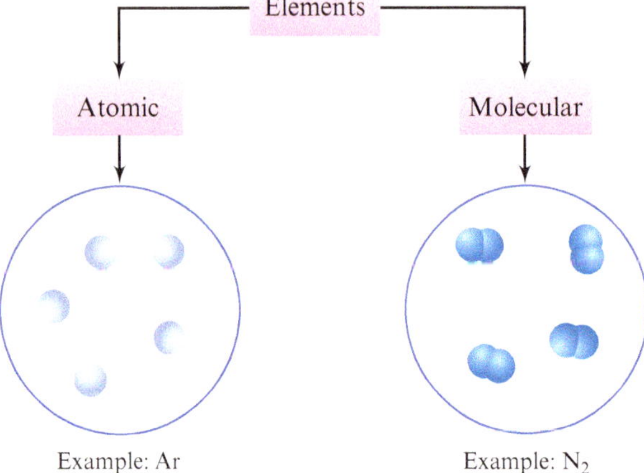

3.3 Compounds and Formulas

- A compound is a substance that contains two or more elements chemically combined in a definite proportion by mass.
- There are two general types of compounds:
  - Molecular—formed of individual molecules composed of atoms
  - Ionic—formed from ions that are either positive or negative
    - Cation—positively charged ion
    - Anion—negatively charged ion
- A chemical formula shows the symbols and the ratios of atoms for the elements in a chemical compound.
- Characteristics of a chemical formula include:
  - It contains symbols of all elements in the compound.
  - The symbol represents one atom of the element.
  - If more than one atom of an element is present, the number of atoms is indicated by a subscript.
  - Parentheses are used to show multiple groups of atoms occurring as a unit in the compound.
  - A formula does not show the arrangement of the atoms in the compound.
- The law of definite composition states that a compound always contains two or more elements combined in a definite proportion by mass.
- The law of multiple proportions states that atoms of two or more elements may combine in different ratios to form more than one compound.

KEY TERMS

compound
molecule
ion
cation
anion
chemical formula
subscripts
natural law
law of definite composition
law of multiple proportions

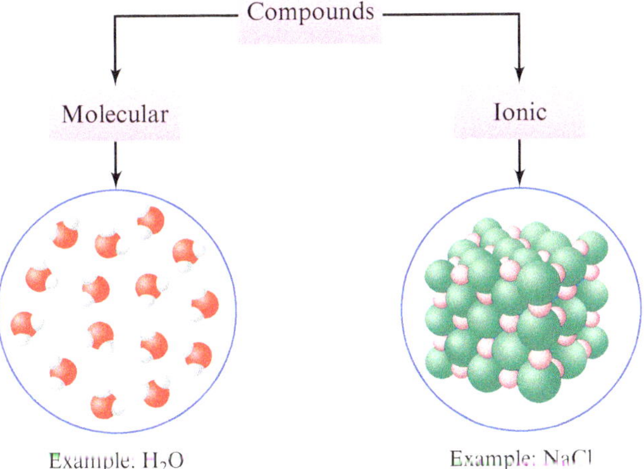

Example: H_2O Example: NaCl

REVIEW QUESTIONS

1. Are there more atoms of silicon or hydrogen in the Earth's crust, seawater, and atmosphere? Use Table 3.1 and the fact that the mass of a silicon atom is about 28 times that of a hydrogen atom.
2. Give the symbols for each of the following elements:
   (a) manganese (e) chlorine
   (b) fluorine (f) vanadium
   (c) sodium (g) zinc
   (d) helium (h) nitrogen
3. Give the names for each of the following elements:
   (a) Fe (e) Be
   (b) Mg (f) Co
   (c) C (g) Ar
   (d) P (h) Hg
4. What does the symbol of an element stand for?
5. List six elements and their symbols in which the first letter of the symbol is different from that of the name. (Table 3.4)
6. Write the names and symbols for the 14 elements that have only one letter as their symbol. (See the periodic table on inside front cover.)
7. Of the six most abundant elements in the human body, how many are metals? nonmetals? metalloids? (Table 3.2)
8. Write the names and formulas for the elements that exist as diatomic molecules. (Table 3.6)
9. Interpret the difference in meaning for each pair:
   (a) CO and Co (c) S_8 and 8 S
   (b) H_2 and 2 H (d) CS and Cs
10. Distinguish between an element and a compound.
11. How many metals are there? nonmetals? metalloids? (Table 3.5)
12. Of the ten most abundant elements in the Earth's crust, seawater, and atmosphere, how many are metals? nonmetals? metalloids? (Table 3.1)
13. Give the names of (a) the solid diatomic nonmetal and (b) the liquid diatomic nonmetal. (Table 3.6)
14. Distinguish between a compound and a mixture.
15. What are the two general types of compounds? How do they differ from each other?
16. What is the basis for distinguishing one compound from another?
17. What is the major difference between a cation and an anion?

PAIRED EXERCISES

Most of the exercises in this chapter are available for assignment via the online homework management program, WileyPLUS (www.wileyplus.com). All exercises with blue numbers have answers in Appendix 5.

1. Which of the following are diatomic molecules?
(a) HCl (b) O_2 (c) N_2O (d) P_4 (e) SiO_2 (f) H_2O_2 (g) CH_4 (h) ClF

2. Which of the following are diatomic molecules?
(a) O_3 (b) H_2O (c) CO_2 (d) HI (e) S_8 (f) Cl_2 (g) CO (h) NH_3

3. What elements are present in each compound?
(a) potassium iodide KI
(b) sodium carbonate Na_2CO_3
(c) aluminum oxide Al_2O_3
(d) calcium bromide $CaBr_2$
(e) acetic acid $HC_2H_3O_2$

4. What elements are present in each compound?
(a) magnesium bromide $MgBr_2$
(b) carbon tetrachloride CCl_4
(c) nitric acid HNO_3
(d) barium sulfate $BaSO_4$
(e) aluminum phosphate $AlPO_4$

5. Write the formula for each compound (the composition is given after each name):
(a) zinc oxide 1 atom Zn, 1 atom O
(b) potassium chlorate 1 atom K, 1 atom Cl, 3 atoms O
(c) sodium hydroxide 1 atom Na, 1 atom O, 1 atom H
(d) ethyl alcohol 2 atoms C, 6 atoms H, 1 atom O

6. Write the formula for each compound (the composition is given after each name):
(a) aluminum bromide 1 atom Al, 3 atoms Br
(b) calcium fluoride 1 atom Ca, 2 atoms F
(c) lead(II) chromate 1 atom Pb, 1 atom Cr, 4 atoms O
(d) benzene 6 atoms C, 6 atoms H

7. Many foods contain interesting compounds that give them their flavors and smells. Some of these compounds are listed below with the chemical composition given after each name. Write the formulas for these compounds.
(a) Allicin gives garlic its flavor (6 carbons, 10 hydrogens, 1 oxygen, and 2 sulfurs).
(b) Capsaicin gives peppers their heat (18 carbons, 27 hydrogens, 1 nitrogen, and 3 oxygens).
(c) Limonene gives oranges their fragrance (10 carbons and 16 hydrogens).

8. Many plants contain interesting compounds that sometimes have medicinal properties. Some of these compounds are listed below with the chemical composition given after each name. Write the formulas for these compounds.
(a) Aescin from horse chestnuts has anti-inflammatory properties (55 carbons, 86 hydrogens, and 24 oxygens).
(b) Proanthocyanidins found in cranberries help to prevent urinary tract infections. Proanthocyanidins are polymers composed of epicatechin units (15 carbons, 14 hydrogens, and 6 oxygens).
(c) Betulinic acid found in the common birch tree is an antimalarial drug (30 carbons, 48 hydrogens, and 3 oxygens).

9. Write the name and number of atoms of each element in each of the following compounds:
(a) Fe_2O_3 (b) $Ca(NO_3)_2$ (c) $Co(C_2H_3O_2)_2$ (d) CH_3COCH_3 (e) K_2CO_3 (f) $Cu_3(PO_4)_2$ (g) C_2H_5OH (h) $Na_2Cr_2O_7$

10. Write the name and number of atoms of each element in each of the following compounds:
(a) $HC_2H_3O_2$ (b) $(NH_4)_3PO_4$ (c) $Mg(HSO_3)_2$ (d) $ZnCl_2$ (e) $NiCO_3$ (f) $KMnO_4$ (g) $CH_3CH_2CH_2CH_3$ (h) $PbCrO_4$

11. How many total atoms are represented in each formula?
(a) $Co(ClO_3)_2$ (b) $(NH_4)_2SO_3$ (c) CH_3CH_2COOH (d) $C_{12}H_{22}O_{11}$

12. How many total atoms are represented in each formula?
(a) CH_3CH_2OH (b) $KAl(SO_4)_2$ (c) $NH_4C_2H_3O_2$ (d) $C_6H_4Cl_2$

13. How many hydrogen atoms are represented in each formula?
(a) $Al(C_2H_3O_2)_3$ (b) $CH_3CH_2CH_2OH$ (c) NH_4OH (d) $C_6H_5CH_2CH_3$

14. How many oxygen atoms are represented in each formula?
(a) $Fe(C_2H_3O_2)_3$ (b) H_3PO_4 (c) $Ba(ClO_3)_2$ (d) $Fe_2(Cr_2O_7)_3$

15. Determine whether each of the following is a pure substance or a mixture:
(a) hot tea (b) beach sand (c) carbon dioxide (d) cement (e) zinc (f) vinegar

16. Determine whether each of the following is a pure substance or a mixture:
(a) dirt (b) salad dressing (c) tungsten (d) dinitrogen monoxide (e) brass (f) egg

17. For Question 15, state whether each pure substance is an element or a compound.

18. For Question 16, state whether each pure substance is an element or a compound.

19. Classify each of the following as an element, a compound, or a mixture:

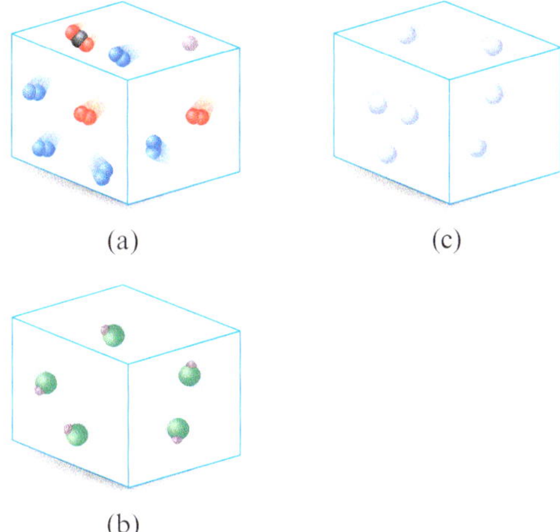

20. Classify each of the following as an element, a compound, or a mixture:

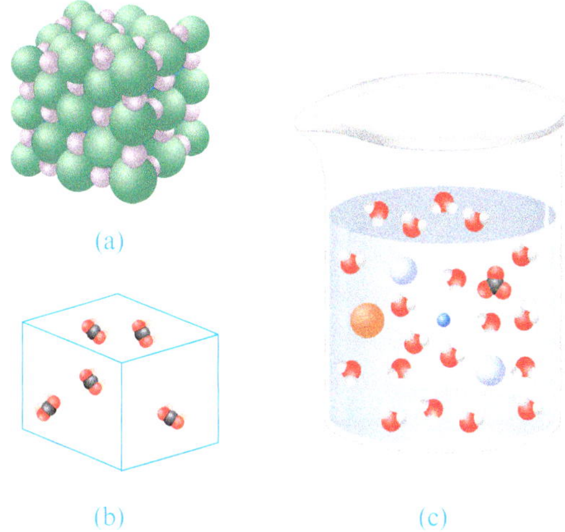

21. Is there a pattern to the location of the gaseous elements on the periodic table? If so, describe it.

22. Is there a pattern to the location of the liquid elements on the periodic table? If so, describe it.

23. What percent of the first 36 elements on the periodic table are metals?

24. What percent of the first 36 elements on the periodic table are solids at room temperature?

25. In Section 3.3, there is a statement about the composition of water. It says that water (H_2O) contains 8 grams of oxygen for every 1 gram of hydrogen. Show why this statement is true.

26. In Section 3.3, there is a statement about the composition of hydrogen peroxide. It says that hydrogen peroxide (H_2O_2) contains 16 grams of oxygen for every 1 gram of hydrogen. Show why this statement is true.

ADDITIONAL EXERCISES

27. You accidentally poured salt into your large grind pepper shaker. This is the only pepper that you have left and you need it to cook a meal, but you don't want the salt to be mixed in. How can you successfully separate these two components?

28. On the periodic table at the front of this book, do you notice anything about the atoms that make up the following ionic compounds: NaCl, KI, and $MgBr_2$? (*Hint*: Look at the position of the atoms in the given compounds on the periodic table.)

29. How many total atoms are present in each of the following compounds?
    (a) CO
    (b) BF_3
    (c) HNO_3
    (d) $KMnO_4$
    (e) $Ca(NO_3)_2$
    (f) $Fe_3(PO_4)_2$

30. The formula for vitamin B_{12} is $C_{63}H_{88}CoN_{14}O_{14}P$.
    (a) How many atoms make up one molecule of vitamin B_{12}?
    (b) What percentage of the total atoms are carbon?
    (c) What fraction of the total atoms are metallic?

31. It has been estimated that there is 4×10^{-4} mg of gold per liter of seawater. At a price of \$19.40/g, what would be the value of the gold in 1 km^3(1×10^{15} cm^3) of the ocean?

32. Calcium dihydrogen phosphate is an important fertilizer. How many atoms of hydrogen are there in ten formula units of $Ca(H_2PO_4)_2$?

33. How many total atoms are there in one molecule of $C_{145}H_{293}O_{168}$?

34. Name the following:
    (a) three elements, all metals, beginning with the letter M
    (b) four elements, all solid nonmetals
    (c) five elements, all solids in the first five rows of the periodic table, whose symbols start with letters different from the element name

35. How would you separate a mixture of sugar and sand and isolate each in its pure form?

36. How many total atoms are there in seven dozen formulas of nitric acid, HNO_3?

37. Make a graph using the following data. Plot the density of air in grams per liter along the x-axis and temperature along the y-axis.

Temperature (°C)	Density (g/L)
0	1.29
10	1.25
20	1.20
40	1.14
80	1.07

(a) What is the relationship between density and temperature according to your graph?
(b) From your plot, find the density of air at these temperatures:
5°C 25°C 70°C

38. These formulas look similar but represent different things.

8 S S_8

Compare and contrast them. How are they alike? How are they different?

39. Write formulas for the following compounds that a colleague read to you:

(a) NA–CL
(b) H2–S–O4
(c) K2–O
(d) Fe2–S3
(e) K3–P–O4
(f) CA (pause) CN taken twice
(g) C6–H12–O6
(h) C2–H5 (pause) OH
(i) CR (pause) NO3 taken three times

40. The abundance of iodine in seawater is 5.0×10^{-8} percent by mass. How many kilograms of seawater must be treated to obtain 1.0 g of iodine?

41. When the tongue detects something sweet, it begins a chain of reactions to tell the body to expect calorie-containing foods. If we consume noncaloric foods containing artificial sweeteners, our bodies expect food and will tell our brains to find more if the sweet food does not contain anything to digest. Some common sweeteners are listed below. Write the name and number of atoms of each element in the sweeteners.

Compound	Common name	Chemical formula
(a) Sucrose	Table sugar	$C_{12}H_{22}O_{11}$
(b) Saccharin	Sweet 'N Low	$C_7H_5O_3NS$
(c) Aspartame	NutraSweet, Equal	$C_{14}H_{18}O_5N_2$
(d) Acesulfame-K	Sunett, Sweet One	$C_4H_4O_3NS$-K
(e) Sucralose	Splenda	$C_{12}H_{19}O_8Cl_3$

42. Should the elemental symbol for cobalt be written as Co or CO? What confusion could arise if the wrong notation were used?

43. Caffeine has the chemical structure illustrated in the image given on the right. Carbon atoms are represented as gray spheres, nitrogen atoms as blue spheres, oxygen atoms as red spheres, and hydrogen atoms as white spheres. Write a chemical formula for caffeine.

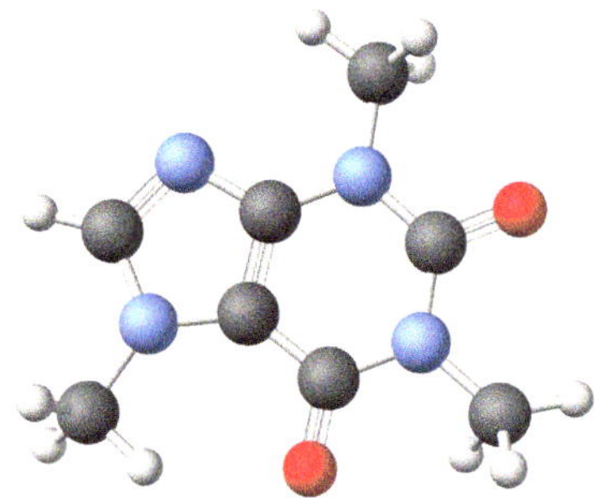

Use these pictures to answer Question 44.

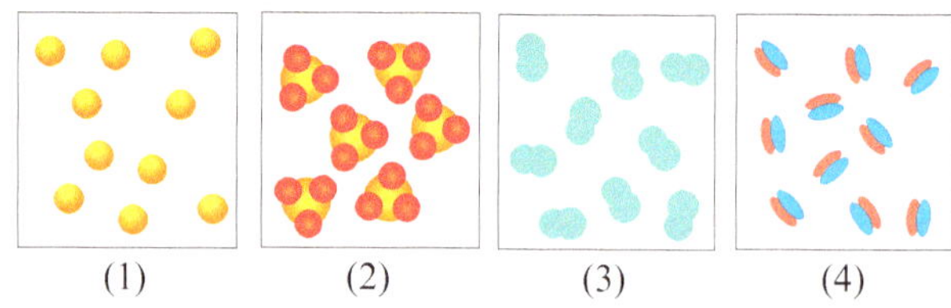

44. (a) Which picture best represents fluorine gas? Why?
(b) Which other elements could that picture also represent?
(c) Which of the pictures could represent SO_3 gas?

CHALLENGE EXERCISES

45. Write the chemical formulas of the neutral compounds that would result if the following ions were combined. Use the Charges of Common Ions table on the inside back cover of the book to help you.

(a) ammonium ion and chloride ion
(b) hydrogen ion and hydrogen sulfate ion
(c) magnesium ion and iodide ion
(d) iron(II) ion and fluoride ion
(e) lead(II) ion and phosphate ion
(f) aluminum ion and oxide ion

46. Write formulas of all the compounds that will form between the first five of the Group 1A (1) and 2A (2) metals and the oxide ion.

47. Unsaturated fats and oils are often used in cooking. For each pair of fatty acids listed below, tell the number of carbon and hydrogen atoms in each molecule.

(a) arachidic acid (saturated) $CH_3(CH_2)_{18}COOH$ and arachidonic acid (unsaturated) $CH_3(CH_2)_4(CH{=}CHCH_2)_4(CH_2)_2COOH$
(b) stearic acid (saturated) $CH_3(CH_2)_{16}COOH$ and linoleic acid (unsaturated) $CH_3(CH_2)_4(CH{=}CHCH_2)_2(CH_2)_6COOH$
(c) What is the ratio of hydrogen to carbon atoms for each pair?
(d) Propose a possible explanation for describing linoleic and stearic acids as being saturated.

48. One cup of Raisin Bran provides 60.% of the U.S. recommended daily allowance (RDA) of iron.

(a) If the cereal provides 11 mg of iron, what is the U.S. RDA for Fe?
(b) When the iron in the cereal is extracted, it is found to be the pure element. What is the volume of iron in a cup of the cereal?

49. Absent-minded Alfred put down a bottle containing silver on the table. When he went to retrieve it, he realized he had forgotten to label the bottle. Unfortunately, there were two full bottles of the same size side-by-side. Alfred realized he had placed a bottle of mercury on the same table last week. State two ways Alfred can determine which bottle contains silver without opening the bottles.

ANSWERS TO PRACTICE EXERCISES

3.1 Cesium, Cs, 1A; Calcium, Ca, 2A; Chromium, Cr, 6B; Cobalt, Co, 8B; Copper, Cu, 1B; Cadmium, Cd, 2B; Carbon, C, 4A; Chlorine, Cl, 7A; Cerium, Ce, 4B; Copernicium, Cn, 2B; Curium, Cm, 8B; Californium, Cf, 1B.

3.2
(a) any of the transition metals
(b) Br and Hg
(c) H, N, O, F, Cl

3.3

gases	H, N, Cl, Ne
liquids	Br, Hg
solids	Na, Ca, S, Fe

3.4

nonmetal	F, Kr, I, S
metal	Na, Cr, Mo, Cu
metalloid	Si, Sb

3.5
(a) K–BR
(b) PB–CL2
(c) CA (pause) CO3
(d) MG (pause) OH taken twice

3.6
(a) KBr
(b) $NaClO_3$
(c) H_2SO_3
(d) $Al_2(CO_3)_3$

4

Properties of Matter

CHAPTER OUTLINE

The world we live in is a kaleidoscope of sights, sounds, smells, and tastes. Our senses help us to describe these objects in our lives. For example, the smell of freshly baked cinnamon rolls creates a mouth-watering desire to gobble down a sample. Just as sights, sounds, smells, and tastes form the properties of the objects around us, each substance in chemistry has its own unique properties that allow us to identify it and predict its interactions.

These interactions produce both physical and chemical changes. When you eat an apple, the ultimate metabolic result is carbon dioxide and water. These same products are achieved by burning logs. Not only does a chemical change occur in these cases, but an energy change occurs as well. Some reactions release energy (as does the apple or the log) whereas others require energy, such as the production of steel or the melting of ice. Over 90% of our current energy comes from chemical reactions.

4.1 Properties of Substances

Compare the physical and chemical properties of a substance.

LEARNING OBJECTIVE

KEY TERMS

properties
physical properties
chemical properties

How do we recognize substances? Each substance has a set of **properties** that is characteristic of that substance and gives it a unique identity. Properties—the "personality traits" of substances—are classified as either physical or chemical. **Physical properties** are the inherent characteristics of a substance that can be determined without altering its composition; they are associated with its physical existence. Common physical properties include color, taste, odor, state of matter (solid, liquid, or gas), density, melting point, and boiling point (FIGURE 4.1). **Chemical properties** describe the ability of a substance to form new substances, either by reaction with other substances or by decomposition.

Martin Dohrn/Science Source Images

FIGURE 4.1 Physical property: The boiling point of water is a physical property. At its boiling point water changes from a liquid to a gas, but the molecules remain water molecules. Although the molecules are farther apart, they are still water.

Let's consider a few of the physical and chemical properties of chlorine. Physically, chlorine is a gas at room temperature about 2.4 times heavier than air. It is greenish yellow in color and has a disagreeable odor. Chemically, chlorine will not burn but will support the combustion of certain other substances. It can be used as a bleaching agent, as a disinfectant for water, and in many chlorinated substances such as refrigerants and insecticides. When chlorine combines with the metal sodium, it forms a salt called sodium chloride (FIGURE 4.2). These properties, among many others, help us characterize and identify chlorine.

CHECK YOUR UNDERSTANDING

4.1 Physical Properties
4.2 Chemical Properties
WileyPLUS

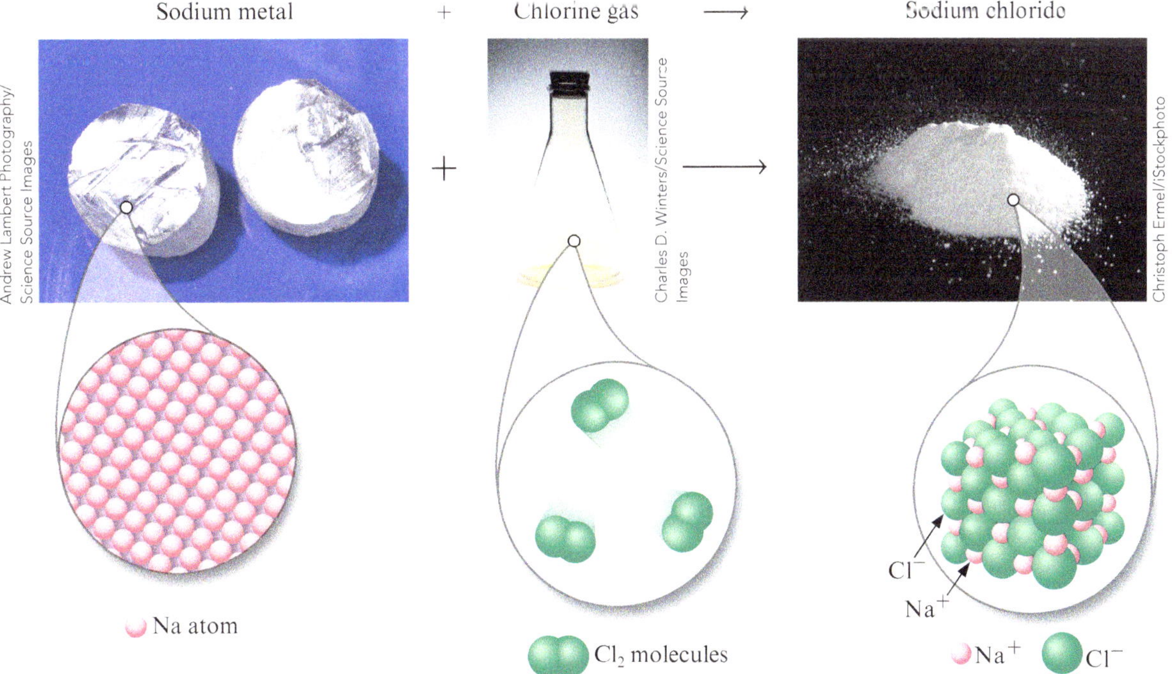

FIGURE 4.2 Chemical property: When sodium metal reacts with chlorine gas, a new substance called sodium chloride forms.

TABLE 4.1 Physical Properties of Selected Substances

Substance	Color	Odor	Physical state	Melting point (°C)	Boiling point (°C)
Chlorine	Greenish yellow	Sharp, suffocating	Gas (20°C)	−101.6	−34.6
Water	Colorless	Odorless	Liquid	0.0	100.0
Sugar	White	Odorless	Solid	—	Decomposes 170–186
Acetic acid	Colorless	Like vinegar	Liquid	16.7	118.0
Nitrogen dioxide	Reddish brown	Sharp, suffocating	Gas	−11.2	21.2
Oxygen	Colorless	Odorless	Gas	−218.4	−183

Many chemists have reference books such as the *Handbook of Chemistry and Physics* to use as a resource.

Substances, then, are recognized and differentiated by their properties. **TABLE 4.1** lists six substances and several of their common physical properties. Information about physical properties, such as that given in Table 4.1, is available in handbooks of chemistry and physics. Scientists don't pretend to know all the answers or to remember voluminous amounts of data, but it is important for them to know where to look for data in the literature and on the Internet.

No two substances have identical physical and chemical properties.

CHEMISTRY *IN ACTION* Making Money

Chemists are heavily involved in the manufacture of our currency. In fact, in a very real way, the money industry depends on chemistry and finding substances with the correct properties. The most common paper currency in the United States is the dollar bill. Chemistry is used to form the ink and paper and in processes used to defeat counterfeiters. The ink used on currency has to do a variety of things. It must be just the right consistency to fill the fine lines of the printing plate and release onto the paper without smearing. The ink must dry almost immediately, since the sheets of currency fly out of the press and into stacks 10,000 sheets tall. The pressure at the bottom of the stack is large, and the ink must not stick to the back of the sheet above it.

The $100 bill has incorporated a color-shifting ink into an image of the liberty bell hidden in the inkwell. If you tilt the bill, you will see the bell change from copper to green making it visible in the inkwell. More recently a security thread has also been incorporated into the bills. This security thread is colored with an ink that will glow with a particular color when exposed to ultraviolet light. The $5 bill glows blue, the $10 orange, the $20 green, the $50 yellow, and the $100 pink. Counterfeiters keep getting more creative and the technology used in fabricating our money continues to improve to stay a step ahead.

The security of our currency also depends on substances in the ink, which is optically variable. This color-changing ink shifts from green to black depending on how the bill is tilted. The ink used for the numbers in the lower right corner on the front of $10, $20, $50, and $100 bills shows this color change.

Once the currency is printed, it has to undergo durability tests. The bills are soaked in different solvents for 24 hours to make sure that they can stand dry cleaning and chemical exposure to household items such as gasoline. Then the bills must pass a washing-machine test to be sure that, for example, your $20 bill is still intact in your pocket after the washer is through with it. Paper currency is a blend of 75% cotton and 25% linen. Last, the bills must pass the crumple test, in which they are rolled up, put in a metal tube, crushed by a plunger, removed, and flattened. This process is repeated up to 50 times.

Once a bill is in circulation, Federal Reserve banks screen it using light to measure wear. If the bill gets too dirty, it is shredded and sent to a landfill. The average $1 bill stays in circulation for only 18 months. No wonder Congress decided in 1997 to revive the dollar coin by creating a new dollar coin to succeed the Susan B. Anthony dollar.

Color-shifting ink on the $100 bill along with a security thread

Sacagawea and presidential dollar coins, which succeeded the Susan B. Anthony dollar, are still not in wide circulation today.

EXAMPLE 4.1

You have three flasks that contain chlorine gas, nitrogen dioxide gas, and oxygen gas. Describe how you would use physical properties to identify the contents in each flask.

SOLUTION

Use Table 4.1 to compare the compounds and note their colors. You will find that chlorine is a greenish yellow color, nitrogen dioxide is reddish brown, and oxygen is colorless. Now you have the information to identify the contents in each flask.

PRACTICE 4.1

You are given three samples of silver compounds: AgCl, AgBr, and AgI. Arrange these silver compounds in the order of their percent silver by mass (highest to lowest). Describe how you arrived at your arrangement. *Hint:* Use the periodic table inside the front cover of your text.

ONLINE LEARNING MODULE
Using Physical Properties to Identify Substances
WileyPLUS

4.2 Physical and Chemical Changes

Compare the physical and chemical changes in a substance.

LEARNING OBJECTIVE

KEY TERMS
physical change
chemical change
chemical equation
reactants
products

Matter can undergo two types of changes, physical and chemical. **Physical changes** are changes in physical properties (such as size, shape, and density) or changes in the state of matter without an accompanying change in composition. The changing of ice into water and water into steam are physical changes from one state of matter into another. No new substances are formed in these physical changes.

When a clean platinum wire is heated in a burner flame, the appearance of the platinum changes from silvery metallic to glowing red. This change is physical because the platinum can be restored to its original metallic appearance by cooling and, more importantly, because the composition of the platinum is not changed by heating and cooling.

In a **chemical change**, new substances are formed that have different properties and composition from the original material. The new substances need not resemble the original material in any way.

When a clean copper wire is heated in a burner flame, the appearance of the copper changes from coppery metallic to glowing red. Unlike the platinum wire, the copper wire is not restored to its original appearance by cooling but instead becomes a black material. This black material is a new substance called copper(II) oxide. It was formed

Sawing wood produces physical change.

Before heating, the wire is 100% copper (1.000 g).

Copper and oxygen from the air combine chemically when the wire is heated.

After heating, the wire is black copper(II) oxide (79.9% copper, 20.1% oxygen) (1.251 g).

Tom Pantages

Tom Pantages

Tom Pantages

Cu atoms

Cu atoms
O_2 molecules

O^{2-} Cu^{2+}

FIGURE 4.3 Chemical change: Forming of copper(II) oxide from copper and oxygen.

by a chemical change when copper combined with oxygen in the air during the heating process. The unheated wire was essentially 100% copper, but the copper(II) oxide is 79.9% copper and 20.1% oxygen. One gram of copper will yield 1.251 g of copper(II) oxide (FIGURE 4.3). The platinum is changed only physically when heated, but the copper is changed both physically and chemically when heated.

Water can be decomposed chemically into hydrogen and oxygen. This is usually accomplished by passing electricity through the water in a process called *electrolysis*. Hydrogen collects at one electrode while oxygen collects at the other (FIGURE 4.4). The composition and the physical appearance of the hydrogen and the oxygen are quite different from those of water. They are both colorless gases, but each behaves differently when a burning wooden splint is placed into the sample: The hydrogen explodes with a pop while the flame

Charles D. Winters/Science Source Images

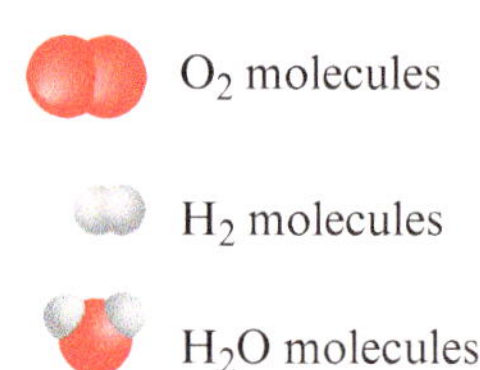

FIGURE 4.4 Electrolysis of water produces hydrogen gas (on the right) and oxygen gas (on the left). Note the ratio of the gases is 2:1.

brightens considerably in the oxygen (oxygen supports and intensifies the combustion of the wood). From these observations, we conclude that a chemical change has taken place.

Chemists have devised a shorthand method for expressing chemical changes in the form of **chemical equations**. The two previous examples of chemical changes can be represented by the following molecular representations, word and symbol equations. In the electrolysis reaction, water decomposes into hydrogen and oxygen when electrolyzed. In the formation reaction, copper plus oxygen when heated produce copper(II) oxide. The arrow means "produces," and it points to the products. The Greek letter delta (Δ) represents heat. The starting substances (water, copper, and oxygen) are called the **reactants**, and the substances produced [hydrogen, oxygen, and copper(II) oxide] are called the **products**. We will learn more about writing chemical equations in later chapters.

Formation of Copper(II) Oxide

Type of equation	Reactants				Products
Word	copper	+	oxygen	$\xrightarrow{\Delta}$	copper(II) oxide
Molecular		+		$\xrightarrow{\Delta}$	
Symbol (formula)	2 Cu	+	O_2	$\xrightarrow{\Delta}$	2 CuO

Electrolysis of Water

Type of equation	Reactants		Products		
Word	water	$\xrightarrow[\text{energy}]{\text{electrical}}$	hydrogen	+	oxygen
Molecular		$\xrightarrow[\text{energy}]{\text{electrical}}$		+	
Symbol (formula)	$2\ H_2O$	$\xrightarrow[\text{energy}]{\text{electrical}}$	$2\ H_2$	+	O_2

Physical change usually accompanies a chemical change. TABLE 4.2 lists some common physical and chemical changes; note that wherever a chemical change occurs, a physical change also occurs. However, wherever a physical change is listed, only a physical change occurs.

TABLE 4.2 Physical or Chemical Changes of Some Common Processes

Process taking place	Type of change	Accompanying observations
Rusting of iron	Chemical	Shiny, bright metal changes to reddish brown rust.
Boiling of water	Physical	Liquid changes to vapor.
Burning of sulfur in air	Chemical	Yellow, solid sulfur changes to gaseous, choking sulfur dioxide.
Boiling of an egg	Chemical	Liquid white and yolk change to solids.
Combustion of gasoline	Chemical	Liquid gasoline burns to gaseous carbon monoxide, carbon dioxide, and water.
Digestion of food	Chemical	Food changes to liquid nutrients and partially solid wastes.
Sawing of wood	Physical	Smaller pieces of wood and sawdust are made from a larger piece of wood.
Burning of wood	Chemical	Wood burns to ashes, gaseous carbon dioxide, and water.
Heating of glass	Physical	Solid becomes pliable during heating, and the glass may change its shape.

ONLINE LEARNING MODULE
Chemical and Physical Changes
➥ WileyPLUS

ENHANCED EXAMPLE 4.2 WileyPLUS

In a distillation, a chemist is boiling pure ethyl alcohol and observes that the boiling point is 78.5°C. Using an apparatus to condense and recover the alcohol vapors back to a liquid, she finds that the boiling point of the condensed liquid also is 78.5°C. Describe whether this experiment is a physical or chemical change.

SOLUTION

The process of changing a liquid to a vapor and back to a liquid is a physical change verified by the fact that both liquids have the same boiling points.

PRACTICE 4.2

When ice melts, is this a physical or chemical change?

4.3 Learning to Solve Problems

LEARNING OBJECTIVE List the basic steps in solving chemistry problems.

Now that we have learned some of the basics of chemistry including names and symbols for the elements, dimensional analysis, and a good amount of chemical terminology, we are ready to begin learning to solve chemistry problems.

One of the great joys of studying chemistry is learning to be a good problem solver. The ability to solve complicated problems is a skill that will help you greatly throughout your life. It is our goal to help you learn to solve problems by thinking through the problems on your own. To this end the basic steps in solving problems are:

Read Read the problem carefully. Determine what is known and what is to be solved for and write them down. It is important to label all factors and measurements with the proper units.

Plan Determine which principles are involved and which unit relationships are needed to solve the problem. You may refer to tables for needed data. Set up the problem in a neat, organized, and logical fashion, making sure all unwanted units cancel. Use the examples as guides for setting up the problem.

Calculate Proceed with the necessary mathematical operations. Make certain that your answer contains the proper number of significant figures.

Check Check the answer to see if it is reasonable.

At first we will help you through this process in each example. Then later in the text as you learn more about solving problems and have lots of practice, we'll help you less. By the end of your course you'll be solving problems on your own.

A few more words about problem solving. Don't allow any formal method of problem solving to limit your use of common sense and intuition. If a problem is clear to you and its solution seems simpler by another method, by all means use it.

If you approach problem solving using this general method, you will be able to learn to solve more and more difficult problems on your own. As you gain confidence you will become an independent, creative problem solver and be able to use this skill wherever you go in "real life."

4.4 Energy

LEARNING OBJECTIVE List the various forms of energy, explain the role of energy in chemical changes, and state the law of conservation of energy.

KEY TERMS
energy
potential energy
kinetic energy
law of conservation of energy

From the early discovery that fire can warm us and cook our food to our discovery that nuclear reactors can be used to produce vast amounts of controlled energy, our technical progress has been directed by our ability to produce, harness, and utilize energy. **Energy** is the capacity of matter to do work. Energy exists in many forms; some of the more familiar forms are mechanical, chemical, electrical, heat, nuclear, and radiant or light energy. Matter can have both potential and kinetic energy.

Potential energy (PE) is stored energy, or energy that an object possesses due to its relative position. For example, a ball located 20 ft above the ground has more potential energy than when located 10 ft above the ground and will bounce higher when allowed to fall. Water backed up behind a dam represents potential energy that can be converted into useful work in the form of electrical or mechanical energy. Gasoline is a source of chemical potential energy. When gasoline burns (combines with oxygen), the heat released is associated with a decrease in potential energy. The new substances formed by burning have less chemical potential energy than the gasoline and oxygen.

Kinetic energy (KE) is energy that matter possesses due to its motion. When the water behind the dam is released and allowed to flow, its potential energy is changed into kinetic energy, which can be used to drive generators and produce electricity. Moving bodies possess kinetic energy. We all know the results when two moving vehicles collide: Their kinetic energy is expended in the crash that occurs. The pressure exerted by a confined gas is due to the kinetic energy of rapidly moving gas particles.

Energy can be converted from one form to another form. Some kinds of energy can be converted to other forms easily and efficiently. For example, mechanical energy can be converted to electrical energy with an electric generator at better than 90% efficiency. On the other hand, solar energy has thus far been directly converted to electrical energy at an efficiency approaching 30%. In chemistry, energy is most frequently released as heat.

B. Yaruin\© Brian Yarvin/The Image Works

The mechanical energy of falling water is converted to electrical energy at the hydroelectric plant at Niagara Falls. As the water falls, potential energy is converted to kinetic energy and turns a turbine to produce electrical energy.

Energy in Chemical Changes

In all chemical changes, matter either absorbs or releases energy. Chemical changes can produce different forms of energy. For example, electrical energy to start automobiles is produced by chemical changes in the lead storage battery. Light energy is produced by the chemical change that occurs in a light stick. Heat and light are released from the combustion of fuels. For example, a rocket can use the reaction between hydrogen and oxygen-producing water vapor to help power it (FIGURE 4.5). All the energy needed for our life processes—breathing, muscle contraction, blood circulation, and so on—is produced by chemical changes occurring within the cells of our bodies.

Conversely, energy is used to cause chemical changes. For example, a chemical change occurs in the electroplating of metals when electrical energy is passed through a salt solution in which the metal is submerged. A chemical change occurs when radiant energy from the sun is used by green plants in the process of photosynthesis. And as we saw, a chemical change occurs when electricity is used to decompose water into hydrogen and oxygen. Chemical changes are often used primarily to produce energy rather than to produce new substances. The heat or thrust generated by the combustion of fuels is more important than the new substances formed.

Steve Nesius/Reuters/Newscom

FIGURE 4.5 The Orion rocket has external fuel tanks which provide the chemicals for the reactions to power the space craft into orbit.

EXAMPLE 4.3

Methane (CH_4) is the major compound in natural gas and is used in our homes for heating and cooking. The combustion of methane in oxygen (air) produces considerable heat energy. The chemical reaction is

$$\underset{\text{methane}}{CH_4} + \underset{\text{oxygen}}{2\,O_2} \rightarrow \underset{\text{carbon dioxide}}{CO_2} + \underset{\text{water}}{2\,H_2O} + \underset{\text{heat}}{802.5\text{ kJ}}$$

If it were possible to make methane from carbon dioxide and water, how much heat energy would it take to make methane? Write the equation.

SOLUTION

From the equation it takes 802.5 kJ to make methane. The chemical equation for the reaction is

$$CO_2 + 2\,H_2O + 802.5\text{ kJ} \rightarrow CH_4 + 2\,O_2$$

ONLINE LEARNING MODULE

Energy

 **WileyPLUS**

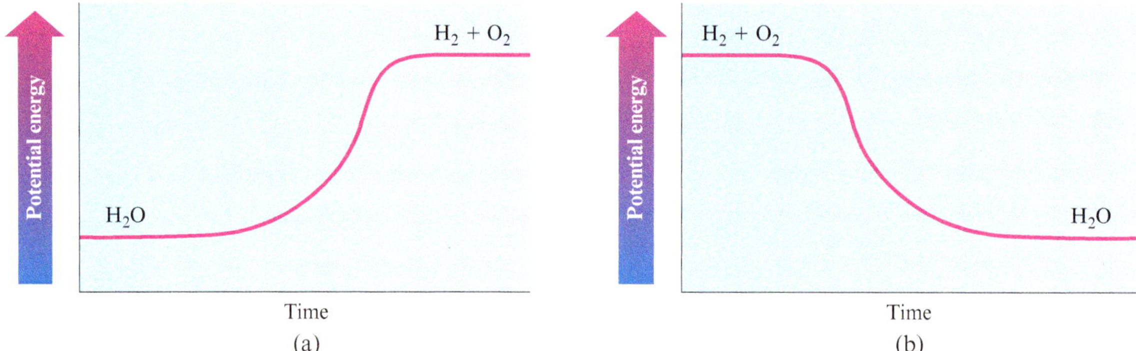

FIGURE 4.6 (a) In electrolysis of water, energy is absorbed by the system, so the products H_2 and O_2 have a higher potential energy. (b) When hydrogen is used as a fuel, energy is released and the product (H_2O) has lower potential energy.

Conservation of Energy

An energy transformation occurs whenever a chemical change occurs (FIGURE 4.6). If energy is absorbed during the change, the products will have more chemical potential energy than the reactants. Conversely, if energy is given off in a chemical change, the products will have less chemical potential energy than the reactants. Water can be decomposed in an electrolytic cell by absorbing electrical energy. The products, hydrogen and oxygen, have a greater chemical potential energy level than that of water (see Figure 4.6a). This potential energy is released in the form of heat and light when the hydrogen and oxygen are burned to form water again (see Figure 4.6b). Thus, energy can be changed from one form to another or from one substance to another and, therefore, is not lost.

The energy changes occurring in many systems have been thoroughly studied. No system has ever been found to acquire energy except at the expense of energy possessed by another system. This is the **law of conservation of energy**: Energy can be neither created nor destroyed, though it can be transformed from one form to another.

4.5 Heat: Quantitative Measurement

LEARNING OBJECTIVE Calculate the amount of heat lost or gained in a given system.

KEY TERMS
joule
calorie
specific heat

The SI-derived unit for energy is the joule (pronounced *jool*, rhyming with *tool*, and abbreviated J). Another unit for heat energy, which has been used for many years, is the calorie (abbreviated cal). The relationship between joules and calories is

$$4.184 \text{ J} = 1 \text{ cal} \quad \text{(exactly)}$$

To give you some idea of the magnitude of these heat units, 4.184 **joules**, or 1 **calorie**, is the quantity of heat energy required to change the temperature of 1 gram of water by 1°C, usually measured from 14.5°C to 15.5°C.

Since joules and calories are rather small units, kilojoules (kJ) and kilocalories (kcal) are used to express heat energy in many chemical processes. The kilocalorie is also known as the nutritional Calorie (spelled with a capital C and abbreviated Cal). In this book, heat energy will be expressed in joules.

The difference in the meanings of the terms *heat* and *temperature* can be seen by this example: Visualize two beakers, A and B. Beaker A contains 100 g of water at 20°C, and beaker B contains 200 g of water also at 20°C. The beakers are heated until the temperature of the water in each reaches 30°C. The temperature of the water in the beakers was raised by exactly the same amount, 10°C. But twice as much heat (8368 J) was required to raise the temperature of the water in beaker B as was required in beaker A (4184 J).

In the middle of the eighteenth century, Joseph Black (1728–1799), a Scottish chemist, was experimenting with the heating of elements. He heated and cooled equal masses

TABLE 4.3 Specific Heat of Selected Substances

Substance	Specific heat (J/g°C)	Specific heat (cal/g°C)
Water	4.184	1.000
Ethyl alcohol	2.138	0.511
Ice	2.059	0.492
Aluminum	0.900	0.215
Iron	0.473	0.113
Copper	0.385	0.0921
Gold	0.131	0.0312
Lead	0.128	0.0305

of iron and lead through the same temperature range. Black noted that much more heat was needed for the iron than for the lead. He had discovered a fundamental property of matter—namely, that every substance has a characteristic heat capacity. Heat capacities may be compared in terms of specific heats. The **specific heat** of a substance is the quantity of heat (lost or gained) required to change the temperature of 1 g of that substance by 1°C. It follows then that the specific heat of liquid water is 4.184 J/g°C (or 1.000 cal/g°C). The specific heat of water is high compared with that of most substances. Aluminum and copper, for example, have specific heats of 0.900 J/g°C and 0.385 J/g°C, respectively (TABLE 4.3). The relation of mass, specific heat, temperature change (Δt), and quantity of heat lost or gained by a system is expressed by this general equation:

$$\begin{pmatrix}\text{mass of}\\ \text{substance}\end{pmatrix}\begin{pmatrix}\text{specific heat}\\ \text{of substance}\end{pmatrix}(\Delta t) = \text{heat}$$

Thus the amount of heat needed to raise the temperature of 200. g of water by 10.0°C can be calculated as follows:

$$(200.\ \cancel{g})\left(\frac{4.184\ \text{J}}{\cancel{g}\cancel{°C}}\right)(10.0\ \cancel{°C}) = 8.37 \times 10^3\ \text{J}$$

The Greek letter Δ is used to mean "change in."

Lowercase t is used for Celsius and upper case T for Kelvin.

Mass is in grams, specific heat is in cal/g°C or J/g°C and Δt is in °C.

CHECK YOUR UNDERSTANDING

4.3 Specific Heat

ONLINE LEARNING MODULE

Specific Heat-1

➡ WileyPLUS

ENHANCED EXAMPLE 4.4 — WileyPLUS

Calculate the specific heat of a solid in J/g°C and cal/g°C if 1638 J raises the temperature of 125 g of the solid from 25.0°C to 52.6°C.

SOLUTION

Read **Knowns:** 125 g of the solid
$\Delta t = 52.6 - 25.0 = 27.6°\text{C}$
Heat = 1638 J

Solving for: specific heat of the solid

Plan Use the equation

$$(\text{mass})(\text{specific heat})(\Delta t) = \text{heat}$$

solving for specific heat

$$\text{specific heat} = \frac{\text{heat}}{\text{g} \times \Delta t}$$

Calculate

$$\text{specific heat} = \frac{1638\ \text{J}}{125\ \text{g} \times 27.6°\text{C}} = 0.475\ \text{J/g°C}$$

Convert joules to calories using 1.00 cal/4.184 J

$$\text{specific heat} = \left(\frac{0.475\ \cancel{J}}{\text{g°C}}\right)\left(\frac{1.000\ \text{cal}}{4.184\ \cancel{J}}\right) = 0.114\ \text{cal/g°C}$$

Check Note that the units in the answer agree with the units for specific heat.

ONLINE LEARNING MODULE
Specific Heat-2
➡ WileyPLUS

ENHANCED EXAMPLE 4.5 WileyPLUS

A sample of a metal with a mass of 212 g is heated to 125.0°C and then dropped into 375 g water at 24.0°C. If the final temperature of the water is 34.2°C, what is the specific heat of the metal? (Assume no heat losses to the surroundings.)

SOLUTION

Read • **Knowns:**

mass (metal) = 212 g
mass (water) = 375 g
Δt (water) = 34.2 − 24.0 = 10.2°C
Δt (metal) = 125.0 − 34.2 = 90.8°C

Solving for: specific heat of the metal

Plan • When the metal enters the water, it begins to cool, losing heat to the water. At the same time, the temperature of the water rises. This continues until the temperature of the metal and the water is equal (34.2°C), after which no net flow of heat occurs.

Heat lost or gained by a system is given by

(mass)(specific heat)(Δt) = energy change (heat)

Calculate • **heat gained by the water = heat lost by the metal**

heat gained by the water = (mass water)(specific heat)(Δt)

$$= (375\ \cancel{g})\left(\frac{4.184\ \text{J}}{\cancel{g}\,\cancel{^\circ\text{C}}}\right)(10.2\ \cancel{^\circ\text{C}}) = 1.60 \times 10^4\ \text{J}$$

heat gained by the water = heat lost by the metal = 1.60×10^4 J

(mass metal)(specific heat metal)(Δt) = 1.60×10^4 J

(212 g)(specific heat metal)(90.8°C) = 1.60×10^4 J

To determine the specific heat of the metal, we rearrange the equation

(mass)(specific heat)(Δt) = heat

solving for the specific heat of the metal

$$\text{specific heat metal} = \frac{1.60 \times 10^4\ \text{J}}{(212\ \text{g})(90.8^\circ\text{C})} = 0.831\ \text{J/g}^\circ\text{C}$$

Check • Note the units in the answer agree with units for specific heat.

PRACTICE 4.3

Calculate the quantity of energy needed to heat 8.0 g of water from 42.0°C to 45.0°C.

PRACTICE 4.4

A 110.0-g sample of metal at 55.5°C raises the temperature of 150.0 g of water from 23.0°C to 25.5°C. Determine the specific heat of the metal in J/g°C.

4.6 Energy in the Real World

LEARNING OBJECTIVE Define a hydrocarbon compound and explain its role in the world's energy supply.

Coal, petroleum, natural gas, and woody plants provide us with a vast resource of energy, all of which is derived from the sun. Plants use the process of photosynthesis to store the sun's energy, and we harvest that energy by burning the plants or using the decay products of the plants. These decay products have been converted over millions of years to fossil fuels. As the plants died and decayed, natural processes changed them into petroleum deposits that we now use in the forms of gasoline and natural gas.

CHEMISTRY *IN ACTION* Popping Popcorn

© EuToch/iStockphoto

Popcorn

What's the secret to getting all those kernels in a bag of microwave popcorn to pop? Food scientists have looked at all sorts of things to improve the poppability of popcorn. In every kernel of popcorn there is starch and a little moisture. As the kernel heats up, the water vaporizes and the pressure builds up in the kernel. Finally, the kernel bursts and the starch expands into the white foam of popped corn.

Researchers in Indiana tested 14 types of popcorn to determine the most poppable. They found a range of 4–47% of unpopped kernels and determined that the kernels that held moisture the best produced the fewest unpopped kernels. Further analysis led them to the conclusion that the determining factor was the pericarp.

The pericarp is the hard casing on the popcorn kernel. It is made of long polymers of cellulose. When the kernel is heated up, the long chains of cellulose line up and make a strong crystalline structure. This strong structure keeps the water vapor inside the kernel, allowing the pressure to rise high enough to pop the kernel.

Getting the most pops comes from the molecular changes resulting from heating the kernel. Energy and chemistry combine to produce a tasty snack!

Although we do not really understand completely how petroleum deposits were formed, they most likely came from the remains of marine organisms living over 500 million years ago. Petroleum is composed of hydrocarbon compounds (which contain only carbon and hydrogen). Carbon is a unique element in that it can produce chains of atoms in a molecule of different lengths. TABLE 4.4 gives the names and formulas of some simple hydrocarbons. Natural gas is associated with petroleum deposits and consists mainly of methane with some ethane, propane, and butane mixed in. Large resources of natural gas are now being made available in the United States through a process called fracking (hydraulic fracturing).

Coal was formed from the remains of plants that were buried and subjected to high pressure and heat over many years. Chemical changes lowered the oxygen and hydrogen content of these plant tissues and produced the black solid we recognize as coal. The energy available from a mass of coal increases as the carbon content increases. Coal is an important energy source in the United States.

TABLE 4.4 Names and Formulas for Simple Hydrocarbons

Names	Formula
Methane	CH_4
Ethane	C_2H_6
Propane	C_3H_8
Butane	C_4H_{10}
Pentane	C_5H_{12}
Hexane	C_6H_{14}
Heptane	C_7H_{16}
Octane	C_8H_{18}
Nonane	C_9H_{20}
Decane	$C_{10}H_{22}$

Alternative energy sources include wind, solar, and biofuels.

© WillSelarep/iStockphoto

© Ashley Cooper/Corbis

© Derek Dammann/iStockphoto

We are currently facing an energy crisis that has caused us to search for new renewable energy sources for our future. We will need to consider economic, supply, and climatic factors to find fuels to serve us as we continue to grow our human population. There are a variety of potential resources: solar, nuclear, biomass, wind, and synthetic fuels. We now are beginning to focus on developing these resources and using them in the most economic way to lower our energy dependence globally.

ONLINE LEARNING MODULE
Chemical Equation for Combustion
WileyPLUS

EXAMPLE 4.6

Much of the energy used in the world is dependent on coal. Large deposits of coal are found all over the world. Coal supplies about 20% of the energy used in the United States. The combustion of coal, which is mainly carbon and reacts with oxygen in the air, produces carbon dioxide and much heat energy. Write the chemical equation for this reaction.

SOLUTION

$$C + O_2 \rightarrow CO_2 + \text{heat (393 kJ)}$$

PRACTICE 4.5

The major compounds in gasoline are hydrocarbons. Without petroleum (from which hydrocarbons are distilled to make gasoline), industry would come to a halt and millions of automobiles would be idled. Devise a general formula (C_xH_y) that represents the hydrocarbons in Table 4.4.

CHAPTER 4 REVIEW

4.1 Properties of Substances

KEY TERMS

properties
physical properties
chemical properties

- Each substance has a set of properties that is characteristic of the substance and gives it a unique identity.
- Physical properties are inherent in the substance and can be determined without altering the composition of the substance:
  - Color
  - Taste
  - Odor
  - State of matter
  - Density
  - Melting point
  - Boiling point
- Chemical properties describe the ability of a substance to interact with other substances to form different substances.

4.2 Physical and Chemical Changes

KEY TERMS

physical change
chemical change
chemical equation
reactants
products

- A physical change is a change in the physical properties or a change in the state of matter for a substance without altering the composition of the substance.
- In a chemical change different substances are formed that have different properties and composition from the original material.
- Chemical changes can be represented by chemical equations:
  - Word equation
  - Molecular equation
  - Symbol (formula) equation

4.3 Learning to Solve Problems

Read Read the problem carefully, determining what is known and what is to be solved for.

Plan Determine what is required to solve the problem and set up the problem in a neat, organized manner using the examples as guides.

Calculate Complete the necessary calculations making sure your answer contains the proper units and significant figures.

Check Check to see if your answer is reasonable.

4.4 Energy

KEY TERMS

energy
potential energy
kinetic energy
law of conservation of energy

- Energy is the capacity to do work.
- Potential energy is energy that results from position or is stored within a substance.
- Kinetic energy is the energy matter possesses as a result of its motion.
- Energy can be converted from one form to another.
- Common forms of energy:
  - Mechanical
  - Chemical
  - Electrical
  - Heat
  - Nuclear
  - Light
- In chemistry, energy is most frequently expressed as heat.
- In all chemical changes matter either absorbs or releases energy.
- Energy can be used to cause chemical change.
- Chemical changes are often used to produce energy rather than new substances.
- Energy can neither be created nor destroyed. It can be transformed from one form to another.
- The energy released or absorbed in a chemical reaction can be summarized in graphical form.

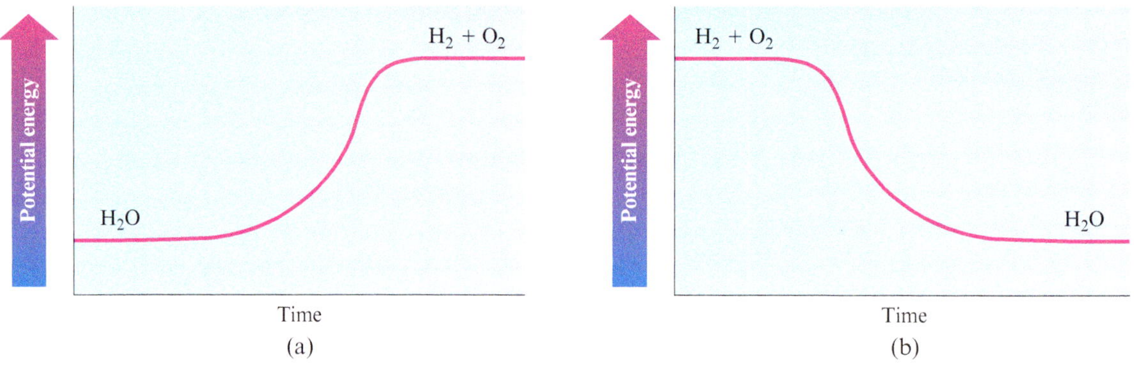

4.5 Heat: Quantitative Measurement

KEY TERMS
joule
calorie
specific heat

- The SI unit for heat is the joule:
  4.184 J = 1 cal
- The calorie is defined as the amount of heat required to change the temperature of 1 gram of water 1°C.
- Every substance has a characteristic heat capacity:
  - The specific heat of a substance is a measure of its heat capacity.
  - The specific heat is the quantity of heat required to change the temperature of 1 gram of the substance by 1°C.
- Heat lost or gained by a system can be calculated by
  heat = (mass of substance)(specific heat of substance)(Δt).

REVIEW QUESTIONS

1. In what physical state does acetic acid exist at 393 K? (Table 4.1)
2. In what physical state does chlorine exist at −65°C? (Table 4.1)
3. Calculate the boiling point of acetic acid in
(a) kelvins
(b) degrees Fahrenheit (Table 4.1)
4. Calculate the melting point of acetic acid in (Table 4.1)
(a) degrees Fahrenheit
(b) kelvins
5. What evidence of chemical change is visible when electricity is run through water? (Figure 4.4)
6. What is the fundamental difference between a chemical change and a physical change?
7. What is the critical first step in solving any chemistry problem?
8. Is the calculation step the last step in solving a chemistry problem? Why or why not?
9. Distinguish between potential and kinetic energy.
10. How is a chemistry calorie (cal) different from a food calorie (Cal)?
11. Which requires more energy to heat from 20°C to 30°C, iron or gold? (See Table 4.3.)
12. Predict which molecule would produce more carbon dioxide when burned: a molecule of ethane or a molecule of octane.
13. What is the common characteristic of all hydrocarbons?
14. Which is the most common element used to produce energy for human use?
15. Why might the combustion of fossil fuels not be the best solution to the planet's energy needs? What are some other energy sources that may become important in the future?

PAIRED EXERCISES

Most of the exercises in this chapter are available for assignment via the online homework management program, WileyPLUS (www.wileyplus.com). All exercises with blue numbers have answers in Appendix 5.

1. Determine whether each of the following represents a physical property or a chemical property:
(a) Vinegar has a pungent odor.
(b) Carbon cannot be decomposed.
(c) Sulfur is a bright yellow solid.
(d) Sodium chloride is a crystalline solid.
(e) Water does not burn.
(f) Mercury is a liquid at 25°C.
(g) Oxygen is not combustible.
(h) Aluminum combines with oxygen to form a protective oxide coating.

2. Determine whether each of the following represents a physical property or a chemical property:
(a) Chlorine gas has a greenish-yellow tint.
(b) The density of water at 4°C is 1.000 g/mL.
(c) Hydrogen gas is very flammable.
(d) Aluminum is a solid at 25°C.
(e) Water is colorless and odorless.
(f) Lemon juice tastes sour.
(g) Gold does not tarnish.
(h) Copper cannot be decomposed.

3. Cite the evidence that indicates that only physical changes occur when a platinum wire is heated in a Bunsen burner flame.

4. Cite the evidence that indicates that both physical and chemical changes occur when a copper wire is heated in a Bunsen burner flame.

5. Identify the reactants and products when a copper wire is heated in air in a Bunsen burner flame.

6. Identify the reactants and products for the electrolysis of water.

7. State whether each of the following represents a chemical change or a physical change:
   (a) A steak is cooked on a grill until well done.
   (b) In the lab, students firepolish the end of a glass rod. The jagged edge of the glass becomes smooth.
   (c) Chlorine bleach is used to remove a coffee stain on a white lab coat.
   (d) When two clear and colorless aqueous salt solutions are mixed together, the solution turns cloudy and yellow.
   (e) One gram of an orange crystalline solid is heated in a test tube, producing a green powdery solid whose volume is 10 times the volume of the original substance.
   (f) In the lab, a student cuts a 20-cm strip of magnesium metal into 1-cm pieces.

8. State whether each of the following represents a chemical change or a physical change:
   (a) A few grams of sucrose (table sugar) are placed in a small beaker of deionized water; the sugar crystals "disappear," and the liquid in the beaker remains clear and colorless.
   (b) A copper statue, over time, turns green.
   (c) When a teaspoon of baking soda (sodium bicarbonate) is placed into a few ounces of vinegar (acetic acid), volumes of bubbles (effervescence) are produced.
   (d) When a few grams of a blue crystalline solid are placed into a beaker of deionized water, the crystals "disappear" and the liquid becomes clear and blue in color.
   (e) In the lab, a student mixes 2 mL of sodium hydroxide with 2 mL of hydrochloric acid in a test tube. He notices that the test tube is very warm to the touch.
   (f) A woman visits a hairdresser and has her hair colored a darker shade of brown. After several weeks the hair, even though washed several times, does not change back to the original color.

9. Are the following examples of potential energy or kinetic energy?
   (a) fan blades spinning
   (b) a bird flying
   (c) sodium hydroxide in a sealed jar
   (d) a person napping
   (e) ocean waves rippling

10. Are the following examples of potential energy or kinetic energy?
   (a) a fish swimming
   (b) a skier at the top of a hill
   (c) a bird's egg in a nest
   (d) water starting to boil
   (e) a leaf unfolding

11. What happens to the kinetic energy of a speeding car when the car is braked to a stop?

12. What energy transformation is responsible for the fiery reentry of a rocket into Earth's atmosphere?

13. Indicate with a plus sign (+) any of these processes that require energy and a negative sign (−) any that release energy.
   (a) arctic ice melting
   (b) starting a car
   (c) flash of lightning
   (d) dry ice changing to vapor
   (e) blowing up a balloon

14. Indicate with a plus sign (+) any of these processes that require energy and a negative sign (−) any that release energy.
   (a) riding a bike
   (b) fireworks bursting
   (c) water evaporating
   (d) tires deflating
   (e) wood burning in a fireplace

15. How many joules of heat are required to heat 125 g of aluminum from 19.0°C to 95.5°C? (Table 4.3)

16. How many joules of heat are required to heat 65 g of lead from 22.0°C to 98.5°C? (Table 4.3)

17. How many joules of heat are required to heat 25.0 g of ethyl alcohol from the prevailing room temperature, 22.5°C, to its boiling point, 78.5°C?

18. How many joules of heat are required to heat 35.0 g of isopropyl alcohol from the prevailing room temperature, 21.2°C, to its boiling point, 82.4°C? (The specific heat of isopropyl alcohol is 2.604 J/g°C.)

19. A 135-g sample of a metal requires 2.50 kJ to change its temperature from 19.5°C to 100.0°C. What is the specific heat of this metal?

20. A 275-g sample of a metal requires 10.75 kJ to change its temperature from 21.2°C to its melting temperature, 327.5°C. What is the specific heat of this metal?

21. A 155-g sample of copper was heated to 150.0°C, then placed into 250.0 g water at 19.8°C. (The specific heat of copper is 0.385 J/g°C.) Calculate the final temperature of the mixture. (Assume no heat is lost to the surroundings.)

22. A 225-g sample of aluminum was heated to 125.5°C, then placed into 500.0 g water at 22.5°C. (The specific heat of aluminum is 0.900 J/g°C). Calculate the final temperature of the mixture. (Assume no heat is lost to the surroundings.)

ADDITIONAL EXERCISES

23. Read the following passage and identify at least two physical and two chemical properties of zeolites.

Zeolites are crystalline solids composed of silicon, aluminum, and oxygen with very porous structures. They are very useful in a variety of applications. Zeolites are formed naturally when volcanic rocks and ash react with alkaline groundwater. They may also be synthesized in the laboratory by allowing aluminosilicate crystals to form around small organic molecules. Zeolites generally have very low density due to their porous nature. They are often used as molecular sieves because they have large pores that can trap molecules of certain sizes and shapes while excluding others. Bandages used by the military sometimes contain zeolites, which absorb water from the blood, thus accelerating the natural clotting process. Zeolites have also replaced phosphates in many detergents, which is an ecological benefit because zeolites do not encourage the growth of algae. Interestingly, the name *zeolite* comes from the fact that zeolites give off water in the form of steam when they are heated; thus, they are called boiling (zeo) stones (lithos).

24. A 110.0-g sample of a gray-colored, unknown, pure metal was heated to 92.0°C and put into a coffee-cup calorimeter containing 75.0 g of water at 21.0°C. When the heated metal was put into the water, the temperature of the water rose to a final temperature of 24.2°C. The specific heat of water is 4.184 J/g°C.
(a) What is the specific heat of the metal?
(b) Is it possible that the metal is either iron or lead? Explain.

25. The specific heat of a human is approximately 3.47 J/g°C. Given this information,
(a) If a 165-lb man eats a candy bar containing 262 Cal, how much will his body temperature increase if all the calories from the candy bar are converted into heat energy? Remember that a food calorie (Cal) is equal to 1 kcal.
(b) If a 165-lb man eats a roll of candy containing 250 Cal, how much will his body temperature increase if all the calories from the candy bar are converted into heat energy?

26. A sample of water at 23.0°C required an input of 1.69×10^4 J of heat to reach its boiling point, 100°C. What was the mass of the water? (Table 4.3)

27. A sample of gold required 3.25×10^4 J of heat to melt it from room temperature, 23.2°C, to its melting point, 1064.4°C. How many pounds of gold were in the sample?

28. If 40.0 kJ of energy are absorbed by 500.0 g of water at 10.0°C, what is the final temperature of the water?

29. A sample of iron was heated to 125.0°C, then placed into 375 g of water at 19.8°C. The temperature of the water rose to 25.6°C. How many grams of iron were in the sample? (Table 4.3)

30. A sample of copper was heated to 275.1°C and placed into 272 g of water at 21.0°C. The temperature of the water rose to 29.7°C. How many grams of copper were in the sample? (Table 4.3)

31. A 100.0-g sample of copper is heated from 10.0°C to 100.0°C.
(a) Determine the number of calories needed.
(b) The same amount of heat is added to 100.0 g of Al at 10.0°C. Which metal gets hotter, the copper or the aluminum?

32. A 500.0-g piece of iron is heated in a flame and dropped into 400.0 g of water at 10.0°C. The temperature of the water rises to 90.0°C. How hot was the iron when it was first removed from the flame?

33. A 20.0-g piece of metal at 203°C is dropped into 100.0 g of water at 25.0°C. The water temperature rose to 29.0°C. Calculate the specific heat of the metal (J/g°C). Assume that all of the heat lost by the metal is transferred to the water and no heat is lost to the surroundings.

34. Assuming no heat loss by the system, what will be the final temperature when 50.0 g of water at 10.0°C are mixed with 10.0 g of water at 50.0°C?

35. Three 500.0-g pans of iron, aluminum, and copper are each used to fry an egg. Which pan fries the egg (105°C) most quickly? Explain.

36. At 6:00 P.M., you put a 300.0-g copper pan containing 800.0 g of water (all at room temperature, which is 25°C) on the stove. The stove supplies 628 J/s. When will the water reach the boiling point? (Assume no heat is lost.)

37. Why does blowing gently across the surface of a cup of hot coffee help to cool it? Why does inserting a spoon into the coffee do the same thing?

38. If you are boiling some potatoes in a pot of water, will they cook faster if the water is boiling vigorously than if the water is only gently boiling? Explain your reasoning.

39. Homogenized whole milk contains 4% butterfat by volume. How many milliliters of fat are there in a glass (250 mL) of milk? How many grams of butterfat ($d = 0.8$ g/mL) are in this glass of milk?

40. Gloves are often worn to protect the hands from being burned when they come in contact with very hot or very cold objects. Gloves are often made of cotton or wool, but many of the newer heat-resistant gloves are made of silicon rubber. The specific heats of these materials are listed below:

Material	Specific heat (J/g°C)
wool felt	1.38
cotton	1.33
paper	1.33
rubber	3.65
silicon rubber	1.46

(a) If a glove with a mass of 99.3 grams composed of cotton increases in temperature by 15.3°F, how much energy was absorbed by the glove?
(b) A glove with a mass of 86.2 grams increases in temperature by 25.9°F when it absorbs 1.71 kJ of energy. Calculate the specific heat of the glove and predict its composition.
(c) If a glove with a mass of 50.0 grams needs to absorb 1.65 kJ of energy, how much will the temperature of the glove increase for each of the materials listed above?
(d) Which is the best material for a heat-resistant glove?
(e) If you were designing a heat-resistant glove, what kind of specific heat would you look for?

41. If solar panels are placed in the Mojave Desert in California that generate 100 megawatts on a 1.3-square-mile site with an average of 7.50 hours of productive daylight each day, how many tons of coal are equivalent to the energy produced in one day by the solar panels? (One ton of coal will produce 26.6 gigajoules of energy, 1 Watt = 1 J/s.)

42. Suppose 25 g of solid sulfur and 35 g of oxygen gas are placed in a sealed container.
(a) Does the container hold a mixture or a compound?
(b) After heating, the container was weighed. From a comparison of the total mass before heating to the total mass after heating, can you tell whether a reaction took place? Explain.
(c) After the container is heated, all the contents are gaseous. Has the density of the container including its contents changed? Explain.

CHALLENGE EXERCISES

43. Suppose a ball is sitting at the top of a hill. At this point, the ball has potential energy. The ball rolls down the hill, so the potential energy is converted into kinetic energy. When the ball reaches the bottom of the hill, it goes halfway up the hill on the other side and stops. If energy is supposed to be conserved, then why doesn't the ball go up the other hill to the same level as it started from?

44. From the following molecular picture, determine whether a physical or a chemical change has occurred. Justify your answer.

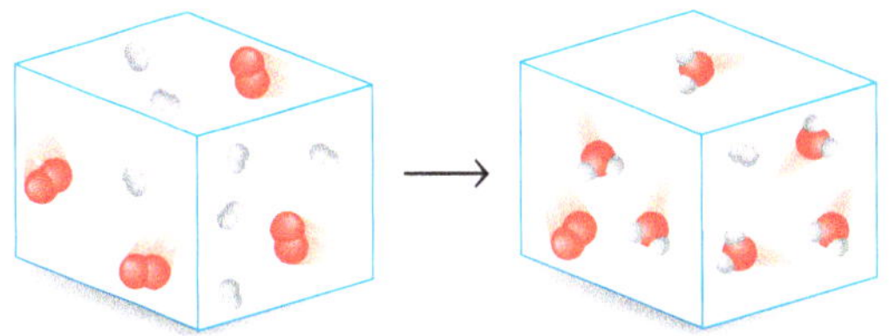

45. From the illustration below (a) describe the change that has occurred and (b) determine whether the change was a physical change or a chemical change. Justify your answer.

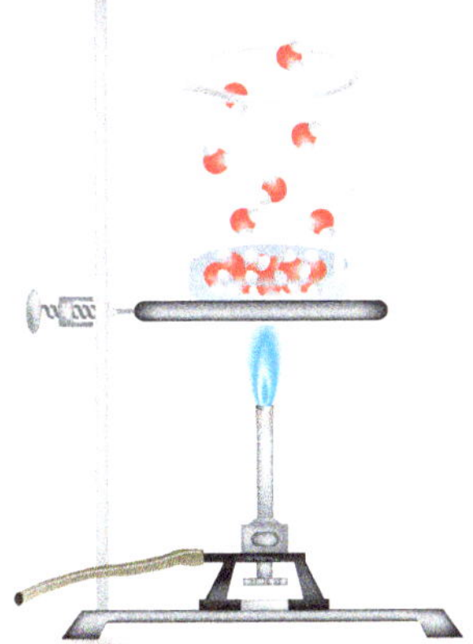

46. Describe the change(s) that you see in the following illustration. Was this a physical or a chemical change?

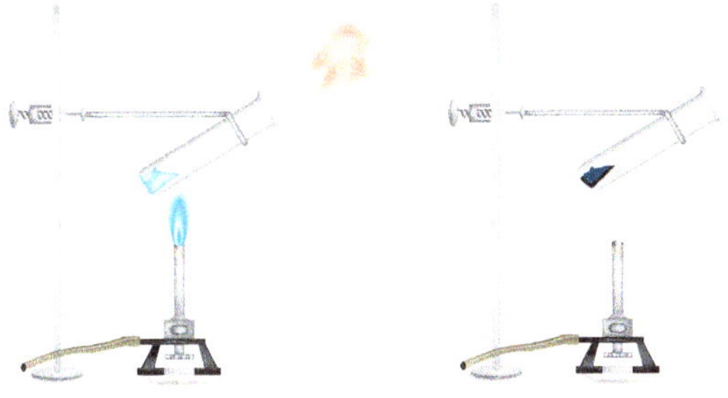

47. Sue and Tim each left a one-quart bowl outside one night. Sue's bowl was full of water and covered, while Tim's was empty and open. The next day there was a huge snowstorm that filled Tim's bowl with snow. The temperature that night went down to 12°F.

(a) Which bowl would require less energy to bring its contents to room temperature (25°C)? Why?

(b) What temperature change (°C) is required to warm the bowls to 25°C?

(c) How much heat (in kJ) is required to raise the temperature of the contents of Sue's bowl to 0°C (without converting the ice to water)?

(d) Did the water in Sue's bowl undergo chemical or physical changes or both?

48. The U.S. military provides personnel on deployment with fuel bars to heat up their meals. A fuel bar provides 300. kJ of energy, and 75% of that energy can be transferred to food. How many ounces of water could be heated by one fuel bar to brew coffee in the field? The initial temperature of the water is 15.0°C and it needs to be heated to boiling at 100.0°C.

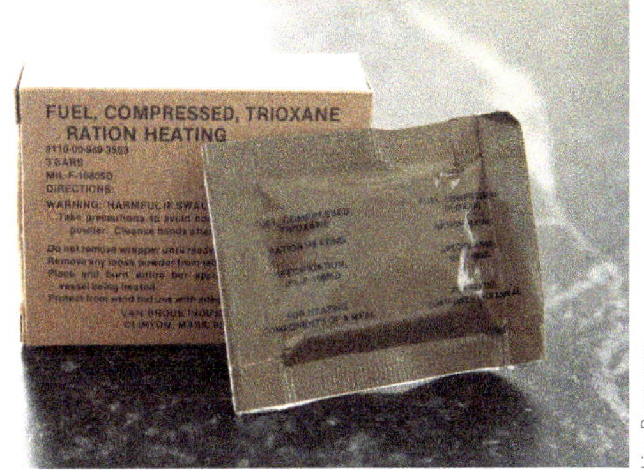

Lisa Passmore

ANSWERS TO PRACTICE EXERCISES

4.1 Each compound has one atom of silver to one atom of the other elements. Using the periodic table, we find the atomic masses of these elements to be Ag (107.9), Cl (35.45), Br (79.90), and I (126.8). Each compound has the same amount of silver, but the percent of silver will be different. The highest percent of silver will be in AgCl, the middle will be in AgBr, and the lowest will be in AgI. So the order is AgCl, AgBr, AgI.

4.2 Physical change

4.3 $1.0 \times 10^2 J = 24 cal$

4.4 0.477 J/g°C

4.5 C_nH_{2n+2} where n is the number of carbons in each formula

5

Early Atomic Theory and Structure

CHAPTER OUTLINE

Pure substances are classified as elements or compounds, but just what makes a substance possess its unique properties? How small a piece of salt will still taste salty? Carbon dioxide puts out fires, is used by plants to produce oxygen, and forms dry ice when solidified. But how small a mass of this material still behaves like carbon dioxide? Substances are in their simplest identifiable form at the atomic, ionic, or molecular level. Further division produces a loss of characteristic properties.

What particles lie within an atom or ion? Can these tiny particles be transferred from atom to atom? Lightning occurs when electrons move to neutralize a charge difference between the clouds and the Earth. How are these tiny particles alike? How do they differ? How far can we continue to divide them? Alchemists began the quest, early chemists laid the foundation, and modern chemists continue to build and expand on models of the atom.

5.1 Dalton's Model of the Atom

Describe Dalton's model of the atom and compare it to the earlier concepts of matter.

LEARNING OBJECTIVE

KEY TERMS

Dalton's atomic model

The structure of matter has long intrigued and engaged us. The earliest models of the atom were developed by the ancient Greek philosophers. About 440 B.C. Empedocles stated that all matter was composed of four "elements"—earth, air, water, and fire. Democritus (about 470–370 B.C.) thought that all forms of matter were composed of tiny indivisible particles, which he called atoms, derived from the Greek word *atomos*, meaning "indivisible." He held that atoms were in constant motion and that they combined with one another in various ways. This hypothesis was not based on scientific observations. Shortly thereafter, Aristotle (384–322 B.C.) opposed the theory of Democritus and instead endorsed and advanced the Empedoclean theory. So strong was the influence of Aristotle that his theory dominated the thinking of scientists and philosophers until the beginning of the seventeenth century.

More than 2000 years after Democritus, the English schoolmaster John Dalton (1766–1844) revived the concept of atoms and proposed an atomic model based on facts and experimental evidence (FIGURE 5.1). His theory, described in a series of papers published from 1803 to 1810, rested on the idea of a different kind of atom for each element. The essence of **Dalton's atomic model** may be summed up as follows:

1. Elements are composed of minute, indivisible particles called atoms.
2. Atoms of the same element are alike in mass and size.
3. Atoms of different elements have different masses and sizes.
4. Chemical compounds are formed by the union of two or more atoms of different elements.
5. Atoms combine to form compounds in simple numerical ratios, such as one to one, one to two, two to three, and so on.
6. Atoms of two elements may combine in different ratios to form more than one compound.

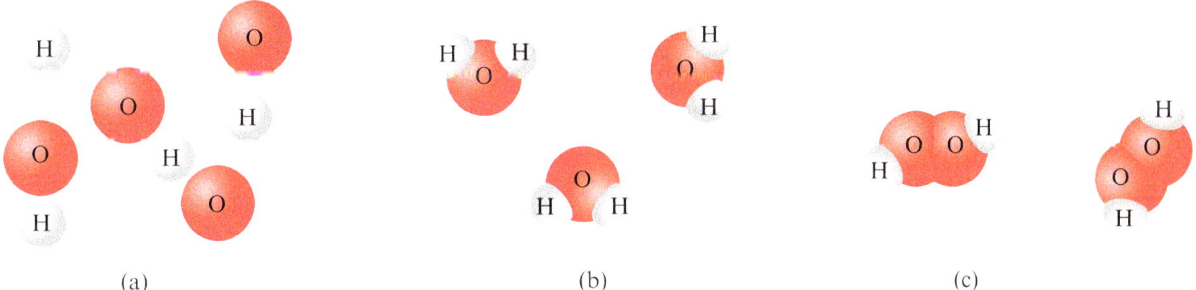

FIGURE 5.1 (a) Dalton's atoms were individual particles, the atoms of each element being alike in mass and size but different in mass and size from other elements; (b) and (c) Dalton's atoms combine in specific ratios to form compounds.

Dalton's atomic model stands as a landmark in the development of chemistry. The major premises of his model are still valid, but some of his statements must be modified or qualified because later investigations have shown that (1) atoms are composed of subatomic particles; (2) not all the atoms of a specific element have the same mass; and (3) atoms, under special circumstances, can be decomposed.

In chemistry we use models (theories) such as Dalton's atomic model to explain the behavior of atoms, molecules, and compounds. Models are modified to explain new information. We frequently learn the most about a system when our models (theories) fail. That is the time when we must rethink our explanation and determine whether we need to modify our model or propose a new or different model to explain the behavior.

CHECK YOUR UNDERSTANDING

5.1 Early Atomic Theory
5.2 Applying Dalton's Model 1
5.3 Applying Dalton's Model 2
5.4 Applying Dalton's Model 3
➡ WileyPLUS

ONLINE LEARNING MODULE
Dalton's Atomic Theory
WileyPLUS

EXAMPLE 5.1

Which statement of Dalton's model of the atom does this series of compounds represent: NO_2, N_2O, N_2O_3, N_2O_5?

SOLUTION

Atoms combine to form compounds in simple numerical ratios. Atoms of two elements combine in different ratios to form more than one compound.

PRACTICE 5.1

List another series of compounds that illustrates the statements in Example 5.1.

5.2 Electric Charge

LEARNING OBJECTIVE Recognize the force between particles and distinguish between a cation and an anion.

You've probably received a shock after walking across a carpeted area on a dry day. You may have also experienced the static electricity associated with combing your hair and have had your clothing cling to you. These phenomena result from an accumulation of *electric charge*. This charge may be transferred from one object to another. The properties of electric charge are as follows:

1. Charge may be of two types, positive and negative.
2. Unlike charges attract (positive attracts negative), and like charges repel (negative repels negative and positive repels positive).
3. Charge may be transferred from one object to another by contact or induction.
4. The less the distance between two charges, the greater the force of attraction between unlike charges (or repulsion between identical charges). The force of attraction (F) can be expressed using the following equation:

$$F = \frac{kq_1q_2}{r^2}$$

where q_1 and q_2 are the charges, r is the distance between the charges, and k is a constant.

CHECK YOUR UNDERSTANDING
5.5 Electric Attraction between Charges
WileyPLUS

Discovery of Ions

English scientist Michael Faraday (1791–1867) made the discovery that certain substances when dissolved in water conduct an electric current. He also noticed that certain compounds decompose into their elements when an electric current is passed through the compound. Atoms of some elements are attracted to the positive electrode, while atoms of other elements are attracted to the negative electrode. Faraday concluded that these atoms are electrically charged. He called them *ions* after the Greek word meaning "wanderer."

Any moving charge is an electric current. The electrical charge must travel through a substance known as a conducting medium. The most familiar conducting media are metals formed into wires.

The Swedish scientist Svante Arrhenius (1859–1927) extended Faraday's work. Arrhenius reasoned that an ion is an atom (or a group of atoms) carrying a positive or negative charge. When a compound such as sodium chloride (NaCl) is melted, it conducts electricity. Water is unnecessary. Arrhenius's explanation of this conductivity was that upon melting, the sodium chloride dissociates, or breaks up, into charged ions, Na^+ and Cl^-. The Na^+ ions move toward the negative electrode (cathode), whereas the Cl^- ions migrate toward the positive electrode (anode). Thus positive ions are called *cations*, and negative ions are called *anions*.

From Faraday's and Arrhenius's work with ions, Irish physicist G. J. Stoney (1826–1911) realized there must be some fundamental unit of electricity associated with atoms. He named this unit the electron in 1891. Unfortunately, he had no means of supporting

Richard Megna/Fundamental Photographs

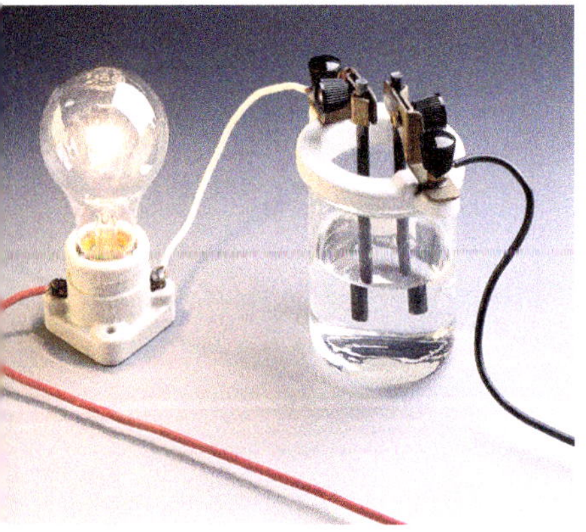

When ions are present in a solution of salt water and an electric current is passed through the solution, the light bulb glows.

his idea with experimental proof. Evidence remained elusive until 1897, when English physicist J. J. Thomson (1856–1940) was able to show experimentally the existence of the electron.

CHECK YOUR UNDERSTANDING

5.6 Cations

5.7 Anions

WileyPLUS

ENHANCED EXAMPLE 5.2 WileyPLUS

What is a cation? What is an anion?

SOLUTION

A cation is a positively charged ion. An anion is a negatively charged ion.

PRACTICE 5.2

Which of the following ions are cations and which are anions?

Na^+, Cl^-, NH_4^+, Ca^{2+}, Br^-, Al^{3+}, NO_3^-, SO_4^{2-}, PO_4^{3-}, K^+, Ag^+, Fe^{2+}, Fe^{3+}, S^{2-}

ONLINE LEARNING MODULE

Anions and Cations

WileyPLUS

5.3 Subatomic Parts of the Atom

Describe the three basic subatomic particles and how they changed Dalton's model of the atom.

LEARNING OBJECTIVE

KEY TERMS

subatomic particles

electron

proton

Thomson model of the atom

neutron

The concept of the atom—a particle so small that until recently it could not be seen even with the most powerful microscope—and the subsequent determination of its structure stand among the greatest creative intellectual human achievements.

Any visible quantity of an element contains a vast number of identical atoms. But when we refer to an atom of an element, we isolate a single atom from the multitude in order to present the element in its simplest form. FIGURE 5.2 shows individual atoms as we can see them today.

What is this tiny particle we call the atom? The diameter of a single atom ranges from 0.1 to 0.5 nanometer ($1 \text{ nm} = 1 \times 10^{-9} \text{ m}$). Hydrogen, the smallest atom, has a diameter of about 0.1 nm. To arrive at some idea of how small an atom is, consider this dot (•), which has a diameter of about 1 mm, or 1×10^6 nm. It would take 10 million hydrogen atoms to form a line of atoms across this dot. As inconceivably small as atoms are, they contain even smaller particles, the **subatomic particles**, including electrons, protons, and neutrons.

The development of atomic theory was helped in large part by the invention of new instruments. For example, the Crookes tube, developed by Sir William Crookes (1832–1919) in 1875, opened the door to the subatomic structure of the atom (FIGURE 5.3). The emissions generated in a Crookes tube are called *cathode rays*. J. J. Thomson demonstrated in 1897 that cathode rays (1) travel in straight lines, (2) are negative in charge, (3) are deflected by electric and magnetic fields, (4) produce sharp shadows, and (5) are capable of moving a small paddle wheel. This was the experimental discovery of the fundamental unit of charge—the electron.

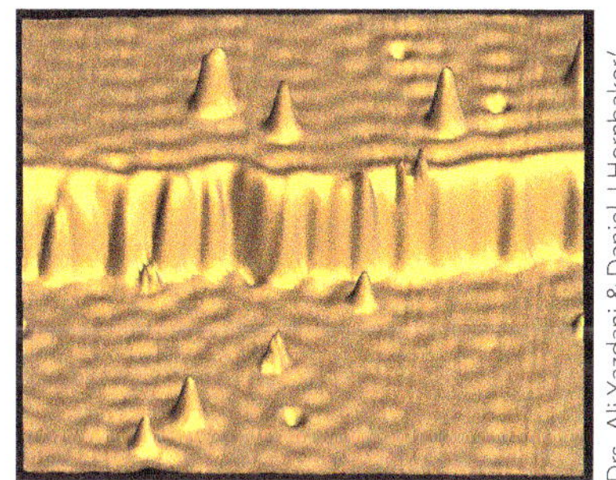

Drs. Ali Yazdani & Daniel J. Hornbaker/Science Source/Photo Researchers, Inc.

FIGURE 5.2 A scanning tunneling microscope shows an array of copper atoms.

CHECK YOUR UNDERSTANDING

5.8 Subatomic Particles

WileyPLUS

Power source (electrical)

(−)

(+)

Metal plate

Metal plate

Stream of electrons (negative particles)

FIGURE 5.3 Cathode ray tube. A stream of electrons passes between electrodes. The fast-moving particles excite the gas inside the tube, creating a greenish glow between the electrodes.

CHECK YOUR UNDERSTANDING

5.9 Charge of a Proton

5.10 Charge of an Electron

WileyPLUS

The **electron** (e^-) is a particle with a negative electrical charge and a mass of 9.110×10^{-28} g. This mass is 1/1837 the mass of a hydrogen atom. Although the actual charge of an electron is known, its value is too cumbersome for practical use and has therefore been assigned a relative electrical charge of −1. The size of an electron has not been determined exactly, but its diameter is believed to be less than 10^{-12} cm.

Protons were first observed by German physicist Eugen Goldstein (1850–1930) in 1886. However, it was Thomson who discovered the nature of the proton. He showed that the proton is a particle, and he calculated its mass to be about 1837 times that of an electron. The **proton** (p) is a particle with actual mass of 1.673×10^{-24} g. Its relative charge (+1) is equal in magnitude, but opposite in sign, to the charge on the electron. The mass of a proton is only very slightly less than that of a hydrogen atom.

Thomson had shown that atoms contain both negatively and positively charged particles. Clearly, the Dalton model of the atom was no longer acceptable. Atoms are not indivisible but are instead composed of smaller parts. Thomson proposed a new model of the atom.

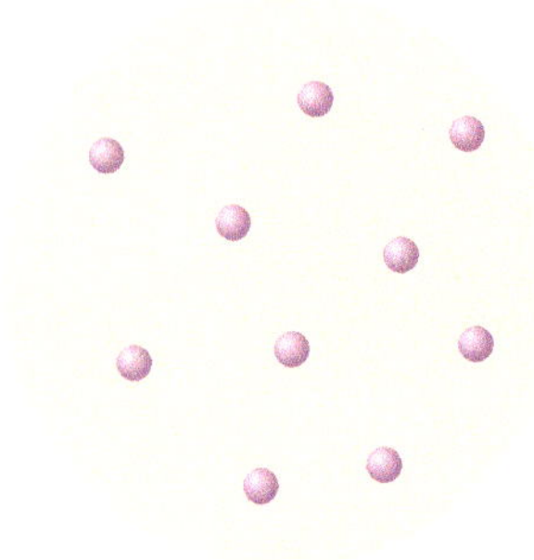

FIGURE 5.4 Thomson model of the atom. In this early model of the atom, negative particles (electrons) were thought to be embedded in a positively charged sphere. It is sometimes called the plum pudding model.

In the **Thomson model of the atom**, the electrons are negatively charged particles embedded in the positively charged atomic sphere (FIGURE 5.4). A neutral atom could become an ion by gaining or losing electrons.

Positive ions were explained by assuming that a neutral atom loses electrons. An atom with a net charge of +1 (for example, Na^+ or Li^+) has lost one electron. An atom with a net charge of +3 (for example, Al^{3+}) has lost three electrons (FIGURE 5.5A).

Negative ions were explained by assuming that additional electrons can be added to atoms. A net charge of −1 (for example, Cl^- or F^-) is produced by the addition of one electron. A net charge of −2 (for example, O^{2-} or S^{2-}) requires the addition of two electrons (FIGURE 5.5B).

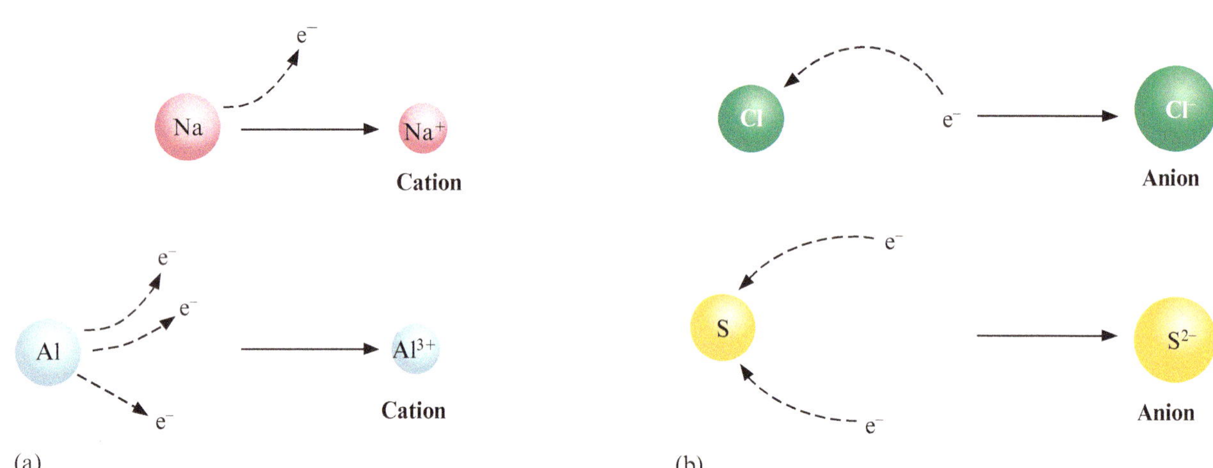

FIGURE 5.5 (a) When one or more electrons are lost from a neutral atom, a cation is formed. (b) When one or more electrons are added to a neutral atom, an anion is formed.

The third major subatomic particle was discovered in 1932 by James Chadwick (1891–1974). This particle, the **neutron** (n), has neither a positive nor a negative charge and has an actual mass (1.675×10^{-24} g) that is only very slightly greater than that of a proton. The properties of these three subatomic particles are summarized in TABLE 5.1.

TABLE 5.1 Electrical Charge and Relative Mass of Electrons, Protons, and Neutrons

Particle	Symbol	Relative electrical charge	Actual mass (g)
Electron	e^-	−1	9.110×10^{-28}
Proton	p	+1	1.673×10^{-24}
Neutron	n	0	1.675×10^{-24}

Nearly all the ordinary chemical properties of matter can be explained in terms of atoms consisting of electrons, protons, and neutrons. The discussion of atomic structure that follows is based on the assumption that atoms contain only these principal subatomic particles. Many other subatomic particles, such as mesons, positrons, neutrinos, and antiprotons, have been discovered, but it is not yet clear whether all these particles are actually present in the atom or whether they are produced by reactions occurring within the nucleus. The fields of atomic and high-energy physics have produced a long list of subatomic particles. Descriptions of the properties of many of these particles are to be found in physics textbooks.

ENHANCED EXAMPLE 5.3 WileyPLUS

The mass of a helium atom is 6.65×10^{-24} g. How many atoms are in a 4.0-g sample of helium?

SOLUTION

$$(4.0 \text{ g})\left(\frac{1 \text{ atom He}}{6.65 \times 10^{-24} \text{g}}\right) = 6.0 \times 10^{23} \text{ atoms He}$$

PRACTICE 5.3

The mass of an atom of hydrogen is 1.673×10^{-24} g. How many atoms are in a 10.0-g sample of hydrogen?

ONLINE LEARNING MODULE

Using Mass to Determine Number of Atoms

➡ WileyPLUS

5.4 The Nuclear Atom

LEARNING OBJECTIVE

Explain how the nuclear model of the atom differs from Dalton's and Thomson's models.

KEY TERMS

nucleus

atomic number

The discovery that positively charged particles are present in atoms came soon after the discovery of radioactivity by Henri Becquerel (1852–1908) in 1896. Radioactive elements spontaneously emit alpha particles, beta particles, and gamma rays from their nuclei (see Chapter 18).

By 1907 Ernest Rutherford (1871–1937) had established that the positively charged alpha particles emitted by certain radioactive elements are ions of the element helium. Rutherford used these alpha particles to establish the nuclear nature of atoms. In experiments performed in 1911, he directed a stream of positively charged helium ions (alpha particles) at a very thin sheet of gold foil (about 1000 atoms thick). See FIGURE 5.6A. He observed that most of the alpha particles passed through the foil with little or no deflection; but a few of the particles were deflected at large angles, and occasionally one even bounced back from the foil (FIGURE 5.6B). It was known that like charges repel each other

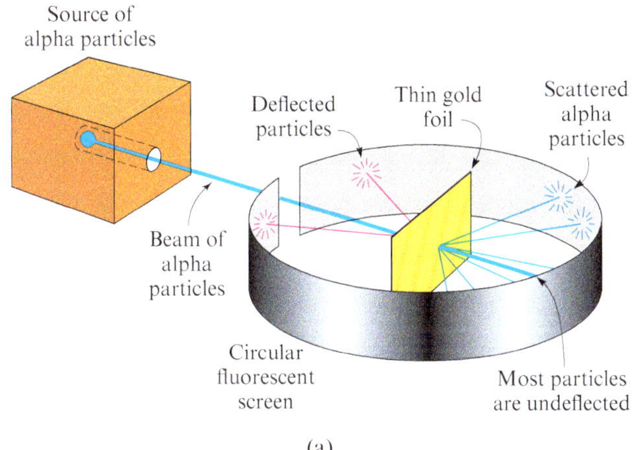

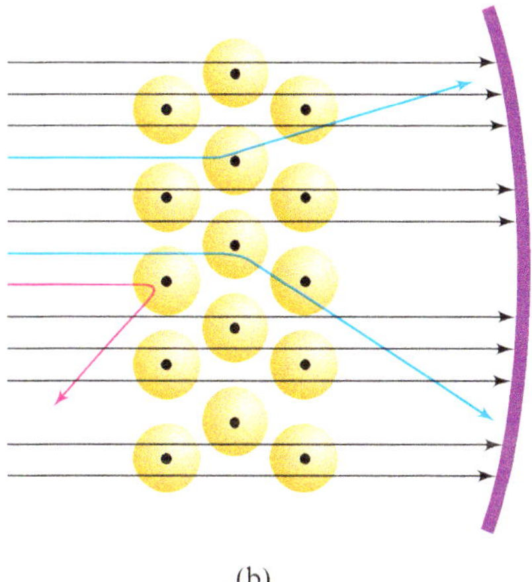

FIGURE 5.6 (a) Rutherford's experiment on alpha-particle scattering, where positive alpha particles (α), emanating from a radioactive source, were directed at a thin gold foil. (b) Deflection (red) and scattering (blue) of the positive alpha particles by the positive nuclei of the gold atoms.

and that an electron with a mass of 1/1837 that of a proton could not possibly have an appreciable effect on the path of an alpha particle, which is about 7350 times more massive than an electron. Rutherford therefore reasoned that each gold atom must contain a positively charged mass occupying a relatively tiny volume and that, when an alpha particle approaches close enough to this positive mass, it is deflected. Rutherford spoke of this positively charged mass as the *nucleus* of the atom. Because alpha particles have relatively high masses, the extent of the deflections (some actually bounced back) indicated to Rutherford that the nucleus is very heavy and dense. (The density of the nucleus of a hydrogen atom is about 10^{12} g/cm^3—about 1 trillion times the density of water.) Because most of the alpha particles passed through the thousand or so gold atoms without any apparent deflection, he further concluded that most of an atom consists of empty space.

CHECK YOUR UNDERSTANDING

5.11 Rutherford's Gold Foil Experiment

5.12 Rutherford's Conclusions

➥ WileyPLUS

When we speak of the mass of an atom, we are referring primarily to the mass of the nucleus. The nucleus contains all the protons and neutrons, which represent more than 99.9% of the total mass of any atom. By way of illustration, the largest number of electrons known to exist in an atom is 118. The mass of even 118 electrons is only about 1/17 of the mass of a single proton or neutron. The mass of an atom therefore is primarily determined by the combined masses of its protons and neutrons.

General Arrangement of Subatomic Particles

The alpha-particle scattering experiments of Rutherford established that the atom contains a dense, positively charged nucleus. The later work of Chadwick demonstrated that the atom contains neutrons, which are particles with mass, but no charge. Rutherford also noted that light, negatively charged electrons are present and offset the positive charges in the nucleus. Based on this experimental evidence, a model of the atom and the location of its subatomic particles was devised in which each atom consists of a **nucleus** surrounded by electrons (FIGURE 5.7). The nucleus contains protons and neutrons but does not contain electrons. In a neutral atom the positive charge of the nucleus (due to protons) is exactly offset by the negative electrons. Because the charge of an electron is equal to, but of opposite sign than, the charge of a proton, a neutral atom must contain exactly the same number of electrons as protons. However, this model of atomic structure provides no information on the arrangement of electrons within the atom.

CHECK YOUR UNDERSTANDING

5.13 Location of the Subatomic Particles

➥ WileyPLUS

A neutral atom contains the same number of protons and electrons.

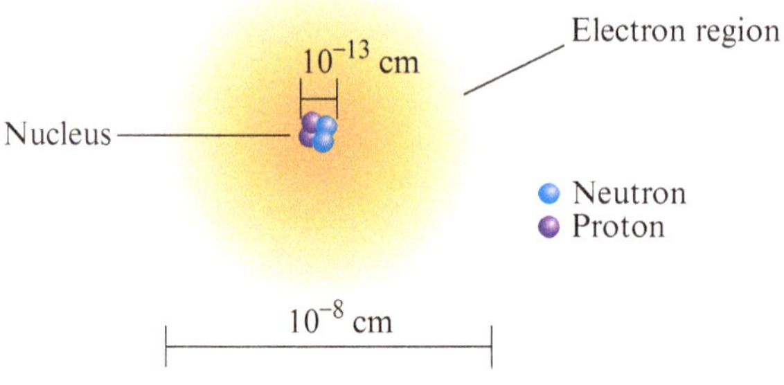

FIGURE 5.7 In the nuclear model of the atom, protons and neutrons are located in the nucleus. The electrons are found in the remainder of the atom (which is mostly empty space because electrons are very tiny).

ONLINE LEARNING MODULE

Atom Structure

➥ WileyPLUS

EXAMPLE 5.4

What are the major subatomic particles and where are they located in an atom?

SOLUTION

Protons, neutrons, and electrons are the three major subatomic particles in an atom. The protons and neutrons are located in the nucleus and the electrons are located outside the nucleus and around the nucleus.

Atomic Numbers of the Elements

The **atomic number** (Z) of an element is the number of protons in the nucleus of an atom of that element. The atomic number determines the identity of an atom. For example, every atom with an atomic number of 1 is a hydrogen atom; it contains one proton in its nucleus. Every atom with an atomic number of 6 is a carbon atom; it contains 6 protons in its nucleus. Every atom with an atomic number of 92 is a uranium atom; it contains 92 protons in its nucleus. The atomic number tells us not only the number of positive charges in the nucleus but also the number of electrons in the neutral atom, since a neutral atom contains the same number of electrons and protons.

$$\text{atomic number} = \text{number of protons in the nucleus}$$

You don't need to memorize the atomic numbers of the elements because a periodic table is usually provided in texts, in laboratories, and on examinations. The atomic numbers of all elements are shown in the periodic table on the inside front cover of this book and are also listed in the table of atomic masses on the inside front endpapers.

PRACTICE 5.4

Look in the periodic table and determine how many protons and electrons are in a Cu atom and an Xe atom.

5.5 Isotopes of the Elements

Define the terms *atomic number*, *mass number*, and *isotope*.

LEARNING OBJECTIVE

KEY TERMS

isotope

mass number

Shortly after Rutherford's conception of the nuclear atom, experiments were performed to determine the masses of individual atoms. These experiments showed that the masses of nearly all atoms were greater than could be accounted for by simply adding up the masses of all the protons and electrons that were known to be present in an atom. This fact led to the concept of the neutron, a particle with no charge but with a mass about the same as that of a proton. Because this particle has no charge, it was very difficult to detect, and the existence of the neutron was not proven experimentally until 1932. All atomic nuclei except that of the simplest hydrogen atom contain neutrons.

All atoms of a given element have the same number of protons. Experimental evidence has shown that, in most cases, all atoms of a given element do not have identical masses. This is because atoms of the same element may have different numbers of neutrons in their nuclei.

Atoms of an element having the same atomic number but different atomic masses are called **isotopes** of that element. Atoms of the various isotopes of an element therefore have the same number of protons and electrons but different numbers of neutrons.

Three isotopes of hydrogen (atomic number 1) are known. Each has one proton in the nucleus and one electron. The first isotope (protium), without a neutron, has a mass number of 1; the second isotope (deuterium), with one neutron in the nucleus, has a mass number of 2; the third isotope (tritium), with two neutrons, has a mass number of 3 (FIGURE 5.8).

The three isotopes of hydrogen may be represented by the symbols $^{1}_{1}H$, $^{2}_{1}H$ and $^{3}_{1}H$, indicating an atomic number of 1 and mass numbers of 1, 2, and 3, respectively. This

An uncut diamond from a Colorado mine. Diamonds are composed of carbon.

$^{1}_{1}H$ Protium | $^{2}_{1}H$ or D Deuterium | $^{3}_{1}H$ or T Tritium

FIGURE 5.8 The isotopes of hydrogen. The number of protons (pink) and neutrons (blue) are shown within the nucleus. The electron (e^-) exists outside the nucleus.

CHECK YOUR UNDERSTANDING
5.14 Isotopes
WileyPLUS

method of representing atoms is called *isotopic notation*. The subscript (Z) is the atomic number; the superscript (A) is the **mass number**, which is the sum of the number of protons and the number of neutrons in the nucleus. The hydrogen isotopes may also be referred to as hydrogen-1, hydrogen-2, and hydrogen-3.

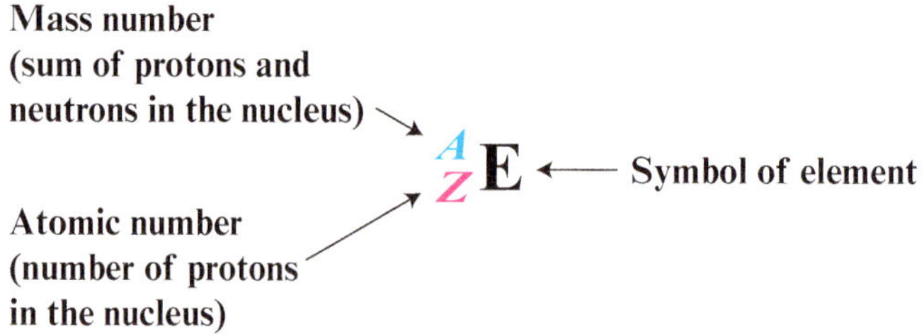

The mass number of an element is the sum of the protons and neutrons in the nucleus.

Most of the elements occur in nature as mixtures of isotopes. However, not all isotopes are stable; some are radioactive and are continuously decomposing to form other elements. For example, of the seven known isotopes of carbon, only two, carbon-12 and carbon-13, are stable. Of the seven known isotopes of oxygen, only three—$^{16}_{8}O$, $^{17}_{8}O$, and $^{18}_{8}O$—are stable. Of the fifteen known isotopes of arsenic, $^{75}_{33}As$ is the only one that is stable.

The relationship between mass number and atomic number is such that if we subtract the atomic number from the mass number of a given isotope, we obtain the number of neutrons in the nucleus of an atom of that isotope. TABLE 5.2 shows this method of determining the number of neutrons. For example, the fluorine atom ($^{19}_{9}F$), atomic number 9, having a mass of 19 contains 10 neutrons:

mass number	−	atomic number	=	number of neutrons
19	−	9	=	10

The atomic masses given in the table on the front endpapers of this book are values accepted by international agreement. You need not memorize atomic masses. In the calculations in this book, the use of atomic masses rounded to four significant figures will give results of sufficient accuracy. (See periodic table.)

Use four significant figures for atomic masses in this text.

TABLE 5.2 Determination of the Number of Neutrons in an Atom by Subtracting Atomic Number from Mass Number

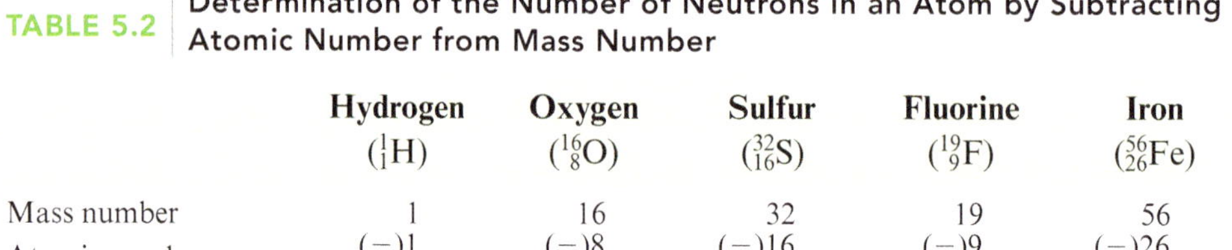

	Hydrogen ($^{1}_{1}H$)	Oxygen ($^{16}_{8}O$)	Sulfur ($^{32}_{16}S$)	Fluorine ($^{19}_{9}F$)	Iron ($^{56}_{26}Fe$)
Mass number	1	16	32	19	56
Atomic number	(−)1	(−)8	(−)16	(−)9	(−)26
Number of neutrons	0	8	16	10	30

CHECK YOUR UNDERSTANDING
5.15 Atomic Number
5.16 Mass Number
WileyPLUS

ONLINE LEARNING MODULE
Isotope Composition
WileyPLUS

ENHANCED EXAMPLE 5.5

WileyPLUS

How many protons, neutrons, and electrons are found in an atom of $^{14}_{6}C$?

SOLUTION

The element is carbon, atomic number 6. The number of protons or electrons equals the atomic number and is 6. The number of neutrons is determined by subtracting the atomic number from the mass number: $14 - 6 = 8$.

PRACTICE 5.5

How many protons, neutrons, and electrons are in each of these isotopes?
(a) $^{16}_{8}O$ (b) $^{80}_{35}Br$ (c) $^{235}_{92}U$ (d) $^{64}_{29}Cu$

PRACTICE 5.6

What are the atomic number and the mass number of the elements that contain
(a) 9 electrons (b) 24 protons and 28 neutrons (c) $^{197}_{79}X$
What are the names of these elements?

CHEMISTRY IN ACTION Isotope Detectives

Scientists are learning to use isotopes to determine the origin of drugs and gems. It turns out that isotope ratios similar to those used in carbon dating can also identify the source of cocaine or the birthplace of emeralds.

Researchers with the Drug Enforcement Agency (DEA) have created a database of the origin of coca leaves that pinpoints the origin of the leaves with a 90% accuracy. Cocaine keeps a chemical signature of the environment where it grew. Isotopes of carbon and nitrogen are found in a particular ratio based on climatic conditions in the growing region. These ratios correctly identified the source of 90% of the samples tested, according to James Ehleringer of the University of Utah, Salt Lake City. This new method can trace drugs a step further back than current techniques, which mainly look at chemicals introduced by processing practices in different locations. This could aid in tracking the original exporters and stopping production at the source.

It turns out that a similar isotopic analysis of oxygen has led researchers in France to be able to track the birthplace of emeralds. Very high quality emeralds have few inclusions (microscopic cavities). Gemologists use these inclusions and the material trapped in them to identify the source of the gem. High-quality gems can now also be identified by using an oxygen isotope ratio. These tests use an ion microscope that blasts a few atoms from the gems' surface (with virtually undetectable damage). The tiny sample is analyzed for its oxygen isotope ratio and then compared to a database from emerald mines around the world. Using the information, gemologists can determine the mine from which the emerald came. Since emeralds from Colombian mines are valued much more highly than those from other countries, this technique can be used to help collectors know just what they are paying for, as well as to identify the history of treasured emeralds.

PjrStudio/Alamy

Raw and cut emeralds.

5.6 Atomic Mass

LEARNING OBJECTIVE Explain the relationship between the atomic mass of an element and the masses of its isotopes.

KEY TERMS
atomic mass unit (u)
atomic mass

The mass of a single atom is far too small to measure on a balance, but fairly precise determinations of the masses of individual atoms can be made with an instrument called a *mass spectrometer*. The mass of a single hydrogen atom is 1.673×10^{-24} g. However, it is neither convenient nor practical to compare the actual masses of atoms expressed in grams; therefore, a table of relative atomic masses using *atomic mass units* was devised. The International Union of Pure and Applied Chemists (IUPAC) recently decided that the abbreviation for atomic mass unit should be u. We will use this unit in this text. (The term *atomic weight* is sometimes used instead of *atomic mass*.) The carbon isotope having six protons and six neutrons and designated carbon-12, or $^{12}_{6}C$, was chosen as the standard for atomic masses. This reference isotope was assigned a value of exactly 12 atomic mass units (u). Thus, 1 **atomic mass unit (u)** is defined as equal to exactly 1/12 of the mass of a carbon-12 atom. The actual mass of a carbon-12 atom is 1.9927×10^{-23} *g* and that of one atomic mass unit is 1.6606×10^{-24} g. In the table of atomic masses, all elements then have values that are relative to the mass assigned to the reference isotope, carbon-12.

$1 \text{ u} = 1.6606 \times 10^{-24}$ g

A table of atomic masses is given on the inside front cover of this book. Hydrogen atoms, with a mass of about 1/12 that of a carbon atom, have an average atomic mass of 1.00794 u on this relative scale. Magnesium atoms, which are about twice as heavy as carbon, have an average mass of 24.305 u. The average atomic mass of oxygen is 15.9994 u.

Since most elements occur as mixtures of isotopes with different masses, the atomic mass determined for an element represents the average relative mass of all the naturally occurring isotopes of that element. The atomic masses of the individual isotopes are approximately whole numbers, because the relative masses of the protons and neutrons are approximately 1.0 u each. Yet we find that the atomic masses given for many of the elements deviate considerably from whole numbers.

For example, the atomic mass of rubidium is 85.4678 u, that of copper is 63.546 u, and that of magnesium is 24.305 u. The deviation of an atomic mass from a whole number is due mainly to the unequal occurrence of the various isotopes of an element.

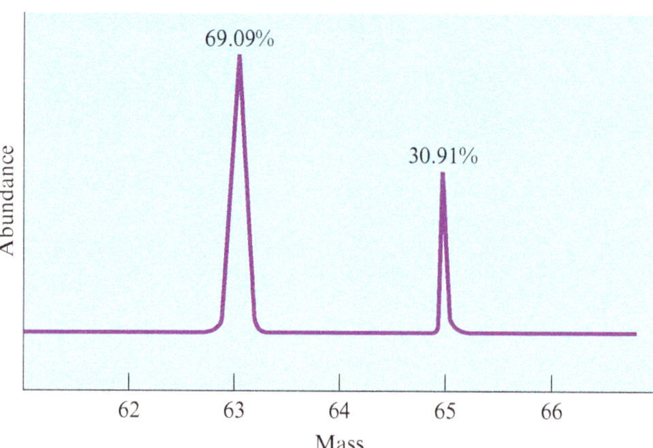

FIGURE 5.9 A typical reading from a mass spectrometer. The two principal isotopes of copper are shown with the abundance (%) given.

The two principal isotopes of copper are $^{63}_{29}Cu$ and $^{65}_{29}Cu$. Copper used in everyday objects and the Liberty Bell contains a mixture of these two isotopes. It is apparent that copper-63 atoms are the more abundant isotope, since the atomic mass of copper, 63.546 u, is closer to 63 u than to 65 u (FIGURE 5.9). The actual values of the copper isotopes observed by mass spectra determination are shown in the following table:

Isotope	Isotopic mass (u)	Abundance (%)	Average atomic mass (u)
$^{63}_{29}Cu$	62.9298	69.09	63.55
$^{65}_{29}Cu$	64.9278	30.91	

© Tetra Images/Getty Images, Inc.

The copper used in casting the Liberty Bell contains a mixture of the isotopes of copper.

The average atomic mass can be calculated by multiplying the atomic mass of each isotope by the fraction of each isotope present and adding the results. The calculation for copper is

$$(62.9298\ \text{u})(0.6909) = 43.48\ \text{u}$$
$$(64.9278\ \text{u})(0.3091) = \underline{20.07\ \text{u}}$$
$$63.55\ \text{u}$$

The average atomic mass is a weighted average of the masses of all the isotopes present in the sample.

The **atomic mass** of an element is the average relative mass of the isotopes of that element compared to the atomic mass of carbon-12 (exactly 12.0000... u).

ENHANCED EXAMPLE 5.6 WileyPLUS

Chlorine is found in nature as two isotopes, $^{37}_{17}Cl$ (24.47%) and $^{35}_{17}Cl$ (75.53%). The atomic masses are 36.96590 and 34.96885 u, respectively. Determine the average atomic mass of chlorine.

SOLUTION

Multiply each mass by its percentage and add the results to find the average:

$$(0.2447)(36.96590\ \text{u}) + (0.7553)(34.96885\ \text{u})$$
$$= 35.4575\ \text{u}$$
$$= 35.46\ \text{u (four significant figures)}$$

ONLINE LEARNING MODULE
Average Atomic Mass
WileyPLUS

PRACTICE 5.7

Silver occurs as two isotopes with atomic masses 106.9041 and 108.9047 u, respectively. The first isotope represents 51.82% and the second 48.18%. Determine the average atomic mass of silver.

CHAPTER 5 REVIEW

KEY TERM

Dalton's atomic model

5.1 Dalton's Model of the Atom

- Greek model of matter:
  - Four elements—earth, air, water, fire
  - Democritus—atoms (indivisible particles) make up matter
  - Aristotle—opposed atomic ideas
- Summary of Dalton's model of the atom:
  - Elements are composed of atoms.
  - Atoms of the same element are alike (mass and size).
  - Atoms of different elements are different in mass and size.
  - Compounds form by the union of two or more atoms of different elements.
  - Atoms form compounds in simple numerical ratios.
  - Atoms of two elements may combine in different ratios to form different compounds.

5.2 Electric Charge

- The properties of electric charge:
  - Charges are one of two types—positive or negative.
  - Unlike charges attract and like charges repel.
  - Charge is transferred from one object to another by contact or by induction.
  - The force of attraction between charges is expressed by

$$F = \frac{kq_1q_2}{r^2}$$

- Michael Faraday discovered electrically charged ions.
- Svante Arrhenius explained that conductivity results from the dissociation of compounds into ions:
  - Cation—positive charge—attracted to negative electrode (cathode)
  - Anion—negative charge—attracted to positive electrode (anode)

KEY TERMS

subatomic particles
electron
proton
Thomson model of the atom
neutron

5.3 Subatomic Parts of the Atom

- Atoms contain smaller subatomic particles:
  - Electron—negative charge, 1/1837 mass of proton
  - Proton—positive charge, 1.673×10^{-24} g
  - Neutron—no charge, 1.675×10^{-24} g
- Thomson model of the atom:
  - Negative electrons are embedded in a positive atomic sphere.
- Number of protons = number of electrons in a neutral atom.
- Ions are formed by losing or gaining electrons.

KEY TERMS

nucleus
atomic number

5.4 The Nuclear Atom

- Rutherford gold foil experiment modified the Thomson model to a nuclear model of the atom:
  - Atoms are composed of a nucleus containing protons and/or neutrons surrounded by electrons, which occupy mostly empty space.
  - Neutral atoms contain equal numbers of protons and electrons.

KEY TERMS

isotope
mass number

5.5 Isotopes of the Elements

- The mass number of an element is the sum of the protons and neutrons in the nucleus.

Mass number (sum of protons and neutrons in the nucleus) → $^{A}_{Z}\mathrm{E}$ ← **Symbol of element**

Atomic number (number of protons in the nucleus)

- The number of neutrons in an atom is determined by
  - mass number − atomic number = number of neutrons
  - $A - Z$ = neutrons

KEY TERMS

atomic mass unit (u)

atomic mass

5.6 Atomic Mass

- The average atomic mass is a weighted average of the masses of all the isotopes present in the sample.

REVIEW QUESTIONS

1. List the main features of Dalton's model of the atom.
2. What elements of Democritus' theory did Dalton incorporate into his model of the atom?
3. How does Dalton's theory explain why chemical formulas are always written with whole-number values?
4. Distinguish between an atom and an ion.
5. Does the force of attraction increase, decrease, or stay the same as two oppositely charged particles approach each other?
6. How do cations and anions differ?
7. A neutron is approximately how many times heavier than an electron?
8. From the chemist's point of view, what are the essential differences among a proton, a neutron, and an electron?
9. What are the atomic numbers of (a) copper, (b) nitrogen, (c) phosphorus, (d) radium, and (e) zinc?
10. What letters are used to designate atomic number and mass number in isotopic notation of atoms?
11. In what ways are isotopes alike? In what ways are they different?
12. Explain why the mass number of an element is always a whole number.

PAIRED EXERCISES

Most of the exercises in this chapter are available for assignment via the online homework management program, WileyPLUS (www.wileyplus.com). All exercises with blue numbers have answers in Appendix 5.

1. Identify the ions in a mixture containing Cu^{2+}, Cu^{2+}, N^{3-}, O^{2-}, Ca^{2+}, Al^{3+}, Te^{3-}, and H^{+} that would be attracted to the positive end of an electric field.
2. Identify the ions in a mixture containing Mg^{2+}, Br^{-}, Cr^{3+}, Ba^{2+}, P^{3-}, Ca^{2+}, Se^{2-}, and Y^{3+} that would be attracted to the anode of a battery.
3. Explain why, in Rutherford's experiments, some alpha particles were scattered at large angles by the gold foil or even bounced back.
4. What experimental evidence led Rutherford to conclude the following?
   (a) The nucleus of the atom contains most of the atomic mass.
   (b) The nucleus of the atom is positively charged.
   (c) The atom consists of mostly empty space.
5. Describe the general arrangement of subatomic particles in the atom.
6. What part of the atom contains practically all its mass?
7. What contribution did these scientists make to atomic models of the atom?
   (a) Dalton (b) Thomson (c) Rutherford
8. Consider the following models of the atom: (a) Dalton, (b) Thomson, (c) Rutherford. How does the location of the electrons in an atom vary? How does the location of the atom's positive matter compare?
9. Explain why the atomic masses of elements are not whole numbers.
10. Is the isotopic mass of a given isotope ever an exact whole number? Is it always? (Consider the masses of $^{12}_{6}C$ and $^{63}_{29}Cu$.)
11. What special names are given to the isotopes of hydrogen?
12. List the similarities and differences in the three isotopes of hydrogen.
13. What is the nuclear composition of the five naturally occurring isotopes of germanium having mass numbers 70, 72, 73, 74, and 76?
14. What is the nuclear composition of the five naturally occurring isotopes of zinc having mass numbers 64, 66, 67, 68, and 70?
15. Identify the element represented by the drawing below and give the isotopic notation for it.

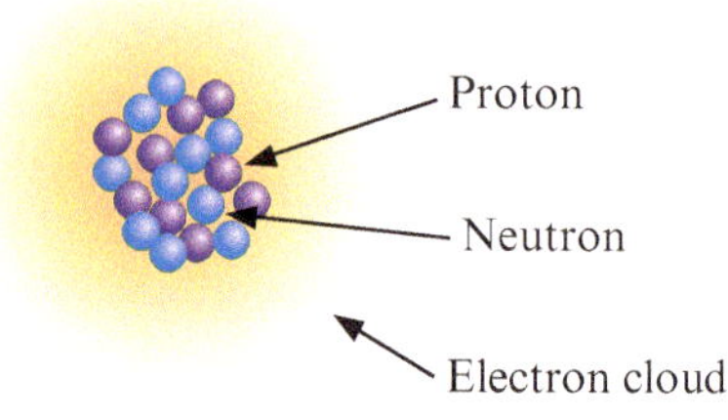

16. Identify the element represented by the drawing below and give the isotopic notation for it.

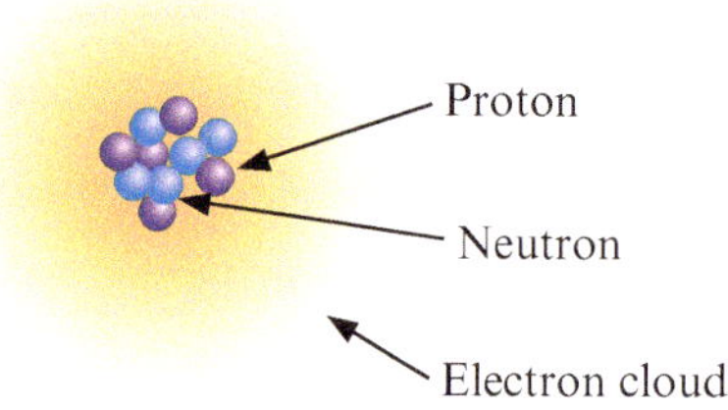

17. Draw a picture representing Rutherford's understanding of the structure of an atom of carbon-12. Show the location of the protons, neutrons, and electron cloud.

19. Write isotopic notation symbols for each of the following:
(a) $Z = 29$, $A = 65$
(b) $Z = 20$, $A = 45$
(c) $Z = 36$, $A = 84$

21. Give the number of protons, neutrons, and electrons in the following isotopes.
(a) ^{63}Cu
(b) ^{32}S
(c) manganese-55 (^{55}Mn)
(d) potassium-39 (^{39}K)

23. An unknown element contains 33 protons, 36 electrons, and has a mass number 76. Answer the following questions:
(a) What is the atomic number of this element?
(b) What are the name and symbol of this element?
(c) How many neutrons does it contain?
(d) What is the charge on this ion? Is it a cation or an anion?
(e) Write the symbolic notation for this element.

25. Nitrogen atoms tend to gain 3 electrons to form nitride anions.
(a) What is the charge on a nitride anion?
(b) What is the average atomic mass of a nitrogen atom in atomic mass units?
(c) Calculate the mass of the 3 electrons gained by a nitride anion in atomic mass units.
(d) Does the mass of a nitrogen atom change significantly when it gains electrons to become an anion?

27. Naturally occurring zirconium exists as five stable isotopes: ^{90}Zr with a mass of 89.905 u (51.45%); ^{91}Zr with a mass of 90.906 u (11.22%); ^{92}Zr with a mass of 91.905 u (17.15%); ^{94}Zr with a mass of 93.906 u (17.38%); and ^{96}Zr with a mass of 95.908 u (2.80%). Calculate the average mass of zirconium.

29. A particular element exists in two stable isotopic forms. One isotope has a mass of 62.9296 u (69.17% abundance). The other isotope has a mass of 64.9278 u. Calculate the average mass of the element and determine its identity.

31. An average dimension for the radius of an atom is 1.0×10^{-8} cm, and the average radius of the nucleus is 1.0×10^{-13} cm. Determine the ratio of atomic volume to nuclear volume. Assume that the atom is spherical $V = (4/3)\pi r^3$ for a sphere].

33. Elemental boron is composed of two isotopes, boron-10 and boron-11.
(a) How many protons and neutrons are found in these two isotopes of boron?
(b) Write the correct isotopic notation for each of these isotopes of boron.
(c) Calculate the mass percent of protons in the nucleus of each of these isotopes.

18. Draw a picture representing Rutherford's understanding of the structure of an atom of nitrogen-14. Show the location of the protons, neutrons, and electron cloud.

20. Write isotopic notation symbols for each of the following:
(a) $Z = 47$, $A = 109$
(b) $Z = 8$, $A = 18$
(c) $Z = 26$, $A = 57$

22. Tell the number of protons, neutrons, and electrons in the following isotopes.
(a) ^{54}Fe
(b) ^{23}Na
(c) bromine-79 (^{79}Br)
(d) phosphorus-31 (^{31}P)

24. An unknown element contains 56 protons, 54 electrons, and has a mass number 135. Answer the following questions:
(a) What is the atomic number of this element?
(b) What are the name and symbol of this element?
(c) How many neutrons does it contain?
(d) What is the charge on this ion? Is it a cation or an anion?
(e) Write the symbolic notation for this element.

26. Calcium atoms tend to lose 2 electrons to form a calcium cation.
(a) What is the charge on a calcium cation?
(b) What is the average atomic mass of a calcium atom in atomic mass units?
(c) Calculate the mass of the 2 electrons lost by a calcium cation in atomic mass units.
(d) Does the mass of a calcium atom change significantly when it loses electrons to become a cation?

28. Naturally occurring titanium exists as five stable isotopes. Four of the isotopes are ^{46}Ti with a mass of 45.953 u (8.0%); ^{47}Ti with a mass of 46.952 u (7.3%); ^{48}Ti with a mass of 47.948 u (73.8%); and ^{49}Ti with a mass of 48.948 u (5.5%). The average mass of an atom of titanium is 47.9 u. Determine the mass of the fifth isotope of titanium.

30. A particular element exists in two stable isotopic forms. One isotope has a mass of 34.9689 u (75.77% abundance). The other isotope has a mass of 36.9659 u. Calculate the average mass of the element and determine its identity.

32. An aluminum atom has an average diameter of about 3.0×10^{-8} cm. The nucleus has a diameter of about 2.0×10^{-13} cm. Calculate the ratio of the atom's diameter to its nucleus.

34. Elemental iridium is composed of two stable isotopes, iridium-191 and iridium-193.
(a) How many protons and neutrons are found in these two isotopes of iridium?
(b) Write the correct isotopic notation for each of these isotopes of iridium.
(c) Calculate the mass percent of protons in the nucleus of each of these isotopes.

ADDITIONAL EXERCISES

35. What experimental evidence supports these statements?
(a) The nucleus of an atom is small.
(b) The atom consists of both positive and negative charges.
(c) The nucleus of the atom is positive.

36. Given the following information, determine which, if any, elements are isotopes.
(a) mass number = 32; number of neutrons = 17
(b) mass number = 32; number of protons = 16
(c) number of protons = 15; number of neutrons = 16

37. The diameter of a silicon atom is 2.34×10^{-8} cm. If the silicon atoms were placed next to each other in a straight line across an 8.5-in. width of paper, how many silicon atoms would be there?

38. How is it possible for there to be more than one kind of atom of the same element?

39. Place each of the following elements in order of increasing number of neutrons (least to most); ^{156}Dy, ^{160}Gd, ^{162}Er, ^{165}Ho. Does this order differ from the elements' positions on the periodic table?

40. Lithium has two stable isotopes, lithium-6 with an atomic mass of 6.0151 u and lithium-7 with an atomic mass of 7.0160 u. Predict which of these two isotopes is more common by looking at the atomic mass from the periodic table. Explain your reasoning.

41. An unknown element Q has two known isotopes: ^{60}Q and ^{63}Q. If the average atomic mass is 61.5 u, what are the relative percentages of the isotopes?

42. The actual mass of one atom of an element is 3.27×10^{-19} mg.
(a) Calculate the atomic mass of the element.
(b) Identify the element, giving the symbol and the name. (There is only one stable isotope of this element.)

43. The mass of an atom of silver is 1.79×10^{-22} g. How many atoms would be in a 0.52-lb sample of silver?

44. Using the periodic table inside the front cover of the book, determine which of the first 20 elements have isotopes that you would expect to have the same number of protons, neutrons, and electrons.

45. Solar winds that send streams of charged particles toward the Earth cause the aurora borealis or the northern lights. As these charged particles hit the upper atmosphere of the Earth or the ionosphere (so named because it is composed of a plasma or gaseous soup of ions), they collide with ions located in this region. These high-energy ions then lose energy in the form of light, which is visible in the northernmost and southernmost regions of the planet as ribbons of light in the night sky. Some of the ions responsible for the colors seen are C^+, O^+, and O^{2+}. Determine the number of protons and electrons in each of these ions.

46. (a) What is an ion?
(b) The average mass of a calcium atom is 40.08 u. Why do we also use 40.08 u as the average mass of a calcium ion (Ca^{2+})?

47. Often the minerals we need in our diet are provided to us in vitamin tablets as cations. Following is a list of ingredients for a vitamin and mineral supplement. The table below highlights some of these minerals and identifies the ion provided for each mineral. Determine the number of protons and electrons in each of these ions.

Mineral supplement	Mineral use	Ion provided	Number of protons	Number of electrons
Calcium carbonate	Bones and teeth	Ca^{2+}		
Iron(II) sulfate	Hemoglobin	Fe^{2+}		
Chromium(III) nitrate	Insulin	Cr^{3+}		
Magnesium sulfate	Bones	Mg^{2+}		
Zinc sulfate	Cellular metabolism	Zn^{2+}		
Potassium iodide	Thyroid function	I^-		

48. Complete the following table with the appropriate data for each isotope given (all are neutral atoms):

Element	Symbol	Atomic number	Mass number	Number of protons	Number of neutrons	Number of electrons
	^{36}Cl					
Gold			197			
		56			79	
					20	18
			58	28		

49. Complete the following table with the appropriate data for each isotope given (all are neutral atoms):

Element	Symbol	Atomic number	Mass number	Number of protons	Number of neutrons	Number of electrons
	^{134}Xe					
Silver			107			
				9	10	
		92			143	92
			41	19		

50. Draw diagrams similar to those shown in Figure 5.5 for the following ions:
(a) O^{2-} (b) Na^+ (c) P^{3-} (d) Ca^{2+}

51. Draw pictures similar to those in Figure 5.8 for the following isotopes:
(a) ^3He and ^4He
(b) ^6Li and ^7Li
(c) ^{10}B and ^{11}B
(d) ^{12}C and ^{13}C

52. How many protons are in one molecule of dichlorine heptoxide? Is it possible to determine precisely how many electrons and neutrons are in a molecule of dichlorine heptoxide? Why or why not?

53. Which subatomic particles are not in the nucleus? What do an ion of Ca and an atom of Ar have in common?

54. What percent of the total mass of one atom of each of the following elements comes from electrons?
(a) aluminum (mass of one atom = 4.480×10^{-23} g)
(b) phosphorus (mass of one atom = 5.143×10^{-23} g)
(c) krypton (mass of one atom = 1.392×10^{-22} g)
(d) platinum (mass of one atom = 3.240×10^{-22} g)

55. What percent of the total mass of one atom of each of the following elements comes from protons?
(a) selenium (mass of one atom = 1.311×10^{-22} g)
(b) xenon (mass of one atom = 2.180×10^{-22} g)
(c) chlorine (mass of one atom = 5.887×10^{-23} g)
(d) barium (mass of one atom = 2.280×10^{-22} g)

56. Figure 5.7 is a representation of the nuclear model of the atom. The area surrounding the nucleus is labeled as the electron region. What is the electron region?

CHALLENGE EXERCISE

57. You have discovered a new element and are trying to determine where on the periodic table it would fit. You decide to do a mass spectrometer analysis of the sample and discover that it contains three isotopes with masses of 270.51 u, 271.23 u, and 269.14 u and relative abundances of 34.07%, 55.12%, and 10.81%, respectively. Sketch the mass spectrometer reading, determine the average atomic mass of the element, estimate its atomic number, and determine its approximate location on the periodic table.

58. Congratulations! You discover a new element you name wyzzlebium (Wz). The average atomic mass of Wz was found to be 303.001 u, and its atomic number is 120.
(a) If the masses of the two isotopes of wyzzlebium are 300.9326 u and 303.9303 u, what is the relative abundance of each isotope?
(b) What are the isotopic notations of the two isotopes? (e.g., $^{A}_{Z}Wz$)
(c) How many neutrons are in one atom of the more abundant isotope?

59. An unidentified atom is found to have an atomic mass 7.18 times that of the carbon-12 isotope.
(a) What is the mass of the unidentified atom?
(b) What are the possible identities of this atom?
(c) Why are you unable to positively identify the element based on the atomic mass and the periodic table?
(d) If the element formed a compound M_2O, where M is the unidentified atom, identify M by writing the isotopic notation for the atom.

60. Scientists such as Dalton, Thomson, and Rutherford proposed important models, which were ultimately challenged by later technology. What do we know to be false in Dalton's atomic model? What was missing in Thomson's model of the atom? What was Rutherford's experiment that led to the current model of the atom?

ANSWERS TO PRACTICE EXERCISES

5.1 CH_4, C_2H_2, C_2H_4, C_2H_6

5.2 cations: Na^+, NH_4^+, Ca^{2+}, Al^{3+}, K^+, Ag^+, Fe^{2+}, Fe^{3+}
anions: Cl^-, Br^-, NO_3^-, S^{2-}, SO_4^{2-}, PO_4^{3-}

5.3 5.98×10^{24} atoms

5.4 Cu, 29 protons and 29 electrons; Xe, 36 protons and 36 electrons

5.5

	protons	neutrons	electrons
(a)	8	8	8
(b)	35	45	35
(c)	92	143	92
(d)	29	35	29

5.6

(a)	atomic number	9
	mass number	19
	name	fluorine
(b)	atomic number	24
	mass number	52
	name	chromium
(c)	atomic number	79
	mass number	179
	name	gold

5.7 107.9 u

6

Nomenclature of Inorganic Compounds

CHAPTER OUTLINE

As children, we begin to communicate with other people in our lives by learning the names of objects around us. As we continue to develop, we learn to speak and use language to complete a wide variety of tasks. As we enter school, we begin to learn of other languages—the languages of mathematics, of other cultures, of computers. In each case, we begin by learning the names of the building blocks and then proceed to more abstract concepts. Chemistry has a language all its own—a whole new way of describing the objects so familiar to us in our daily lives. For example, the seashell shown above is formed from the chemical calcium carbonate, commonly called limestone. It is the same chemical used in many calcium supplements for our diets. Once we learn the language, we can begin to understand the macroscopic and microscopic world of chemistry.

6.1 Common and Systematic Names

LEARNING OBJECTIVE

Distinguish between the common and systematic names of chemical substances.

Chemical nomenclature is the system of names that chemists use to identify compounds. When a new substance is formulated, it must be named in order to distinguish it from all other substances (FIGURE 6.1). In this chapter, we will restrict our discussion to the nomenclature of inorganic compounds—compounds that do not generally contain carbon.

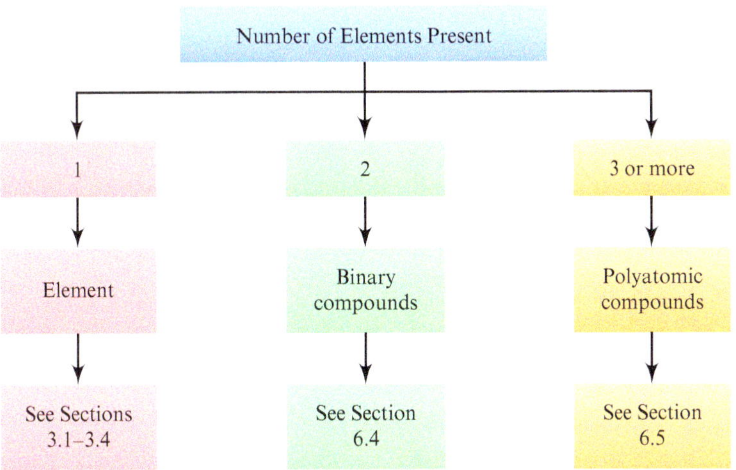

FIGURE 6.1 Where to find rules for naming inorganic substances in this book.

Common names are arbitrary names that are not based on the chemical composition of compounds. Before chemistry was systematized, a substance was given a name that generally associated it with one of its outstanding physical or chemical properties. For example, *quicksilver* is a common name for mercury, and nitrous oxide (N_2O), used as an anesthetic in dentistry, has been called *laughing gas* because it induces laughter when inhaled. Water and ammonia are also common names because neither provides any information about the chemical composition of the compounds. If every substance were assigned a common name, the amount of memorization required to learn over 50 million names would be astronomical.

Common names have distinct limitations, but they remain in frequent use. Common names continue to be used because the systematic name is too long or too technical for everyday use. For example, calcium oxide (CaO) is called *lime* by plasterers; photographers refer to *hypo* rather than sodium thiosulfate ($Na_2S_2O_3$); and nutritionists use the name *vitamin* D_3, instead of 9,10-secocholesta-5,7,10(19)-trien-3-ß-ol ($C_{27}H_{44}O$). TABLE 6.1 lists the common names, formulas, and systematic names of some familiar substances.

Water (H_2O) and ammonia (NH_3) are almost always referred to by their common names.

Chemists prefer systematic names that precisely identify the chemical composition of chemical compounds. The system for inorganic nomenclature was devised by the International Union of Pure and Applied Chemistry (IUPAC), which was founded in 1921. The IUPAC meets regularly and constantly reviews and updates the system.

6.2 Elements and Ions

LEARNING OBJECTIVE

Discuss the formation, charge, and naming of simple ions.

In Chapter 3, we studied the names and symbols for the elements as well as their location on the periodic table. In Chapter 5, we investigated the composition of the atom and learned that all atoms are composed of protons, electrons, and neutrons; that a particular element is defined by the number of protons it contains; and that atoms are uncharged because they contain equal numbers of protons and electrons.

The formula for most elements is simply the symbol of the element. In chemical reactions or mixtures, an element behaves as though it were a collection of individual particles. A small number of elements have formulas that are not single atoms at normal temperatures. Seven of the elements are *diatomic* molecules—that is, two atoms bonded together

TABLE 6.1 Common Names, Formulas, and Chemical Names of Familiar Substances

Common names	Formula	Chemical names	Common names	Formula	Chemical names
Acetylene	C_2H_2	ethyne	Gypsum	$CaSO_4 \cdot 2\ H_2O$	calcium sulfate dihydrate
Lime	CaO	calcium oxide	Grain alcohol	C_2H_5OH	ethanol, ethyl alcohol
Slaked lime	$Ca(OH)_2$	calcium hydroxide	Hypo	$Na_2S_2O_3$	sodium thiosulfate
Water	H_2O	water	Nitrous oxide	N_2O	dinitrogen monoxide
Galena	PbS	lead(II) sulfide	Lye, caustic soda	$NaOH$	sodium hydroxide
Alumina	Al_2O_3	aluminum oxide	Milk of magnesia	$Mg(OH)_2$	magnesium hydroxide
Baking soda	$NaHCO_3$	sodium hydrogen carbonate	Muriatic acid	HCl	hydrochloric acid
Cane or beet sugar	$C_{12}H_{22}O_{11}$	sucrose	Plaster of paris	$CaSO_4 \cdot {}^1/_2\ H_2O$	calcium sulfate hemihydrate
Borax	$Na_2B_4O_7 \cdot 10\ H_2O$	sodium tetraborate decahydrate	Potash	K_2CO_3	potassium carbonate
			Pyrite (fool's gold)	FeS_2	iron(II) sufide
Brimstone	S	sulfur	Quicksilver	Hg	mercury
Calcite, marble, limestone	$CaCO_3$	calcium carbonate	Saltpeter (chile)	$NaNO_3$	sodium nitrate
			Table salt	$NaCl$	sodium chloride
Cream of tartar	$KHC_4H_4O_6$	potassium hydrogen tartrate	Vinegar	$HC_2H_3O_2$	acetic acid
Epsom salts	$MgSO_4 \cdot 7\ H_2O$	magnesium sulfate heptahydrate	Washing soda	$Na_2CO_3 \cdot 10\ H_2O$	sodium carbonate decahydrate
			Wood alcohol	CH_3OH	methanol, methyl alcohol

to form a molecule. These diatomic elements are hydrogen, H_2, oxygen, O_2, nitrogen, N_2, fluorine, F_2, chlorine, Cl_2, bromine, Br_2, and iodine, I_2. Three other elements that are commonly polyatomic are S_8, sulfur, Se_8, selenium, and P_4, phosphorus.

Elements occurring as polyatomic molecules

Hydrogen	H_2	Chlorine	Cl_2	Sulfur	S_8
Oxygen	O_2	Fluorine	F_2	Phosphorus	P_4
Nitrogen	N_2	Bromine	Br_2	Selenium	Se_8
		Iodine	I_2		

We have learned that a charged particle, known as an *ion*, can be produced by adding or removing one or more electrons from a neutral atom. For example, potassium atoms contain 19 protons and 19 electrons. To make a potassium ion, we remove one electron, leaving 19 protons and only 18 electrons. This gives an ion with a positive one (+1) charge:

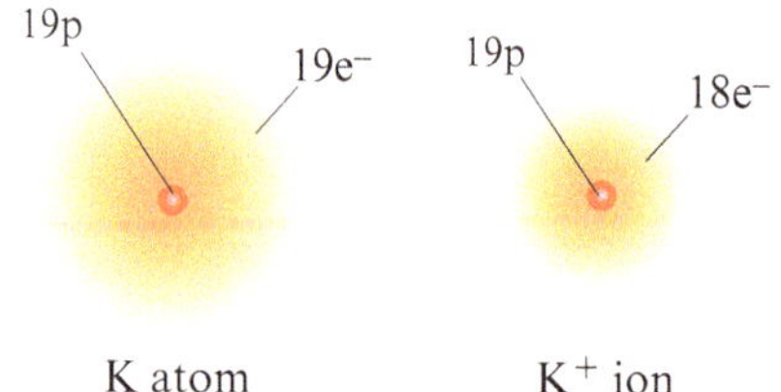

Written in the form of an equation, $K \rightarrow K^+ + e^-$. A positive ion is called a *cation*. Any neutral atom that *loses* an electron will form a cation. Sometimes an atom may lose one electron, as in the potassium example. Other atoms may lose more than one electron:

$$Mg \rightarrow Mg^{2+} + 2e^-$$

or

$$Al \rightarrow Al^{3+} + 3e^-$$

Cations are named the same as their parent atoms, as shown here:

Atom		Ion	
K	potassium	K^+	potassium ion
Mg	magnesium	Mg^{2+}	magnesium ion
Al	aluminum	Al^{3+}	aluminum ion

CHECK YOUR UNDERSTANDING

6.1 Cations

6.2 Naming Cations

WileyPLUS

Ions can also be formed by adding electrons to a neutral atom. For example, the chlorine atom contains 17 protons and 17 electrons. The equal number of positive charges and negative charges results in a net charge of zero for the atom. If one electron is added to the chlorine atom, it now contains 17 protons and 18 electrons, resulting in a net charge of negative one (−1) on the ion:

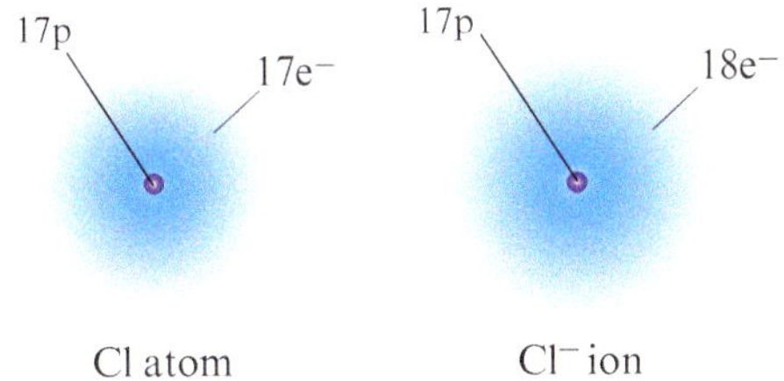

CHEMISTRY *IN ACTION* What's in a Name?

When a scientist discovered a new element in the early days of chemistry, he or she had the honor of naming it. Now researchers must submit their choices for a name to an international committee called the International Union of Pure and Applied Chemistry (IUPAC) before the name can be placed on the periodic table. Since 1997 the IUPAC has decided on names for 11 new elements from 104 through 118. The new elements are called rutherfordium (Rf), dubnium (Db), seaborgium (Sg), bohrium (Bh), hessium (Hs), meitnerium (Mt), darmstadtium (Ds), roentgenium (Rg), copernicium (Cn), flerovium (Fv), and livermorium (Lv).

The new names are compromises among the choices presented by different research teams. Some names given to the elements recognize places where much of the research to find these elements is done. The Russians gained recognition for work done at a laboratory in Dubna, the Germans for their laboratories in Darmstadt and Hesse, and the Americans for the Lawrence Livermore National Laboratory (LLNL).

Other names of elements recognize the contributions of great scientists and researchers such as the American Glenn Seaborg, the first living scientist to have an element named after him. The British recognized Ernest Rutherford, who discovered the atom's nucleus. The Germans recognized Lise Meitner who co-discovered atomic fission. Both the Germans and the Russians won recognition for Niels Bohr whose model of the atom led to modern ideas about atomic structure. Wilhelm Roentgen was honored for discovering X-rays and winning the first Nobel Prize in physics. Most recently, the astronomer Copernicus was honored for changing our model of the solar system to a heliocentric model and to highlight the link between astronomy and nuclear science. The Russians honored Georgy Flerov, who developed the Soviet nuclear program, with the name for element 114.

Elements 113, 115, 117, and 118 do not yet have official names and are currently shown on the periodic table using three-letter symbols representing their numbers.

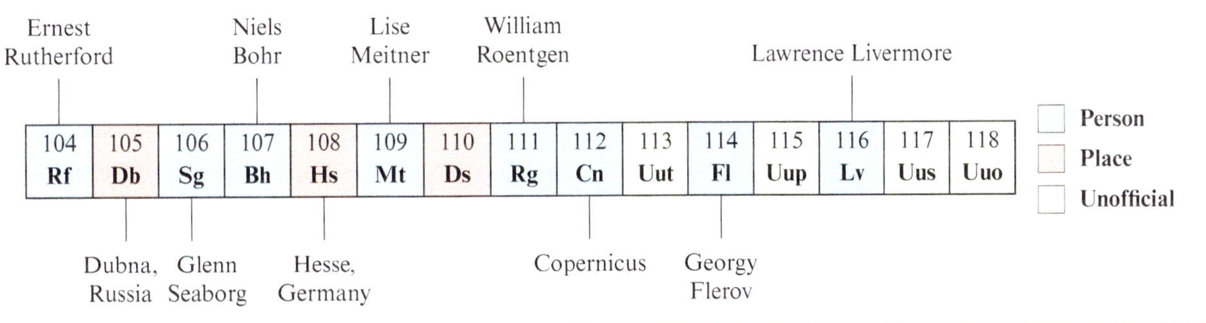

In a chemical equation, this process is summarized as $Cl + e^- \rightarrow Cl^-$. A negative ion is called an *anion*. Any neutral atom that *gains* an electron will form an anion. Atoms may gain more than one electron to form anions with different charges:

$$O + 2e^- \rightarrow O^{2-}$$

$$N + 3e^- \rightarrow N^{3-}$$

CHECK YOUR UNDERSTANDING

6.3 Anions

WileyPLUS

Anions are named differently from cations. To name an anion consisting of only one element, use the stem of the parent element name and change the ending to *-ide*. For example, the Cl^- ion is named by using the stem *chlor-* from chlorine and adding *-ide* to form chloride ion. Here are some examples:

Symbol	Name of atom	Ion	Name of ion
F	fluorine	F^-	fluoride ion
Br	bromine	Br^-	bromide ion
Cl	chlorine	Cl^-	chloride ion
I	iodine	I^-	iodide ion
O	oxygen	O^{2-}	oxide ion
N	nitrogen	N^{3-}	nitride ion

Ions are formed by adding or removing electrons from an atom. Atoms do not form ions on their own. Most often ions are formed when metals combine with nonmetals.

The charge on an ion can often be predicted from the position of the element on the periodic table. FIGURE 6.2 shows the charges of selected ions from several groups on the periodic table. Notice that all the metals and hydrogen in the far left column (Group 1A(1)) are (1+), all those in the next column (Group 2A(2)) are (2+), and the metals in the next tall column (Group 3A(13)) form (3+) ions. The elements in the lower center part of the table are called *transition metals*. These elements tend to form cations with various positive charges. There is no easy way to predict the charges on these cations. All metals lose electrons to form positive ions.

CHECK YOUR UNDERSTANDING

6.4 Making an Anion

6.5 Making a Cation

WileyPLUS

In contrast, the nonmetals form anions by gaining electrons. On the right side of the periodic table in Figure 6.2, you can see that the atoms in Group 7A(17) form (1−) ions. The nonmetals in Group 6A(16) form (2−) ions. It's important to learn the charges on the ions shown in Figure 6.2 and their relationship to the group number at the top of the column. For nontransition metals the charge is equal to the group number. For nonmetals, the charge is equal to the group number minus 8. We will learn more about why these ions carry their particular charges later in the course.

1 1A	2 2A	3	4	5	6	7	8	9	10	11	12	13 3A	14 4A	15 5A	16 6A	17 7A	18 8A
H^+																	
Li^+	Be^{2+}													N^{3-}	O^{2-}	F^-	
Na^+	Mg^{2+}											Al^{3+}		P^{3-}	S^{2-}	Cl^-	
K^+	Ca^{2+}				Cr^{2+} Cr^{3+}		Fe^{2+} Fe^{3+}			Cu^+ Cu^{2+}	Zn^{2+}					Br^-	
Rb^+	Sr^{2+}			Transition metals						Ag^+	Cd^{2+}					I^-	
Cs^+	Ba^{2+}																

FIGURE 6.2 Charges of selected ions in the periodic table.

ENHANCED EXAMPLE 6.1 WileyPLUS

What is the general basis for naming a negative ion in a binary compound?

SOLUTION

Establish an identifying stem from the name of the negative element; then add the suffix *-ide* to that stem. For example, the stem for chlorine is *chlor-*. Add the suffix *-ide* to get the name of the ion, *chloride*. The stem for oxygen is *ox-*. Add the suffix *-ide* to get the name of the ion, *oxide*.

PRACTICE 6.1

Derive the names of these ions when used as the negative ion in a binary compound: P, N, B, C.

CHECK YOUR UNDERSTANDING

6.6 Naming Anions

➡ WileyPLUS

6.3 Writing Formulas from Names of Ionic Compounds

LEARNING OBJECTIVE

Write the chemical formula for an ionic compound from the name of the compound.

CHECK YOUR UNDERSTANDING

6.7 Charge on an Alkali Metal Ion

6.8 Charge on an Alkaline Earth Metal Ion

6.9 Charge on a Halogen Ion

6.10 Charge on a Group 5A (15) Ion

6.11 Charge on a Group 6A (16) Ion

➡ WileyPLUS

In Chapters 3 and 5, we learned that compounds can be composed of ions. These substances are called *ionic compounds* and will conduct electricity when dissolved in water. An excellent example of an ionic compound is ordinary table salt composed of crystals of sodium chloride. When dissolved in water, sodium chloride conducts electricity very well, as shown in FIGURE 6.3.

A chemical compound must have a net charge of zero. If it contains ions, the charges on the ions must add up to zero in the formula for the compound. This is relatively easy in the case of sodium chloride. The sodium ion (1+) and the chloride ion (1−) add to zero, resulting in the formula NaCl. Now consider an ionic compound containing calcium (Ca^{2+}) and fluoride (F^-) ions. How can we write a formula with a net charge of zero? To do this we need one Ca^{2+} and two F^- ions. The correct formula is CaF_2. The subscript 2 indicates that two fluoride ions are needed for each calcium ion. Aluminum oxide is a bit more complicated because it consists of Al^{3+} and O^{2-} ions. Since 6 is the least common multiple of 3 and 2, we have $2(3+) + 3(2-) = 0$, or a formula containing 2 Al^{3+} ions and 3 O^{2-} ions for Al_2O_3. Here are a few more examples of formula writing for ionic compounds:

Name of compound	Ions	Least common multiple	Sum of charges on ions	Formula
Sodium bromide	Na^+, Br^-	1	$(+1) + (-1) = 0$	NaBr
Potassium sulfide	K^+, S^{2-}	2	$2(+1) + (-2) = 0$	K_2S
Zinc sulfate	Zn^{2+}, SO_4^{2-}	2	$(+2) + (-2) = 0$	$ZnSO_4$
Ammonium phosphate	NH_4^+, PO_4^{3-}	3	$3(+1) + (-3) = 0$	$(NH_4)_3PO_4$
Aluminum chromate	Al^{3+}, CrO_4^{2-}	6	$2(+3) + 3(-2) = 0$	$Al_2(CrO_4)_3$

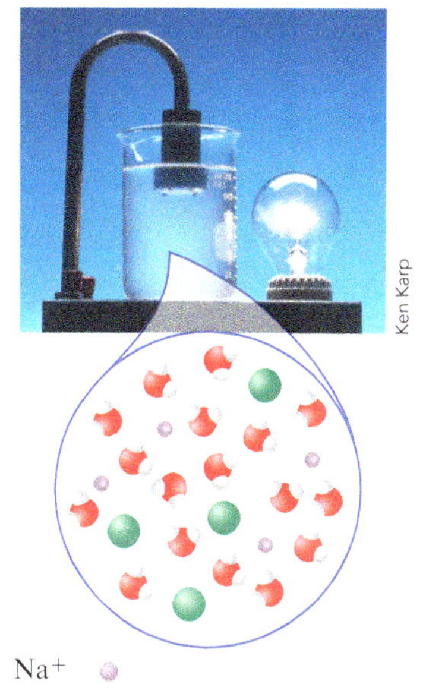

FIGURE 6.3 A solution of salt water contains Na^+ and Cl^- ions in addition to water molecules. The ions cause the solution to conduct electricity, lighting the bulb.

RULES FOR WRITING FORMULAS FOR IONIC COMPOUNDS

1. Write the formula for the metal ion followed by the formula for the nonmetal ion.
2. Combine the smallest numbers of each ion needed to give the charge sum equal to zero.
3. Write the formula for the compound as the symbol for the metal and nonmetal, each followed by a subscript of the number determined in 2.

ONLINE LEARNING MODULE

Writing the Formula of an Ionic Compound

➥ WileyPLUS

ENHANCED EXAMPLE 6.2

WileyPLUS

Write formulas for (a) calcium chloride, (b) magnesium oxide, and (c) barium phosphide.

SOLUTION

(a) Use the following steps for calcium chloride.

1. From the name, we know that calcium chloride is composed of calcium and chloride ions. First write down the formulas of these ions:

$$Ca^{2+} \qquad Cl^-$$

2. To write the formula of the compound, combine the smallest numbers of Ca^{2+} and Cl^- ions to give the charge sum equal to zero. In this case the lowest common multiple of the charges is 2:

$$(Ca^{2+}) + 2(Cl^-) = 0$$
$$(2+) + 2(1-) = 0$$

3. Therefore, the formula is $CaCl_2$.

(b) Use the same procedure for magnesium oxide:

1. From the name, we know that magnesium oxide is composed of magnesium and oxide ions. First write down the formulas of these ions:

$$Mg^{2+} \qquad O^{2-}$$

2. To write the formula of the compound, combine the smallest numbers of Mg^{2+} and O^{2-} ions to give the charge sum equal to zero:

$$(Mg^{2+}) + (O^{2-}) = 0$$
$$(2+) + (2-) = 0$$

3. The formula is MgO.

(c) Use the same procedure for barium phosphide:

1. From the name, we know that barium phosphide is composed of barium and phosphide ions. First write down the formulas of these ions:

$$Ba^{2+} \qquad P^{3-}$$

2. To write the formula of this compound, combine the smallest numbers of Ba^{2+} and P^{3-} ions to give the charge sum equal to zero. In this case the lowest common multiple of the charges is 6:

$$3(Ba^{2+}) + 2(P^{3-}) = 0$$
$$3(2+) + 2(3-) = 0$$

3. The formula is Ba_3P_2.

PRACTICE 6.2

Write formulas for compounds containing the following ions:

(a) K^+ and F^-
(b) Ca^{2+} and Br^-
(c) Mg^{2+} and N^{3-}
(d) Na^+ and S^{2-}
(e) Ba^{2+} and O^{2-}
(f) As^{5+} and CO_3^{2-}

6.4 Naming Binary Compounds

Name binary ionic and nonionic compounds.

LEARNING OBJECTIVE

KEY TERMS

binary compounds
Stock System

Binary compounds contain only two different elements. Many binary compounds are formed when a metal combines with a nonmetal to form a *binary ionic compound*. The metal loses one or more electrons to become a cation, while the nonmetal gains one or more electrons to become an anion. The cation is written first in the formula, followed by the anion.

Binary Ionic Compounds Containing a Metal Forming Only One Type of Cation

The chemical name is composed of the name of the metal followed by the name of the nonmetal, which has been modified to an identifying stem plus the suffix *-ide*.

For example, sodium chloride, NaCl, is composed of one atom each of sodium and chlorine. The name of the metal, sodium, is written first and is not modified. The second part of the name is derived from the nonmetal, chlorine, by using the stem *chlor-* and adding the ending *-ide*; it is named *chloride*. The compound name is sodium chloride.

	NaCl
Elements:	Sodium (metal)
	Chlorine (nonmetal) name modified to the stem *chlor-* + *-ide*
Name of compound:	Sodium chloride

Stems of the more common negative-ion-forming elements are shown in TABLE 6.2.

TABLE 6.2 Examples of Elements Forming Anions

Symbol	Element	Stem	Anion name
Br	bromine	brom	bromide
Cl	chlorine	chlor	chloride
F	fluorine	fluor	fluoride
H	hydrogen	hydr	hydride
I	iodine	iod	iodide
N	nitrogen	nitr	nitride
O	oxygen	ox	oxide
P	phosphorus	phosph	phosphide
S	sulfur	sulf	sulfide

RULES FOR NAMING BINARY IONIC COMPOUNDS OF METAL FORMING ONE TYPE OF CATION

1. Write the name of the cation.
2. Write the stem for the anion and add the suffix *-ide*.

Compounds may contain more than one atom of the same element, but as long as they contain only two different elements and only one compound of these two elements exists, the name follows the rules for binary compounds:

Examples:

$CaBr_2$	Mg_3N_2	Li_2O
calcium bromide	magnesium nitride	lithium oxide

Some of the compounds with names ending in *-ide* are shown in TABLE 6.3.

TABLE 6.3 Examples of Compounds with Names Ending in *-ide*

Formula	Name	Formula	Name
$AlCl_3$	aluminum chloride	BaS	barium sulfide
Al_2O_3	aluminum oxide	LiI	lithium iodide
CaC_2	calcium carbide	$MgBr_2$	magnesium bromide
HCl	hydrogen chloride	NaH	sodium hydride
HI	hydrogen iodide	Na_2O	sodium oxide

CHECK YOUR UNDERSTANDING

6.12 Naming Binary Ionic Compounds

➡ WileyPLUS

ENHANCED EXAMPLE 6.3 WileyPLUS

Name the compound CaF_2.

SOLUTION

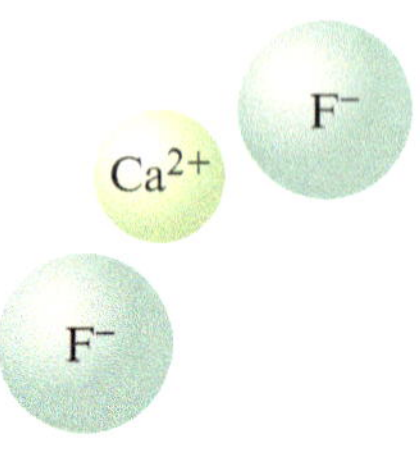

From the formula it is a two-element compound and follows the rules for binary compounds.

1. The compound is composed of a metal, Ca, and a nonmetal, F. Elements in the 2A(2) column of the periodic table form only one type of cation. Thus, we name the positive part of the compound *calcium*.
2. Modify the name of the second element to the stem *fluor-* and add the binary ending *-ide* to form the name of the negative part, *fluoride*.

Therefore, the name of the compound is *calcium fluoride*.

PRACTICE 6.3

Write formulas for these compounds:

(a) strontium chloride
(b) potassium iodide
(c) aluminum nitride
(d) calcium sulfide
(e) sodium oxide

Binary Ionic Compounds Containing a Metal That Can Form Two or More Types of Cations

The metals in the center of the periodic table (including the transition metals) often form more than one type of cation. For example, iron can form Fe^{2+} and Fe^{3+} ions, and copper can form Cu^{+} and Cu^{2+} ions. This can be confusing when you are naming compounds. For example, copper chloride could be $CuCl_2$ or CuCl. To resolve this difficulty the IUPAC devised a system, known as the **Stock System**, to name these compounds. This system is currently recognized as the official system to name these compounds, although another older system is sometimes used. In the Stock System, when a compound contains a metal that can form more than one type of cation, the charge on the cation of the metal is designated by a Roman numeral placed in parentheses immediately following the name of the metal. The negative element is treated in the usual manner for binary compounds:

Cation charge	+1	+2	+3	+4	+5
Roman numeral	(I)	(II)	(III)	(IV)	(V)

Examples:

$FeCl_2$	iron(II) chloride	Fe^{2+}
$FeCl_3$	iron(III) chloride	Fe^{3+}
CuCl	copper(I) chloride	Cu^{+}
$CuCl_2$	copper(II) chloride	Cu^{2+}

The Roman numeral is part of the cation name.

The fact that $FeCl_2$ has two chloride ions, each with a -1 charge, establishes that the charge on Fe is $+2$. To distinguish between the two iron chlorides, $FeCl_2$ is named iron(II) chloride and $FeCl_3$ is named iron(III) chloride:

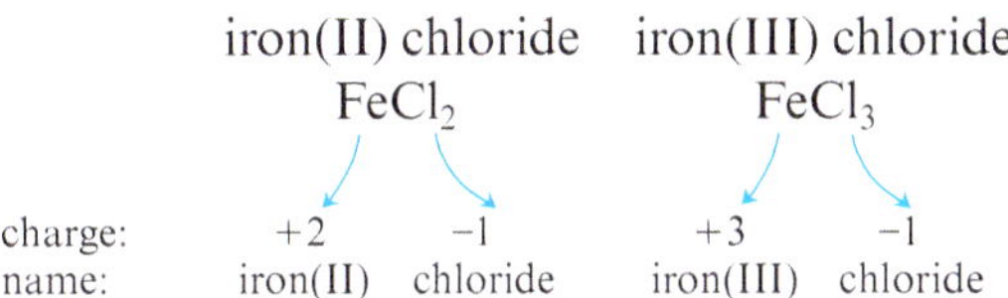

When a metal forms only one possible cation, we need not distinguish one cation from another, so Roman numerals are not needed. Thus we do not say calcium(II) chloride for $CaCl_2$, but rather calcium chloride, since the charge of calcium is understood to be $+2$.

RULES FOR NAMING BINARY IONIC COMPOUNDS OF METAL FORMING TWO OR MORE TYPES OF CATIONS (STOCK SYSTEM)

1. Write the name of the cation.
2. Write the charge on the cation as a Roman numeral in parentheses.
3. Write the stem of the anion and add the suffix *-ide*.

In classical nomenclature, when the metallic ion has only two cation types, the name of the metal (usually the Latin name) is modified with the suffixes *-ous* and *-ic* to distinguish between the two. The lower-charge cation is given the *-ous* ending, and the higher one, the *-ic* ending.

Examples:

$FeCl_2$	ferrous chloride	Fe^{2+}	(lower-charge cation)
$FeCl_3$	ferric chloride	Fe^{3+}	(higher-charge cation)
CuCl	cuprous chloride	Cu^{+}	(lower-charge cation)
$CuCl_2$	cupric chloride	Cu^{2+}	(higher-charge cation)

Some of the common metals that have more than one type of cation are shown in TABLE 6.4.

TABLE 6.4 Names and Charges of Some Common Metal Ions That Have More Than One Type of Cation

Formula	Stock System name	Classical name	Formula	Stock System name	Classical name
Cu^{1+}	copper(I)	cuprous	Sn^{4+}	tin(IV)	stannic
Cu^{2+}	copper(II)	cupric	Pb^{2+}	lead(II)	plumbous
Hg^{1+} $(Hg_2)^{2+}$	mercury(I)	mercurous	Pb^{4+}	lead(IV)	plumbic
Hg^{2+}	mercury(II)	mercuric	As^{3+}	arsenic(III)	arsenous
Fe^{2+}	iron(II)	ferrous	As^{5+}	arsenic(V)	arsenic
Fe^{3+}	iron(III)	ferric	Ti^{3+}	titanium(III)	titanous
Sn^{2+}	tin(II)	stannous	Ti^{4+}	titanium(IV)	titanic

CHECK YOUR UNDERSTANDING
6.13 Elements Forming Multiple Ions
➡ WileyPLUS

CHECK YOUR UNDERSTANDING
6.14 Using Suffixes to Identify Charge
➡ WileyPLUS

Notice that the *ous–ic* naming system does not give the charge of the cation of an element but merely indicates that at least two types of cations exist. The Stock System avoids any possible uncertainty by clearly stating the charge on the cation.

In this book we will use mainly the Stock System.

ONLINE LEARNING MODULE
Naming an Ionic Compound Containing a Metal with Two or More Types of Cations
➡ WileyPLUS

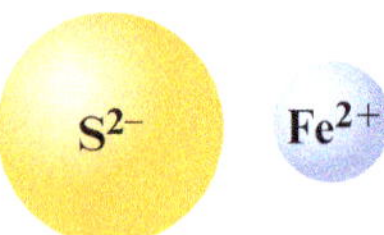

ENHANCED EXAMPLE 6.4 WileyPLUS

Name the compound FeS.

SOLUTION

This compound follows the rules for a binary compound.

1, 2. It is a compound of Fe, a metal, and S, a nonmetal, and Fe is a transition metal that has more than one type of cation. In sulfides, the charge on the S is −2. Therefore, the charge on Fe must be +2, and the name of the positive part of the compound is *iron(II)*.

3. We have already determined that the name of the negative part of the compound will be *sulfide*.

The name of FeS is *iron(II) sulfide*.

PRACTICE 6.4

Write the name for each of the following compounds using the Stock System:
(a) PbI_2 (b) SnF_4 (c) Fe_2O_3 (d) CuO

PRACTICE 6.5

Write formulas for the following compounds:
(a) tin(IV) chromate (c) tin(II) fluoride
(b) chromium(III) bromide (d) copper(I) oxide

Binary Compounds Containing Two Nonmetals

Compounds between nonmetals are molecular, not ionic. Therefore, a different system for naming them is used. In a compound formed between two nonmetals, the element that occurs first in this series is written and named first:

Si, B, P, H, C, S, I, Br, N, Cl, O, F

The name of the second element retains the *-ide* ending as though it were an anion. A Latin or Greek prefix (*mono-*, *di-*, *tri-*, and so on) is attached to the name of each element to indicate the number of atoms of that element in the molecule. The prefix *mono-* is rarely used for naming the first element. Some common prefixes and their numerical equivalences are as shown in the margin table.

Prefix	Number
mono	1
di	2
tri	3
tetra	4
penta	5
hexa	6
hepta	7
octa	8
nona	9
deca	10

RULES FOR NAMING BINARY COMPOUNDS CONTAINING TWO NONMETALS

1. Write the name for the first element using a prefix if there is more than one atom of this element.
2. Write the stem of the second element and add the suffix *-ide*. Use a prefix to indicate the number of atoms for the second element.

Here are some examples of compounds that illustrate this system:

N_2O_3

(di) nitrogen — Indicates two nitrogen atoms
(tri) oxide — Indicates three oxygen atoms

CO	carbon monoxide	N_2O	dinitrogen monoxide
CO_2	carbon dioxide	N_2O_4	dinitrogen tetroxide
PCl_3	phosphorus trichloride	NO	nitrogen monoxide
SO_2	sulfur dioxide	N_2O_3	dinitrogen trioxide
P_2O_5	diphosphorus pentoxide	S_2Cl_2	disulfur dichloride
CCl_4	carbon tetrachloride	S_2F_{10}	disulfur decafluoride

CHECK YOUR UNDERSTANDING
6.15 Using Prefixes in Naming Compounds 1
6.16 Using Prefixes in Naming Compounds 2
6.17 Using Prefixes in Naming Compounds 3
➡ WileyPLUS

These examples illustrate that we sometimes drop the final *o* (mono) or *a* (penta) of the prefix when the second element is oxygen. This avoids creating a name that is awkward to pronounce. For example, CO is carbon monoxide instead of carbon monooxide.

ENHANCED EXAMPLE 6.5 **WileyPLUS**

Name the compound PCl_5.

SOLUTION

1. Phosphorus and chlorine are nonmetals, so the rules for naming binary compounds containing two nonmetals apply. Phosphorus is named first. Therefore, the compound is a chloride.
2. No prefix is needed for phosphorus because each molecule has only one atom of phosphorus. The prefix *penta-* is used with chloride to indicate the five chlorine atoms. (PCl_3 is also a known compound.)

The name for PCl_5 is *phosphorus pentachloride*.

ONLINE LEARNING MODULE
Naming a Binary Compound Containing Two Nonmetals
➡ WileyPLUS

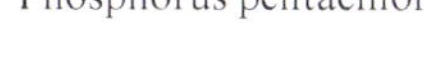

Phosphorus pentachloride

PRACTICE 6.6

Name these compounds:

(a) Cl_2O (b) SO_2 (c) CBr_4 (d) N_2O_5 (e) NH_3 (f) ICl_3

PRACTICE 6.7

Name these compounds:

(a) KBr (b) Ca_3N_2 (c) SO_3 (d) SnF_2 (e) $CuCl_2$ (f) N_2O_4

6.5 Naming Compounds Containing Polyatomic Ions

LEARNING OBJECTIVE

Recognize names, formulas, and charges of polyatomic ions, name compounds containing polyatomic ions, and write formulas from names of compounds containing polyatomic ions.

KEY TERM
polyatomic ion

A **polyatomic ion** is an ion that contains two or more elements. Compounds containing polyatomic ions are composed of three or more elements and usually consist of one or more cations combined with a negative polyatomic ion. In general, naming compounds containing polyatomic ions is similar to naming binary compounds. The cation is named first, followed by the name for the negative polyatomic ion.

To name these compounds, you must learn to recognize the common polyatomic ions (TABLE 6.5) and know their charges. Consider the formula $KMnO_4$. You must be able to recognize that it consists of two parts: $K\ MnO_4$. These parts are composed of a K^+ ion

CHECK YOUR UNDERSTANDING
6.18 Multiple Atom Ions
➡ WileyPLUS

TABLE 6.5 Names, Formulas, and Charges of Some Common Polyatomic Ions

Name	Formula	Charge	Name	Formula	Charge
Acetate	$C_2H_3O_2^-$	−1	Cyanide	CN^-	−1
Ammonium	NH_4^+	+1	Dichromate	$Cr_2O_7^{2-}$	−2
Arsenate	AsO_4^{3-}	−3	Hydroxide	OH^-	−1
Hydrogen carbonate	HCO_3^-	−1	Nitrate	NO_3^-	−1
Hydrogen sulfate	HSO_4^-	−1	Nitrite	NO_2^-	−1
Bromate	BrO_3^-	−1	Permanganate	MnO_4^-	−1
Carbonate	CO_3^{2-}	−2	Phosphate	PO_4^{3-}	−3
Chlorate	ClO_3^-	−1	Sulfate	SO_4^{2-}	−2
Chromate	CrO_4^{2-}	−2	Sulfite	SO_3^{2-}	−2

More ions are listed on the back endpapers.

Potassium permanganate crystals are dark purple.

and a MnO_4^- ion. The correct name for this compound is potassium permanganate. Many polyatomic ions that contain oxygen are called *oxy-anions* and generally have the suffix *-ate* or *-ite*. Unfortunately, the suffix doesn't indicate the number of oxygen atoms present. The *-ate* form contains more oxygen atoms than the *-ite* form. Examples include sulfate (SO_4^{2-}), sulfite (SO_3^{2-}), nitrate (NO_3^-), and nitrite (NO_2^-).

RULES FOR NAMING COMPOUNDS CONTAINING POLYATOMIC IONS

1. Write the name of the cation.
2. Write the name of the anion.

CHECK YOUR UNDERSTANDING

6.19 Naming Polyatomic Ions 1

6.20 Naming Polyatomic Ions 2

6.21 Naming Polyatomic Ions 3

WileyPLUS

Only four of the common negatively charged polyatomic ions do not use the *ate–ite* system. These exceptions are hydroxide (OH^-), hydrogen sulfide (HS^-), peroxide (O_2^{2-}), and cyanide (CN^-). Care must be taken with these, as their endings can easily be confused with the *-ide* ending for binary compounds (Section 6.4).

There are three common positively charged polyatomic ions as well—the ammonium, the mercury(I) (Hg_2^{2+}), and the hydronium (H_3O^+) ions. The ammonium ion (NH_4^+) is frequently found in polyatomic compounds, whereas the hydronium ion (H_3O^+) is usually associated with aqueous solutions of acids (Chapter 15).

ENHANCED EXAMPLE 6.6

WileyPLUS

What is the minimum number of different elements in a compound that contains a polyatomic ion?

SOLUTION

Most polyatomic ions are anions and contain two different elements. Adding a cation to a polyatomic ion to form a compound means the minimum is three different elements in a polyatomic compound.

PRACTICE 6.8

Name these compounds:

(a) $NaNO_3$ (b) $Ca_3(PO_4)_2$ (c) KOH (d) Li_2CO_3 (e) $NaClO_3$

Inorganic compounds are also formed from more than three elements (TABLE 6.6). In these cases, one or more of the ions is often a polyatomic ion. Once you have learned to recognize the polyatomic ions, naming these compounds follows the patterns we have already learned. First identify the ions. Name the cations in the order given and follow them with the names of the anions. Study the following examples:

Compound	Ions	Name
$NaHCO_3$	Na^+; HCO_3^-	sodium hydrogen carbonate
NaHS	Na^+; HS^-	sodium hydrogen sulfide
$MgNH_4PO_4$	Mg^{2+}; NH_4^+; PO_4^{3-}	magnesium ammonium phosphate
$NaKSO_4$	Na^+; K^+; SO_4^{2-}	sodium potassium sulfate

PRACTICE 6.9

Name these compounds:

(a) $KHCO_3$ (b) $NaHC_2O_4$ (c) $BaNH_4PO_4$ (d) $NaAl(SO_4)_2$

TABLE 6.6 Names of Selected Compounds That Contain More Than One Kind of Positive Ion

Formula	Name of compound
$KHSO_4$	potassium hydrogen sulfate
$Ca(HSO_3)_2$	calcium hydrogen sulfite
NH_4HS	ammonium hydrogen sulfide
$MgNH_4PO_4$	magnesium ammonium phosphate
NaH_2PO_4	sodium dihydrogen phosphate
Na_2HPO_4	sodium hydrogen phosphate
KHC_2O_4	potassium hydrogen oxalate
$KAl(SO_4)_2$	potassium aluminum sulfate
$Al(HCO_3)_3$	aluminum hydrogen carbonate

6.6 Acids

LEARNING OBJECTIVE

Use the rules to name an acid from its formula and to write the formula for an acid from its name.

While we will learn much more about acids later (see Chapter 15), it is helpful to be able to recognize and name common acids both in the laboratory and in class. The simplest way to recognize many acids is to know that acid formulas often begin with hydrogen.

Binary Acids

Certain binary hydrogen compounds, when dissolved in water, form solutions that have *acid* properties. Because of this property, these compounds are given acid names in addition to their regular *-ide* names. For example, HCl is a gas and is called *hydrogen chloride*, but its water solution is known as *hydrochloric acid*. Binary acids are composed of hydrogen and one other nonmetallic element. However, not all binary hydrogen compounds are acids. To express the formula of a binary acid, it's customary to write the symbol of hydrogen first, followed by the symbol of the second element (e.g., HCl, HBr, or H_2S). When we see formulas such as CH_4 or NH_3, we understand that these compounds are not normally considered to be acids.

To name a binary acid, place the prefix *hydro-* in front of, and the suffix *-ic* after, the stem of the nonmetal name. Then add the word *acid*:

	HCl	H_2S
Examples:	*Hydro*-chlor-*ic acid*	*Hydro*-sulfur-*ic acid*
	(hydrochloric acid)	(hydrosulfuric acid)

RULES FOR NAMING BINARY ACIDS

1. Write the prefix *hydro-* followed by the stem of the second element and add the suffix *-ic*.
2. Add the word *acid*.

ENHANCED EXAMPLE 6.7 WileyPLUS

Name the acid HI.

SOLUTION

The compound follows the Rules for Naming Binary Acids.

1. Write the prefix *hydro-* and the stem of the second element with the suffix *-ic*: *hydroiodic*.
2. Add the word *acid*: hydroiodic *acid*.

PRACTICE 6.10

Name these binary acids:

(a) HCl (b) HBr

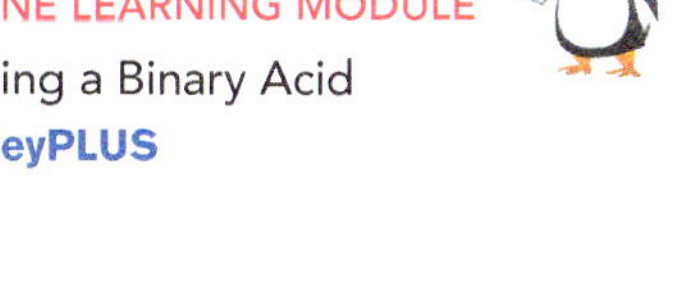

ONLINE LEARNING MODULE
Naming a Binary Acid
➡ WileyPLUS

Acids are hydrogen-containing substances that liberate hydrogen ions when dissolved in water. The same formula is often used to express binary hydrogen compounds, such as HCl, regardless of whether or not they are dissolved in water. TABLE 6.7 shows several examples of binary acids.

TABLE 6.7 Names and Formulas of Selected Binary Acids

Formula	Acid name
HF	Hydrofluoric acid
HCl	Hydrochloric acid
HBr	Hydrobromic acid
HI	Hydroiodic acid
H_2S	Hydrosulfuric acid
H_2Se	Hydroselenic acid

The process for naming binary compounds is summarized in FIGURE 6.4.

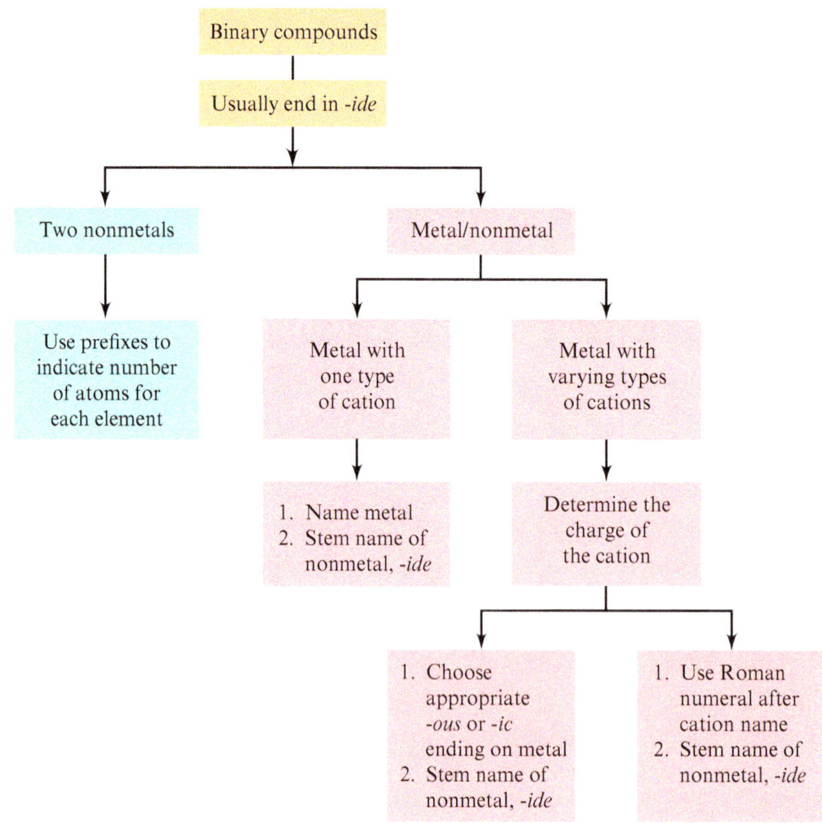

FIGURE 6.4 Flow diagram for naming binary compounds.

Naming Oxy-Acids

Inorganic compounds containing hydrogen, oxygen, and one other element are called *oxy-acids*. The first step in naming these acids is to determine that the compound in question is really an oxy-acid. The keys to identification are (1) hydrogen is the first element in the compound's formula and (2) the second part of the formula consists of a polyatomic ion containing oxygen.

Hydrogen in an oxy-acid is not specifically designated in the acid name. The presence of hydrogen in the compound is indicated by the use of the word *acid* in the name of the substance. To determine the particular type of acid, the polyatomic ion following hydrogen must be examined. The name of the polyatomic ion is modified in the following manner: (1) *-ate* changes to an *-ic* ending; (2) *-ite* changes to an *-ous* ending (TABLE 6.8). The compound with the *-ic* ending contains more oxygen than the one with the *-ous* ending.

TABLE 6.8 Comparison of Acid and Anion Names for Selected Oxy-Acids

Acid	Anion	Acid	Anion	Acid	Anion
H_2SO_4	SO_4^{2-}	H_2CO_3	CO_3^{2-}	HIO_3	IO_3^-
Sulfuric acid	Sulfate ion	Carbonic acid	Carbonate ion	Iodic acid	Iodate ion
H_2SO_3	SO_3^{2-}	H_3BO_3	BO_3^{3-}	$HC_2H_3O_2$	$C_2H_3O_2^-$
Sulfurous acid	Sulfite ion	Boric acid	Borate ion	Acetic acid	Acetate ion
HNO_3	NO_3^-	H_3PO_4	PO_4^{3-}	$H_2C_2O_4$	$C_2O_4^{2-}$
Nitric acid	Nitrate ion	Phosphoric acid	Phosphate ion	Oxalic acid	Oxalate ion
HNO_2	NO_2^-	H_3PO_3	PO_3^{3-}	$HBrO_3$	BrO_3^-
Nitrous acid	Nitrite ion	Phosphorous acid	Phosphite ion	Bromic acid	Bromate ion

ONLINE LEARNING MODULE
Naming an Oxy-Acid
➡ WileyPLUS

ENHANCED EXAMPLE 6.8 WileyPLUS

Name the acids H_2SO_4 and H_2SO_3.

SOLUTION

The names are written by modifying the name of the polyatomic ion in the acid.

For H_2SO_4 the polyatomic ion is the SO_4^{2-} sulfate ion. The *-ate* ending changes to an *-ic* ending in the acid. The stem of the sulfate ion is *sulfur*, so the name for the compound is *sulfuric* and the word *acid* is added to show the presence of hydrogen in the compound: *sulfuric acid*.

For H_2SO_3 the polyatomic ion is the SO_3^{2-} sulfite ion. The *-ite* ending changes to an *-ous* ending in the acid, *sulfurous*, and the word *acid* is added to show the presence of hydrogen in the compound: *sulfurous acid*.

PRACTICE 6.11

Name these polyatomic acids:
(a) H_2CO_3 (b) $HC_2H_3O_2$ (c) HNO_3 (d) HNO_2

Some elements form more than two different polyatomic ions containing oxygen. To name these ions, prefixes are used in addition to a suffix. To indicate more oxygen than in the *-ate* form we add the prefix *per-*, which is a short form of *hyper-* meaning "more." The prefix *hypo-*, meaning "less" (oxygen in this case), is used for the ion containing less oxygen than the *-ite* form. An example of this system is shown for the oxy-acids containing chlorine and oxygen in TABLE 6.9. The prefixes are also used with other similar ions, such as iodate (IO_3^-) and bromate (BrO_3^-).

TABLE 6.9 Oxy-Acids of Chlorine

Acid formula	Anion formula	Anion name	Acid name
$HClO$	ClO^-	*hypo*chlor*ite*	*hypo*chlor*ous* acid
$HClO_2$	ClO_2^-	chlor*ite*	chlor*ous* acid
$HClO_3$	ClO_3^-	chlor*ate*	chlor*ic* acid
$HClO_4$	ClO_4^-	*per*chlor*ate*	*per*chlor*ic* acid

We have now looked at ways of naming a variety of inorganic compounds—binary compounds consisting of a metal and a nonmetal and of two nonmetals, acids, and polyatomic compounds (FIGURE 6.5). These compounds are just a small part of the classified chemical compounds. Most of the remaining classes are in the broad field of organic chemistry under such categories as hydrocarbons, alcohols, ethers, aldehydes, ketones, phenols, and carboxylic acids.

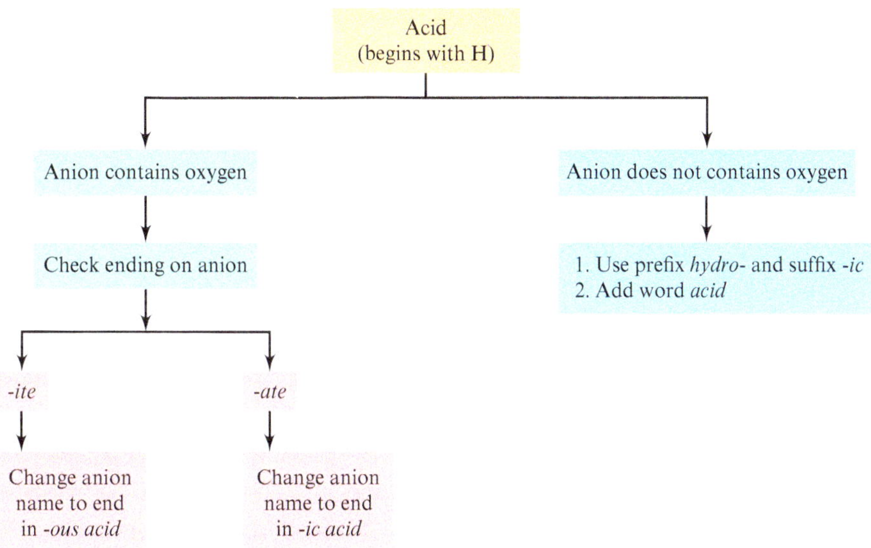

FIGURE 6.5 Flow diagram for naming acids.

CHAPTER 6 REVIEW

6.1 Common and Systematic Names

- Common names are arbitrary and do not describe the chemical composition of compounds.
  - Examples—ammonia, water
- A standard system of nomenclature was devised and is maintained by the IUPAC.

6.2 Elements and Ions

- Diatomic elements—H_2, O_2, N_2, F_2, Cl_2, Br_2, I_2.
- Polyatomic elements—P_4, S_8.
- Ions form through the gain or loss of electrons from neutral atoms.
- The charge on ions can often be predicted from the periodic table.

6.3 Writing Formulas from Names of Ionic Compounds

- Compounds must have a net charge of zero.
- To write a formula from a name:
  - Identify and write the symbols for the elements in the compound.
  - Combine the smallest number of ions required to produce a net charge of zero.

6.4 Naming Binary Compounds

KEY TERMS

binary compounds
Stock System

- The rules for naming binary compounds can be summarized in the following chart.

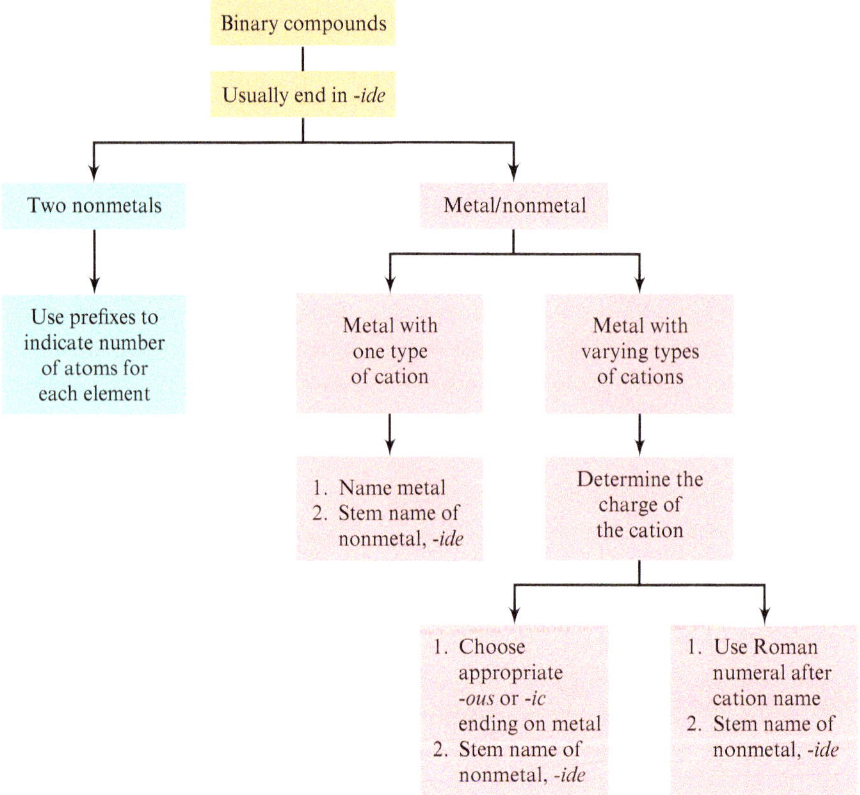

KEY TERM

polyatomic ion

6.5 Naming Compounds Containing Polyatomic Ions

- The rules for naming these compounds:
  1. Write the name of the cation.
  2. Write the name of the anion.

6.6 Acids

- The rules for naming acids can be summarized in the chart that follows:

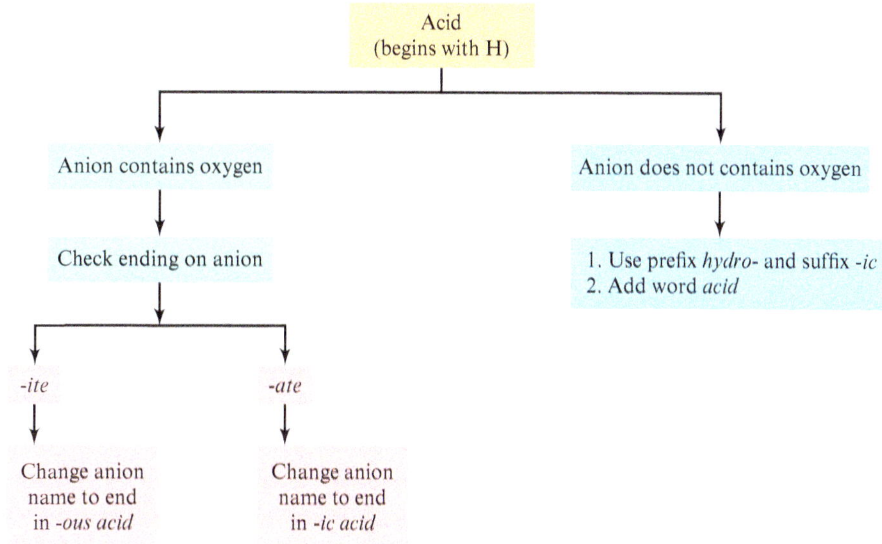

REVIEW QUESTIONS

1. List two compounds almost always referred to by their common names. What would be their systematic names?
2. What must happen in order to convert an atom into an anion? Into a cation?
3. Does the fact that two elements combine in a one-to-one atomic ratio mean that the charges on their ions are both 1? Explain.
4. When naming a binary ionic compound, how do you determine which element to name first?
5. Use the common ion table on the back endpapers of your text to determine the formulas for compounds composed of the following ions:
   (a) potassium and sulfide
   (b) cobalt(II) and bromate
   (c) ammonium and nitrate
   (d) hydrogen and phosphate
   (e) iron(III) and oxide
   (f) magnesium and hydroxide
6. Explain why P_2O_5 is named dinitrogen pentoxide, using prefixes, but Al_2O_3 is named aluminum oxide, without prefixes.
7. Explain why $FeCl_2$ is named iron(II) chloride, using a Roman numeral, and $BaCl_2$ is named barium chloride, using no Roman numeral. Why does neither use prefixes?
8. The names of the nitrogen oxides are nitrogen dioxide (NO_2) and dinitrogen monoxide (N_2O). Why is the *mono-* prefix used only on the second molecule?
9. What is the advantage of using the Stock System for naming metals with variable charge instead of using the common name?
10. Write formulas for the compounds formed when a nickel(II) ion is combined with
    (a) sulfate (f) acetate
    (b) phosphide (g) dichromate
    (c) chromate (h) bromide
    (d) hydroxide (i) nitrate
    (e) iodite (j) hypochlorite
    (Use the common ion table on the back endpapers of your text.)
11. Write the names and formulas for the four oxy-acids containing (a) bromine, (b) iodine.
12. How do you differentiate between different anions formed from a polyprotic acid?

PAIRED EXERCISES

Most of the exercises in this chapter are available for assignment via the online homework management program, WileyPLUS (www.wileyplus.com). All exercises with blue numbers have answers in Appendix 5.

1. Write the formula of the compound that will be formed between these elements:
(a) Ba and S (b) Cs and P (c) Li and Br (d) Mg and N (e) Ca and I (f) H and Cl

2. Write the formula of the compound that will be formed between these elements:
(a) Al and S (b) H and F (c) K and N (d) Sr and O (e) Cs and P (f) Al and Cl

3. Write formulas for the following cations:
(a) potassium (b) ammonium (c) copper(I) (d) titanium(IV) (e) nickel(III) (f) cesium (g) mercury(II) (h) calcium (i) lead(II) (j) zinc (k) silver (l) hydrogen (m) tin(II) (n) iron(III)

4. Write formulas for the following anions:
(a) fluoride (b) acetate (c) iodide (d) carbonate (e) sulfide (f) nitrate (g) phosphide (h) oxide (i) dichromate (j) hydrogen carbonate (k) phosphate (l) sulfate (m) nitride (n) chloride

5. Write the systematic names for the following:
(a) baking soda ($NaHCO_3$) (b) quicksilver (Hg) (c) lime (CaO) (d) vinegar ($HC_2H_3O_2$) (e) Epsom salts ($MgSO_4 \cdot 7H_2O$) (f) lye (NaOH)

6. Write the systematic names for the following:
(a) hypo ($Na_2S_2O_3$) (b) laughing gas (N_2O) (c) alumina (Al_2O_3) (d) table salt (NaCl) (e) milk of magnesia ($Mg(OH)_2$) (f) galena (PbS)

7. Complete the table, filling in each box with the proper formula.

Cations	Anions				
	Br^-	O^{2-}	NO_3^-	PO_4^{3-}	CO_3^{2-}
K^+	KBr				
Mg^{2+}					
Al^{3+}					
Zn^{2+}				$Zn_3(PO_4)_2$	
H^+					

8. Complete the table, filling in each box with the proper formula.

Cations	Anions				
	SO_4^{2-}	OH^-	AsO_4^{3-}	$C_2H_3O_2^-$	CrO_4^{2-}
NH_4^+			$(NH_4)_3AsO_4$		
Ca^{2+}					
Fe^{3+}	$Fe_2(SO_4)_3$				
Ag^+					
Cu^{2+}					

9. Write the names of each of the compounds formed in Question 7.

10. Write the names of each of the compounds formed in Question 8.

11. Write formulas for each of the following binary compounds, all of which are composed of nonmetals:
(a) diphosphorus pentoxide (b) carbon dioxide (c) tribromine octoxide (d) sulfur hexachloride (e) carbon tetrachloride (f) dichlorine heptoxide (g) boron trifluoride (h) tetranitrogen hexasulfide

12. Write formulas for each of the following binary compounds, all of which are composed of a metal and nonmetal:
(a) potassium nitride (b) barium oxide (c) iron(II) oxide (d) strontium phosphide (e) calcium nitride (f) cesium bromide (g) manganese(III) iodide (h) sodium selenide

13. Name each of the following binary compounds, all of which are composed of a metal and a nonmetal:
(a) BaO (b) K_2S (c) $CaCl_2$ (d) Cs_2S (e) Al_2O_3 (f) $CaBr_2$ (g) SrI_2 (h) Mg_3N_2

14. Name each of the following binary compounds, all of which are composed of nonmetals:
(a) PBr_5 (b) I_4O_9 (c) N_2S_5 (d) S_2F_{10} (e) $SiCl_4$ (f) ClO_2 (g) P_4S_7 (h) IF_6

15. Name these compounds by the Stock System (IUPAC):
(a) $CuCl_2$ (b) $FeCl_2$ (c) $Fe(NO_3)_2$ (d) $FeCl_3$ (e) SnF_2 (f) VPO_4

16. Write formulas for these compounds:
(a) tin(IV) bromide (b) copper(I) sulfate (c) nickel(II) borate (d) mercury(II) nitrite (e) cobalt(III) carbonate (f) iron(II) acetate

17. Write formulas for these acids:
(a) hydrochloric acid (b) chloric acid (c) nitric acid (d) carbonic acid (e) sulfurous acid (f) phosphoric acid

18. Write formulas for these acids:
(a) acetic acid (b) hydrofluoric acid (c) hydrosulfuric acid (d) boric acid (e) nitrous acid (f) hypochlorous acid

19. Name these acids:
(a) HNO_2 (b) H_2SO_4 (c) $H_2C_2O_4$ (d) HBr (e) H_3PO_3 (f) $HC_2H_3O_2$ (g) HF (h) $HBrO_3$ (i) HIO_4

20. Name these acids:
(a) H_3PO_4 (b) H_2CO_3 (c) HIO_3 (d) HCl (e) HClO (f) HNO_3 (g) HI (h) $HClO_4$ (i) H_2SO_3

21. Write formulas for these compounds:
(a) silver sulfite
(b) cobalt(II) bromide
(c) tin(II) hydroxide
(d) aluminum sulfate
(e) lead(II) chloride
(f) ammonium carbonate
(g) chromium(III) oxide
(h) copper(II) chloride
(i) potassium permanganate
(j) arsenic(V) sulfite
(k) sodium peroxide
(l) iron(II) sulfate
(m) potassium dichromate
(n) bismuth(III) chromate

22. Write formulas for these compounds:
(a) sodium chromate
(b) magnesium hydride
(c) nickel(II) acetate
(d) calcium chlorate
(e) magnesium bromate
(f) potassium dihydrogen phosphate
(g) manganese(II) hydroxide
(h) cobalt(II) hydrogen carbonate
(i) sodium hypochlorite
(j) barium perchlorate
(k) chromium(III) sulfite
(l) antimony(III) sulfate
(m) sodium oxalate
(n) potassium thiocyanate

23. Write the name of each compound:
(a) $ZnSO_4$
(b) Hg_2S
(c) $CuCO_3$
(d) $Cd(NO_3)_2$
(e) $Al(C_2H_3O_2)_3$
(f) CoF_2
(g) $Cr(ClO_3)_3$
(h) Ag_3PO_4
(i) MnS
(j) $BaCrO_4$

24. Write the name of each compound:
(a) $Ca(HSO_4)_2$
(b) $As_2(SO_3)_3$
(c) $Sn(NO_2)_2$
(d) CuI
(e) $KHCO_3$
(f) $BiAsO_4$
(g) $(NH_4)_2CO_3$
(h) $(NH_4)_2HPO_4$
(i) $NaClO$
(j) $KMnO_4$

25. Write the chemical formulas for these substances:
(a) slaked lime
(b) fool's gold
(c) washing soda
(d) calcite
(e) cane sugar
(f) borax
(g) wood alcohol
(h) acetylene

26. Write the chemical formulas for these substances:
(a) grain alcohol
(b) cream of tartar
(c) gypsum
(d) brimstone
(e) muriatic acid
(f) plaster of paris
(g) lye
(h) laughing gas

27. The early Romans discovered lead in mines for other valuable metals such as silver. They discovered rich fields of galena or lead sulfide in Derbyshire, which is located in what is now known as the United Kingdom, and began to mine it for use in their aqueducts and water pipes. What are the chemical formulas and correct IUPAC names of the two lead sulfides that could exist? Which of these compounds is given the common name galena?

28. Compounds of lead have been used throughout the ages before anyone realized lead's toxicity.
(a) Flake white was a white pigment used by Renaissance artists, which was composed of a combination of lead(II) carbonate and lead(II) hydroxide. What are the correct chemical formulas for these two compounds?
(b) Lead(II) oxide has been added to glass to make crystal that glistens beautifully. When wine is stored in a crystal decanter, the lead slowly leaches out making lead(II) acetate, which is also known as the sugar of lead. It gives the wine a very sweet flavor but is unfortunately also very toxic. Write the correct chemical formulas for these two compounds.

ADDITIONAL EXERCISES

29. Write equations similar to those found in Section 6.2 for the formation of
(a) potassium ion
(b) iodide ion
(c) bromide ion
(d) iron(II) ion
(e) calcium ion
(f) oxide ion

30. Name each of the following polyatomic ions:

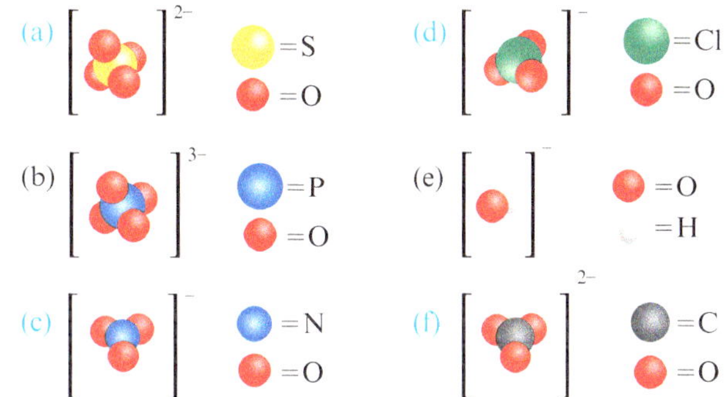

31. Write formulas for all possible compounds formed between the calcium ion and the anions shown in Question 30.

32. Write formulas for all possible compounds formed between the potassium ion and the anions shown in Question 30.

33. Write the formula and name for each of the following compounds:

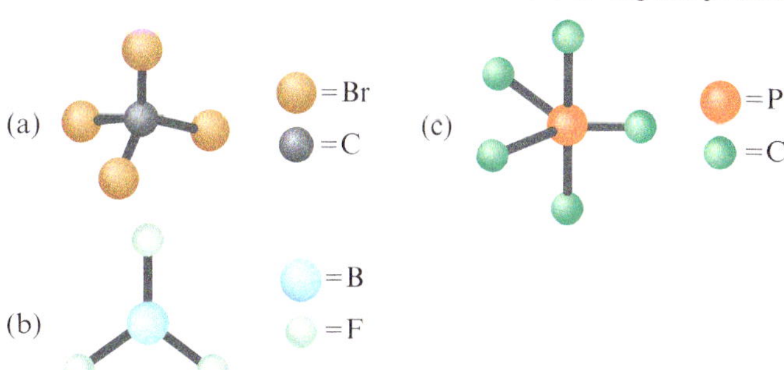

34. Write the formula and name for the following compound:

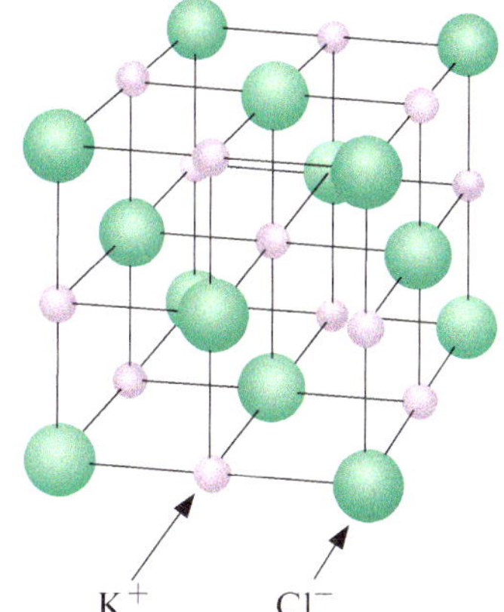

35. State how each of the following is used in naming inorganic compounds: *ide*, *ous*, *ic*, *hypo*, *per*, *ite*, *ate*, Roman numerals.

36. Several excerpts from newspaper articles follow. Tell the chemical formula of the chemical spilled and indicate whether this spill should raise concern in the local area.
(a) A mistake at the Columbia Water and Light Department's West Ash Street pumping station sent an estimated 7200 gallons of calcium-carbonate-laden water into Harmony Creek early yesterday, forcing crews to spend hours cleaning up the mess.
(b) A tanker truck carrying 6000 gallons of acetic acid overturned on the I-5 freeway in San Diego, California.
(c) A university in New Hampshire was discovered dumping dihydrogen oxide into the local sewer system.

37. How many of each type of subatomic particle (protons and electrons) is in
(a) an atom of tin?
(b) a Sn^{2+} ion?
(c) a Sn^{4+} ion?

38. The compound X_2Y_3 is a stable solid. What ionic charge do you expect for X and Y? Explain.

39. The ferricyanide ion has the formula $Fe(CN)_6^{3-}$. Write the formula for the compounds that ferricyanide would form with the cations of elements 3, 13, and 30.

40. Compare and contrast the formulas of
(a) nitride with nitrite
(b) nitrite with nitrate
(c) nitrous acid with nitric acid

41. For each of the following compounds, indicate what is wrong with the name and why. If possible, fix the name.
(a) iron hydroxide
(b) dipotassium dichromium heptoxide
(c) sulfur oxide

42. Sulfur dioxide is a gas formed as a by-product of burning coal. Sulfur trioxide is a significant contributor to acid rain. Does the existence of these two substances violate the law of multiple proportions? Explain.

CHALLENGE EXERCISE

43. An unidentified metal forms an ionic compound with phosphate. The metal forms a 2+ cation. If the minimum ratio of protons in the metal to the phosphorus is 6:5, what metal is it? (*Hint:* First write the formula for the ionic compound formed with phosphate anion.)

44. After studying chemistry, you should be able to recognize more of the substances listed on consumer products. A list of ingredients for dog food follows:

Chicken By-Product Meal (Natural source of Chondroitin Sulfate and Glucosamine), Corn Meal, Ground Whole Grain Sorghum, Ground Whole Grain Barley, Fish Meal (source of fish oil), Chicken, Chicken Fat (preserved with mixed Tocopherols, a source of Vitamin E), Dried Beet Pulp, Chicken Flavor, Dried Egg Product, Potassium Chloride, Brewers Dried Yeast, Salt, Sodium Hexametaphosphate, Fructooligosaccharides, Fish Oil (preserved with mixed Tocopherols, a source of Vitamin E), Calcium Carbonate, Flax Meal, Choline Chloride, Minerals (Ferrous Sulfate, Zinc Oxide, Manganese Sulfate, Copper Sulfate, Manganous Oxide, Potassium Iodide, Cobalt Carbonate), Vitamin E Supplement, Dried Chicken Cartilage (Natural source of Chondroitin Sulfate and Glucosamine), DL-Methionine, Vitamins (Ascorbic Acid, Vitamin A Acetate, Calcium Pantothenate, Biotin, Thiamine Mononitrate (source of Vitamin B1), Vitamin B12 Supplement, Niacin, Riboflavin Supplement (source of Vitamin B2), Inositol, Pyridoxine Hydrochloride (source of Vitamin B6), Vitamin D3 Supplement, Folic Acid), Beta-Carotene, L-Carnitine, Marigold, Citric Acid, Rosemary Extract.

Many of the substances in this bag of dog food are ionic compounds that you should be able to recognize. The manufacturers of this dog food did not completely identify some of the compounds. Write the chemical formula of the following compounds found in this ingredient list.
(a) potassium chloride
(b) calcium carbonate
(c) ferrous sulfate
(d) zinc oxide
(e) manganese sulfate
(f) copper sulfate
(g) manganous oxide
(h) potassium iodide
(i) cobalt carbonate
(j) sodium chloride

ANSWERS TO PRACTICE EXERCISES

6.1 phosphorus stem is *phosph-* + *-ide* = phosphide; nitrogen stem is *nitr-* + *-ide* = nitride; boron stem is *bor-* + *-ide* = boride; carbon stem is *carb-* + *-ide* = carbide

6.2 (a) KF; (b) $CaBr_2$; (c) Mg_3N_2; (d) Na_2S; (e) BaO; (f) $As_2(CO_3)_5$

6.3 (a) $SrCl_2$; (b) KI; (c) AlN; (d) CaS; (e) Na_2O

6.4 (a) lead(II) iodide; (b) tin(IV) fluoride; (c) iron(III) oxide; (d) copper(II) oxide

6.5 (a) $Sn(CrO_4)_2$; (b) $CrBr_3$; (c) SnF_2; (d) Cu_2O

6.6 (a) dichlorine monoxide; (b) sulfur dioxide; (c) carbon tetrabromide; (d) dinitrogen pentoxide; (e) ammonia; (f) iodine trichloride

6.7 (a) potassium bromide; (b) calcium nitride; (c) sulfur trioxide; (d) tin(II) fluoride; (e) copper(II) chloride; (f) dinitrogen tetroxide

6.8 (a) sodium nitrate; (b) calcium phosphate; (c) potassium hydroxide; (d) lithium carbonate; (e) sodium chlorate

6.9 (a) potassium hydrogen carbonate; (b) sodium hydrogen oxalate; (c) barium ammonium phosphate; (d) sodium aluminum sulfate

6.10 (a) hydrochloric acid; (b) hydrobromic acid

6.11 (a) carbonic acid; (b) acetic acid; (c) nitric acid; (d) nitrous acid

7

Quantitative Composition of Compounds

© Hemis/Alamy

CHAPTER OUTLINE

Cereals, cleaning products, and pain remedies all list their ingredients on the package label. The ingredients are listed in order from most to least, but the amounts are rarely given. However, it is precisely these amounts that give products their desired properties and distinguish them from the competition. Understandably, manufacturers carefully regulate the amounts of ingredients to maintain quality and hopefully their customers' loyalty. The black pearls in the photo above are made of layers of calcium carbonate. They can be measured by counting or weighing.

The composition of compounds is an important concept in chemistry. Determining numerical relationships among the elements in compounds and measuring exact quantities of particles are fundamental tasks that chemists routinely perform in their daily work.

7.1 The Mole

LEARNING OBJECTIVE Apply the concepts of the mole, molar mass, and Avogadro's number to solve chemistry problems.

KEY TERMS
Avogadro's number
mole
molar mass

Oranges can be "counted" by weighing them in the store.

The atom is an incredibly tiny object. Its mass is far too small to measure on an ordinary balance. In Chapter 5 (Section 5.6), we learned to compare atoms using a table of atomic mass units. These units are valuable when we compare the masses of individual atoms (mentally), but they have no practical use in the laboratory. The mass in grams for an "average" carbon atom (atomic mass 12.01 u) would be 2.00×10^{-23} g, which is much too tiny for the best laboratory balance.

So how can we confidently measure masses for these very tiny atoms? We increase the number of atoms in a sample until we have an amount large enough to measure on a laboratory balance. The problem then is how to count our sample of atoms.

Consider for a moment the produce in a supermarket. Frequently, apples and oranges are sorted by size and then sold by weight, not by the piece of fruit. The grocer is counting by weighing. To do this, he needs to know the mass of an "average" apple (235 g) and the mass of an "average" orange (186 g). Now suppose he has an order from the local college for 275 apples and 350 oranges. It would take a long time to count and package this order. The grocer can quickly count fruit by weighing. For example,

$$(275 \text{ apples})\left(\frac{235 \text{ g}}{\text{apple}}\right) = 6.46 \times 10^4 \text{ g} = 64.6 \text{ kg}$$

$$(350 \text{ oranges})\left(\frac{186 \text{ g}}{\text{orange}}\right) = 6.51 \times 10^4 \text{ g} = 65.1 \text{ kg}$$

He can now weigh 64.6 kg of apples and 65.1 kg of oranges and pack them without actually counting them. Manufacturers and suppliers often count by weighing. Other examples of counting by weighing include nuts, bolts, and candy.

Chemists also count atoms by weighing. We know the "average" masses of atoms, so we can count atoms by defining a unit to represent a larger number of atoms. Chemists have chosen the mole (mol) as the unit for counting atoms. The mole is a unit for counting just as a dozen or a ream or a gross is used to count:

$$1 \text{ dozen} = 12 \text{ objects}$$

$$1 \text{ ream} = 500 \text{ objects}$$

$$1 \text{ gross} = 144 \text{ objects}$$

$$1 \text{ mole} = 6.022 \times 10^{23} \text{ objects}$$

Note that we use a unit only when it is appropriate. A dozen eggs is practical in our kitchen, a gross might be practical for a restaurant, but a ream of eggs would not be very practical. Chemists can't use dozens, grosses, or reams because atoms are so tiny that a dozen, gross, or ream of atoms still couldn't be measured in the laboratory.

Units of measurement need to be appropriate for the object being measured.

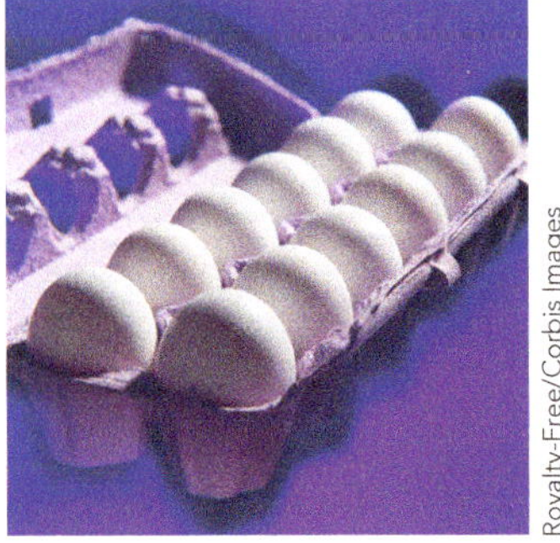
Eggs are measured by the dozen.

Paper is measured by the ream (500 sheets).

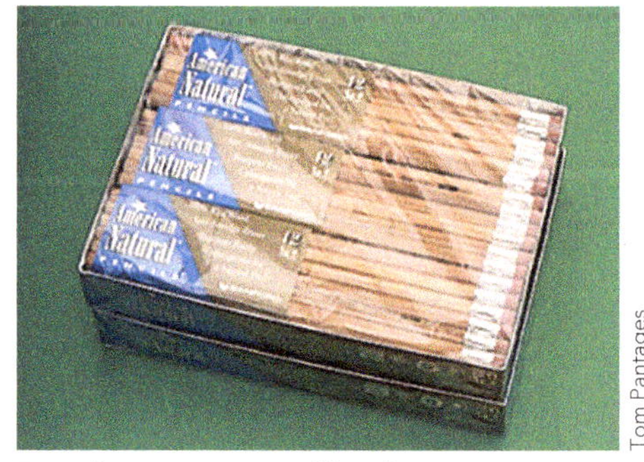

Pencils are measured by the gross (144).

The number represented by 1 mol, 6.022×10^{23}, is called **Avogadro's number**, in honor of Amedeo Avogadro (1776–1856), who investigated several quantitative aspects in chemistry. It's difficult to imagine how large Avogadro's number really is, but this example may help: If 10,000 people started to count Avogadro's number, and each counted at the rate of 100 numbers per minute each minute of the day, it would take them over 1 trillion (10^{12}) years to count the total number. So even the tiniest amount of matter contains extremely large numbers of atoms.

Symbol name	C (carbon)	Al (aluminum)	Pb (lead)
Element			
Average atomic mass	12.01 u	26.98 u	207.2 u
Mass of sample	12.01 g	26.98 g	207.2 g
Number of atoms in sample	6.022×10^{23} atoms	6.022×10^{23} atoms	6.022×10^{23} atoms

Avogadro's number has been experimentally determined by several methods. How does it relate to atomic mass units? Remember that the atomic mass for an element is the average relative mass of all the isotopes for the element. The atomic mass (expressed in grams) of 1 mole of any element contains the same number of particles (Avogadro's number) as there are in exactly 12 g of ^{12}C. Thus, 1 **mole** of anything is the amount of the substance that contains the same number of items as there are atoms in exactly 12 g of ^{12}C.

Remember that ^{12}C is the reference isotope for atomic masses.

$$1 \text{ mole} = 6.022 \times 10^{23} \text{ items}$$

From the definition of mole, we can see that the atomic mass in grams of any element contains 1 mol of atoms. The term *mole* is so commonplace in chemistry that chemists use it as freely as the words *atom* and *molecule*. A mole of atoms, molecules, ions, or electrons represents Avogadro's number of these particles. If we can speak of a mole of atoms, we can also speak of a mole of molecules, a mole of electrons, or a mole of ions, understanding that in each case we mean 6.022×10^{23} particles:

The abbreviation for mole is mol, both singular and plural.

$$1 \text{ mol of atoms} = 6.022 \times 10^{23} \text{ atoms}$$
$$1 \text{ mol of molecules} = 6.022 \times 10^{23} \text{ molecules}$$
$$1 \text{ mol of ions} = 6.022 \times 10^{23} \text{ ions}$$

PRACTICE 7.1

How many atoms are in 1.000 mole of the following?

(a) Fe (b) H_2 (c) H_2SO_4

The atomic mass of an element in grams contains Avogadro's number of atoms and is defined as the **molar mass** of that element. To determine molar mass, we change the units of the atomic mass (found in the periodic table) from atomic mass units to grams. For example, sulfur has an atomic mass of 32.07 u, so 1 mol of sulfur has a molar mass of 32.07 g and contains 6.022×10^{23} atoms of sulfur. Here are some other examples:

CHECK YOUR UNDERSTANDING

7.1 Units for Atoms, Molecules, and Ions

WileyPLUS

Element	Atomic mass	Molar mass	Number of atoms
H	1.008 u	1.008 g	6.022×10^{23}
Mg	24.31 u	24.31 g	6.022×10^{23}
Na	22.99 u	22.99 g	6.022×10^{23}

In this text molar masses of elements are given to four significant figures.

To summarize:

1. The atomic mass expressed in grams is the *molar mass* of an element. It is different for each element. In this text, molar masses are expressed to four significant figures.

$$1 \text{ molar mass} = \text{atomic mass of an element in grams}$$

2. One mole of any element contains Avogadro's number of atoms.

$$1 \text{ mol of atoms} = 6.022 \times 10^{23} \text{ atoms}$$

CHECK YOUR UNDERSTANDING

7.2 Mass of a Mole

7.3 Molecules of Carbon Dioxide

WileyPLUS

We can use these relationships to make conversions between number of atoms, mass, and moles, as shown in the following examples.

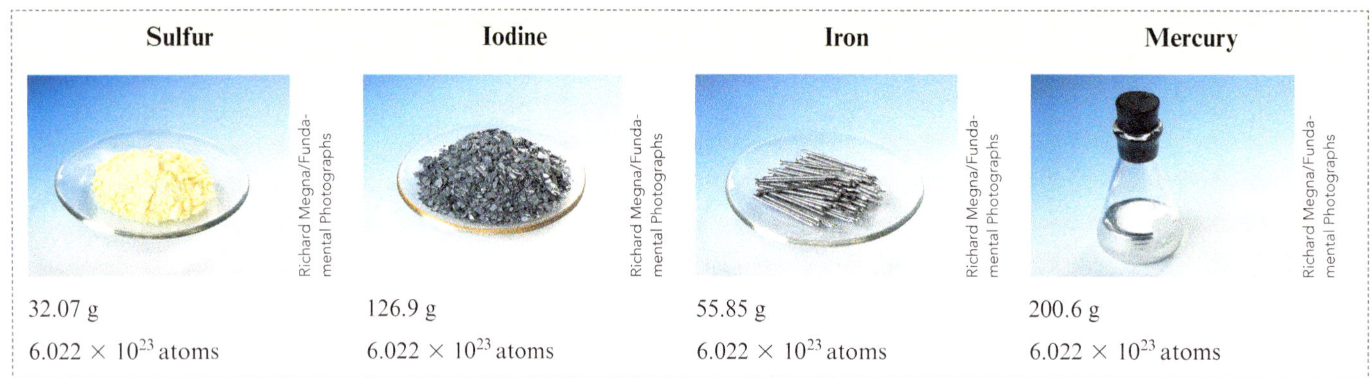

ONLINE LEARNING MODULE

Changing Grams to Moles

WileyPLUS

ENHANCED EXAMPLE 7.1

WileyPLUS

How many moles of iron does 25.0 g of iron (Fe) represent?

SOLUTION

Plan • **Solution map:** g Fe → moles Fe

1 mole Fe = 55.85 g (from periodic table)

The conversion factor needed is

$$\frac{1 \text{ mole Fe}}{55.85 \text{ g Fe}} \quad \text{or} \quad \frac{55.85 \text{ g Fe}}{1 \text{ mole Fe}}$$

Calculate • $(25.0 \cancel{\text{g Fe}})\left(\frac{1 \text{ mole Fe}}{55.85 \cancel{\text{g Fe}}}\right) = 0.448 \text{ mol Fe}$

Check • Since 25.0 g has three significant figures, the number of significant figures allowed in the answer is three. The answer is reasonable since 0.448 mol is about ½ mole and 25.0 g is about ½ of 55.85 g (one mole Fe).

ONLINE LEARNING MODULE

Determining the Number of Atoms in a Sample

WileyPLUS

ENHANCED EXAMPLE 7.2

WileyPLUS

How many magnesium atoms are contained in 5.00 g Mg?

SOLUTION

Plan • **Solution map:** g Mg → atoms Mg

1 mol Mg = 24.31 g (from periodic table)

The conversion factor needed is

$$\frac{6.022 \times 10^{23} \text{ atoms Mg}}{24.31 \text{ g Mg}} \quad \text{or} \quad \frac{24.31 \text{ g Mg}}{6.022 \times 10^{23} \text{ atoms Mg}}$$

Calculate • $(5.00 \cancel{\text{g Mg}})\left(\frac{6.022 \times 10^{23} \text{ atoms Mg}}{24.31 \cancel{\text{g Mg}}}\right) = 1.24 \times 10^{23} \text{ atoms Mg}$

Check • Since 5.00 g has three significant figures, the number of significant figures allowed in the answer is three.

ENHANCED EXAMPLE 7.3 WileyPLUS

What is the mass, in grams, of one atom of carbon (C)?

SOLUTION

Plan **Solution map:** atoms C → grams C

1 mol C = 12.01 g (from periodic table)

The conversion factor needed is

$$\frac{6.022 \times 10^{23}\ \text{atoms C}}{12.01\ \text{g C}} \quad \text{or} \quad \frac{12.01\ \text{g C}}{6.022 \times 10^{23}\ \text{atoms C}}$$

Calculate

$$(1\ \cancel{\text{atom C}})\left(\frac{12.01\ \text{g C}}{6.022 \times 10^{23}\ \cancel{\text{atoms C}}}\right) = 1.994 \times 10^{-23}\ \text{g C}$$

Check Since 1 atom is an exact number, the number of significant figures allowed in the answer is four based on the number of significant figures in the molar mass of C.

ENHANCED EXAMPLE 7.4 WileyPLUS

ONLINE LEARNING MODULE
Converting Atoms to Mass
➡ WileyPLUS

What is the mass of 3.01×10^{23} atoms of sodium (Na)?

SOLUTION

Plan **Solution map:** atoms Na → grams Na

1 mol Na = 22.99 g (from periodic table)

The conversion factor needed is

$$\frac{6.022 \times 10^{23}\ \text{atoms Na}}{22.99\ \text{g Na}} \quad \text{or} \quad \frac{22.99\ \text{g Na}}{6.022 \times 10^{23}\ \text{atoms Na}}$$

Calculate

$$(3.01 \times 10^{23}\ \cancel{\text{atoms Na}})\left(\frac{22.99\ \text{g Na}}{6.022 \times 10^{23}\ \cancel{\text{atoms Na}}}\right) = 11.5\ \text{g Na}$$

Check Since 3.01×10^{23} has three significant figures, the answer (mass) should have three as well.

ENHANCED EXAMPLE 7.5 WileyPLUS

What is the mass of 0.252 mol of copper (Cu)?

SOLUTION

Plan **Solution map:** mol Cu → grams Cu

1 mol Cu = 63.55 g (from periodic table)

The conversion factor needed is

$$\frac{1\ \text{mol Cu}}{63.55\ \text{g Cu}} \quad \text{or} \quad \frac{63.55\ \text{g Cu}}{1\ \text{mol Cu}}$$

Calculate

$$(0.252\ \cancel{\text{mol Cu}})\left(\frac{63.55\ \text{g Cu}}{1\ \cancel{\text{mol Cu}}}\right) = 16.0\ \text{g Cu}$$

Check The number of significant figures in both the question and answer is three.

ONLINE LEARNING MODULE
Finding Molecules in Moles of a Diatomic Substance
WileyPLUS

ENHANCED EXAMPLE 7.6 WileyPLUS

How many oxygen atoms are present in 1.00 mol of oxygen molecules?

SOLUTION

Plan Oxygen is a diatomic molecule with the formula O_2. Therefore, a molecule of oxygen contains 2 oxygen atoms: $\frac{2 \text{ atoms O}}{1 \text{ molecule } O_2}$.

Solution map: moles O_2 → molecules O_2 → atoms O

The conversion factors needed are

$$\frac{6.022 \times 10^{23} \text{ molecules } O_2}{1 \text{ mole } O_2} \quad \text{and} \quad \frac{2 \text{ atoms O}}{1 \text{ molecule } O_2}$$

Calculate $$(1.00 \cancel{\text{mol } O_2})\left(\frac{6.022 \times 10^{23} \cancel{\text{molecules } O_2}}{1 \cancel{\text{mole } O_2}}\right)\left(\frac{2 \text{ atoms O}}{1 \cancel{\text{molecule } O_2}}\right)$$
$$= 1.20 \times 10^{24} \text{ atoms O}$$

PRACTICE 7.2

What is the mass of 2.50 mol of helium (He)?

PRACTICE 7.3

How many atoms are present in 0.025 mol of iron?

7.2 Molar Mass of Compounds

LEARNING OBJECTIVE Calculate the molar mass of a compound.

A formula unit is indicated by the formula, for example, Mg, MgS, H_2O, NaCl.

One mole of a compound contains Avogadro's number of *formula units* of that compound. The terms *molecular weight, molecular mass, formula weight*, and *formula mass* have been used in the past to refer to the mass of 1 mol of a compound. However, the term *molar mass* is more inclusive because it can be used for all types of compounds.

If the formula of a compound is known, its molar mass can be determined by adding the molar masses of all the atoms in the formula. If more than one atom of any element is present, its mass must be added as many times as it appears in the compound.

ENHANCED EXAMPLE 7.7 WileyPLUS

The formula for water is H_2O. What is its molar mass?

SOLUTION

Read **Known:** H_2O

Plan The formula contains 1 atom O and 2 atoms H. We need to use the atomic masses from the inside front cover of the text.

Calculate
$$\begin{aligned} O &= 16.00 \text{ g} &&= 16.00 \text{ g} \\ 2\text{ H} &= 2(1.008 \text{ g}) &&= \underline{\ 2.016 \text{ g}\ } \\ &&& \ 18.02 \text{ g} = \text{molar mass of } H_2O \end{aligned}$$

ENHANCED EXAMPLE 7.8 WileyPLUS

Calculate the molar mass of calcium hydroxide, $Ca(OH)_2$.

SOLUTION

Read • **Known:** $Ca(OH)_2$

Plan • The formula contains 1 atom Ca and 2 atoms each of O and H. We need to use the atomic masses from the inside front cover of the text.

Calculate •

$$
\begin{aligned}
1\ \text{Ca} &= 1(40.08\ \text{g}) = 40.08\ \text{g} \\
2\ \text{O} &= 2(16.00\ \text{g}) = 32.00\ \text{g} \\
2\ \text{H} &= 2(1.008\ \text{g}) = \underline{2.016\ \text{g}} \\
& \qquad 74.10\ \text{g} = \text{molar mass of } Ca(OH)_2
\end{aligned}
$$

PRACTICE 7.4

Calculate the molar mass of KNO_3.

ONLINE LEARNING MODULE
Molar Mass
➡ WileyPLUS

In this text we round all molar masses to four significant figures, although you may need to use a different number of significant figures for other work (in the lab).

The mass of 1 mol of a compound contains Avogadro's number of formula units. Consider the compound hydrogen chloride (HCl). One atom of H combines with one atom of Cl to form HCl. When 1 mol of H (1.008 g of H or 6.022×10^{23} H atoms) combines with 1 mol of Cl (35.45 g of Cl or 6.022×10^{23} Cl atoms), 1 mol of HCl (36.46 g of HCl or 6.022×10^{23} HCl molecules) is produced. These relationships are summarized in the following table:

CHECK YOUR UNDERSTANDING
7.4 Molar Mass and Significant Figures
➡ WileyPLUS

H	Cl	HCl
6.022×10^{23} H *atoms*	6.022×10^{23} Cl *atoms*	6.022×10^{23} HCl *molecules*
1 mol H *atoms*	1 mol Cl *atoms*	1 mol HCl *molecules*
1.008 g H	35.45 g Cl	36.46 g HCl
1 molar mass H *atoms*	1 molar mass Cl *atoms*	1 molar mass HCl *molecules*

CHECK YOUR UNDERSTANDING
7.5 Atoms and Molar Mass
➡ WileyPLUS

In dealing with diatomic elements (H_2, O_2, N_2, F_2, Cl_2, Br_2, and I_2), we must take special care to distinguish between a mole of atoms and a mole of molecules. For example, consider 1 mol of oxygen molecules, which has a mass of 32.00 g. This quantity is equal

Tom Pantages

A mole of table salt (in front of the salt shaker) and a mole of water (in the plastic container) have different sizes, but both contain Avogadro's number of formula units.

to 2 mol of oxygen atoms. Remember that 1 mol represents Avogadro's number of the particular chemical entity that is under consideration:

Formula units are often used in place of molecules for substances that contain ions.

$$1 \text{ mol } H_2O = 18.02 \text{ g } H_2O = 6.022 \times 10^{23} \text{ molecules}$$
$$1 \text{ mol NaCl} = 58.44 \text{ g NaCl} = 6.022 \times 10^{23} \text{ formula units}$$
$$1 \text{ mol } H_2 = 2.016 \text{ g } H_2 = 6.022 \times 10^{23} \text{ molecules}$$
$$1 \text{ mol } HNO_3 = 63.02 \text{ g } HNO_3 = 6.022 \times 10^{23} \text{ molecules}$$
$$1 \text{ mol } K_2SO_4 = 174.3 \text{ g } K_2SO_4 = 6.022 \times 10^{23} \text{ formula units}$$
$$1 \text{ mol} = 6.022 \times 10^{23} \text{ formula units or molecules}$$
$$= 1 \text{ molar mass of a compound}$$

CHECK YOUR UNDERSTANDING

7.6 Molecules and Formula Units

WileyPLUS

ENHANCED EXAMPLE 7.9 WileyPLUS

What is the mass of 1 mol of sulfuric acid (H_2SO_4)?

SOLUTION

Read • **Known:** H_2SO_4

Plan • We need to use the atomic masses from the inside front cover of the text.

Calculate •

$$2 \text{ H} = 2(1.008 \text{ g}) = 2.016 \text{ g}$$
$$1 \text{ S} = 32.07 \text{ g} = 32.07 \text{ g}$$
$$4 \text{ O} = 4(16.00 \text{ g}) = \underline{64.00 \text{ g}}$$
$$98.09 \text{ g} = \text{mass of 1 mol of } H_2SO_4$$

ENHANCED EXAMPLE 7.10 WileyPLUS

How many moles of sodium hydroxide (NaOH) are there in 1.00 kg of sodium hydroxide?

SOLUTION

Read • **Known:** 1.00 kg NaOH

Plan • We need to use the atomic masses from the inside front cover of the text to find the molar mass of NaOH.

$$\overset{\text{Na}}{22.99 \text{ g}} + \overset{\text{O}}{16.00 \text{ g}} + \overset{\text{H}}{1.008 \text{ g}} = 40.00 \text{ g NaOH} = \text{molar mass NaOH}$$

Solution map: kg NaOH → g NaOH → mol NaOH

Create the appropriate conversion factor by placing the unit desired in the numerator and the unit to be eliminated in the denominator.

We need two conversion factors: $\dfrac{1000 \text{ g}}{1 \text{ kg}}$ and $\dfrac{1 \text{ mol NaOH}}{40.00 \text{ g NaOH}}$.

Calculate • $$(1.00 \cancel{\text{kg NaOH}})\left(\frac{1000 \cancel{\text{g NaOH}}}{1 \cancel{\text{kg NaOH}}}\right)\left(\frac{1 \text{ mol NaOH}}{40.00 \cancel{\text{g NaOH}}}\right)$$
$$= 25.00 \text{ mol NaOH}$$

Therefore, 1.00 kg NaOH = 25.0 mol NaOH.

ENHANCED EXAMPLE 7.11 WileyPLUS

What is the mass of 5.00 mol water?

SOLUTION

Read • **Known:** 5.00 mol H_2O

Plan We need to use the atomic masses from the inside front cover of the text to find the molar mass of H_2O.

Solution map: mol H_2O → g H_2O

$$\frac{18.02 \text{ g H}_2\text{O}}{1 \text{ mol H}_2\text{O}} \text{ (conversion factor)}$$

Calculate $(5.00 \cancel{\text{mol H}_2\text{O}})\left(\frac{18.02 \text{ g H}_2\text{O}}{1 \cancel{\text{mol H}_2\text{O}}}\right) = 90.1 \text{ g H}_2\text{O}$

Therefore, 5.00 mol H_2O = 90.1 g H_2O.

ENHANCED EXAMPLE 7.12 WileyPLUS

ONLINE LEARNING MODULE

Converting Mass to Molecules

WileyPLUS

How many molecules of hydrogen chloride (HCl) are there in 25.0 g of hydrogen chloride?

SOLUTION

Read **Known:** 25.0 g HCl

Plan We need to use the atomic masses from the inside front cover of the text to find the molar mass of HCl.

$$\overset{\text{H}}{1.008 \text{ g}} + \overset{\text{Cl}}{35.45 \text{ g}} = 36.46 \text{ g} = \text{molar mass HCl}$$

Solution map: g HCl → mol HCl → molecules HCl

$$\frac{1 \text{ mol HCl}}{36.46 \text{ g HCl}} \text{ and } \frac{6.022 \times 10^{23} \text{ molecules HCl}}{1 \text{ mol HCl}} \text{ (conversion factors)}$$

Calculate $(25.0 \cancel{\text{g HCl}})\left(\frac{1 \cancel{\text{mol HCl}}}{36.46 \cancel{\text{g HCl}}}\right)\left(\frac{6.022 \times 10^{23} \text{ molecules HCl}}{1 \cancel{\text{mol HCl}}}\right)$

$$= 4.13 \times 10^{23} \text{ molecules HCl}$$

PRACTICE 7.5

What is the mass of 0.150 mol of Na_2SO_4?

PRACTICE 7.6

How many moles and molecules are there in 500.0 g of $HC_2H_3O_2$?

7.3 Percent Composition of Compounds

LEARNING OBJECTIVE

Calculate the percent composition of a compound from its chemical formula and from experimental data.

KEY TERM

percent composition of a compound

Percent means parts per 100 parts. Just as each piece of pie is a percent of the whole pie, each element in a compound is a percent of the whole compound. The **percent composition of a compound** is the *mass percent* of each element in the compound. The molar mass represents the total mass, or 100%, of the compound. Thus, the percent composition of water, H_2O, is 11.19% H and 88.79% O by mass. According to the law of definite composition, the percent composition must be the same no matter what size sample is taken.

The percent composition of a compound can be determined (1) from knowing its formula or (2) from experimental data.

CHECK YOUR UNDERSTANDING

7.7 Percent Composition

WileyPLUS

Percent Composition from Formula

If the formula of a compound is known, a two-step process is needed to determine the percent composition:

PROBLEM-SOLVING STRATEGY

For Calculating Percent Composition from Formula

1. Calculate the molar mass (Section 7.2).
2. Divide the total mass of each element in the formula by the molar mass and multiply by 100. This gives the percent composition:

$$\frac{\text{total mass of the element}}{\text{molar mass}} \times 100 = \text{percent of the element}$$

ENHANCED EXAMPLE 7.13

WileyPLUS

Calculate the percent composition of sodium chloride (NaCl).

SOLUTION

Read • **Known:** NaCl

Plan • Use the Problem-Solving Strategy for Calculating Percent Composition from Formula.

Calculate • 1. We need to use the atomic masses from the inside front cover of the text to find the molar mass of NaCl.

$$\overset{\text{Na}}{22.99\text{ g}} + \overset{\text{Cl}}{35.45\text{ g}} = 58.44\text{ g} = \text{molar mass NaCl}$$

2. Calculate the percent composition for each element.

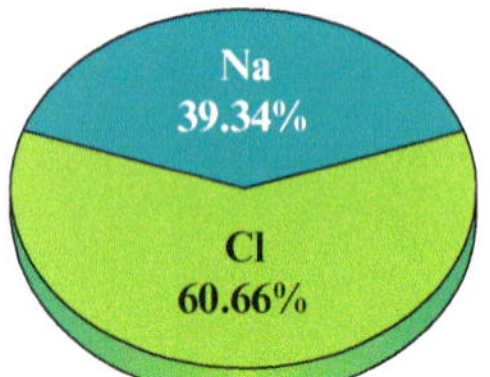

$$\text{Na:}\quad \left(\frac{22.99\text{ g Na}}{58.44\text{ g}}\right)(100) = 39.34\%\text{ Na}$$

$$\text{Cl:}\quad \left(\frac{35.45\text{ g Cl}}{58.44\text{ g}}\right)(100) = \frac{60.66\%\text{ Cl}}{100.00\%\text{ total}}$$

NOTE In any two-component system, if the percent of one component is known, the other is automatically defined by difference; that is, if Na is 39.43%, then Cl is 100% − 39.34% = 60.66%. However, the calculation of the percent should be carried out for each component, since this provides a check against possible error. The percent composition should add to 100% ± 0.2%.

EXAMPLE 7.14

Calculate the percent composition of potassium chloride (KCl).

SOLUTION

Read • **Known:** KCl

Plan • Use the Problem-Solving Strategy for Calculating Percent Composition from Formula.

Calculate • 1. We need to use the atomic masses from the inside front cover of the text to find the molar mass of KCl.

$$\overset{\text{K}}{39.10\text{ g}} + \overset{\text{Cl}}{35.45\text{ g}} = 74.55\text{ g} = \text{molar mass KCl}$$

2. Calculate the percent composition for each element.

$$\text{K:}\quad \left(\frac{39.10\text{ g K}}{74.55\text{ g}}\right)(100) = 52.45\%\text{ K}$$

$$\text{Cl:}\quad \left(\frac{35.45\text{ g Cl}}{74.55\text{ g}}\right)(100) = \underline{47.55\%\text{ Cl}}$$

$$100.00\%\text{ total}$$

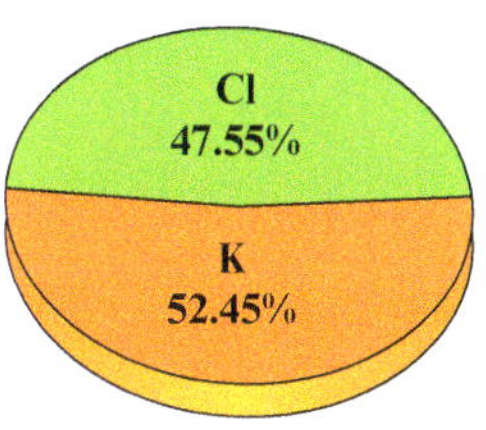

Comparing the results calculated for NaCl and for KCl, we see that NaCl contains a higher percentage of Cl by mass, although each compound has a one-to-one atom ratio of Cl to Na and Cl to K. The reason for this mass percent difference is that Na and K do not have the same atomic masses.

It is important to realize that when we compare *1 mol of NaCl with 1 mol of KCl*, each quantity contains the same number of Cl atoms—namely, 1 mol of Cl atoms. However, if we compare *equal masses* of NaCl and KCl, there will be more Cl atoms in the mass of NaCl, since NaCl has a higher mass percent of Cl.

1 mol NaCl contains	**100.0 g NaCl contains**	**1 mol KCl contains**	**100.0 g KCl contains**
1 mol Na	39.34 g Na	1 mol K	52.45 g K
1 mol Cl	60.66 g Cl	1 mol Cl	47.55 g Cl
	60.66% Cl		47.55% Cl

ENHANCED EXAMPLE 7.15 **WileyPLUS**

ONLINE LEARNING MODULE

Percent Composition

➡ WileyPLUS

Calculate the percent composition of potassium sulfate (K_2SO_4).

SOLUTION

Read • **Known:** K_2SO_4

Plan • Use the Problem-Solving Strategy for Calculating Percent Composition from Formula.

Calculate • 1. We need to use the atomic masses from the inside front cover of the text to find the molar mass of K_2SO_4.

$$\overset{\text{K}}{2(39.10)\text{ g}} + \overset{\text{S}}{32.07\text{ g}} + \overset{\text{O}}{4(16.00)} = 174.3\text{ g} = \text{molar mass } K_2SO_4$$

2. Calculate the percent composition for each element.

$$\text{K:}\quad \left(\frac{78.20\text{ g K}}{174.3\text{ g}}\right)(100) = 44.87\%\text{ K}$$

$$\text{S:}\quad \left(\frac{32.07\text{ g S}}{174.3\text{ g}}\right)(100) = 18.40\%\text{ S}$$

$$\text{O:}\quad \left(\frac{64.00\text{ g O}}{174.3\text{ g}}\right)(100) = \underline{36.72\%\text{ O}}$$

$$99.99\%\text{ total}$$

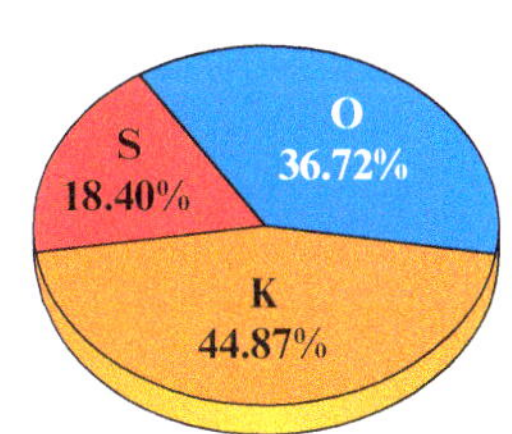

PRACTICE 7.7

Calculate the percent composition of $Ca(NO_3)_2$.

PRACTICE 7.8

Calculate the percent composition of K_2CrO_4.

CHEMISTRY IN ACTION Feeling the Molecular Heat?

Habanero chilis are very hot because they contain a high concentration of capsaicin.

Humans are able to detect the presence of some molecules at very low concentrations. You certainly know this if you have ever eaten chili peppers. When cooking with peppers, you may feel a burning sensation on your skin or in your eyes. The molecule responsible for this burning sensation is capsaicin, which we are able to detect in even very small amounts. How is it that we are able to detect such small amounts of capsaicin? The answer lies in the fact that even a very small mass of this substance still contains many molecules. In a 1-picogram sample of capsaicin there are over a trillion molecules!

How exactly can we detect these molecules at such low concentrations? David Julius, a biologist at UCSF, has been studying this question for over 20 years and has discovered protein receptors that are able to sense the presence of capsaicin, which we perceive as heat. These proteins are found in nerve fibers throughout the skin and tongue. These proteins seem to function by opening a channel or gate when they come in contact with the capsaicin from the chili pepper. Charged particles or ions flowing through the channel then trigger an electrical signal that sends a message to the brain. Interestingly, this same protein senses both high temperatures and hot peppers. This may be the reason that the physiological response to a hot room and eating chili peppers is often the same. Furthermore, once your skin or tongue has been exposed to capsaicin, it will become even more sensitive to other heat stimuli because this protein tends to amplify the signal it is sending to the brain as it detects additional heat or peppers. For this reason, consuming a hot beverage after eating chili peppers will seem to burn your tongue more than usual.

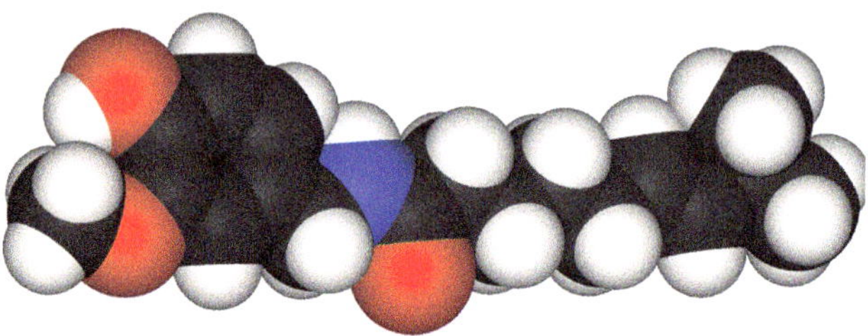
Capsaicin, the molecule responsible for the heat in chili peppers.

Percent Composition from Experimental Data

If experimental data is given, a two-step process is needed to determine the percent composition:

PROBLEM-SOLVING STRATEGY

For Percent Composition from Experimental Data

1. Calculate the mass of the compound formed.
2. Divide the mass of each element by the total mass of the compound and multiply by 100.

ENHANCED EXAMPLE 7.16 WileyPLUS

Zinc oxide is a compound with many uses from preventing sunburn to acting as a pigment in white paint. When heated in air, 1.63 g of zinc (Zn) combines with 0.40 g of oxygen (O_2) to form zinc oxide. Calculate the percent composition of the compound formed.

SOLUTION

Read **Knowns:** 1.63 g Zn

0.40 g O_2

Plan Use the Problem-Solving Strategy for Calculating Percent Composition from Experimental Data.

Calculate 1. We need to calculate the mass of the product formed.

$$1.63 \text{ g Zn} + 0.40 \text{ g } O_2 = 2.03 \text{ g product}$$

2. Calculate the percent for each element.

$$\text{Zn: } \left(\frac{1.63 \text{ g Zn}}{2.03 \text{ g}}\right)(100) = 80.3\% \text{ Zn}$$

$$\text{O: } \left(\frac{0.40 \text{ g O}}{2.03 \text{ g}}\right)(100) = \underline{20.0\% \text{ O}}$$

$$100.3\% \text{ total}$$

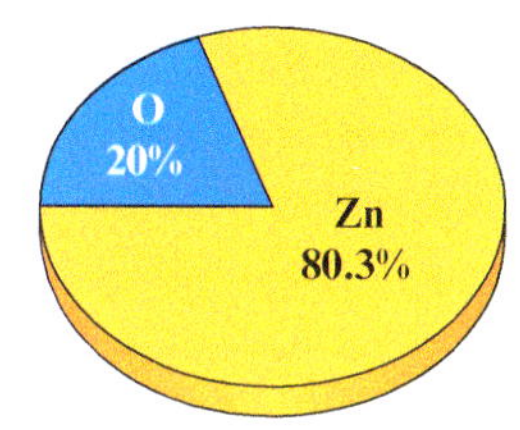

PRACTICE 7.9

Aluminum chloride is formed by reacting 13.43 g aluminum with 53.18 g chlorine. What is the percent composition of the compound?

7.4 Calculating Empirical Formulas

Determine the empirical formula for a compound from its percent composition.

LEARNING OBJECTIVE

KEY TERM
empirical formula

The **empirical formula**, or *simplest formula*, gives the smallest whole-number ratio of atoms present in a compound. This formula gives the relative number of atoms of each element in the compound.

We can establish empirical formulas because (1) individual atoms in a compound are combined in whole-number ratios and (2) each element has a specific atomic mass.

To calculate an empirical formula, we need to know (1) the elements that are combined, (2) their atomic masses, and (3) the ratio by mass or percentage in which they are combined. If elements A and B form a compound, we may represent the empirical formula as A_xB_y, where x and y are small whole numbers that represent the atoms of A and B. To write the empirical formula, we must determine x and y:

CHECK YOUR UNDERSTANDING
7.8 Empirical Formula
WileyPLUS

PROBLEM-SOLVING STRATEGY

For Calculating an Empirical Formula

1. Assume a definite starting quantity (usually 100.0 g) of the compound, if not given, and express the mass of each element in grams.
2. Convert the grams of each element into moles using each element's molar mass. This conversion gives the number of moles of atoms of each element in the quantity assumed in Step 1. At this point, these numbers will usually not be whole numbers.
3. Divide each value obtained in Step 2 by the smallest of these values. If the numbers obtained are whole numbers, use them as subscripts and write the empirical formula. If the numbers obtained are not whole numbers, go on to Step 4.
4. Multiply the values obtained in Step 3 by the smallest number that will convert them to whole numbers. Use these whole numbers as the subscripts in the empirical formula. For example, if the ratio of A to B is 1.0 : 1.5, multiply both numbers by 2 to obtain a ratio of 2:3. The empirical formula then is A_2B_3.

Some common fractions and their decimal equivalents are

$\frac{1}{4} = 0.25$
$\frac{1}{3} = 0.333 \ldots$
$\frac{2}{3} = 0.666 \ldots$
$\frac{1}{2} = 0.5$
$\frac{3}{4} = 0.75$

Multiply the decimal equivalent by the number in the denominator of the fraction to get a whole number: 4(0.75) = 3.

ONLINE LEARNING MODULE
Empirical Formula from Percent Composition
WileyPLUS

In many of these calculations, results will vary somewhat from an exact whole number; this can be due to experimental errors in obtaining the data or from rounding off numbers. Calculations that vary by no more than ±0.1 from a whole number usually are rounded off to the nearest whole number. Deviations greater than about 0.1 unit usually mean that the calculated ratios need to be multiplied by a factor to make them all whole numbers. For example, an atom ratio of 1 : 1.33 should be multiplied by 3 to make the ratio 3 : 4. Let's do a few examples to see how it works.

ENHANCED EXAMPLE 7.17

WileyPLUS

Calculate the empirical formula of a compound containing 11.19% hydrogen (H) and 88.79% oxygen (O).

SOLUTION

Read • **Knowns:** 11.19% H
88.79% O

Plan • Use the Problem-Solving Strategy for Calculating an Empirical Formula.

Calculate • 1. Assume 100.0 g of material. We know that the percent of each element equals the grams of each element:

11.19 g H

88.79 g O

2. Convert grams of each element to moles:

$$\text{H:}\ (11.19\ \text{g})\left(\frac{1\ \text{mol H atoms}}{1.008\ \text{g H}}\right) = 11.10\ \text{mol H atoms}$$

$$\text{O:}\ (88.79\ \text{g})\left(\frac{1\ \text{mol O atoms}}{16.00\ \text{g O}}\right) = 5.549\ \text{mol O atoms}$$

The formula could be expressed as $H_{11.10}O_{5.549}$. However, it's customary to use the smallest whole-number ratio of atoms.

3. Change the number of moles to whole numbers by dividing by the smallest number.

$$\text{H} = \frac{11.10\ \text{mol}}{5.549\ \text{mol}} = 2.000$$

$$\text{O} = \frac{5.549\ \text{mol}}{5.549\ \text{mol}} = 1.000$$

The simplest ratio of H to O is 2:1.

Empirical formula = H_2O

ENHANCED EXAMPLE 7.18

WileyPLUS

The analysis of a salt shows that it contains 56.58% potassium (K), 8.68% carbon (C), and 34.73% oxygen (O). Calculate the empirical formula for this substance.

SOLUTION

Read • **Knowns:** 56.58% K
8.68% C
34.73% O

Plan • Use the Problem-Solving Strategy for Calculating an Empirical Formula.

Calculate 1. Assume 100.0 g of material. We know that the percent of each element equals the grams of each element:

56.58 g K
8.68 g C
34.73 g O

2. Convert the grams of each element to moles:

$$\text{K:} \quad (56.58\ \cancel{\text{g K}})\left(\frac{1\ \text{mol K atoms}}{39.10\ \cancel{\text{g K}}}\right) = 1.447\ \text{mol K atoms}$$

$$\text{C:} \quad (8.68\ \cancel{\text{g C}})\left(\frac{1\ \text{mol C atoms}}{12.01\ \cancel{\text{g C}}}\right) = 0.723\ \text{mol C atoms}$$

$$\text{O:} \quad (34.73\ \cancel{\text{g O}})\left(\frac{1\ \text{mol O atoms}}{16.00\ \cancel{\text{g O}}}\right) = 2.171\ \text{mol O atoms}$$

3. Change the number of moles to whole numbers by dividing by the smallest number.

$$\text{K} = \frac{1.447\ \text{mol}}{0.723\ \text{mol}} = 2.00$$

$$\text{C} = \frac{0.723\ \text{mol}}{0.723\ \text{mol}} = 1.00$$

$$\text{O} = \frac{2.171\ \text{mol}}{0.723\ \text{mol}} = 3.00$$

The simplest ratio of K:C:O is 2:1:3.

Empirical formula = K_2CO_3

ENHANCED EXAMPLE 7.19

WileyPLUS

ONLINE LEARNING MODULE

Empirical Formula from Experimental Data

➥ **WileyPLUS**

A sulfide of iron was formed by combining 2.233 g of iron (Fe) with 1.926 g of sulfur (S). What is the empirical formula for the compound?

SOLUTION

Read **Knowns:** 2.233 g Fe
1.926 g S

Plan Use the Problem-Solving Strategy for Calculating an Empirical Formula.

Calculate 1. The mass of each element is known, so we use it directly.

2. Convert grams of each element to moles.

$$\text{Fe:} \quad (2.233\ \cancel{\text{g Fe}})\left(\frac{1\ \text{mol Fe atoms}}{55.85\ \cancel{\text{g Fe}}}\right) = 0.03998\ \text{mol Fe atoms}$$

$$\text{S:} \quad (1.926\ \cancel{\text{g S}})\left(\frac{1\ \text{mol S atoms}}{32.07\ \cancel{\text{g S}}}\right) = 0.06006\ \text{mol S atoms}$$

3. Change the number of moles to whole numbers by dividing by the smallest number.

$$\text{Fe} = \frac{0.03998\ \text{mol}}{0.03998\ \text{mol}} = 1.000$$

$$\text{S} = \frac{0.06006\ \text{mol}}{0.03998\ \text{mol}} = 1.502$$

4. We still have not reached a ratio that gives whole numbers in the formula, so we multiply by a number that will give us whole numbers.

$$\text{Fe:} \quad (1.000)2 = 2.000$$
$$\text{S:} \quad (1.502)2 = 3.004$$

$$\text{Empirical formula} = Fe_2S_3$$

PRACTICE 7.10

Calculate the empirical formula of a compound containing 52.14% C, 13.12% H, and 34.73% O.

PRACTICE 7.11

Calculate the empirical formula of a compound that contains 43.7% phosphorus and 56.3% O by mass.

7.5 Calculating the Molecular Formula from the Empirical Formula

LEARNING OBJECTIVE Compare an empirical formula to a molecular formula and calculate a molecular formula from the empirical formula of the compound and its molar mass.

KEY TERM
molecular formula

The **molecular formula** is the true formula, representing the total number of atoms of each element present in one molecule of a compound. It is entirely possible that two or more substances will have the same percent composition, yet be distinctly different compounds. For example, acetylene (C_2H_2) is a common gas used in welding; benzene (C_6H_6) is an important solvent obtained from coal tar and is used in the synthesis of styrene and nylon. Both acetylene and benzene contain 92.3% C and 7.7% H. The smallest ratio of C and H corresponding to these percentages is CH (1:1). Therefore, the *empirical* formula for both acetylene and benzene is CH, even though the *molecular* formulas are C_2H_2 and C_6H_6, respectively. Often the molecular formula is the same as the empirical formula. If the molecular formula is not the same, it will be an integral (whole-number) multiple of the empirical formula. For example,

$$CH = \text{empirical formula}$$
$$(CH)_2 = C_2H_2 = \text{acetylene} \quad \text{(molecular formula)}$$
$$(CH)_6 = C_6H_6 = \text{benzene} \quad \text{(molecular formula)}$$

CHECK YOUR UNDERSTANDING
7.9 Molecular Formula
WileyPLUS

TABLE 7.1 compares the formulas of these substances. TABLE 7.2 shows empirical and molecular formula relationships of other compounds.

TABLE 7.1 Molecular Formulas of Two Compounds Having an Empirical Formula with a 1:1 Ratio of Carbon and Hydrogen Atoms

	Composition		
Formula	**% C**	**% H**	**Molar mass**
CH (empirical)	92.3	7.7	13.02 (empirical)
C_2H_2 (acetylene)	92.3	7.7	26.04 (2 × 13.02)
C_6H_6 (benzene)	92.3	7.7	78.12 (6 × 13.02)

TABLE 7.2 Some Empirical and Molecular Formulas

Substance	Empirical formula	Molecular formula	Substance	Empirical formula	Molecular formula
Acetylene	CH	C_2H_2	Diborane	BH_3	B_2H_6
Benzene	CH	C_6H_6	Hydrazine	NH_2	N_2H_4
Ethylene	CH_2	C_2H_4	Hydrogen	H	H_2
Formaldehyde	CH_2O	CH_2O	Chlorine	Cl	Cl_2
Acetic acid	CH_2O	$C_2H_4O_2$	Bromine	Br	Br_2
Glucose	CH_2O	$C_6H_{12}O_6$	Oxygen	O	O_2
Hydrogen chloride	HCl	HCl	Nitrogen	N	N_2
Carbon dioxide	CO_2	CO_2	Iodine	I	I_2

The molecular formula can be calculated from the empirical formula if the molar mass is known. The molecular formula will be equal either to the empirical formula or some multiple of it. For example, if the empirical formula of a compound of hydrogen and fluorine is HF, the molecular formula can be expressed as $(HF)_n$, where $n = 1, 2, 3, 4, \ldots$. This n means that the molecular formula could be HF, H_2F_2, H_3F_3, H_4F_4, and so on. To determine the molecular formula, we must evaluate n:

$$n = \frac{\text{molar mass}}{\text{mass of empirical formula}} = \text{number of empirical formula units}$$

What we actually calculate is the number of units of the empirical formula contained in the molecular formula.

ENHANCED EXAMPLE 7.20 **WileyPLUS**

A compound of nitrogen and oxygen with a molar mass of 92.00 g was found to have an empirical formula of NO_2. What is its molecular formula?

SOLUTION

Read **Knowns:** NO_2 = empirical formula
Molar mass = 92.00 g

Plan Let n be the number of NO_2 units in a molecule; then the molecular formula is $(NO_2)_n$. Find the molar mass of an NO_2 unit and determine the numbers of units needed for the molar mass of the compound.

Calculate The molar mass of NO_2 is 14.01 g + 2(16.00 g) = 46.01 g.

$$n = \frac{92.00\text{ g}}{46.01\text{ g}} = 2 \text{ (empirical formula units)}$$

The molecular formula is $(NO_2)_2$ or N_2O_4.

N_2O_4

ONLINE LEARNING MODULE
Molecular Formula
WileyPLUS

ENHANCED EXAMPLE 7.21

WileyPLUS

Propylene is a compound frequently polymerized to make polypropylene, which is used for rope, laundry bags, blankets, carpets, and textile fibers. Propylene has a molar mass of 42.08 g and contains 14.3% H and 85.7% C. What is its molecular formula?

SOLUTION

Read • **Knowns:** Molar mass = 42.08 g
14.3% H
85.7% C

Plan • First find the empirical formula using the Problem-Solving Strategy for Calculating an Empirical Formula. Then use n to determine the molecular formula from the empirical formula.

Calculate • 1. Assume 100.0 g of material. We know that the percent of each element equals the grams of each element:

14.3 g H

85.7 g C

2. Convert the grams of each element to moles:

$$\text{H:}\quad (14.3\ \cancel{\text{g H}})\left(\frac{1\ \text{mol H atoms}}{1.008\ \cancel{\text{g H}}}\right) = 14.2\ \text{mol H atoms}$$

$$\text{C:}\quad (85.7\ \cancel{\text{g C}})\left(\frac{1\ \text{mol C atoms}}{12.01\ \cancel{\text{g C}}}\right) = 7.14\ \text{mol C atoms}$$

3. Change the number of moles to whole numbers by dividing by the smallest number.

$$\text{H} = \frac{14.2\ \text{mol}}{7.14\ \text{mol}} = 1.99$$

$$\text{C} = \frac{7.14\ \text{mol}}{7.14\ \text{mol}} = 1.00$$

Empirical formula = CH_2

We determine the molecular formula from the empirical formula and the molar mass:

Molecular formula = $(CH_2)_n$

CH_2 molar mass = 12.01 g + 2(1.008 g) = 14.03 g

$$n = \frac{42.08\ \text{g}}{14.03\ \text{g}} = 3\ \text{(empirical formula units)}$$

The molecular formula is $(CH_2)_3 = C_3H_6$.

When propylene is polymerized into polypropylene, it can be used for rope and other recreational products.

C_3H_6

PRACTICE 7.12

Calculate the empirical and molecular formulas of a compound that contains 80.0% C, 20.0% H, and has a molar mass of 30.00 g.

CHAPTER 7 REVIEW

KEY TERMS
Avogadro's number
mole
molar mass

7.1 The Mole

- We count atoms by weighing them since they are so tiny.
- 1 mole = 6.022×10^{23} items.
- Avogadro's number is 6.022×10^{23}.

7.2 Molar Mass of Compounds

- One mole of a compound contains Avogadro's number of formula units of that compound.
- The mass (grams) of one mole of a compound is the molar mass.
- Molar mass is determined by adding the molar masses of all the atoms in a formula.
- Molar masses are given to four significant figures in this text.

KEY TERM
percent composition of a compound

7.3 Percent Composition of Compounds

- To determine the percent composition from a formula:
  - Calculate the molar mass.
  - For each element in the formula

$$\frac{\text{total mass of the element}}{\text{molar mass of the compound}} \times 100 = \text{percent of the element}$$

- To determine percent composition from experimental data:
  - Calculate the mass of the compound formed.
  - For each element in the formula

$$\frac{\text{mass of the element}}{\text{mass of the compound formed}} \times 100 = \text{percent of the element}$$

KEY TERM
empirical formula

7.4 Calculating Empirical Formulas

- The empirical formula is the simplest formula giving the smallest whole-number ratio of atoms present in a compound.
- To determine the empirical formula for a compound you need to know:
  - The elements that are combined
  - Their atomic masses
  - The ratio of masses or percentage in which they are combined
- Empirical formulas are represented in the form A_xB_y. To determine the empirical formula of a compound:
  - Assume a starting quantity (100.0 g is a good choice).
  - Convert mass (g) to moles for each element.
  - Divide each element's moles by the smallest number of moles.
  - If the ratios are whole numbers, use them as subscripts and write the empirical formula.
  - If the ratios are not whole numbers, multiply them all by the smallest number, which will convert them to whole numbers.
  - Use the whole numbers to write the empirical formula.

KEY TERM
molecular formula

7.5 Calculating the Molecular Formula from the Empirical Formula

- The molecular formula is the true formula representing the total number of atoms of each element present in one molecule of the compound.
- Two or more substances may have the same empirical formulas but different molecular formulas.
- The molecular formula is calculated from the empirical formula when the molar mass is known:

$$n = \frac{\text{molar mass}}{\text{mass of empirical formula}} = \text{number of empirical formula units}$$

  - The molecular formula is $(A_xB_y)_n$.

REVIEW QUESTIONS

1. What is a mole?
2. Which would have a higher mass: a mole of K atoms or a mole of Au atoms?
3. Which would contain more atoms: a mole of K atoms or a mole of Au atoms?
4. Which would contain more electrons: a mole of K atoms or a mole of Au atoms?
5. What is the numerical value of Avogadro's number?
6. What is the relationship between Avogadro's number and the mole?
7. Which has the greater number of oxygen atoms, 1 mole of oxygen gas (O_2) or 1 mole of ozone gas (O_3)?
8. What is molar mass?
9. If the atomic mass scale had been defined differently, with an atom of $^{12}_{6}C$ defined as a mass of 50 u, would this have any effect on the value of Avogadro's number? Explain.
10. Complete these statements, supplying the proper quantity.
    (a) A mole of O atoms contains _____ atoms.
    (b) A mole of O_2 molecules contains _____ molecules.
    (c) A mole of O_2 molecules contains _____ atoms.
    (d) A mole of O atoms has a mass of _____ grams.
    (e) A mole of O_2 molecules has a mass of _____ grams.
11. How many molecules are present in 1 molar mass of sulfuric acid (H_2SO_4)? How many atoms are present?
12. What are the two types of information that can be used to calculate the percent composition of a compound?
13. A compound is composed of only nitrogen and oxygen. If the compound contains 43.8% nitrogen, what is the percent oxygen in the compound?
14. Write the steps used to calculate percent composition of an element in a compound.
15. In calculating the empirical formula of a compound from its percent composition, why do we choose to start with 100.0 g of the compound?
16. Which of the following compounds have the same empirical formula? What is the empirical formula? CH_4, C_2H_4, C_4H_6, C_4H_{10}, C_8H_{12}, C_8H_{18}
17. What is the difference between an empirical formula and a molecular formula?
18. What data beyond the percent composition are generally used to determine the molecular formula of a compound?
19. What information does the ratio $\frac{\text{molar mass}}{\text{mass of empirical formula}}$ provide?

PAIRED EXERCISES

Most of the exercises in this chapter are available for assignment via the online homework management program, WileyPLUS (www.wileyplus.com). All exercises with blue numbers have answers in Appendix 5.

1. Determine the molar masses of these compounds:
   (a) KBr
   (b) Na_2SO_4
   (c) $Pb(NO_3)_2$
   (d) C_2H_5OH
   (e) $HC_2H_3O_2$
   (f) Fe_3O_4
   (g) $C_{12}H_{22}O_{11}$
   (h) $Al_2(SO_4)_3$
   (i) $(NH_4)_2HPO_4$

2. Determine the molar masses of these compounds:
   (a) NaOH
   (b) Ag_2CO_3
   (c) Cr_2O_3
   (d) $(NH_4)_2CO_3$
   (e) $Mg(HCO_3)_2$
   (f) C_6H_5COOH
   (g) $C_6H_{12}O_6$
   (h) $K_4Fe(CN)_6$
   (i) $BaCl_2 \cdot 2H_2O$

3. How many moles of atoms are contained in the following?
   (a) 22.5 g Zn
   (b) 0.688 g Mg
   (c) 4.5×10^{22} atoms Cu
   (d) 382 g Co
   (e) 0.055 g Sn
   (f) 8.5×10^{24} molecules N_2

4. How many moles of atoms are contained in the following?
   (a) 25.0 g NaOH
   (b) 44.0 g Br_2
   (c) 0.684 g $MgCl_2$
   (d) 14.8 g CH_3OH
   (e) 2.88 g Na_2SO_4
   (f) 4.20 lb ZnI_2

5. Calculate the number of grams in each of the following:
   (a) 0.550 mol Au
   (b) 15.8 mol H_2O
   (c) 12.5 mol Cl_2
   (d) 3.15 mol NH_4NO_3

6. Calculate the number of grams in each of the following:
   (a) 4.25×10^{-4} mol H_2SO_4
   (b) 4.5×10^{22} molecules CCl_4
   (c) 0.00255 mol Ti
   (d) 1.5×10^{16} atoms S

7. How many molecules are contained in each of the following?
   (a) 2.5 mol S_8
   (b) 7.35 mol NH_3
   (c) 17.5 g C_2H_5OH
   (d) 225 g Cl_2

8. How many molecules are contained in each of the following?
   (a) 9.6 mol C_2H_4
   (b) 2.76 mol N_2O
   (c) 23.2 g CH_3OH
   (d) 32.7 g CCl_4

9. The Japanese symbol for an atom has been written on a piece of copper metal using 99 iron atoms. Calculate the mass of the iron used to create this symbol.

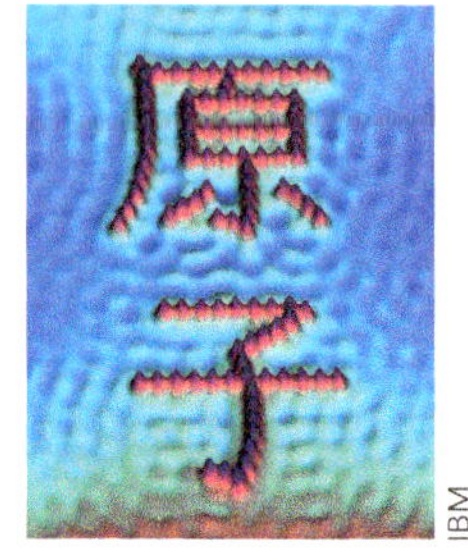

IBM

10. The starship *Enterprise* has been drawn on a piece of copper metal using 69 carbon monoxide molecules. Calculate the mass of the carbon monoxide used to create this image.

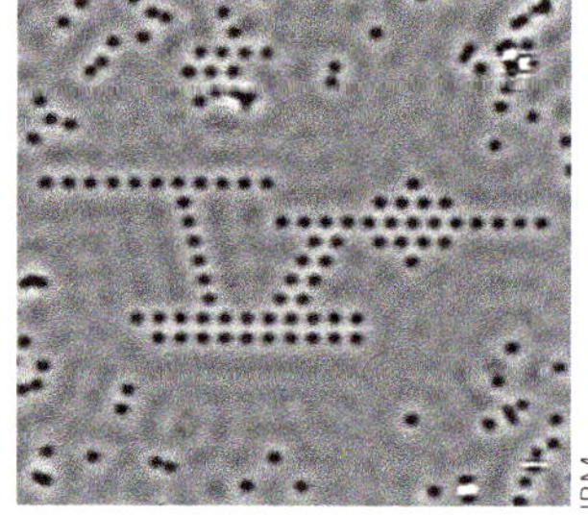
IBM

11. How many atoms are contained in each of the following?
    (a) 25 molecules P_2O_5
    (b) 3.62 mol O_2
    (c) 12.2 mol CS_2
    (d) 1.25 g Na
    (e) 2.7 g CO_2
    (f) 0.25 g CH_4

12. How many atoms are contained in each of the following?
    (a) 2 molecules CH_3COOH
    (b) 0.75 mol C_2H_6
    (c) 25 mol H_2O
    (d) 92.5 g Au
    (e) 75 g PCl_3
    (f) 15 g $C_6H_{12}O_6$

13. Many companies are working to develop cups, straws, and even nail polish that will be able to detect a variety of drugs that may impair a person's ability to make good decisions. One of these drugs is rohypnol, also known as the date rape drug. This molecule is similar to the one found in Valium, a drug prescribed to treat anxiety disorders, seizures, and other medical conditions. If a nail polish will change color when it comes in contact with 50.0 mg of rohypnol ($C_{16}H_{12}FN_3O_3$), how many molecules of rohypnol are required to change the color of the nail polish?

14. Receptor proteins that allow us to sense heat have been isolated as well as molecules that cause us to perceive heat such as hot pepper and wasabi. If such proteins are able to detect as little as 20.0 μg of capsaicin, the chemical responsible for the burn in peppers, how many molecules of capsaicin ($C_{18}H_{27}NO_3$) are required before the tongue can detect it?

15. Calculate the mass in grams of each of the following:
(a) 1 atom He
(b) 15 atoms C
(c) 4 molecules N_2O_5
(d) 11 molecules $C_6H_5NH_2$

16. Calculate the mass in grams of each of the following:
(a) 1 atom Xe
(b) 22 atoms Cl
(c) 9 molecules CH_3COOH
(d) 15 molecules $C_4H_4O_2(NH_2)_2$

17. Make the following conversions:
(a) 25 kg CO_2 to mol CO_2
(b) 5 atoms Pb to mol Pb
(c) 6 mol O_2 to atoms oxygen
(d) 25 molecules P_4 to g P_4

18. Make the following conversions:
(a) 275 atoms W to mol W
(b) 95 mol H_2O to kg H_2O
(c) 12 molecules SO_2 to g SO_2
(d) 25 mol Cl_2 to atoms chlorine

19. 1 molecule of tetraphosphorus decoxide contains
(a) how many moles?
(b) how many grams?
(c) how many phosphorus atoms?
(d) how many oxygen atoms?
(e) how many total atoms?

20. 125 grams of disulfur decafluoride contain
(a) how many moles?
(b) how many molecules?
(c) how many total atoms?
(d) how many atoms of sulfur?
(e) how many atoms of fluorine?

21. How many atoms of hydrogen are contained in each of the following?
(a) 25 molecules $C_6H_5CH_3$
(b) 3.5 mol H_2CO_3
(c) 36 g CH_3CH_2OH

22. How many atoms of hydrogen are contained in each of the following?
(a) 23 molecules CH_3CH_2COOH
(b) 7.4 mol H_3PO_4
(c) 57 g $C_6H_5ONH_2$

23. Cosmone is a molecule used by fragrance manufacturers to provide a rich and elegant musky essence to many perfumes. Cosmone has the molecular formula $C_{15}H_{26}O$.
(a) Calculate the molar mass of cosmone.
(b) Calculate the mass of 3.82 moles of cosmone.
(c) Calculate the number of molecules of cosmone in a sample containing 8.36×10^{-4} mol cosmone.
(d) Calculate the number of moles of carbon in a 8.35-mol sample of cosmone.
(e) Calculate the mass of oxygen in a 4.29-g sample of cosmone.
(f) Calculate the mass of a sample of cosmone that contains 6.58×10^{19} atoms of hydrogen.
(g) Calculate the mass of one molecule of cosmone.
(h) Calculate the number of atoms of carbon in 8.00-g of cosmone.

Mary Evans/Library of Congress/Age

24. 4-Unadecanolide is a molecule used by fragrance manufacturers to provide a strong peach essence to many personal care products. It has the molecular formula $C_{11}H_{20}O_2$.
(a) Calculate the molar mass of 4-unadecanolide.
(b) Calculate the mass of 8.13 moles of 4-unadecanolide.
(c) Calculate the number of molecules of 4-unadecanolide in a sample containing 5.77×10^{-12} mol of 4-unadecanolide.
(d) Calculate the number of moles of hydrogen in a 35.7-mol sample of 4-unadecanolide.
(e) Calculate the mass of carbon in a 85.1-g sample of 4-unadecanolide.
(f) Calculate the mass of a sample of 4-unadecanolide that contains 9.47×10^{19} atoms of oxygen.
(g) Calculate the mass of one molecule of 4-unadecanolide.
(h) Calculate the number of atoms of carbon in 4.50 g of 4-unadecanolide.

Mediablitzimages / Alamy

25. Calculate the number of
(a) grams of silver in 25.0 g AgBr
(b) grams of nitrogen in 6.34 mol $(NH_4)_3PO_4$
(c) grams of oxygen in 8.45×10^{22} molecules SO_3

26. Calculate the number of
(a) grams of chlorine in 5.0 g $PbCl_2$
(b) grams of hydrogen in 4.50 mol H_2SO_4
(c) grams of hydrogen in 5.45×10^{22} molecules NH_3

27. Calculate the percent composition by mass of these compounds:
(a) NaBr
(b) $KHCO_3$
(c) $FeCl_3$
(d) $SiCl_4$
(e) $Al_2(SO_4)_3$
(f) $AgNO_3$

28. Calculate the percent composition by mass of these compounds:
(a) $ZnCl_2$
(b) $NH_4C_2H_3O_2$
(c) MgP_2O_7
(d) $(NH_4)_2SO_4$
(e) $Fe(NO_3)_3$
(f) ICl_3

29. Calculate the percent of iron in the following compounds:
(a) FeO
(b) Fe_2O_3
(c) Fe_3O_4
(d) $K_4Fe(CN)_6$

30. Which of the following chlorides has the highest and which has the lowest percentage of chlorine, by mass, in its formula?
(a) KCl
(b) $BaCl_2$
(c) $SiCl_4$
(d) LiCl

31. A 73.16-g sample of an interesting barium silicide reported to have superconducting properties was found to contain 33.62 g barium and the remainder silicon. Calculate the percent composition of the compound.

32. One of the compounds that give garlic its characteristic odor is ajoene. A 7.52-g sample of this compound contains 3.09 g sulfur, 0.453 g hydrogen, 0.513 g oxygen, and the remainder is carbon. Calculate the percent composition of ajoene.

33. Examine the following formulas. Which compound has the
(a) higher percent by mass of hydrogen: H_2O or H_2O_2?
(b) lower percent by mass of nitrogen: NO or N_2O_3?
(c) higher percent by mass of oxygen: NO_2 or N_2O_4?

Check your answers by calculation if you wish.

34. Examine the following formulas. Which compound has the
(a) lower percent by mass of chlorine: $NaClO_3$ or $KClO_3$?
(b) higher percent by mass of sulfur: $KHSO_4$ or K_2SO_4?
(c) lower percent by mass of chromium: Na_2CrO_4 or $Na_2Cr_2O_7$?

Check your answers by calculation if you wish.

35. Calculate the empirical formula of each compound from the percent compositions given:
(a) 63.6% N, 36.4% O
(b) 46.7% N, 53.3% O
(c) 25.9% N, 74.1% O
(d) 43.4% Na, 11.3% C, 45.3% O
(e) 18.8% Na, 29.0% Cl, 52.3% O
(f) 72.02% Mn, 27.98% O

36. Calculate the empirical formula of each compound from the percent compositions given:
(a) 64.1% Cu, 35.9% Cl
(b) 47.2% Cu, 52.8% Cl
(c) 51.9% Cr, 48.1% S
(d) 55.3% K, 14.6% P, 30.1% O
(e) 38.9% Ba, 29.4% Cr, 31.7% O
(f) 3.99% P, 82.3% Br, 13.7% Cl

37. Determine the empirical formula for each of the following, using the given masses:
(a) a compound containing 26.08 g zinc, 4.79 g carbon, and 19.14 g oxygen
(b) a 150.0-g sample of a compound containing 57.66 g carbon, 7.26 g hydrogen, and the remainder, chlorine
(c) a 75.0-g sample of an oxide of vanadium, containing 42.0 g V and the remainder, oxygen
(d) a compound containing 67.35 g nickel, 43.46 g oxygen, and 23.69 g phosphorus

38. Determine the empirical formula for each of the following, using the given masses:
(a) a compound containing 55.08 g carbon, 3.85 g hydrogen, and 61.07 g bromine
(b) a 65.2-g sample of a compound containing 36.8 g silver, 12.1 g chlorine, and the remainder, oxygen
(c) a 25.25-g sample of a sulfide of vanadium, containing 12.99 g V, and the remainder, sulfur
(d) a compound containing 38.0 g zinc and 12.0 g phosphorus

39. A 15.267-g sample of a sulfide of gold was found to contain 12.272 g gold, and the remainder, sulfur. Calculate the empirical formula of this sulfide of gold.

40. A 10.724-g sample of an oxide of titanium was found to contain 7.143 g titanium, and the remainder, oxygen. Calculate the empirical formula of this oxide of titanium.

41. Lenthionine, a naturally occurring sulfur compound, gives shitake mushrooms their interesting flavor. This compound is composed of carbon, sulfur, and hydrogen. A 5.000-g sample of lenthionine contains 0.6375 g carbon and 0.1070 g hydrogen, and the remainder is sulfur. The molar mass of lenthionine is 188.4 g/mol. Determine the empirical formula for lenthionine.

42. A 5.276-g sample of an unknown compound was found to contain 3.989 g of mercury and the remainder, chlorine. Calculate the empirical formula of this chloride of mercury.

43. Traumatic acid is a plant hormone that induces injured cells to begin to divide and repair the trauma they have experienced. Traumatic acid contains 63.13% carbon, 8.830% hydrogen, and 28.03% oxygen. Its molar mass is 228 g/mol. Determine the empirical and molecular formulas of traumatic acid.

44. Dixanthogen, an herbicide, is composed of 29.73% carbon, 4.16% hydrogen, 13.20% oxygen, and 52.91% sulfur. It has a molar mass of 242.4 g/mol. Determine the empirical and the molecular formulas.

45. Determine the empirical formula of florhydral, a fragrance molecule with a fresh, natural aroma often used in fabric detergents. It is composed of 86.65% C, 4.48% H, and 8.88% O.

46. Determine the empirical formula of jasvanol, a fragrance molecule with a floral aromatic jasmine note with a hint of fruit. It is composed of 81.02% C, 11.79% H, and 7.20% O.

47. Ethanedioic acid, a compound that is present in many vegetables, has a molar mass of 90.04 g/mol and a composition of 26.7% C, 2.2% H, and 71.1% O. What is the molecular formula for this substance?

48. Butyric acid is a compound that is present in butter. It has a molar mass of 88.11 g/mol and is composed of 54.5% C, 9.2% H, and 36.3% O. What is the molecular formula for this substance?

49. Calculate the percent composition, determine the molecular formula and the empirical formula for the nitrogen-oxygen compound that results when 12.04 g of nitrogen are reacted with enough oxygen to produce 39.54 g of product. The molar mass of the product is 92.02 g.

50. Calculate the percent composition, determine the molecular formula and the empirical formula for the carbon-hydrogen-oxygen compound that results when 30.21 g of carbon, 40.24 g of oxygen, and 5.08 g of hydrogen are reacted to produce a product with a molar mass of 180.18 g.

51. The compound XYZ_3 has a molar mass of 100.09 g and a percent composition (by mass) of 40.04% X, 12.00% Y, and 47.96% Z. What is the formula of the compound?

52. The compound $X_2(YZ_3)_3$ has a molar mass of 282.23 g and a percent composition (by mass) of 19.12% X, 29.86% Y, and 51.02% Z. What is the formula of the compound?

ADDITIONAL EXERCISES

53. White phosphorus is one of several forms of phosphorus and exists as a waxy solid consisting of P_4 molecules. How many atoms are present in 0.350 mol of P_4?
54. How many grams of sodium contain the same number of atoms as 10.0 g of potassium?
55. One molecule of an unknown compound is found to have a mass of 3.27×10^{-22} g. What is the molar mass of this compound?
56. How many molecules of sugar are in a 5-lb bag of sugar? The formula for table sugar or sucrose is $C_{12}H_{22}O_{11}$.
57. If a stack of 500 sheets of paper is 4.60 cm high, what will be the height, in meters, of a stack of Avogadro's number of sheets of paper?
58. There are about 7.0 billion (7.0×10^9) people on Earth. If exactly 1 mol of dollars were distributed equally among these people, how many dollars would each person receive?
59. If 20. drops of water equal 1.0 mL (1.0 cm^3),
    (a) how many drops of water are there in a cubic mile of water?
    (b) what would be the volume in cubic miles of a mole of drops of water?
60. Silver has a density of 10.5 g/cm^3. If 1.00 mol of silver were shaped into a cube,
    (a) what would be the volume of the cube?
    (b) what would be the length of one side of the cube?
61. Given 1.00-g samples of each of the compounds CO_2, O_2, H_2O, and CH_3OH,
    (a) which sample will contain the largest number of molecules?
    (b) which sample will contain the largest number of atoms? Show proof for your answers.
62. How many grams of Fe_2S_3 will contain a total number of atoms equal to Avogadro's number?
63. How many grams of calcium must be combined with 1 g phosphorus to form the compound Ca_3P_2?
64. An iron ore contains 5% $FeSO_4$ by mass. How many grams of iron could be obtained from 1.0 ton of this ore?
65. How many grams of lithium will combine with 20.0 g of sulfur to form the compound Li_2S?
66. Calculate the percentage of
    (a) mercury in $HgCO_3$
    (b) oxygen in $Ca(ClO_3)_2$
    (c) nitrogen in $C_{10}H_{14}N_2$ (nicotine)
    (d) Mg in $C_{55}H_{72}MgN_4O_5$ (chlorophyll)
67. Zinc and sulfur react to form zinc sulfide, ZnS. If we mix 19.5 g of zinc and 9.40 g of sulfur, have we added sufficient sulfur to fully react all the zinc? Show evidence for your answer.
68. Vomitoxin is produced by some fungi that grow on wheat and barley. It derives its name from the fact that it causes pigs that eat contaminated wheat to vomit. The chemical formula for vomitoxin is $C_{15}H_{20}O_6$. What is the percent composition of this compound?
69. Diphenhydramine hydrochloride, a drug used commonly as an antihistamine, has the formula $C_{17}H_{21}NO \cdot HCl$. What is the percent composition of each element in this compound?
70. What is the percent composition of each element in sucrose, $C_{12}H_{22}O_{11}$?
71. Aspirin is well known as a pain reliever (analgesic) and as a fever reducer (antipyretic). It has a molar mass of 180.2 g/mol and a composition of 60.0% C, 4.48% H, and 35.5% O. Calculate the molecular formula of aspirin.
72. How many grams of oxygen are contained in 8.50 g of $Al_2(SO_4)_3$?
73. Gallium arsenide is one of the newer materials used to make semiconductor chips for use in supercomputers. Its composition is 48.2% Ga and 51.8% As. What is the empirical formula?
74. Calcium tartrate is used as a preservative for certain foods and as an antacid. It contains 25.5% C, 2.1% H, 21.3% Ca, and 51.0% O. What is the empirical formula for calcium tartrate?
75. The compositions of four different compounds of carbon and chlorine follow. Determine the empirical formula and the molecular formula for each compound.

	Percent C	Percent Cl	Molar mass (g)
(a)	7.79	92.21	153.8
(b)	10.13	89.87	236.7
(c)	25.26	74.74	284.8
(d)	11.25	88.75	319.6

76. How many years is a mole of seconds?
77. A normal penny has a mass of about 2.5 g. If we assume the penny to be pure copper (which means the penny is very old since newer pennies are a mixture of copper and zinc), how many atoms of copper does it contain?
78. What would be the mass (in grams) of one thousand trillion molecules of glycerin ($C_3H_8O_3$)?
79. If we assume that there are 7.0 billion people on the Earth, how many moles of people is this?
80. An experimental catalyst used to make polymers has the following composition: Co, 23.3%; Mo, 25.3%; and Cl, 51.4%. What is the empirical formula for this compound?
81. If a student weighs 18 g of aluminum and needs twice as many atoms of magnesium as she has of aluminum, how many grams of Mg does she need?
82. If 10.0 g of an unknown compound composed of carbon, hydrogen, and nitrogen contains 17.7% N and 3.8×10^{23} atoms of hydrogen, what is its empirical formula?
83. A substance whose formula is A_2O (A is a mystery element) is 60.0% A and 40.0% O. Identify the element A.
84. For the following compounds whose molecular formulas are given, indicate the empirical formula:
    (a) $C_6H_{12}O_6$ glucose
    (b) C_8H_{18} octane
    (c) $C_3H_6O_3$ lactic acid
    (d) $C_{25}H_{52}$ paraffin
    (e) $C_{12}H_4Cl_4O_2$ dioxin (a powerful poison)
85. Copper is often added to paint for oceangoing ships and boats to discourage the growth of marine organisms on their hulls. Unfortunately the copper dissolves into the water and is toxic to the marine life in harbors where these boats are docked. When copper(II) ions reach a concentration of 9.0 microgram/L, the growth of some marine life slows. How many copper ions are found in a liter of water containing 9.0 micrograms of Cu^{2+}/L?
86. Molecules of oxygen gas have been very difficult to find in space. Recently the European Space Agency's Herschel space observatory has found molecular oxygen in the Orion star-forming complex. It is present at very low concentrations, 1 molecule of oxygen per every million (1.0×10^6) molecules of hydrogen, but it does exist. How many molecules of oxygen would exist in a 3.0-g sample of hydrogen gas?

87. Scientists from the European Space Agency have been studying geysers located around the southern pole of one of Saturn's moons. These geysers send 250.0 kg of water into the atmosphere every second. Scientists trying to discover what happens to this water have found that it forms a very thin vapor ring around Saturn. How many moles of water are shot into the atmosphere in a Saturn day? (One Saturn day is equal to 10 hours and 45 minutes.)

88. Researchers from the University of Cape Town in South Africa have discovered that penguins are able to detect a chemical released by plankton as they are being consumed by schools of fish. By following this scent they are able to find these schools of fish and feast on them. The chemical they smell is composed of 38.65% carbon, 9.74% hydrogen, and 51.61% sulfur. Determine the empirical formula of this compound.

CHALLENGE EXERCISES

89. The compound $A(BC)_3$ has a molar mass of 78.01 g and a percent composition of 34.59% A, 61.53% B, and 3.88% C. Determine the identity of the elements A, B, and C. What is the percent composition by mass of the compound A_2B_3?

90. A 2.500-g sample of an unknown compound containing only C, H, and O was burned in O_2. The products were 4.776 g of CO_2 and 2.934 g of H_2O.

(a) What is the percent composition of the original compound?

(b) What is the empirical formula of the compound?

91. Researchers at Anna Gudmundsdottir's laboratory at the University of Cincinnati have been studying extremely reactive chemicals known as radicals. One of the interesting phenomena they have discovered is that these radicals can be chemically attached to fragrance molecules, effectively tethering them to a solution. When light strikes these tethered molecules, the fragrance is released. This property would allow us to produce perfumes, cleansers, and other consumer products that release fragrance only when exposed to light. If limonene, $C_{10}H_{16}$, the molecule that gives fruits their citrus scent, were able to be tethered to one of these radicals and every photon of light would release one molecule of limonene, calculate the time in seconds required to release 1.00 picogram of limonene if ambient light releases 2.64×10^{18} photons/sec.

92. A new mineral, putnisite, has been discovered in western Australia. The chemistry of this mineral is unique, and it is found to have an empirical formula of $SrCa_4Cr_8(CO_3)_8(SO_4)(OH)_{16} \cdot 25H_2O$. It is a hydrate like opals, and the $\cdot 25H_2O$ means that every mole of the substance incorporates 25 moles of water into its crystal structure.

(a) Calculate the molar mass of this hydrate.

(b) How many atoms of carbon are there in 1.00 mole of putnisite?

(c) How many atoms of oxygen are there in 1.00 mole of putnisite?

(d) What mass of putnisite contains 1.00 mol of calcium?

(e) How many grams of sulfur are found in 1 kg of putnisite?

Dr. Peter Elliott, University of Adelaide

93. The ingredient list for chocolate chip cookies is shown below.

Ingredients

- 2.0 cups all-purpose flour
- 1.0 teaspoon baking soda
- 1.0 teaspoon salt
- 1.0 cup butter
- 0.25 cup granulated sugar
- 1.3 cups packed brown sugar
- 1.0 teaspoon vanilla extract
- 2 large eggs
- 2.0 cups chocolate chips

© miir/iStockphoto

(a) One cup of flour has a mass of 125 g and contains 32% starch. Find the number of starch molecules (molar mass 1.8×10^5 g/mol) in this flour.

(b) One teaspoon of baking soda contains 11 g NaHCO3. Caluclate the number of formula units of baking soda in the recipe.

(c) One cup of brown sugar contains 220 g sucrose ($C_{12}H_{22}O_{11}$). Calculate the number of sucrose molecules in one batch of cookies.

(d) One large egg contains 6.3 g protein (primarily ovalbumin). The molar mass of ovalbumin is 4.5×10^3 g/mol. Determine the mass of ovalbumin in a batch of chocolate chip cookies.

(e) One teaspoon of vanilla extract contains 23 mg of vanillin ($C_8H_8O_3$). Calculate the mass of vanilla in one teaspoon.

(f) Butter is 81% fat and 1 cup of butter is 227 g. The molar mass of this fat is 8.0×10^2 g/mol. Determine the mass of the fat in the butter used in a batch of these cookies.

94. Compound X requires 104 g of O_2 to produce 2 moles of CO_2 and 2.5 moles of H_2O.

(a) What is the empirical formula for X?

(b) What additional information would you need to determine the molecular formula for X?

ANSWERS TO PRACTICE EXERCISES

7.1 (a) 6.022×10^{23} atoms Fe
(b) 1.204×10^{24} atoms H
(c) 4.215×10^{24} atoms

7.2 10.0 g helium

7.3 1.5×10^{22} atoms

7.4 101.1 g KNO_3

7.5 21.3 g Na_2SO_4

7.6 8.326 mol and 5.014×10^{24} molecules $HC_2H_3O_2$

7.7 24.42% Ca; 17.07% N; 58.50% O

7.8 40.27% K; 26.78% Cr; 32.96% O

7.9 20.16% Al; 79.84% Cl

7.10 C_2H_6O

7.11 P_2O_5

7.12 The empirical formula is CH_3; the molecular formula is C_2H_6.

8

Chemical Equations

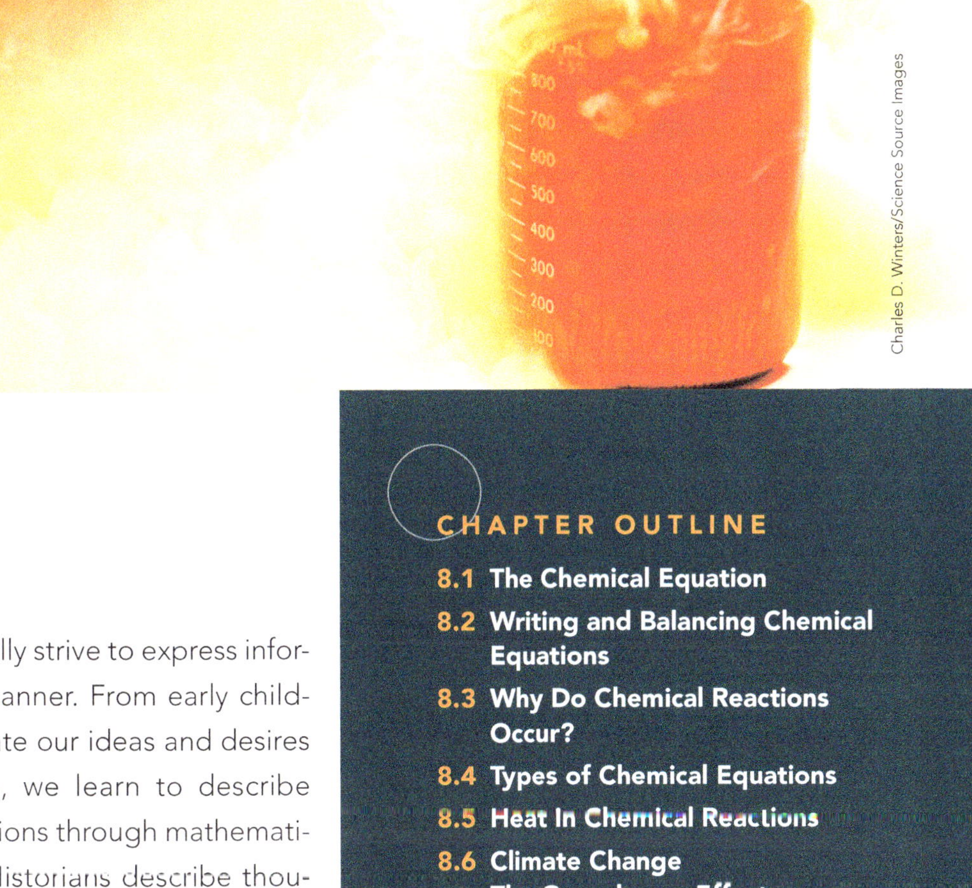

Charles D. Winters/Science Source Images

CHAPTER OUTLINE

In the world today, we continually strive to express information in a concise, useful manner. From early childhood, we are taught to translate our ideas and desires into sentences. In mathematics, we learn to describe numerical relationships and situations through mathematical expressions and equations. Historians describe thousands of years of history in 500-page textbooks. Filmmakers translate entire events, such as the Olympics, into a few hours of entertainment.

Chemists use chemical equations to describe reactions they observe in the laboratory or in nature. For example, as shown in the photo, flames and sparks result when aluminum foil is dropped into liquid bromine. Chemical equations provide us with the means to (1) summarize the reaction, (2) display the substances that are reacting, (3) show the products, and (4) indicate the amounts of all component substances in a reaction.

8.1 The Chemical Equation

LEARNING OBJECTIVE Describe the information present in a chemical equation.

KEY TERMS
reactants
products
chemical equation
law of conservation of mass

Chemical reactions always involve change. Atoms, molecules, or ions rearrange to form different substances, sometimes in a spectacular manner. For example, the thermite reaction is a reaction between aluminum metal and iron(III) oxide, which produces molten iron and aluminum oxide. The substances entering the reaction are called the **reactants**, and the substances formed are called the **products**. In our example,

reactants	aluminum
	iron(III) oxide
products	iron
	aluminum oxide

Charles D. Winters/Science Source

Molten iron falls from the reaction container during the thermite reaction.

During reactions, chemical bonds are broken and new bonds are formed. The reactants and products may be present as solids, liquids, gases, or in solution.

In a chemical reaction atoms are neither created nor destroyed. All atoms present in the reactants must also be present in the products.

A **chemical equation** is a shorthand expression for a chemical change or reaction. A chemical equation uses the chemical symbols and formulas of the reactants and products and other symbolic terms to represent a chemical reaction. The equations are written according to this general format:

1. Reactants are separated from products by an arrow ($\longrightarrow$) that indicates the direction of the reaction. The reactants are placed to the left and the products to the right of the arrow. A plus sign (+) is placed between reactants and between products when needed.

$$\underbrace{Al + Fe_2O_3}_{\text{reactants}} \longrightarrow \underbrace{Fe + Al_2O_3}_{\text{products}}$$

2. Coefficients (whole numbers) are placed in front of substances to balance the equation and to indicate the number of units (atoms, molecules, moles, ions) of each substance reacting or being produced. When no number is shown, it is understood that one unit of the substance is indicated.

$$2\,Al + Fe_2O_3 \longrightarrow 2\,Fe + Al_2O_3$$

3. Conditions required to carry out the reaction may, if desired, be placed above or below the arrow or equality sign. For example, a delta sign placed over the arrow ($\xrightarrow{\Delta}$) indicates that heat is supplied to the reaction.

$$2\,Al + Fe_2O_3 \xrightarrow{\Delta} 2\,Fe + Al_2O_3$$

4. The physical state of a substance is indicated by the following symbols: (*s*) for solid state; (*l*) for liquid state; (*g*) for gaseous state; and (*aq*) for substances in aqueous solution. States are not always given in chemical equations.

$$2\,Al(s) + Fe_2O_3(s) \xrightarrow{\Delta} Fe(l) + Al_2O_3(s)$$

Symbols commonly used in chemical equations are given in **TABLE 8.1**.

TABLE 8.1 Symbols Commonly Used in Chemical Equations

Symbol	Meaning
+	Plus or added to (placed between substances)
$\longrightarrow$	Yields; produces (points to products)
(*s*)	Solid state (written after a substance)
(*l*)	Liquid state (written after a substance)
(*g*)	Gaseous state (written after a substance)
(*aq*)	Aqueous solution (substance dissolved in water)
Δ	Heat is added (when written above or below arrow)

EXAMPLE 8.1

Classify each formula in the chemical reactions below as a reactant or a product.

(a) $MnO_2 + 4\ HCl \longrightarrow MnCl_2 + Cl_2 + 2\ H_2O$

(b) $Na_2SO_3 + H_2SO_4 \longrightarrow Na_2SO_4 + SO_2 + H_2O$

SOLUTION

(a) Reactants: MnO_2 and HCl
Products: $MnCl_2$, Cl_2, and H_2O

(b) Reactants: Na_2SO_3, and H_2SO_4
Products: Na_2SO_4, SO_2, and H_2O

PRACTICE 8.1

How is a substance in a chemical equation identified as a solid, a gas, or in a water solution?

ONLINE LEARNING MODULE
Analyzing Chemical Equations
WileyPLUS

Conservation of Mass

The **law of conservation of mass** states that no change is observed in the total mass of the substances involved in a chemical change. This law, tested by extensive laboratory experimentation, is the basis for the quantitative mass relationships among reactants and products.

The decomposition of water into hydrogen and oxygen illustrates this law. Thus, 100.0 g of water decompose into 11.2 g of hydrogen and 88.8 g of oxygen:

water	$\longrightarrow$	hydrogen	+	oxygen
100.0 g		11.2 g		88.8 g
100.0 g	$\longrightarrow$		100.0 g	
reactant			products	

In a chemical reaction **mass of reactants = mass of products**

CHECK YOUR UNDERSTANDING
8.1 Law of Conservation of Mass
WileyPLUS

8.2 Writing and Balancing Chemical Equations

Write and balance chemical equations.

LEARNING OBJECTIVE

KEY TERMS
balanced equation

To represent the quantitative relationships of a reaction, the chemical equation must be balanced. A **balanced equation** contains the same number of each kind of atom on each side of the equation. The balanced equation therefore obeys the law of conservation of mass.

Every chemistry student must learn to *balance* equations. Many equations are balanced by trial and error, but care and attention to detail are still required. The way to balance an equation is to adjust the number of atoms of each element so that they are the same on each side of the equation, but a correct formula is never changed in order to balance an equation. The general procedure for balancing equations is as follows:

Correct formulas are not changed to balance an equation.

PROBLEM-SOLVING STRATEGY

For Writing and Balancing a Chemical Equation

1. Identify the reaction. Write a description or word equation for the reaction. For example, let's consider mercury(II) oxide decomposing into mercury and oxygen.

$$\text{mercury(II) oxide} \xrightarrow{\Delta} \text{mercury} + \text{oxygen}$$

Study this procedure carefully and refer to it when you work examples.

2. **Write the unbalanced (skeleton) equation.** Make sure that the formula for each substance is correct and that reactants are written to the left and products to the right of the arrow. For our example,

$$HgO \xrightarrow{\Delta} Hg + O_2$$

The correct formulas must be known or determined from the periodic table, lists of ions, or experimental data.

3. **Balance the equation.** Use the following process as necessary:

(a) Count and compare the number of atoms of each element on each side of the equation and determine those that must be balanced:

Hg is balanced (1 on each side)

O needs to be balanced (1 on reactant side, 2 on product side)

Leave elements that are in two or more formulas (on the same side of the equation) unbalanced until just before balancing hydrogen and oxygen.

(b) Balance each element, one at a time, by placing whole numbers (coefficients) in front of the formulas containing the unbalanced element. It is usually best to balance metals first, then nonmetals, then hydrogen and oxygen. Select the smallest coefficients that will give the same number of atoms of the element on each side. A coefficient placed in front of a formula multiplies every atom in the formula by that number (e.g., 2 H_2SO_4 means two units of sulfuric acid and also means four H atoms, two S atoms, and eight O atoms). Place a 2 in front of HgO to balance O:

$$2\,HgO \xrightarrow{\Delta} Hg + O_2$$

(c) Check all other elements after each individual element is balanced to see whether, in balancing one element, other elements have become unbalanced. Make adjustments as needed. Now Hg is not balanced. To adjust this, we write a 2 in front of Hg:

$$2\,HgO \xrightarrow{\Delta} 2\,Hg + O_2 \qquad \text{(balanced)}$$

(d) Do a final check, making sure that each element and/or polyatomic ion is balanced and that the smallest possible set of whole-number coefficients has been used:

$$2\,HgO \xrightarrow{\Delta} 2\,Hg + O_2 \qquad \text{(correct form)}$$

$$4\,HgO \xrightarrow{\Delta} 4\,Hg + 2\,O_2 \qquad \text{(incorrect form)}$$

Not all chemical equations can be balanced by the simple method of inspection just described. The following examples show *stepwise* sequences leading to balanced equations. Study each one carefully.

ENHANCED EXAMPLE 8.2 WileyPLUS

Write the balanced equation for the reaction that takes place when magnesium metal is burned in air to produce magnesium oxide.

SOLUTION

Use the Problem-Solving Strategy for Writing and Balancing a Chemical Equation.

Elements shown in color are balanced.

1. *Word equation:*

$$\text{magnesium} + \text{oxygen} \longrightarrow \text{magnesium oxide}$$

reactants (R) product (P)

R = reactant
P = product

2. *Skeleton equation:*

$$Mg + O_2 \longrightarrow MgO \qquad \text{(unbalanced)}$$

3. *Balance:*
   (a) Mg is balanced.
   Oxygen is not balanced. Two O atoms appear on the left side and one on the right side.
   (b) Place the coefficient 2 in front of MgO to balance oxygen:

$$Mg + O_2 \longrightarrow 2\ MgO \quad \text{(unbalanced)}$$

   (c) Now Mg is not balanced. One Mg atom appears on the left side and two on the right side. Place a 2 in front of Mg:

$$2\ Mg + O_2 \longrightarrow 2\ MgO \quad \text{(balanced)}$$

   (d) *Check:* Each side has two Mg and two O atoms.

(a)	R	1 Mg	2 O
	P	1 Mg	1 O
(b)	R	1 Mg	2 O
	P	2 Mg	2 O
(c)	R	2 Mg	2 O
	P	2 Mg	2 O

ENHANCED EXAMPLE 8.3

WileyPLUS

When methane, CH_4, undergoes complete combustion, it reacts with oxygen to produce carbon dioxide and water. Write the balanced equation for this reaction.

ONLINE LEARNING MODULE

Balancing a Combustion Equation.

➡ WileyPLUS

SOLUTION

Use the Problem-Solving Strategy for Writing and Balancing a Chemical Equation.

1. *Word equation:*

$$\text{methane} + \text{oxygen} \longrightarrow \text{carbon dioxide} + \text{water}$$

2. *Skeleton equation:*

$$CH_4 + O_2 \longrightarrow CO_2 + H_2O \quad \text{(unbalanced)}$$

3. *Balance:*
   (a) Carbon is balanced.
   Hydrogen and oxygen are not balanced.
   (b) Balance H atoms by placing a 2 in front of H_2O:

$$CH_4 + O_2 \longrightarrow CO_2 + 2\ H_2O \quad \text{(unbalanced)}$$

   Each side of the equation has four H atoms; oxygen is still not balanced. Place a 2 in front of O_2 to balance the oxygen atoms:

$$CH_4 + 2\ O_2 \longrightarrow CO_2 + 2\ H_2O \quad \text{(balanced)}$$

   (c) The other atoms remain balanced.
   (d) *Check:* The equation is correctly balanced; it has one C, four O, and four H atoms on each side.

(a)	R	1 C	4 H	2 O
	P	1 C	2 H	3 O
(b)	R	1 C	4 H	2 O
	P	1 C	4 H	4 O
(c)	R	1 C	4 H	4 O
	P	1 C	4 H	4 O

ENHANCED EXAMPLE 8.4

WileyPLUS

Oxygen and potassium chloride are formed by heating potassium chlorate. Write a balanced equation for this reaction.

SOLUTION

Use the Problem-Solving Strategy for Writing and Balancing a Chemical Equation.

1. *Word equation:*

$$\text{potassium chlorate} \xrightarrow{\Delta} \text{potassium chloride} + \text{oxygen}$$

2. *Skeleton equation:*

$$KClO_3 \xrightarrow{\Delta} KCl + O_2 \quad \text{(unbalanced)}$$

(a)	R	1 K	1 Cl	3 O
	P	1 K	1 Cl	2 O
(b)	R	2 K	2 Cl	6 O
	P	1 K	1 Cl	6 O
(c)	R	2 K	2 Cl	6 O
	P	2 K	2 Cl	6 O

3. *Balance:*
   (a) Potassium and chlorine are balanced.
   Oxygen is unbalanced (three O atoms on the left and two on the right side).
   (b) How many oxygen atoms are needed? The subscripts of oxygen (3 and 2) in $KClO_3$ and O_2 have a least common multiple of 6. Therefore, coefficients for $KClO_3$ and O_2 are needed to get six O atoms on each side. Place a 2 in front of $KClO_3$ and a 3 in front of O_2 to balance oxygen:
   $$2\ KClO_3 \xrightarrow{\Delta} KCl + 3\ O_2 \qquad \text{(unbalanced)}$$
   (c) Now K and Cl are not balanced. Place a 2 in front of KCl, which balances both K and Cl at the same time:
   $$2\ KClO_3 \xrightarrow{\Delta} 2\ KCl + 3\ O_2 \qquad \text{(balanced)}$$
   (d) *Check:* Each side now contains two K, two Cl, and six O atoms.

ONLINE LEARNING MODULE
Balancing a Chemical Equation
➡ WileyPLUS

EXAMPLE 8.5

Silver nitrate reacts with hydrogen sulfide to produce silver sulfide and nitric acid. Write a balanced equation for this reaction.

SOLUTION

Use the Problem-Solving Strategy for Writing and Balancing a Chemical Equation.

1. *Word equation:*
   silver nitrate + hydrogen sulfide ⟶ silver sulfide + nitric acid
2. *Skeleton equation:*
   $$AgNO_3 + H_2S \longrightarrow Ag_2S + HNO_3 \qquad \text{(unbalanced)}$$
3. *Balance:*
   (a) Ag and H are unbalanced.
   (b) Place a 2 in front of $AgNO_3$ to balance Ag:
   $$2\ AgNO_3 + H_2S \longrightarrow Ag_2S + HNO_3 \qquad \text{(unbalanced)}$$
   (c) H and NO_3^- are still unbalanced. Balance by placing a 2 in front of HNO_3:
   $$2\ AgNO_3 + H_2S \longrightarrow Ag_2S + 2\ HNO_3 \qquad \text{(balanced)}$$
   In this example, N and O atoms are balanced by balancing the NO_3^- ion as a unit.
   (d) The other atoms remain balanced.
   (e) *Check:* Each side has two Ag, two H, and one S atom. Also, each side has two NO_3^- ions.

(a)	R	1 Ag	2 H	1 S	1 NO_3
	P	2 Ag	1 H	1 S	1 NO_3
(b)	R	2 Ag	2 H	1 S	2 NO_3
	P	2 Ag	1 H	1 S	1 NO_3
(c)	R	2 Ag	2 H	1 S	2 NO_3
	P	2 Ag	2 H	1 S	2 NO_3

ONLINE LEARNING MODULE
Writing and Balancing Chemical Equations
➡ WileyPLUS

EXAMPLE 8.6

When aluminum hydroxide is mixed with sulfuric acid, the products are aluminum sulfate and water. Write a balanced equation for this reaction.

SOLUTION

Use the Problem-Solving Strategy for Writing and Balancing a Chemical Equation.

1. *Word equation:*
   aluminum hydroxide + sulfuric acid ⟶ aluminum sulfate + water

2. *Skeleton equation:*

$$Al(OH)_3 + H_2SO_4 \longrightarrow Al_2(SO_4)_3 + H_2O \qquad \text{(unbalanced)}$$

3. *Balance:*

(a) All elements are unbalanced.

(b) Balance Al by placing a 2 in front of $Al(OH)_3$. Treat the unbalanced SO_4^{2-} ion as a unit and balance by placing a 3 in front of H_2SO_4:

$$2\,Al(OH)_3 + 3\,H_2SO_4 \longrightarrow Al_2(SO_4)_3 + H_2O \qquad \text{(unbalanced)}$$

Balance the unbalanced H and O by placing a 6 in front of H_2O:

$$2\,Al(OH)_3 + 3\,H_2SO_4 \longrightarrow Al_2(SO_4)_3 + 6\,H_2O \qquad \text{(balanced)}$$

(c) The other atoms remain balanced.

(d) *Check:* Each side has 2 Al, 12 H, 3 S, and 18 O atoms.

(a)	R	1 Al	1 SO_4	3 O	5 H
	P	2 Al	3 SO_4	1 O	2 H
(b)	R	2 Al	3 SO_4	6 O	12 H
	P	2 Al	3 SO_4	1 O	2 H
(c)	R	2 Al	3 SO_4	6 O	12 H
	P	2 Al	3 SO_4	6 O	12 H

ENHANCED EXAMPLE 8.7

WileyPLUS

When the fuel in a butane gas stove undergoes complete combustion, it reacts with oxygen to form carbon dioxide and water. Write the balanced equation for this reaction.

Butane C_4H_{10}

SOLUTION

Use the Problem-Solving Strategy for Writing and Balancing a Chemical Equation.

1. *Word equation:*

$$\text{butane} + \text{oxygen} \longrightarrow \text{carbon dioxide} + \text{water}$$

2. *Skeleton equation:*

$$C_4H_{10} + O_2 \longrightarrow CO_2 + H_2O \qquad \text{(unbalanced)}$$

3. *Balance:*

(a) All elements are unbalanced.

(b) Balance C by placing a 4 in front of CO_2:

$$C_4H_{10} + O_2 \longrightarrow 4\,CO_2 + H_2O \qquad \text{(unbalanced)}$$

Balance H by placing a 5 in front of H_2O:

$$C_4H_{10} + O_2 \longrightarrow 4\,CO_2 + 5\,H_2O \qquad \text{(unbalanced)}$$

Oxygen remains unbalanced. The oxygen atoms on the right side are fixed because 4 CO_2 and 5 H_2O are derived from the single C_4H_{10} molecule on the left. When we try to balance the O atoms, we find that there is no whole number that can be placed in front of O_2 to bring about a balance, so we double the coefficients of each substance and then balance the oxygen:

$$2\,C_4H_{10} + 13\,O_2 \longrightarrow 8\,CO_2 + 10\,H_2O \qquad \text{(balanced)}$$

(c) The other atoms remain balanced.

(d) *Check:* Each side now has 8 C, 20 H, and 26 O atoms.

(a)	R	4 C	10 H	2 O
	P	1 C	2 H	3 O
(b)	R	4 C	10 H	2 O
	P	4 C	10 H	13 O
(c)	R	8 C	20 H	26 O
	P	8 C	20 H	26 O

PRACTICE 8.2

Write a balanced formula equation for:

$$\text{aluminum} + \text{oxygen} \longrightarrow \text{aluminum oxide}$$

PRACTICE 8.3

Write a balanced formula equation for:

$$\text{magnesium hydroxide} + \text{phosphoric acid} \longrightarrow \text{magnesium phosphate} + \text{water}$$

Information in a Chemical Equation

Depending on the particular context in which it is used, a formula can have different meanings. A formula can refer to an individual chemical entity (atom, ion, molecule, or formula unit) or to a mole of that chemical entity. For example, the formula H_2O can mean any of the following:

- 2 H atoms and 1 O atom
- 1 molecule of water
- 1 mol of water
- 6.022×10^{23} molecules of water
- 18.02 g of water

Formulas used in equations can represent units of individual chemical entities or moles, the latter being more commonly used. For example, in the reaction of hydrogen and oxygen to form water,

We generally use moles in equations because molecules are so small.

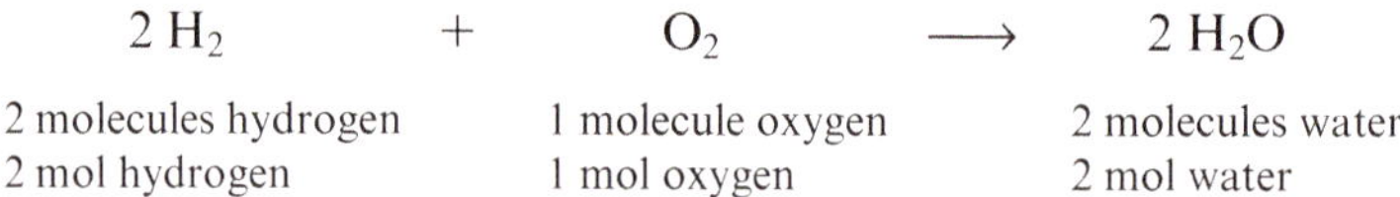

$$2\,H_2 \quad + \quad O_2 \quad \longrightarrow \quad 2\,H_2O$$

$2\,H_2$	O_2	$2\,H_2O$
2 molecules hydrogen	1 molecule oxygen	2 molecules water
2 mol hydrogen	1 mol oxygen	2 mol water

As indicated earlier, a chemical equation is a shorthand description of a chemical reaction. Interpretation of a balanced equation gives us the following information:

1. What the reactants are and what the products are
2. The formulas of the reactants and products
3. The number of molecules or formula units of reactants and products in the reaction
4. The number of atoms of each element involved in the reaction
5. The number of moles of each substance

CHECK YOUR UNDERSTANDING

8.2 Information in a Chemical Equation

WileyPLUS

Consider the reaction that occurs when propane gas (C_3H_8) is burned in air; the products are carbon dioxide (CO_2) and water (H_2O). The balanced equation and its interpretation are as follows:

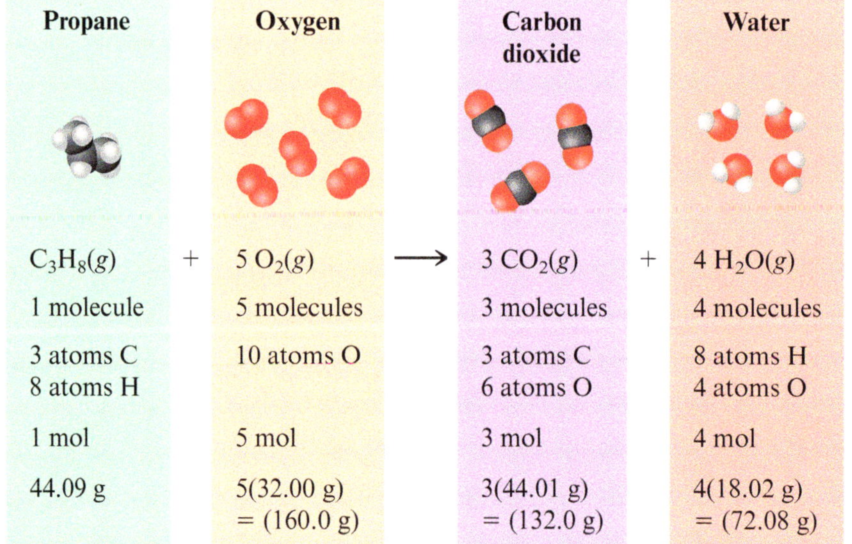

Propane		Oxygen		Carbon dioxide		Water
$C_3H_8(g)$	+	$5\,O_2(g)$	$\longrightarrow$	$3\,CO_2(g)$	+	$4\,H_2O(g)$
1 molecule		5 molecules		3 molecules		4 molecules
3 atoms C 8 atoms H		10 atoms O		3 atoms C 6 atoms O		8 atoms H 4 atoms O
1 mol		5 mol		3 mol		4 mol
44.09 g		5(32.00 g) = (160.0 g)		3(44.01 g) = (132.0 g)		4(18.02 g) = (72.08 g)

PRACTICE 8.4

Consider the reaction that occurs when hydrogen gas reacts with chlorine gas to produce gaseous hydrogen chloride:

(a) Write a word equation for this reaction.
(b) Write a balanced formula equation including the state for each substance.
(c) Label each reactant and product to show the relative amounts of each substance.
  (1) number of molecules (3) number of moles
  (2) number of atoms (4) mass
(d) What mass of HCl would be produced if you reacted 2 mol hydrogen gas with 2 mol chlorine gas?

The quantities involved in chemical reactions are important when working in industry or the laboratory. We will study the relationship among quantities of reactants and products in the next chapter.

8.3 Why Do Chemical Reactions Occur?

LEARNING OBJECTIVE

Determine the factors that cause reactions to occur.

Chemical equations represent chemical changes or reactions. But how can we predict whether or not a chemical reaction will occur? Chemists have learned over time to see observations in reactants that seem to predict when reactants will become products. These "driving forces" that indicate reactants will change into products are

- Formation of a solid
- Formation of a gas
- Formation of water
- Transfer of electrons

If any of these things occur when chemicals are placed together, a chemical reaction is probable. Therefore, to predict whether a reaction will occur we need to consider these "driving forces." We will consider these forces as we learn more about chemical reactions and become better at predicting the outcome of mixing reactants.

8.4 Types of Chemical Equations

LEARNING OBJECTIVE

Give examples of a combination reaction, decomposition reaction, single-displacement reaction, and double-displacement reaction.

KEY TERMS
synthesis reaction
combustion reaction
decomposition reaction
single-displacement reaction
double-displacement reaction

Reactions are classified into types to assist in writing equations and in predicting other reactions. Many chemical reactions fit one or another of the four principal reaction types that we discuss in the following paragraphs. Reactions are also classified as oxidation–reduction. Special methods are used to balance complex oxidation–reduction equations. (See Chapter 17.)

Synthesis (Combination) Reaction

In a **synthesis reaction**, two reactants combine to give one product. Synthesis is one of the most important activities of chemistry. Producing new compounds including new medicines and synthetic compounds such as plastics have greatly influenced our lives. The general form of the equation is

$$A + B \longrightarrow AB$$

in which A and B are either elements or compounds and AB is a compound. The formula of the compound in many cases can be determined from knowledge of the ionic charges

CHECK YOUR UNDERSTANDING
8.3 Synthesis Reactions
WileyPLUS

Charles D. Winters/Science Source Images

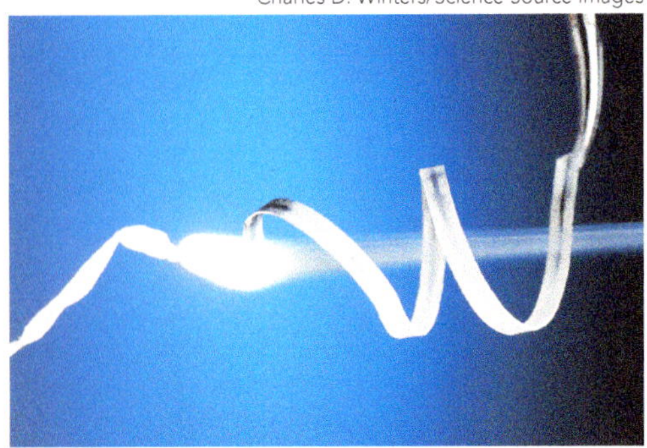

A strip of Mg burning in air produces MgO (left of the flame).

of the reactants in their combined states. Some reactions begin with elements such as the following:

$$2\,Mg(s) + O_2(g) \longrightarrow 2\,MgO(s)$$
$$S(s) + O_2(g) \longrightarrow SO_2(g)$$
$$N_2(g) + O_2(g) \longrightarrow 2\,NO_2(g)$$
$$2\,Na(s) + Cl_2(g) \longrightarrow 2\,NaCl(s)$$

Other reactions begin with compounds that are changed into a new compound:

$$Na_2O(s) + H_2O(l) \longrightarrow 2\,NaOH(aq)$$
$$SO_3(g) + H_2O(l) \longrightarrow H_2SO_4(aq)$$

Many of these reactions can also be classified as oxidation–reduction reactions as we shall see in future chapters.

CHEMISTRY *IN ACTION* CO Poisoning—A Silent Killer

Carbon monoxide (CO) is one possible product when carbon-containing fuels such as coal, natural gas, propane, heating oil, and gasoline are burned. For example, when methane, the main component of natural gas, burns, the reaction is

$$CH_4(g) + 2\,O_2(g) \longrightarrow 2\,H_2O(g) + CO_2(g)$$

However, if not enough oxygen is available, carbon monoxide will be produced instead of carbon dioxide:

$$2\,CH_4(g) + 3\,O_2(g) \longrightarrow 4\,H_2O(g) + 2\,CO(g)$$

Carbon monoxide is colorless and has no flavor or smell. It often can poison several people at once, since it can build up undetected in enclosed spaces. The symptoms of carbon monoxide poisoning are so easily missed that people often don't realize that they have been poisoned. Common symptoms are headache, ringing in ears, nausea, dizziness, weakness, confusion, and drowsiness. A low concentration of CO in the air may go completely unnoticed while the toxicity level increases in the blood.

Carbon monoxide in the blood is poisonous because of its ability to bond to a hemoglobin molecule. Both CO and O_2 bond to the same place on the hemoglobin molecule. The CO bonds more tightly than O_2. We use hemoglobin to transport oxygen to the various tissues from the lungs. Unfortunately, CO also bonds to hemoglobin, forming a molecule called carboxyhemoglobin. In fact, hemoglobin prefers CO to O_2. Once a CO is bound to a hemoglobin molecule, it cannot carry an oxygen molecule. So, until the CO is released, that molecule of hemoglobin is lost as an O_2 carrier. If you breathe normal air (about 20% oxygen) containing 0.1% CO, in just one hour the CO bonds to 50% of the hemoglobin molecules. To keep CO from being a silent killer, there are three possible actions:

1. Immediate treatment of victims to restore the hemoglobin
2. Detection of CO levels before poisoning occurs
3. Conversion of CO to CO_2 to eliminate the threat

If a person already has CO poisoning, the treatment must focus on getting CO released from the hemoglobin. This can be done by giving the victim oxygen or by placing the victim in a hyperbaric chamber (at about 3 atm pressure). These treatments reduce the time required for carboxyhemoglobin to release CO from between 4 to 6 hours to 30 to 90 minutes. This treatment also supplies more oxygen to the system to help keep the victim's brain functioning.

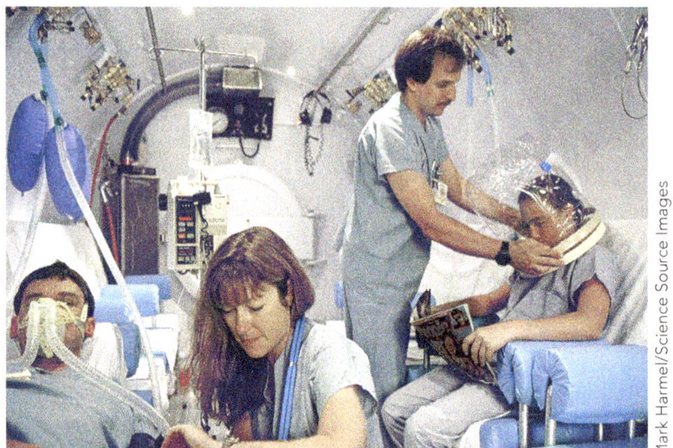

Mark Harmel/Science Source Images

People in a hyperbaric chamber.

A CO detector can sound an alarm before the CO level becomes toxic in a home. Current CO detectors sound an alarm when levels increase rapidly or when lower concentrations are present over a long time. Inside a CO detector is a beam of infrared light that shines on a chromophore (a substance that turns darker with increasing CO levels). If the infrared light passing through the chromophore is too low, the alarm sounds. Typically, these alarms ring when CO levels reach 70 ppm, or greater than 30 ppm for 30 days.

The final option is to convert CO to CO_2 before toxicity occurs. David Schreyer from NASA's Langley Research Center has devised a catalyst (tin hydroxide with small amounts of Pd) that speeds up the conversion of CO to CO_2 at room temperature. He and his colleagues have adapted their catalyst to home ventilation systems to eliminate even low levels of CO before it has a chance to build up. Using all of these approaches can help to eliminate 500–10,000 cases of CO poisoning each year.

Combustion Reaction

In a **combustion reaction** a substance reacts with oxygen to form oxides. The most common combustion reactions are between hydrocarbons (compounds composed of hydrogen and carbon) and oxygen to form carbon dioxide and water. These reactions are commonly used to generate energy from fossil fuels. Here are some examples of combustion reactions of hydrocarbons that may be familiar to you:

Methane or CH_4 is the fuel most often used in gas ovens and stoves. It is also the fuel generally used to power Bunsen burners in the chemistry laboratory.

$$CH_4(g) + 2\,O_2(g) \longrightarrow CO_2(g) + 2\,H_2O(g)$$

(Notice that the water is generally vaporized.)

Octane or C_8H_{18} is one of the major components of the gasoline used in automobiles.

$$2\,C_8H_{18}(l) + 25\,O_2(g) \longrightarrow 16\,CO_2(g) + 18\,H_2O(g)$$

Ethanol or C_2H_5OH is sometimes added to gasoline as an alternative to fossil fuels. It burns more cleanly than hydrocarbons.

$$C_2H_5OH(l) + 3\,O_2(g) \longrightarrow 2\,CO_2(g) + 3\,H_2O(g)$$

(Remember to account for the oxygen in the ethanol when balancing the equation.)

Notice that one of the products of the combustion of carbon-based fuels is carbon dioxide, which contributes to climate change as described in Section 8.5.

Combustion reactions are not limited to hydrogen or carbon compounds. Two by-products of the combustion of fossil fuels are oxides of nitrogen and sulfur. Often fossil fuels contain sulfur, which can also react with the oxygen when it is exposed to the high temperatures in a gasoline engine. Atmospheric nitrogen will form an oxide as well under these conditions. The reactions shown here could be classified as synthesis reactions or combustion reactions.

$$S(s) + O_2(g) \longrightarrow SO_2(g)$$

$$N_2(g) + 2\,O_2(g) \longrightarrow 2\,NO_2(g)$$

Decomposition Reaction

In a **decomposition reaction**, a single substance is decomposed, or broken down, to give two or more different substances. This reaction may be considered the reverse of combination. The starting material must be a compound, and the products may be elements or compounds. The general form of the equation is

$$AB \longrightarrow A + B$$

Predicting the products of a decomposition reaction can be difficult and requires an understanding of each individual reaction. Heating oxygen-containing compounds often results in decomposition. Here are some reactions that fall into this category:

CHECK YOUR UNDERSTANDING
8.4 Decomposition Reactions
➡ WileyPLUS

$$2\,HgO(s) \xrightarrow{\Delta} 2\,Hg(l) + O_2(g)$$

$$2\,PbO_2(s) \xrightarrow{\Delta} 2\,PbO(s) + O_2(g)$$

$$CaCO_3(s) \xrightarrow{\Delta} CaO(s) + CO_2(g)$$

$$2\,NaHCO_3(s) \xrightarrow{\Delta} Na_2CO_3(s) + H_2O(g) + CO_2(g)$$

$$2\,KClO_3(s) \xrightarrow{\Delta} 2\,KCl(s) + 3\,O_2(g)$$

$$2\,NaNO_3(s) \xrightarrow{\Delta} 2\,NaNO_2(s) + O_2(g)$$

$$2\,H_2O_2(l) \xrightarrow{\Delta} 2\,H_2O(l) + O_2(g)$$

Note that a driving force in many of these reactions is the production of a gas.

Charles D. Winters/Science Source Images

When pieces of zinc metal are placed in hydrochloric acid, hydrogen bubbles form immediately.

CHECK YOUR UNDERSTANDING

8.5 Single-Displacement Reactions

➡ WileyPLUS

TABLE 8.2 Activity Series

	Metals	Halogens
↑ increasing activity	K	F_2
	Ca	Cl_2
	Na	Br_2
	Mg	I_2
	Al	
	Zn	
	Fe	
	Ni	
	Sn	
	Pb	
	H	
	Cu	
	Ag	
	Hg	
	Au	

Single-Displacement Reaction

In a **single-displacement reaction**, one element reacts with a compound to replace one of the elements of that compound, yielding a different element and a different compound. The general forms of the equation follow.

If A is a *metal*, A will replace B to form AC, provided that A is a more reactive metal than B.

$$A + BC \longrightarrow B + AC$$

If A is a *halogen*, it will replace C to form BA, provided that A is a more reactive halogen than C.

$$A + BC \longrightarrow C + BA$$

A brief activity series of selected metals (and hydrogen) and halogens is shown in TABLE 8.2. This series is listed in descending order of chemical activity, with the most active metals and halogens at the top. Many chemical reactions can be predicted from an activity series because the atoms of any element in the series will replace the atoms of those elements below it. For example, zinc metal will replace hydrogen from a hydrochloric acid solution. But copper metal, which is below hydrogen on the list and thus less reactive than hydrogen, will not replace hydrogen from a hydrochloric acid solution. Here are some reactions that fall into this category:

1. reactions with metals as A:

$$Zn(s) + 2\,HCl(aq) \longrightarrow H_2(g) + ZnCl_2(aq)$$
$$2\,Al(s) + 3\,H_2SO_4(aq) \longrightarrow 3\,H_2(g) + Al_2(SO_4)_3(aq)$$
$$2\,Na(s) + 2\,H_2O(l) \longrightarrow H_2(g) + 2\,NaOH(aq)$$
$$Ca(s) + 2\,H_2O(l) \longrightarrow H_2(g) + Ca(OH)_2(aq)$$
$$3\,Fe(s) + \underset{\text{steam}}{4\,H_2O(g)} \longrightarrow 4\,H_2(g) + Fe_3O_4(s)$$
$$Fe(s) + CuSO_4(aq) \longrightarrow Cu(s) + FeSO_4(aq)$$
$$Cu(s) + 2\,AgNO_3(aq) \longrightarrow 2\,Ag(s) + Cu(NO_3)_2(aq)$$

2. reactions with a halogen as A:

$$Cl_2(g) + 2\,NaBr(aq) \longrightarrow Br_2(l) + 2\,NaCl(aq)$$
$$Cl_2(g) + 2\,KI(aq) \longrightarrow I_2(s) + 2\,KCl(aq)$$

A common chemical reaction is the displacement of hydrogen from water or acids. This reaction is a good illustration of the relative reactivity of metals and the use of the activity series. For example,

- K, Ca, and Na displace hydrogen from cold water, steam (H_2O), and acids.
- Mg, Al, Zn, and Fe displace hydrogen from steam and acids.
- Ni, Sn, and Pb displace hydrogen only from acids.
- Cu, Ag, Hg, and Au do not displace hydrogen.

ENHANCED EXAMPLE 8.8 WileyPLUS

Will a reaction occur between (a) nickel metal and hydrochloric acid and (b) tin metal and a solution of aluminum chloride? Write balanced equations for the reactions.

SOLUTION

(a) Nickel is more reactive than hydrogen, so it will displace hydrogen from hydrochloric acid. The products are hydrogen gas and a salt of Ni^{2+} and Cl^- ions:

$$Ni(s) + 2\,HCl(aq) \longrightarrow H_2(g) + NiCl_2(aq)$$

(b) According to the activity series, tin is less reactive than aluminum, so no reaction will occur:

$$Sn(s) + AlCl_3(aq) \longrightarrow \text{no reaction}$$

PRACTICE 8.5

Write balanced equations for these reactions:
(a) iron metal and a solution of magnesium chloride
(b) zinc metal and a solution of lead(II) nitrate

Double-Displacement Reaction

In a **double-displacement reaction**, two compounds exchange partners with each other to produce two different compounds. The general form of the equation is

$$AB + CD \longrightarrow AD + CB$$

This reaction can be thought of as an exchange of positive and negative groups, in which A combines with D and C combines with B. In writing formulas for the products, we must account for the charges of the combining groups.

It's also possible to write an equation in the form of a double-displacement reaction when a reaction has not occurred. For example, when solutions of sodium chloride and potassium nitrate are mixed, the following equation can be written:

$$NaCl(aq) + KNO_3(aq) \longrightarrow NaNO_3(aq) + KCl(aq)$$

When the procedure is carried out, no physical changes are observed, indicating that no chemical reaction has taken place.

A double-displacement reaction is accompanied by evidence of reaction such as:

1. The evolution of heat
2. The formation of an insoluble precipitate
3. The production of gas bubbles

CHECK YOUR UNDERSTANDING
8.6 Double-Displacement Reactions
WileyPLUS

Let's look at some of these reactions more closely: The production of a molecule of water from an H^+ and an OH^- ion is accompanied by a *release of heat*, which can be detected by touching the reaction container or by using a thermometer.

$$HCl(aq) + NaOH(aq) \longrightarrow NaCl(aq) + H_2O(l) + \text{heat}$$

$$H_2SO_4(aq) + Ba(OH)_2(aq) \longrightarrow BaSO_4(s) + 2\ H_2O(l) + \text{heat}$$

$$CuO(s) + 2\ HNO_3(aq) \longrightarrow Cu(NO_3)_2(aq) + H_2O(l) + \text{heat}$$

$$CaO(s) + 2\ HCl(aq) \longrightarrow CaCl_2(aq) + H_2O(l) + \text{heat}$$

Formation of a solid The solubilities of the products can be determined by consulting the solubility table in Appendix 4 to see whether one or both of the products are insoluble in water. A solid is indicated by placing (*s*) after its formula in the equation.

$$BaCl_2(aq) + 2\ AgNO_3(aq) \longrightarrow 2\ AgCl(s) + Ba(NO_3)_2(aq)$$

$$Pb(NO_3)_2(aq) + 2\ KI(aq) \longrightarrow PbI_2(s) + 2\ KNO_3(aq)$$

Richard Megna/Fundamental Photographs

A double-displacement reaction results from pouring a clear, colorless solution of $Pb(NO_3)_2$ into a clear, colorless solution of KI, forming a yellow solid of PbI_2. $Pb(NO_3)_2 + 2KI \longrightarrow PbI_2 + 2KNO_3$.

Formation of a gas A gas such as HCl or H_2S may be produced directly, as in these two examples:

$$H_2SO_4(aq) + NaCl(s) \longrightarrow NaHSO_4(s) + HCl(g)$$

$$2\ HCl(aq) + ZnS(s) \longrightarrow ZnCl_2(aq) + H_2S(g)$$

A gas can also be produced indirectly. Some unstable compounds formed in a double-displacement reaction, such as H_2CO_3, H_2SO_3, and NH_4OH, will decompose to form water and a gas:

$$2\,HCl(aq) + Na_2CO_3(aq) \longrightarrow 2\,NaCl(aq) + H_2CO_3(aq) \longrightarrow 2\,NaCl(aq) + H_2O(l) + CO_2(g)$$

$$2\,HNO_3(aq) + K_2SO_3(aq) \longrightarrow 2\,KNO_3(aq) + H_2SO_3(aq) \longrightarrow 2\,KNO_3(aq) + H_2O(l) + SO_2(g)$$

$$NH_4Cl(aq) + NaOH(aq) \longrightarrow NaCl(aq) + NH_4OH(aq) \longrightarrow NaCl(aq) + H_2O(l) + NH_3(g)$$

ONLINE LEARNING MODULE
Writing a Double-Displacement Chemical Equation
WileyPLUS

ENHANCED EXAMPLE 8.9

WileyPLUS

Write the equation for the reaction between aqueous solutions of hydrobromic acid and potassium hydroxide.

SOLUTION

First we write the formulas for the reactants. They are HBr and KOH. Now rewrite the equation using the formulas for the known substances:

$$HBr(aq) + KOH(aq) \longrightarrow$$

In this reaction, which is a double-displacement type, the H^+ from the acid combines with the OH^- from the base to form water. The ionic compound must be composed of the other two ions, K^+ and Br^-. We determine the formula of the ionic compound to be KBr from the fact that K is a +1 cation and Br is a −1 anion. The final balanced equation is

$$HBr(aq) + KOH(aq) \longrightarrow KBr(aq) + H_2O(l)$$

ENHANCED EXAMPLE 8.10

WileyPLUS

Complete and balance the equation for the reaction between aqueous solutions of barium chloride and sodium sulfate.

SOLUTION

First determine the formulas for the reactants. They are $BaCl_2$ and Na_2SO_4. Both substances are ionic compounds. Since both substances are compounds, the reaction will be of the double-displacement type. Start writing the equation with the reactants:

$$BaCl_2(aq) + Na_2SO_4(aq) \longrightarrow$$

If the reaction is double-displacement, Ba^{2+} will be written combined with SO_4^{2-}, and Na^+ with Cl^- as the products. The balanced equation is

$$BaCl_2(aq) + Na_2SO_4(aq) \longrightarrow BaSO_4(s) + 2\,NaCl(aq)$$

The final step is to determine the nature of the products, which controls whether or not the reaction will take place. If both products are soluble, we have a mixture of all the ions in solution. But if an insoluble precipitate is formed, the reaction will definitely occur. We know from experience that NaCl is fairly soluble in water, but what about $BaSO_4$? Consulting the solubility table in Appendix 4 we see that $BaSO_4$ is insoluble in water, so it will be a precipitate in the reaction. Thus, the reaction will occur, forming a precipitate. The equation is

$$BaCl_2(aq) + Na_2SO_4(aq) \longrightarrow BaSO_4(s) + 2\,NaCl(aq)$$

Richard Megna/Fundamental Photographs

When barium chloride is poured into a solution of sodium sulfate, a white precipitate of barium sulfate forms.

PRACTICE 8.6

Complete and balance the equations for these reactions in water:

(a) potassium phosphate + barium chloride
(b) hydrochloric acid + nickel carbonate
(c) ammonium chloride + sodium nitrate

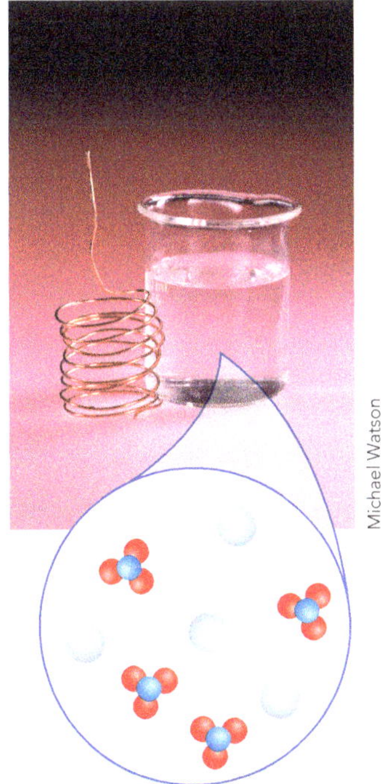

A solution of silver nitrate contains Ag^+ ions and NO_3^- ions.

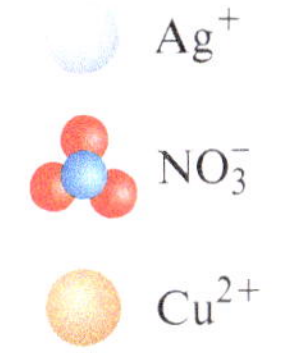

A copper wire is placed in a solution of silver nitrate. After 24 hours, crystals of silver are seen hanging on the copper wire and the solution has turned blue, indicating copper ions are present there.

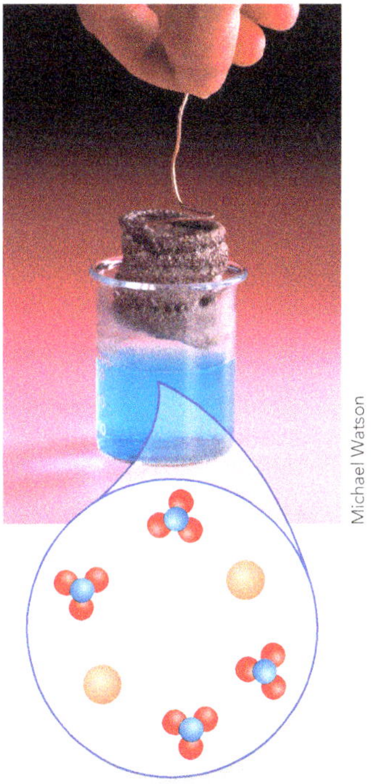

The silver metal clings to the copper wire, and the blue solution contains Cu^{2+} and NO_3^- ions.

FIGURE 8.1 Reaction of a copper wire and silver nitrate solution.

Some of the reactions you attempt may fail because the substances are not reactive or because the proper conditions for reaction are not present. For example, mercury(II) oxide does not decompose until it is heated; magnesium does not burn in air or oxygen until the temperature reaches a certain point. When silver is placed in a solution of copper(II) sulfate, no reaction occurs. When copper wire is placed in a solution of silver nitrate, a single-displacement reaction takes place because copper is a more reactive metal than silver (**FIGURE 8.1**).

$$Cu(s) + 2\ AgNO_3(aq) \longrightarrow 2Ag(s) + Cu(NO_3)_2\,(aq)$$

The successful prediction of the products of a reaction is not always easy. The ability to predict products correctly comes with knowledge and experience. Although you may not be able to predict many reactions at this point, as you continue to experiment you will find that reactions can be categorized and that prediction of the products becomes easier, if not always certain.

CHECK YOUR UNDERSTANDING

8.7 Identifying the Type of Reaction 1

8.8 Identifying the Type of Reaction 2

8.9 Identifying the Type of Reaction 3

8.10 Identifying the Type of Reaction 4

WileyPLUS

8.5 Heat in Chemical Reactions

LEARNING OBJECTIVE

Explain the following terms and how they relate to a chemical reaction: exothermic reaction, endothermic reaction, heat of reaction, and activation energy.

KEY TERMS

exothermic reactions
endothermic reactions
heat of reaction
hydrocarbons
activation energy

Energy changes always accompany chemical reactions. One reason reactions occur is that the products attain a lower, more stable energy state than the reactants. When the reaction leads to a more stable state, energy is released to the surroundings as heat and/or work. When a solution of a base is neutralized by the addition of an acid, the liberation of heat energy is signaled by an immediate rise in the temperature of the solution. When an automobile engine burns gasoline, heat is certainly liberated; at the same time, part of the liberated energy does the work of moving the automobile.

Reactions are either exothermic or endothermic. **Exothermic reactions** liberate heat; **endothermic reactions** absorb heat. In an exothermic reaction, heat is a product and may be written on the right side of the equation for the reaction. In an endothermic reaction, heat can be regarded as a reactant and is written on the left side of the equation. Here are two examples:

$$H_2(g) + Cl_2(g) \longrightarrow 2\,HCl(g) + 185\ kJ \quad (\textit{exothermic})$$

$$N_2(g) + O_2(g) + 181\ kJ \longrightarrow 2\,NO(g) \quad (\textit{endothermic})$$

CHECK YOUR UNDERSTANDING

8.11 Endothermic Reactions

8.12 Exothermic Reactions

WileyPLUS

The quantity of heat produced by a reaction is known as the **heat of reaction**. The units used can be kilojoules or kilocalories. Consider the reaction represented by this equation:

$$C(s) + O_2(g) \longrightarrow CO_2(g) + 393\ kJ$$

When the heat released is expressed as part of the equation, the substances are expressed in units of moles. Thus, when 1 mol (12.01 g) of C combines with 1 mol (32.00 g) of O_2, 1 mol (44.01 g) of CO_2 is formed and 393 kJ of heat are released. In this reaction, as in many others, the heat energy is more useful than the chemical products.

Aside from relatively small amounts of energy from nuclear processes, the sun is the major provider of energy for life on Earth. The sun maintains the temperature necessary for life and also supplies light energy for the endothermic photosynthetic reactions of green plants. In photosynthesis, carbon dioxide and water are converted to free oxygen and glucose:

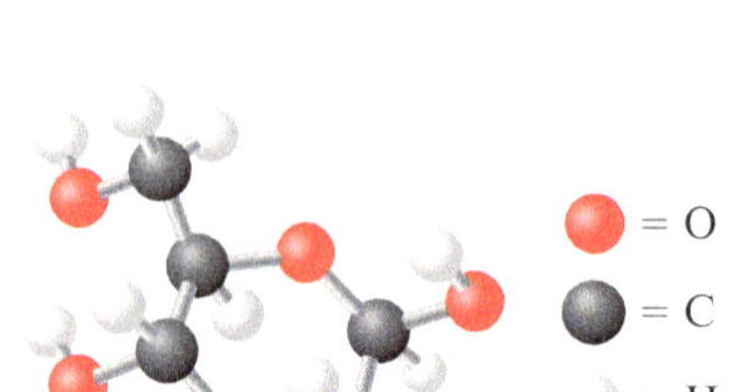

Glucose
$C_6H_{12}O_6$

$$6\,CO_2 + 6\,H_2O + 2519\ kJ \longrightarrow \underset{\text{glucose}}{C_6H_{12}O_6} + 6\,O_2$$

Nearly all of the chemical energy used by living organisms is obtained from glucose or compounds derived from glucose.

This cornfield is a good example of the endothermic reactions happening through photosynthesis in plants.

Paul Damien/National Geographic Stock

The major source of energy for modern technology is fossil fuel—coal, petroleum, and natural gas. The energy is obtained from the combustion (burning) of these fuels, which are converted to carbon dioxide and water. Fossil fuels are mixtures of **hydrocarbons**, compounds containing only hydrogen and carbon.

Natural gas is primarily methane, CH_4. Petroleum is a mixture of hydrocarbons (compounds of carbon and hydrogen). Liquefied petroleum gas (LPG) is a mixture of propane (C_3H_8) and butane (C_4H_{10}).

Here are some examples:

$$CH_4(g) + 2\,O_2(g) \longrightarrow CO_2(g) + 2\,H_2O(g) + 890\text{ kJ}$$

$$C_3H_8(g) + 5\,O_2(g) \longrightarrow 3\,CO_2(g) + 4\,H_2O(g) + 2200\text{ kJ}$$

The combustion of these fuels releases a tremendous amount of energy, but reactions won't occur to a significant extent at ordinary temperatures. A spark or a flame must be present before methane will ignite. The amount of energy that must be supplied to start a chemical reaction is called the **activation energy**. In an exothermic reaction, once this activation energy is provided, enough energy is then generated to keep the reaction going.

Be careful not to confuse an exothermic reaction that requires heat (activation energy) to get it *started* with an endothermic process that requires energy to keep it going. The combustion of magnesium, for example, is highly exothermic, yet magnesium must be heated to a fairly high temperature in air before combustion begins. Once started, however, the combustion reaction goes very vigorously until either the magnesium or the available supply of oxygen is exhausted. The electrolytic decomposition of water to hydrogen and oxygen is highly endothermic. If the electric current is shut off when this process is going on, the reaction stops instantly. The relative energy levels of reactants and products in exothermic and endothermic processes are presented graphically in **FIGURES 8.2** and **8.3**.

Combustion of Mg inside a block of dry ice (CO_2) makes a glowing lantern.

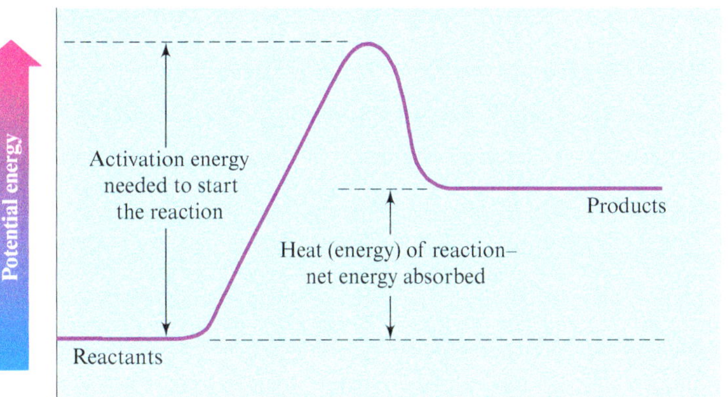

FIGURE 8.2 An endothermic process occurs when a cold pack containing an ampule of solid NH_4Cl is broken, releasing the crystals into water in the pack. The dissolving of NH_4Cl in water is endothermic, producing a salt solution that is cooler than the surroundings. This process is represented graphically to show the energy changes between the reactants and the products.

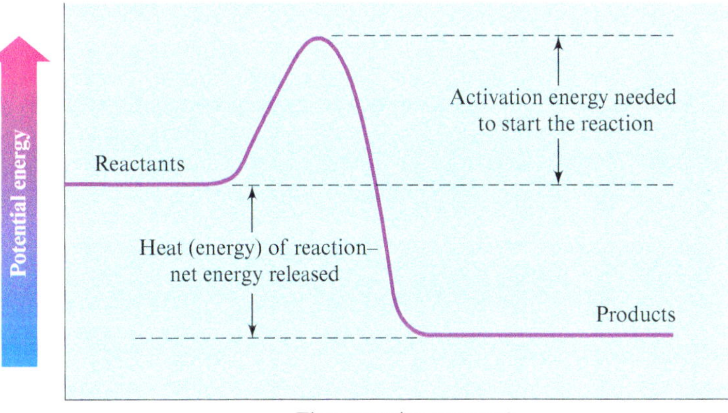

FIGURE 8.3 Exothermic reaction between $KClO_3$ and sugar. A sample of $KClO_3$ and sugar is well mixed and placed on a fireproof pad. Several drops of concentrated H_2SO_4 are used to ignite the mixture.

CHECK YOUR UNDERSTANDING

8.13 Activation Energy and Heat of Reaction

➡ WileyPLUS

Examples of endothermic and exothermic processes can be easily demonstrated. In Figure 8.2, the products are at a higher potential energy than the reactants. Energy has therefore been absorbed, and the reaction is endothermic. An endothermic reaction takes place when you apply a cold pack to an injury. When a cold pack is activated, ammonium chloride (NH_4Cl) dissolves in water. For example, temperature changes from 24.5°C to 18.1°C result when 10 g of NH_4Cl are added to 100 mL of water. Energy, in the form of heat, is taken from the immediate surroundings (water), causing the salt solution to become cooler.

ONLINE LEARNING MODULE

Heat in Chemical Reactions

➡ WileyPLUS

EXAMPLE 8.11

Label the graph in Figure 8.2 showing the specific reactants and products as well as the heat of reaction for the reaction.

SOLUTION

For Figure 8.2 the chemical equation is $NH_4Cl(s) + H_2O(l) \longrightarrow NH_4^+(aq) + Cl^-(aq) + H_2O(l)$. The reactants are $NH_4Cl(s) + H_2O(l)$ and the products are $NH_4^+(aq) + Cl^-(aq) + H_2O(l)$. The graph is labeled and shown here:

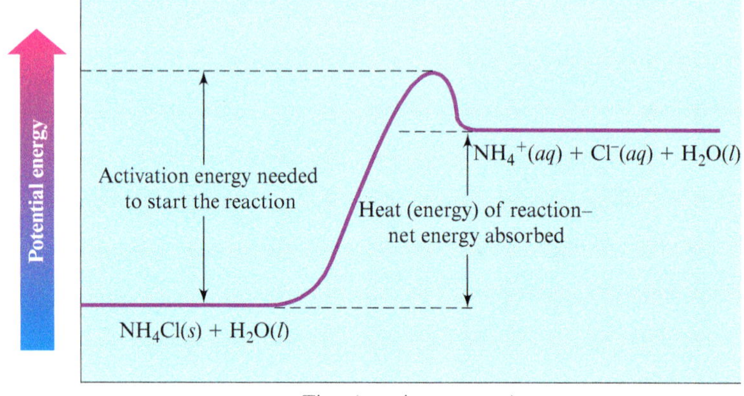

PRACTICE 8.7

For each of the following reactions, draw a reaction graph (like those in Figure 8.2 or 8.3) and label the parts of the graph with the information from the chemical equation.

(a) $2\ Na + Cl_2 \longrightarrow 2\ NaCl + 822\ kJ$

(b) $2\ BrF_3 + 300.8\ kJ/mol \longrightarrow Br_2 + 3\ F_2$

In Figure 8.3, the products are at a lower potential energy than the reactants. Energy (heat) is given off, producing an exothermic reaction. Here, potassium chlorate ($KClO_3$) and sugar are mixed and placed into a pile. A drop of concentrated sulfuric acid is added, creating a spectacular exothermic reaction.

8.6 Climate Change: The Greenhouse Effect

LEARNING OBJECTIVE Discuss the possible causes and results of the greenhouse effect.

Fossil fuels, derived from coal and petroleum, provide the energy we use to power our industries, heat and light our homes and workplaces, and run our cars. As these fuels are burned, they produce carbon dioxide and water, releasing over 50 billion tons of carbon dioxide into our atmosphere each year.

The concentration of CO_2 has been monitored by scientists since 1958. Each week a scientist walks outdoors at a particular location and, carefully holding his breath, opens a basketball-sized glass sphere and collects 5 liters of the atmosphere. At these same remote sites all over the globe (and at dozens of others), instruments sniff the air adding more measurements of the atmosphere to the data. These data are summarized in FIGURE 8.4. Several trends have been discovered in these data.

1. In the Northern Hemisphere the concentration of CO_2 rises and falls 7 ppm over a year, peaking in May and dropping as plants use the CO_2 to produce growth till October, when a second peak occurs as a result of the fallen leaves decaying.
2. The 7-ppm annual variation is superimposed on an average increase in the concentration of CO_2. Our current average is more than 380 ppm, or an increase of more than 65 ppm over the last 50 years. It continues to rise about 2 ppm annually.

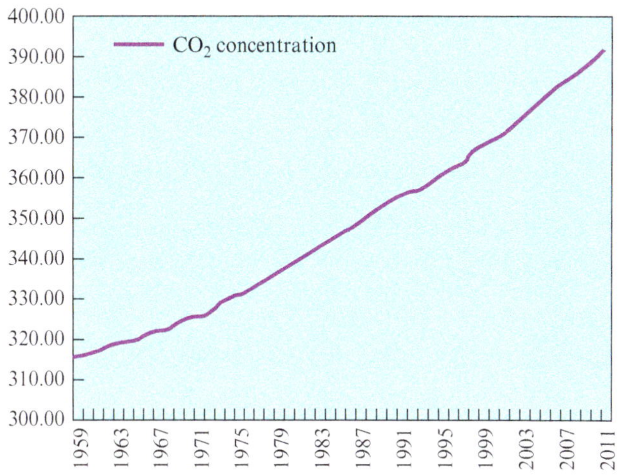

FIGURE 8.4 Concentration of CO_2 in the atmosphere.

Source: Data from the National Oceanic and Atmospheric Administration, http://www.esrl.noaa.gov/gmd/ccgg/trends/#mol-full, February 28, 2012.

Carbon dioxide is a minor component in our atmosphere and is not usually considered to be a pollutant. The concern expressed by scientists arises from the dramatic increase occurring in the Earth's atmosphere. Without the influence of humans in the environment, the exchange of carbon dioxide between plants and animals would be relatively balanced. Our continued use of fossil fuels led to an increase of 7.4% in carbon dioxide between 1900 and 1970 and an additional 3.5% increase during the 1980s. Continued increases were observed in the 1990s.

Besides our growing consumption of fossil fuels, other factors contribute to increased carbon dioxide levels in the atmosphere: Rain forests are being destroyed by cutting and burning to make room for increased population and agricultural needs. Carbon dioxide is added to the atmosphere during the burning, and the loss of trees diminishes the uptake of carbon dioxide by plants.

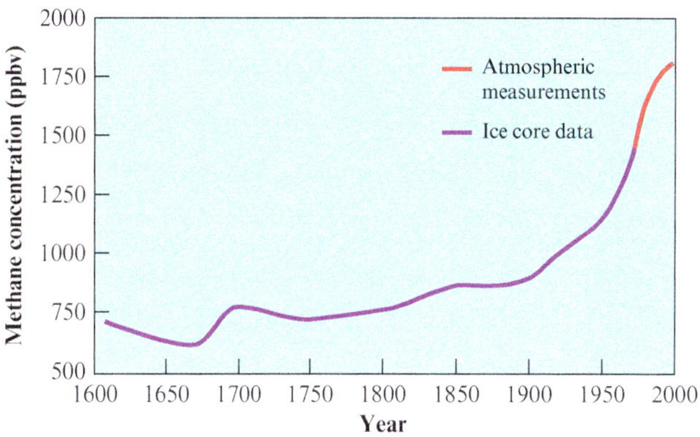

FIGURE 8.5 Concentration of methane in the atmosphere.

About half of all the carbon dioxide released into our atmosphere each year remains there, thus increasing its concentration. The other half is absorbed by plants during photosynthesis or is dissolved in the ocean to form hydrogen carbonates and carbonates.

Methane is another important greenhouse gas. Its concentration in the atmosphere has also increased significantly since the 1850s, as shown in FIGURE 8.5. Methane is produced by cows, termites, agriculture, and anaerobic bacteria. Coal mining and oil wells also release methane into the atmosphere. The greenhouse effect of methane is 20 times more than that of CO_2, but there is less methane in the atmosphere.

Carbon dioxide and other greenhouse gases, such as methane and water, act to warm our atmosphere by trapping heat near the surface of the Earth. Solar radiation strikes the Earth and warms the surface. The warmed surface then reradiates this energy as heat. The greenhouse gases absorb some of this heat energy from the surface, which then warms our atmosphere (FIGURE 8.6). A similar principle is illustrated in a greenhouse, where sunlight comes through the glass yet heat cannot escape.

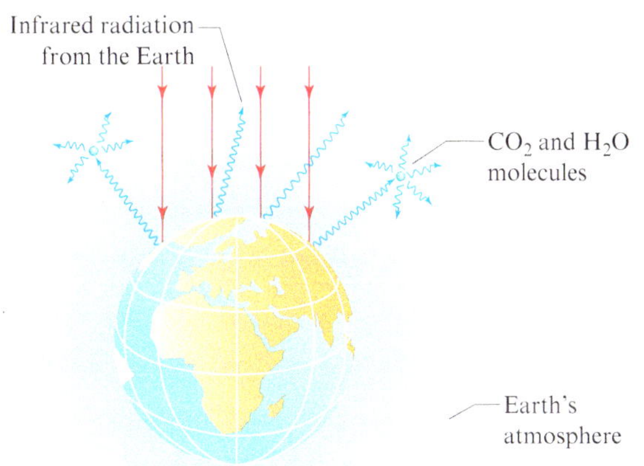

FIGURE 8.6 Visible light from the sun (red arrows) penetrates the atmosphere and strikes the Earth. Part of this light is changed to infrared radiation. Molecules of CO_2 and H_2O in the atmosphere absorb infrared radiation, acting like the glass of a greenhouse by trapping the energy.

CHEMISTRY *IN ACTION* Decreasing Carbon Footprints

In order to slow down climate change, we need to stop the increase of carbon dioxide (CO_2) in the air. We can, of course, continue to try to limit our dependence on fossil fuels. But scientists are also working to capture carbon dioxide emissions. Capturing CO_2 presents two challenges: how to remove the CO_2 from the air and what to do with it once it has been collected.

One way of capturing and storing carbon dioxide is to let nature work. Plants capture 505 million tons of carbon dioxide in North America through photosynthesis by storing it as carbon in woody plants and trees or as organic material in the soil and wetlands or sediments in rivers and lakes. The difficulty with this method is that as the decay process continues in an oxygen-rich environment, the carbon is returned to the atmosphere as CO_2. Ning Zing, from the University of Maryland, suggests that we could prevent the release of this CO_2 from forests by burying wood in an oxygen-poor environment. The carbon would then remain in the ground and not in the atmosphere. He notes that it is possible to accomplish this process at a cost of $14 per ton of wood, but that would mean collecting wood and burying it in trenches, which is not an environment-friendly method of capture and storage. Other scientists have suggested seeding the oceans to increase the algae and phytoplankton. This would also pull CO_2 from the atmosphere, but many scientists think that this method would not have much effect on climate since the oceans could only capture a small fraction of the CO_2 emitted by humans.

Coal and petroleum each account for about 40% of global CO_2 emissions. Coal is the resource that is the greatest threat to our climate since it produces more CO_2 per unit of energy than any other fossil fuel. Each year coal-fired power plants produce 8 billion tons of CO_2 (containing 2.2 billion tons of C). Future coal plants (planned in the next 25 years) will generate 660 billion tons of CO_2. This is 25% more than all the CO_2 that humans have produced by burning coal since 1751. How can we reduce these numbers? The current goal is to capture 90% of the CO_2 emissions from a power plant but only increase the cost of electricity by 20%. Scientists are developing materials that act as CO_2 sponges. These materials, called zeolites, are very porous and have large surface areas where CO_2 molecules can attach themselves.

Steven Raniszewski/Design Pics/Newscom

Zeolite (a basaltic volcanic rock) could be used as a storage place for carbon dioxide from power plants.

The question here is that once the CO_2 is captured, how can it be stored? Many scientists propose to lock it underground or in deep ocean. Since CO_2 is a liquid at high pressure (as in ocean depths below 500 meters) and does not mix very well with water, it could be stored as large pool on the ocean floor. This storage method raises a number of environmental concerns. In some places there are saline aquifers underground. The water from these aquifers is not drinkable, and so some of them might be a suitable storage place for CO_2. Another possibility is using volcanic rock as a storage site. Tests indicate that liquid CO_2 will react with basalt to produce minerals such as calcium carbonate. These deposits contain many holes and cracks where the CO_2 might be injected. It is thought that in the United States these saline aquifers and geologic formations could store 150 years' worth of power plant emissions.

We are now looking seriously at ways to slow down the rise in atmospheric CO_2 and to decrease our carbon footprints without losing our fossil fuels as one of our energy sources.

The air in the greenhouse warms, producing a climate considerably different than that in nature.

The long-term effects of climate change are still a matter of debate. One consideration is whether the polar ice caps will melt; this would cause a rise in sea level and lead to major flooding on the coasts of our continents. Further effects could include shifts in rainfall patterns, producing droughts and extreme seasonal change in such major agricultural regions as California. To reverse these trends will require major efforts in the following areas:

- The development of new energy sources to cut our dependence on fossil fuels
- An end to deforestation worldwide
- Intense efforts to improve conservation

On an individual basis each of us plays a significant role. Recycling, switching to more fuel-efficient cars, and using energy-efficient appliances, heaters, and air conditioners all would result in decreased energy consumption.

CHAPTER 8 REVIEW

8.1 The Chemical Equation

KEY TERMS

reactants
products
chemical equation
law of conservation of mass

- A chemical equation is shorthand for expressing a chemical change or reaction.
- In a chemical reaction atoms are neither created nor destroyed.
- All atoms in the reactants must be present in the products.

8.2 Writing and Balancing Chemical Equations

KEY TERM

balanced equation

- To balance a chemical equation:
  - Identify the reaction.
  - Write the unbalanced (skeleton) equation.
  - Balance the equation:
    - Count the number of atoms of each element on each side and determine which need to be balanced.
    - Balance each element (one at a time) by placing whole numbers (coefficients) in front of the formulas containing the unbalanced element:
      - Begin with metals, then nonmetals, and then H and O.
    - Check the other elements to see if they have become unbalanced in the process of balancing the chosen element. If so, rebalance as needed.
    - Do a final check to make sure all elements are balanced.
- The following information can be found in a chemical equation:
  - Identity of reactants and products
  - Formulas for reactants and products
  - Number of formula units for reactants and products
  - Number of atoms of each element in the reaction
  - Number of moles of each substance

8.3 Why Do Chemical Reactions Occur?

- Formation of a solid
- Formation of a gas
- Formation of water
- Transfer of electrons

8.4 Types of Chemical Equations

KEY TERMS

synthesis reaction
combustion reaction
decomposition reaction
single-displacement reaction
double-displacement reaction

- Synthesis reactions — $A + B \longrightarrow AB$
- Decomposition reactions — $AB \longrightarrow A + B$
- Single-displacement reactions:
  - In which A is a metal — $A + BC \longrightarrow B + AC$
  - In which A is a halogen — $A + BC \longrightarrow C + BA$
- Double-displacement reactions — $AB + CD \longrightarrow AD + CB$
- Evidence for a chemical reaction:
  - Evolution of heat
  - Formation of an insoluble precipitate
  - Production of a gas

KEY TERMS

exothermic reactions
endothermic reactions
heat of reaction
hydrocarbons
activation energy

8.5 Heat in Chemical Reactions

- Exothermic reactions release heat.

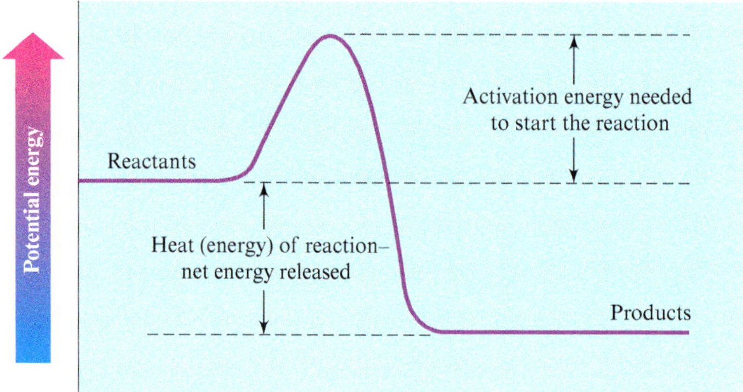

- Endothermic reactions absorb heat.

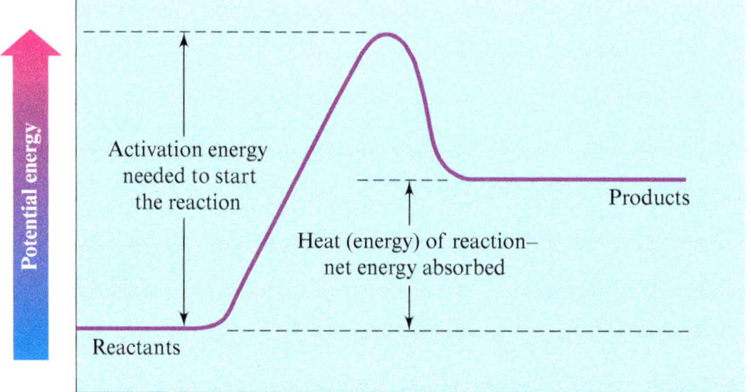

- The amount of heat released or absorbed in a chemical reaction is called the heat of reaction:
  - It can be written as a reactant or product in the chemical equation.
  - Units are joules (J) or kilojoules (kJ).
- The major source of energy for modern technology is hydrocarbon combustion.
- To initiate a chemical reaction, activation energy is required:
  - In exothermic reactions, this energy is returned and more is released, which allows the reaction to continue on its own.
  - In endothermic reactions, energy must be added to start and continue to be added to sustain the reaction.

8.6 Climate Change: The Greenhouse Effect

- Carbon dioxide levels are now increasing on our planet every year.
- Carbon dioxide and other greenhouse gases warm the atmosphere by trapping heat near the surface of the Earth.

REVIEW QUESTIONS

1. What is represented by the numbers (coefficients) that are placed in front of the formulas in a balanced equation?
2. What is meant by the physical state of a substance? What symbols are used to represent these physical states, and what does each symbol mean?
3. What is the purpose of balancing chemical equations?
4. In a balanced chemical equation:
   (a) Are atoms conserved?
   (b) Are molecules conserved?
   (c) Are moles conserved?
   Explain your answers briefly.
5. In Section 8.2, there are small charts in color along the margins. What information is represented in those charts?
6. Why is it incorrect to balance chemical reactions by changing the subscripts on the reactants and products?
7. What is a combustion reaction? How can you identify whether a reaction is a combustion reaction?
8. What information does the activity series shown in Table 8.2 give?
9. What major types of chemical reactions are studied in this chapter?
10. What three observations indicate a chemical reaction has taken place?
11. Explain how endothermic reactions differ from exothermic reactions.
12. Why does an exothermic reaction require heat to occur?
13. Name several greenhouse gases.
14. What factors are responsible for the annual rise and fall in carbon dioxide levels?

PAIRED EXERCISES

Most of the exercises in this chapter are available for assignment via the online homework management program, WileyPLUS (www.wileyplus.com). All exercises with blue numbers have answers in Appendix 5.

1. Classify the following as an endothermic or exothermic reaction:
   (a) freezing water
   (b) the reaction inside an ice pack
   (c) burning wood
   (d) combustion of Mg in dry ice
   (e) melting ice

2. Classify the following as an endothermic or exothermic reaction:
   (a) making popcorn in a microwave oven
   (b) a burning match
   (c) boiling water
   (d) burning rocket fuel
   (e) the reaction inside a heat pack

3. Balance each of the following equations. Classify each reaction as synthesis, decomposition, single-displacement, or double-displacement.
   (a) $H_2 + O_2 \longrightarrow H_2O$
   (b) $N_2H_4(l) \longrightarrow NH_3(g) + N_2(g)$
   (c) $H_2SO_4 + NaOH \longrightarrow H_2O + Na_2SO_4$
   (d) $Al_2(CO_3)_3 \xrightarrow{\Delta} Al_2O_3 + CO_2$
   (e) $NH_4I + Cl_2 \longrightarrow NH_4Cl + I_2$

4. Balance each of the following equations. Classify each reaction as synthesis, decomposition, single-displacement, or double-displacement.
   (a) $H_2 + Br_2 \longrightarrow HBr$
   (b) $BaO_2(s) + H_2SO_4(aq) \longrightarrow BaSO_4(s) + H_2O_2(aq)$
   (c) $Ba(ClO_3)_2 \xrightarrow{\Delta} BaCl_2 + O_2$
   (d) $CrCl_3 + AgNO_3 \longrightarrow Cr(NO_3)_3 + AgCl$
   (e) $H_2O_2 \longrightarrow H_2O + O_2$

5. Balance each of the following reactions and identify them as synthesis, decomposition, single-displacement, or double-displacement reactions.
   (a) $Na_2CrO_4 + CaCl_2 \longrightarrow CaCrO_4 + NaCl$
   (b) $Al + CuCl_2 \longrightarrow Cu + AlCl_3$
   (c) 

   (d) 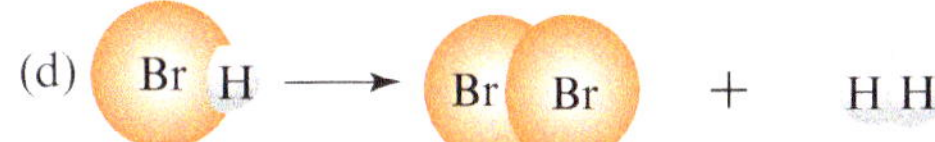

   (e) 

6. Balance each of the following reactions and identify them as synthesis, decomposition, single-displacement, or double-displacement reactions.
   (a) $FeCl_3 + Li_2CO_3 \longrightarrow LiCl + Fe_2(CO_3)_3$
   (b) $CdCl_2 + TiCl_3 \longrightarrow Ti + CdCl_2$
   (c)

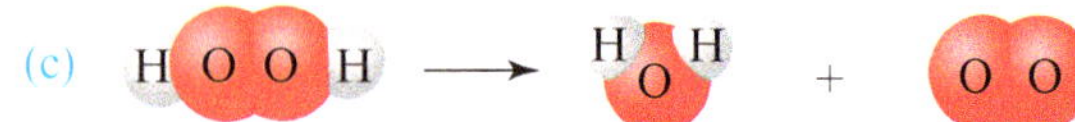

   (d)
   (e)

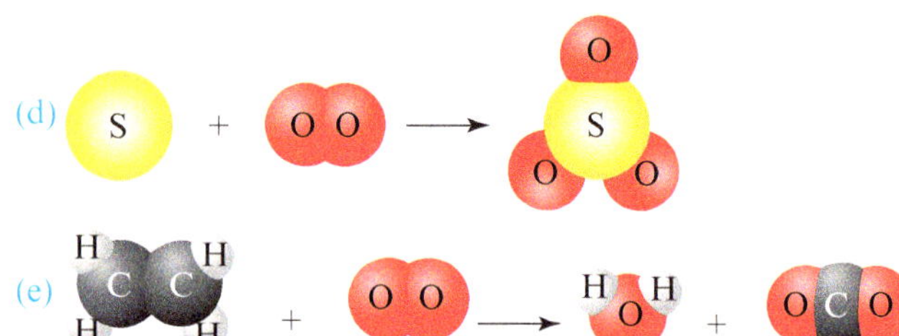

7. Complete and balance the reaction for the complete combustion of the following fuels:
   (a) propane, C_3H_8
   (b) decane, $C_{10}H_{22}$
   (c) hexane, C_6H_{14}
   (d) butyl alcohol, C_4H_9OH

8. Complete and balance the reaction for the complete combustion of the following fuels:
   (a) butane, C_4H_{10}
   (b) paraffin wax, $C_{34}H_{64}$
   (c) acetylene, C_2H_2
   (d) methanol, CH_3OH

9. Balance the following equations:
   (a) $MnO_2 + CO \longrightarrow Mn_2O_3 + CO_2$
   (b) $Cu_2O(s) + C(s) \longrightarrow Cu(s) + CO(g)$
   (c) $C_3H_5(NO_3)_3 \longrightarrow CO_2 + H_2O + N_2 + O_2$
   (d) $FeS + O_2 \longrightarrow Fe_2O_3 + SO_2$
   (e) $Cu(NO_3)_2 \longrightarrow CuO + NO_2 + O_2$
   (f) $NO_2 + H_2O \longrightarrow HNO_3 + NO$
   (g) $Fe(s) + S(l) \longrightarrow Fe_2S_3(s)$
   (h) $HCN + O_2 \longrightarrow N_2 + CO_2 + H_2O$
   (i) $B_5H_9 + O_2 \longrightarrow B_2O_3 + H_2O$

10. Balance the following equations:
   (a) $SO_2 + O_2 \longrightarrow SO_3$
   (b) $Li_2O(s) + H_2O(l) \longrightarrow LiOH(aq)$
   (c) $Na + H_2O \longrightarrow NaOH + H_2$
   (d) $AgNO_3 + Ni \longrightarrow Ni(NO_3)_2 + Ag$
   (e) $Bi_2S_3 + HCl \longrightarrow BiCl_3 + H_2S$
   (f) $PbO_2 \xrightarrow{\Delta} PbO + O_2$
   (g) $Hg_2(C_2H_3O_2)_2(aq) + KCl(aq) \longrightarrow Hg_2Cl_2(s) + KC_2H_3O_2(aq)$
   (h) $KI + Br_2 \longrightarrow KBr + I_2$
   (i) $K_3PO_4 + BaCl_2 \longrightarrow KCl + Ba_3(PO_4)_2$

11. Change these word equations into formula equations and balance them. Be sure to use the proper symbols to indicate the state of each substance, as given.
   (a) Magnesium metal is placed into hydrobromic acid solution, forming hydrogen gas and aqueous magnesium bromide.
   (b) When heated, solid calcium chlorate decomposes into calcium chloride solid, releasing oxygen gas.
   (c) Lithium metal reacts with oxygen gas to form solid lithium oxide.
   (d) Solutions of barium bromate and sodium phosphate combine to form solid barium phosphate and aqueous sodium bromate.
   (e) Solutions of acetic acid and sodium carbonate are mixed together, forming a solution of sodium acetate, along with carbon dioxide gas and liquid water.
   (f) Solutions of silver nitrate and aluminum iodide are mixed together, forming solid silver iodide and aqueous aluminum nitrate.

12. Change these word equations into formula equations and balance them. Be sure to use the proper symbols to indicate the state of each substance, as given.
   (a) Upon heating, solid magnesium carbonate decomposes into solid magnesium oxide and carbon dioxide gas.
   (b) Solid calcium hydroxide reacts with aqueous chloric acid to form a solution of calcium chlorate along with liquid water.
   (c) Solutions of iron(III) sulfate and sodium hydroxide are mixed together, forming solid iron(III) hydroxide and a solution of sodium sulfate.
   (d) Zinc metal is placed into a solution of acetic acid, producing hydrogen gas and aqueous zinc acetate.
   (e) Gaseous sulfur trioxide reacts with liquid water to form a solution of sulfuric acid.
   (f) Solutions of sodium carbonate and cobalt(II) chloride react to form solid cobalt(II) carbonate and a solution of sodium chloride.

13. For each of the following reactions, predict the products, converting each to a balanced formula equation:
 (a) Aqueous solutions of sulfuric acid and sodium hydroxide are mixed together. (Heat is released during the reaction.)
 (b) Aqueous solutions of lead(II) nitrate and potassium bromide are mixed together. (The solution turns cloudy white during the reaction.)
 (c) Aqueous solutions of ammonium chloride and silver nitrate are mixed together. (The solution turns cloudy white during the reaction.)
 (d) Solid calcium carbonate is mixed with acetic acid. (Bubbles of gas are formed during the reaction.)

14. For each of the following reactions, predict the products, converting each to a balanced formula equation:
 (a) Aqueous solutions of copper(II) sulfate and potassium hydroxide are mixed together. (The solution turns cloudy and light blue during the reaction.)
 (b) Aqueous solutions of phosphoric acid and sodium hydroxide are mixed together. (Heat is produced during the reaction.)
 (c) Solid sodium bicarbonate is mixed with phosphoric acid. (Bubbles of gas are formed during the reaction.)
 (d) Aqueous solutions of aluminum chloride and lead(II) nitrate are mixed together. (The solution turns cloudy white during the reaction.)

15. Use the activity series to predict which of the following reactions will occur. Complete and balance the equations. If no reactions occurs, write "no reaction" as the product.
 (a) $Ca(s) + H_2O(l) \longrightarrow$
 (b) $Br_2(l) + KI(aq) \longrightarrow$
 (c) $Cu(s) + HCl(aq) \longrightarrow$
 (d) $Al(s) + H_2SO_4(aq) \longrightarrow$

16. Use the activity series to predict which of the following reactions will occur. Complete and balance the equations. If no reaction occurs, write "no reaction" as the product.
 (a) $Cu(s) + NiCl_2(aq) \longrightarrow$
 (b) $Rb(s) + H_2O(l) \longrightarrow$
 (c) $I_2(s) + CaCl_2(aq) \longrightarrow$
 (d) $Mg(s) + Al(NO_3)_3(aq) \longrightarrow$

17. A sample of manganese is placed into a solution of lead(II) chloride and the manganese dissolves. A precipitate of elemental lead is observed to form. Another sample of manganese is placed into a solution of barium chloride and no reaction occurs.
 (a) Write the equation for the reaction that takes place between lead(II) chloride and manganese. (Manganese will form an ion with a +2 charge.)
 (b) Which element is more active, lead or manganese?
 (c) Which element is more active, manganese or barium?
 (d) Would you expect a reaction to occur if you placed a sample of lead metal into a solution of barium chloride? Why or why not? If you expect a reaction to occur, write and balance the equation for the reaction and state what you would expect to see in the reaction mixture.

18. A sample of aluminum is placed into a solution of nickel(II) chloride and the aluminum dissolves. A precipitate of elemental nickel is observed to form. A drop of liquid mercury is placed into a solution of nickel(II) chloride and no reaction occurs.
 (a) Write the equation for the reaction that takes place between nickel(II) chloride and aluminum.
 (b) Which element is more active, nickel or aluminum?
 (c) Which element is more active, mercury or nickel?
 (d) Would you expect a reaction to occur if you placed a sample of nickel metal into a solution of mercury(II) chloride? Why or why not? If you expect a reaction to occur, write and balance the equation for the reaction and state what you would expect to see in the reaction mixture.

19. Complete and balance the equations for each of the following reactions. All yield products.
 (a) $Sr(s) + H_2O(l) \longrightarrow$
 (b) $BaCl_2(aq) + AgNO_3(aq) \longrightarrow$
 (c) $Mg(s) + ZnBr_2(aq) \longrightarrow$
 (d) $K(s) + Cl_2(g) \longrightarrow$

20. Complete and balance the equations for each of the following reactions. All yield products.
 (a) $Li_2O(s) + H_2O(l) \longrightarrow$
 (b) $Na_2SO_4(aq) + Pb(NO_3)_2(aq) \longrightarrow$
 (c) $Zn(s) + CuSO_4(aq) \longrightarrow$
 (d) $Al(s) + O_2(g) \longrightarrow$

21. Complete and balance the equations for these reactions. All reactions yield products.
 (a) $Ba + O_2 \longrightarrow$
 (b) $NaHCO_3 \xrightarrow{\Delta} Na_2CO_3 +$
 (c) $Ni + CuSO_4 \longrightarrow$
 (d) $MgO + HCl \longrightarrow$
 (e) $H_3PO_4 + KOH \longrightarrow$

22. Complete and balance the equations for these reactions. All reactions yield products.
 (a) $C + O_2 \longrightarrow$
 (b) $Al(ClO_3)_3 \xrightarrow{\Delta} O_2 +$
 (c) $CuBr_2 + Cl_2 \longrightarrow$
 (d) $SbCl_3 + (NH_4)_2S \longrightarrow$
 (e) $NaNO_3 \longrightarrow NaNO_2 +$

23. Interpret these chemical reactions in terms of the number of moles of each reactant and product:
 (a) $MgBr_2 + 2\,AgNO_3 \longrightarrow Mg(NO_3)_2 + 2\,AgBr$
 (b) $N_2 + 3\,H_2 \longrightarrow 2\,NH_3$
 (c) $2\,C_3H_7OH + 9\,O_2 \longrightarrow 6\,CO_2 + 8\,H_2O$

24. Interpret these equations in terms of the relative number of moles of each substance involved and indicate whether the reaction is exothermic or endothermic:
 (a) $2\,Na + Cl_2 \longrightarrow 2\,NaCl + 822\text{ kJ}$
 (b) $PCl_5 + 92.9\text{ kJ} \longrightarrow PCl_3 + Cl_2$
 (c) $S(s) + 2\,CO(g) \longrightarrow SO_2(g) + 2\,C(s) + 76\text{ kJ}$

25. Write balanced equations for each of these reactions, including the heat term:
 (a) Solid mercury(II) oxide decomposes into liquid mercury and oxygen gas upon the absorption of 90.8 kJ for each mole of mercury(II) oxide decomposed.
 (b) Hydrogen gas reacts with oxygen gas to form liquid water. The reaction produces 285.8 kJ of heat for each mole of water formed.

26. Write balanced equations for each of these reactions, including the heat term:
 (a) Calcium metal reacts with water to produce calcium hydroxide and hydrogen gas, releasing 635.1 kJ of heat for every mole of calcium that reacts.
 (b) Bromine trifluoride decomposes into bromine and fluorine upon the absorption of 300.8 kJ of heat for each mole of bromine trifluoride decomposed.

27. Determine what reactants would form the given products. Give a balanced equation for each and classify the reaction as a synthesis, decomposition, single-displacement, or double-displacement reaction.
(a) $AgCl(s) + O_2(g)$ (c) $ZnCl_2(s)$
(b) $H_2(g) + FeSO_4(aq)$ (d) $KBr(aq) + H_2O(l)$

28. Determine what reactants would form the given products. Give a balanced equation for each and classify the reaction as a synthesis, decomposition, single-displacement, or double-displacement reaction.
(a) $Pb(s) + Ni(NO_3)_2(aq)$ (c) $Hg(l) + O_2(g)$
(b) $Mg(OH)_2(s)$ (d) $PbCO_3(s) + NH_4Cl(aq)$

ADDITIONAL EXERCISES

29. An aqueous mixture of the nitrate salts of sodium, calcium, and silver is mixed with ammonium salts of chloride, sulfate, and carbonate. Write balanced equations for any combination that will form a precipitate.

30. Balance this equation, using the smallest possible whole numbers. Then determine how many atoms of oxygen appear on each side of the equation:

$$P_4O_{10} + HClO_4 \longrightarrow Cl_2O_7 + H_3PO_4$$

31. Suppose that in a balanced equation the term 5 $Ni_3(PO_4)_2$ appears.
(a) How many atoms of nickel are represented?
(b) How many atoms of phosphorus are represented?
(c) How many atoms of oxygen are represented?
(d) How many atoms of all kinds are represented?

32. In a lab experiment, a student places a few stones, composed mainly of calcium carbonate, into a beaker and then adds some acetic acid. The student observes that effervescence occurs immediately. Explain this observation by writing a balanced formula equation.

33. Make a drawing to show the combustion reaction of one molecule of methane, CH_4.

34. A variety of iron-containing ores can be converted into iron and steel in a blast furnace. The reactions that take place are outlined below. Write balanced chemical equations for each reaction.
(a) Pure carbon is reacted with oxygen gas to form carbon dioxide.
(b) Carbon dioxide is reacted with more carbon to form monoxide.
(c) Carbon monoxide reacts with the iron ore to form elemental iron and carbon dioxide. There are two main types of ore that may be used: hematite (Fe_2O_3) and magnetite (Fe_3O_4).

35. The box at the left represents the reactant atoms in blue (B) and molecules in green (G_2). After reacting, these reactants form the compound pictured in the box at the right. Write a balanced chemical reaction for this process.

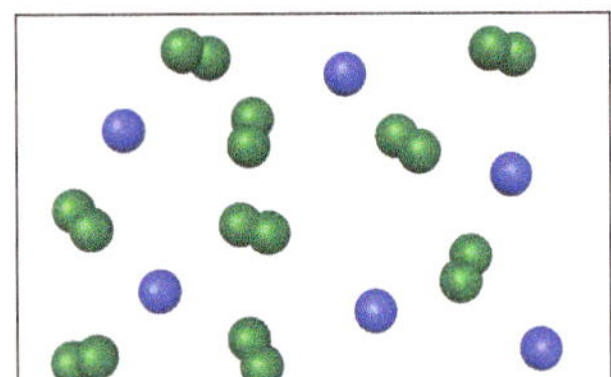

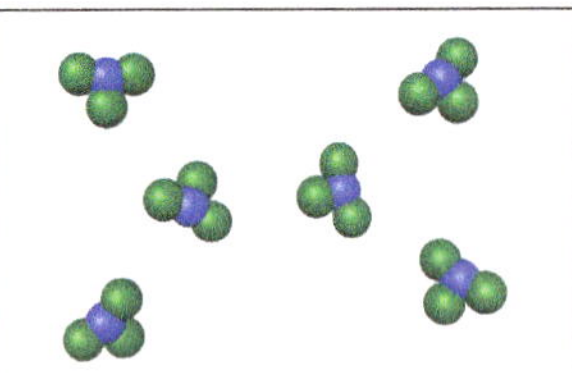

36. Students in a lab exercise were asked to choose among a sample of different metals to react with a solution of nickel(II) chloride. Which of the following metals should they choose? Write balanced equations for all that will react and explain why each will or will not react.
(a) copper (d) lead
(b) zinc (e) calcium
(c) aluminum

37. A student does an experiment to determine where titanium metal should be placed on the activity series chart. He places newly cleaned pieces of titanium into solutions of nickel(II) nitrate, lead(II) nitrate, and magnesium nitrate. He finds that the titanium reacts with the nickel(II) nitrate and lead(II) nitrate solutions, but not with the magnesium nitrate solution. From this information, place titanium in the activity series in a position relative to these ions.

38. Predict the products of each of the following combination reactions in words. Then change each equation into a balanced formula equation.
(a) Cesium reacts with oxygen.
(b) Aluminum metal reacts with sulfur.
(c) Sulfur trioxide reacts with water.
(d) Sodium oxide reacts with water.

39. Automobile designers are currently working on cars that will run on hydrogen. Hydrogen-fueled cars have the advantage that they produce only water as a waste product. In these cars, hydrogen gas reacts with oxygen gas to form gaseous water.
(a) Write a balanced chemical equation for this reaction.
(b) There are many ways to generate hydrogen to fuel these cars. The most commonly used method of making hydrogen gas is to react methane gas (CH_4) with gaseous water at 1000°C to form gaseous carbon monoxide and hydrogen gas. The carbon monoxide gas can then be reacted with more gaseous water to produce more hydrogen gas and carbon dioxide gas. Write balanced chemical equations for these two reactions.
(c) One of the products of burning fossil fuels to provide energy is carbon dioxide. This carbon dioxide may be increasing the temperature of the planet. Methane can also be used to provide energy by reacting it with oxygen gas to form carbon dioxide and water. Write the balanced chemical equation for this reaction.
(d) Compare the number of molecules of carbon dioxide produced out of one molecule of methane by burning it directly to produce energy and by converting it to hydrogen gas and using this hydrogen gas to produce energy. Is there an advantage to using hydrogen gas as the fuel source? Explain.

40. Many paint pigments are composed of ionic compounds. For each of the pigments listed below, write a complete balanced equation to synthesize the pigment using the reagents listed in parentheses. Write the correct IUPAC name of each pigment.
(a) chrome yellow, $PbCrO_4$ ($Pb(NO_3)_2(aq) + K_2CrO_4(aq)$)
(b) cadmium yellow, CdS ($CdCl_2(aq) + Li_2S(aq)$)
(c) white lead, $PbCO_3$ ($Pb(C_2H_3O_2)_2(aq) + H_2CO_3(aq)$)
(d) vermillion, HgS ($Hg(NO_3)_2(aq) + Na_2S(aq)$)
(e) titanium black, Ti_2O_3 ($Ti(s) + O_2(g)$)
(f) iron oxide red, Fe_2O_3 ($Fe(s) + O_2(g)$)

41. Predict the products of each of the following decomposition reactions in words. Then change each equation into a balanced formula equation.
(a) Zinc oxide decomposes upon heating strongly.
(b) Tin(IV) oxide decomposes into two products when heated.
(c) Sodium carbonate decomposes when heated.
(d) Magnesium chlorate decomposes when heated.

42. Predict the products of each of the following single-displacement reactions in words. Then change each equation into a balanced formula equation. All reactions occur in aqueous solution.
   (a) Magnesium metal reacts with hydrochloric acid.
   (b) Sodium bromide reacts with chlorine.
   (c) Zinc metal reacts with iron(III) nitrate.
   (d) Aluminum reacts with copper(II) sulfate.
43. Predict the products of each of the following double-displacement reactions in words. Then change each equation into a balanced formula equation. All reactants are in aqueous solution.
   (a) Ammonium phosphate reacts with barium nitrate.
   (b) Sodium sulfide reacts with lead(II) acetate.
   (c) Copper(II) sulfate reacts with calcium chlorate.
   (d) Barium hydroxide reacts with oxalic acid.
   (e) Phosphoric acid reacts with potassium hydroxide.
   (f) Sulfuric acid reacts with sodium carbonate.
44. Predict which of the following double-displacement reactions will occur. For those that do, write a balanced formula equation. For any reaction that does not occur, write "no reaction" as the product. All reactants are in aqueous solution.
   (a) potassium sulfate and barium acetate
   (b) sulfuric acid and lithium hydroxide
   (c) ammonium phosphate and sodium bromide
   (d) calcium iodide and silver nitrate
   (e) nitric acid and strontium hydroxide
   (f) cesium nitrate and calcium hydroxide
45. Write balanced equations for each of the following combustion reactions:

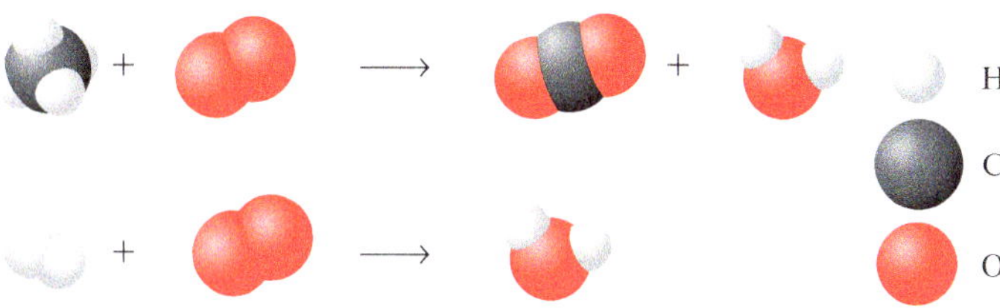

46. Write balanced equations for the complete combustion of the following hydrocarbons:
   (a) CH_4 (b) C_3H_6 (c) $C_6H_5CH_3$
47. List the factors that contribute to an increase in carbon dioxide in our atmosphere.
48. List three gases considered to be greenhouse gases. Explain why they are given this name.
49. How can the effects of climate change be reduced?
50. What happens to carbon dioxide released into our atmosphere?
51. In the Northern Hemisphere, the concentration of CO_2 peaks twice a year. When do these peaks occur, and what causes them?

CHALLENGE EXERCISE

52. You are given a solution of the following cations in water: Ag^+, Co^{2+}, Ba^{2+}, Zn^{2+}, and Sn^{2+}. You want to separate the ions out *one at a time* by precipitation using the following reagents: NaF, NaI, Na_2SO_4, and NaCl.
   (a) In what order would you add the reagents to ensure that *only one cation* is precipitating out at a time? Use the solubility table in Appendix 4 to help you. (*Note:* A precipitate that is slightly soluble in water is still considered to precipitate out of the solution.)
   (b) Why are the anionic reagents listed above all sodium salts?
53. Compound A decomposes at room temperature in an exothermic reaction, while compound B requires heating before it will decompose in an endothermic reaction.
   (a) Draw reaction profiles (potential energy vs. reaction progress) for both reactions.
   (b) The decomposition of 0.500 mol of $NaHCO_3$ to form sodium carbonate, water, and carbon dioxide requires 85.5 kJ of heat. How many grams of water could be collected, and how many kJ of heat are absorbed when 24.0 g of CO_2 are produced?
   (c) Could $NaHCO_3$ be compound A or compound B? Explain your reasoning.
54. When solutions containing mercury(II) nitrate and potassium iodide were mixed, a red-orange precipitate resulted.
   (a) What type of reaction occurred?
   (b) What are the name and formula for the solid product?

ANSWERS TO PRACTICE EXERCISES

8.1 To show that a substance is a solid, write (*s*) after the formula of the substance. When the substance is a gas, write (*g*) after the formula, and when the substance is in an aqueous solution write (*aq*) after the formula.

8.2 $4\ Al + 3\ O_2 \longrightarrow 2\ Al_2O_3$

8.3 $3\ Mg(OH)_2 + 2\ H_3PO_4 \longrightarrow Mg_3(PO_4)_2 + 6\ H_2O$

8.4 (a) hydrogen gas + chlorine gas ⟶ hydrogen chloride gas

(b) $H_2(g) + Cl_2(g) \longrightarrow 2\ HCl(g)$

(c)

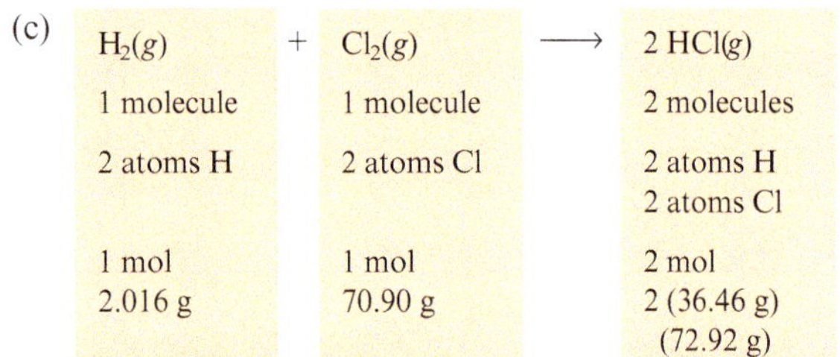

$H_2(g)$	+	$Cl_2(g)$	⟶	$2\ HCl(g)$
1 molecule		1 molecule		2 molecules
2 atoms H		2 atoms Cl		2 atoms H 2 atoms Cl
1 mol 2.016 g		1 mol 70.90 g		2 mol 2 (36.46 g) (72.92 g)

(d) $2\ mol\ H_2 + 2\ mol\ Cl_2 \longrightarrow 4\ mol\ HCl(145.8\ g)$

8.5 (a) $Fe + MgCl_2 \longrightarrow$ no reaction

(b) $Zn(s) + Pb(NO_3)_2(aq) \longrightarrow Pb(s) + Zn(NO_3)_2(aq)$

8.6 (a) $2\ K_3PO_4(aq) + 3\ BaCl_2(aq) \longrightarrow Ba_3(PO_4)_2(s) + 6\ KCl(aq)$

(b) $2\ HCl(aq) + NiCO_3(aq) \longrightarrow NiCl_2(aq) + H_2O(l) + CO_2(g)$

(c) $NH_4Cl(aq) + NaNO_3(aq) \longrightarrow$ no reaction

8.7 (a)

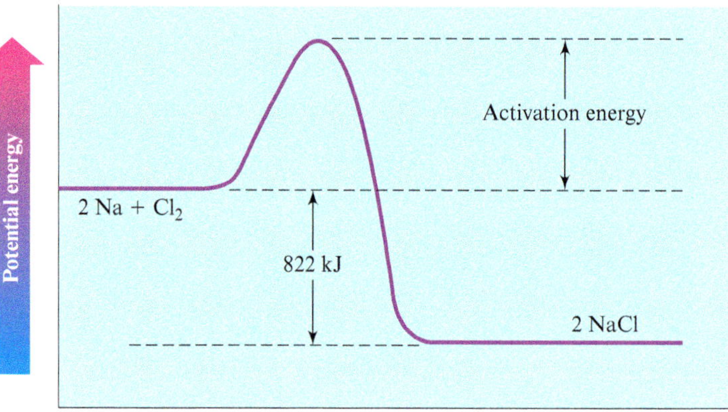

(b)

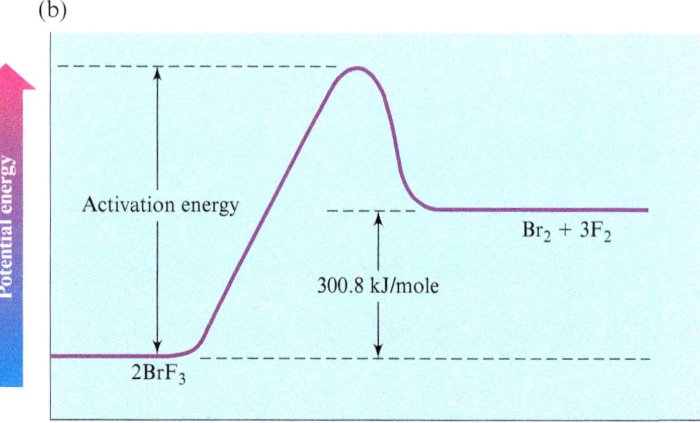

9 Calculations from Chemical Equations

CHAPTER OUTLINE

9.1 Introduction to Stoichiometry
9.2 Mole–Mole Calculations
9.3 Mole–Mass Calculations
9.4 Mass–Mass Calculations
9.5 Limiting Reactant and Yield Calculations

The old adage "waste not, want not" is appropriate in our daily life and in the laboratory. Determining correct amounts comes into play in almost all professions. A seamstress determines the amount of material, lining, and trim necessary to produce a gown for her client by relying on a pattern or her own experience to guide the selection. A carpet layer determines the correct amount of carpet and padding necessary to recarpet a customer's house by calculating the floor area. The IRS determines the correct deduction for federal income taxes from your paycheck based on your expected annual income.

In chemistry, calculations from chemical equations are important too. These same calculations and amounts are used frequently in medicine and pharmacy. Accurate calculations of dose and accurate measurement of chemicals are required in order for a pharmacist to dispense the correct dosage of medicine to her patients.

9.1 Introduction to Stoichiometry

LEARNING OBJECTIVE Define stoichiometry and describe the strategy required to solve problems based on chemical equations.

KEY TERMS
stoichiometry
molar mass
mole ratio

We often need to calculate the amount of a substance that is either produced from, or needed to react with, a given quantity of another substance. The area of chemistry that deals with quantitative relationships among reactants and products is known as **stoichiometry** (*stoy-key-ah-meh-tree*). Solving problems in stoichiometry requires the use of *moles* in the form of *mole ratios.*

The chemist also finds it necessary to calculate amounts of products or reactants by using a balanced chemical equation. With these calculations, the chemist can control the amount of product by scaling the reaction up or down to fit the needs of the laboratory and can thereby minimize waste or excess materials formed during the reaction.

A Short Review

Molar mass The sum of the atomic masses of all the atoms in an element or compound is called **molar mass**. The term *molar mass* also applies to the mass of a mole of any formula unit—atoms, molecules, or ions; it is the atomic mass of an atom or the sum of the atomic masses in a molecule or an ion (in grams).

Relationship between molecule and mole. A molecule is the smallest unit of a molecular substance (e.g., Br_2), and a mole is Avogadro's number (6.022×10^{23}) of molecules of that substance. A mole of bromine (Br_2) has the same number of molecules as a mole of carbon dioxide, a mole of water, or a mole of any other molecular substance. When we relate molecules to molar mass, 1 molar mass is equivalent to 1 mol, or 6.022×10^{23} molecules.

The term *mole* also refers to any chemical species. It represents a quantity (6.022×10^{23} particles) and may be applied to atoms, ions, electrons, and formula units of nonmolecular substances. In other words,

A mole of water, salt, and any gas all have the same number of particles (6.022×10^{23}).

Richard Megna/Fundamental Photographs

$$1 \text{ mole} = \begin{cases} 6.022 \times 10^{23} \text{ molecules} \\ 6.022 \times 10^{23} \text{ formula units} \\ 6.022 \times 10^{23} \text{ atoms} \\ 6.022 \times 10^{23} \text{ ions} \end{cases}$$

Other useful mole relationships are

$$\text{molar mass} = \frac{\text{grams of a substance}}{\text{number of moles of the substance}}$$

$$\text{molar mass} = \frac{\text{grams of a monatomic element}}{\text{number of moles of the element}}$$

$$\text{number of moles} = \frac{\text{number of molecules}}{6.022 \times 10^{23} \text{ molecules/mole}}$$

Balanced equations. When using chemical equations for calculations of mole–mass–volume relationships between reactants and products, the equations must be balanced. *Remember:* The number in front of a formula in a balanced chemical equation represents the number of moles of that substance in the chemical reaction.

A **mole ratio** is a ratio between the number of moles of any two species involved in a chemical reaction. For example, in the reaction

$$\underset{2\text{ mol}}{2\,H_2} + \underset{1\text{ mol}}{O_2} \longrightarrow \underset{2\text{ mol}}{2\,H_2O}$$

six mole ratios can be written:

$$\frac{2\text{ mol }H_2}{1\text{ mol }O_2} \qquad \frac{2\text{ mol }H_2}{2\text{ mol }H_2O} \qquad \frac{1\text{ mol }O_2}{2\text{ mol }H_2O}$$

$$\frac{1\text{ mol }O_2}{2\text{ mol }H_2} \qquad \frac{2\text{ mol }H_2O}{2\text{ mol }H_2} \qquad \frac{2\text{ mol }H_2O}{1\text{ mol }O_2}$$

We use the mole ratio to convert the number of moles of one substance to the corresponding number of moles of another substance in a chemical reaction. For example, if we want to calculate the number of moles of H_2O that can be obtained from 4.0 mol of O_2, we use the mole ratio 2 mol H_2O/1 mol O_2:

$$(4.0\ \cancel{\text{mol } O_2})\left(\frac{2\text{ mol } H_2O}{1\ \cancel{\text{mol } O_2}}\right) = 8.0\text{ mol } H_2O$$

EXAMPLE 9.1

Use the following equation $CO_2 + 4\,H_2 \rightarrow CH_4 + 2\,H_2O$ to write the mole ratio needed to calculate

(a) the number of moles of water that can be produced from 3 moles of CO_2.
(b) the number of moles of hydrogen required to make 3 moles of water.

SOLUTION

In order to write the mole ratio, we need to determine the relationship between the desired substance and the starting substance.

(a) In this case the desired substance is water. The moles of water belong in the numerator of the mole ratio. The starting substance is CO_2. The moles of CO_2 then belong in the denominator of the mole ratio. In the mole ratio the coefficients of the balanced equation are used. Therefore, the mole ratio is $\frac{2\text{ mol } H_2O}{1\text{ mol } CO_2}$.

(b) In this case the desired substance is hydrogen. The moles of hydrogen belong in the numerator of the mole ratio. The starting substance is water. The moles of water then belong in the denominator of the mole ratio. In the mole ratio the coefficients of the balanced equation are used. Therefore, the mole ratio is $\frac{4\text{ mol } H_2}{2\text{ mol } H_2O}$.

PRACTICE 9.1

Use the equation $2\,KClO_3 \rightarrow 2\,KCl + 3\,O_2$ to write the mole ratio needed to calculate

(a) moles of KCl produced from 3.5 moles of $KClO_3$.
(b) moles of KCl produced when 4.5 moles of O_2 are produced.

CHECK YOUR UNDERSTANDING

9.1 Information from a Balanced Chemical Equation
9.2 Mole Ratios 1
9.3 Mole Ratios 2
➡ WileyPLUS

ONLINE LEARNING MODULE
Identifying Mole Ratios
➡ WileyPLUS

PROBLEM-SOLVING STRATEGY

For Stoichiometry Problems

1. Convert the quantity of starting substance to moles (if it is not given in moles).
2. Convert the moles of starting substance to moles of desired substance.
3. Convert the moles of desired substance to the units specified in the problem.

Write and balance the equation before you begin the problem.

Like balancing chemical equations, making stoichiometric calculations requires practice. Several worked examples follow. Study this material and practice on the problems at the end of this chapter.

PROBLEM-SOLVING STRATEGY

For Stoichiometry Problems

Use a balanced equation.

1. **Determine the number of moles of starting substance.** Identify the starting substance from the data given in the problem statement. Convert the quantity of the starting substance to moles, if it is not already done:

$$\text{moles} = (\text{grams})\left(\frac{1\text{ mole}}{\text{molar mass}}\right)$$

2. **Determine the mole ratio of the desired substance to the starting substance.** The number of moles of each substance in the balanced equation is indicated by the coefficient in front of each substance. Use these coefficients to set up the mole ratio:

As in all problems with units, the desired quantity is in the numerator, and the quantity to be eliminated is in the denominator.

$$\text{mole ratio} = \frac{\text{moles of desired substance in the equation}}{\text{moles of starting substance in the equation}}$$

Multiply the number of moles of starting substance (from Step 1) by the mole ratio to obtain the number of moles of desired substance:

Units of moles of starting substance cancel in the numerator and denominator.

$$\text{moles of desired substance} = \underbrace{\left(\text{moles of starting substance}\right)}_{\text{From Step 1}} \left(\frac{\text{moles of desired substance in the balanced equation}}{\text{moles of starting substance in the balanced equation}}\right)$$

3. **Calculate the desired substance in the units specified in the problem.** If the answer is to be in moles, the calculation is complete. If units other than moles are wanted, multiply the moles of the desired substance (from Step 2) by the appropriate factor to convert moles to the units required. For example, if grams of the desired substance are wanted:

$$\text{grams} = \underbrace{(\text{moles})}_{\text{From Step 2}} \left(\frac{\text{molar mass}}{1\ \text{mol}}\right)$$

If moles ⟶ atoms, use $\dfrac{6.022 \times 10^{23}\ \text{atoms}}{1\ \text{mol}}$

If moles ⟶ molecules, use $\dfrac{6.022 \times 10^{23}\ \text{molecules}}{1\ \text{mol}}$

The steps for converting the mass of a starting substance A to either the mass, atoms, or molecules of desired substance B are summarized in FIGURE 9.1.

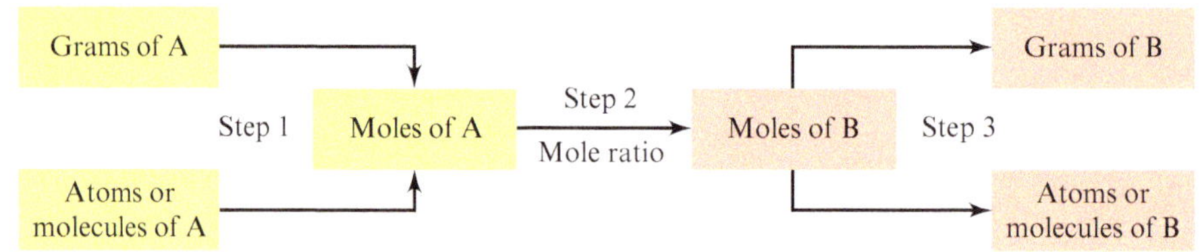

FIGURE 9.1 Steps for converting starting substance A to mass, atoms, or molecules of desired substance B.

9.2 Mole–Mole Calculations

LEARNING OBJECTIVE Solve problems in which the reactants and products are both in moles.

Let's solve stoichiometric problems for mole–mole calculations. The quantity of starting substance is given in moles, and the quantity of desired substance is requested in moles.

ONLINE LEARNING MODULE
Mole–Mole Calculations
WileyPLUS

ENHANCED EXAMPLE 9.2 WileyPLUS

How many moles of carbon dioxide will be produced by the complete reaction of 2.0 mol of glucose ($C_6H_{12}O_6$) according to the following equation?

$$\underset{1\ \text{mol}}{C_6H_{12}O_6} + \underset{6\ \text{mol}}{6\,O_2} \rightarrow \underset{6\ \text{mol}}{6\,CO_2} + \underset{6\ \text{mol}}{6\,H_2O}$$

SOLUTION

Read **Knowns:** 2.0 mol of glucose
$C_6H_{12}O_6 + 6\,O_2 \rightarrow 6\,CO_2 + 6\,H_2O$

Solving for: moles of carbon dioxide

Plan The balanced equation states that 6 mol CO_2 will be produced from 1 mol $C_6H_{12}O_6$. Even though we can readily see that 12 mol CO_2 will be formed, let's use the Problem-Solving Strategy for Stoichiometry Problems.

Solution map: moles of $C_6H_{12}O_6$ → moles of CO_2

The mole ratio from the balanced equation is $\frac{6 \text{ mol } CO_2}{1 \text{ mol } C_6H_{12}O_6}$.

The mole ratio is exact and does not affect the number of significant figures in the answer.

Calculate $(2.0 \cancel{\text{mol } C_6H_{12}O_6})\left(\frac{6 \text{ mol } CO_2}{1 \cancel{\text{mol } C_6H_{12}O_6}}\right) = 12 \text{ mol } CO_2$

Check Note that the moles $C_6H_{12}O_6$ cancel and that the answer agrees with our reasoned answer.

ENHANCED EXAMPLE 9.3

WileyPLUS

How many moles of ammonia can be produced from 8.00 mol of hydrogen reacting with nitrogen? The balanced equation is

$$3\ H_2 + N_2 \rightarrow 2\ NH_3$$

CHECK YOUR UNDERSTANDING

9.4 Mole–Mole Calculations

➥ WileyPLUS

SOLUTION

Read **Knowns:** 8.00 mol of hydrogen

$3\ H_2 + N_2 \rightarrow 2\ NH_3$

Solving for: moles of ammonia

Ammonia
NH_3

Plan The balanced equation states that we get 2 mol NH_3 for every 3 mol H_2 that react. Let's use the Problem-Solving Strategy for Stoichiometry Problems.

Solution map: moles of H_2 → moles of NH_3

The mole ratio from the balanced equation is $\frac{2 \text{ mol } NH_3}{3 \text{ mol } H_2}$.

Calculate $(8.00 \cancel{\text{mol } H_2})\left(\frac{2 \text{ mol } NH_3}{3 \cancel{\text{mol } H_2}}\right) = 5.33 \text{ mol } NH_3$

Check Note that the moles H_2 cancel and that the answer is less than the starting 8.00 mol, which makes sense since our mole ratio is 2:3.

ENHANCED EXAMPLE 9.4

WileyPLUS

Given the balanced equation

$$\underset{1\text{ mol}}{K_2Cr_2O_7} + \underset{6\text{ mol}}{6\ KI} + 7\ H_2SO_4 \rightarrow Cr_2(SO_4)_3 + 4\ K_2SO_4 + \underset{3\text{ mol}}{3\ I_2} + 7\ H_2O$$

Calculate (a) the number of moles of potassium dichromate ($K_2Cr_2O_7$) that will react with 2.0 mol of potassium iodide (KI) and (b) the number of moles of iodine (I_2) that will be produced from 2.0 mol of potassium iodide.

SOLUTION

Read **Knowns:**

$$K_2Cr_2O_7 + 6\ KI + 7\ H_2SO_4 \rightarrow Cr_2(SO_4)_3 + 4\ K_2SO_4 + 3\ I_2 + 7\ H_2O$$

(a) 2.0 mol KI
(b) 2.0 mol KI

(a) **Solving for:** moles of $K_2Cr_2O_7$

Plan Use the Problem-Solving Strategy for Stoichiometry Problems.

Solution map: moles of KI → moles of $K_2Cr_2O_7$

The mole ratio from the balanced equation is $\frac{1 \text{ mol } K_2Cr_2O_7}{6 \text{ mol KI}}$.

Calculate $(2.0 \text{ mol KI})\left(\frac{1 \text{ mol } K_2Cr_2O_7}{6 \text{ mol KI}}\right) = 0.33 \text{ mol } K_2Cr_2O_7$

Check Note that the moles KI cancel and that the answer is less than the starting 2.0 mol, which makes sense since our mole ratio is 1:6.

(b) **Solving for:** moles of I_2

Plan Use the Problem-Solving Strategy for Stoichiometry Problems.

Solution map: mole of KI → moles of I_2

The mole ratio from the balanced equation is $\frac{3 \text{ mol } I_2}{6 \text{ mol KI}}$.

Calculate $(2.0 \text{ mol KI})\left(\frac{3 \text{ mol } I_2}{6 \text{ mol KI}}\right) = 1.0 \text{ mol } I_2$

Check Note that the moles KI cancel and that the answer is half the starting 2.0 mol, which makes sense since our mole ratio is 1:2.

ONLINE LEARNING MODULE

Mole–Molecule Calculations

WileyPLUS

ENHANCED EXAMPLE 9.5 WileyPLUS

How many molecules of water can be produced by reacting 0.010 mol of oxygen with hydrogen? The balanced equation is $2\,H_2 + O_2 \rightarrow 2\,H_2O$.

SOLUTION

Read **Knowns:** $0.010 \text{ mol } O_2$

$2\,H_2 + O_2 \rightarrow 2\,H_2O$

Solving for: molecules H_2O

Plan Use the Problem-Solving Strategy for Stoichiometry Problems.

Solution map: moles of O_2 → moles of H_2O → molecules of H_2O

The mole ratio from the balanced equation is $\frac{2 \text{ mol } H_2O}{1 \text{ mol } O_2}$.

Calculate $(0.010 \text{ mol } O_2)\left(\frac{2 \text{ mol } H_2O}{1 \text{ mol } O_2}\right) = 0.020 \text{ mol } H_2O$

Conversion factor for moles H_2O → molecules H_2O

$$\frac{6.022 \times 10^{23} \text{ molecules}}{1 \text{ mol}}$$

$$(0.020 \text{ mol } H_2O)\left(\frac{6.022 \times 10^{23} \text{ molecules}}{1 \text{ mol}}\right) = 1.2 \times 10^{22} \text{ molecules } H_2O$$

NOTE: 0.020 mol is still quite a large number of water molecules.

PRACTICE 9.2

How many moles of aluminum oxide will be produced from 0.50 mol of oxygen?

$$4\ Al + 3\ O_2 \longrightarrow 2\ Al_2O_3$$

PRACTICE 9.3

How many moles of aluminum hydroxide are required to produce 22.0 mol of water?

$$2\ Al(OH)_3 + 3\ H_2SO_4 \longrightarrow Al_2(SO_4)_3 + 6\ H_2O$$

9.3 Mole–Mass Calculations

LEARNING OBJECTIVE

Solve problems in which mass is given and the answer is to be determined in moles or the moles are given and mass is to be determined.

The object of this type of problem is to calculate the mass of one substance that reacts with or is produced from a given number of moles of another substance in a chemical reaction. If the mass of the starting substance is given, we need to convert it to moles. We use the mole ratio to convert moles of starting substance to moles of desired substance. We can then change moles of desired substance to mass.

ENHANCED EXAMPLE 9.6 WileyPLUS

ONLINE LEARNING MODULE
Mass–Mole Calculations
➡ WileyPLUS

What mass of hydrogen can be produced by reacting 6.0 mol of aluminum with hydrochloric acid? The balanced equation is $2\ Al(s) + 6\ HCl(aq) \rightarrow 2\ AlCl_3(aq) + 3\ H_2(g)$.

SOLUTION

Read **Knowns:** 6.0 mol Al

$$2\ Al(s) + 6\ HCl(aq) \rightarrow 2\ AlCl_3(aq) + 3\ H_2(g)$$

Solving for: mass H_2

Plan Use the Problem-Solving Strategy for Stoichiometry Problems.

Solution map: moles of Al → moles of H_2 → grams H_2

The mole ratio from the balanced equation is $\dfrac{3\text{ mol } H_2}{2\text{ mol Al}}$.

Calculate

$$(6.0\ \cancel{\text{mol Al}})\left(\frac{3\text{ mol } H_2}{2\ \cancel{\text{mol Al}}}\right) = 9.0\text{ mol } H_2$$

Conversion factor for moles H_2 → grams H_2 $\quad \dfrac{2.016\text{ g } H_2}{1\text{ mol } H_2}$

$$(9.0\ \cancel{\text{mol } H_2})\left(\frac{2.016\text{ g } H_2}{1\ \cancel{\text{mol } H_2}}\right) = 18\text{ g } H_2$$

NOTE: You can keep the answer for the first calculation (9.0 mol H_2) on your calculator and continue the calculation to the answer of 18 g H_2 in one continuous calculation as shown below:

moles of Al → moles of H_2 → grams H_2

$$(6.0\ \cancel{\text{mol Al}})\left(\frac{3\ \cancel{\text{mol } H_2}}{2\ \cancel{\text{mol Al}}}\right)\left(\frac{2.016\text{ g } H_2}{1\ \cancel{\text{mol } H_2}}\right) = 18\text{ g } H_2$$

ENHANCED EXAMPLE 9.7 WileyPLUS

How many moles of water can be produced by burning 325 g of octane (C_8H_{18})? The balanced equation is $2\ C_8H_{18}(l) + 25\ O_2(g) \rightarrow 16\ CO_2(g) + 18\ H_2O(g)$.

SOLUTION

Read **Knowns:** 325 g C_8H_{18}

$2\ C_8H_{18}(l) + 25\ O_2(g) \rightarrow 16\ CO_2(g) + 18\ H_2O(g)$

Solving for: mol H_2O

Plan Use the Problem-Solving Strategy for Stoichiometry Problems.

Solution map: grams of C_8H_{18} → moles of C_8H_{18} → moles of H_2O

Conversion factor for grams C_8H_{18} → moles C_8H_{18} $\dfrac{1\ \text{mol}\ C_8H_{18}}{114.2\ \text{g}\ C_8H_{18}}$

The mole ratio from the balanced equation is $\dfrac{18\ \text{mol}\ H_2O}{2\ \text{mol}\ C_8H_{18}}$.

Octane
C_8H_{18}

Calculate $(325\ \cancel{\text{g}\ C_8H_{18}})\left(\dfrac{1\ \text{mol}\ C_8H_{18}}{114.2\ \cancel{\text{g}\ C_8H_{18}}}\right) = 2.85\ \text{mol}\ C_8H_{18}$

$$(2.85\ \cancel{\text{mol}\ C_8H_{18}})\left(\dfrac{18\ \text{mol}\ H_2O}{2\ \cancel{\text{mol}\ C_8H_{18}}}\right) = 25.7\ \text{mol}\ H_2O$$

Check On your calculator in continuous calculation you might get 25.6 mol H_2O for your answer. The difference in the last digit is due to rounding off at different points in the calculation. You should check with your instructor to find out the appropriate rules for your course. In our examples we round at the end of each step to show the proper number of significant figures for each calculation.

PRACTICE 9.4

How many moles of potassium chloride and oxygen can be produced from 100.0 g of potassium chlorate? The balanced equation is:

$$2\ KClO_3 \longrightarrow 2\ KCl + 3\ O_2$$

PRACTICE 9.5

How many grams of silver nitrate are required to produce 0.25 mol of silver sulfide? The balanced equation is:

$$2\ AgNO_3 + H_2S \longrightarrow Ag_2S + 2\ HNO_3$$

9.4 Mass–Mass Calculations

LEARNING OBJECTIVE Solve problems in which mass is given and the answer is to be determined as mass.

Solving mass–mass stoichiometry problems requires all the steps of the mole-ratio method. The mass of starting substance is converted to moles. The mole ratio is then used to determine moles of desired substance, which, in turn, is converted to mass of desired substance.

CHECK YOUR UNDERSTANDING
9.5 Mass–Mass Calculations
9.6 Converting from Grams to Moles
WileyPLUS

ONLINE LEARNING MODULE
Mass–Mass Calculations
WileyPLUS

ENHANCED EXAMPLE 9.8 WileyPLUS

What mass of carbon dioxide is produced by the complete combustion of 100. g of the hydrocarbon pentane, C_5H_{12}? The balanced equation is $C_5H_{12}(l) + 8\ O_2(g) \rightarrow 5\ CO_2(g) + 6\ H_2O(g)$.

SOLUTION

Read **Knowns:** 100. g C_5H_{12}

$C_5H_{12}(l) + 8\ O_2(g) \rightarrow 5\ CO_2(g) + 6\ H_2O(g)$

Solving for: mass CO_2

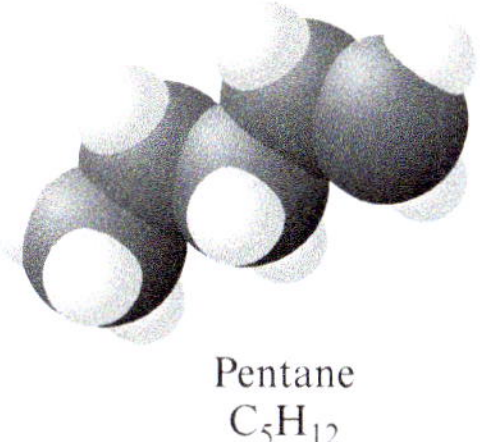
Pentane
C_5H_{12}

Plan Use the Problem-Solving Strategy for Stoichiometry Problems.

Solution map: grams of C_5H_{12} → moles of C_5H_{12} → moles CO_2 → g CO_2

Conversion factor for grams C_5H_{12} → moles C_5H_{12} $\dfrac{1 \text{ mol } C_5H_{12}}{72.15 \text{ g } C_5H_{12}}$

The mole ratio from the balanced equation is $\dfrac{5 \text{ mol } CO_2}{1 \text{ mol } C_5H_{12}}$.

Conversion factor for moles CO_2 → g CO_2 $\dfrac{44.01 \text{ g } CO_2}{1 \text{ mol } CO_2}$

Calculate grams of C_5H_{12} → moles of C_5H_{12}

$$(100. \text{ g } C_5H_{12})\left(\frac{1 \text{ mol } C_5H_{12}}{72.15 \text{ g } C_5H_{12}}\right) = 1.39 \text{ mol } C_5H_{12}$$

moles of C_5H_{12} → moles CO_2

$$(1.39 \text{ mol } C_5H_{12})\left(\frac{5 \text{ mol } CO_2}{1 \text{ mol } C_5H_{12}}\right) = 6.95 \text{ mol } CO_2$$

moles of CO_2 → g CO_2

$$(6.95 \text{ mol } CO_2)\left(\frac{44.01 \text{ g } CO_2}{1 \text{ mol } CO_2}\right) = 306 \text{ g } CO_2$$

ENHANCED EXAMPLE 9.9

WileyPLUS

How many grams of nitric acid (HNO_3) are required to produce 8.75 g of dinitrogen monoxide (N_2O) according to the following equation?

$$4 \text{ Zn}(s) + 10 \text{ HNO}_3(aq) \rightarrow 4 \text{ Zn}(\text{NO}_3)_2(aq) + \text{N}_2\text{O}(g) + 5 \text{ H}_2\text{O}(l)$$

SOLUTION

Read **Knowns:** 8.75 g N_2O

$$4 \text{ Zn}(s) + 10 \text{ HNO}_3(aq) \rightarrow 4 \text{ Zn}(\text{NO}_3)_2(aq) + \text{N}_2\text{O}(g) + 5 \text{ H}_2\text{O}(l)$$

Solving for: g HNO_3

Plan Use the Problem-Solving Strategy for Stoichiometry Problems.

Solution map: grams of N_2O → moles N_2O → moles HNO_3 → g HNO_3

Conversion factor for grams N_2O → moles N_2O $\dfrac{1 \text{ mol } N_2O}{44.02 \text{ g } N_2O}$

The mole ratio from the balanced equation is $\dfrac{10 \text{ mol } HNO_3}{1 \text{ mol } N_2O}$.

Conversion factor for moles HNO_3 → g HNO_3 $\dfrac{63.02 \text{ g } HNO_3}{1 \text{ mol } HNO_3}$

Calculate grams of N_2O → moles of N_2O

$$(8.75 \text{ g } N_2O)\left(\frac{1 \text{ mol } N_2O}{44.02 \text{ g } N_2O}\right) = 0.199 \text{ mol } N_2O$$

moles N_2O → moles HNO_3

$$(0.199 \text{ mol } N_2O)\left(\frac{10 \text{ mol } HNO_3}{1 \text{ mol } N_2O}\right) = 1.99 \text{ mol } HNO_3$$

moles HNO_3 → g HNO_3

$$(1.99 \text{ mol } HNO_3)\left(\frac{63.02 \text{ g } HNO_3}{1 \text{ mol } HNO_3}\right) = 125 \text{ g } HNO_3$$

PRACTICE 9.6

How many grams of chromium(III) chloride are required to produce 75.0 g of silver chloride? The balanced equation is:

$$CrCl_3 + 3\,AgNO_3 \longrightarrow Cr(NO_3)_3 + 3\,AgCl$$

PRACTICE 9.7

What mass of water is produced by the complete combustion of 225.0 g of butane (C_4H_{10})? The balanced equation is:

$$2\,C_4H_{10} + 13\,O_2 \longrightarrow 8\,CO_2 + 10\,H_2O$$

9.5 Limiting Reactant and Yield Calculations

LEARNING OBJECTIVE Solve problems involving limiting reactants and yield.

KEY TERMS
limiting reactant
theoretical yield
actual yield
percent yield

In many chemical processes, the quantities of the reactants used are such that one reactant is in excess. The amount of the product(s) formed in such a case depends on the reactant that is not in excess. This reactant is called the **limiting reactant**—it limits the amount of product that can be formed.

Consider the case illustrated in FIGURE 9.2. How many bicycles can be assembled from the parts shown? The limiting part in this case is the number of pedal assemblies; only three bicycles can be built because there are only three pedal assemblies. The wheels and frames are parts in excess.

Let's consider a chemical example at the molecular level in which seven molecules of H_2 are combined with four molecules of Cl_2 (FIGURE 9.3 before reaction). How many molecules of HCl can be produced according to this reaction?

$$H_2 + Cl_2 \longrightarrow 2\,HCl$$

CHECK YOUR UNDERSTANDING
9.7 Identifying Limiting Reactant Problems
WileyPLUS

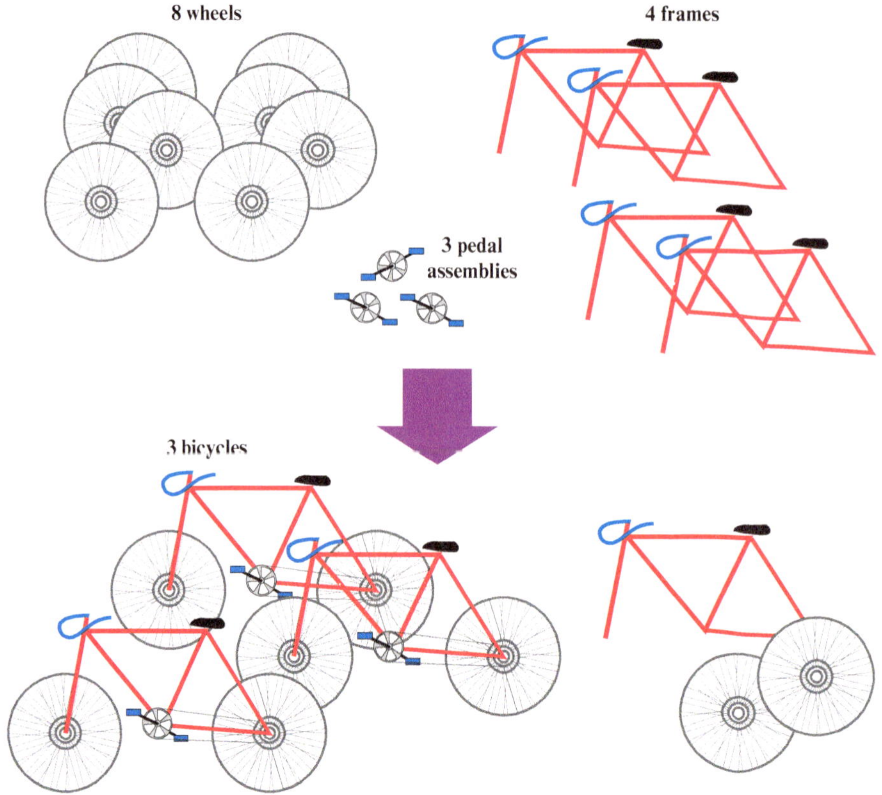

FIGURE 9.2 The number of bicycles that can be built from these parts is determined by the "limiting reactant" (the pedal assemblies).

CHEMISTRY IN ACTION A Shrinking Technology

The microchip has revolutionized the field of electronics. Engineers at Bell Laboratories, Massachusetts Institute of Technology, the University of California, and Stanford University are racing to produce parts for tiny machines and robots. New techniques now produce gears smaller than a grain of sand and motors lighter than a speck of dust.

To produce ever smaller computers, calculators, and even microbots (microsized robots), precise quantities of chemicals in exact proportions are required. The secret to producing minute circuits is to print the entire circuit or blueprint at one time. Computers are used to draw a chip.

These nanorobots attach themselves to red blood cells.

This image is then transferred onto a pattern, or mask, with details finer than a human hair. In a process similar to photography, light is then shined through the mask onto a silicon-coated surface. The areas created on the silicon exhibit high or low resistance to chemical etching. Chemicals then etch away the silicon.

Micromachinery is produced in the same way. First a thin layer of silicon dioxide is applied (sacrificial material); then a layer of polysilicon is carefully applied (structural material). Next, a mask is applied, and the whole structure is covered with plasma (excited gas). The plasma acts as a tiny sandblaster, removing everything the mask doesn't protect. This process is repeated as the entire machine is constructed. When the entire assembly is complete, the whole machine is placed in hydrofluoric acid, which dissolves all the sacrificial material and permits the various parts of the machine to move.

To turn these micromachines into true microbots, current research is focusing on locomotion and sensing imaging systems. Possible uses for these microbots include "smart" pills, which could contain sensors or drug reservoirs (currently used in birth control). Tiny pumps, once inside the body, will dispense the proper amount of medication at precisely the correct site. These microbots are currently in production for the treatment of diabetes (to release insulin).

If the molecules of H_2 and Cl_2 are taken apart and recombined as HCl (Figure 9.3 after reaction), we see that eight molecules of HCl can be formed before we run out of Cl_2. Therefore, the Cl_2 is the limiting reactant and H_2 is in excess—three molecules of H_2 remain unreacted.

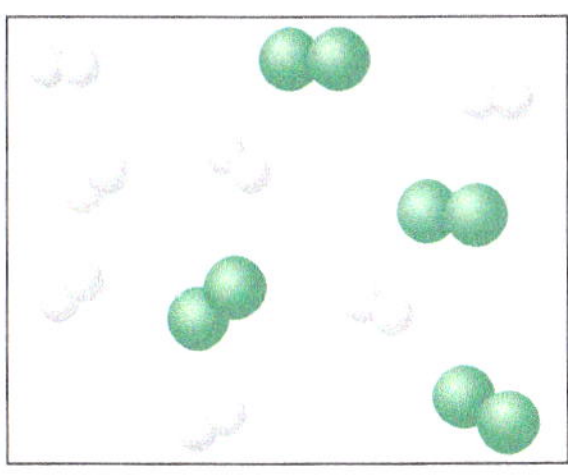

Before reaction

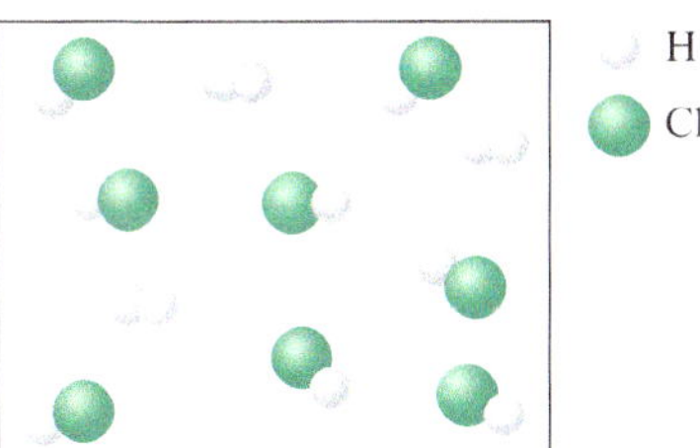

After reaction

FIGURE 9.3 Reaction of H_2 with Cl_2. Eight molecules of HCl are formed, and three molecules of H_2 remain. The limiting reactant is Cl_2.

When problem statements give the amounts of two reactants, one of them is usually a limiting reactant. We can identify the limiting reactant using the following method:

CHECK YOUR UNDERSTANDING

9.8 Limiting Reactant Calculation

➡ WileyPLUS

PROBLEM-SOLVING STRATEGY

For Limiting Reactant Problems

1. Calculate the amount of product (moles or grams, as needed) formed from each reactant.
2. Determine which reactant is limiting. (The reactant that gives the least amount of product is the limiting reactant; the other reactant is in excess.)
3. Once we know the limiting reactant, the amount of product formed can be determined. It is the amount determined by the limiting reactant.
4. If we need to know how much of the other reactant remains, we calculate the amount of the other reactant required to react with the limiting reactant, then subtract this amount from the starting quantity of the reactant. This gives the amount of that substance that remains unreacted.

ONLINE LEARNING MODULE

Limiting Reactant

➡ WileyPLUS

ENHANCED EXAMPLE 9.10 WileyPLUS

How many moles of HCl can be produced by reacting 4.0 mol H_2 and 3.5 mol Cl_2? Which compound is the limiting reactant? The equation is $H_2(g) + Cl_2(g) \longrightarrow 2\ HCl(g)$.

SOLUTION

Read • **Knowns:** 4.0 mol H_2

3.5 mol Cl_2

$H_2(g) + Cl_2(g) \rightarrow 2\ HCl(g)$

Solving for: mol HCl

Plan • Use the Problem-Solving Strategy for Limiting Reactant Problems.

Calculate the mol HCl formed from each reactant.

The mole ratios from the balanced equation are

$$\frac{2 \text{ mol HCl}}{1 \text{ mol } H_2} \qquad \frac{2 \text{ mol HCl}}{1 \text{ mol } Cl_2}$$

Calculate • moles H_2 → moles HCl

$$(4.0 \cancel{\text{mol } H_2}) = \frac{2 \text{ mol HCl}}{1 \cancel{\text{mol } H_2}} = 8.0 \text{ mol HCl}$$

moles of Cl_2 → moles HCl

$$(3.5 \cancel{\text{mol } Cl_2}) = \frac{2 \text{ mol HCl}}{1 \cancel{\text{mol } Cl_2}} = 7.0 \text{ mol HCl}$$

Determine the limiting reactant. The limiting reactant is Cl_2 because it produces less HCl than H_2. The H_2 is in excess. The yield of HCl is 7.0 mol.

Check • In this reaction we can decide on the limiting reactant by inspection. From the equation you can see that 1 mol H_2 reacts with 1 mol Cl_2. Therefore, when we react 4.0 mol H_2 with 3.5 mol Cl_2, Cl_2 is the reactant that limits the amount of HCl produced, since it is present in smaller amount.

ENHANCED EXAMPLE 9.11 WileyPLUS

How many moles of Fe_3O_4 can be obtained by reacting 16.8 g Fe with 10.0 g H_2O? Which substance is the limiting reactant? Which substance is in excess?

$$3\ Fe(s) + 4\ H_2O(g) \xrightarrow{\Delta} Fe_3O_4(s) + 4\ H_2(g)$$

SOLUTION

Read • **Knowns:** 16.5 g Fe

10.0 g H_2O

$3\ Fe(s) + 4\ H_2O(g) \xrightarrow{\Delta} Fe_3O_4(s) + 4\ H_2(g)$

Solving for: mol Fe_3O_4

Plan • Use the Problem-Solving Strategy for Limiting Reactant Problems.

Calculate the mol Fe_3O_4 formed from each reactant:

g reactant → mol reactant → mol Fe_3O_4

Calculate • For Fe

$$(16.8 \cancel{\text{g Fe}})\left(\frac{1 \cancel{\text{mol Fe}}}{55.85 \cancel{\text{g Fe}}}\right)\left(\frac{1 \text{ mol } Fe_3O_4}{3 \cancel{\text{mol Fe}}}\right) = 0.100 \text{ mol } Fe_3O_4$$

For H_2O

$$(10.0\ \cancel{g\ H_2O})\left(\frac{1\ \cancel{mol\ H_2O}}{18.02\ \cancel{g\ H_2O}}\right)\left(\frac{1\ mol\ Fe_3O_4}{4\ \cancel{mol\ H_2O}}\right) = 0.139\ mol\ Fe_3O_4$$

Determine the limiting reactant. The limiting reactant is Fe because it produces less Fe_3O_4 than H_2O. The H_2O is in excess. The yield of Fe_3O_4 is 0.100 mol.

ENHANCED EXAMPLE 9.12

WileyPLUS

How many grams of silver bromide (AgBr) can be formed when solutions containing 50.0 g of $MgBr_2$ and 100.0 g of $AgNO_3$ are mixed together? How many grams of excess reactant remain unreacted?

SOLUTION

Read **Knowns:** 50.0 g $MgBr_2$
100.0 g $AgNO_3$

$$MgBr_2(aq) + 2\ AgNO_3(aq) \rightarrow 2\ AgBr(s) + Mg(NO_3)_2(aq)$$

Solving for: g AgBr

Plan Use the Problem-Solving Strategy for Limiting Reactant Problems.
Calculate the g AgBr formed from each reactant:

g reactant → mol reactant → mol AgBr → g AgBr

Calculate For $MgBr_2$

$$(50.0\ \cancel{g\ MgBr_2})\left(\frac{1\ \cancel{mol\ MgBr_2}}{184.1\ \cancel{g\ MgBr_2}}\right)\left(\frac{2\ \cancel{mol\ AgBr}}{1\ \cancel{mol\ MgBr_2}}\right)\left(\frac{187.8\ g\ AgBr}{1\ \cancel{mol\ AgBr}}\right)$$
$$= 102\ g\ AgBr$$

For $AgNO_3$

$$(100.0\ \cancel{g\ AgNO_3})\left(\frac{1\ \cancel{mol\ AgNO_3}}{169.9\ \cancel{g\ AgNO_3}}\right)\left(\frac{2\ \cancel{mol\ AgBr}}{2\ \cancel{mol\ AgNO_3}}\right)\left(\frac{187.8\ g\ AgBr}{1\ \cancel{mol\ AgBr}}\right)$$
$$= 110.5\ g\ AgBr$$

Determine the limiting reactant. The limiting reactant is $MgBr_2$ because it produces less AgBr than $AgNO_3$. The $AgNO_3$ is in excess. The yield of AgBr is 102 g.

Calculate the mass of unreacted $AgNO_3$. We must first determine the g $AgNO_3$ that react with 50.0 g $MgBr_2$.

g $MgBr_2$ → mol $MgBr_2$ → mol $AgNO_3$ → g $AgNO_3$

$$(50.0\ \cancel{g\ MgBr_2})\left(\frac{1\ \cancel{mol\ MgBr_2}}{184.1\ \cancel{g\ MgBr_2}}\right)\left(\frac{2\ \cancel{mol\ AgNO_3}}{1\ \cancel{mol\ MgBr_2}}\right)\left(\frac{169.9\ g\ AgNO_3}{1\ \cancel{mol\ AgNO_3}}\right)$$
$$= 92.3\ g\ AgNO_3$$

$$\text{Unreacted } AgNO_3 = 100.0\ g\ AgNO_3 - 92.3\ g\ AgNO_3 = 7.7\ g\ AgNO_3$$

The final mixture will contain:

102 g AgBr(*s*) 7.7 g $AgNO_3$

PRACTICE 9.8

How many grams of hydrogen chloride can be produced from 0.490 g of hydrogen and 50.0 g of chlorine? The balanced equation is:

$$H_2(g) + Cl_2(g) \longrightarrow 2\ HCl(g)$$

PRACTICE 9.9

How many grams of barium sulfate will be formed from 200.0 g of barium nitrate and 100.0 g of sodium sulfate? The balanced equation is:

$$Ba(NO_3)_2(aq) + Na_2SO_4(aq) \longrightarrow BaSO_4(s) + 2\ NaNO_3(aq)$$

The quantities of the products we have been calculating from chemical equations represent the maximum yield (100%) of product according to the reaction represented by the equation. Many reactions, especially those involving organic substances, fail to give a 100% yield of product. The main reasons for this failure are the side reactions that give products other than the main product and the fact that many reactions are reversible. In addition, some product may be lost in handling and transferring from one vessel to another. The **theoretical yield** of a reaction is the calculated amount of product that can be obtained from a given amount of reactant, according to the chemical equation. The **actual yield** is the amount of product that we finally obtain.

CHECK YOUR UNDERSTANDING
9.9 Reason for a Low Yield
WileyPLUS

CHECK YOUR UNDERSTANDING
9.10 Percent Yield
WileyPLUS

The **percent yield** is the ratio of the actual yield to the theoretical yield multiplied by 100. Both the theoretical and the actual yields must have the same units to obtain a percent:

$$\frac{\text{actual yield}}{\text{theoretical yield}} \times 100 = \text{percent yield}$$

For example, if the theoretical yield calculated for a reaction is 14.8 g, and the amount of product obtained is 9.25 g, the percent yield is

Round off as appropriate for your particular course.

$$\text{percent yield} = \left(\frac{9.25\ \text{g}}{14.8\ \text{g}}\right)(100) = 62.5\%\ \text{yield}$$

ONLINE LEARNING MODULE
Percent Yield
WileyPLUS

ENHANCED EXAMPLE 9.13 WileyPLUS

Carbon tetrachloride (CCl_4) was prepared by reacting 100. g of carbon disulfide and 100. g of chlorine. Calculate the percent yield if 65.0 g of CCl_4 were obtained from the reaction $CS_2 + 3\ Cl_2 \rightarrow CCl_4 + S_2Cl_2$.

SOLUTION

Read • **Knowns:** 100. g CS_2
100. g Cl_2
65.0 g CCl_4 obtained
$CS_2 + 3\ Cl_2 \rightarrow CCl_4 + S_2Cl_2$

Solving for: % yield of CCl_4

Plan • Use the equation for percent yield:

$$\frac{\text{actual yield}}{\text{theoretical yield}} \times 100 = \text{percent yield}$$

Determine the theoretical yield. Calculate the g CCl_4 formed from each reactant.

$$\text{g reactant} \rightarrow \text{mol reactant} \rightarrow \text{mol } CCl_4 \rightarrow \text{g } CCl_4$$

Calculate • For CS_2

$$(100.\ \text{g } CS_2)\left(\frac{1\ \text{mol } CS_2}{76.15\ \text{g } CS_2}\right)\left(\frac{1\ \text{mol } CCl_4}{1\ \text{mol } CS_2}\right)\left(\frac{153.8\ \text{g } CCl_4}{1\ \text{mol } CCl_4}\right) = 202\ \text{g } CCl_4$$

For Cl_2

$$(100.\ \text{g Cl}_2)\left(\frac{1\ \text{mol Cl}_2}{70.90\ \text{g Cl}_2}\right)\left(\frac{1\ \text{mol CCl}_4}{3\ \text{mol Cl}_2}\right)\left(\frac{153.8\ \text{g CCl}_4}{1\ \text{mol CCl}_4}\right)$$
$$= 72.3\ \text{g CCl}_4$$

Determine the limiting reactant. The limiting reactant is Cl_2 because it produces less CCl_4 than CS_2. The CS_2 is in excess. The theoretical yield of CCl_4 is 72.3 g.

Calculate the percent yield:

$$\frac{\text{actual yield}}{\text{theoretical yield}} \times 100 = \left(\frac{65.0\ \text{g}}{72.3\ \text{g}}\right)(100) = 89.9\%\ \text{yield}$$

ENHANCED EXAMPLE 9.14

WileyPLUS

Silver bromide was prepared by reacting 200.0 g of magnesium bromide and an adequate amount of silver nitrate. Calculate the percent yield if 375.0 g of silver bromide were obtained from the reaction:

$$MgBr_2 + 2\ AgNO_3 \rightarrow Mg(NO_3)_2 + 2\ AgBr$$

SOLUTION

Read **Knowns:** 200.0 g $MgBr_2$

375.0 g AgBr

$MgBr_2 + 2\ AgNO_3 \rightarrow Mg(NO_3)_2 + 2\ AgBr$

Solving for: % yield AgBr

Plan Use the equation for percent yield:

$$\frac{\text{actual yield}}{\text{theoretical yield}} \times 100 = \text{percent yield}$$

Determine the theoretical yield. Calculate the grams of AgBr formed:

$$\text{g MgBr}_2 \rightarrow \text{mol MgBr}_2 \rightarrow \text{mol AgBr} \rightarrow \text{g AgBr}$$

Calculate $$(200.0\ \text{g MgBr}_2)\left(\frac{1\ \text{mol MgBr}_2}{184.1\ \text{g MgBr}_2}\right)\left(\frac{2\ \text{mol AgBr}}{1\ \text{mol MgBr}_2}\right)\left(\frac{187.8\ \text{g AgBr}}{1\ \text{mol AgBr}}\right)$$
$$= 408.0\ \text{g AgBr}$$

The theoretical yield is 408.0 g AgBr.

Calculate the percent yield:

$$\frac{\text{actual yield}}{\text{theoretical yield}} \times 100 = \left(\frac{375.0\ \text{g}}{408.0\ \text{g}}\right)(100) = 91.91\%\ \text{yield}$$

PRACTICE 9.10

Aluminum oxide was prepared by heating 225 g of chromium(II) oxide with 125 g of aluminum. Calculate the percent yield if 100.0 g of aluminum oxide were obtained. The balanced equation is:

$$2\ Al + 3\ CrO \longrightarrow Al_2O_3 + 3\ Cr$$

CHAPTER 9 REVIEW

9.1 Introduction to Stoichiometry

KEY TERMS

stoichiometry
molar mass
mole ratio

- Solving stoichiometry problems requires the use of moles and mole ratios.
- To solve a stoichiometry problem:
  - Convert the quantity of starting substance to moles.
  - Convert the moles of starting substance to moles of desired substance.
  - Convert the moles of desired substance to the appropriate unit.

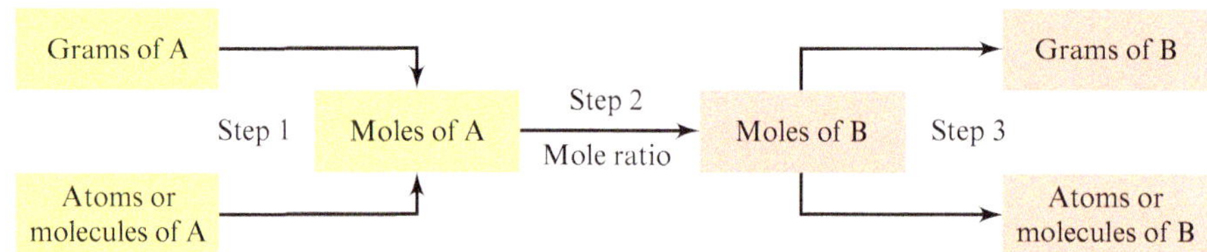

9.2 Mole–Mole Calculations

- If the starting substance is given in moles and the unit specified for the answer is moles:
  - Convert the moles of starting substance to moles of desired substance.

9.3 Mole–Mass Calculations

- If the starting substance is given in moles and the unit specified for the answer is grams:
  - Convert the moles of starting substance to moles of desired substance.
  - Convert the moles of desired substance to grams.

9.4 Mass–Mass Calculations

- If the starting substance is given in grams and the unit specified for the answer is grams:
  - Convert the quantity of starting substance to moles.
  - Convert the moles of starting substance to moles of desired substance.
  - Convert the moles of desired substance to grams.

9.5 Limiting Reactant and Yield Calculations

KEY TERMS

limiting reactant
theoretical yield
actual yield
percent yield

- To identify the limiting reactant in a reaction:
  - Calculate the amount of product formed from each reactant.
  - Determine the limiting reactant by selecting the one that gives the least amount of product.
- To determine the actual amount of product formed in a limiting reactant situation:
  - Use the result calculated that is the least amount of product.
- To determine the amount of the other reactants required to react with the limiting reactant:
  - Calculate the amount of the other reactant needed to react with the limiting reactant.
  - Subtract this amount from the original amount of the other reactant to find the amount of excess reactant (unreacted).
- The theoretical yield of a chemical reaction is the calculated amount of product from a given amount of reactant (the limiting reactant).
- The actual yield is the amount of product actually obtained experimentally.
- The percent yield is

$$\frac{\text{actual yield}}{\text{theoretical yield}} \times 100 = \text{percent yield}$$

REVIEW QUESTIONS

1. What is a mole ratio?
2. What piece of information is needed to convert grams of a compound to moles of the same compound?
3. Phosphine (PH_3) can be prepared by the reaction of calcium phosphide, Ca_3P_2:

$$Ca_3P_2 + 6\,H_2O \longrightarrow 3\,Ca(OH)_2 + 2\,PH_3$$

Based on this equation, which of the following statements are correct? Show evidence to support your answer.
(a) One mole of Ca_3P_2 produces 2 mol of PH_3.
(b) One gram of Ca_3P_2 produces 2 g of PH_3.
(c) Three moles of $Ca(OH)_2$ are produced for each 2 mol of PH_3 produced.
(d) The mole ratio between phosphine and calcium phosphide is

$$\frac{2 \text{ mol } PH_3}{1 \text{ mol } Ca_3P_2}$$

(e) When 2.0 mol of Ca_3P_2 and 3.0 mol of H_2O react, 4.0 mol of PH_3 can be formed.
(f) When 2.0 mol of Ca_3P_2 and 15.0 mol of H_2O react, 6.0 mol of $Ca(OH)_2$ can be formed.
(g) When 200. g of Ca_3P_2 and 100. g of H_2O react, Ca_3P_2 is the limiting reactant.
(h) When 200. g of Ca_3P_2 and 100. g of H_2O react, the theoretical yield of PH_3 is 57.4 g.

4. The equation representing the reaction used for the commercial preparation of hydrogen cyanide is

$$2\,CH_4 + 3\,O_2 + 2\,NH_3 \longrightarrow 2\,HCN + 6\,H_2O$$

Based on this equation, which of the following statements are correct? Rewrite incorrect statements to make them correct.
(a) Three moles of O_2 are required for 2 mol of NH_3.
(b) Twelve moles of HCN are produced for every 16 mol of O_2 that react.
(c) The mole ratio between H_2O and CH_4 is $\frac{6 \text{ mol } H_2O}{2 \text{ mol } CH_4}$
(d) When 12 mol of HCN are produced, 4 mol of H_2O will be formed.
(e) When 10 mol of CH_4, 10 mol of O_2, and 10 mol of NH_3 are mixed and reacted, O_2 is the limiting reactant.
(f) When 3 mol each of CH_4, O_2, and NH_3 are mixed and reacted, 3 mol of HCN will be produced.

5. What information beyond the mole ratio is needed to convert from moles of reactant to grams of product?
6. Draw a flowchart showing the conversions necessary to determine grams of ammonia formed from 10 g of hydrogen gas.

$$3\,H_2 + 2\,N_2 \longrightarrow 2\,NH_3$$

7. What is the difference between the theoretical and the actual yield of a chemical reaction?
8. How is the percent yield of a reaction calculated?

PAIRED EXERCISES

Most of the exercises in this chapter are available for assignment via the online homework management program, WileyPLUS (www.wileyplus.com). All exercises with blue numbers have answers in Appendix 5.

1. Calculate the number of moles in these quantities:
(a) 25.0 g KNO_3
(b) 56 millimol NaOH
(c) 5.4×10^2 g $(NH_4)_2C_2O_4$
(d) 16.8 mL H_2SO_4 solution (d = 1.727 g/mL, 80.0% H_2SO_4 by mass)

2. Calculate the number of moles in these quantities:
(a) 2.10 kg $NaHCO_3$
(b) 525 mg $ZnCl_2$
(c) 9.8×10^{24} molecules CO_2
(d) 250 mL ethyl alcohol, C_2H_5OH (d = 0.789 g/mL)

3. Calculate the number of grams in these quantities:
(a) 2.55 mol $Fe(OH)_3$
(b) 125 kg $CaCO_3$
(c) 10.5 mol NH_3
(d) 72 millimol HCl
(e) 500.0 mL of liquid Br_2 (d = 3.119 g/mL)

4. Calculate the number of grams in these quantities:
(a) 0.00844 mol $NiSO_4$
(b) 0.0600 mol $HC_2H_3O_2$
(c) 0.725 mol Bi_2S_3
(d) 4.50×10^{21} molecules glucose, $C_6H_{12}O_6$
(e) 75 mL K_2CrO_4 solution (d = 1.175 g/mL, 20.0% K_2CrO_4 by mass)

5. Which contains the larger number of molecules, 10.0 g H_2O or 10.0 g H_2O_2? Show evidence for your answer.

6. Which contains the larger numbers of molecules, 25.0 g HCl or 85.0 g $C_6H_{12}O_6$? Show evidence for your answer.

7. Balance the equation for the synthesis of sucrose
$CO_2 + H_2O \longrightarrow C_{12}H_{22}O_{11} + O_2$
and set up the mole ratio of
(a) CO_2 to H_2O
(b) H_2O to $C_{12}H_{22}O_{11}$
(c) O_2 to CO_2
(d) $C_{12}H_{22}O_{11}$ to CO_2
(e) H_2O to O_2
(f) O_2 to $C_{12}H_{22}O_{11}$

8. Balance the equation for the combustion of butanol
$C_4H_9OH + O_2 \longrightarrow CO_2 + H_2O$
and set up the mole ratio of
(a) O_2 to C_4H_9OH
(b) H_2O to O_2
(c) CO_2 to H_2O
(d) C_4H_9OH to CO_2
(e) H_2O to C_4H_9OH
(f) CO_2 to O_2

9. Given the unbalanced equation

$$CO_2\ H_2 \longrightarrow CH_4\ H_2O$$

(a) How many moles of water can be produced from 25 moles of carbon dioxide?
(b) How many moles of CH_4 will be produced along with 12 moles of water?

10. Given the unbalanced equation

$$H_2SO_4 + NaOH \longrightarrow Na_2SO_4 + H_2O$$

(a) How many moles of NaOH will react with 17 mol of H_2SO_4?
(b) How many moles of Na_2SO_4 will be produced when 21 mol of NaOH react?

11. Given the equation

$$MnO_2(s) + HCl(aq) \longrightarrow Cl_2(g) + MnCl_2(aq) + H_2O(l)$$
(unbalanced)

(a) How many moles of HCl will react with 1.05 mol of MnO_2?
(b) How many moles of $MnCl_2$ will be produced when 1.25 mol of H_2O are formed?
(c) How many grams of Cl_2 will be produced when 3.28 mol of MnO_2 react with excess HCl?
(d) How many moles of HCl are required to produce 15.0 kg of $MnCl_2$?

12. Given the equation

$$Al_4C_3 + 12\,H_2O \longrightarrow 4\,Al(OH)_3 + 3\,CH_4$$

(a) How many moles of water are needed to react with 100. g of Al_4C_3?
(b) How many moles of $Al(OH)_3$ will be produced when 0.600 mol of CH_4 is formed?
(c) How many moles of CH_4 will be formed by the reaction of 275 grams of Al_4C_3 with excess water?
(d) How many grams of water are required to produce 4.22 moles of $Al(OH)_3$?

13. Carbonates react with acids to form a salt, water, and carbon dioxide gas. When 50.0 g of calcium carbonate are reacted with sufficient hydrochloric acid, how many grams of calcium chloride will be produced? (Write a balanced equation first.)

14. Certain metals can displace hydrogen from acids to produce hydrogen gas and a salt. When 25.2 g of aluminum metal are placed into hydrobromic acid, how many grams of aluminum bromide will be produced? (Write the balanced equation first.)

15. In a blast furnace, iron(III) oxide reacts with coke (carbon) to produce molten iron and carbon monoxide:

$$Fe_2O_3 + 3\,C \xrightarrow{\Delta} 2\,Fe + 3\,CO$$

How many kilograms of iron would be formed from 125 kg of Fe_2O_3?

16. How many grams of steam and iron must react to produce 375 g of magnetic iron oxide, Fe_3O_4?

$$3\,Fe(s) + 4\,H_2O(g) \xrightarrow{\Delta} Fe_3O_4(s) + 4\,H_2(g)$$

17. Polychlorinated biphenyls (PCBs), which were formerly used in the manufacture of electrical transformers, are environmental and health hazards. They break down very slowly in the environment. The decomposition of PCBs can be represented by the equation

$$2\,C_{12}H_4Cl_6 + 23\,O_2 + 2\,H_2O \longrightarrow 24\,CO_2 + 12\,HCl$$

(a) How many moles of water are needed to react 10.0 mol O_2?
(b) How many grams of HCl are produced when 15.2 mol H_2O react?
(c) How many moles of CO_2 are produced when 76.5 g HCl are produced?
(d) How many grams of $C_{12}H_4Cl_6$ are reacted when 100.25 g CO_2 are produced?
(e) How many grams of HCl are produced when 2.5 kg $C_{12}H_4Cl_6$ react?

18. Mercury can be isolated from its ore by the reaction

$$4\,HgS + 4\,CaO \longrightarrow 4\,Hg + 3\,CaS + CaSO_4$$

(a) How many moles of CaS are produced from 2.5 mol CaO?
(b) How many grams of Hg are produced along with 9.75 mol of $CaSO_4$?
(c) How many moles of CaO are required to react with 97.25 g of HgS?
(d) How many grams of Hg are produced from 87.6 g of HgS?
(e) How many grams of CaS are produced along with 9.25 kg of Hg?

19. Draw the molecules of product(s) formed for each of the following reactions; then determine which substance is the limiting reactant:

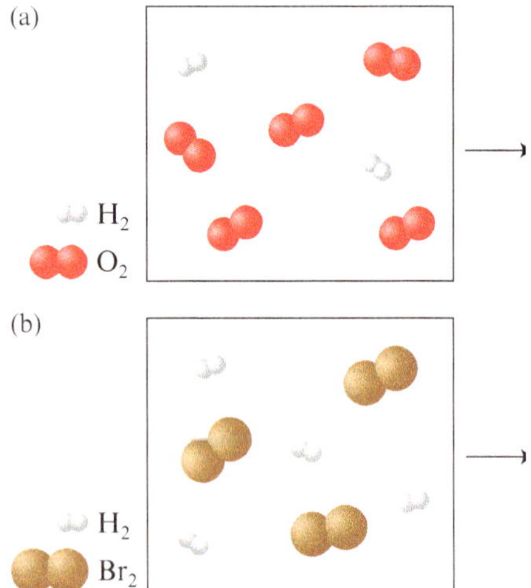

20. Draw the molecules of product(s) formed for the following reactions; then determine which substance is the limiting reactant:

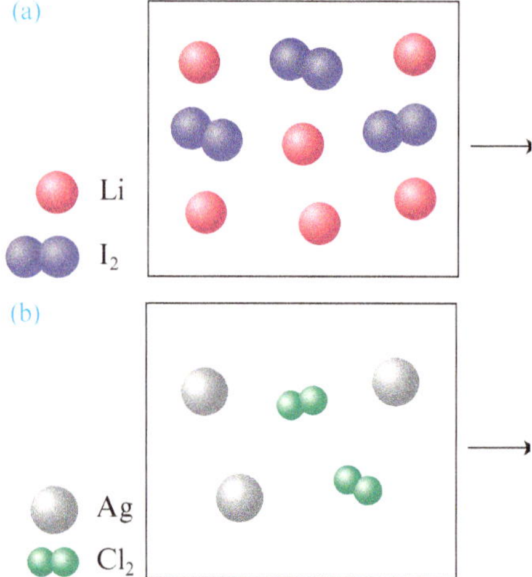

21. Draw pictures similar to those in Question 19 and determine which is the limiting reactant in the following reactions:
(a) Eight atoms of potassium react with five molecules of chlorine to produce potassium chloride.
(b) Ten atoms of aluminum react with three molecules of oxygen to produce aluminum oxide.

22. Draw pictures similar to those in Question 20 and determine which is the limiting reactant in the following reactions:
(a) Eight molecules of nitrogen react with six molecules of oxygen to produce nitrogen dioxide.
(b) Fifteen atoms of iron react with twelve molecules of water to produce magnetic iron oxide (Fe_3O_4) and hydrogen.

23. In the following equations, determine which reactant is the limiting reactant and which reactant is in excess. The amounts used are given below each reactant. Show evidence for your answers.

(a) $KOH + HNO_3 \longrightarrow KNO_3 + H_2O$
16.0 g 12.0 g

(b) $2\ NaOH + H_2SO_4 \longrightarrow Na_2SO_4 + H_2O$
10.0 g 10.0 g

24. In the following equations, determine which reactant is the limiting reactant and which reactant is in excess. The amounts used are given below each reactant. Show evidence for your answers.

(a) $2\ Bi(NO_3)_3 + 3\ H_2S \longrightarrow Bi_2S_3 + 6\ HNO_3$
50.0 g 6.00 g

(b) $3\ Fe + 4\ H_2O \longrightarrow Fe_3O_4 + 4\ H_2$
40.0 g 16.0 g

25. Aluminum sulfate has many uses in industry. It can be prepared by reacting aluminum hydroxide and sulfuric acid:

$$2\ Al(OH)_3 + 3\ H_2SO_4 \longrightarrow Al_2(SO_4)_3 + 6\ H_2O$$

(a) When 35.0 g of $Al(OH)_3$ and 35.0 g of H_2SO_4 are reacted, how many moles of $Al_2(SO_4)_3$ can be produced? Which substance is the limiting reactant?
(b) When 45.0 g of H_2SO_4 and 25.0 g of $Al(OH)_3$ are reacted, how many grams of $Al_2(SO_4)_3$ are produced? Which reactant is in excess?
(c) In a reaction vessel, 2.5 mol of $Al(OH)_3$ and 5.5 mol of H_2SO_4 react. Determine which substances will be present in the container when the reaction goes to completion and how many moles of each.

26. Phosphorous trichloride, a precursor for the manufacture of certain pesticides and herbicides, can be produced by the reaction of phosphorous and chlorine:

$$P_4 + 6\ Cl_2 \longrightarrow 4\ PCl_3$$

(a) When 20.5 g of P_4 and 20.5 g of Cl_2 are reacted, how many moles of PCl_3 can be produced?
(b) When 55 g of Cl_2 are reacted with 25 g of P_4, how many grams of PCl_3 can be produced? Which reactant is in excess?
(c) In a reaction vessel, 15 mol of P_4 and 35 mol of Cl_2 react. Determine which substances will be present in the container after the reaction goes to completion and how many moles of each substance is present.

27. When a certain nonmetal whose formula is X_8 burns in air, XO_3 forms. Write a balanced equation for this reaction. If 120.0 g of oxygen gas are consumed completely, along with 80.0 g of X_8, identify the element X.

28. When a particular metal X reacts with HCl, the resulting products are XCl_2 and H_2, Write and balance the equation. When 78.5 g of the metal react completely, 2.42 g of hydrogen gas result. Identify the element X.

29. Elemental silicon can be produced by the reduction of silicon dioxide, or sand, with carbon:

$$SiO_2 + 2\ C \longrightarrow Si + 2\ CO$$

When 35.0 kg of SiO_2 reacts with 25.3 kg of carbon, and 14.4 kg of silicon is recovered, what is the percent yield for the reaction?

30. Chromium metal can be produced by the reduction of Cr_2O_3 with elemental silicon:

$$2\ Cr_2O_3 + 3\ Si \longrightarrow 4\ Cr + 3\ SiO_2$$

If 350.0 grams of Cr_2O_3 are reacted with 235.0 grams of elemental silicon and 213.2 grams of chromium metal are recovered, what is the percent yield?

31. In a laboratory experiment, 27.5 g of Cu were reacted with 125 g of HNO_3:

$$3\ Cu + 8\ HNO_3 \longrightarrow 3\ Cu(NO_3)_2 + 4\ H_2O + 2\ NO$$

(a) How many grams of $Cu(NO_3)_2$ can be theoretically produced?
(b) How many grams of the excess reactant remain?
(c) If the percent yield of $Cu(NO_3)_2$ is 87.3%, what is the actual yield in grams?

32. In aqueous solution, metal oxides can react with acids to form a salt and water:

$$Fe_2O_3(s) + 6\ HCl(aq) \longrightarrow 2\ FeCl_3(aq) + 3\ H_2O(l)$$

(a) How many moles of each product will be formed when 35 g of Fe_2O_3 react with 35 g of HCl?
(b) From part (a), how many grams of the excess reactant remain?
(c) If the percent yield of $FeCl_3$ is 92.5%, what is the actual yield in grams?

ADDITIONAL EXERCISES

33. A tool set contains 6 wrenches, 4 screwdrivers, and 2 pliers. The manufacturer has 1000 pliers, 2000 screwdrivers, and 3000 wrenches in stock. Can an order for 600 tool sets be filled? Explain briefly.

34. Write a balanced equation for the fermentation of glucose, $C_6H_{12}O_6$, into ethyl alcohol, C_2H_5OH and CO_2. Determine how many liters of ethyl alcohol ($d = 0.789$ g/mL) can be produced from 575 lb of glucose.

35. For mass–mass calculation, why is it necessary to convert the grams of starting material to moles of starting material, then determine the moles of product from the moles of starting material, and convert the moles of product to grams of product? Why can't you calculate the grams of product directly from the grams of starting material?

36. Oxygen masks for producing O_2 in emergency situations contain potassium superoxide (KO_2). It reacts according to this equation:

$$4\ KO_2 + 2\ H_2O + 4\ CO_2 \longrightarrow 4\ KHCO_3 + 3\ O_2$$

(a) If a person wearing such a mask exhales 0.85 g of CO_2 every minute, how many moles of KO_2 are consumed in 10.0 minutes?
(b) How many grams of oxygen are produced in 1.0 hour?

37. In the presence of concentrated H_2SO_4, sucrose, $C_{12}H_{22}O_{11}$, undergoes dehydration, forming carbon and water:

$$C_{12}H_{22}O_{11} \xrightarrow{H_2SO_4} 12\ C + 11\ H_2O$$

(a) How many grams of carbon and how many grams of water can be produced from 2.0 lb of sucrose?
(b) If 25.2 g sucrose are dehydrated, how many grams of liquid water with a density of 0.994 g/mL can be collected?

38. Alfred Nobel extended Sobrero's work on nitroglycerin and discovered that when mixed with diatomaceous earth it would form a stable mixture. This mixture is now known as dynamite. The reaction of nitroglycerin is

$$4\ C_3H_5(NO_3)_3(l) \longrightarrow 12\ CO_2(g) + 10\ H_2O(g) + 6\ N_2(g) + O_2(g)$$

(a) Calculate the number of grams of each product that are produced by the decomposition of 45.0 g of nitroglycerin.
(b) Calculate the total number of moles of gas produced.

39. The methyl alcohol (CH_3OH) used in alcohol burners combines with oxygen gas to form carbon dioxide and water. How many grams of oxygen are required to burn 60.0 mL of methyl alcohol ($d = 0.787$ g/mL)?

40. Hydrazine (N_2H_4) and hydrogen peroxide (H_2O_2) have been used as rocket propellants. They react according to the following equation:

$$7\ H_2O_2 + N_2H_4 \longrightarrow 2\ HNO_3 + 8\ H_2O$$

(a) When 75 kg of hydrazine react, how many grams of nitric acid can be formed?
(b) When 250 L of hydrogen peroxide ($d = 1.41$ g/mL) react, how many grams of water can be formed?
(c) How many grams of hydrazine will be required to react with 725 g hydrogen peroxide?
(d) How many grams of water can be produced when 750 g hydrazine combine with 125 g hydrogen peroxide?
(e) How many grams of the excess reactant in part (d) are left unreacted?

41. Chlorine gas can be prepared according to the reaction

$$16\ HCl + 2\ KMnO_4 \longrightarrow 5\ Cl_2 + 2\ KCl + 2\ MnCl_2 + 8\ H_2O$$

(a) How many moles of $MnCl_2$ can be produced when 25 g $KMnO_4$ are mixed with 85 g HCl?
(b) How many grams of water will be produced when 75 g KCl are produced?
(c) What is the percent yield of Cl_2 if 150 g HCl are reacted, producing 75 g of Cl_2?
(d) When 25 g HCl react with 25 g $KMnO_4$, how many grams of Cl_2 can be produced?
(e) How many grams of the excess reactant in part (d) are left unreacted?

42. Silver tarnishes in the presence of hydrogen sulfide (which smells like rotten eggs) and oxygen because of the reaction

$$4\ Ag + 2\ H_2S + O_2 \longrightarrow 2\ Ag_2S + 2\ H_2O$$

(a) How many grams of silver sulfide can be formed from a mixture of 1.1 g Ag, 0.14 g H_2S, and 0.080 g O_2?
(b) How many more grams of H_2S would be needed to completely react all of the Ag?

43. Humans long ago discovered that they could make soap by cooking fat with ashes from a fire. This happens because the ash contains potassium hydroxide that reacts with the triglycerides in the fat to form soap. The chemical reaction for this process is

$$\underset{\text{(triglyceride)}}{C_3H_5(C_{15}H_{31}CO_2)_3} + 3\ KOH \longrightarrow C_3H_5(OH)_3 + \underset{\text{(soap)}}{3\ C_{15}H_{31}CO_2K}$$

If 500.0 g of this triglyceride are mixed with 500.0 g of potassium hydroxide and 361.7 g of soap are produced, what is the percent yield for the reaction?

44. Group 2A metal oxides react with water to produce hydroxides. When 35.55 g of CaO are placed into 125 mL water ($d = 1.000$ g/mL), how many grams of $Ca(OH)_2$ can be produced?

45. When a solution of lead(II) nitrate is mixed with a solution of sodium chromate, a yellow precipitate forms.

(a) Write the balanced equation for the reaction, including the states of all the substances.
(b) When 26.41 g of lead(II) nitrate are mixed with 18.33 g of sodium chromate, what is the percent yield of the solid if 21.23 g are recovered?

46. Ethyl alcohol burns, producing carbon dioxide and water:

$$C_2H_5OH + 3\ O_2 \longrightarrow 2\ CO_2 + 3\ H_2O$$

(a) If 2.5 mol of C_2H_5OH react with 7.5 mol of O_2 in a closed container, what substances and how many moles of each will be present in the container when the reaction has gone to completion?
(b) When 225 g of C_2H_5OH are burned, how many grams of CO_2 and how many grams of H_2O will be produced?

47. After 180.0 g of zinc were dropped into a beaker of hydrochloric acid and the reaction ceased, 35 g of unreacted zinc remained in the beaker:

$$Zn + HCl \longrightarrow ZnCl_2 + H_2$$

(a) How many grams of hydrogen gas were produced?
(b) How many grams of HCl were reacted?
(c) How many more grams of HCl would be required to completely react with the original sample of zinc?

48. Use this equation to answer (a) and (b):

$$Fe(s) + CuSO_4(aq) \longrightarrow Cu(s) + FeSO_4(aq)$$

(a) When 2.0 mol of Fe and 3.0 mol of $CuSO_4$ are reacted, what substances will be present when the reaction is over? How many moles of each substance are present?
(b) When 20.0 g of Fe and 40.0 g of $CuSO_4$ are reacted, what substances will be present when the reaction is over? How many grams of each substance are present?

49. Methyl alcohol (CH_3OH) is made by reacting carbon monoxide and hydrogen in the presence of certain metal oxide catalysts. How much alcohol can be obtained by reacting 40.0 g of CO and 10.0 g of H_2? How many grams of excess reactant remain unreacted?

$$CO(g) + 2\ H_2(g) \longrightarrow CH_3OH(l)$$

50. Ethyl alcohol (C_2H_5OH), also called grain alcohol, can be made by the fermentation of glucose:

$$\underset{\text{glucose}}{C_6H_{12}O_6} \longrightarrow \underset{\text{ethyl alcohol}}{2\ C_2H_5OH} + 2\ CO_2$$

If an 84.6% yield of ethyl alcohol is obtained,
(a) what mass of ethyl alcohol will be produced from 750 g of glucose?
(b) what mass of glucose should be used to produce 475 g of C_2H_5OH?

51. Both $CaCl_2$ and $MgCl_2$ react with $AgNO_3$ to precipitate AgCl. When solutions containing equal masses of $CaCl_2$ and $MgCl_2$ are reacted with $AgNO_3$, which salt solution will produce the larger amount of AgCl? Show proof.

52. An astronaut excretes about 2500 g of water a day. If lithium oxide (Li_2O) is used in the spaceship to absorb this water, how many kilograms of Li_2O must be carried for a 30-day space trip for three astronauts?

$$Li_2O + H_2O \longrightarrow 2\ LiOH$$

53. Much commercial hydrochloric acid is prepared by the reaction of concentrated sulfuric acid with sodium chloride:

$$H_2SO_4 + 2\ NaCl \longrightarrow Na_2SO_4 + 2\ HCl$$

How many kilograms of concentrated H_2SO_4, 96% H_2SO_4 by mass, are required to produce 20.0 L of concentrated hydrochloric acid ($d = 1.20$ g/mL, 42.0% HCl by mass)?

54. Silicon carbide or carborundum is a widely used ceramic material in car brakes and bullet-proof vests because of its hardness. Carborundum is found naturally in the mineral moissanite, but the demand for it exceeds the supply. For this reason it is generally synthesized in the laboratory. The reaction is

$$SiO_2 + 3\,C \longrightarrow SiC + 2\,CO$$

How many kg of silicon carbide can be formed by the reaction of 250.0 kg of sand (SiO_2) with excess carbon?

55. Three chemical reactions that lead to the formation of sulfuric acid are

$$S + O_2 \longrightarrow SO_2$$

$$2\,SO_2 + O_2 \longrightarrow 2\,SO_3$$

$$SO_3 + H_2O \longrightarrow H_2SO_4$$

Starting with 100.0 g of sulfur, how many grams of sulfuric acid will be formed, assuming a 10% loss in each step? What is the percent yield of H_2SO_4?

56. A 10.00-g mixture of $NaHCO_3$ and Na_2CO_3 was heated and yielded 0.0357 mol H_2O and 0.1091 mol CO_2. Calculate the percent composition of the mixture:

$$NaHCO_3 \longrightarrow Na_2O + CO_2 + H_2O$$

$$Na_2CO_3 \longrightarrow Na_2O + CO_2$$

57. Draw a picture showing the reaction of 2 molecules CH_4 with 3 molecules of O_2 to form water and carbon dioxide gas. Also show any unreacted CH_4 and O_2. Does one of the reactants limit the amount of products formed? If so, which one?

CHALLENGE EXERCISES

58. Sugar can decompose by the following reaction:

$$C_6H_{12}O_6 \longrightarrow 2\,C_2H_5OH + 2\,CO_2(g)$$

(a) If 25.0 g of $C_6H_{12}O_6$ were used, and only 11.2 g of C_2H_5OH were produced, how much reactant was left over and what volume of gas was produced? (*Note:* One mole of gas occupies 24.0 L of space at room temperature.)
(b) What was the yield of the reaction?
(c) What type of reaction took place?

59. When solutions containing 25 g each of lead(II) nitrate and potassium iodide were mixed, a yellow precipitate resulted.
(a) What type of reaction occurred?
(b) What are the name and formula for the solid product?
(c) If after filtration and drying, the solid product weighed 7.66 g, what was the percent yield for the reaction?

60. Consider the following unbalanced reaction:

$$XNO_3 + CaCl_2 \longrightarrow XCl + Ca(NO_3)_2$$

(a) If 30.8 g of $CaCl_2$ produced 79.6 g of XCl, what is X?
(b) Would X be able to displace hydrogen from an acid?

61. When 12.82 g of a mixture of $KClO_3$ and NaCl are heated strongly, the $KClO_3$ reacts according to this equation:

$$2\,KClO_3(s) \longrightarrow 2\,KCl(s) + 3\,O_2(g)$$

The NaCl does not undergo any reaction. After the heating, the mass of the residue (KCl and NaCl) is 9.45 g. Assuming that all the loss of mass represents loss of oxygen gas, calculate the percent of $KClO_3$ in the original mixture.

62. Gastric juice contains about 3.0 g HCl per liter. If a person produces about 2.5 L of gastric juice per day, how many antacid tablets, each containing 400 mg of $Al(OH)_3$, are needed to neutralize all the HCl produced in one day?

$$Al(OH)_3(s) + 3\,HCl(aq) \longrightarrow AlCl_3(aq) + 3\,H_2O(l)$$

63. Phosphoric acid, H_3PO_4, can be synthesized from phosphorus, oxygen, and water according to these two reactions:

$$4\,P + 5\,O_2 \longrightarrow P_4O_{10}$$

$$P_4O_{10} + 6\,H_2O \longrightarrow 4\,H_3PO_4$$

Starting with 20.0 g P, 30.0 g O_2, and 15.0 g H_2O, what is the mass of phosphoric acid that can be formed?

64. You have just finished your first year of college in South Carolina and decide to take a cross-country trip to visit California. You begin your drive west, but upon arriving in Kansas you discover to your dismay that your funds have dwindled. You meet up with the manager of a feed mill who has just lost his formulations chemist and needs help determining how many bags of feed supplement he can produce with the oats, molasses, alfalfa, and apples that are arriving on Monday. Luckily you just finished your chemistry class and remember learning about limiting reactants. If you can use these ideas to solve the following problem, your money worries may be over! Here is the problem: A new customer just ordered a new formulation of supplement containing 25.0 lb of oats, 17.5 lb of molasses, 30.0 lb of alfalfa, and 28.5 lb of apples per bag. If four train cars of materials are due in the plant on Monday containing the components listed below, how much money will you make for the remainder of your trip if you earn $2.25 for every bag of supplement produced?

Raw materials

1 boxcar of oats containing 4.20 tons of oats
1 tanker of molasses containing 2490 gallons of molasses (molasses has a density of 1.46 g/mL)
250.0 crates of apples, each containing 12.5 kg of apples
350.0 bales of alfalfa, each weighing 62.8 kg

ANSWERS TO PRACTICE EXERCISES

9.1 (a) $\dfrac{2 \text{ mol KCl}}{2 \text{ mol KClO}_3}$ (b) $\dfrac{2 \text{ mol KCl}}{3 \text{ mol O}_2}$
9.2 0.33 mol Al_2O_3
9.3 7.33 mol $Al(OH)_3$
9.4 0.8157 mol KCl, 1.223 mol O_2
9.5 85 g $AgNO_3$
9.6 27.6 g $CrCl_3$
9.7 348.6 g H_2O
9.8 17.7 g HCl
9.9 164.3 g $BaSO_4$
9.10 89.1% yield

10

Modern Atomic Theory and the Periodic Table

CHAPTER OUTLINE

How do we go about studying an object that is too small to see? Think back to that birthday present you could look at but not yet open. Judging from the wrapping and size of the box was not very useful, but shaking, turning, and lifting the package all gave indirect clues to its contents. After all your experiments were done, you could make a fairly good guess about the contents. But was your guess correct? The only way to know for sure would be to open the package.

Chemists have the same dilemma when they study the atom. Atoms are so very small that it isn't possible to use our normal senses to describe them. We are essentially working in the dark with this package we call the atom. As seen in fireworks displays, atoms can emit light under the right conditions. The light emitted can give us clues as to the electron structure of the atom. Additionally, our improvements in instruments (X-ray machines and scanning tunneling microscopes) and measuring devices (spectrophotometers and magnetic resonance imaging, MRI) as well as in our mathematical skills are bringing us closer to revealing the secrets of the atom.

10.1 Electromagnetic Radiation

List the three basic characteristics of electromagnetic radiation.

LEARNING OBJECTIVE

KEY TERMS

wavelength
frequency
speed
photons

In the last 200 years, vast amounts of data have been accumulated to support atomic theory. When atoms were originally suggested by the early Greeks, no physical evidence existed to support their ideas. Early chemists did a variety of experiments, which culminated in Dalton's model of the atom. Because of the limitations of Dalton's model, modifications were proposed first by Thomson and then by Rutherford, which eventually led to our modern concept of the nuclear atom. These early models of the atom work reasonably well—in fact, we continue to use them to visualize a variety of chemical concepts. There remain questions that these models cannot answer, including an explanation of how atomic structure relates to the periodic table. In this chapter, we will present our modern model of the atom; we will see how it varies from and improves upon earlier atomic models.

Electromagnetic Radiation

Scientists have studied energy and light for centuries, and several models have been proposed to explain how energy is transferred from place to place. One way energy travels through space is by *electromagnetic radiation*. Examples of electromagnetic radiation include light from the sun, X-rays in your dentist's office, microwaves from your microwave oven, radio and television waves, and radiant heat from your fireplace. While these examples seem quite different, they are all similar in some important ways. Each shows wavelike behavior, and all travel at the same speed in a vacuum (3.00×10^8 m/s).

The study of wave behavior is a topic for another course, but we need some basic terminology to understand atoms. Waves have three basic characteristics: wavelength, frequency, and speed. **Wavelength** (lambda, λ) is the distance between consecutive peaks (or troughs) in a wave, as shown in FIGURE 10.1. **Frequency** (nu, ν) tells how many waves pass a particular point per second. **Speed** (v) tells how fast a wave moves through space.

Tony Freeman/PhotoEdit

Surfers judge the wavelength, frequency, and speed of waves to get the best ride.

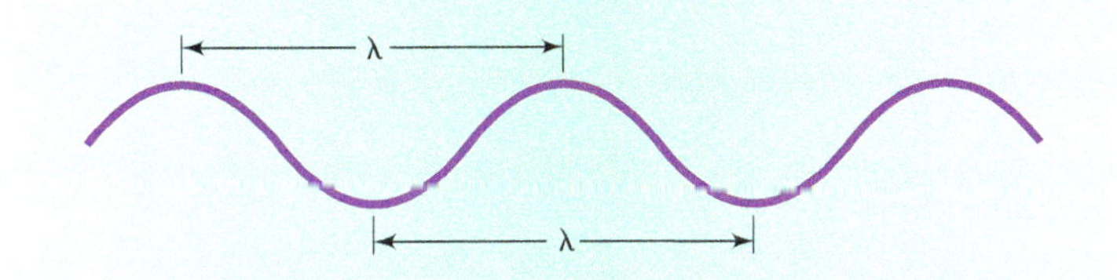

FIGURE 10.1 The wavelength of this wave is shown by λ. It can be measured from peak to peak or trough to trough.

Light is one form of electromagnetic radiation and is usually classified by its wavelength, as shown in FIGURE 10.2. Visible light, as you can see, is only a tiny part of the electromagnetic spectrum. Some examples of electromagnetic radiation involved in energy transfer outside the visible region are hot coals in your backyard grill, which transfer infrared radiation to cook your food, and microwaves, which transfer energy to water molecules in the food, causing them to move more quickly and thus raise the temperature of the food.

CHECK YOUR UNDERSTANDING

10.1 Wavelength, Frequency, and Speed

WileyPLUS

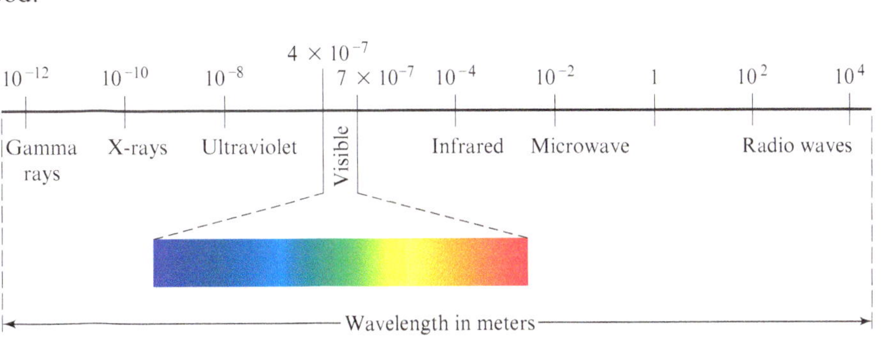

FIGURE 10.2 The electromagnetic spectrum.

CHEMISTRY *IN ACTION* You Light Up My Life

Birds in the parrot family have an unusual way to attract their mates—their feathers glow in the dark! This phenomenon is called fluorescence. It results from the absorption of ultraviolet (UV) light, which is then reemitted at longer wavelengths that both birds and people can see. In everyday life this happens in a fluorescent bulb or in the many glow-in-the-dark products such as light sticks.

Kathleen Arnold from the University of Glasgow, Scotland, discovered that the feathers of parrots that fluoresced were only those used in display or those shown off during courtship. She decided to experiment using budgerigars with their natural colors. The researchers offered birds a choice of two companion birds, which were smeared with petroleum jelly. One of the potential companions also had a UV blocker in the petroleum jelly. The birds clearly preferred companions without the UV blocker, leading the researchers to conclude that the parrots prefer to court radiant partners. The researchers also tested same-sex companions and discovered they did not prefer radiant companions.

Perhaps we may find that the glow of candlelight really does add to the radiance of romance. Ultraviolet light certainly does for parrots!

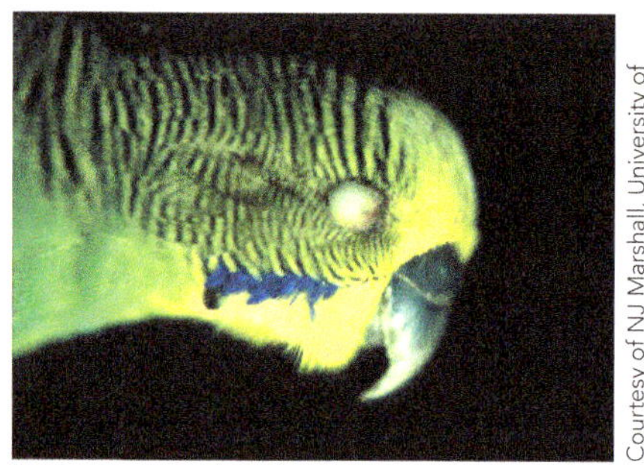

Courtesy of NJ Marshall, University of Queensland

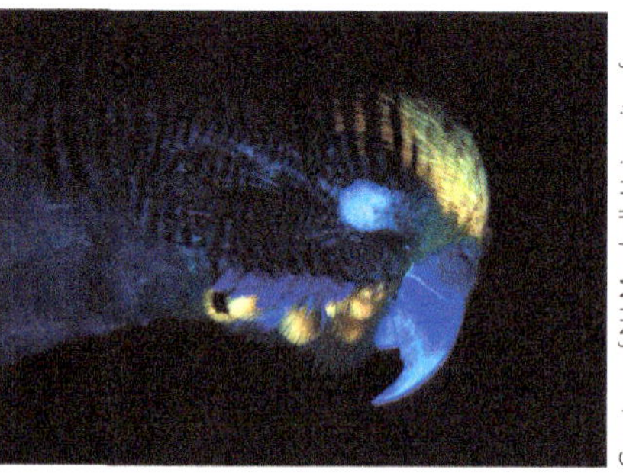

Courtesy of NJ Marshall, University of Queensland

Budgerigars under normal light and under UV light showing the glow used to attract mates.

CHECK YOUR UNDERSTANDING

10.2 Photons

WileyPLUS

We have evidence for the wavelike nature of light. We also know that a beam of light behaves like a stream of tiny packets of energy called **photons**. So what is light exactly? Is it a particle? Is it a wave? Scientists have agreed to explain the properties of electromagnetic radiation by using both wave and particle properties. Neither explanation is ideal, but currently these are our best models.

ONLINE LEARNING MODULE

Wavelength

WileyPLUS

EXAMPLE 10.1

What is a wavelength?

SOLUTION

A wavelength is the distance between consecutive peaks or troughs of a wave. The symbol used for wavelength is lambda (λ).

PRACTICE 10.1

Draw a wave function and indicate the distance of two wavelengths (2 lambda).

10.2 The Bohr Atom

LEARNING OBJECTIVE

Explain the relationship between the line spectrum and the quantized energy levels of an electron in an atom.

KEY TERMS

line spectrum
quanta
ground state
orbital

As scientists struggled to understand the properties of electromagnetic radiation, evidence began to accumulate that atoms could radiate light. At high temperatures, or when subjected to high voltages, elements in the gaseous state give off colored light. Brightly colored neon signs illustrate this property of matter very well. When the light emitted by a gas is passed through a prism or diffraction grating, a set of brightly

FIGURE 10.3 Line spectrum of hydrogen. Each line corresponds to the wavelength of the energy emitted when the electron of a hydrogen atom, which has absorbed energy, falls back to a lower principal energy level.

colored lines called a **line spectrum** results (FIGURE 10.3). These colored lines indicate that the light is being emitted only at certain wavelengths, or frequencies, that correspond to specific colors. Each element possesses a unique set of these spectral lines that is different from the sets of all the other elements.

In 1912–1913, while studying the line spectrum of hydrogen, Niels Bohr (1885–1962), a Danish physicist, made a significant contribution to the rapidly growing knowledge of atomic structure. His research led him to believe that electrons exist in specific regions at various distances from the nucleus. He also visualized the electrons as revolving in orbits around the nucleus, like planets rotating around the sun, as shown in FIGURE 10.4.

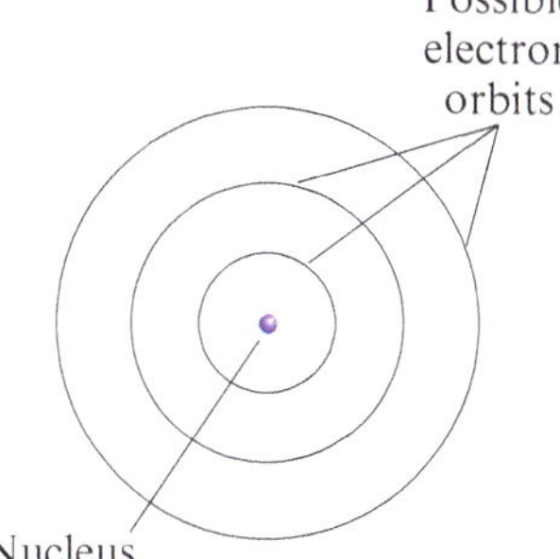

FIGURE 10.4 The Bohr model of the hydrogen atom described the electron revolving in certain allowed circular orbits around the nucleus.

Bohr's first paper in this field dealt with the hydrogen atom, which he described as a single electron revolving in an orbit about a relatively heavy nucleus. He applied the concept of energy quanta, proposed in 1900 by the German physicist Max Planck (1858–1947), to the observed line spectrum of hydrogen. Planck stated that energy is never emitted in a continuous stream but only in small, discrete packets called **quanta** (Latin, *quantus*, "how much"). From this, Bohr theorized that electrons have several possible energies corresponding to several possible orbits at different distances from the nucleus. Therefore, an electron has to be in one specific energy level; it cannot exist between energy levels. In other words, the energy of the electron is said to be quantized. Bohr also stated that when a hydrogen atom absorbed one or more quanta of energy, its electron would "jump" to a higher energy level.

Bohr was able to account for spectral lines of hydrogen this way. A number of energy levels are available, the lowest of which is called the **ground state**. When electrons are located in these lowest energy orbitals, the atom is said to be in the ground state. Any other electron arrangement would be called an excited state of the atom. When an electron falls from a high energy level to a lower one (say, from the fourth to the second), a quantum of energy is emitted as light at a specific frequency, or wavelength (FIGURE 10.5). This light corresponds to one of the lines visible in the hydrogen spectrum (Figure 10.3). Several lines are visible in this spectrum, each one corresponding to a specific electron energy-level shift within the hydrogen atom.

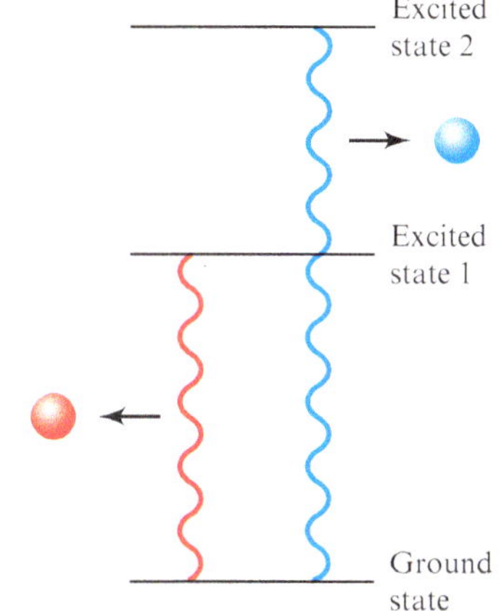

FIGURE 10.5 When an excited electron returns to the ground state, energy emitted as a photon is released. The color (wavelength) of the light is determined by the difference in energy between the two states (excited and ground).

EXAMPLE 10.2

What does the line spectrum of a hydrogen atom illustrate?

SOLUTION

Each line shows the frequency of light energy emitted when an electron that has absorbed energy falls from a higher energy level to a lower energy level.

PRACTICE 10.2

What is a photon?

ONLINE LEARNING MODULE

Atomic Spectra

➡ WileyPLUS

The chemical properties of an element and its position in the periodic table depend on electron behavior within the atoms. In turn, much of our knowledge of the behavior of electrons within atoms is based on spectroscopy. Niels Bohr contributed a great deal to our knowledge of atomic structure by (1) suggesting quantized energy levels for electrons and (2) showing that spectral lines result from the radiation of small increments of energy (Planck's quanta) when electrons shift from one energy level to another. Bohr's

CHECK YOUR UNDERSTANDING

10.3 The Ground State of an Atom

➡ WileyPLUS

calculations succeeded very well in correlating the experimentally observed spectral lines with electron energy levels for the hydrogen atom. However, Bohr's methods of calculation did not succeed for heavier atoms. More theoretical work on atomic structure was needed.

In 1924, the French physicist Louis de Broglie (1892–1957) suggested a surprising hypothesis: All objects have wave properties. Louis de Broglie used sophisticated mathematics to show that the wave properties for an object of ordinary size, such as a baseball, are too small to be observed. But for smaller objects, such as an electron, the wave properties become significant. Other scientists confirmed de Broglie's hypothesis, showing that electrons do exhibit wave properties. In 1926, Erwin Schrödinger (1887–1961), an Austrian physicist, created a mathematical model that described electrons as waves. Using Schrödinger's wave mechanics, we can determine the *probability* of finding an electron in a certain region around the nucleus of the atom.

FIGURE 10.6 An orbital for a hydrogen atom. The intensity of the dots shows that the electron spends more time closer to the nucleus.

This treatment of the atom led to a new branch of physics called *wave mechanics* or *quantum mechanics*, which forms the basis for our modern understanding of atomic structure. Although the wave-mechanical description of the atom is mathematical, it can be translated, at least in part, into a visual model. It is important to recognize that we cannot locate an electron precisely within an atom; however, it is clear that electrons are not revolving around the nucleus in orbits as Bohr postulated. The electrons are instead found in *orbitals*. An **orbital** is pictured in FIGURE 10.6 as a region in space around the nucleus where there is a high probability of finding a given electron.

10.3 Energy Levels of Electrons

LEARNING OBJECTIVE Describe the principal energy levels, sublevels, and orbitals of an atom.

KEY TERMS
- principal energy levels
- sublevels
- spin
- Pauli exclusion principle

One of the ideas Bohr contributed to the modern concept of the atom was that the energy of the electron is quantized—that is, the electron is restricted to only certain allowed energies. The wave-mechanical model of the atom also predicts discrete **principal energy levels** within the atom. These energy levels are designated by the letter n, where n is a positive integer (FIGURE 10.7). The lowest principal energy level corresponds to $n = 1$, the next to $n = 2$, and so on. As n increases, the energy of the electron increases, and the electron is found on average farther from the nucleus.

Each principal energy level is divided into **sublevels**, which are illustrated in FIGURE 10.8. The first principal energy level has one sublevel. The second principal energy level has two sublevels, the third energy level has three sublevels, and so on. Each of these sublevels contains spaces for electrons called orbitals.

In each sublevel the electrons are found within specified orbitals (s, p, d, f). Let's consider each principal energy level in turn. The first principal energy level ($n = 1$) has one

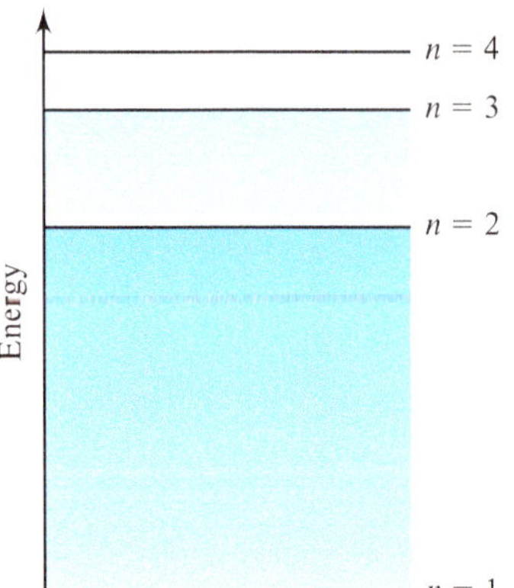

FIGURE 10.7 The first four principal energy levels in the hydrogen atom. Each level is assigned a principal quantum number n.

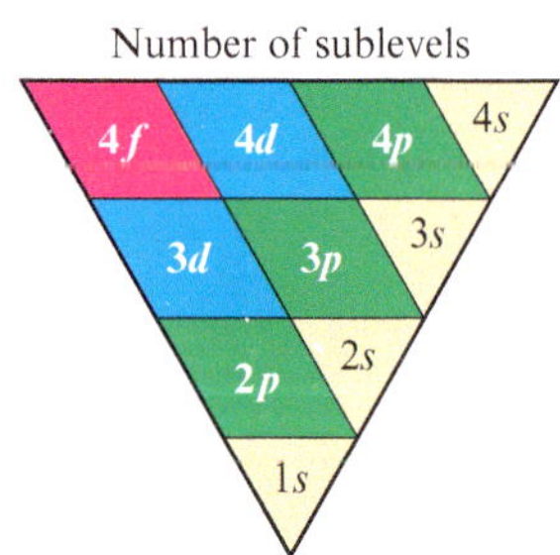

FIGURE 10.8 The types of orbitals on each of the first four principal energy levels.

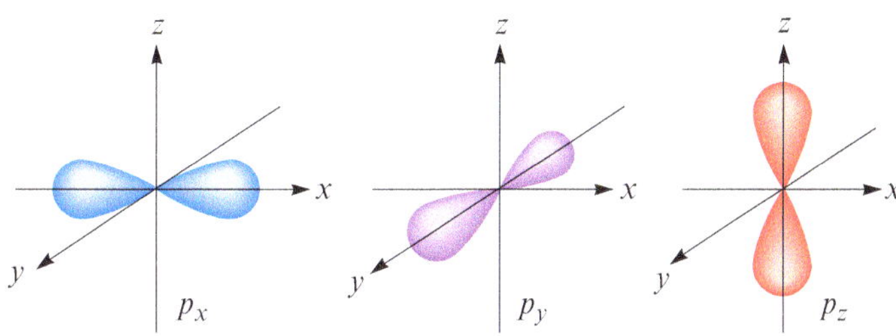

FIGURE 10.9 Perspective representation of the p_x, p_y, and p_z atomic orbitals.

sublevel or type of orbital. It is spherical in shape and is designated as 1*s*. It is important to understand what the spherical shape of the 1*s* orbital means. The electron does *not* move around on the surface of the sphere, but rather the surface encloses a space where there is a 90% probability where the electron may be found. It might help to consider these orbital shapes in the same way we consider the atmosphere. There is no distinct dividing line between the atmosphere and "space." The boundary is quite fuzzy. The same is true for atomic orbitals. Each has a region of highest density roughly corresponding to its shape. The probability of finding the electron outside this region drops rapidly but never quite reaches zero. Scientists often speak of orbitals as electron "clouds" to emphasize the fuzzy nature of their boundaries.

How many electrons can fit into a 1*s* orbital? To answer this question, we need to consider one more property of electrons. This property is called **spin**. Each electron appears to be spinning on an axis, like a globe. It can only spin in two directions. We represent this spin with an arrow: ↑ or ↓. In order to occupy the same orbital, electrons must have *opposite* spins. That is, two electrons with the same spin cannot occupy the same orbital. This gives us the answer to our question: An atomic orbital can hold a maximum of two electrons, which must have opposite spins. This rule is called the **Pauli exclusion principle**. The first principal energy level contains one type of orbital (1*s*) that holds a maximum of two electrons.

What happens with the second principal energy level ($n = 2$)? Here we find two sublevels, 2*s* and 2*p*. Like 1*s* in the first principal energy level, the 2*s* orbital is spherical in shape but is larger in size and higher in energy. It also holds a maximum of two electrons. The second type of orbital is designated by 2*p*. The 2*p* sublevel consists of three orbitals: $2p_x$, $2p_y$, and $2p_z$. The shape of *p* orbitals is quite different from the *s* orbitals, as shown in FIGURE 10.9.

CHECK YOUR UNDERSTANDING
10.4 Pauli Exclusion Principle
10.5 Electrons in an *s* Sublevel
WileyPLUS

Each *p* orbital has two "lobes." Remember, the space enclosed by these surfaces represents the regions of probability for finding the electrons 90% of the time. There are three separate *p* orbitals, each oriented in a different direction, and each *p* orbital can hold a maximum of two electrons. Thus the total number of electrons that can reside in all three *p* orbitals is six. To summarize our model, the first principal energy level of an atom has a 1*s* orbital. The second principal energy level has a 2*s* and three 2*p* orbitals labeled $2p_x$, $2p_y$, and $2p_z$, as shown in FIGURE 10.10.

Notice that there is a correspondence between the energy level and number of sublevels.

CHECK YOUR UNDERSTANDING
10.6 Orbitals in a *p* Sublevel
WileyPLUS

The third principal energy level has three sublevels labeled 3*s*, 3*p*, and 3*d*. The 3*s* orbital is spherical and larger than the 1*s* and 2*s* orbitals. The $3p_x$, $3p_y$, and $3p_z$ orbitals are shaped

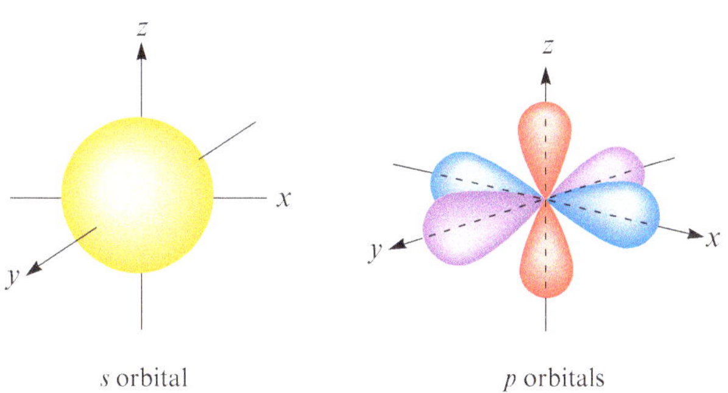

FIGURE 10.10 Orbitals on the second principal energy level are one 2*s* and three 2*p* orbitals.

CHEMISTRY *IN ACTION* Atomic Clocks

Imagine a clock that keeps time to within 1 second over a million years. The National Institute of Standards and Technology in Boulder, Colorado, has an atomic clock that does just that—a little better than your average alarm, grandfather, or cuckoo clock! This atomic clock serves as the international standard for time and frequency. How does it work?

Within the case are several layers of magnetic shielding. In the heart of the clock is a small oven that heats cesium metal to release cesium atoms, which are collected into a narrow beam (1 mm wide). The beam of atoms passes down a long evacuated tube while being excited by a laser until all the cesium atoms are in the same electron state.

The atoms then enter another chamber filled with reflecting microwaves. The frequency of the microwaves (9,192,631,770 cycles per second) is exactly the same frequency required to excite a cesium atom from its ground state to the next higher energy level. These excited cesium atoms then release electromagnetic radiation in a process known as fluorescence. Electronic circuits maintain the microwave frequency at precisely the right level to keep the cesium atoms moving from one level to the next. One second is equal to 9,192,631,770 of these vibrations. The clock is set to this frequency and can keep accurate time for over a million years.

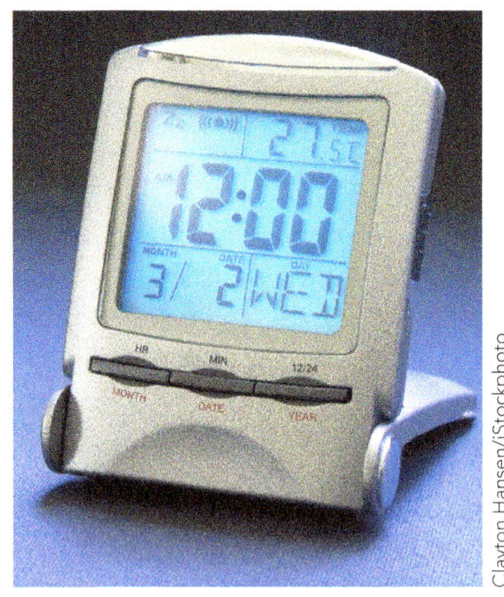

This clock automatically updates itself by comparing time with an atomic clock by radio signal.

CHECK YOUR UNDERSTANDING
10.7 Electrons in a *d* Sublevel
WileyPLUS

like those of the second level, only larger. The five 3*d* orbitals have the shapes shown in FIGURE 10.11. You don't need to memorize these shapes, but notice that they look different from the *s* or *p* orbitals.

Each time a new principal energy level is added, we also add a new sublevel. This makes sense because each energy level corresponds to a larger average distance from the nucleus, which provides more room on each level for new sublevels containing more orbitals.

The pattern continues with the fourth principal energy level. It has 4*s*, 4*p*, 4*d*, and 4*f* orbitals. There are one 4*s*, three 4*p*, five 4*d*, and seven 4*f* orbitals. The shapes of the *s*, *p*, and *d* orbitals are the same as those for lower levels, only larger. We will not consider the shapes of the *f* orbitals. Remember that for all *s*, *p*, *d*, and *f* orbitals, the maximum number of electrons per orbital is two. We summarize each principal energy level:

$n = 1$	1*s*			
$n = 2$	2*s*	2*p* 2*p* 2*p*		
$n = 3$	3*s*	3*p* 3*p* 3*p*	3*d* 3*d* 3*d* 3*d* 3*d*	
$n = 4$	4*s*	4*p* 4*p* 4*p*	4*d* 4*d* 4*d* 4*d* 4*d*	4*f* 4*f* 4*f* 4*f* 4*f* 4*f* 4*f*

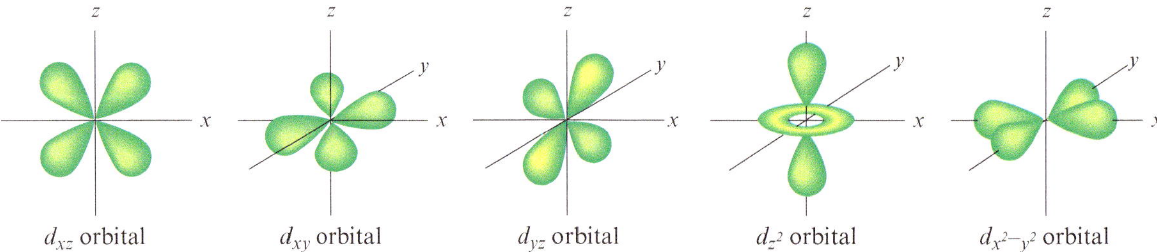

FIGURE 10.11 The five *d* orbitals are found in the third principal energy level along with one 3*s* orbital and three 3*p* orbitals.

EXAMPLE 10.3

Energy levels of the atoms are designated as principal energy levels and energy sublevels.
(a) What labels are used for these energy levels?
(b) How many electrons are in each sublevel type?

SOLUTION

(a) The letter n is used to designate the principal energy levels. The letters s, p, d, and f are used to designate the energy sublevels.
(b) According to the periodic table, there are seven (7) principal energy levels and $2s$, $6p$, $10d$, and $14f$ energy sublevels.

PRACTICE 10.3

(a) How many electrons can occupy a single subenergy level?
(b) What is the maximum number of electrons that can occupy the third principal energy level? List them.

FIGURE 10.12 The modern concept of a hydrogen atom consists of a proton and an electron in an s orbital. The shaded area represents a region where the electron may be found with 90% probability.

The hydrogen atom consists of a nucleus (containing one proton) and one electron occupying a region outside of the nucleus. In its ground state, the electron occupies a $1s$ orbital, but by absorbing energy the electron can become *excited* and move to a higher energy level.

The hydrogen atom can be represented as shown in FIGURE 10.12. The diameter of the nucleus is about 10^{-13} cm, and the diameter of the electron orbital is about 10^{-8} cm. The diameter of the electron cloud of a hydrogen atom is about 100,000 times greater than the diameter of the nucleus.

10.4 Atomic Structures of the First 18 Elements

Use the guidelines to write electron configurations.

LEARNING OBJECTIVE

KEY TERMS
electron configuration
orbital diagram
valence electrons

We have seen that hydrogen has one electron that can occupy a variety of orbitals in different principal energy levels. Now let's consider the structure of atoms with more than one electron. Because all atoms contain orbitals similar to those found in hydrogen, we can describe the structures of atoms beyond hydrogen by systematically placing electrons in these hydrogen-like orbitals. We use the following guidelines:

1. No more than two electrons can occupy one orbital.
2. Electrons occupy the lowest energy orbitals available. They enter a higher energy orbital only when the lower orbitals are filled. For the atoms beyond hydrogen, orbital energies vary as $s < p < d < f$ for a given value of n.
3. Each orbital in a sublevel is occupied by a single electron before a second electron enters. For example, all three p orbitals must contain one electron before a second electron enters a p orbital.

We can use several methods to represent the atomic structures of atoms, depending on what we are trying to illustrate. When we want to show both the nuclear makeup and the electron structure of each principal energy level (without orbital detail), we can use a diagram such as FIGURE 10.13.

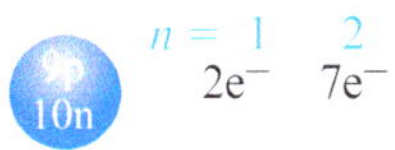

Fluorine atom

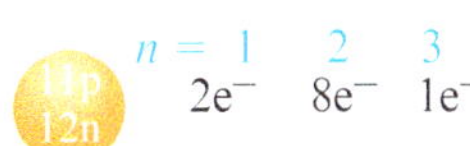

Sodium atom

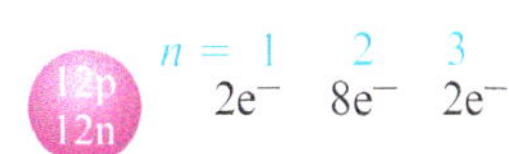

Magnesium atom

FIGURE 10.13 Atomic structure diagrams of fluorine, sodium, and magnesium atoms. The number of protons and neutrons is shown in the nucleus. The number of electrons is shown in each principal energy level outside the nucleus.

Often we are interested in showing the arrangement of the electrons in an atom in their orbitals. There are two ways to do this. The first method is called the **electron configuration**. In this method, we list each type of orbital, showing the number of electrons in it as an exponent. An electron configuration is read as follows:

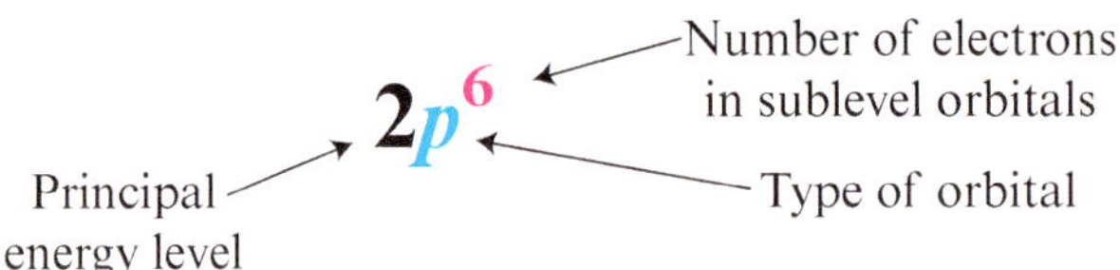

We can also represent this configuration with an **orbital diagram** in which boxes represent the orbitals (containing small arrows indicating the electrons). When the orbital contains one electron, an arrow, pointing upward (↑), is placed in the box. A second arrow, pointing downward (↓), indicates the second electron in that orbital.

Let's consider each of the first 18 elements on the periodic table in turn. The order of filling for the orbitals in these elements is 1*s*, 2*s*, 2*p*, 3*s*, 3*p*, and 4*s*. Hydrogen, the first element, has only one electron. The electron will be in the 1*s* orbital because this is the most favorable position (where it will have the greatest attraction for the nucleus). Both representations are shown here:

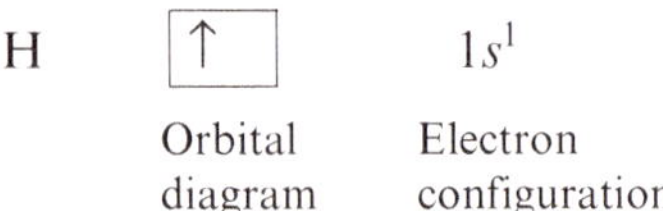

Helium, with two electrons, can be shown as

He [↑↓] $1s^2$

Orbital diagram — Electron configuration

The first energy level, which can hold a maximum of two electrons, is now full. An atom with three electrons will have its third electron in the second principal energy level. Thus, in lithium (atomic number 3), the first two electrons are in the 1*s* orbital, and the third electron is in the 2*s* orbital of the second energy level.

Lithium has the following structure:

Li [↑↓] (1*s*) [↑] (2*s*) $1s^2 2s^1$

All four electrons of beryllium are *s* electrons:

Be [↑↓] (1*s*) [↑↓] (2*s*) $1s^2 2s^2$

The next six elements illustrate the filling of the *p* orbitals. Boron has the first *p* electron. Because *p* orbitals all have the same energy, it doesn't matter which of these orbitals fills first:

B [↑↓] (1*s*) [↑↓] (2*s*) [↑][ ][ ] (2*p*) $1s^2 2s^2 2p^1$

Carbon is the sixth element. It has two electrons in the 1*s* orbital, two electrons in the 2*s* orbital, and two electrons to place in the 2*p* orbitals. Because it is more difficult for the *p* electrons to pair up than to occupy a second *p* orbital, the second *p* electron is located in a different *p* orbital. We could show this by writing $2p_x^1 2p_y^1$, but we usually write it as $2p^2$; it is *understood* that the electrons are in different *p* orbitals. The spins on these electrons are alike, for reasons we will not explain here.

C [↑↓] (1*s*) [↑↓] (2*s*) [↑][↑][ ] (2*p*) $1s^2 2s^2 2p^2$

Nitrogen has seven electrons. They occupy the 1*s*, 2*s*, and 2*p* orbitals. The third *p* electron in nitrogen is still unpaired and is found in the $2p_z$ orbital:

	1*s*	2*s*	2*p*	
N	↑↓	↑↓	↑ ↑ ↑	$1s^22s^22p^3$

Oxygen is the eighth element. It has two electrons in both the 1*s* and 2*s* orbitals and four electrons in the 2*p* orbitals. One of the 2*p* orbitals is now occupied by a second electron, which has a spin opposite the electron already in that orbital:

	1*s*	2*s*	2*p*	
O	↑↓	↑↓	↑↓ ↑ ↑	$1s^22s^22p^4$

The next two elements are fluorine with nine electrons and neon with ten electrons:

	1*s*	2*s*	2*p*	
F	↑↓	↑↓	↑↓ ↑↓ ↑	$1s^22s^22p^5$
Ne	↑↓	↑↓	↑↓ ↑↓ ↑↓	$1s^22s^22p^6$

With neon, the first and second energy levels are filled as shown in **TABLE 10.1**. The second energy level can hold a maximum of eight electrons, $2s^22p^6$.

Sodium, element 11, has two electrons in the first energy level and eight electrons in the second energy level, with the remaining electron occupying the 3*s* orbital in the third energy level:

	1*s*	2*s*	2*p*	3*s*	
Na	↑↓	↑↓	↑↓ ↑↓ ↑↓	↑	$1s^22s^22p^63s^1$

TABLE 10.1 | **Orbital Filling for the First Ten Elements***

Atomic number	Element	Orbitals 1*s*	2*s*	2*p*	Electron configuration
1	H	↑			$1s^1$
2	He	↑↓			$1s^2$
3	Li	↑↓	↑		$1s^22s^1$
4	Be	↑↓	↑↓		$1s^22s^2$
5	B	↑↓	↑↓	↑	$1s^22s^22p^1$
6	C	↑↓	↑↓	↑ ↑	$1s^22s^22p^2$
7	N	↑↓	↑↓	↑ ↑ ↑	$1s^22s^22p^3$
8	O	↑↓	↑↓	↑↓ ↑ ↑	$1s^22s^22p^4$
9	F	↑↓	↑↓	↑↓ ↑↓ ↑	$1s^22s^22p^5$
10	Ne	↑↓	↑↓	↑↓ ↑↓ ↑↓	$1s^22s^22p^6$

*Boxes represent the orbitals grouped by sublevel. Electrons are shown by arrows.

TABLE 10.2 | **Orbital Diagrams and Electron Configurations for Elements 11–18**

Atomic number	Element	1s	2s	2p	3s	3p	Electron configuration
11	Na	↑↓	↑↓	↑↓ ↑↓ ↑↓	↑		$1s^2 2s^2 2p^6 3s^1$
12	Mg	↑↓	↑↓	↑↓ ↑↓ ↑↓	↑↓		$1s^2 2s^2 2p^6 3s^2$
13	Al	↑↓	↑↓	↑↓ ↑↓ ↑↓	↑↓	↑ _ _	$1s^2 2s^2 2p^6 3s^2 3p^1$
14	Si	↑↓	↑↓	↑↓ ↑↓ ↑↓	↑↓	↑ ↑ _	$1s^2 2s^2 2p^6 3s^2 3p^2$
15	P	↑↓	↑↓	↑↓ ↑↓ ↑↓	↑↓	↑ ↑ ↑	$1s^2 2s^2 2p^6 3s^2 3p^3$
16	S	↑↓	↑↓	↑↓ ↑↓ ↑↓	↑↓	↑↓ ↑ ↑	$1s^2 2s^2 2p^6 3s^2 3p^4$
17	Cl	↑↓	↑↓	↑↓ ↑↓ ↑↓	↑↓	↑↓ ↑↓ ↑	$1s^2 2s^2 2p^6 3s^2 3p^5$
18	Ar	↑↓	↑↓	↑↓ ↑↓ ↑↓	↑↓	↑↓ ↑↓ ↑↓	$1s^2 2s^2 2p^6 3s^2 3p^6$

The number of the column in 1A–7A of the periodic table gives the number of valence electrons of the elements in those columns.

Magnesium (12), aluminum (13), silicon (14), phosphorus (15), sulfur (16), chlorine (17), and argon (18) follow in order. TABLE 10.2 summarizes the filling of the orbitals for elements 11–18.

The electrons in the outermost (highest) energy level of an atom are called the **valence electrons**. For example, oxygen, which has the electron configuration of $1s^2 2s^2 2p^4$, has electrons in the first and second energy levels. Therefore, the second principal energy level is the valence level for oxygen. The $2s$ and $2p$ electrons are the valence electrons. Valence electrons are involved in bonding atoms together to form compounds and are of particular interest to chemists, as we will see in Chapter 11.

CHECK YOUR UNDERSTANDING

10.8 Valence Electrons

➡ WileyPLUS

ONLINE LEARNING MODULE

Electron Configurations of Valence Electrons

➡ WileyPLUS

ENHANCED EXAMPLE 10.4 WileyPLUS

What is the electron configuration for the valence electrons in magnesium?

SOLUTION

Magnesium has the electron configuration of $1s^2 2s^2 2p^6 3s^2$. The electrons are in the first, second, and third energy levels. This means the valence electrons are in the third energy level. The $3s$ electrons are the valence electrons for magnesium. The electron configuration for the valence electrons is $3s^2$.

PRACTICE 10.4

Give the electron configuration for the valence electrons in the following elements.

(a) B (b) N (c) Na (d) Cl

10.5 Electron Structures and the Periodic Table

LEARNING OBJECTIVE

Describe how the electron configurations of the atoms relate to their position on the periodic table and write electron configurations for elements based on their position on the periodic table.

KEY TERMS

period

groups or families (of elements)

representative elements

transition elements

We have seen how the electrons are assigned for the atoms of elements 1–18. How do the electron structures of these atoms relate to their position on the periodic table? To answer this question, we need to look at the periodic table more closely.

The periodic table represents the efforts of chemists to organize the elements logically. Chemists of the early nineteenth century had sufficient knowledge of the properties of

CHEMISTRY *IN ACTION* Collecting the Elements

Theodore Gray loves collecting samples of elements. He had an idea some years ago to design a conference table for his company, Wolfram Research in Champaign, Illinois. He built the table in the shape of the periodic table and made little boxes below each element to house a sample of the element. As he was collecting the element samples, he began to realize that he needed to keep track of where each sample had come from before he forgot. So he began to keep notes on each element sample, where he obtained it, what it was, and why it was there. Gray says, "Once I had collected three or four dozen such descriptions, I thought I had better put them on a website, because that's what you do when you have anything these days—you put it on your own website." His element collection is ever expanding and contains more than 1400 samples.

Theodore Gray's websites, http://www.theodoregray.com/PeriodicTable and periodictable.com, include elemental data and many stories that are both entertaining and very distracting. Gray also has created video for each element from all angles in stunning high-resolution photography, formed into video loops. He used to take all the photos himself, but the project has grown so big he now has an assistant and utilizes Mathematica, a computer program from Wolfram Research, to help him. He has also created a beautiful periodic table poster from the element images.

Take a couple of minutes and spend some time with Gray on his websites enjoying the photos and stories of the elements. He has some truly amazing examples of the elements.

© Mike Walker Photography

Theodore Gray and his periodic table conference table containing samples of each element.

elements to recognize similarities among groups of elements. In 1869, Dimitri Mendeleev (1834–1907) of Russia and Lothar Meyer (1830–1895) of Germany independently published periodic arrangements of the elements based on increasing atomic masses. Mendeleev's arrangement is the precursor to the modern periodic table, and his name is associated with it. The modern periodic table is shown on the inside front cover of this book.

Each horizontal row in the periodic table is called a **period**, as shown in FIGURE 10.14. There are seven periods of elements. The number of each period corresponds to

Group number; Noble gases

Key: 9 — Atomic number; F — Symbol

Period	1 1A	2 2A	3 3B	4 4B	5 5B	6 6B	7 7B	8 8B	9 8B	10 8B	11 1B	12 2B	13 3A	14 4A	15 5A	16 6A	17 7A	18 8A
1	1 H																	2 He
2	3 Li	4 Be											5 B	6 C	7 N	8 O	9 F	10 Ne
3	11 Na	12 Mg											13 Al	14 Si	15 P	16 S	17 Cl	18 Ar
4	19 K	20 Ca	21 Sc	22 Ti	23 V	24 Cr	25 Mn	26 Fe	27 Co	28 Ni	29 Cu	30 Zn	31 Ga	32 Ge	33 As	34 Se	35 Br	36 Kr
5	37 Rb	38 Sr	39 Y	40 Zr	41 Nb	42 Mo	43 Tc	44 Ru	45 Rh	46 Pd	47 Ag	48 Cd	49 In	50 Sn	51 Sb	52 Te	53 I	54 Xe
6	55 Cs	56 Ba	57–71 La–Lu	72 Hf	73 Ta	74 W	75 Re	76 Os	77 Ir	78 Pt	79 Au	80 Hg	81 Tl	82 Pb	83 Bi	84 Po	85 At	86 Rn
7	87 Fr	88 Ra	89–103 Ac–Lr	104 Rf	105 Db	106 Sg	107 Bh	108 Hs	109 Mt	110 Ds	111 Rg	112 Cn	113 Uut	114 Fl	115 Uup	116 Lv	117 Uus	118 Uuo

FIGURE 10.14 The periodic table of the elements.

FIGURE 10.15 Valence electron configurations for the first 18 elements.

1 1A							Noble gases 18
1 **H** $1s^1$	2 2A	3 3A	14 4A	15 5A	16 6A	17 7A	2 **He** $1s^2$
3 **Li** $2s^1$	4 **Be** $2s^2$	5 **B** $2s^22p^1$	6 **C** $2s^22p^2$	7 **N** $2s^22p^3$	8 **O** $2s^22p^4$	9 **F** $2s^22p^5$	10 **Ne** $2s^22p^6$
11 **Na** $3s^1$	12 **Mg** $3s^2$	13 **Al** $3s^23p^1$	14 **Si** $3s^23p^2$	15 **P** $3s^23p^3$	16 **S** $3s^23p^4$	17 **Cl** $3s^23p^5$	18 **Ar** $3s^23p^6$

CHECK YOUR UNDERSTANDING

10.9 Horizontal Rows on the Periodic Table

10.10 Vertical Rows on the Periodic Table

➡ **WileyPLUS**

the outermost energy level that contains electrons for elements in that period. Those in Period 1 contain electrons only in energy level 1, while those in Period 2 contain electrons in levels 1 and 2. In Period 3, electrons are found in levels 1, 2, 3, and so on.

Elements that behave in a similar manner are found in **groups** or **families**. These form the vertical columns on the periodic table. Two systems exist for numbering the groups. In one system, the columns are numbered from left to right using the numbers 1–18. The A groups are known as the **representative elements**. The B groups are called the **transition elements**. In this book we will focus on the representative elements. The groups (columns) of the periodic table often have family names. For example, the group on the far right side of the periodic table (He, Ne, Ar, Kr, Xe, and Rn) is called the *noble gases*. Group 1A(1) is called the *alkali metals*, Group 2A(2) the *alkaline earth metals*, and Group 7A(17) the *halogens*.

Both numbering systems are shown in the periodic table on the inside front cover.

How is the structure of the periodic table related to the atomic structures of the elements? We've just seen that the periods of the periodic table are associated with the energy level of the outermost electrons of the atoms in that period. Look at the valence electron configurations of the elements we have just examined (FIGURE 10.15). Do you see a pattern? The valence electron configuration for the elements in each column is the same. The chemical behavior and properties of elements in a particular family must therefore be associated with the electron configuration of the elements. The number for the principal energy level is different. This is expected since each new period is associated with a different energy level for the valence electrons.

The electron configurations for elements beyond these first 18 become long and tedious to write. We often abbreviate the electron configuration using the following notation:

$$\text{Na} \quad [\text{Ne}]3s^1$$

Look carefully at Figure 10.15 and you will see that the p orbitals are full at the noble gases. By placing the symbol for the noble gas in square brackets, we can abbreviate the complete electron configuration and focus our attention on the valence electrons (the electrons we will be interested in when we discuss bonding in Chapter 11). To write the abbreviated electron configuration for any element, go back to the previous noble gas and place its symbol in square brackets. Then list the valence electrons. Here are some examples:

B	$1s^22s^22p^1$	[He] $2s^22p^1$
Cl	$1s^22s^22p^63s^23p^5$	[Ne] $3s^23p^5$
Na	$1s^22s^22p^63s^1$	[Ne] $3s^1$

The sequence for filling the orbitals is exactly as we would expect up through the $3p$ orbitals. The third energy level might be expected to fill with $3d$ electrons before electrons enter the $4s$ orbital, but this is not the case. The behavior and properties of the next two elements, potassium (19) and calcium (20), are very similar to the elements in Groups 1A(1) and 2A(2), respectively. They clearly belong in these groups. The other elements in Group 1A(1) and Group 2A(2) have electron configurations that indicate valence electrons in the *s* orbitals. For example, since the electron configuration is connected to the element's properties, we should place the last electrons for potassium and calcium in the 4*s* orbital. Their electron configurations are

K	$1s^22s^22p^63s^23p^64s^1$	or	$[Ar]\,4s^1$
Ca	$1s^22s^22p^63s^23p^64s^2$	or	$[Ar]\,4s^2$

PRACTICE 10.5

Write the abbreviated electron configuration for the following elements:
(a) Br (b) Sr (c) Ba (d) Te

Elements 21–30 belong to the elements known as *transition elements*. Electrons are placed in the 3*d* orbitals for each of these elements. When the 3*d* orbitals are full, the electrons fill the 4*p* orbitals to complete the fourth period. Let's consider the overall relationship between orbital filling and the periodic table. FIGURE 10.16 illustrates the type of orbital filling and its location on the periodic table. The tall columns on the table (labeled 1A(1) – 7A(17), and noble gases) are often called the *representative elements*. Valence electrons in these elements occupy *s* and *p* orbitals. The period number corresponds to the energy level of the valence electrons. The elements in the center of the periodic table (shown in) are the transition elements where the *d* orbitals are being filled. Notice that the number for the *d* orbitals is one less than the period number. The two rows shown at the bottom of the table in Figure 10.16 are called the *inner transition elements* or the *lanthanide* and *actinide series*. The last electrons in these elements are placed in the *f* orbitals. The number for the *f* orbitals is always two less than that of the *s* and *p* orbitals. A periodic table is almost always available to you, so if you understand the relationship between the orbitals and the periodic table, you can write the electron configuration for any element. There are several minor variations to these rules, but we won't concern ourselves with them in this course.

CHECK YOUR UNDERSTANDING

10.11 Periodic Table - Elements in Group 7A(17)

10.12 Periodic Table - Elements in Groups 1B through 8B (3–12)

WileyPLUS

FIGURE 10.16 Arrangement of elements according to the electron sublevel being filled.

ONLINE LEARNING MODULE

Electron Configurations of Elements

➥ WileyPLUS

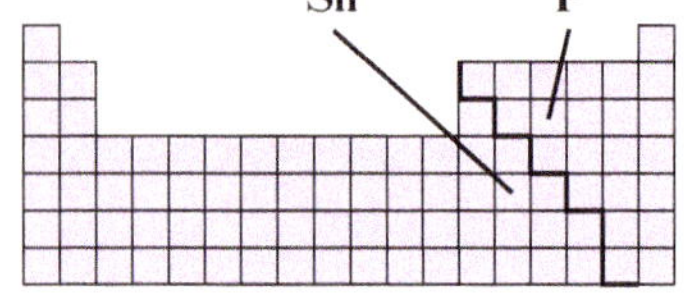

ENHANCED EXAMPLE 10.5

WileyPLUS

Use the periodic table to write the electron configuration for phosphorus and tin.

SOLUTION

Phosphorus is element 15 and is located in Period 3, Group 5A. The electron configuration must have a full first and second energy level:

$$\text{P} \quad 1s^22s^22p^63s^23p^3 \quad \text{or} \quad [\text{Ne}]3s^23p^3$$

You can determine the electron configuration by looking across the period and counting the element blocks.

Tin is element 50 in Period 5, Group 4A(14), two places after the transition metals. It must have two electrons in the 5*p* series. Its electron configuration is

$$\text{Sn} \quad 1s^22s^22p^63s^23p^64s^23d^{10}4p^65s^24d^{10}5p^2 \quad \text{or} \quad [\text{Kr}]5s^24d^{10}5p^2$$

Notice that the *d* series of electrons always has a principal energy level number one less than its period.

PRACTICE 10.6

Use the periodic table to write the complete electron configuration for (a) O, (b) Ca, and (c) Ti.

The early chemists classified the elements based only on their observed properties, but modern atomic theory gives us an explanation for why the properties of elements vary periodically. For example, as we "build" atoms by filling orbitals with electrons, the same orbitals occur on each energy level. This means that the same electron configuration reappears regularly for each level. Groups of elements show similar chemical properties because of the similarity of these outermost electron configurations.

In FIGURE 10.17, only the electron configuration of the outermost electrons is given. This periodic table illustrates these important points:

1. The number of the period corresponds with the highest energy level occupied by electrons in that period.

Period	Group number 1 1A	2 2A	3 3B	4 4B	5 5B	6 6B	7 7B	8 8B	9 8B	10 8B	11 1B	12 2B	13 3A	14 4A	15 5A	16 6A	17 7A	Noble gases 18 8A
1	1 **H** $1s^1$																	2 **He** $1s^2$
2	3 **Li** $2s^1$	4 **Be** $2s^2$											5 **B** $2s^22p^1$	6 **C** $2s^22p^2$	7 **N** $2s^22p^3$	8 **O** $2s^22p^4$	9 **F** $2s^22p^5$	10 **Ne** $2s^22p^6$
3	11 **Na** $3s^1$	12 **Mg** $3s^2$											13 **Al** $3s^23p^1$	14 **Si** $3s^23p^2$	15 **P** $3s^23p^3$	16 **S** $3s^23p^4$	17 **Cl** $3s^23p^5$	18 **Ar** $3s^23p^6$
4	19 **K** $4s^1$	20 **Ca** $4s^2$	21 **Sc** $4s^23d^1$	22 **Ti** $4s^23d^2$	23 **V** $4s^23d^3$	24 **Cr** $4s^13d^5$	25 **Mn** $4s^23d^5$	26 **Fe** $4s^23d^6$	27 **Co** $4s^23d^7$	28 **Ni** $4s^23d^8$	29 **Cu** $4s^13d^{10}$	30 **Zn** $4s^23d^{10}$	31 **Ga** $4s^24p^1$	32 **Ge** $4s^24p^2$	33 **As** $4s^24p^3$	34 **Se** $4s^24p^4$	35 **Br** $4s^24p^5$	36 **Kr** $4s^24p^6$
5	37 **Rb** $5s^1$	38 **Sr** $5s^2$	39 **Y** $5s^24d^1$	40 **Zr** $5s^24d^2$	41 **Nb** $5s^14d^4$	42 **Mo** $5s^14d^5$	43 **Tc** $5s^14d^6$	44 **Ru** $5s^14d^7$	45 **Rh** $5s^14d^8$	46 **Pd** $5s^04d^{10}$	47 **Ag** $5s^14d^{10}$	48 **Cd** $5s^24d^{10}$	49 **In** $5s^25p^1$	50 **Sn** $5s^25p^2$	51 **Sb** $5s^25p^3$	52 **Te** $5s^25p^4$	53 **I** $5s^25p^5$	54 **Xe** $5s^25p^6$
6	55 **Cs** $6s^1$	56 **Ba** $6s^2$	57 **La** $6s^25d^1$	72 **Hf** $6s^25d^2$	73 **Ta** $6s^25d^3$	74 **W** $6s^25d^4$	75 **Re** $6s^25d^5$	76 **Os** $6s^25d^6$	77 **Ir** $6s^25d^7$	78 **Pt** $6s^15d^9$	79 **Au** $6s^15d^{10}$	80 **Hg** $6s^25d^{10}$	81 **Tl** $6s^26p^1$	82 **Pb** $6s^26p^2$	83 **Bi** $6s^26p^3$	84 **Po** $6s^26p^4$	85 **At** $6s^26p^5$	86 **Rn** $6s^26p^6$
7	87 **Fr** $7s^1$	88 **Ra** $7s^2$	89 **Ac** $7s^26d^1$	104 **Rf** $7s^26d^2$	105 **Db** $7s^26d^3$	106 **Sg** $7s^26d^4$	107 **Bh** $7s^26d^5$	108 **Hs** $7s^26d^6$	109 **Mt** $7s^26d^7$	110 **Ds** $7s^16d^9$	111 **Rg** $7s^16d^{10}$	112 **Cn** $7s^25d^{10}$	113 **Uut** $7s^27p^1$	114 **Fl** $7s^27p^2$	115 **Uup** $7s^27p^3$	116 **Lv** $7s^27p^4$	117 **Uus** $7s^27p^5$	118 **Uuo** $7s^27p^6$

FIGURE 10.17 Outermost electron configurations.

2. The group numbers for the representative elements are equal to the total number of outermost electrons in the atoms of the group. For example, elements in Group 7A(17) always have the electron configuration ns^2np^5. The *d* and *f* electrons are always in a lower energy level than the highest principal energy level and so are not considered as outermost (valence) electrons.
3. The elements of a family have the same outermost electron configuration except that the electrons are in different principal energy levels.
4. The elements within each of the *s*, *p*, *d*, *f* blocks are filling the *s*, *p*, *d*, *f* orbitals, as shown in Figure 10.17.
5. Within the transition elements, some discrepancies in the order of filling occur. (Explanation of these discrepancies and similar ones in the inner transition elements is beyond the scope of this book.)

EXAMPLE 10.6

Write the complete electron configuration of a zinc atom and a rubidium atom.

SOLUTION

The atomic number of zinc is 30; it therefore has 30 protons and 30 electrons in a neutral atom. Using Figure 10.14, we see that the electron configuration of a zinc atom is $1s^22s^22p^63s^23p^64s^23d^{10}$. Check by adding the superscripts, which should equal 30.

The atomic number of rubidium is 37; therefore, it has 37 protons and 37 electrons in a neutral atom. With a little practice using a periodic table, you can write the electron configuration directly. The electron configuration of a rubidium atom is $1s^22s^22p^63s^23p^64s^23d^{10}4p^65s^1$. Check by adding the superscripts, which should equal 37.

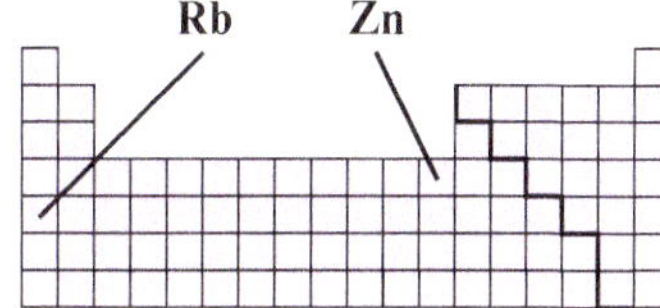

PRACTICE 10.7

Write the complete electron configuration for a gallium atom and a lead atom.

CHAPTER 10 REVIEW

10.1 Electromagnetic Radiation

KEY TERMS

wavelength
frequency
speed
photons

- Atomic theory has changed over the past 200 years:
  - Early Greek ideas
  - Dalton's model
  - Thomson's model
  - Rutherford's model
  - Modern atomic theory
- Basic wave characteristics include
  - Wavelength (λ)
  - Frequency (ν)
  - Speed (v)
- Light is a form of electromagnetic radiation.
- Light can also be considered to be composed of energy packets called photons, which behave like particles.

10.2 The Bohr Atom

KEY TERMS

line spectrum
quanta
ground state
orbital

- The study of atomic spectra led Bohr to propose that
  - Electrons are found in quantized energy levels in an atom.
  - Spectral lines result from the radiation of quanta of energy when the electron moves from a higher level to a lower level.
- The chemical properties of an element and its position on the periodic table depend on electrons.
- Louis de Broglie suggested that all objects have wave properties.
- Schrödinger created a mathematical model to describe electrons as waves.
  - Electrons are located in orbitals, or regions of probability, around the nucleus of an atom.

10.3 Energy Levels of Electrons

KEY TERMS

principal energy levels
sublevels
spin
Pauli exclusion principle

- Modern atomic theory predicts that
  - Electrons are found in discrete principal energy levels (n = 1, 2, 3 . . .).
    - Energy levels contain sublevels.

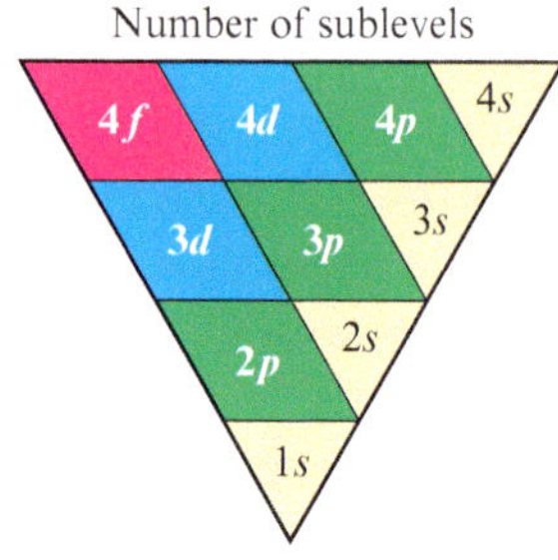

- Two electrons fit into each orbital but must have opposite spin to do so.

10.4 Atomic Structures of the First 18 Elements

KEY TERMS

electron configuration
orbital diagram
valence electrons

- Guidelines for writing electron configurations:
  - Not more than two electrons per orbital
  - Electrons fill lowest energy levels first:
    - $s < p < d < f$ for a given value of n
  - Orbitals on a given sublevel are each filled with a single electron before pairing of electrons begins to occur
- For the representative elements, only electrons in the outermost energy level (valence electrons) are involved in bonding.

10.5 Electron Structures and the Periodic Table

KEY TERMS

period
groups or families (of elements)
representative elements
transition elements

- Elements in horizontal rows on the periodic table contain elements whose valence electrons (*s* and *p*) are generally on the same energy level as the number of the row.
- Elements that are chemically similar are arranged in columns (groups) on the periodic table.
- The valence electron configurations of elements in a group or family are the same, but they are located in different principal energy levels.

REVIEW QUESTIONS

1. Define the terms *wavelength* and *frequency*. What symbols do chemists generally use to represent these quantities?
2. What is the wavelength range for visible light? Which has a longer wavelength, red light or blue light?
3. What is the name given to a packet of energy?
4. What is an orbital?
5. What is meant when we say the electron structure of an atom is in its ground state?
6. What is the major difference between an orbital and a Bohr orbit?
7. Explain how and why Bohr's model of the atom was modified to include the cloud model of the atom.
8. How do 1*s* and 2*s* orbitals differ? How are they alike?
9. What letters are used to designate the types of orbitals?
10. List the following orbitals in order of increasing energy: 2*s*, 2*p*, 4*s*, 1*s*, 3*d*, 3*p*, 4*p*, 3*s*.
11. How many *s* electrons, *p* electrons, and *d* electrons are possible in any energy level?
12. Sketch the *s*, p_x, p_y, and p_z orbitals.
13. Under what conditions can a second electron enter an orbital already containing one electron?
14. What is a valence shell?
15. What are valence electrons, and why are they important?
16. A lanthanide element has the designation $4f^3$ in its electron structure. What is the significance of the 4, the *f*, and the 3?
17. Of the elements Ir, Pb, Xe, Zr, and Ag, which ones are not contained in the groups of representative elements?
18. From the standpoint of electron structure, what do the elements in the *p* block have in common?
19. Write symbols for the elements with atomic numbers 6, 7, 8, 15, and 33. Which of those elements have something in common? Explain.
20. Write the symbols of the first three elements that have six electrons in their outermost energy level.
21. What is the greatest number of elements to be found in any period? Which periods have this number?
22. From the standpoint of energy level, how does the placement of the last electron in the Group A elements differ from that of the Group B elements?
23. Find the places in the periodic table where elements are not in proper sequence according to atomic mass. (See inside of front cover for periodic table.)
24. What are the names of the two scientists who independently published results that have led to the establishment of the periodic table?
25. Which scientist is credited with being the author of the modern periodic table?

PAIRED EXERCISES

Most of the exercises in this chapter are available for assignment via the online homework management program, WileyPLUS (www.wileyplus.com). All exercises with blue numbers have answers in Appendix 5.

1. How many total and how many valence electrons are in each of the following neutral atoms?
   (a) lithium (c) calcium
   (b) magnesium (d) fluorine
2. How many total and how many valence electrons are in each of the following neutral atoms?
   (a) sodium (c) phosphorus
   (b) arsenic (d) aluminum
3. Write the complete electron configuration for each of the following elements:
   (a) scandium (c) bromine
   (b) rubidium (d) sulfur
4. Write the complete electron configuration for each of the following elements:
   (a) manganese (c) gallium
   (b) krypton (d) boron
5. Explain how the spectral lines of hydrogen occur.
6. Explain how Bohr used the data from the hydrogen spectrum to support his model of the atom.
7. How many orbitals exist in the fourth principal energy level? What are they, and in what periods can they be found?
8. How many total electrons can be present in the third principal energy level? In what orbitals are they found?
9. Write orbital diagrams for these elements:
   (a) N (b) Cl (c) Zn (d) Zr (e) I
10. Write orbital diagrams for these elements:
    (a) Si (b) S (c) Ar (d) V (e) P

11. For each of the orbital diagrams given, write out the corresponding electron configurations.

(a) O [↑↓] [↑↓] [↑↓|↑|↑]

(b) Ca [↑↓] [↑↓] [↑↓|↑↓|↑↓] [↑↓] [↑↓|↑↓|↑↓] [↑↓]

(c) Ar [↑↓] [↑↓] [↑↓|↑↓|↑↓] [↑↓] [↑↓|↑↓|↑↓]

(d) Br [↑↓] [↑↓] [↑↓|↑↓|↑↓] [↑↓] [↑↓|↑↓|↑↓] [↑↓] [↑↓|↑↓|↑↓|↑↓|↑↓] [↑↓|↑↓|↑]

(e) Fe [↑↓] [↑↓] [↑↓|↑↓|↑↓] [↑↓] [↑↓|↑↓|↑↓] [↑↓] [↑↓|↑|↑|↑|↑]

12. For each of the orbital diagrams given, write out the corresponding electron configurations.

(a) Li [↑↓] [↑]

(b) P [↑↓] [↑↓] [↑↓|↑↓|↑↓] [↑↓] [↑|↑|↑]

(c) Zn [↑↓] [↑↓] [↑↓|↑↓|↑↓] [↑↓] [↑↓|↑↓|↑↓] [↑↓] [↑↓|↑↓|↑↓|↑↓|↑↓]

(d) Na [↑↓] [↑↓] [↑↓|↑↓|↑↓] [↑]

(e) K [↑↓] [↑↓] [↑↓|↑↓|↑↓] [↑↓] [↑↓|↑↓|↑↓] [↑]

13. Identify the correct orbital diagrams below. For the incorrect diagrams redraw the correct orbital diagram.

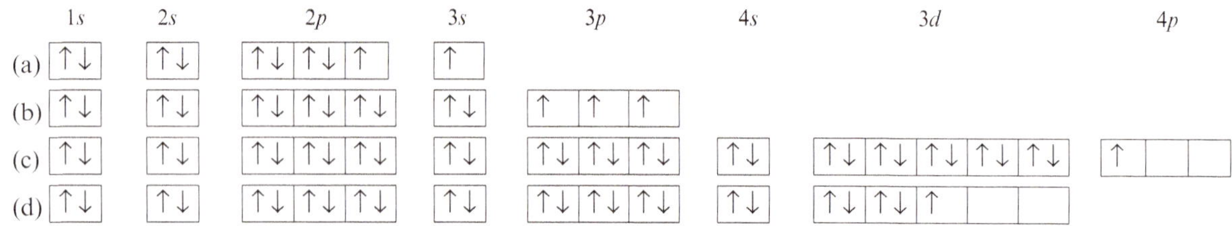

14. Identify the correct orbital diagrams below. For the incorrect diagrams redraw the correct orbital diagram.

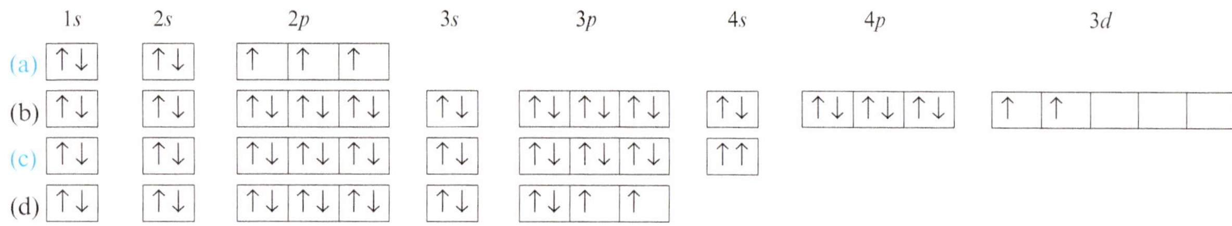

15. Identify the element represented by each of the orbital diagrams in Problem 13.

16. Identify the element represented by each of the orbital diagrams in Problem 14.

17. If a *p* sublevel has five electrons, which orbitals will they occupy? Draw the sublevel using arrows to represent electrons and show spin based on the direction of the arrow.

18. If a *d* sublevel has seven electrons, which orbitals will they occupy? Draw the sublevel using arrows to represent electrons and show spin based on the direction of the arrow.

19. Which elements have these electron configurations? Give both the symbol and the name:
(a) $1s^22s^22p^5$
(b) $1s^22s^22p^63s^1$
(c) $[Ne]3s^23p^4$
(d) $[Ar]4s^23d^8$

20. Which of the elements have these electron configurations? Give both the symbol and the name:
(a) $1s^22s^22p^1$
(b) $1s^22s^22p^63s^23p^2$
(c) $[Xe]6s^24f^{14}5d^{10}6p^2$
(d) $[Kr]5s^24d^{10}5p^4$

21. Identify the element and draw orbital diagrams for elements with these atomic numbers:
(a) 22 (b) 18 (c) 33 (d) 35 (e) 25

22. Identify the element and draw orbital diagrams for elements with these atomic numbers:
(a) 15 (b) 30 (c) 20 (d) 34 (e) 19

23. For each of the electron configurations given, write the corresponding orbital diagrams.
(a) F $1s^22s^22p^5$
(b) S $1s^22s^22p^63s^23p^4$
(c) Co $1s^22s^22p^63s^23p^64s^23d^7$
(d) Kr $1s^22s^22p^63s^23p^64s^23d^{10}4p^6$
(e) Ru $1s^22s^22p^63s^23p^64s^23d^{10}4p^65s^24d^6$

24. For each of the electron configuration given, write the corresponding orbital diagrams.
(a) Cl $1s^22s^22p^63s^23p^5$
(b) Mg $1s^22s^22p^63s^2$
(c) Ni $1s^22s^22p^63s^23p^64s^23d^8$
(d) Cu $1s^22s^22p^63s^23p^64s^23d^9$
(e) Ba $1s^22s^22p^63s^23p^64s^23d^{10}4p^65s^24d^{10}5p^66s^2$

25. Identify these elements from their atomic structure diagrams:

(a) (16p, 16n) $2e^-$ $8e^-$ $6e^-$

(b) (28p, 32n) $2e^-$ $8e^-$ $16e^-$ $2e^-$

26. Diagram the atomic structures (as you see in Exercise 25) for these elements:

(a) $^{27}_{13}Al$

(b) $^{48}_{22}Ti$

27. Why is the eleventh electron of the sodium atom located in the third energy level rather than in the second energy level?

28. Why is the last electron in potassium located in the fourth energy level rather than in the third energy level?

29. What electron structure do the noble gases have in common?

30. What is unique about the noble gases from an electron point of view?

31. How are elements in a period related to one another?

32. How are elements in a group related to one another?

33. How many valence electrons do each of the following elements have?
(a) C (b) S (c) K (d) I (e) B

34. How many valence electrons do each of the following elements have?
(a) N (b) P (c) O (d) Ba (e) Al

35. What do the electron structures of the alkali metals have in common?

36. Why would you expect elements zinc, cadmium, and mercury to be in the same chemical family?

37. Pick the electron structures that represent elements in the same chemical family:
(a) $1s^22s^1$
(b) $1s^22s^22p^4$
(c) $1s^22s^22p^2$
(d) $1s^22s^22p^63s^23p^4$
(e) $1s^22s^22p^63s^23p^6$
(f) $1s^22s^22p^63s^23p^64s^2$
(g) $1s^22s^22p^63s^23p^64s^1$
(h) $1s^22s^22p^63s^23p^64s^23d^1$

38. Pick the electron structures that represent elements in the same chemical family:
(a) $[He]2s^22p^6$
(b) $[Ne]3s^1$
(c) $[Ne]3s^2$
(d) $[Ne]3s^23p^3$
(e) $[Ar]4s^23d^{10}$
(f) $[Ar]4s^23d^{10}4p^6$
(g) $[Ar]4s^23d^5$
(h) $[Kr]5s^24d^{10}$

39. In the periodic table, calcium, element 20, is surrounded by elements 12, 19, 21, and 38. Which of these have physical and chemical properties most resembling calcium?

40. In the periodic table, phosphorus, element 15, is surrounded by elements 14, 7, 16, and 33. Which of these have physical and chemical properties most resembling phosphorus?

41. Classify the following elements as metals, nonmetals, or metalloids (review Chapter 3 if you need help):
(a) potassium
(b) plutonium
(c) sulfur
(d) antimony

42. Classify the following elements as metals, nonmetals, or metalloids (review Chapter 3 if you need help):
(a) iodine
(b) tungsten
(c) molybdenum
(d) germanium

43. Look for antimony in the periodic table.
(a) Write the complete electron configuration for an atom of antimony.
(b) Write the shorthand electron configuration for an atom of antimony.
(c) How many valence electrons does antimony have?
(d) Underline the valence electrons in the shorthand electron configuration.

44. Look for selenium in the periodic table.
(a) Write the complete electron configuration for an atom of selenium.
(b) Write the shorthand electron configuration for an atom of selenium.
(c) How many valence electrons does selenium have?
(d) Underline the valence electrons in the shorthand electron configuration.

45. In which period and group does an electron first appear in an *f* orbital?

46. In which period and group does an electron first appear in a *d* orbital?

47. How many electrons occur in the valence level of Group 7A(17) and 7B(7) elements? Why are they different?

48. How many electrons occur in the valence level of Group 3A(13) and 3B(3) elements? Why are they different?

49. Determine the identity of the element that contains exactly
(a) three $4p$ electrons in the ground state
(b) seven $3d$ electrons in the ground state
(c) one $2s$ electrons in the ground state
(d) five $3p$ electrons in the ground state

50. Determine the identity of the element that contains exactly
(a) two $6p$ electrons in the ground state
(b) five $4f$ electrons in the ground state
(c) one $4p$ electrons in the ground state
(d) seven $5d$ electrons in the ground state

ADDITIONAL EXERCISES

51. Using only the periodic table, explain how the valence energy level and the number of valence electrons could be determined.

52. Using only a periodic table, write the full electron configuration for each of the following elements. Circle the valence electrons.
(a) Mg (b) P (c) K (d) F (e) Se (f) N

53. Which of the following would have the same number of valence electrons?
(a) Na^+ (b) O (c) Li (d) F^- (e) Ne

54. Name the group in which each of the following elements appear:
(a) $1s^22s^22p^5$
(b) $1s^22s^22p^63s^23p^64s^2$
(c) $1s^22s^22p^63s^1$
(d) $1s^22s^22p^63s^23p^64s^23d^{10}4p^6$
(e) $1s^2$
(f) $1s^22s^22p^63s^23p^64s^23d^{10}4p^65s^1$

55. Chromium is a lustrous silver-colored metal that has been used to prevent corrosion for centuries. Bronze swords and other weapons discovered in burial pits from the Qin Dynasty were coated with chromium and had not corroded at all since their entombing. Today many items are coated with a layer of chromium as a decorative and protective covering.

O. Louis Mazzatenta/NG Image Collection

(a) Sometimes the actual electron configurations of the elements differ from those predicted by the periodic table. The experimentally determined electron configuration for chromium is $1s^22s^22p^63s^23p^64s^13d^5$. Is this the electron configuration you would predict based on the periodic table? If not, what is the configuration predicted by the periodic table for chromium?

(b) Chromium has a density of 7.19 g/cm^3. How many atoms of chromium are contained in a 5.00-cm^3 sample of chromium?

(c) If the radius of a chromium atom is 1.40×10^{-8} cm, what is the volume $\left(V = \frac{4}{3}\pi r^3\right)$ of a single chromium atom?

(d) How many chromium atoms occupy a volume of 5.00 cm^3?

56. Anyone watching fireworks enjoys the beautiful colors produced. The colors are produced by adding different elements to the mix of fuels in the fireworks. Red is produced by adding calcium, yellow by adding sodium, and blue by adding copper. Explain why adding different elements might give fireworks their characteristic colors.

57. There has long been speculation that Wolfgang Amadeus Mozart may have been poisoned by a rival. There is some evidence that he may have been poisoned by an antimony compound. This compound would have been prescribed by his doctor, however, suggesting that there was no foul play. What is the electron configuration of antimony?

58. A dog is suspected of attacking a runner in the park. This runner was wearing a pearlescent lip gloss containing bismuth hypochlorite, a compound often used in cosmetics to give this pearlescent effect. Upon analyzing a residue on the dog's fur, this compound is detected giving additional support to the victim's claim. Write the complete and shorthand electron configurations for bismuth.

59. Write the electron configuration of each of the following neutral atoms. (You may need to refer to Figure 3.2.)

(a) the four most abundant elements in the Earth's crust, seawater, and air

(b) the five most abundant elements in the human body

60. Give the maximum number of electrons that can reside in the following:

(a) a *p* orbital

(b) a *d* sublevel

(c) the third principal energy level

(d) an *s* orbital

(e) an *f* sublevel

61. Give the names of each of the following elements based on the information given:

(a) the second element in Period 3

(b) [Ne]$3s^23p^3$

(c) [↑↓] [↑↓] [↑↓|↑↓|↑↓] [↑↓] [↑↓|↑↓|↑↓]
$\quad$ 1*s* $\quad$ 2*s* $\quad$ 2*p* $\quad$ 3*s* $\quad$ 3*p*

62. Why does the emission spectrum for nitrogen reveal many more spectral lines than that for hydrogen?

63. List the first element on the periodic table that satisfies each of these conditions:

(a) a completed set of *p* orbitals

(b) two 4*p* electrons

(c) seven valence electrons

(d) three unpaired electrons

64. In Group 7A(17), the first two elements are gases, the next one is a liquid, and the fourth one is a solid. What do they have in common that causes them to be in the same group?

65. Explain why each of the following electron configurations for a *p* sublevel is disallowed.

(a) [↓↓↓] [↓] [↓]

(b) [↓↑] [↓↑] [ ]

66. How is the line spectra of an atom produced? (Answer this on an atomic level.)

67. What is meant by the term *ground state*? Write the electron configuration of an atom of carbon in the ground state. Write one possible electron configuration of a carbon atom in an excited state (i.e., not in the ground state).

68. In which groups are the transition elements located?

69. How do the electron structures of the transition elements differ from those of the representative elements?

70. The atomic numbers of the noble gases are 2, 10, 18, 36, 54 and 86. Without looking at a periodic table, determine the atomic numbers for the elements that have five electrons in their valence shell.

71. What is the family name for

(a) Group 1A(1)?

(b) Group 2A(2)?

(c) Group 7A(17)?

72. What sublevel is being filled in

(a) Period 3, Group 3A(13) to 7A(17)?

(b) Period 5, transition elements?

(c) the lanthanide series?

73. Classify each of the following as a noble gas, a representative element, or a transition metal. Also indicate whether the element is a metal, nonmetal, or metalloid.

(a) Na (d) Ra
(b) N (e) As
(c) Mo (f) Ne

74. If element 36 is a noble gas, in which groups would you expect elements 35 and 37 to occur?

75. Write a paragraph describing the general features of the periodic table.

76. What is the relationship between two elements if
(a) one of them has 10 electrons, 10 protons, and 10 neutrons and the other has 10 electrons, 10 protons, and 12 neutrons?
(b) one of them has 23 electrons, 23 protons, and 27 neutrons and the other has 24 electrons, 24 protons, and 26 neutrons?

77. Is there any pattern for the location of gases on the periodic table? for the location of liquids? for the location of solids?

CHALLENGE EXERCISES

78. A valence electron in an atom of sulfur is excited by heating a sample. The electron jumps from the *s* orbital to the *p* orbital. What is the electron configuration of the excited sulfur atom, and what would the orbital diagram look like?

79. Element 87 is in Group 1A(1), Period 7. In how many principal energy levels are electrons located? Describe its outermost energy level.

80. Use the periodic table to explain why metals tend to lose electrons and nonmetals tend to gain electrons.

81. Show how the periodic table helps determine the expected electron configuration for any element.

82. An alkaline earth metal, M, combines with a halide, X. Will the resulting compound be ionic or covalent? Why? What is the Lewis structure for the compound?

83. "All electrons in atoms with even atomic numbers are paired." Is this statement true or false? Explain your answer using an example.

ANSWERS TO PRACTICE EXERCISES

10.1

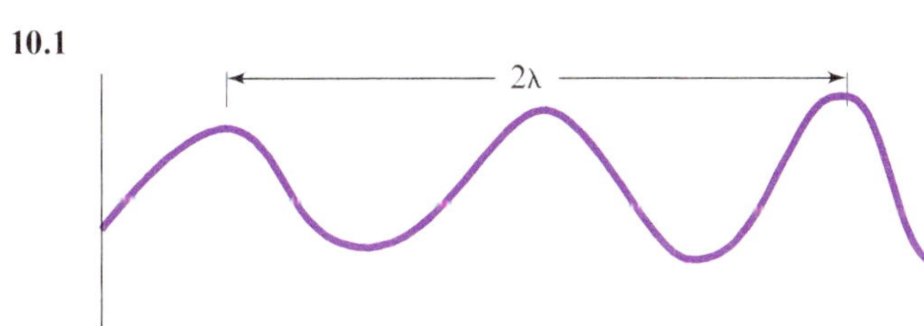

10.2 A photon is a little packet of electromagnetic radiation energy and has the property of both a particle and a wave. It travels at the speed of light, 3.00×10^8 meters per second.

10.3 (a) Two electrons. Wolfgang Pauli's exclusion principle.
(b) $1s^22s^22p^63s^23p^6$ for a total of 18 electrons.

10.4 (a) $2s^22p^1$ (c) $3s^1$
(b) $2s^22p^3$ (d) $3s^23p^5$

10.5 (a) $[Ar]4s^24p^5$ (c) $[Xe]6s^2$
(b) $[Kr]5s^2$ (d) $[Kr]5s^25p^4$

10.6 (a) O $1s^22s^22p^4$
(b) Ca $1s^22s^22p^63s^23p^64s^2$
(c) Ti $1s^22s^22p^63s^23p^64s^23d^2$

10.7 Ga, $1s^22s^22p^63s^23p^64s^23d^{10}4p^1$
Pb, $1s^22s^22p^63s^23p^64s^23d^{10}4p^65s^24d^{10}5p^66s^24f^{14}5d^{10}6p^2$

11

Chemical Bonds: The Formation of Compounds from Atoms

Pasieka/Photo Researchers, Inc.

CHAPTER OUTLINE

For centuries we've been aware that certain metals cling to a magnet. High-speed levitation trains are heralded to be the wave of the future. How do they function? In each case, forces of attraction and repulsion are at work.

Human interactions also suggest that "opposites attract" and "likes repel." Attractions draw us into friendships and significant relationships, whereas repulsive forces may produce debate and antagonism. We form and break apart interpersonal bonds throughout our lives.

In chemistry, we also see this phenomenon. Substances form chemical bonds as a result of electrical attractions. These bonds provide the tremendous diversity of compounds found in nature. The photograph above shows a crystal formed by molecules of tartaric acid, a molecule found in baking powder and other food additives. The atoms in tartaric acid bond together in a very specific orientation to form the shape of the molecule and produce this beautiful pattern. This chapter is one of the most significant and useful chapters in the book—chemical bonding between atoms. This is what chemistry is really all about. Study it carefully.

11.1 Periodic Trends in Atomic Properties

LEARNING OBJECTIVE

Discuss the atomic trends for metals and nonmetals, atomic radius, and ionization energy.

KEY TERM

ionization energy

Although atomic theory and electron configuration help us understand the arrangement and behavior of the elements, it's important to remember that the design of the periodic table is based on observing properties of the elements. Before we use the concept of atomic structure to explain how and why atoms combine to form compounds, we need to understand the characteristic properties of the elements and the trends that occur in these properties on the periodic table. These trends allow us to use the periodic table to accurately predict properties and reactions of a wide variety of substances.

Metals and Nonmetals

In Section 3.5, we classified elements as metals, nonmetals, or metalloids. The heavy stair-step line beginning at boron and running diagonally down the periodic table separates the elements into metals and nonmetals. Metals are usually lustrous, malleable, and good conductors of heat and electricity. Nonmetals are just the opposite—nonlustrous, brittle, and poor conductors. Metalloids are found bordering the heavy diagonal line and may have properties of both metals and nonmetals.

Most elements are classified as metals (FIGURE 11.1). Metals are found on the left side of the stair-step line, while the nonmetals are located toward the upper right of the table. Note that hydrogen does not fit into the division of metals and nonmetals. It displays nonmetallic properties under normal conditions, even though it has only one outermost electron like the alkali metals. Hydrogen is considered to be a unique element.

It is the chemical properties of metals and nonmetals that interest us most. Metals tend to lose electrons and form positive ions, while nonmetals tend to gain electrons and form negative ions. When a metal reacts with a nonmetal, electrons are often transferred from the metal to the nonmetal.

Metals
Metalloids
Nonmetals

1	2	3	4	5	6	7	8	9	10	11	12	13	14	15	16	17	18
1 H																	2 He
3 Li	4 Be											5 B	6 C	7 N	8 O	9 F	10 Ne
11 Na	12 Mg											13 Al	14 Si	15 P	16 S	17 Cl	18 Ar
19 K	20 Ca	21 Sc	22 Ti	23 V	24 Cr	25 Mn	26 Fe	27 Co	28 Ni	29 Cu	30 Zn	31 Ga	32 Ge	33 As	34 Se	35 Br	36 Kr
37 Rb	38 Sr	39 Y	40 Zr	41 Nb	42 Mo	43 Tc	44 Ru	45 Rh	46 Pd	47 Ag	48 Cd	49 In	50 Sn	51 Sb	52 Te	53 I	54 Xe
55 Cs	56 Ba	57 La*	72 Hf	73 Ta	74 W	75 Re	76 Os	77 Ir	78 Pt	79 Au	80 Hg	81 Tl	82 Pb	83 Bi	84 Po	85 At	86 Rn
87 Fr	88 Ra	89 Ac†	104 Rf	105 Db	106 Sg	107 Bh	108 Hs	109 Mt	110 Ds	111 Rg	112 Cn	113 Uut	114 Fl	115 Uup	116 Lv	117 Uus	118 Uuo

*	58 Ce	59 Pr	60 Nd	61 Pm	62 Sm	63 Eu	64 Gd	65 Tb	66 Dy	67 Ho	68 Er	69 Tm	70 Yb	71 Lu
†	90 Th	91 Pa	92 U	93 Np	94 Pu	95 Am	96 Cm	97 Bk	98 Cf	99 Es	100 Fm	101 Md	102 No	103 Lr

FIGURE 11.1 The elements are classified as metals, nonmetals, and metalloids.

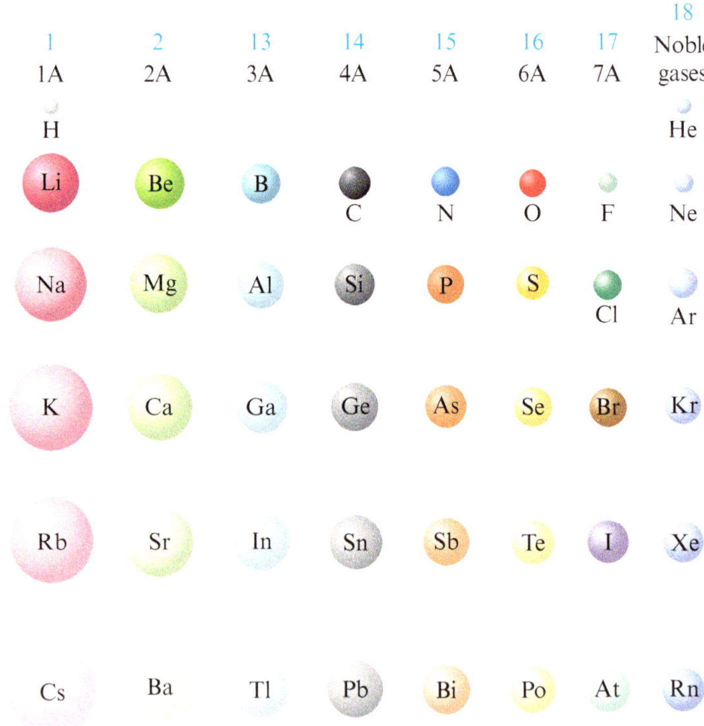

FIGURE 11.2 Relative atomic radii for the representative elements. Atomic radius decreases across a period and increases down a group in the periodic table.

Atomic Radius

The relative radii of the representative elements are shown in FIGURE 11.2. Notice that the radii of the atoms tend to increase down each group and that they tend to decrease from left to right across a period.

The increase in radius down a group can be understood if we consider the electron structure of the atoms. For each step down a group, an additional energy level is added to the atom. The average distance from the nucleus to the outside edge of the atom must increase as each new energy level is added. The atoms get bigger as electrons are placed in these new higher-energy levels.

Understanding the decrease in atomic radius across a period requires more thought, however. As we move from left to right across a period, electrons within the same block are being added to the same principal energy level. Within a given energy level, we expect the orbitals to have about the same size. We would then expect the atoms to be about the same size across the period. But each time an electron is added, a proton is added to the nucleus as well. The increase in positive charge (in the nucleus) pulls the electrons closer to the nucleus, which results in a gradual decrease in atomic radius across a period.

CHECK YOUR UNDERSTANDING
11.1 Atomic Radius
WileyPLUS

Ionization Energy

The **ionization energy** of an atom is the energy required to remove an electron from the atom. For example,

$$Na + \text{ionization energy} \longrightarrow Na^{+} + e^{-}$$

The first ionization energy is the amount of energy required to remove the first electron from an atom, the second is the amount required to remove the second electron from that atom, and so on.

TABLE 11.1 gives the ionization energies for the removal of one to five electrons from several elements. The table shows that even higher amounts of energy are needed to remove the second, third, fourth, and fifth electrons. This makes sense because removing electrons leaves fewer electrons attracted to the same positive charge in the nucleus. The data in Table 11.1 also show that an extra-large ionization energy (blue) is needed when

CHECK YOUR UNDERSTANDING
11.2 Ionization Energy 1
WileyPLUS

TABLE 11.1 | Ionization Energies for Selected Elements*

Element	Required amounts of energy (kJ/mol) 1st e^-	2nd e^-	3rd e^-	4th e^-	5th e^-
H	1,314				
He	2,372	5,247			
Li	520	7,297	11,810		
Be	900	1,757	14,845	21,000	
B	800	2,430	3,659	25,020	32,810
C	1,088	2,352	4,619	6,222	37,800
Ne	2,080	3,962	6,276	9,376	12,190
Na	496	4,565	6,912	9,540	13,355

*Values are expressed in kilojoules per mole, showing energies required to remove 1 to 5 electrons per atom. Blue type indicates the energy needed to remove an electron from a noble gas electron structure.

an electron is removed from a noble gas–like structure, clearly showing the stability of the electron structure of the noble gases.

First ionization energies have been experimentally determined for most elements. FIGURE 11.3 plots these energies for representative elements in the first four periods. Note these important points:

1. Ionization energy in Group A elements decreases from top to bottom in a group. For example, in Group 1A(1) the ionization energy changes from 520 kJ/mol for Li to 419 kJ/mol for K.
2. Ionization energy gradually increases from left to right across a period. Noble gases have a relatively high value, confirming the nonreactive nature of these elements.

Metals don't behave in exactly the same manner. Some metals give up electrons much more easily than others. In the alkali metal family, cesium gives up its 6*s* electron much more easily than the metal lithium gives up its 2*s* electron. This makes sense when we consider that the size of the atoms increases down the group. The distance between the nucleus and the outer electrons increases, and the ionization energy decreases. The most chemically active metals are located at the lower left of the periodic table.

CHECK YOUR UNDERSTANDING

11.3 Ionization Energy 2

WileyPLUS

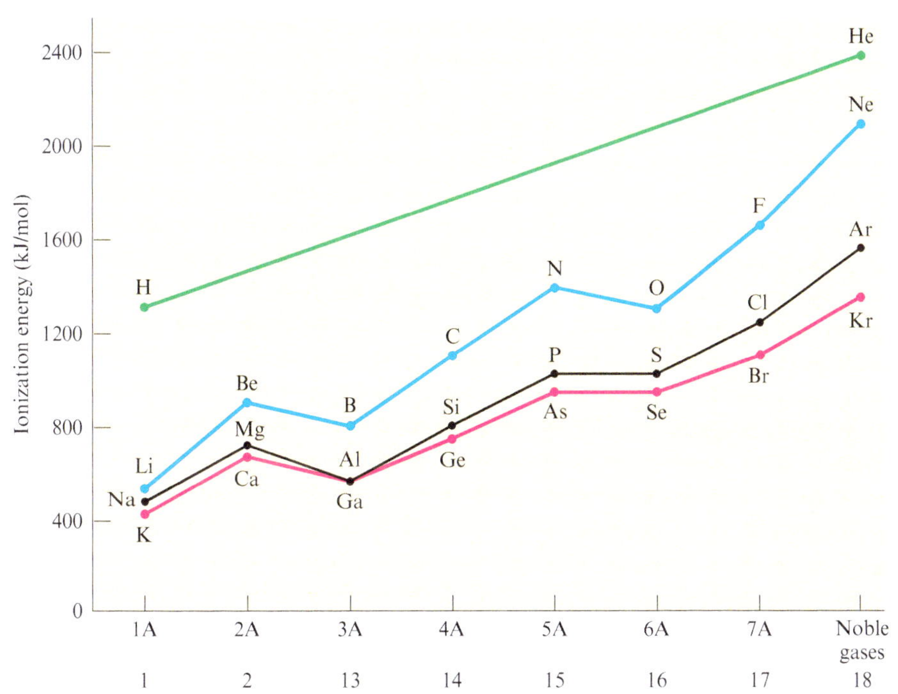

FIGURE 11.3 Periodic relationship of the first ionization energy for representative elements in the first four periods.

Nonmetals have relatively large ionization energies compared to metals. Nonmetals tend to gain electrons and form anions. Since the nonmetals are located at the right side of the periodic table, it is not surprising that ionization energies tend to increase from left to right across a period. The most active nonmetals are found in the *upper* right corner of the periodic table (excluding the noble gases).

EXAMPLE 11.1

Why is there such a large increase in the ionization energy to remove the fifth electron from a carbon atom? (See Table 11.1.)

SOLUTION

Two main factors apply here. (1) The remaining carbon atom has six protons attracting only two electrons, and (2) these two electrons constitute the stable noble gas helium. Both factors make it much more difficult to withdraw the fifth electron.

PRACTICE 11.1

(a) What are the trends in the first ionization energy for the groups of elements and for the periods of elements in the periodic table?
(b) Consider your answer to part (a) and state why this is the case for the groups of elements.

11.2 Lewis Structures of Atoms

LEARNING OBJECTIVE Draw the Lewis structure for a given atom.

KEY TERM
Lewis structure

Metals tend to form cations (positively charged ions) and nonmetals form anions (negatively charged ions) in order to attain a stable valence electron structure. For many elements this stable valence level contains eight electrons (two *s* and six *p*), identical to the valence electron configuration of the noble gases. Atoms undergo rearrangements of electron structure to lower their chemical potential energy (or to become more stable). These rearrangements are accomplished by losing, gaining, or sharing electrons with other atoms. For example, a hydrogen atom could accept a second electron and attain an electron structure the same as the noble gas helium. A fluorine atom could gain an electron and attain an electron structure like neon. A sodium atom could lose one electron to attain an electron structure like neon.

The valence electrons in the outermost energy level of an atom are responsible for the electron activity that occurs to form chemical bonds. The **Lewis structure** of an atom is a representation that shows the valence electrons for that atom. American chemist Gilbert N. Lewis (1875–1946) proposed using the symbol for the element and dots for electrons. The number of dots placed around the symbol equals the number of *s* and *p* electrons in the outermost energy level of the atom. Paired dots represent paired electrons; unpaired dots represent unpaired electrons. For example, **H·** is the Lewis symbol for a hydrogen atom, $1s^1$; **:Ḃ** is the Lewis symbol for a boron atom, with valence electrons $2s^22p^1$. In the case of boron, the symbol B represents the boron nucleus and the $1s^2$ electrons; the dots represent only the $2s^22p^1$ electrons.

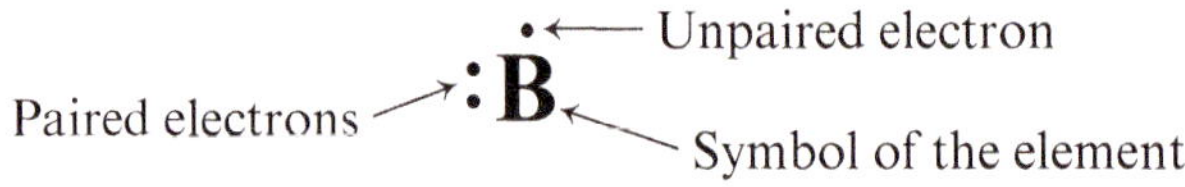

The Lewis method is used not only because of its simplicity of expression but also because much of the chemistry of the atom is directly associated with the electrons in the outermost energy level. FIGURE 11.4 shows Lewis structures for the elements hydrogen through calcium.

CHECK YOUR UNDERSTANDING
11.4 Lewis Structures for Atoms
WileyPLUS

1 1A	2 2A	13 3A	14 4A	15 5A	16 6A	17 7A	18 Noble gases
H·							He:
Li·	Be:	:B	:C·	:N·	·O:	:F:	:Ne:
Na·	Mg:	:Al	:Si·	:P·	·S:	:Cl:	:Ar:
K·	Ca:						

FIGURE 11.4 Lewis structures of the first 20 elements. Dots represent electrons in the outermost *s* and *p* energy levels only.

ENHANCED EXAMPLE 11.2 WileyPLUS

Write the Lewis structure for a phosphorus atom.

SOLUTION

First establish the electron structure for a phosphorus atom, which is $1s^2 2s^2 2p^6 3s^2 3p^3$. Note that there are five electrons in the outermost energy level; they are $3s^2 3p^3$. Write the symbol for phosphorus and place the five electrons as dots around it.

The $3s^2$ electrons are paired and are represented by the paired dots. The $3p^3$ electrons, which are unpaired, are represented by the single dots.

ONLINE LEARNING MODULE
Lewis Structures of Atoms
➡ WileyPLUS

A quick way to determine the correct number of dots (electrons) for a Lewis structure is to use the group number. For the A groups on the periodic table, the group number is the same as the number of electrons in the Lewis structure.

PRACTICE 11.2

Write the Lewis structure for the following elements:
(a) N (b) Al (c) Sr (d) Br

11.3 The Ionic Bond: Transfer of Electrons from One Atom to Another

LEARNING OBJECTIVE

Discuss the formation of an ionic bond and the chemical change that results from the bond.

KEY TERMS
ionic bond

The chemistry of many elements, especially the representative ones, is to attain an outer electron structure like that of the chemically stable noble gases. With the exception of helium, this stable structure consists of eight electrons in the outermost energy level (TABLE 11.2).

TABLE 11.2 Arrangement of Electrons in the Noble Gases*

Noble gas	Symbol	Electron structure n = 1	2	3	4	5	6
Helium	He	$1s^2$					
Neon	Ne	$1s^2$	$2s^2 2p^6$				
Argon	Ar	$1s^2$	$2s^2 2p^6$	$3s^2 3p^6$			
Krypton	Kr	$1s^2$	$2s^2 2p^6$	$3s^2 3p^6 3d^{10}$	$4s^2 4p^6$		
Xenon	Xe	$1s^2$	$2s^2 2p^6$	$3s^2 3p^6 3d^{10}$	$4s^2 4p^6 4d^{10}$	$5s^2 5p^6$	
Radon	Rn	$1s^2$	$2s^2 2p^6$	$3s^2 3p^6 3d^{10}$	$4s^2 4p^6 4d^{10} 4f^{14}$	$5s^2 5p^6 5d^{10}$	$6s^2 6p^6$

*Each gas except helium has eight electrons in its outermost energy level.

Let's look at the electron structures of sodium and chlorine to see how each element can attain a structure of 8 electrons in its outermost energy level. A sodium atom has 11 electrons: 2 in the first energy level, 8 in the second energy level, and 1 in the third energy level. A chlorine atom has 17 electrons: 2 in the first energy level, 8 in the second energy level, and 7 in the third energy level. If a sodium atom transfers or loses its 3*s* electron, its third energy level becomes vacant, and it becomes a sodium ion with an electron configuration identical to that of the noble gas neon. This process requires energy:

11^+ $11e^-$ ⟶ 11^+ $10e^-$ + $1e^-$

Na atom ($1s^22s^22p^63s^1$) Na$^+$ ion ($1s^22s^22p^6$)

An atom that has lost or gained electrons will have a positive or negative charge, depending on which particles (protons or electrons) are in excess. Remember that a charged particle or group of particles is called an *ion*.

By losing a negatively charged electron, the sodium atom becomes a positively charged particle known as a sodium ion. The charge, +1, results because the nucleus still contains 11 positively charged protons, and the electron orbitals contain only 10 negatively charged electrons. The charge is indicated by a plus sign (+) and is written as a superscript after the symbol of the element (Na^+).

CHECK YOUR UNDERSTANDING
11.5 Alkaline Earth Ion Charges
➡ WileyPLUS

A chlorine atom with seven electrons in the third energy level needs one electron to pair up with its one unpaired 3*p* electron to attain the stable outer electron structure of argon. By gaining one electron, the chlorine atom becomes a chloride ion (Cl^-), a negatively charged particle containing 17 protons and 18 electrons. This process releases energy:

CHECK YOUR UNDERSTANDING
11.6 Group 7A(17) Ion Charges
➡ WileyPLUS

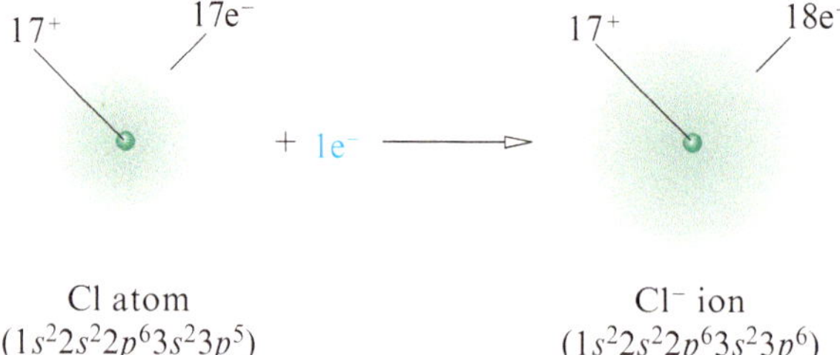

Consider sodium and chlorine atoms reacting with each other. The 3*s* electron from the sodium atom transfers to the half-filled 3*p* orbital in the chlorine atom to form a positive sodium ion and a negative chloride ion. The compound sodium chloride results because the Na^+ and Cl^- ions are strongly attracted to each other by their opposite electrostatic charges. The force holding the oppositely charged ions together is called an ionic bond:

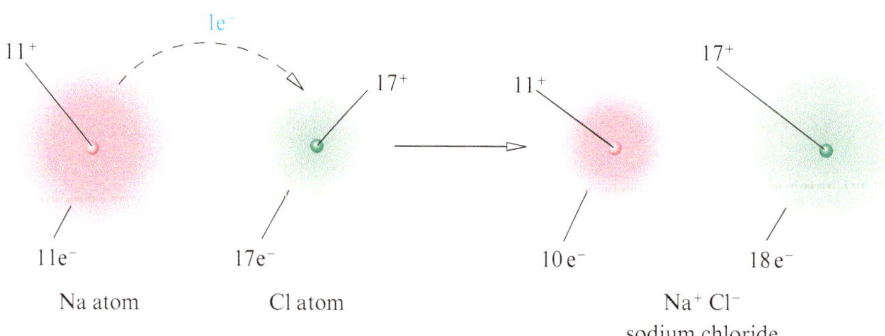

The Lewis representation of sodium chloride formation is

$$Na\cdot + \cdot\ddot{\underset{\cdot\cdot}{Cl}}: \longrightarrow [Na]^+ [:\ddot{\underset{\cdot\cdot}{Cl}}:]^-$$

The chemical reaction between sodium and chlorine is a very vigorous one, producing considerable heat in addition to the salt formed. When energy is released in a chemical

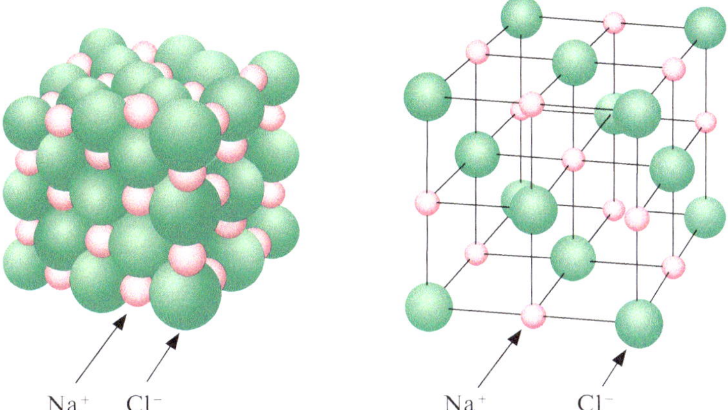

FIGURE 11.5 Sodium chloride crystal. Diagram represents a small fragment of sodium chloride, which forms cubic crystals. Each sodium ion is surrounded by six chloride ions, and each chloride ion is surrounded by six sodium ions. The tiny NaCl crystals show the cubic crystal structure of salt.

These tiny NaCl crystals show the cubic structure illustrated in Figure 11.5.

reaction, the products are more stable than the reactants. Note that in NaCl both atoms attain a noble gas electron structure.

Sodium chloride is made up of cubic crystals in which each sodium ion is surrounded by six chloride ions and each chloride ion by six sodium ions, except at the crystal surface. A visible crystal is a regularly arranged aggregate of millions of these ions, but the ratio of sodium to chloride ions is 1:1, hence the formula NaCl. The cubic crystalline lattice arrangement of sodium chloride is shown in FIGURE 11.5.

FIGURE 11.6 contrasts the relative sizes of sodium and chlorine atoms with those of their ions. The sodium ion is smaller than the atom due primarily to two factors: (1) The sodium atom has lost its outermost electron, thereby reducing its size; and (2) the 10 remaining electrons are now attracted by 11 protons and are thus drawn closer to the nucleus. Conversely, the chloride ion is larger than the atom because (1) it has 18 electrons but only 17 protons and (2) the nuclear attraction on each electron is thereby decreased, allowing the chlorine atom to expand as it forms an ion.

Remember: A cation is always smaller than its parent atom, whereas an anion is always larger than its parent atom.

We've seen that when sodium reacts with chlorine, each atom becomes an ion. Sodium chloride, like all ionic substances, is held together by the attraction existing between positive and negative charges. An **ionic bond** is the attraction between oppositely charged ions.

Ionic bonds are formed whenever one or more electrons are transferred from one atom to another. Metals, which have relatively little attraction for their valence electrons, tend to form ionic bonds when they combine with nonmetals.

It's important to recognize that substances with ionic bonds do not exist as molecules. In sodium chloride, for example, the bond does not exist solely between a single sodium ion and a single chloride ion. Each sodium ion in the crystal attracts six near-neighbor negative chloride ions; in turn, each negative chloride ion attracts six near-neighbor positive sodium ions (see Figure 11.5).

A metal will usually have one, two, or three electrons in its outer energy level. In reacting, metal atoms characteristically lose these electrons, attain the electron structure of a noble gas, and become positive ions. A nonmetal, on the other hand, is only a few electrons short of having a noble gas electron structure in its outer energy level and thus has a tendency to gain electrons. In reacting with metals, nonmetal atoms characteristically gain one to four electrons, attain the electron structure of a noble gas, and become negative ions. The ions

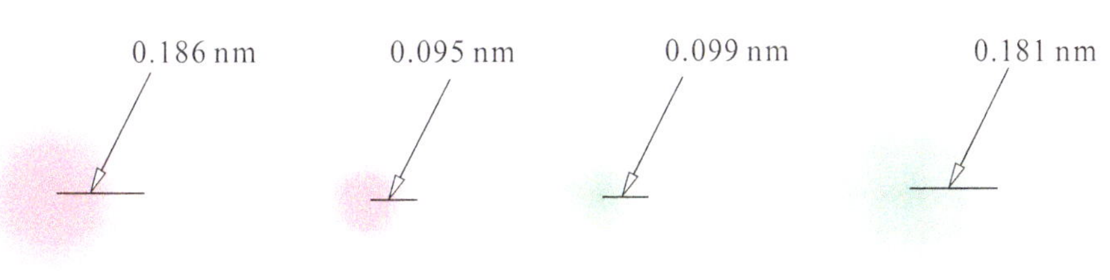

FIGURE 11.6 Relative radii of sodium and chlorine atoms and their ions.

TABLE 11.3 Change in Atomic Radii (nm) of Selected Metals and Nonmetals*

Atomic radius		Ionic radius		Atomic radius		Ionic radius	
Li	0.152	Li^+	0.060	F	0.071	F^-	0.136
Na	0.186	Na^+	0.095	Cl	0.099	Cl^-	0.181
K	0.227	K^+	0.133	Br	0.114	Br^-	0.195
Mg	0.160	Mg^{2+}	0.065	O	0.074	O^{2-}	0.140
Al	0.143	Al^{3+}	0.050	S	0.103	S^{2-}	0.184

*The metals shown lose electrons to become positive ions. The nonmetals gain electrons to become negative ions.

formed by loss of electrons are much smaller than the corresponding metal atoms; the ions formed by gaining electrons are larger than the corresponding nonmetal atoms. The dimensions of the atomic and ionic radii of several metals and nonmetals are given in TABLE 11.3.

PRACTICE 11.3

What noble gas structure is formed when an atom of each of these metals loses all its valence electrons? Write the formula for the metal ion formed.

(a) K (b) Mg (c) Al (d) Ba

Study the following examples. Note the loss and gain of electrons between atoms; also note that the ions in each compound have a noble gas electron structure.

ONLINE LEARNING MODULE
Ionic Bonding-1
WileyPLUS

EXAMPLE 11.3

Explain how magnesium and chlorine combine to form magnesium chloride, $MgCl_2$.

SOLUTION

A magnesium atom of electron structure $1s^22s^22p^63s^2$ must lose two electrons or gain six to reach a stable electron structure. If magnesium reacts with chlorine and each chlorine atom can accept only one electron, two chlorine atoms will be needed for the two electrons from each magnesium atom. The compound formed will contain one magnesium ion and two chloride ions. The magnesium atom, having lost two electrons, becomes a magnesium ion with a +2 charge. Each chloride ion will have a −1 charge. The transfer of electrons from a magnesium atom to two chlorine atoms is shown in the following illustration:

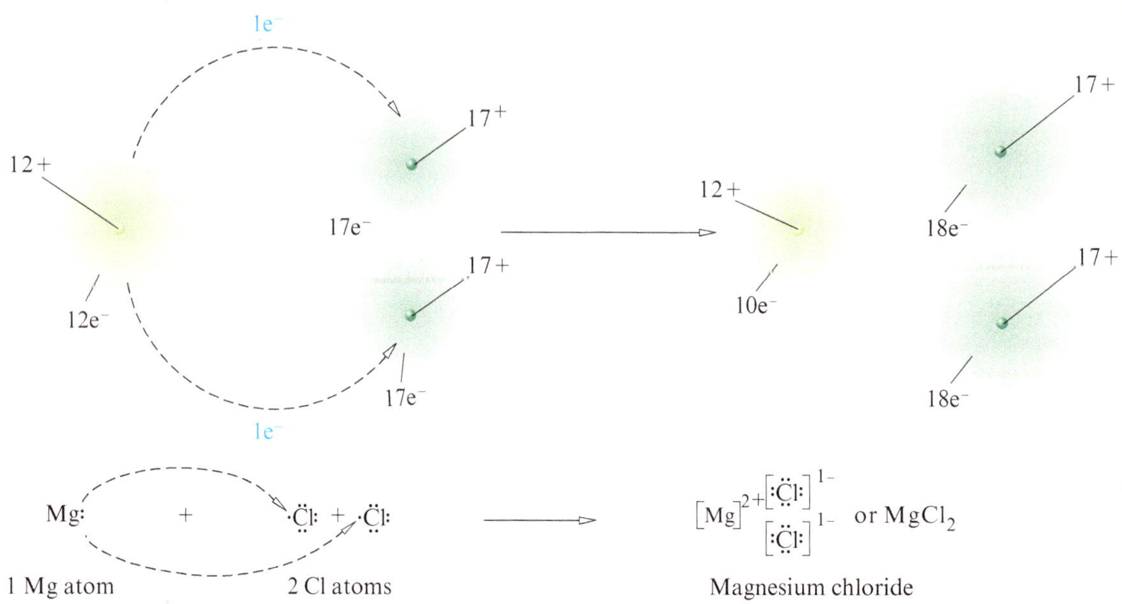

EXAMPLE 11.4

Explain the formation of sodium fluoride (NaF) from its elements.

SOLUTION

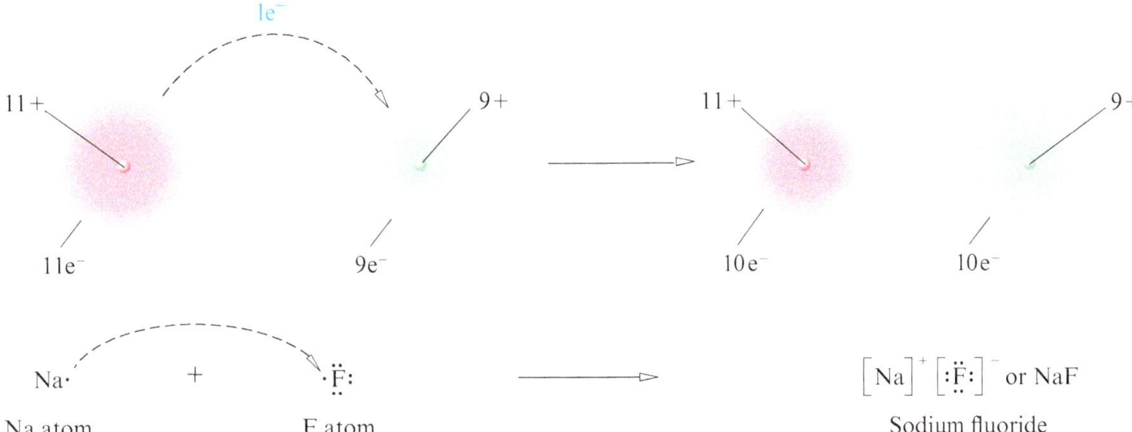

The fluorine atom, with seven electrons in its outer energy level, behaves similarly to the chlorine atom.

EXAMPLE 11.5

Explain the formation of aluminum fluoride (AlF_3) from its elements.

SOLUTION

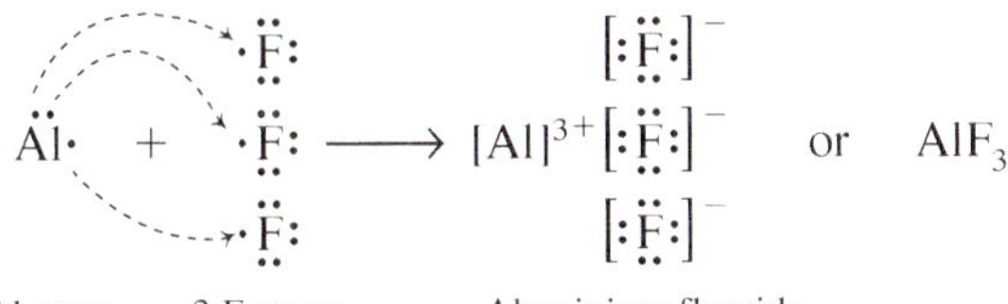

Each fluorine atom can accept only one electron. Therefore, three fluorine atoms are needed to combine with the three valence electrons of one aluminum atom. The aluminum atom has lost three electrons to become an aluminum ion (Al^{3+}) with a +3 charge.

ONLINE LEARNING MODULE
Ionic Bonding-2
WileyPLUS

EXAMPLE 11.6

Explain the formation of magnesium oxide (MgO) from its elements.

SOLUTION

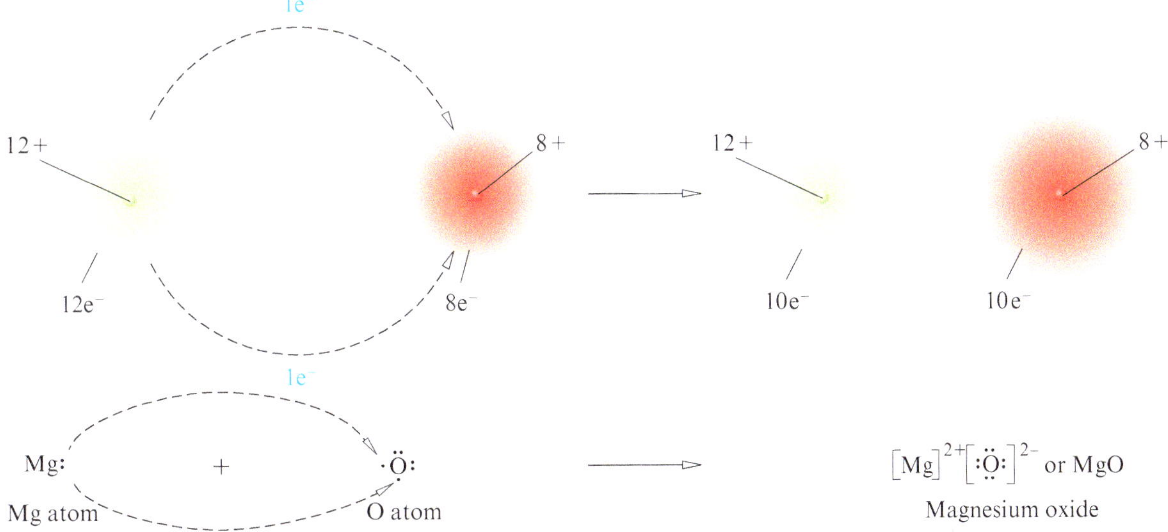

The magnesium atom, with two electrons in the outer energy level, exactly fills the need of one oxygen atom for two electrons. The resulting compound has a ratio of one magnesium atom to one oxygen atom. The oxygen (oxide) ion has a −2 charge, having gained two electrons. In combining with oxygen, magnesium behaves the same way as when it combines with chlorine—it loses two electrons.

EXAMPLE 11.7

Explain the formation of sodium sulfide (Na_2S) from its elements.

SOLUTION

$$Na\cdot + Na\cdot + \cdot\ddot{S}: \longrightarrow [Na]^+ [Na]^+ [:\ddot{S}:]^{2-} \quad \text{or} \quad Na_2S$$

2 Na atoms 1 S atom Sodium sulfide

Two sodium atoms supply the electrons that one sulfur atom needs to make eight in its outer energy level.

EXAMPLE 11.8

Explain the formation of aluminum oxide (Al_2O_3) from its elements.

SOLUTION

$$2\,\dot{Al}\cdot + 3\,\cdot\ddot{O}: \longrightarrow [Al]^{3+} [Al]^{3+} [:\ddot{O}:]^{2-} [:\ddot{O}:]^{2-} [:\ddot{O}:]^{2-} \quad \text{or} \quad Al_2O_3$$

2 Al atoms 3 O atoms Aluminum oxide

One oxygen atom, needing two electrons, cannot accommodate the three electrons from one aluminum atom. One aluminum atom falls one electron short of the four electrons needed by two oxygen atoms. A ratio of two atoms of aluminum to three atoms of oxygen, involving the transfer of six electrons (two to each oxygen atom), gives each atom a stable electron configuration.

Note that in each of these examples, outer energy levels containing eight electrons were formed in all the negative ions. This formation resulted from the pairing of all the *s* and *p* electrons in these outer energy levels.

11.4 Predicting Formulas of Ionic Compounds

LEARNING OBJECTIVE Predict the formulas of ionic compounds from their position on the periodic table.

In previous examples, we learned that when a metal and a nonmetal react to form an ionic compound, the metal loses one or more electrons to the nonmetal. In Chapter 6, where we learned to name compounds and write formulas, we saw that Group 1A(1) metals always form +1 cations, whereas Group 2A(2) elements form +2 cations. Group 7A(17) elements form +1 anions and Group 6A(16) elements form +2 anions.

It stands to reason, then, that this pattern is directly related to the stability of the noble gas configuration. Metals lose electrons to attain the electron configuration of a noble gas (the previous one on the periodic table). A nonmetal forms an ion by gaining enough electrons to achieve the electron configuration of the noble gas following it on the periodic table. These observations lead us to an important chemical principle:

In almost all stable chemical compounds of representative elements, each atom attains a noble gas electron configuration. This concept forms the basis for our understanding of chemical bonding.

We can apply this principle in predicting the formulas of ionic compounds. To predict the formula of an ionic compound, we must recognize that chemical compounds are always electrically neutral. In addition, the metal will lose electrons to achieve a noble gas configuration and the nonmetal will gain electrons to achieve a noble gas configuration. Consider the compound formed between barium and sulfur. Barium has two valence electrons, whereas sulfur has six valence electrons:

$$\text{Ba} \quad [\text{Xe}]6s^2 \qquad \text{S} \quad [\text{Ne}]3s^2 3p^4$$

If barium loses two electrons, it will achieve the configuration of xenon. By gaining two electrons, sulfur achieves the configuration of argon. Consequently, a pair of electrons is transferred between atoms. Now we have Ba^{2+} and S^{2-}. Since compounds are electrically neutral, there must be a ratio of one Ba to one S, giving the formula BaS.

The same principle works for many other cases. Since the key lies in the electron configuration, the periodic table can be used to extend the prediction even further. Because of similar electron structures, the elements in a family generally form compounds with the same atomic ratios. In general, if we know the atomic ratio of a particular compound—say, NaCl—we can predict the atomic ratios and formulas of the other alkali metal chlorides. These formulas are LiCl, KCl, RbCl, CsCl, and FrCl (TABLE 11.4).

Similarly, if we know that the formula of the oxide of hydrogen is H_2O, we can predict that the formula of the sulfide will be H_2S, because sulfur has the same valence electron structure as oxygen. Recognize, however, that these are only predictions; it doesn't necessarily follow that every element in a group will behave like the others or even that a predicted compound will actually exist. For example, knowing the formulas for potassium chlorate, bromate, and iodate to be $KClO_3$, $KBrO_3$, and KIO_3, we can correctly predict the corresponding sodium compounds to have the formulas $NaClO_3$, $NaBrO_3$, and $NaIO_3$. Fluorine belongs to the same family of elements (Group 7A(17)) as chlorine, bromine, and iodine, so it would appear that fluorine should combine with potassium and sodium to give fluorates with the formulas KFO_3 and $NaFO_3$. However, potassium and sodium fluorates are not known to exist.

In the discussion in this section, we refer only to representative metals [Groups 1A(1), 2A(2), and 3A(13)]. The transition metals (Group B) show more complicated behavior (they form multiple ions), and their formulas are not as easily predicted.

CHECK YOUR UNDERSTANDING

11.7 Alkali Metals and Oxides

11.8 Alkaline Earth Metals and Nitrides

WileyPLUS

TABLE 11.4 Formulas of Compounds Formed by Alkali Metals

Lewis structure	Oxides	Chlorides	Bromides	Sulfates
Li·	Li_2O	LiCl	LiBr	Li_2SO_4
Na·	Na_2O	NaCl	NaBr	Na_2SO_4
K·	K_2O	KCl	KBr	K_2SO_4
Rb·	Rb_2O	RbCl	RbBr	Rb_2SO_4
Cs·	Cs_2O	CsCl	CsBr	Cs_2SO_4

ENHANCED EXAMPLE 11.9 WileyPLUS

The formula for calcium sulfide is CaS and that for lithium phosphide is Li_3P. Predict formulas for (a) magnesium sulfide, (b) potassium phosphide, and (c) magnesium selenide.

SOLUTION

(a) Look up calcium and magnesium in the periodic table; they are both in Group 2A(2). The formula for calcium sulfide is CaS, so it's reasonable to predict that the formula for magnesium sulfide is MgS.

(b) Find lithium and potassium in the periodic table; they are in Group 1A(1). Since the formula for lithium phosphide is Li_3P, it's reasonable to predict that K_3P is the formula for potassium phosphide.

(c) Find selenium in the periodic table; it is in Group 6A(16) just below sulfur. Therefore, it's reasonable to assume that selenium forms selenide in the same way that sulfur forms sulfide. Since MgS was the predicted formula for magnesium sulfide in part (a), we can reasonably assume that the formula for magnesium selenide is MgSe.

PRACTICE 11.4

The formula for sodium oxide is Na_2O. Predict the formula for

(a) sodium sulfide (b) rubidium oxide

PRACTICE 11.5

The formula for barium phosphide is Ba_3P_2. Predict the formula for

(a) magnesium nitride (b) barium arsenide

11.5 The Covalent Bond: Sharing Electrons

LEARNING OBJECTIVE Draw the electron structure of a covalent bond.

KEY TERM
covalent bond
polar covalent bond

Some atoms do not transfer electrons from one atom to another to form ions. Instead they form a chemical bond by sharing pairs of electrons between them. A **covalent bond** consists of a pair of electrons shared between two atoms. This bonding concept was introduced in 1916 by G. N. Lewis. In the millions of known compounds, the covalent bond is the predominant chemical bond.

True molecules exist in substances in which the atoms are covalently bonded. It is proper to refer to molecules of such substances as hydrogen, chlorine, hydrogen chloride, carbon dioxide, water, or sugar (FIGURE 11.7). These substances contain only covalent bonds and exist as aggregates of molecules. We don't use the term *molecule* when talking about ionically bonded compounds, such as sodium chloride, because such substances exist as large aggregates of positive and negative ions, not as molecules (Figure 11.7).

A study of the hydrogen molecule gives us an insight into the nature of the covalent bond and its formation. The formation of a hydrogen molecule (H_2) involves the overlapping and pairing of 1*s* electron orbitals from two hydrogen atoms, shown in FIGURE 11.8. Each atom contributes one electron of the pair that is shared jointly by two hydrogen nuclei. The orbital of the electrons now includes both hydrogen nuclei, but probability factors show that the most likely place to find the electrons (the point of highest electron density) is between the two nuclei. The

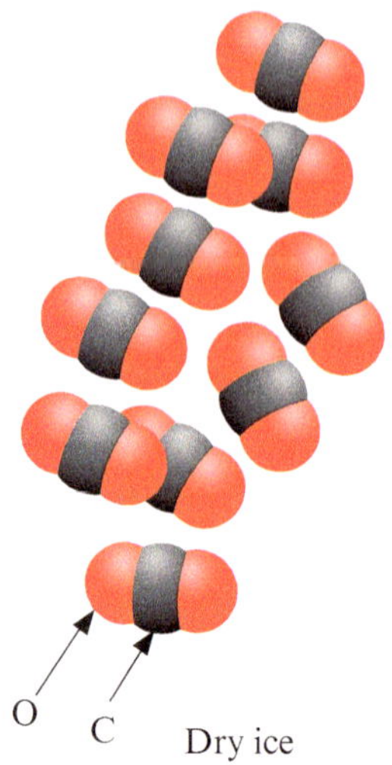

Na^+ Cl^-

Sodium chloride

FIGURE 11.7 Solid carbon dioxide (dry ice) is composed of individual covalently bonded molecules of CO_2 closely packed together. Table salt is a large aggregate of Na^+ and Cl^- ions instead of molecules.

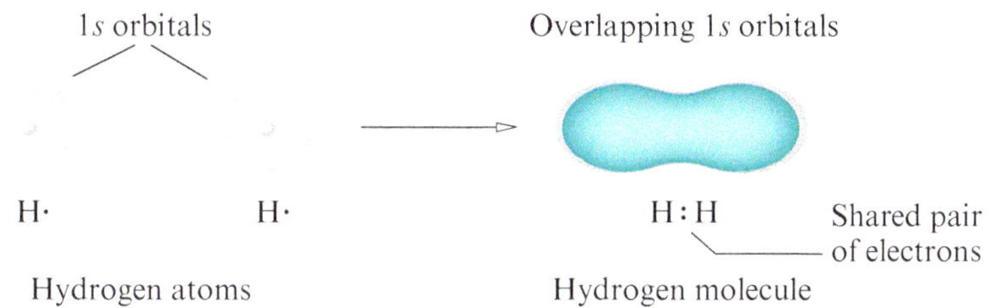

FIGURE 11.8 The formation of a hydrogen molecule from two hydrogen atoms. The two 1s orbitals overlap, forming the H_2 molecule. In this molecule the two electrons are shared between the atoms, forming a covalent bond.

two nuclei are shielded from each other by the pair of electrons, allowing the two nuclei to be drawn very close to each other.

The formula for chlorine gas is Cl_2. When the two atoms of chlorine combine to form this molecule, the electrons must interact in a manner similar to that shown in the hydrogen example. Each chlorine atom would be more stable with eight electrons in its outer energy level. But chlorine atoms are identical, and neither is able to pull an electron away from the other. What happens is this: The unpaired $3p$ electron orbital of one chlorine atom overlaps the unpaired $3p$ electron orbital of the other atom, resulting in a pair of electrons that is mutually shared between the two atoms. Each atom furnishes one of the pair of shared electrons. Thus, each atom attains a stable structure of eight electrons by sharing an electron pair with the other atom. The pairing of the p electrons and the formation of a chlorine molecule are illustrated in FIGURE 11.9. Neither chlorine atom has a positive or negative charge because both contain the same number of protons and have equal attraction for the pair of electrons being shared. Other examples of molecules in which electrons are equally shared between two atoms are hydrogen (H_2), oxygen (O_2), nitrogen (N_2), fluorine (F_2), bromine (Br_2), and iodine (I_2). Note that more than one pair of electrons may be shared between atoms:

CHECK YOUR UNDERSTANDING

11.9 Ionic and Covalent Compounds

➡ WileyPLUS

H:H	:F̤̈:F̤̈:	:B̤̈r:B̤̈r:	:Ï̤:Ï̤:	:Ö::Ö:	:N:::N:
hydrogen	fluorine	bromine	iodine	oxygen	nitrogen

The Lewis structure given for oxygen does not adequately account for all the properties of the oxygen molecule. Other theories explaining the bonding in oxygen molecules have been advanced, but they are complex and beyond the scope of this book.

In writing structures, we commonly replace the pair of dots used to represent a shared pair of electrons with a dash (—). One dash represents a single bond; two dashes, a double bond; and three dashes, a triple bond. The six structures just shown may be written thus:

H—H :F̤̈—F̤̈: :B̤̈r—B̤̈r: :Ï̤—Ï̤: :Ö=Ö: :N≡N:

Tom Pantages

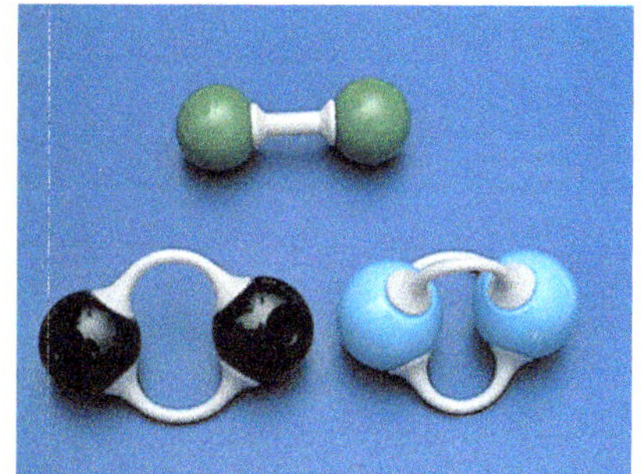

Molecular models for F_2 (green, single bond), O_2 (black, double bond), and N_2 (blue, triple bond).

The ionic bond and the covalent bond represent two extremes. In ionic bonding the atoms are so different that electrons are transferred between them, forming a charged pair of ions. In covalent bonding, two identical atoms share electrons equally. The bond is the mutual attraction of the two nuclei for the shared electrons. Between these extremes lie many cases in which the atoms are not different enough for a transfer of electrons but are different enough that the electron pair cannot be shared equally. This unequal sharing of electrons results in the formation of a **polar covalent bond**.

Remember: A dash represents a shared pair of electrons.

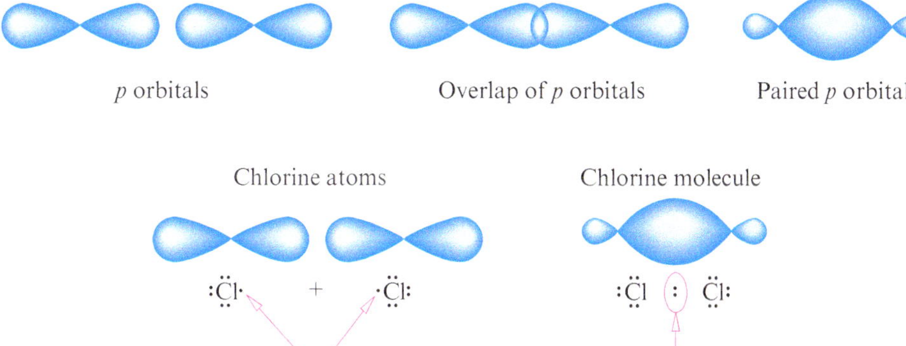

FIGURE 11.9 Pairing of p electrons in the formation of a chlorine molecule.

CHECK YOUR UNDERSTANDING

11.10 Covalent Bonds

WileyPLUS

EXAMPLE 11.10

What is the basic concept of a covalent bond?

SOLUTION

The basic concept of a covalent bond is the sharing of one or more pairs of electrons between two atoms to form a covalent bond between the two atoms.

PRACTICE 11.6

The atoms in the two molecules HBr and C_2H_6 are bonded by covalent bonds. Draw Lewis structures showing these covalent bonds.

11.6 Electronegativity

LEARNING OBJECTIVE Explain how electronegativities of component atoms in a molecule determine the polarity of the molecule.

KEY TERM

electronegativity
nonpolar covalent bond
dipole

When two *different* kinds of atoms share a pair of electrons, a bond forms in which electrons are shared unequally. One atom assumes a partial positive charge and the other a partial negative charge with respect to each other. This difference in charge occurs because the two atoms exert unequal attraction for the pair of shared electrons. The attractive force that an atom of an element has for shared electrons in a molecule or polyatomic ion is known as its **electronegativity**. Elements differ in their electronegativities. For example, both hydrogen and chlorine need one electron to form stable electron configurations. They share a pair of electrons in hydrogen chloride (HCl). Chlorine is more electronegative and therefore has a greater attraction for the shared electrons than does hydrogen. As a result, the pair of electrons is displaced toward the chlorine atom, giving it a partial negative charge and leaving the hydrogen atom with a partial positive charge. Note that the electron is not transferred entirely to the chlorine atom (as in the case of sodium chloride) and that no ions are formed. The entire molecule, HCl, is electrically neutral. A partial charge is usually indicated by the Greek letter delta, δ. Thus, a partial positive charge is represented by δ+ and a partial negative charge by δ−.

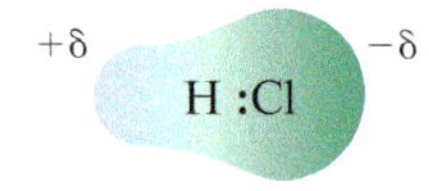

δ+ H:Cl: δ−

Hydrogen chloride

The pair of shared electrons in HCl is closer to the more electronegative chlorine atom than to the hydrogen atom, giving chlorine a partial negative charge with respect to the hydrogen atom.

A scale of relative electronegativities, in which the most electronegative element, fluorine, is assigned a value of 4.0, was developed by the Nobel Laureate (1954 and 1962) Linus Pauling (1901–1994). TABLE 11.5 shows that the relative electronegativity of the nonmetals is high and that of the metals is low. These electronegativities indicate that atoms of metals have a greater tendency to lose electrons than do atoms of nonmetals and that nonmetals have a greater tendency to gain electrons than do metals. The higher the electronegativity value, the greater the attraction for electrons. Note that electronegativity generally increases from left to right across a period and decreases down a group for the representative elements. The highest electronegativity is 4.0 for fluorine, and the lowest is 0.7 for francium and cesium. It's important to remember that the higher the electronegativity, the more strongly an atom attracts electrons.

CHECK YOUR UNDERSTANDING

11.11 Electronegativity

WileyPLUS

EXAMPLE 11.11

What are the electronegativity trends for the groups of representative elements and the periods of representative elements in the periodic table?

SOLUTION

In general, electronegativity decreases from top to bottom for a group of representative elements and increases from left to right across a period of representative elements.

TABLE 11.5 Three-Dimensional Representation of Electronegativity

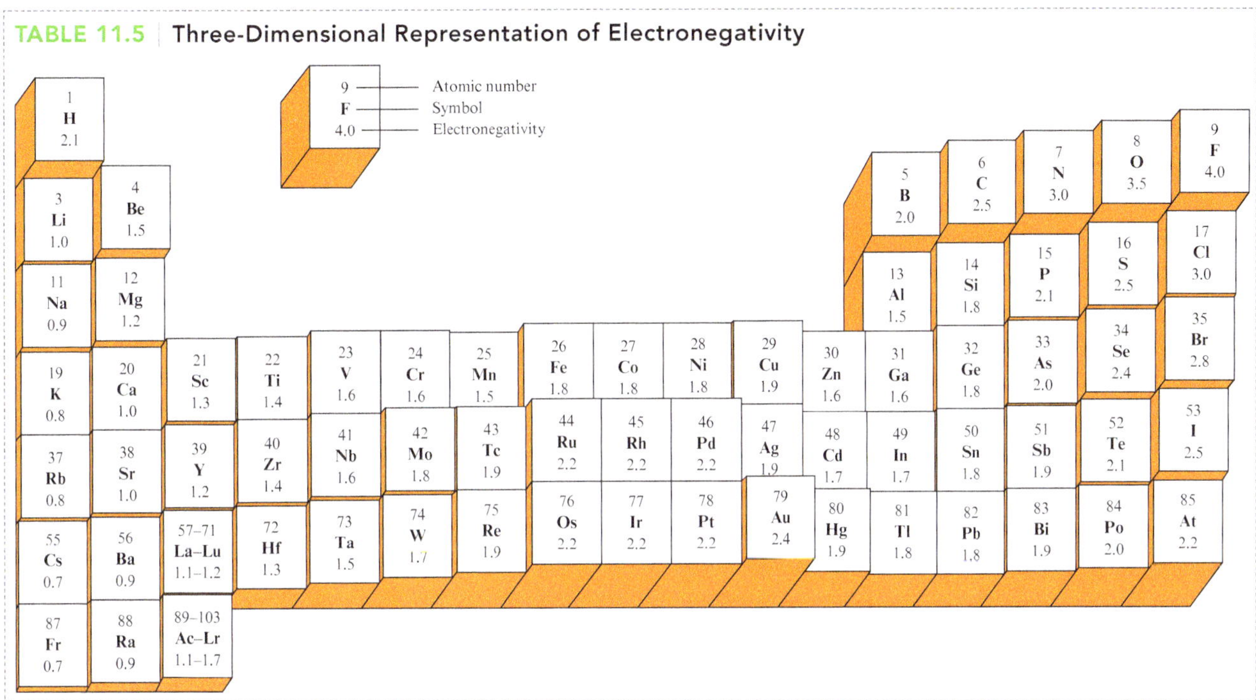

PRACTICE 11.7

Explain the term *electronegativity* and how it relates to the elements in the periodic table.

The polarity of a bond is determined by the difference in electronegativity values of the atoms forming the bond (FIGURE 11.10). If the electronegativities are the same, the bond is **nonpolar covalent** and the electrons are shared equally. If the atoms have greatly different electronegativities, the bond is very *polar*. At the extreme, one or more electrons are actually transferred and an ionic bond results.

CHECK YOUR UNDERSTANDING

11.12 Electronegativity Trends

➡ WileyPLUS

A **dipole** is a molecule that is electrically asymmetrical, causing it to be oppositely charged at two points. A dipole is often written as (+ −). A hydrogen chloride molecule is polar and behaves as a small dipole. The HCl dipole may be written as H⟼Cl. The arrow points toward the negative end of the dipole. Molecules of H_2O, HBr, and ICl are polar:

H⟼Cl   H⟼Br   I⟼Cl   H⟼O⟻H

How do we know whether a bond between two atoms is ionic or covalent? The difference in electronegativity between the two atoms determines the character of the bond formed between them. As the difference in electronegativity increases, the polarity of the bond (or percent ionic character) increases.

In general, if the electronegativity difference between two bonded atoms is greater than 1.7–1.9, the bond will be more ionic than covalent.

If the electronegativity difference is greater than 2.0, the bond is strongly ionic. If the electronegativity difference is less than 1.5, the bond is strongly covalent.

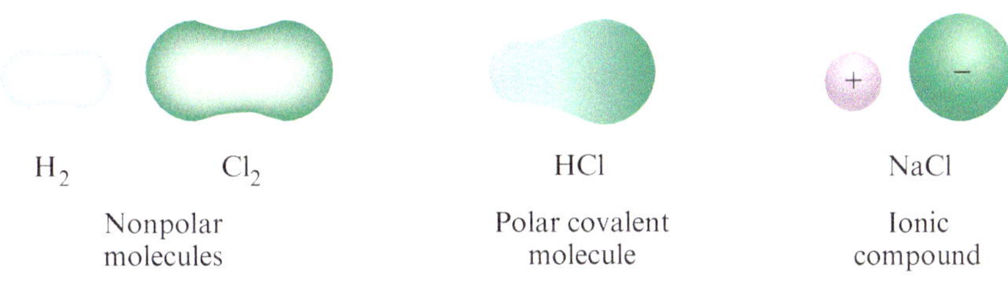

FIGURE 11.10 Nonpolar, polar covalent, and ionic compounds.

CHEMISTRY IN ACTION Trans-forming Fats

Trans fats are virtually everywhere in the American diet. These fats remain when vegetable oils are converted into solid substances that are used in many processed foods. In fact, beginning January 1, 2006, manufacturers were required to label products to show their *trans* fat content. So just what is a *trans* fat?

Different categories of fat can be identified from the pattern of bonds and hydrogen atoms in the molecule. Fatty acids are one component of fats and contain long chains of carbon atoms with hydrogen atoms bonded to some or all of the carbon atoms. Unsaturated fats (such as corn or soybean oil) contain double bonds between some of the carbon atoms in the chains. A carbon that is doubly bonded to another carbon usually also bonds to a hydrogen atom. If the fatty acid has only one double bond, it is monounsaturated (such as olive oil). If the fatty acid contains more than one double bond, it is polyunsaturated. All fat with no double bonds is saturated—all the carbon atoms have the maximum possible hydrogen atoms bonded to them.

Trans fats contain a particular kind of unsaturated fatty acid. When there is a double bond between carbon atoms, the molecule bends in one of two ways: the *cis* or the *trans* direction. In *cis* configuration, the carbon chain on both sides of the double bond bends in to the same side of the double bond (see structure). In the *trans* configuration, the chain on either side of the double bond bends toward opposite sides of the double bond (see structure). Most *trans* fats come from processing oils for prepared foods and from solid fats such as margarine.

When an oil is converted into a solid fat, some of the double bonds are converted to single bonds by adding hydrogen (hydrogenation). This process is easier at *cis* double bonds and, therefore, the remaining double bonds are mainly in the *trans* configuration. These *trans* fatty acids tend to stack together, forming a solid easier than the *cis* forms. Studies have linked diets high in *trans* fats to poor health, high cholesterol, heart disease, and diabetes. Food producers are working on ways to lower the *trans* fat content of foods. Gary List at USDA in Peoria, Illinois, has used high-pressure hydrogen gas on soybean oil at 140°–170°C to hydrogenate the oil, producing a soft margarine containing 5%–6% *trans* fat instead of the ~40% from the standard hydrogenation techniques. We are seeing many more products that qualify for labels of 0 g *trans* fat.

H H
C=C O
$CH_3-(CH_2)_7$ $(CH_2)_7-C$
OH

cis fatty acid

O
H $(CH_2)_7-C$
C=C OH
$CH_3-(CH_2)_7$ H

trans fatty acid

PRACTICE 11.8

Which of these compounds would you predict to be ionic and which would be covalent?

(a) $SrCl_2$
(b) PCl_3
(c) NH_3
(d) RbBr
(e) LiCl
(f) CS_2

Care must be taken to distinguish between polar bonds and polar molecules. A covalent bond between different kinds of atoms is always polar. But a molecule containing different kinds of atoms may or may not be polar, depending on its shape or geometry. Molecules of HF, HCl, HBr, HI, and ICl are all polar because each contains a single polar bond. However, CO_2, CH_4, and CCl_4 are nonpolar molecules despite the fact that all three contain polar bonds. The carbon dioxide molecule O=C=O is nonpolar because the carbon–oxygen dipoles cancel each other by acting in opposite directions.

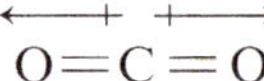

O=C=O

dipoles in equal and opposite directions

Carbon tetrachloride (CCl_4) is nonpolar because the four C—Cl polar bonds are identical, and since these bonds emanate from the center to the corners of a tetrahedron in the molecule, their polarities cancel one another. Methane has the same molecular structure and is also nonpolar. We will discuss the shapes of molecules later in this chapter.

Spacefilling molecular model CCl_4 (top) and methane (CH_4).

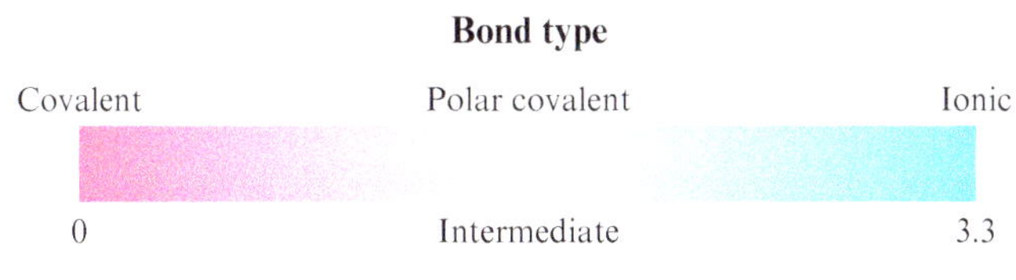

FIGURE 11.11 Relating bond type to electronegativity difference between atoms.

We have said that water is a polar molecule. If the atoms in water were linear like those in carbon dioxide, the two O—H dipoles would cancel each other, and the molecule would be nonpolar. However, water is definitely polar and has a nonlinear (bent) structure with an angle of 105° between the two O—H bonds.

The relationships among types of bonds are summarized in FIGURE 11.11. It is important to realize that bonding is a continuum; that is, the difference between ionic and covalent is a gradual change.

CHECK YOUR UNDERSTANDING

11.13 Polar Covalent Molecules

➡ WileyPLUS

11.7 Lewis Structures of Compounds

LEARNING OBJECTIVE

Draw the Lewis structure of a covalent compound.

As we have seen, Lewis structures are a convenient way of showing the covalent bonds in many molecules or ions of the representative elements. In writing Lewis structures, the most important consideration for forming a stable compound is that the atoms attain a noble gas configuration.

The most difficult part of writing Lewis structures is determining the arrangement of the atoms in a molecule or an ion. In simple molecules with more than two atoms, one atom will be the central atom surrounded by the other atoms. Thus, Cl_2O has two possible arrangements, Cl—Cl—O or Cl—O—Cl. Usually, but not always, the single atom in the formula (except H) will be the central atom.

Although Lewis structures for many molecules and ions can be written by inspection, the following procedure is helpful for learning to write them:

PROBLEM-SOLVING STRATEGY

Writing a Lewis Structure

1. Obtain the total number of valence electrons to be used in the structure by adding the number of valence electrons in all the atoms in the molecule or ion. If you are writing the structure of an ion, add one electron for each negative charge or subtract one electron for each positive charge on the ion.
2. Write the skeletal arrangement of the atoms and connect them with a single covalent bond (two dots or one dash). Hydrogen, which contains only one bonding electron, can form only one covalent bond. Oxygen atoms are not normally bonded to each other, except in compounds known to be peroxides. Oxygen atoms normally have a maximum of two covalent bonds (two single bonds or one double bond).
3. Subtract two electrons for each single bond you used in Step 2 from the total number of electrons calculated in Step 1. This gives you the net number of electrons available for completing the structure.
4. Distribute pairs of electrons (pairs of dots) around each atom (except hydrogen) to give each atom a noble gas structure.
5. If there are not enough electrons to give these atoms eight electrons, change single bonds between atoms to double or triple bonds by shifting unbonded pairs of electrons as needed. Check to see that each atom has a noble gas electron structure (two electrons for hydrogen and eight for the others). A double bond counts as four electrons for each atom to which it is bonded.

Remember: The number of valence electrons of Group A elements is the same as their group number in the periodic table.

ONLINE LEARNING MODULE

Valence Electrons

➡ WileyPLUS

ENHANCED EXAMPLE 11.12

WileyPLUS

How many valence electrons are in each of these atoms: Cl, H, C, O, N, S, P, I?

SOLUTION

You can look at the periodic table to determine the electron structure, or, if the element is in Group A of the periodic table, the number of valence electrons is equal to the A group number:

Atom	Group	Valence electrons
Cl	7A(17)	7
H	1A(1)	1
C	4A(14)	4
O	6A(16)	6
N	5A(15)	5
S	6A(16)	6
P	5A(15)	5
I	7A(17)	7

ONLINE LEARNING MODULE

Writing a Lewis Structure for Water

➡ WileyPLUS

ENHANCED EXAMPLE 11.13

WileyPLUS

Use the Writing a Lewis Structure Problem-Solving Strategy for water (H_2O).

SOLUTION

1. The total number of valence electrons is eight, two from the two hydrogen atoms and six from the oxygen atom.
2. The two hydrogen atoms are connected to the oxygen atom. Write the skeletal structure:

   H O
     H   or   H O H

   Place two dots between the hydrogen and oxygen atoms to form the covalent bonds:

   H:O
      ⠆
      H   or   H:O:H

3. Subtract the four electrons used in Step 2 from eight to obtain four electrons yet to be used.
4. Distribute the four electrons in pairs around the oxygen atom. Hydrogen atoms cannot accommodate any more electrons:

   H—Ö:
       |
       H   or   H—Ö—H (with a lone pair below O)

These arrangements are Lewis structures because each atom has a noble gas electron structure. Note that the shape of the molecule is not shown by the Lewis structure.

ONLINE LEARNING MODULE

Writing a Lewis Structure for Methane

➡ WileyPLUS

ENHANCED EXAMPLE 11.14

WileyPLUS

Use the Writing a Lewis Structure Problem-Solving Strategy for a molecule of methane (CH_4).

SOLUTION

1. The total number of valence electrons is eight, one from each hydrogen atom and four from the carbon atom.

2. The skeletal structure contains four H atoms around a central C atom. Place two electrons between the C and each H.

```
  H          H
H C H      H:C:H
  H          H
```

3. Subtract the eight electrons used in Step 2 from eight (obtained in Step 1) to obtain zero electrons yet to be placed. Therefore, the Lewis structure must be as written in Step 2:

```
                  H
  H               |
H:C:H    or   H — C — H
  H               |
                  H
```

EXAMPLE 11.15

Use the Writing a Lewis Structure Problem-Solving Strategy for a molecule of carbon tetrachloride (CCl_4).

SOLUTION

1. The total number of valence electrons to be used is 32, 4 from the carbon atom and 7 from each of the four chlorine atoms.
2. The skeletal structure contains the four Cl atoms around a central C atom. Place 2 electrons between the C and each Cl:

```
   Cl          Cl
Cl C Cl     Cl:C:Cl
   Cl          Cl
```

3. Subtract the 8 electrons used in Step 2 from 32 (obtained in Step 1) to obtain 24 electrons yet to be placed.
4. Distribute the 24 electrons (12 pairs) around the Cl atoms so that each Cl atom has 8 electrons around it:

```
                          ..
     ..                  :Cl:
    :Cl:                   |
 .. .. ..              ..  |  ..
:Cl:C:Cl:    or       :Cl— C —Cl:
 .. .. ..              ..  |  ..
    :Cl:                   |
     ..                  :Cl:
                          ..
```

This arrangement is the Lewis structure; CCl_4 contains four covalent bonds.

EXAMPLE 11.16

ONLINE LEARNING MODULE
Writing a Lewis Structure for Carbon Dioxide
➡ WileyPLUS

Use the Writing a Lewis Structure Problem-Solving Strategy for CO_2.

SOLUTION

1. The total number of valence electrons is 16, 4 from the C atom and 6 from each O atom.
2. The two O atoms are bonded to a central C atom. Write the skeletal structure and place 2 electrons between the C and each O atom.

```
O : C : O
```

3. Subtract the 4 electrons used in Step 2 from 16 (found in Step 1) to obtain 12 electrons yet to be placed.

4. Distribute the 12 electrons (six pairs) around the C and O atoms. Several possibilities exist:

$$\underset{\text{I}}{:\ddot{\underset{\cdot\cdot}{O}}:C:\ddot{\underset{\cdot\cdot}{O}}:} \qquad \underset{\text{II}}{:\ddot{O}:\underset{\cdot\cdot}{C}:\ddot{\underset{\cdot\cdot}{O}}:} \qquad \underset{\text{III}}{:\ddot{O}:\ddot{\underset{\cdot\cdot}{C}}:\underset{\cdot\cdot}{O}:}$$

5. Not all the atoms have 8 electrons around them (noble gas structure). Remove one pair of unbonded electrons from each O atom in structure I and place one pair between each O and the C atom, forming two double bonds:

$$:\ddot{O}::C::\ddot{O}: \quad \text{or} \quad :\ddot{O}=C=\ddot{O}:$$

Each atom now has 8 electrons around it. The carbon is sharing four pairs of electrons, and each oxygen is sharing two pairs. These bonds are known as double bonds because each involves sharing two pairs of electrons.

PRACTICE 11.9

Write the Lewis structures for the following:
(a) PBr_3 (b) $CHCl_3$ (c) HF (d) H_2CO (e) N_2

CHEMISTRY *IN ACTION* Strong Enough to Stop a Bullet?

What do color-changing pens, bullet-resistant vests, and calculators have in common? The chemicals that make each of them work are liquid crystals. These chemicals find numerous applications; you are probably most familiar with liquid crystal displays (LCDs) and color-changing products, but these chemicals are also used to make superstrong synthetic fibers.

Molecules in a normal crystal remain in an orderly arrangement, but in a liquid crystal the molecules can flow *and* maintain an orderly arrangement at the same time. Liquid crystal molecules are linear and polar. Since the atoms tend to lie in a relatively straight line, the molecules are generally much longer than they are wide. These polar molecules are attracted to each other and are able to line up in an orderly fashion without solidifying.

Liquid crystals with twisted arrangements of molecules give us novelty color-changing products. In these liquid crystals the molecules lie side by side in a nearly flat layer. The next layer is similar but at an angle to the one below. The closely packed flat layers have a special effect on light. As the light strikes the surface, some of it is reflected from the top layer and some from lower layers. When the same wavelength is reflected from many layers, we see a color. (This is similar to the rainbow of colors formed by oil in a puddle on the street or the film of a soap bubble.) As the temperature is increased, the molecules move faster, causing a change in the angle and the space between the layers. This results in a color change in the reflected light. Different compounds change color within different temperature ranges, allowing a variety of practical and amusing applications.

Liquid crystal (nematic) molecules that lie parallel to one another are used to manufacture very strong synthetic fibers. Perhaps the best example of these liquid crystals is Kevlar, a synthetic fiber used in bullet-resistant vests, canoes, and parts of the space shuttle. Kevlar is a synthetic polymer, like nylon or polyester, that gains strength by passing through a liquid crystal state during its manufacture.

In a typical polymer, the long molecular chains are jumbled together, somewhat like spaghetti. The strength of the material is limited by the disorderly arrangement. The trick is to get the molecules to line up parallel to each other. Once the giant molecules have been synthesized, they are dissolved in sulfuric acid. At the proper concentration the molecules align, and the solution is forced through tiny holes in a nozzle and further aligned. The sulfuric acid is removed in a water bath, thereby forming solid fibers in near-perfect alignment. One strand of Kevlar is stronger than an equal-sized strand of steel. It has a much lower density as well, making it a material of choice in bullet-resistant vests.

© Haku Nellies/iStockphoto

Kevlar is used to make protective vests for police.

Although many compounds attain a noble gas structure in covalent bonding, there are numerous exceptions. Sometimes it's impossible to write a structure in which each atom has 8 electrons around it. For example, in BF_3 the boron atom has only 6 electrons around it, and in SF_6 the sulfur atom has 12 electrons around it.

Although there are exceptions, many molecules can be described using Lewis structures where each atom has a noble gas electron configuration. This is a useful model for understanding chemistry.

11.8 Complex Lewis Structures

LEARNING OBJECTIVE

Draw the resonance structures for a polyatomic ion.

KEY TERMS

resonance structure

Most Lewis structures give bonding pictures that are consistent with experimental information on bond strength and length. There are some molecules and polyatomic ions for which no single Lewis structure consistent with all characteristics and bonding information can be written. For example, consider the nitrate ion, NO_3^-. To write a Lewis structure for this polyatomic ion, we use the following steps.

1. The total number of valence electrons is 24, 5 from the nitrogen atom, 6 from each oxygen atom, and 1 from the −1 charge.
2. The three O atoms are bonded to a central N atom. Write the skeletal structure and place two electrons between each pair of atoms. Since we have an extra electron in this ion, resulting in a −1 charge, we enclose the group of atoms in square brackets and add a −1 charge as shown.

$$\left[\begin{array}{c} \text{O} \\ \text{O:}\overset{\cdot\cdot}{\text{N}}\text{:O} \end{array}\right]^-$$

3. Subtract the 6 electrons used in Step 2 from 24 (found in Step 1) to obtain 18 electrons yet to be placed.
4. Distribute the 18 electrons around the N and O atoms:

$$\begin{array}{c} :\ddot{\text{O}} \leftarrow \text{electron deficient} \\ :\ddot{\underset{\cdot\cdot}{\text{O}}}:\ddot{\underset{\cdot\cdot}{\text{N}}}:\ddot{\underset{\cdot\cdot}{\text{O}}}: \end{array}$$

5. One pair of electrons is still needed to give all the N and O atoms a noble gas structure. Move the unbonded pair of electrons from the N atom and place it between the N and the electron-deficient O atom, making a double bond.

$$\left[\begin{array}{c} :\ddot{\text{O}} \\ \| \\ :\underset{\cdot\cdot}{\ddot{\text{O}}}-\text{N}-\underset{\cdot\cdot}{\ddot{\text{O}}}: \end{array}\right]^- \text{ or } \left[\begin{array}{c} :\ddot{\text{O}}: \\ | \\ :\underset{\cdot\cdot}{\ddot{\text{O}}}-\text{N}=\ddot{\text{O}}: \end{array}\right]^- \text{ or } \left[\begin{array}{c} :\ddot{\text{O}}: \\ | \\ \underset{\cdot\cdot}{\ddot{\text{O}}}=\text{N}-\underset{\cdot\cdot}{\ddot{\text{O}}}: \end{array}\right]^-$$

Are these all valid Lewis structures? Yes, so there really are three possible Lewis structures for NO_3^-.

A molecule or ion that has multiple correct Lewis structures shows *resonance*. Each of these Lewis structures is called a **resonance structure**. In this book, however, we will not be concerned with how to choose the correct resonance structure for a molecule or ion. Therefore, any of the possible resonance structures may be used to represent the ion or molecule.

ENHANCED EXAMPLE 11.17 WileyPLUS

Write the Lewis structure for a carbonate ion (CO_3^{2-}).

SOLUTION

1. These four atoms have 22 valence electrons plus 2 electrons from the −2 charge, which makes 24 electrons to be placed.

ONLINE LEARNING MODULE

Writing a Lewis Structure for a Polyatomic Ion

➡ WileyPLUS

2. In the carbonate ion, the carbon is the central atom surrounded by the three oxygen atoms. Write the skeletal structure and place 2 electrons (or a single line) between each C and O:

3. Subtract the 6 electrons used in Step 2 from 24 (from Step 1) to give 18 electrons yet to be placed.

4. Distribute the 18 electrons around the three oxygen atoms and indicate that the carbonate ion has a −2 charge:

   The difficulty with this structure is that the carbon atom has only 6 electrons around it instead of a noble gas octet.

5. Move one of the nonbonding pairs of electrons from one of the oxygens and place it between the carbon and the oxygen. Three Lewis structures are possible:

   or   or

PRACTICE 11.10

Write the Lewis structure for each of the following:

(a) NH_3 (b) H_3O^+ (c) NH_4^+ (d) HCO_3^-

11.9 Compounds Containing Polyatomic Ions

LEARNING OBJECTIVE Describe a compound that contains both ionic and covalent bonds.

A polyatomic ion is a stable group of atoms that has either a positive or a negative charge and behaves as a single unit in many chemical reactions. Sodium carbonate, Na_2CO_3, contains two sodium ions and a carbonate ion. The carbonate ion (CO_3^{2-}) is a polyatomic ion composed of one carbon atom and three oxygen atoms and has a charge of −2. One carbon and three oxygen atoms have a total of 22 electrons in their outer energy levels. The carbonate ion contains 24 outer electrons and therefore has a charge of −2. In this case, the two additional electrons come from the two sodium atoms, which are now sodium ions:

$[Na]^+$ $[Na]^+$

Sodium carbonate

Carbonate ion

Sodium carbonate has both ionic and covalent bonds. Ionic bonds exist between each of the sodium ions and the carbonate ion. Covalent bonds are present between the carbon and oxygen atoms within the carbonate ion. One important difference between the ionic and covalent bonds in this compound can be demonstrated by dissolving sodium carbonate in water. It dissolves in water, forming three charged particles—two sodium ions and one carbonate ion—per formula unit of Na_2CO_3:

$$\underset{\text{sodium carbonate}}{Na_2CO_3(s)} \xrightarrow{\text{water}} \underset{\text{sodium ions}}{2\,Na^+(aq)} + \underset{\text{carbonate ion}}{CO_3^{2-}(aq)}$$

The CO_3^{2-} ion remains as a unit, held together by covalent bonds; but where the bonds are ionic, dissociation of the ions takes place. Do not think, however, that polyatomic ions are so stable that they cannot be altered. Chemical reactions by which polyatomic ions can be changed to other substances do exist.

PRACTICE 11.11

Write Lewis structures for the calcium compounds of the following anions: nitrate and sulfate.

11.10 Molecular Shape

LEARNING OBJECTIVE

Determine the shape of a compound by using the VSEPR method.

KEY TERMS

linear structure
trigonal planar structure
tetrahedral structure
bent structure

So far in our discussion of bonding we have used Lewis structures to represent valence electrons in molecules and ions, but they don't indicate anything regarding the molecular or geometric shape of a molecule. The three-dimensional arrangement of the atoms within a molecule is a significant feature in understanding molecular interactions. Let's consider several examples illustrated in FIGURE 11.12.

Water is known to have the geometric shape known as "bent" or "V-shaped." Carbon dioxide exhibits a linear shape. BF_3 forms a third molecular shape called *trigonal planar* since all the atoms lie in one plane in a triangular arrangement. One of the more common molecular shapes is the tetrahedron, illustrated by the molecule methane (CH_4).

How do we predict the geometric shape of a molecule? We will now study a model developed to assist in making predictions from the Lewis structure.

The Valence Shell Electron Pair Repulsion (VSEPR) Model

The chemical properties of a substance are closely related to the structure of its molecules. A change in a single site on a large biomolecule can make a difference in whether or not a particular reaction occurs.

Instrumental analysis can be used to determine exact spatial arrangements of atoms. Quite often, though, we only need to be able to predict the approximate structure of a

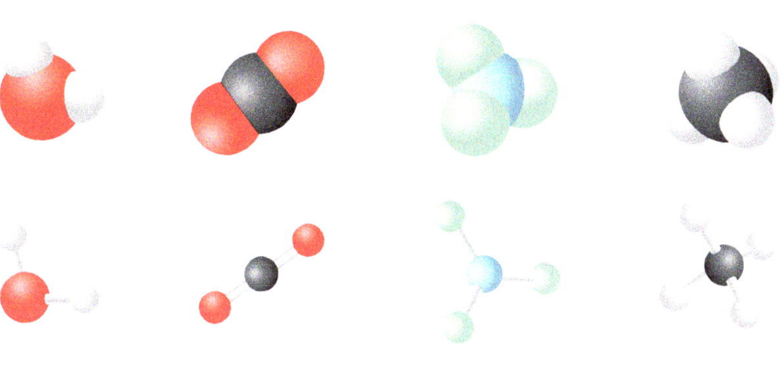

Water H_2O (V-shaped) | Carbon dioxide CO_2 (linear shape) | Boron trifluoride BF_3 (trigonal planar) | Methane CH_4 (tetrahedral)

FIGURE 11.12 Geometric shapes of common molecules. Each molecule is shown as a ball-and-stick model (showing the bonds) and as a spacefilling model (showing the shape).

molecule. A relatively simple model has been developed to allow us to make predictions of shape from Lewis structures.

Nonbonding pairs of electrons on outer atoms are not shown here, so you can focus your attention on the shapes, not the electron arrangement.

The VSEPR model is based on the idea that electron pairs will repel each other electrically and will seek to minimize this repulsion. To accomplish this minimization, the electron pairs will be arranged around a central atom as far apart as possible. Consider $BeCl_2$, a molecule with only two pairs of electrons surrounding the central atom. These electrons are arranged 180° apart for maximum separation:

180°

Cl — Be — Cl

Note that neither beryllium nor boron have an octet in these molecules. There are no additional electrons on the central atoms.

This molecular structure can now be labeled as a **linear structure**. When only two pairs of electrons surround a central atom, they should be placed 180° apart to give a linear structure. Although we saw in chapter 6 that a metal bonded to a non-metal tends to form an ionic compound, the small difference in electronegativity ($3.0 - 1.5 = 1.5$) means that there is significant covalent character to this bond.

What occurs when there are only three pairs of electrons around the central atom? Consider the BF_3 molecule. The greatest separation of electron pairs occurs when the angles between atoms are 120°:

F 120° F
B
120° 120°
F

This arrangement of atoms is flat (planar) and, as noted earlier, is called **trigonal planar structure**. When three pairs of electrons surround an atom, they should be placed 120° apart to show the trigonal planar structure.

Now consider the most common situation (CH_4), with four pairs of electrons on the central carbon atom. In this case the central atom exhibits a noble gas electron structure. What arrangement best minimizes the electron pair repulsions? At first, it seems that an obvious choice is a 90° angle with all the atoms in a single plane:

H
90° 90°
H — C — H
90° 90°
H

However, we must consider that molecules are three-dimensional. This concept results in a structure in which the electron pairs are actually 109.5° apart:

H 109.5° H
C
H H

CHECK YOUR UNDERSTANDING

11.14 The Valence Shell Electron Pair Repulsion (VSEPR) Model and Molecular Shape

WileyPLUS

In this diagram the wedged line seems to protrude from the page, whereas the dashed line recedes. Two representations of this arrangement, known as **tetrahedral structure**, are illustrated in FIGURE 11.13. When four pairs of electrons surround a central atom, they should be placed 109.5° apart to give them a tetrahedral structure.

The VSEPR model is based on the premise that we are counting electron pairs. It's quite possible that one or more of these electron pairs may be nonbonding (lone) pairs. What happens to the molecular structure in these cases?

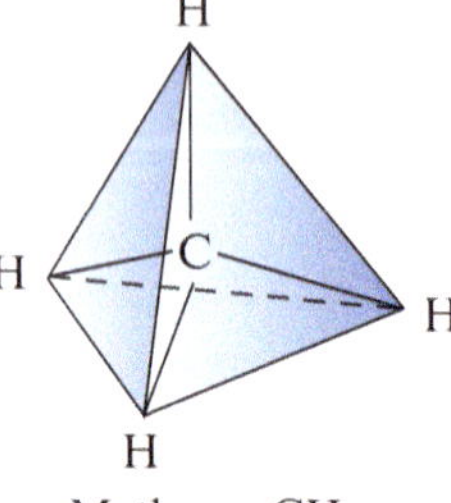

Methane, CH_4

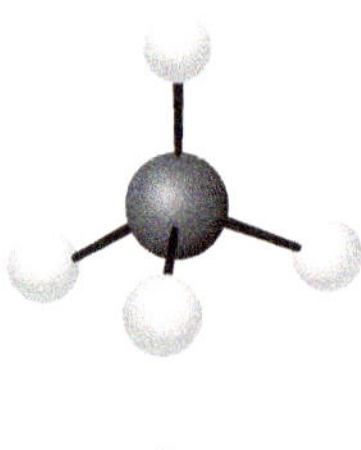
CH_4

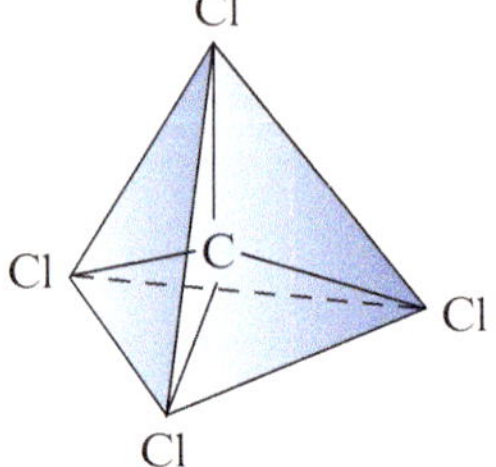

Carbon tetrachloride, CCl_4

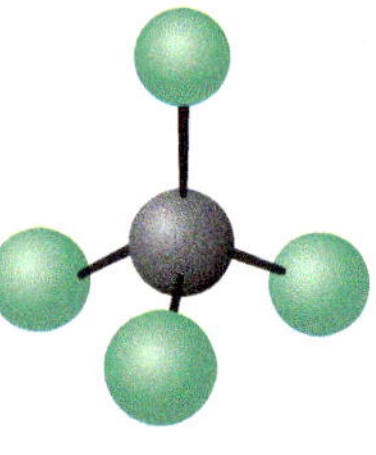
CCl_4

FIGURE 11.13 Ball-and-stick models of methane and carbon tetrachloride. Methane and carbon tetrachloride are nonpolar molecules because their polar bonds cancel each other in the tetrahedral arrangement of their atoms. The carbon atoms are located in the centers of the tetrahedrons.

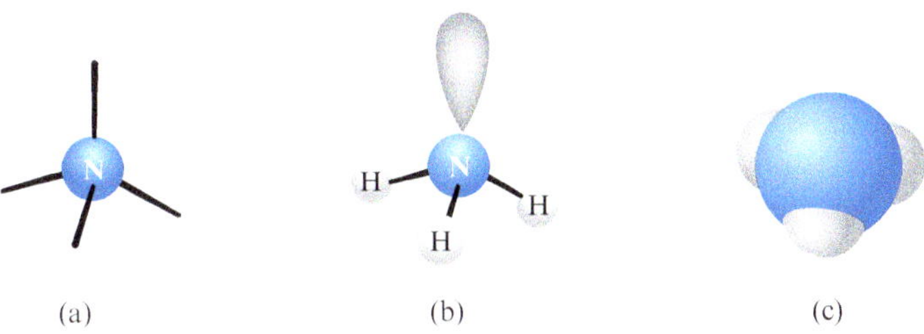

FIGURE 11.14 (a) The tetrahedral arrangement of electron pairs around the N atom in the NH_3 molecule. (b) Three pairs are shared and one is unshared. (c) The NH_3 molecule is pyramidal.

Consider the ammonia molecule. First we draw the Lewis structure to determine the number of electron pairs around the central atom:

H:N̈:H
   Ḧ

Since there are four pairs of electrons, the arrangement of electrons around the central atom will be tetrahedral (FIGURE 11.14a). However, only three of the pairs are bonded to another atom, so the molecule itself is pyramidal. It is important to understand that the placement of the electron pairs determines the shape but the name for the molecule is determined by the position of the atoms themselves. Therefore, ammonia is pyramidal. See Figure 11.14c.

CHECK YOUR UNDERSTANDING

11.15 Electron Pair Arrangement - Central Atom Bonded to Three Atoms, and a Lone Pair

WileyPLUS

Now consider the effect of two unbonded pairs of electrons in the water molecule. The Lewis structure for water is

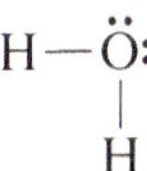

The four electron pairs indicate that a tetrahedral electron arrangement is necessary (FIGURE 11.15a). The molecule is not called tetrahedral because two of the electron pairs are unbonded pairs. The water molecule displays a **bent structure**.

Let's summarize the VSEPR model.

The arrangement of electron pairs around an atom determines its shape, but we name the shape of molecules by the position of the atoms.

PROBLEM-SOLVING STRATEGY

Determining Molecular Shape Using VSEPR

1. Draw the Lewis structure for the molecule.
2. Count the electron pairs around the central atom and arrange them to minimize repulsions (as far apart as possible). This determines the electron pair arrangement.
3. Determine the positions of the atoms.
4. Name the molecular structure from the position of the atoms.

CHECK YOUR UNDERSTANDING

11.16 Electron Pair Arrangement - Central Atom Bonded to Two Atoms, and Two Lone Pairs

WileyPLUS

It is important to recognize that the placement of the electron pairs determines the structure but the name of the molecular structure is determined by the position of the atoms. TABLE 11.6 shows the results of this process. Note that when the number of electron pairs is the same as the number of atoms bonded to the central atom, the electron pair arrangement and the molecular structure are the same. But when the number of atoms and the number of electron pairs are not the same, the molecular structure is different from the electron pair arrangement. This is illustrated when the number of electron pairs is four (a tetrahedral arrangement) in Table 11.6.

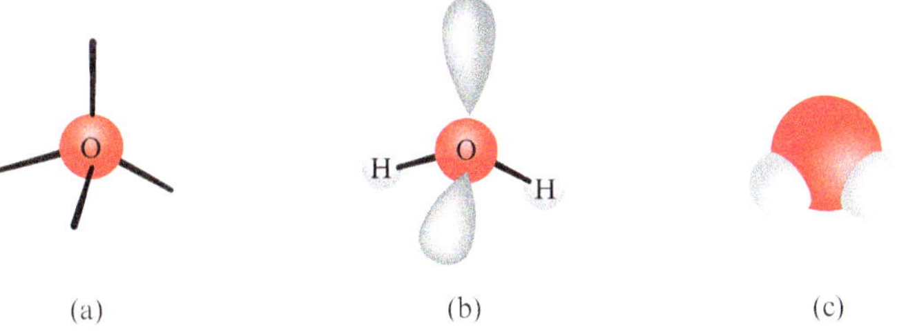

FIGURE 11.15 (a) The tetrahedral arrangement of the four electron pairs around oxygen in the H_2O molecule. (b) Two of the pairs are shared and two are unshared. (c) The H_2O molecule is bent.

TABLE 11.6 Arrangement of Electron Pairs and Molecular Structure

Number of electron pairs	Electron pair arrangement	Ball-and-stick model	Bonds	Molecular structure	Molecular structure model
2	Linear	180°	2	Linear	
3	Trigonal planar	120°	3	Trigonal planar	
4	Tetrahedral	109.5°	4	Tetrahedral	
4	Tetrahedral	109.5°	3	Pyramidal	
4	Tetrahedral	109.5°	2	Bent	

CHECK YOUR UNDERSTANDING

11.17 Electron Pair Arrangement - Central Atom Bonded to Three Atoms, and No Lone Pairs

WileyPLUS

ONLINE LEARNING MODULE

Using VSEPR to Predict Molecular Shape

WileyPLUS

ENHANCED EXAMPLE 11.18

WileyPLUS

Predict the molecular shape for these molecules: H_2S, CCl_4, $AlCl_3$.

SOLUTION

1. Draw the Lewis structure.
2. Count the electron pairs around the central atom and determine the electron arrangement that will minimize repulsions.
3. Determine the positions of the atoms and name the shape of the molecule.

Molecule	Lewis structure	Number of electron pairs	Electron pair arrangement	Molecular shape
H_2S	H:S̤̈:H	4	tetrahedral	bent
CCl_4	:C̤̈l: :C̤̈l:C̤̈:C̤̈l: :C̤̈l:	4	tetrahedral	tetrahedral
$AlCl_3$	:C̤̈l: :C̤̈l:Al:C̤̈l:	3	trigonal planar	trigonal planar

PRACTICE 11.12

Predict the shape for CF_4, NF_3, and BeI_2.

CHAPTER 11 REVIEW

11.1 Periodic Trends in Atomic Properties

KEY TERM
ionization energy

- Metals and nonmetals
- Atomic radius:
  - Increases down a group
  - Decreases across a row
- Ionization energy:
  - Energy required to remove an electron from an atom
  - Decreases down a group
  - Increases across a row

11.2 Lewis Structures of Atoms

KEY TERM
Lewis structure

- A Lewis structure is a representation of the atom where the symbol represents the element and dots around the symbol represent the valence electrons.
- To determine a Lewis structure for representative elements, use the group number as the number of electrons to place around the symbol for the element.

11.3 The Ionic Bond: Transfer of Electrons from One Atom to Another

KEY TERM
ionic bond

- The goal of bonding is to achieve stability:
  - For representative elements, this stability can be achieved by attaining a valence electron structure of a noble gas.
- In an ionic bond stability is attained by transferring an electron from one atom to another:
  - The atom that loses an electron becomes a cation:
    - Positive ions are smaller than their parent atoms.
    - Metals tend to form cations.
  - The atom gaining an electron becomes an anion:
    - Negative ions are larger than their parent atoms.
    - Nonmetals tend to form anions.
  - Ionic compounds do not exist as molecules:
    - Ions are attracted by multiple ions of the opposite charge to form a crystalline structure.

11.4 Predicting Formulas of Ionic Compounds

- Chemical compounds are always electrically neutral.
- Metals lose electrons and nonmetals gain electrons to form compounds.
- Stability is achieved (for representative elements) by attaining a noble gas electron configuration.

11.5 The Covalent Bond: Sharing Electrons

KEY TERMS
covalent bond
polar covalent bond

- Covalent bonds are formed when two atoms share a pair of electrons between them:
  - This is the predominant type of bonding in compounds.
  - True molecules exist in covalent compounds.
  - Overlap of orbitals forms a covalent bond.
- Unequal sharing of electrons results in a polar covalent bond.

11.6 Electronegativity

KEY TERMS
electronegativity
nonpolar covalent bond
dipole

- Electronegativity is the attractive force an atom has for shared electrons in a molecule or polyatomic ion.
- Electrons spend more time closer to the more electronegative atom in a bond forming a polar bond.
- The polarity of a bond is determined by the electronegativity difference between the atoms involved in the bond:
  - The greater the difference, the more polar the bond is.
  - At the extremes:
    - Large differences result in ionic bonds.
    - Tiny differences (or no difference) result(s) in a nonpolar covalent bond.

- A molecule that is electrically asymmetrical has a dipole, resulting in charged areas within the molecule.

+δ H :Cl −δ

δ+ δ−
H : Cl :

hydrogen chloride

- If the electronegativity difference between two bonded atoms is greater than 1.7–1.9, the bond will be more ionic than covalent.
- Polar bonds do not always result in polar molecules.

11.7 Lewis Structures of Compounds

Problem-Solving Strategy

Writing a Lewis Structure

1. Obtain the total number of valence electrons to be used in the structure by adding the number of valence electrons in all the atoms in the molecule or ion. If you are writing the structure of an ion, add one electron for each negative charge or subtract one electron for each positive charge on the ion.
2. Write the skeletal arrangement of the atoms and connect them with a single covalent bond (two dots or one dash). Hydrogen, which contains only one bonding electron, can form only one covalent bond. Oxygen atoms are not normally bonded to each other, except in compounds known to be peroxides. Oxygen atoms normally have a maximum of two covalent bonds (two single bonds or one double bond).
3. Subtract two electrons for each single bond you used in Step 2 from the total number of electrons calculated in Step 1. This gives you the net number of electrons available for completing the structure.
4. Distribute pairs of electrons (pairs of dots) around each atom (except hydrogen) to give each atom a noble gas structure.
5. If there are not enough electrons to give these atoms eight electrons, change single bonds between atoms to double or triple bonds by shifting unbonded pairs of electrons as needed. Check to see that each atom has a noble gas electron structure (two electrons for hydrogen and eight for the others). A double bond counts as four electrons for each atom to which it is bonded.

11.8 Complex Lewis Structures

KEY TERM
resonance structure

- When a single unique Lewis structure cannot be drawn for a molecule, resonance structures (multiple Lewis structures) are used to represent the molecule.

11.9 Compounds Containing Polyatomic Ions

- Polyatomic ions behave like a single unit in many chemical reactions.
- The bonds within a polyatomic ion are covalent.

11.10 Molecular Shape

KEY TERMS
linear structure
trigonal planar structure
tetrahedral structure
bent structure

- Lewis structures do not indicate the shape of a molecule.

PROBLEM-SOLVING STRATEGY

Determining Molecular Shape Using VSEPR

1. Draw the Lewis structure for the molecule.
2. Count the electron pairs around the central atom and arrange them to minimize repulsions (as far apart as possible). This determines the electron pair arrangement.
3. Determine the positions of the atoms.
4. Name the molecular structure from the position of the atoms.

REVIEW QUESTIONS

1. Rank these elements according to the radii of their atoms, from smallest to largest: Na, Mg, Cl, K, and Rb. (Figure 11.2)
2. Explain why much more ionization energy is required to remove the first electron from neon than from sodium. (Table 11.1)
3. Explain the large increase in ionization energy needed to remove the third electron from beryllium compared with that needed for the second electron. (Table 11.1)
4. Does the first ionization energy increase or decrease from top to bottom in the periodic table for the alkali metal family? Explain. (Figure 11.3)
5. Does the first ionization energy increase or decrease from top to bottom in the periodic table for the noble gas family? Explain. (Figure 11.3)
6. Explain the reason that an atom of helium has a much higher first ionization energy than does an atom of hydrogen. (Table 11.1)
7. Why is there such a large increase in the energy required to remove a second electron from an atom of lithium? (Table 11.1)
8. Which element has the larger radius?
   (a) Li or Be
   (b) K or Rb
   (c) Al or P
   (d) F or Cl
   (e) O or Se
   (f) As or Kr
9. In Groups 1A(1)–7A(18), which element in each group has the smallest atomic radius? (Figure 11.2)
10. Why does the atomic size increase in going down any family of the periodic table?
11. Explain why potassium usually forms a K^+ ion but not a K^{2+} ion.
12. Why does an aluminum ion have a +3 charge?
13. Why are only valence electrons represented in a Lewis structure?
14. All the atoms within each Group A family of elements can be represented by the same Lewis structure. Complete the following table, expressing the Lewis structure for each group. (Use E to represent the elements.) (Figure 11.4)

Group	1A(1)	2A(2)	3A(13)	4A(14)	5A(15)	6A(16)	7A(17)
	E·						

15. Draw the Lewis structure for Cs, Ba, Tl, Pb, Po, At, and Rn. How do these structures correlate with the group in which each element occurs?
16. What are valence electrons?
17. Why do metals tend to lose electrons and nonmetals tend to gain electrons when forming ionic bonds?
18. State whether the elements in each group gain or lose electrons in order to achieve a noble gas configuration. Explain.
   (a) Group 1A(1)
   (b) Group 2A(2)
   (c) Group 6A(16)
   (d) Group 7A(17)
19. What is the overall charge on an ionic compound?
20. Which family of elements tends to form ionic compounds with a 1 : 1 ratio of cation to anion with sulfur?
21. If the formula for calcium phosphate is $Ca_3(PO_4)_2$, predict the formulas of magnesium phosphate, beryllium phosphate, strontium phosphate, and barium phosphate.
22. Explain why magnesium tends to lose two electrons when forming an ionic compound.
23. Why is the term *molecule* used to describe covalent compounds but not ionic compounds?
24. How many shared electrons make up a single covalent bond? What is the maximum number of covalent bonds that can be formed between any two atoms?
25. Explain in simple terms how a chemist describes the formation of a covalent bond.
26. What does the dash between atoms represent in a Lewis electron dot structure?
27. Are all molecules that contain polar bonds polar molecules? Explain.
28. In a polar covalent bond, how do you determine which atom has a partial negative charge ($\delta-$) and which has a partial positive charge ($\delta+$)?
29. In which general areas of the periodic table are the elements with (a) the highest and (b) the lowest electronegativities located?
30. What is the purpose of a Lewis structure?
31. In a Lewis structure, what do the dots represent and what do the lines represent?
32. Can there be more than one correct Lewis structure for a compound? Explain.
33. If a molecule has more than one correct Lewis structure, what is the term used to describe these structures?
34. When drawing a Lewis structure for an ion, how is the charge represented?
35. Write two examples of molecules that have both ionic and covalent bonds.
36. What is the difference between electron pair arrangement and molecular shape?

PAIRED EXERCISES

Most of the exercises in this chapter are available for assignment via the online homework management program, WileyPLUS (www.wileyplus.com). All exercises with blue numbers have answers in Appendix 5.

1. An atom of potassium and a K^+ ion are drawn below. Identify each and explain your choice.

1 2

2. An atom of chlorine and a Cl^- ion are drawn below. Identify each and explain your choice.

1 2

3. Which one in each pair has the larger radius? Explain.
(a) a calcium atom or a calcium ion
(b) a chlorine atom or a chloride ion
(c) a magnesium ion or an aluminum ion
(d) a sodium atom or a silicon atom
(e) a potassium ion or a bromide ion

4. Which one in each pair has the larger radius? Explain.
(a) Fe^{2+} or Fe^{3+}
(b) a potassium atom or a potassium ion
(c) a sodium ion or a chloride ion
(d) a strontium atom or an iodine atom
(e) a rubidium ion or a strontium ion

5. When an atom forms a cation, its size changes. Explain how and why its size changes.

6. When an atom forms an anion, its size changes. Explain how and why its size changes.

7. Define the term *ionization energy*. Describe how the ionization energy of the elements changes as you move across and down the periodic table.

8. Define the term *electronegativity*. Describe how the electronegativity of the elements changes as you move across and down the periodic table.

9. Using the table of electronegativity values (Table 11.5), indicate which element is more positive and which is more negative in these compounds:
(a) H_2O
(b) RbCl
(c) NH_3
(d) PbS
(e) PF_3
(f) CH_4

10. Using the table of electronegativity values (Table 11.5), indicate which element is more positive and which is more negative in these compounds:
(a) HCl
(b) SO_2
(c) CCl_4
(d) IBr
(e) CsI
(f) OF_2

11. Classify the bond between these pairs of elements as principally ionic or principally covalent (use Table 11.5):
(a) sulfur and oxygen
(b) barium and nitrogen
(c) potassium and bromine
(d) carbon and chlorine

12. Classify the bond between these pairs of elements as principally ionic or principally covalent (use Table 11.5):
(a) sodium and oxygen
(b) nitrogen and hydrogen
(c) oxygen and hydrogen
(d) phosphorus and chlorine

13. Write an equation representing each of the following:
(a) the change of a magnesium atom to a magnesium ion
(b) the change of a bromine atom to a bromide ion

14. Write an equation representing each of the following:
(a) the change of a potassium atom to a potassium ion
(b) the change of a sulfur atom to a sulfide ion

15. Use Lewis structures to show the electron transfer that enables these ionic compounds to form:
(a) Li_2O
(b) K_3N

16. Use Lewis structures to show the electron transfer that enables these ionic compounds to form:
(a) K_2S
(b) Ca_3N_2

17. State the number of valence electrons in an atom of each of the following elements:
(a) selenium
(b) phosphorus
(c) bromine
(d) magnesium
(e) helium
(f) arsenic

18. State the number of valence electrons in an atom of each of the following elements:
(a) lead
(b) lithium
(c) oxygen
(d) cesium
(e) gallium
(f) argon

19. How many electrons must be gained or lost for the following to achieve a noble gas electron configuration?
(a) a potassium atom
(b) an aluminum ion
(c) a bromine atom
(d) a selenium atom

20. How many electrons must be gained or lost for the following to achieve a noble gas electron structure?
(a) a sulfur atom
(b) a calcium atom
(c) a nitrogen atom
(d) an iodide ion

21. Answer the following questions for an atom of vanadium.
(a) Write the complete electron configuration.
(b) Using boxes to represent orbitals and arrows to represent electrons, draw a picture to show the electrons in the highest occupied *d* sublevel.
(c) Which electrons are lost when a manganese ion forms a vanadium(II) ion?
(d) Write the complete electron configuration for a V^{2+} ion.

22. Answer the following questions for an atom of manganese.
(a) Write the complete electron configuration.
(b) Using boxes to represent orbitals and arrows to represent electrons, draw a picture to show the electrons in the highest occupied *d* sublevel.
(c) Which electrons are lost when a manganese ion forms a manganese(II) ion?
(d) Write the complete electron configuration for a Mn^{2+} ion.

23. Determine whether the following atoms will form an ionic compound or a molecular compound, and give the formula of the compound.
(a) sodium and chlorine
(b) carbon and 4 hydrogen
(c) magnesium and bromine
(d) 2 bromine
(e) carbon and 2 oxygen

24. Determine whether each of the following atoms will form a nonpolar covalent compound or a polar covalent compound, and give the formula of the compound.
(a) 2 oxygen
(b) hydrogen and bromine
(c) oxygen and 2 hydrogen
(d) 2 iodine

25. Let E be any representative element. Following the pattern in the table, write formulas for the hydrogen and oxygen compounds of the following:
(a) Na (c) Al
(b) Ca (d) Sn

Group						
1A(1)	2A(2)	3A(13)	4A(14)	5A(15)	6A(16)	7A(17)
EH	EH_2	EH_3	EH_4	EH_3	H_2E	HE
E_2O	EO	E_2O_3	EO_2	E_2O_5	EO_3	E_2O_7

26. Let E be any representative element. Following the pattern in the table, write formulas for the hydrogen and oxygen compounds of the following:
(a) Sb (c) Cl
(b) Se (d) C

Group						
1A(1)	2A(2)	3A(13)	4A(14)	5A(15)	6A(16)	7A(17)
EH	EH_2	EH_3	EH_4	EH_3	H_2E	HE
E_2O	EO	E_2O_3	EO_2	E_2O_5	EO_3	E_2O_7

27. The formula for sodium sulfate is Na_2SO_4. Write the names and formulas for the other alkali metal sulfates.

28. The formula for calcium bromide is $CaBr_2$. Write the names and formulas for the other alkaline earth metal bromides.

29. Write Lewis structures for the following:
(a) Na (b) Br^- (c) O^{2-}

30. Write Lewis structures for the following:
(a) Ga (b) Ga^{3+} (c) Ca^{2+}

31. Classify the bonding in each compound as ionic or covalent:
(a) H_2O (c) MgO
(b) NaCl (d) Br_2

32. Classify the bonding in each compound as ionic or covalent:
(a) HCl (c) NH_3
(b) $BaCl_2$ (d) SO_2

33. Predict the type of bond that would be formed between the following pairs of atoms:
(a) P and I (b) N and S (c) Br and I

34. Predict the type of bond that would be formed between the following pairs of atoms:
(a) H and Si (b) O and F (c) Si and Br

35. Draw Lewis structures for the following:
(a) H_2 (b) N_2 (c) Cl_2

36. Draw Lewis structures for the following:
(a) O_2 (b) Br_2 (c) I_2

37. Draw Lewis structures for the following:
(a) NCl_3 (c) C_2H_6
(b) H_2CO_3 (d) $NaNO_3$

38. Draw Lewis structures for the following:
(a) H_2S (c) NH_3
(b) CS_2 (d) NH_4Cl

39. Draw Lewis structures for the following:
(a) Ba^{2+} (d) CN^-
(b) Al^{3+} (e) HCO_3^-
(c) SO_3^{2-}

40. Draw Lewis structures for the following:
(a) I^- (d) ClO_3^-
(b) S^{2-} (e) NO_3^-
(c) CO_3^{2-}

41. Draw Lewis structures showing the possible resonance structures for each of the following molecules:
(a) SO_2 (sulfur is central atom)
(b) N_2O (nitrogen is central atom)

42. Draw Lewis structures showing the possible resonance structures for each of the following ions:
(a) NO_2^- (nitrogen is central atom)
(b) HCO_2^- (carbon is central atom)

43. Classify each of the following molecules as polar or nonpolar:
(a) CH_3Cl (c) OF_2
(b) Cl_2 (d) PBr_3

44. Classify each of the following molecules as polar or nonpolar:
(a) H_2 (c) CH_3OH
(b) NI_3 (d) CS_2

45. Give the number and arrangement of the electron pairs around the central atom:
(a) C in CCl_4 (c) Al in AlH_3
(b) S in H_2S

46. Give the number and arrangement of the electron pairs around the central atom:
(a) Ga in $GaCl_3$ (c) Cl in ClO_3^-
(b) N in NF_3

47. Use VSEPR theory to predict the structure of these polyatomic ions:
(a) sulfate ion (c) periodate ion
(b) chlorate ion

48. Use VSEPR theory to predict the structure of these polyatomic ions:
(a) ammonium ion (c) phosphate ion
(b) sulfite ion

49. An imaginary molecule is drawn below. Predict the electron pair arrangement and molecular structure of the indicated atoms. (Note: The structures may be drawn inaccurately.)

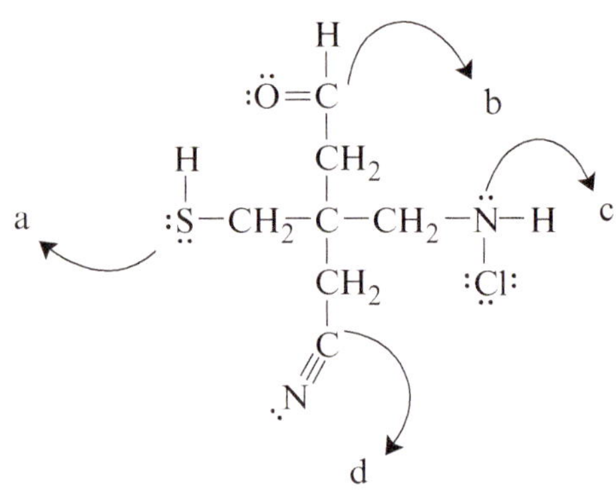

50. An imaginary molecule is drawn below. Predict the electron pair arrangement and molecular structure of the indicated atoms. (Note: The structures may be drawn inaccurately.)

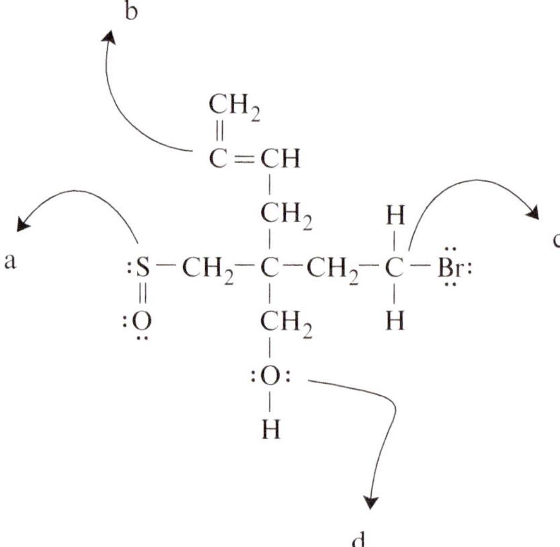

51. Use VSEPR theory to predict the shape of these molecules:
(a) SiH_4 (b) PH_3 (c) SeF_2

52. Use VSEPR theory to predict the shape of these molecules:
(a) SiF_4 (b) OF_2 (c) Cl_2O

53. Element X reacts with sodium to form the compound Na_2X and is in the second period on the periodic table. Identify this element.

54. Element Y reacts with oxygen to form the compound Y_2O and has the lowest ionization energy of any Period 4 element on the periodic table. Identify this element.

ADDITIONAL EXERCISES

55. Atoms of sodium, cesium, and argon are drawn below. Identify each and explain your choice.

56. Spheres representing Br^-, Se^{2-}, Rb^+, Sr^{2+}, and Kr are drawn below. Identify each and explain your choice.

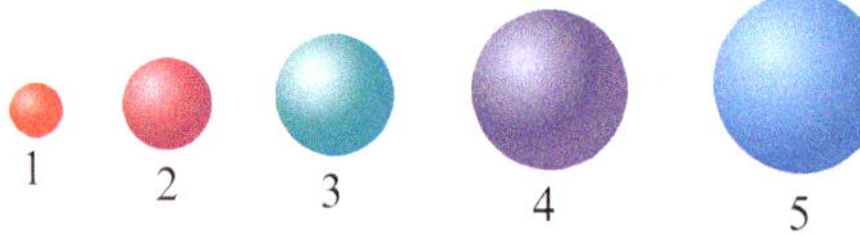

57. The first ionization energy (IE) of potassium is lower than the first IE for calcium, but the second IE of calcium is lower than the second IE of potassium. Use an electron configuration or size argument to explain this trend in ionization energies.

58. Chlorine has a very large first ionization energy, yet it forms a chloride ion relatively easily. Explain.

59. Three particles have the same electron configuration. One is a cation of an alkali metal, one is an anion of the halide in the third period, and the third particle is an atom of a noble gas. What are the identities of the three particles (including charges)? Which particle should have the smallest atomic/ionic radius, which should have the largest, and why?

60. Write Lewis structures for hydrazine (N_2H_4) and hydrazoic acid (HN_3).

61. Draw Lewis structures and give the shape of each of the following compounds:
(a) NO_2^-
(b) SO_4^{2-}
(c) $SOCl_2$
(d) Cl_2O

62. Draw Lewis structures for each of the following compounds:
(a) ethane (C_2H_6)
(b) ethylene (C_2H_4)
(c) acetylene (C_2H_2)

63. Using the periodic table, identify each element given the description:
(a) the most electronegative element of the alkaline earth metals
(b) the noble gas with the highest first ionization energy
(c) the element in Period 4 with the lowest first ionization energy
(d) the element with the highest electronegativity
(e) the alkali metal with the largest radius
(f) the element in Period 2 with the smallest radius

64. Choose the element that fits each description:
(a) the higher electronegativity Cl or Br
(b) the higher first ionization energy O or S
(c) the lower electronegativity Ca or C

65. Why do you think there are no electronegativity values given for the noble gases in Table 11.5?

66. When one electron is removed from an atom of Li, it has two left. Helium atoms also have two electrons. Why is more energy required to remove the second electron from Li than to remove the first from He?

67. Group 1B(11) elements (see the periodic table on the inside cover of your book) have one electron in their outer energy level, as do Group 1A(1) elements. Would you expect them to form compounds such as CuCl, AgCl, and AuCl? Explain.

68. The formula for lead(II) bromide is $PbBr_2$. Predict formulas for tin(II) and germanium(II) bromides.

69. Why is it not proper to speak of sodium chloride molecules?

70. What is a covalent bond? How does it differ from an ionic bond?

71. Briefly comment on the structure Na:$\ddot{\underset{\cdot\cdot}{O}}$:Na for the compound Na_2O.
72. What are the four most electronegative elements?
73. Rank these elements from highest electronegativity to lowest: Mg, S, F, H, O, Cs.
74. Is it possible for a molecule to be nonpolar even though it contains polar covalent bonds? Explain.
75. Why is CO_2 a nonpolar molecule, whereas CO is a polar molecule?
76. Rank the following bonds in order of increasing polarity:
   (a) NaO, CO, NO
   (b) CO, SiO, GeO
   (c) BO, BS, BSe
77. Estimate the bond angle between atoms in these molecules:
   (a) H_2S
   (b) NH_3
   (c) NH_4^+
   (d) $SiCl_4$
78. Why is the Lewis structure of AlF_3 not written as

$$
\begin{array}{ccc}
 & & :\ddot{F}: \\
 & & / \\
:\ddot{\underset{\cdot\cdot}{F}} & - & Al \\
 & & \backslash \\
 & & :\underset{\cdot\cdot}{F}:
\end{array}
$$

   What is the correct Lewis structure, and which electrons are shown in a Lewis structure?
79. Why does carbon have a maximum of four covalent bonds?
80. Draw the Lewis structure of the atom whose electron configuration is $1s^22s^22p^63s^23p^64s^23d^{10}4p^5$. Would you expect this atom to form an ionic, nonpolar covalent, or polar covalent bond with sulfur?
81. Consider the two molecules BF_3 and NF_3. Compare and contrast them in terms of the following:
   (a) valence-level orbitals on the central atom that are used for bonding
   (b) shape of the molecule
   (c) number of lone electron pairs on the central atom
   (d) type and number of bonds found in the molecule
82. With respect to electronegativity, why is fluorine such an important atom? What combination of atoms on the periodic table results in the most ionic bond?
83. Why does the Lewis structure of each element in a given group of representative elements on the periodic table have the same number of dots?
84. A sample of an air pollutant composed of sulfur and oxygen was found to contain 1.40 g sulfur and 2.10 g oxygen. What is the empirical formula for this compound? Draw a Lewis structure to represent it.
85. A dry-cleaning fluid composed of carbon and chlorine was found to have the composition 14.5% carbon and 85.5% chlorine. Its known molar mass is 166 g/mol. Draw a Lewis structure to represent the compound.
86. Identify each of the following drawings as representing ionic compounds, covalent compounds, or compounds composed of both ionic and covalent portions.

(a)

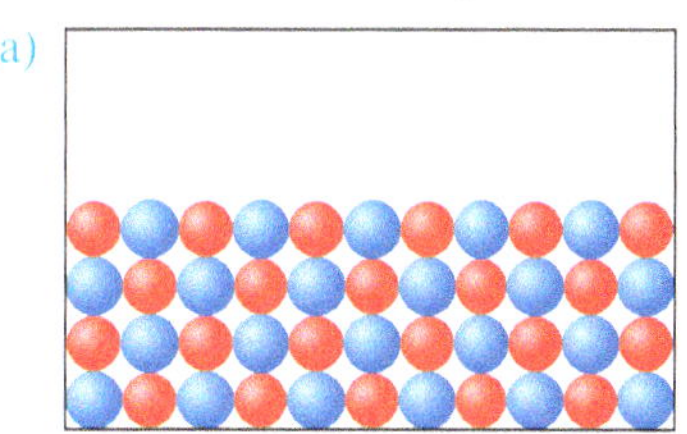

(b)

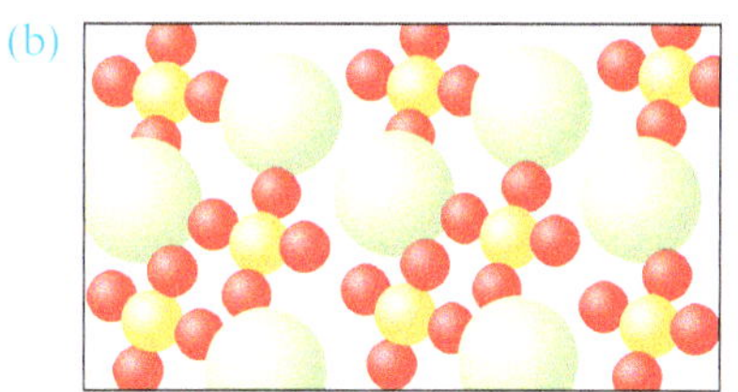

(c)

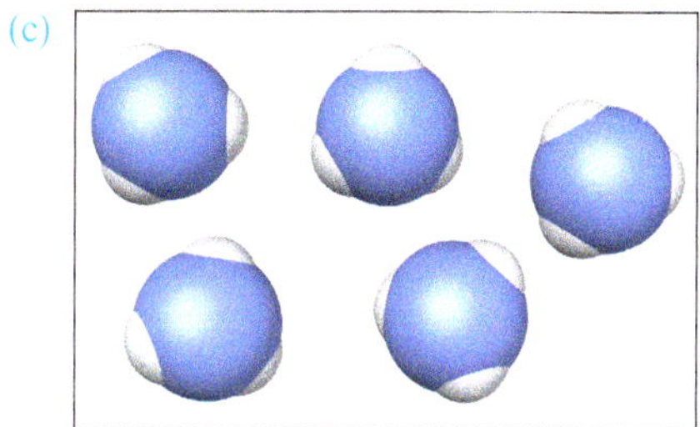

(d)

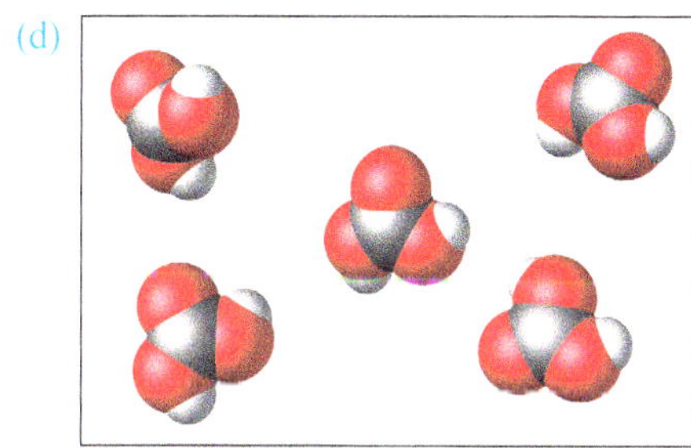

87. Identify each of the following compounds as ionic, covalent, or both. Give the correct IUPAC name for each compound.
   (a) Na_3P
   (b) NH_4I
   (c) SO_2
   (d) H_2S
   (e) $Cu(NO_3)_2$
   (f) MgO

CHALLENGE EXERCISES

88. Draw the following Lewis structures, showing any full charges where they exist on ions:
   (a) H_2SO_4
   (b) $NaNO_3$
   (c) K_2CO_3
   (d) HCN
   (e) $Al_2(SO_4)_3$
   (f) CH_3COOH
89. The first ionization energy for lithium is 520 kJ/mol. How much energy would be required to change 25 g of lithium atoms to lithium ions? (Refer to Table 11.1.)
90. What is the total amount of energy required to remove the first two electrons from 15 moles of sodium atoms? (Refer to Table 11.1.)
91. Sulfur trioxide gas reacts with water to form sulfuric acid. Draw the Lewis structures of each substance to represent the reaction.
92. Compounds found in garlic have been shown to have significant biological effects including immune system enhancement, anticancer activity, and cardiovascular activity. Garlic clearly

seems to have more value than as just a great tasting addition to spaghetti! Some of the compounds in garlic that are currently being studied are shown in the following figures.

```
 H     H    HH    H   O
  \     \  /  \  /    ||
   C     C      C     C—O—H
H/  \\C/  \S/    \C/
      |    |      | \H
      H    O      N
                 / \
                H   H
```

Alliin

```
 H    H   H   O    H
  \    \ /    |    |
H—C==C   C   S    C==C—H
     |    \S/  \C/     |
     H         / \     H
              H   H
```

Allicin (gives garlic its characteristic aroma)

Redraw these molecules adding in dots to represent any lone pairs present on atoms. Draw a circle around atoms with tetrahedral geometry, draw a square around atoms with trigonal planar geometry, draw a pentagon around atoms with a trigonal pyramidal geometry, and draw a hexagon around atoms with a bent geometry. Draw a triangle around any atoms that have an unexpected number of bonds.

93. Two of the most complex organic compounds found outside of our solar system have been detected in a massive star-forming region near the heart of the Milky Way known as Sagittarius B2. The skeleton structures of these two compounds, ethyl formate (the compound that gives raspberries their flavor) and *n*-propyl cyanide, are drawn below. Draw the complete Lewis electron dot structures for these two compounds, identify the electron pair and molecular geometry around each atom, and predict the bond angles.

```
  H  H       O              H  H  H
  |  |       |              |  |  |
H—C—C—O—C—H              H—C—C—C—C—H
  |  |                      |  |  |
  H  H                      H  H  H
```

Ethyl formate   *n*-propyl cyanide

94. Have you ever dropped your cell phone? If so, you have had the opportunity to read all of your texts through a haze of glass shards as you try not to cut your finger on the broken pieces. Cell phone manufacturers are working very hard to create the next generation of glass that will resist breakage. One of the techniques used is to exchange the smaller Na ions for larger K ions in the glass, which compresses the glass making it stronger. The radii of these ions are given in Table 11.3. Calculate the volume of a sodium ion and a potassium ion and determine how many sodium ions will fit in the space vacated by a potassium ion. (Assume these ions are spheres with volume $= \frac{4}{3}\pi r^3$.)

95. Both NCl_3 and BF_3 have a central atom bonded to three other atoms, yet one is pyramidal and the other is trigonal planar. Explain.

ANSWERS TO PRACTICE EXERCISES

11.1 (a) Ionization energy decreases from top to bottom for a group of elements and increases from left to right across a period of elements.

(b) For a group of elements the outer shell of electrons is increasingly further away from the nucleus as you go from top to bottom of the group. Thus, they are not held as strongly by the positive nucleus and are easier to ionize.

11.2 (a) :N̈· (with dots: :N·, one dot above and below)
(b) :Ȧl
(c) Sr:
(d) :B̤̈r·

11.3 (a) Ar; K^+
(b) Ne; Mg^{2+}
(c) Ne; Al^{3+}
(d) Xe; Ba^{2+}

11.4 (a) Na_2S
(b) Rb_2O

11.5 (a) Mg_3N_2
(b) Ba_3As_2

11.6 H:B̤̈r: or H—Br

```
  H H               H  H
  ¨ ¨               |  |
H:C:C:H    or    H—C—C—H
  ¨ ¨               |  |
  H H               H  H
```

11.7 Electronegativity of an atom gives information about the strength that the atom has for attracting a pair of shared electrons in a covalent bond. The higher the electronegativity value, the greater the attraction for a pair of shared electrons. Consequently, when two different atoms with different electronegativities share a pair of electrons, one atom becomes partially negative and the other atom becomes partially positive due to the attractive forces for the pair of electrons. The bond formed is said to be polar covalent.

11.8 ionic: (a), (d), (e)

covalent: (b), (c), (f)

11.9 (a)
```
  :Br:
:P:Br:
  :Br:
```
(b)
```
    H
:Cl:C:Cl:
   :Cl:
```
(c) H:F̤̈:

(d)
```
  :O:
H:C:H
```
(e) :N:::N:

11.10 (a)
```
H:N:H
  H
```
(b)
```
[H:O:H]+
   H
```
(c)
```
[  H  ]+
[H:N:H]
[  H  ]
```
(d)
```
[   :O:   ]−
[H:O:C::O:]
```

11.11

```
        [   :O:   ]2−    [  :O:  ]−          [  :O:  ]−
[Ca]2+  [    |    ]      [   |   ]   [Ca]2+  [   ||  ]
        [ O=S=O   ]      [   N   ]           [   N   ]
        [    |    ]      [ O/ \O ]           [ O/ \O ]
        [   :O:   ]
```

11.12 CF_4, tetrahedral; NF_3, pyramidal; BeI_2, linear

12

The Gaseous State of Matter

Kees Van Den Berg/Science Source Images

CHAPTER OUTLINE

Our atmosphere is composed of a mixture of gases, including nitrogen, oxygen, argon, carbon dioxide, ozone, and trace amounts of others. These gases are essential to life, yet they can also create hazards for us. For example, ozone surrounds the Earth at high altitudes and protects us from harmful ultraviolet rays, but it also destroys rubber and plastics. We require air to live, yet scuba divers must be concerned about oxygen poisoning and the "bends." The gases in the air can be warmed inside a hot-air balloon and expand, lifting us to experience the freedom of flight.

In chemistry, the study of the behavior of gases allows us to understand our atmosphere and the effects that gases have on our lives.

12.1 Properties of Gases

LEARNING OBJECTIVE Explain atmospheric pressure and how it is measured. Be able to convert among the various units of pressure.

KEY TERMS
pressure
atmospheric pressure
barometer
1 atmosphere

Gases are the least dense and most mobile of the three states of matter (FIGURE 12.1). A solid has a rigid structure, and its particles remain in essentially fixed positions. When a solid absorbs sufficient heat, it melts and changes into a liquid. Melting occurs because the molecules (or ions) have absorbed enough energy to break out of the rigid structure of the solid. The molecules or ions in the liquid are more energetic than they were in the solid, as indicated by their increased mobility. Molecules in the liquid state cling to one another. When the liquid absorbs additional heat, the more energetic molecules break away from the liquid surface and go into the gaseous state—the most mobile state of matter. Gas molecules move at very high velocities and have high kinetic energy. The average velocity of hydrogen molecules at 0°C is over 1600 meters (1 mile) per second. Mixtures of gases are uniformly distributed within the container in which they are confined.

The same quantity of a substance occupies a much greater volume as a gas than it does as a liquid or a solid (FIGURE 12.2). For example, 1 mol of water (18.02 g) has a volume of 18 mL at 4°C. This same amount of liquid water would occupy about 22,400 mL in the gaseous state—more than a 1200-fold increase in volume. We may assume from this difference in volume that (1) gas molecules are relatively far apart, (2) gases can be greatly compressed, and (3) the volume occupied by a gas is mostly empty space.

FIGURE 12.2 A mole of water occupies 18 mL as a liquid but would fill this box (22.4 L) as a gas at the same temperature.

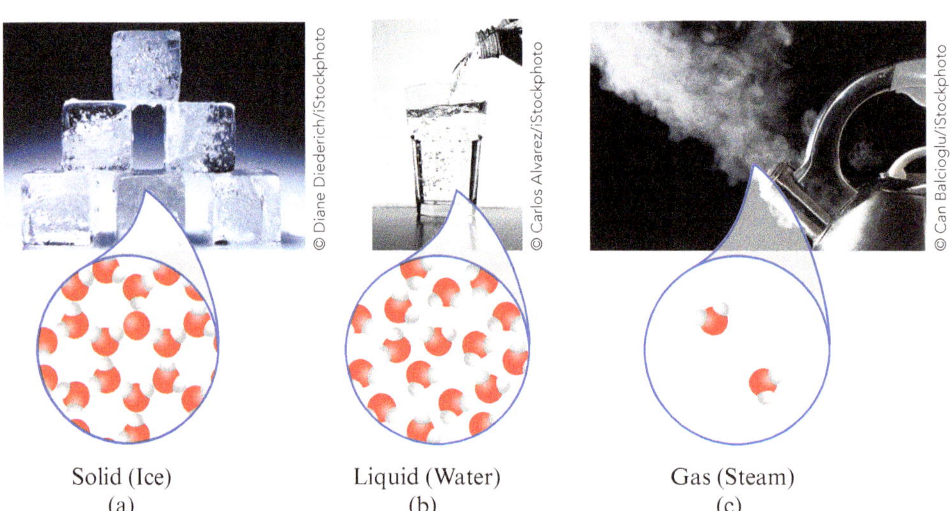

FIGURE 12.1 The three states of matter. (a) Solid—water molecules are held together rigidly and are very close to each other. (b) Liquid—water molecules are close together but are free to move around and slide over each other. (c) Gas—water molecules are far apart and move freely and randomly.

FIGURE 12.3 The pressure resulting from the collisions of gas molecules with the walls of the balloon keeps the balloon inflated.

Measuring the Pressure of a Gas

Pressure is defined as force per unit area. When a rubber balloon is inflated with air, it stretches and maintains its larger size because the pressure on the inside is greater than that on the outside. Pressure results from the collisions of gas molecules with the walls of the balloon (FIGURE 12.3). When the gas is released, the force or pressure of the air escaping from the small neck propels the balloon in a rapid, irregular flight. If the balloon is inflated until it bursts, the gas escaping all at once causes an explosive noise.

The effects of pressure are also observed in the mixture of gases in the atmosphere, which is composed of about 78% nitrogen, 21% oxygen, 1% argon, and other minor constituents by volume (TABLE 12.1). The outer boundary of the atmosphere is not known

TABLE 12.1 Average Composition of Dry Air

Gas	Percent by volume	Gas	Percent by volume
N_2	78.08	He	0.0005
O_2	20.95	CH_4	0.0002
Ar	0.93	Kr	0.0001
CO_2	0.033	Xe, H_2, and N_2O	Trace
Ne	0.0018		

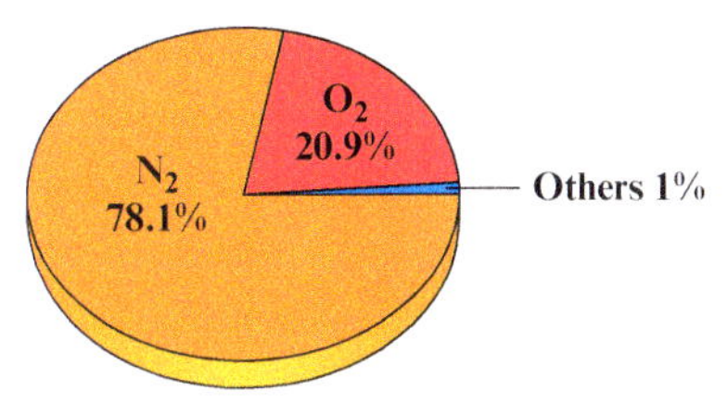

precisely, but more than 99% of the atmosphere is below an altitude of 32 km (20 miles). Thus, the concentration of gas molecules in the atmosphere decreases with altitude, and at about 6.4 km (4 miles) the amount of oxygen is insufficient to sustain human life. The gases in the atmosphere exert a pressure known as **atmospheric pressure**. The pressure exerted by a gas depends on the number of molecules of gas present, the temperature, and the volume in which the gas is confined. Gravitational forces hold the atmosphere relatively close to Earth and prevent air molecules from flying off into space. Thus, the atmospheric pressure at any point is due to the mass of the atmosphere pressing downward at that point.

The pressure of the gases in the atmosphere can be measured with a **barometer**. A mercury barometer may be prepared by completely filling a long tube with pure, dry mercury and inverting the open end into an open dish of mercury. If the tube is longer than 760 mm, the mercury level will drop to a point at which the column of mercury in the tube is just supported by the pressure of the atmosphere. If the tube is properly prepared, a vacuum will exist above the mercury column. The weight of mercury, per unit area, is equal to the pressure of the atmosphere. The column of mercury is supported by the pressure of the atmosphere, and the height of the column is a measure of this pressure (FIGURE 12.4). The mercury barometer was invented in 1643 by the Italian physicist E. Torricelli (1608–1647), for whom the unit of pressure *torr* was named.

Air pressure is measured and expressed in many units. The standard atmospheric pressure, or simply **1 atmosphere** (atm), is the pressure exerted by a column of mercury 760 mm high at a temperature of 0°C. The normal pressure of the atmosphere at sea level is 1 atm, or 760 torr, or 760 mm Hg. The SI unit for pressure is the pascal (Pa), where 1 atm = 101,325 Pa, or 101.3 kPa. Other units for expressing pressure are inches of mercury, centimeters of mercury, the millibar (mbar), and pounds per square inch ($lb/in.^2$ or psi). The values of these units equivalent to 1 atm are summarized in TABLE 12.2.

TABLE 12.2 Pressure Units Equivalent to 1 Atmosphere

1 atm
760 torr
760 mm Hg
76 cm Hg
101.325 kPa
1013 mbar
29.9 in. Hg
14.7 $lb/in.^2$

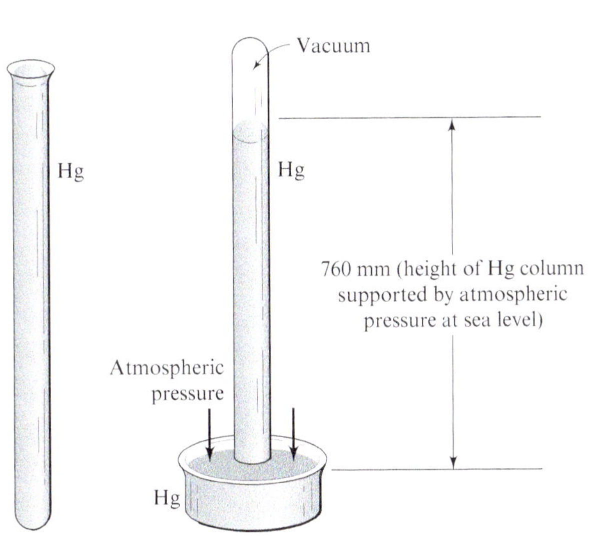

FIGURE 12.4 Preparation of a mercury barometer. The full tube of mercury at the left is inverted and placed in a dish of mercury.

CHECK YOUR UNDERSTANDING

12.1 Atmospheric Pressure Conversions – Hg to Torr

12.2 Atmospheric Pressure Conversions – Hg to Atmospheres

WileyPLUS

Atmospheric pressure varies with altitude. The average pressure at Denver, Colorado, 1.61 km (1 mile) above sea level, is 630 torr (0.83 atm). Atmospheric pressure is 380 torr (0.5 atm) at about 5.5 km (3.4 miles) altitude.

Pressure is often measured by reading the height of a mercury column in millimeters on a barometer. Thus pressure may be recorded as mm Hg (torr). In problems dealing with gases, it is necessary to make conversions among the various pressure units. Since atm, torr, and mm Hg are common pressure units, we give examples involving all three of these units:

$$1 \text{ atm} = 760 \text{ torr} = 760 \text{ mm Hg}$$

ONLINE LEARNING MODULE

Pressure Units

WileyPLUS

ENHANCED EXAMPLE 12.1 WileyPLUS

The average atmospheric pressure at Walnut, California, is 740. mm Hg. Calculate this pressure in (a) torr and (b) atmospheres.

SOLUTION

Let's use conversion factors that relate one unit of pressure to another.

(a) To convert mm Hg to torr, use the conversion factor

760 torr/760 mm Hg (1 torr/1 mm Hg):

$$(740. \cancel{\text{mm Hg}})\left(\frac{1 \text{ torr}}{1 \cancel{\text{mm Hg}}}\right) = 740. \text{ torr}$$

(b) To convert mm Hg to atm, use the conversion factor

1 atm/760. mm Hg:

$$(740. \cancel{\text{mm Hg}})\left(\frac{1 \text{ atm}}{760. \cancel{\text{mm Hg}}}\right) = 0.974 \text{ atm}$$

PRACTICE 12.1

A barometer reads 1.12 atm. Calculate the corresponding pressure in (a) torr, (b) mm Hg, and (c) kilopascals.

Pressure Dependence on the Number of Molecules and the Temperature

Pressure is produced by gas molecules colliding with the walls of a container. At a specific temperature and volume, the number of collisions depends on the number of gas molecules present. The number of collisions can be increased by increasing the number of gas molecules present. If we double the number of molecules, the frequency of collisions and the pressure should double. We find, for an ideal gas, that this doubling is actually what happens. When the temperature and mass are kept constant, the pressure is directly proportional to the number of moles or molecules of gas present. FIGURE 12.5 illustrates this concept.

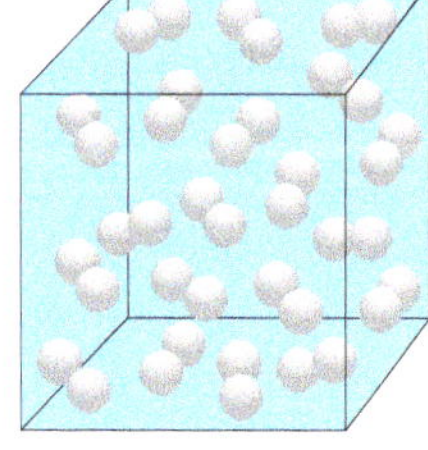

2 mol H_2
$P = 2$ atm

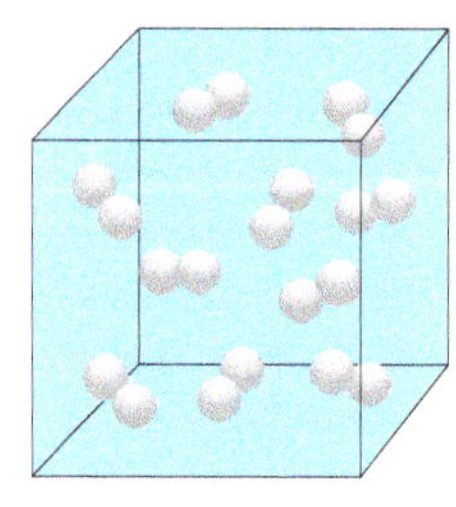

1 mol H_2
$P = 1$ atm

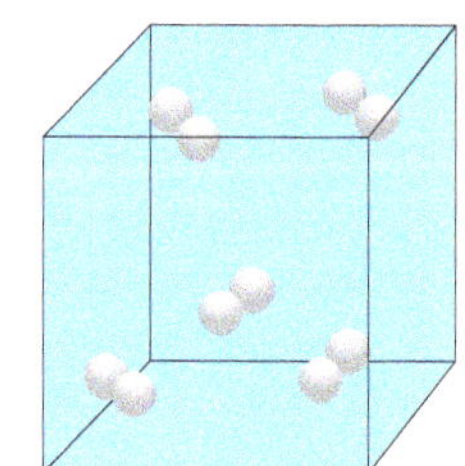

0.5 mol H_2
$P = 0.5$ atm

FIGURE 12.5 The pressure exerted by a gas is directly proportional to the number of molecules present. In each case shown, the volume is 22.4 L and the temperature is 0°C.

CHEMISTRY *IN ACTION* What the Nose Knows

The sense of smell is one of the most intriguing senses. The air around us is filled with traces of the chemicals composing our world. Much work has gone into finding ways to identify the chemicals present in the air. Developing methods to identify chemicals at low concentrations has tremendous benefits.

Better Coffee? Kenneth Suslick at the University of Illinois, Urbana-Champaign, and his colleagues have developed an electronic nose. Molecules placed in polymer films change color when exposed to the aromatic compounds found in coffee. By comparing the relative amounts of each of these compounds to the amounts found in an exceptional cup of coffee, coffee roasters are able to easily identify problems with their coffee beans and thereby ensure a consistently good cup of coffee.

No more stinky feet? If you have been flying recently, you have probably been required to remove your shoes. If researchers at the University of Illinois are successful, you may no longer have to remove your shoes at the airport. A handheld sensor is able to quickly and accurately detect the presence of a commonly used explosive in shoes.

The doctor nose you're sick! Many human diseases produce volatile compounds that are exhaled with every breath you take. Physicians have been able to diagnose diabetes simply by sniffing the breath of their patients for the odor of acetone. While most diseases do not produce easily detectable odors, physicians can take advantage of technology to detect these compounds at the low concentrations in which they are produced.

Scientists can now analyze breath for a number of chemicals specific to diseases. Some tests to detect disease are already approved and on the market (see table). Other breath tests are in development for conditions such as lung, liver, and breast cancer, heart failure, and obesity.

Breath testing poses challenges as a diagnostic tool. Some of the molecules in the breath come from the air and are simply breathed in and released. Others come from substances in the blood taken in through the skin or by eating or are the products of metabolism. Still others come from bacteria that reside in our systems. To analyze the mixture requires knowledge of what "healthy" breath contains and how "sick breath" is different. Once researchers have this figured out for a particular disease and have tested their results for accuracy, breath analysis will become more common as a diagnostic tool.

Testing devices are already shrinking in size and price, and they are noninvasive and quick to use with fast results. Someday soon you may be able to blow into a small device the size of a thumb drive and learn about the health of many of your systems electronically.

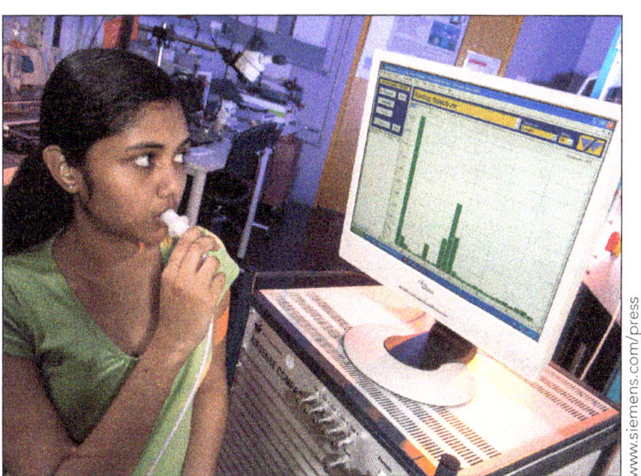

www.siemens.com/press

Condition	Molecule used for diagnosis
Heart transplant rejection	Alkanes and methylalkanes
Irritable Bowel Syndrome	Hydrogen
Diabetes	Acetone
Asthma	Nitric oxide
Stomach ulcers	Carbon isotope in CO_2
Liver disease	CO_2 levels

A good example of this molecule–pressure relationship may be observed in an ordinary cylinder of compressed gas equipped with a pressure gauge. When the valve is opened, gas escapes from the cylinder. The volume of the cylinder is constant, and the decrease in quantity (moles) of gas is registered by a drop in pressure indicated on the gauge.

The pressure of a gas in a fixed volume also varies with temperature. When the temperature is increased, the kinetic energy of the molecules increases, causing more frequent and more energetic collisions of the molecules with the walls of the container. This increase in collision frequency and energy results in a pressure increase (FIGURE 12.6).

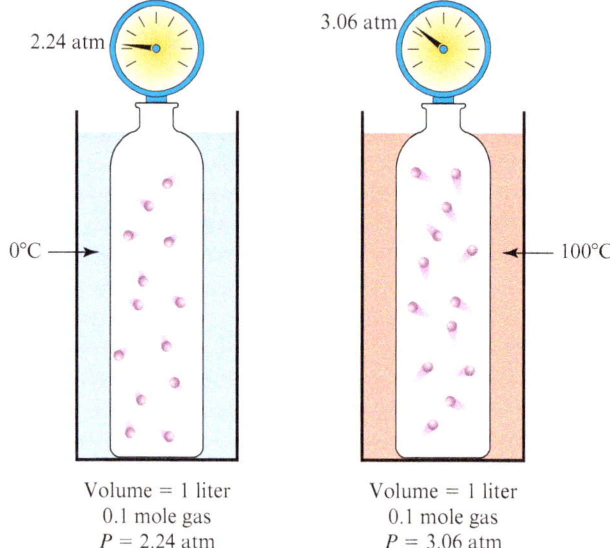

FIGURE 12.6 The pressure of a gas in a fixed volume increases with increasing temperature. The increased pressure is due to more frequent and more energetic collisions of the gas molecules with the walls of the container at the higher temperature.

12.2 Boyle's Law

LEARNING OBJECTIVE Use Boyle's law to calculate changes in pressure or volume of a sample of gas at a constant temperature.

KEY TERM
Boyle's law

Through a series of experiments, Robert Boyle (1627–1691) determined the relationship between the pressure (P) and volume (V) of a particular quantity of a gas. This relationship of P and V is known as **Boyle's law**:

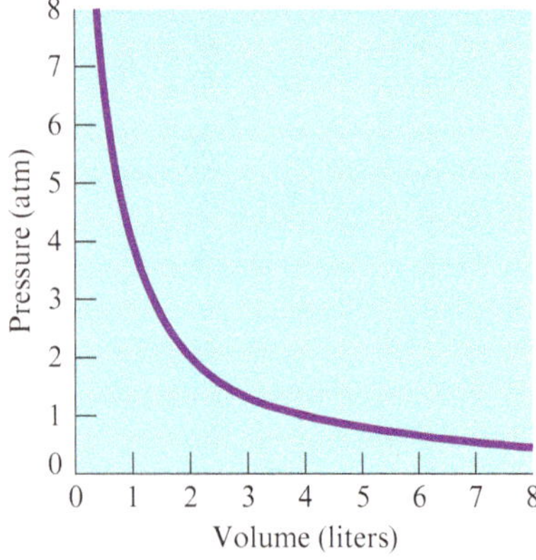

FIGURE 12.7 Graph of pressure versus volume showing the inverse PV relationship of an ideal gas.

> At constant temperature (T), the volume (V) of a fixed mass of a gas is inversely proportional to the pressure (P), which may be expressed as
>
> $$V \propto \frac{1}{P} \quad \text{or} \quad P_1V_1 = P_2V_2$$

This law says that the volume varies ($\propto$) inversely with the pressure at constant mass and temperature. When the pressure on a gas is increased, its volume will decrease, and vice versa. The inverse relationship of pressure and volume is graphed in FIGURE 12.7.

When Boyle doubled the pressure on a specific quantity of a gas, keeping the temperature constant, the volume was reduced to one-half the original volume; when he tripled the pressure on the system, the new volume was one-third the original volume; and so on. His work showed that the product of volume and pressure is constant if the temperature is not changed:

$$PV = \text{constant} \quad \text{or} \quad PV = k \quad \text{(mass and temperature are constant)}$$

Let's demonstrate this law using a cylinder with a movable piston so that the volume of gas inside the cylinder may be varied by changing the external pressure (FIGURE 12.8). Assume that the temperature and the number of gas molecules do not change. We start with a volume of 1000 mL and a pressure of 1 atm. When we change the pressure to 2 atm, the gas molecules are crowded closer together, and the volume is reduced to 500 mL. When we increase the pressure to 4 atm, the volume becomes 250 mL.

Note that the product of the pressure times the volume is the same number in each case, verifying Boyle's law. We may then say that

$$P_1V_1 = P_2V_2$$

where P_1V_1 is the pressure–volume product at one set of conditions, and P_2V_2 is the product at another set of conditions. In each case, the new volume may be calculated by

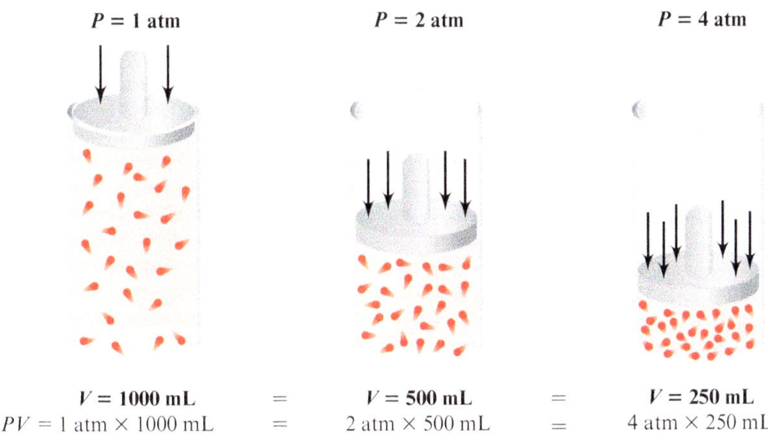

FIGURE 12.8 The effect of pressure on the volume of a gas.

multiplying the starting volume by a ratio of the two pressures involved.

$$V_2 = V_1\left(\frac{P_1}{P_2}\right)$$

Of course, the ratio of pressures used must reflect the direction in which the volume should change. When the pressure is changed from 1 atm to 2 atm, the ratio to be used is 1 atm/2 atm. Now we can verify the volumes shown in Figure 12.8:

1. Starting volume, 1000 mL; pressure change, 1 atm $\longrightarrow$ 2 atm ($P_1 \longrightarrow P_2$)

$$(1000\ \text{mL})\left(\frac{1\ \cancel{\text{atm}}}{2\ \cancel{\text{atm}}}\right) = 500\ \text{mL}$$

2. Starting volume, 1000 mL; pressure change, 1 atm $\longrightarrow$ 4 atm ($P_1 \longrightarrow P_2$)

$$(1000\ \text{mL})\left(\frac{1\ \cancel{\text{atm}}}{4\ \cancel{\text{atm}}}\right) = 250\ \text{mL}$$

3. Starting volume, 500 mL; pressure change, 2 atm $\longrightarrow$ 4 atm ($P_1 \longrightarrow P_2$)

$$(500\ \text{mL})\left(\frac{2\ \cancel{\text{atm}}}{4\ \cancel{\text{atm}}}\right) = 250\ \text{mL}$$

In summary, a change in the volume of a gas due to a change in pressure can be calculated by multiplying the original volume by a ratio of the two pressures. If the pressure is increased, the ratio should have the smaller pressure in the numerator and the larger pressure in the denominator. If the pressure is decreased, the larger pressure should be in the numerator and the smaller pressure in the denominator:

$$\text{new volume} = \text{original volume} \times \text{ratio of pressures} = V_1\left(\frac{P_1}{P_2}\right)$$

We use Boyle's law in the following examples. If no mention is made of temperature, assume that it remains constant.

CHECK YOUR UNDERSTANDING

12.3 Boyle's Law Predictions Based on Pressure Changes

12.4 Boyle's Law Predictions Based on Volume Changes

WileyPLUS

ONLINE LEARNING MODULE

Boyle's Law

WileyPLUS

ENHANCED EXAMPLE 12.2 WileyPLUS

What volume will 2.50 L of a gas occupy if the pressure is changed from 760. mm Hg to 630. mm Hg?

SOLUTION

Read • **Knowns:** $V_1 = 2.50\text{ L}$

$P_1 = 760.\text{ mm Hg} \qquad P_2 = 630.\text{ mm Hg}$

Solving for: new volume (V_2)

Plan • Since the problem is concerned only with pressure and volume, Boyle's law can be used: $P_1V_1 = P_2V_2$

Solution map: pressure decreases → volume increases

$P_1V_1 = P_2V_2$ solving for V_2 gives $V_2 = \dfrac{P_1V_1}{P_2}$

Calculate • $$V_2 = \frac{(2.50\text{ L})(760.\text{ }\cancel{\text{mm Hg}})}{630.\text{ }\cancel{\text{mm Hg}}} = 3.20\text{ L}$$

Another way to think about this is to use a ratio of pressures that will result in an increase in volume.

$$V_2 = (2.50\text{ L})\left(\frac{760.\text{ }\cancel{\text{mm Hg}}}{630.\text{ }\cancel{\text{mm Hg}}}\right) = 3.20\text{ L}$$

Note this produces the same equation as we used above to solve the problem.

Check • The final volume is larger than the original as predicted by Boyle's law and the solution map.

ENHANCED EXAMPLE 12.3 WileyPLUS

A given mass of hydrogen occupies 40.0 L at 700. torr. What volume will it occupy at 5.00 atm pressure?

SOLUTION

Read • **Knowns:** $P_1 = 700.\text{ torr} = 700.\text{ torr}\left(\dfrac{1\text{ atm}}{760\text{ torr}}\right) = 0.921\text{ atm}$

$P_2 = 5.00\text{ atm}$

$V_1 = 40.0\text{ L}$

Solving for: new volume (V_2)

Plan • Since the problem is concerned only with pressure and volume, Boyle's law can be used: $P_1V_1 = P_2V_2$

Solution map: pressure increases → volume decreases

$P_1V_1 = P_2V_2$ solving for V_2 gives $V_2 = \dfrac{P_1V_1}{P_2}$

Calculate • $$V_2 = \frac{(0.921\text{ }\cancel{\text{atm}})(40.0\text{ L})}{5.00\text{ }\cancel{\text{atm}}} = 7.37\text{ L}$$

We can also use a ratio of pressures to solve this problem.

$$V_2 = (40.0\text{ L})\left(\frac{0.921\text{ }\cancel{\text{atm}}}{5.00\text{ }\cancel{\text{atm}}}\right) = 7.37\text{ L}$$

Check • The final volume is smaller than the original as predicted by Boyle's law and the solution map.

ENHANCED EXAMPLE 12.4 WileyPLUS

A gas occupies a volume of 200. mL at 400. torr pressure. To what pressure must the gas be subjected in order to change the volume to 75.0 mL?

SOLUTION

Read • **Knowns:** $P_1 = 400.\ \text{torr}$ $V_1 = 200.\ \text{mL}$

$V_2 = 75.0\ \text{mL}$

Solving for: new pressure (P_2)

Plan • Since the problem is concerned only with pressure and volume, Boyle's law can be used: $P_1V_1 = P_2V_2$

Solution map: volume decreases → pressure increases

Since $P_1V_1 = P_2V_2$ solving for P_2 gives $P_2 = \frac{P_1V_1}{V_2}$

Calculate • $$P_2 = \frac{(400.\ \text{torr})(200.\ \cancel{\text{mL}})}{75.0\ \cancel{\text{mL}}} = 1.07 \times 10^3\ \text{torr}$$

We can also think about this by using a ratio of volumes that will result in an increase in pressure.

$$V_2 = (400.\ \text{torr})\left(\frac{200.\ \cancel{\text{mL}}}{75.0\ \cancel{\text{mL}}}\right) = 1.07 \times 10^3\ \text{torr}$$

This result is the same as the one used above to solve the problem.

Check • The final pressure is larger than the original as predicted by Boyle's law and the solution map.

PRACTICE 12.2

A gas occupies a volume of 3.86 L at 0.750 atm. At what pressure will the volume be 4.86 L?

12.3 Charles' Law

LEARNING OBJECTIVE Use Charles' law to calculate changes in temperature or volume of a sample of gas at constant pressure.

KEY TERMS
absolute zero
Charles' law

The effect of temperature on the volume of a gas was observed in about 1787 by the French physicist J. A. C. Charles (1746–1823). Charles found that various gases expanded by the same fractional amount when they underwent the same change in temperature. Later it was found that if a given volume of any gas initially at 0°C was cooled by 1°C, the volume decreased by $\frac{1}{273}$; if cooled by 2°C, it decreased by $\frac{2}{273}$; if cooled by 20°C, by $\frac{20}{273}$; and so on. Since each degree of cooling reduced the volume by $\frac{1}{273}$, it was apparent that any quantity of any gas would have zero volume if it could be cooled to −273°C. Of course, no real gas can be cooled to −273°C for the simple reason that it would liquefy before that temperature is reached. However, −273°C (more precisely, −273.15°C) is referred to as **absolute zero**; this temperature is the zero point on the Kelvin (absolute) temperature scale—the temperature at which the volume of an ideal, or perfect, gas would become zero.

The volume–temperature relationship for methane is shown graphically in FIGURE 12.9. Experimental data show the graph to be a straight line that, when extrapolated, crosses the temperature axis at −273.15°C, or absolute zero. This is characteristic for all gases.

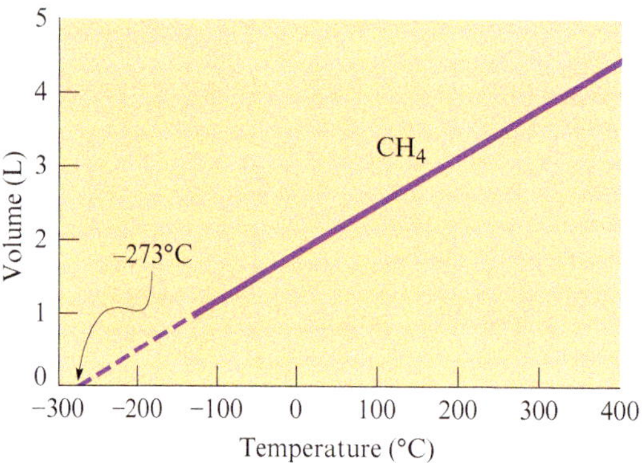

FIGURE 12.9 Volume–temperature relationship of methane (CH_4). Extrapolated portion of the graph is shown by the broken line.

In modern form, **Charles' law** is as follows:

At *constant pressure* the volume of a fixed mass of any gas is directly proportional to the absolute temperature, which may be expressed as

$$V \propto T \qquad \text{or} \qquad \frac{V_1}{T_1} = \frac{V_2}{T_2}$$

A capital T is usually used for absolute temperature (K) and a small t for °C.

Mathematically, this states that the volume of a gas varies directly with the absolute temperature when the pressure remains constant. In equation form, Charles' law may be written as

$$V = kT \qquad \text{or} \qquad \frac{V}{T} = k \qquad \text{(at constant pressure)}$$

where k is a constant for a fixed mass of the gas. If the absolute temperature of a gas is doubled, the volume will double.

To illustrate, let's return to the gas cylinder with the movable or free-floating piston (FIGURE 12.10). Assume that the cylinder labeled (a) contains a quantity of gas and the pressure on it is 1 atm. When the gas is heated, the molecules move faster, and their kinetic energy increases. This action should increase the number of collisions per unit of time and therefore increase the pressure. However, the increased internal pressure will cause the piston to rise to a level at which the internal and external pressures again equal 1 atm, as we see in cylinder (b). The net result is an increase in volume due to an increase in temperature.

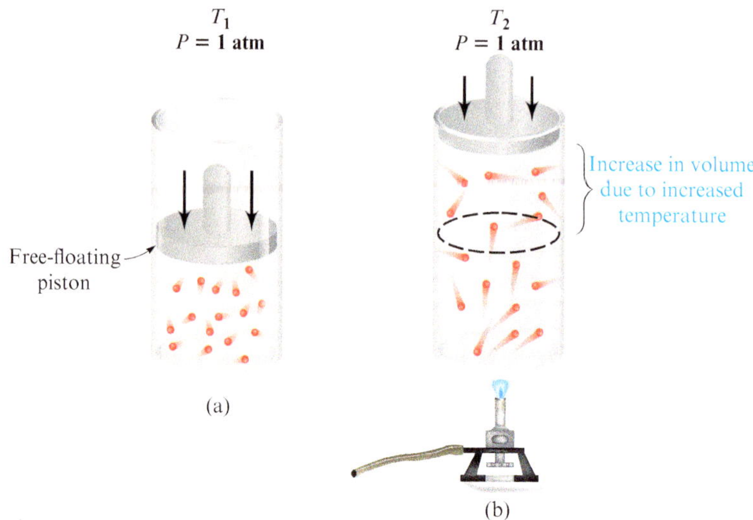

FIGURE 12.10 The effect of temperature on the volume of a gas. The gas in cylinder (a) is heated from T_1 to T_2. With the external pressure constant at 1 atm, the free-floating piston rises, resulting in an increased volume, shown in cylinder (b).

Another equation often used to relate the volume of a gas at two different temperatures is

$$\frac{V_1}{T_1} = \frac{V_2}{T_2} \qquad \text{(constant P)}$$

where V_1 and T_1 are one set of conditions and V_2 and T_2 are another set of conditions.

A simple experiment showing the variation of the volume of a gas with temperature is illustrated in FIGURE 12.11. A balloon is placed in a beaker, and liquid N_2 is poured over it. The volume is reduced, as shown by the collapse of the balloon; when the balloon is removed from the liquid N_2, the gas expands as it warms back to room temperature and the balloon increases in size.

CHECK YOUR UNDERSTANDING
12.5 Charles' Law and Temperature Changes
12.6 Charles' Law and Volume Changes
➡ WileyPLUS

Charles D. Winters/Science Source Images

(a)

Charles D Winters/Photo Researchers Inc.

(b)

Charles D. Winters/Science SourceImages

(c)

FIGURE 12.11 The air-filled balloons (a) are being placed into the beaker containing liquid nitrogen. Notice that the blue and red balloons on the left in (a) have been placed in the beaker in (b) and are much smaller. In (c) all of the balloons now fit inside the beaker.

ENHANCED EXAMPLE 12.5

WileyPLUS

ONLINE LEARNING MODULE
Charles' Law
➡ WileyPLUS

Three liters (3.00L) of hydrogen at −20.°C are allowed to warm to a room temperature of 27°C. What is the volume at room temperature if the pressure remains constant?

SOLUTION

Read **Knowns:** $V_1 = 3.00\text{ L}$ $\quad T_1 = -20.°\text{C} + 273 = 253\text{ K}$

$T_2 = 27°\text{C} + 273 = 300.\text{ K}$

Solving for: new volume (V_2)

Remember: Temperature must be changed to Kelvin in gas law problems. Note that we use 273 to convert instead of 273.15 since our original measurements are to the nearest degree.

Plan Since the problem is concerned only with temperature and volume, Charles' law can be used: $\frac{V_1}{T_1} = \frac{V_2}{T_2}$

Solution map: temperature increases → volume increases

Since $\frac{V_1}{T_1} = \frac{V_2}{T_2}$, $\quad V_2 = \frac{V_1 T_2}{T_1}$

Calculate $$V_2 = \frac{(3.00\text{ L})(300.\text{ K})}{253\text{ K}} = 3.56\text{ L}$$

Note the solution map also produces the same result by using a ratio of temperatures that will result in an increase in volume.

$$V_2 = (3.00\text{ L})\left(\frac{300.\text{ K}}{253\text{ K}}\right) = 3.56\text{ L}$$

Check The final volume is larger than the original as predicted by Charles' law and the solution map.

ENHANCED EXAMPLE 12.6 WileyPLUS

If 20.0 L of oxygen are cooled from 100.°C to 0.°C, what is the new volume?

SOLUTION

Read • Since no mention is made of pressure, we assume the pressure does not change. Remember that temperature must be in kelvins.

Knowns: $V_1 = 20.0\text{ L}$ $T_1 = 100.°\text{C} = 373\text{ K}$

$T_2 = 0.°\text{C} = 273\text{ K}$

Solving for: new volume (V_2)

Plan • Since the problem is concerned only with temperature and volume, Charles' law can be used: $\frac{V_1}{T_1} = \frac{V_2}{T_2}$

Solution map: temperature decreases → volume decreases

Since $\frac{V_1}{T_1} = \frac{V_2}{T_2}$, $V_2 = \frac{V_1 T_2}{T_1}$

Calculate • $V_2 = \frac{(20.0\text{ L})(273\text{ K})}{373\text{ K}} = 14.6\text{ L}$

Or we can use the solution map to multiply the original volume by a ratio of temperatures that will result in a decrease in volume.

$$V_2 = (20.0\text{ L})\left(\frac{273\text{ K}}{373\text{ K}}\right) = 14.6\text{ L}$$

Check • The final volume is smaller than the original as predicted by Charles' law and the solution map.

PRACTICE 12.3

A 4.50-L container of nitrogen gas at 28.0°C is heated to 56.0°C. Assuming that the volume of the container can vary, what is the new volume of the gas?

12.4 Avogadro's Law

LEARNING OBJECTIVE Solve problems using the relationship among moles, mass, and volume of gases.

KEY TERMS
Gay-Lussac's law of combining volumes
Avogadro's law

Early in the nineteenth century, French chemist, J.L. Gay-Lussac (1778–1850) studied the volume relationships of reacting gases. His results, published in 1809, were summarized in a statement known as **Gay-Lussac's law of combining volumes**:

> When measured at the same temperature and pressure, the ratios of the volumes of reacting gases are small whole numbers.

Thus, H_2 and O_2 combine to form water vapor in a volume ratio of 2:1 (FIGURE 12.12); H_2 and Cl_2 react to form HCl in a volume ratio of 1:1, and H_2 and N_2 react to form NH_3 in a volume ratio of 3:1.

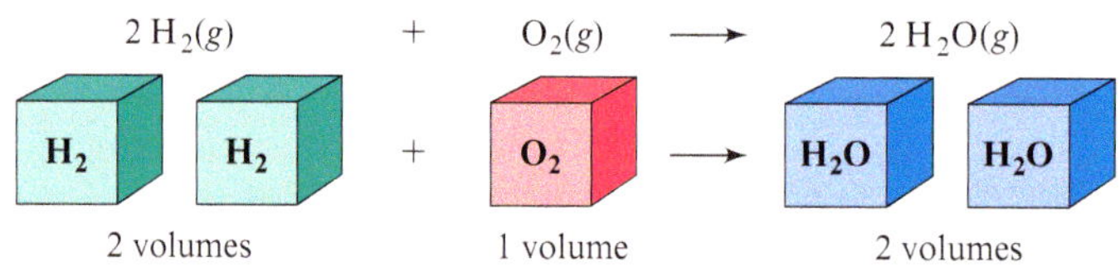

FIGURE 12.12 Gay-Lussac's law of combining volumes of gases applied to the reaction of hydrogen and oxygen. When measured at the same temperature and pressure, hydrogen and oxygen react in a volume ratio of 2:1.

Two years later, in 1811, Amedeo Avogadro (1776–1856) used the law of combining volumes of gases to make a simple but significant and far-reaching generalization concerning gases. **Avogadro's law** states:

> Equal volumes of different gases at the same temperature and pressure contain the same number of molecules.

This law was a real breakthrough in understanding the nature of gases.

1. It offered a rational explanation of Gay-Lussac's law of combining volumes of gases and indicated the diatomic nature of such elemental gases as hydrogen, chlorine, and oxygen.
2. It provided a method for determining the molar masses of gases and for comparing the densities of gases of known molar mass (see Section 12.8).
3. It afforded a firm foundation for the development of the kinetic-molecular theory.

CHECK YOUR UNDERSTANDING
12.7 Moles of a Gas and Volume
➡ WileyPLUS

By Avogadro's law, equal volumes of hydrogen and chlorine at the same temperature and pressure contain the same number of molecules. On a volume basis, hydrogen and chlorine react thus (FIGURE 12.13):

hydrogen	+	chlorine	⟶	hydrogen chloride
1 volume		1 volume		2 volumes

Therefore, hydrogen molecules react with chlorine molecules in a 1:1 ratio. Since two volumes of hydrogen chloride are produced, one molecule of hydrogen and one molecule of chlorine must produce two molecules of hydrogen chloride. Therefore, each hydrogen molecule and each chlorine molecule must consist of two atoms. The coefficients of the balanced equation for the reaction give the correct ratios for volumes, molecules, and moles of reactants and products:

H_2	+	Cl_2	⟶	2HCl
1 volume		1 volume		2 volumes
1 molecule		1 molecule		2 molecules
1 mol		1 mol		2 mol

By like reasoning, oxygen molecules also must contain at least two atoms because one volume of oxygen reacts with two volumes of hydrogen to produce two volumes of water vapor.

The volume of a gas depends on the temperature, the pressure, and the number of gas molecules. Different gases at the same temperature have the same average kinetic energy. Hence, if two different gases are at the same temperature, occupy equal volumes, and exhibit equal pressures, each gas must contain the same number of molecules. This statement is true because systems with identical PVT properties can be produced only by equal numbers of molecules having the same average kinetic energy.

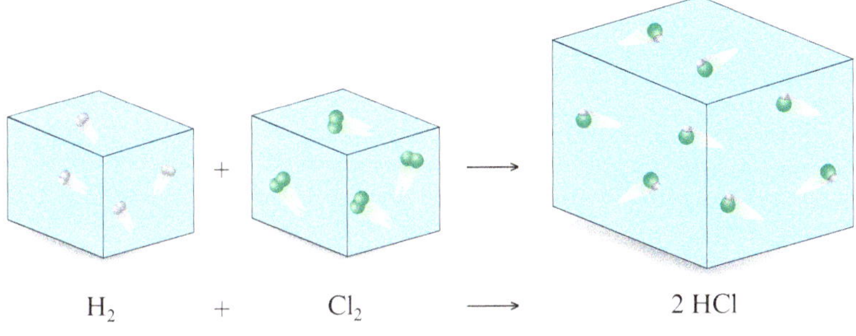

FIGURE 12.13 Avogadro's law proved the concept of diatomic molecules for hydrogen and chlorine.

12.5 Combined Gas Laws

LEARNING OBJECTIVE Use the combined gas law to calculate changes in pressure, temperature, or volume of a sample of gas.

KEY TERMS
standard conditions
standard temperature and pressure (STP)
molar volume

To compare volumes of gases, common reference points of temperature and pressure were selected and called **standard conditions** or **standard temperature and pressure** (abbreviated **STP**). Standard temperature is 273.15 K (0°C), and standard pressure is 1 atm (760 torr, 760 mm Hg, or 101.325 kPa). For purposes of comparison, volumes of gases are often changed to STP conditions:

In this text we'll use 273 K for temperature conversions and calculations. Check with your instructor for rules in your class.

$$\text{standard temperature} = 273.15\text{ K or }0.00°\text{C}$$

$$\text{standard pressure} = 1\text{ atm or }760\text{ torr, }760\text{ mm Hg, or }101.325\text{ kPa}$$

Remember that you determine the correct ratios by thinking about the final result. For example, an increase in pressure should decrease the volume, and so on.

When temperature and pressure change at the same time, the new volume may be calculated by multiplying the initial volume by the correct ratios of both pressure and temperature as follows:

$$\text{final volume} = (\text{initial volume})\begin{pmatrix}\text{ratio of}\\\text{pressures}\end{pmatrix}\begin{pmatrix}\text{ratio of}\\\text{temperatures}\end{pmatrix}$$

The P, V, and T relationships for a given mass of any gas, in fact, may be expressed as a single equation, $PV/T = k$. For problem solving, this equation is usually written

$$\frac{P_1V_1}{T_1} = \frac{P_2V_2}{T_2}$$

where P_1, V_1, and T_1 are the initial conditions and P_2, V_2, and T_2 are the final conditions.

This equation can be solved for any one of the six variables and is useful in dealing with the pressure–volume–temperature relationships of gases. Note what happens to the combined gas law when one of the variables is constant:

- T constant $\rightarrow P_1V_1 = P_2V_2$ Boyle's law
- P constant $\rightarrow \dfrac{V_1}{T_1} = \dfrac{V_2}{T_2}$ Charles' law

CHECK YOUR UNDERSTANDING
12.8 Standard Temperature and Pressure (STP)
12.9 Moles of a Gas and Volume at STP
12.10 Temperature and Pressure Relationship
WileyPLUS

ONLINE LEARNING MODULE
Combined Gas Law
WileyPLUS

ENHANCED EXAMPLE 12.7 WileyPLUS

Given 20.0 L of ammonia gas at 5°C and 730. torr, calculate the volume at 50.°C and 800. torr.

SOLUTION

Read Remember that temperature must be in kelvins.

Knowns:
$P_1 = 730.\text{ torr}$ $P_2 = 800.\text{ torr}$
$T_1 = 5°\text{C} = 278\text{ K}$ $T_2 = 50.°\text{C} = 323\text{ K}$
$V_1 = 20.0\text{ L}$

Solving for: new volume (V_2)

Plan Since P, V, and T are all changing, we must use the combined gas law:

$$\frac{P_1V_1}{T_1} = \frac{P_2V_2}{T_2}$$

$$V_2 = \frac{V_1P_1T_2}{P_2T_1}$$

Calculate $V_2 = \frac{(20.0\ \text{L})(730.\ \cancel{\text{torr}})(323\ \cancel{\text{K}})}{(800.\ \cancel{\text{torr}})(278\ \cancel{\text{K}})} = 21.2\ \text{L}$

Check The number of significant figures should be three, and it is, as indicated by the significant figures, in the knowns.

ENHANCED EXAMPLE 12.8

WileyPLUS

To what temperature (°C) must 10.0 L of nitrogen at 25°C and 700. torr be heated in order to have a volume of 15.0 L and a pressure of 760. torr?

SOLUTION

Read Remember that temperature must be in kelvins.

Knowns: $P_1 = 700.\ \text{torr}$ $\quad P_2 = 760.\ \text{torr}$

$V_1 = 10.0\ \text{L}$ $\quad V_2 = 15.0\ \text{L}$

$T_1 = 25°\text{C} = 298\ \text{K}$

Solving for: new temperature (T_2)

Plan Since P, V, and T are all changing, we use the combined gas law:

$$\frac{P_1V_1}{T_1} = \frac{P_2V_2}{T_2}$$

$$T_2 = \frac{T_1P_2V_2}{P_1V_1}$$

Calculate $T_2 = \frac{(298\ \text{K})(760.\ \cancel{\text{torr}})(15.0\ \cancel{\text{L}})}{(700.\ \cancel{\text{torr}})(10.0\ \cancel{\text{L}})} = 485\ \text{K}$

Since the problem asks for °C, we must convert our answer:

$485\ \text{K} - 273 = 212°\text{C}$

ENHANCED EXAMPLE 12.9

WileyPLUS

The volume of a gas-filled balloon is 50.0 L at 20.°C and 742 torr. What volume will it occupy at standard temperature and pressure (760. torr and 273 K)?

SOLUTION

Read Remember that temperature must be in kelvins.

Knowns: $P_1 = 742\ \text{torr}$ $\quad P_2 = 760.\ \text{torr}$ (standard pressure)

$T_1 = 20.°\text{C} = 293\ \text{K}$ $\quad T_2 = 273\ \text{K}$ (standard temperature)

$V_1 = 50.0\ \text{L}$

Solving for: new volume (V_2)

Plan Since P, V, and T are all involved, we use the combined gas law:

$$\frac{P_1V_1}{T_1} = \frac{P_2V_2}{T_2}$$

$$V_2 = \frac{P_1V_1T_2}{P_2T_1}$$

Calculate $V_2 = \frac{(742\ \cancel{\text{torr}})(50.0\ \text{L})(273\ \cancel{\text{K}})}{(760.\ \cancel{\text{torr}})(293\ \cancel{\text{K}})} = 45.5\ \text{L}$

TABLE 12.3 | Mass of a Molar Volume (22.4 L) of Various Gases

Gas	Mass of 22.4 L
Xe	131.3 g
O_2	32.00 g
CO_2	44.01 g
N_2	28.02 g
NH_3	17.03 g
Ar	39.95 g
H_2S	34.09 g
H_2	2.016 g
SO_2	64.07 g
HCl	36.46 g
CH_4	16.04 g
Cl_2	71.90 g

PRACTICE 12.4

15.00 L of gas at 45.0°C and 800. torr are heated to 400.°C, and the pressure is changed to 300. torr. What is the new volume?

PRACTICE 12.5

To what temperature must 5.00 L of oxygen at 50.°C and 600. torr be heated in order to have a volume of 10.0 L and a pressure of 800. torr?

Mole–Mass–Volume Relationships of Gases

Because a mole contains 6.022×10^{23} molecules (Avogadro's number), a mole of any gas will have the same volume as a mole of any other gas at the same temperature and pressure. It has been experimentally determined that the volume occupied by a mole of any gas is 22.4 L at STP. This volume, 22.4 L, is known as the **molar volume** of a gas. The molar volume is a cube about 28.2 cm (11.1 in.) on a side. The molar masses of several gases, each occupying 22.4 L at STP, are shown in TABLE 12.3 and FIGURE 12.14:

One mole of a gas occupies 22.4 L at STP.

As with many constants, the molar volume is known more exactly to be 22.414 L. We use 22.4 L in our calculations, since the extra figures don't often affect the result, given the other measurements in the calculation.

The molar volume is useful for determining the molar mass of a gas or of substances that can be easily vaporized. If the mass and the volume of a gas at STP are known, we can calculate its molar mass. For example, 1.00 L of pure oxygen at STP has a mass of 1.429 g. The molar mass of oxygen may be calculated by multiplying the mass of 1.00 L by 22.4 L/mol:

$$\left(\frac{1.429\ \text{g}}{1.00\ \cancel{\text{L}}}\right)\left(\frac{22.4\ \cancel{\text{L}}}{1\ \text{mol}}\right) = 32.0\ \text{g/mol} \qquad \text{(molar mass)}$$

If the mass and volume are at other than standard conditions, we change the volume to STP and then calculate the molar mass.

The molar volume, 22.4 L/mol, is used as a conversion factor to convert grams per liter to grams per mole (molar mass) and also to convert liters to moles. The two conversion factors are

$$\frac{22.4\ \text{L}}{1\ \text{mol}} \quad \text{and} \quad \frac{1\ \text{mol}}{22.4\ \text{L}}$$

Standard conditions apply only to pressure, temperature, and volume. Mass is not affected.

These conversions must be done at STP except under certain special circumstances.

ONLINE LEARNING MODULE
Calculating Molar Mass at STP
➡ WileyPLUS

ENHANCED EXAMPLE 12.10 WileyPLUS

If 2.00 L of a gas measured at STP has a mass of 3.23 g, what is the molar mass of the gas?

SOLUTION

Read • **Knowns:** $V = 2.00$ L
$T = 273$ K
$P = 1.00$ atm
$m = 3.23$ g

Plan • The unit for molar mass is g/mol.
Solution map: g/L → g/mol
1 mol = 22.4 L, so the conversion factor is 22.4 L/1 mol.

Calculate • $\left(\frac{3.23\ \text{g}}{2.00\ \cancel{\text{L}}}\right)\left(\frac{22.4\ \cancel{\text{L}}}{1\ \text{mol}}\right) = 36.2$ g/mol (molar mass)

ENHANCED EXAMPLE 12.11 WileyPLUS

Measured at 40.°C and 630. torr, the mass of 691 mL of diethyl ether is 1.65 g. Calculate the molar mass of diethyl ether.

Diethyl ether

SOLUTION

Read To find the molar mass we must first convert to STP.

Knowns: $P_1 = 630.\text{ torr}$ $P_2 = 760\text{ torr}$

$T_1 = 40.°\text{C} = 313\text{ K}$ $T_2 = 273\text{ K}$

$V_1 = 691\text{ mL}$

Plan Since P, V, and T are all changing, we must use the combined gas law:

$$\frac{P_1V_1}{T_1} = \frac{P_2V_2}{T_2}$$

$$V_2 = \frac{V_1P_1T_2}{P_2T_1}$$

Calculate $$V_2 = \frac{(691\text{ mL})(630.\text{ torr})(273\text{ K})}{(760.\text{ torr})(313\text{ K})} = 500.\text{ mL} = 0.500\text{ L (at STP)}$$

$$\left(\frac{1.65\text{ g}}{0.500\text{ L}}\right)\left(\frac{22.4\text{ L}}{1\text{ mol}}\right) = 73.9\text{ g/mol}$$

PRACTICE 12.6

A gas with a mass of 86 g occupies 5.00 L at 25°C and 3.00 atm pressure. What is the molar mass of the gas?

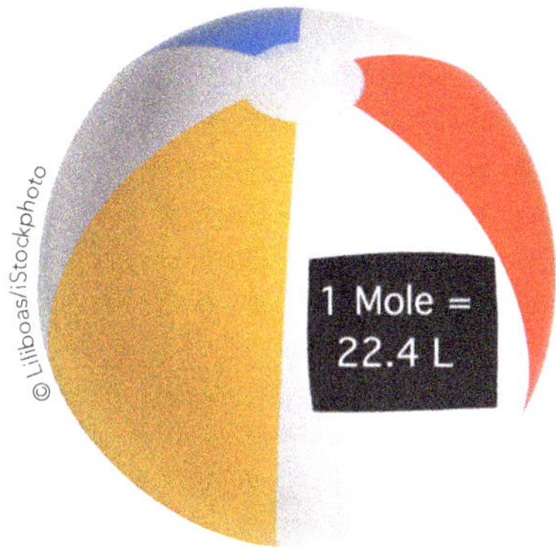

FIGURE 12.14 One mole of a gas occupies 22.4 L at STP.

12.6 Ideal Gas Law

Use the ideal gas law to solve problems involving pressure, volume, temperature, and number of moles.

LEARNING OBJECTIVE

KEY TERMS
ideal gas law
kinetic molecular theory (KMT)
ideal gas

Now that we have considered all of the relationships used in gas calculations, we can simplify our work by combining the relationships into a single equation describing a gas. The four variables that characterize a gas are:

- volume (V)
- pressure (P)
- absolute temperature (T)
- number of molecules of moles (n)

Combining these variables, we obtain

$$V \propto \frac{nT}{P} \quad \text{or} \quad V = \frac{nRT}{P}$$

where R is a proportionality constant known as the *ideal gas constant*. The equation is commonly written as

$$PV = nRT$$

and is known as the **ideal gas law**. This law summarizes in a single expression what we have considered in our earlier discussions. The value and units of R depend on the units of P, V, and T.

CHECK YOUR UNDERSTANDING

12.11 Formula Used for Ideal Gas Law

WileyPLUS

We can calculate one value of R by taking 1 mol of a gas at STP conditions. Solve the equation for R:

$$R = \frac{PV}{nT} = \frac{(1\ \text{atm})(22.4\ \text{L})}{(1\ \text{mol})(273\ \text{K})} = 0.0821 \frac{\text{L} \cdot \text{atm}}{\text{mol} \cdot \text{K}}$$

CHECK YOUR UNDERSTANDING

12.12 Units Used for Ideal Gas Law Calculations

WileyPLUS

The units of R in this case are liter · atmospheres (L · atm) per mole kelvin (mol · K). When the value of $R = 0.0821$ L · atm/mol · K, P is in atmospheres, n is in moles, V is in liters, and T is in kelvins.

The ideal gas equation can be used to calculate any one of the four variables when the other three are known.

ONLINE LEARNING MODULE

Ideal Gas Law

WileyPLUS

ENHANCED EXAMPLE 12.12 WileyPLUS

What pressure will be exerted by 0.400 mol of a gas in a 5.00-L container at 17°C?

SOLUTION

Read • **Knowns:** $V = 5.00$ L

$T = 17°\text{C} = 290.\ \text{K}$

$n = 0.400$ mol

Plan • We need to use the ideal gas law:

$$PV = nRT$$

$$P = \frac{nRT}{V}$$

Calculate •
$$P = \frac{(0.400\ \cancel{\text{mol}})\left(0.0821 \frac{\cancel{\text{L}} \cdot \text{atm}}{\cancel{\text{mol}} \cdot \cancel{\text{K}}}\right)(290.\ \cancel{\text{K}})}{5.00\ \cancel{\text{L}}} = 1.90\ \text{atm}$$

ENHANCED EXAMPLE 12.13 WileyPLUS

How many moles of oxygen gas are in a 50.0-L tank at 22°C if the pressure gauge reads 2000. lb/in.^{2}?

SOLUTION

1 atm = 14.7 lb/in.^{2} from Table 12.2.

Read • **Knowns:**
$$P = \left(\frac{2000.\ \cancel{\text{lb}}}{\cancel{\text{in.}^2}}\right)\left(\frac{1\ \text{atm}}{14.7 \frac{\cancel{\text{lb}}}{\cancel{\text{in.}^2}}}\right) = 136.1\ \text{atm}$$

$V = 50.0$ L

$T = 22°\text{C} = 295$ K

Plan • We need to use the ideal gas law:

$$PV = nRT$$

$$n = \frac{PV}{RT}$$

Calculate •
$$n = \frac{(136.1\ \cancel{\text{atm}})(50.0\ \cancel{\text{L}})}{\left(0.0821 \frac{\cancel{\text{L}} \cdot \cancel{\text{atm}}}{\text{mol} \cdot \cancel{\text{K}}}\right)(295\ \cancel{\text{K}})} = 281\ \text{mol}\ O_2$$

PRACTICE 12.7

A 23.8-L cylinder contains oxygen gas at 20.0°C and 732 torr. How many moles of oxygen are in the cylinder?

The molar mass of a gaseous substance can also be determined using the ideal gas law. Since molar mass = g/mol, it follows that mol = g/molar mass. Using M for molar mass and g for grams, we can substitute g/M for n (moles) in the ideal gas law to get

$$PV = \frac{g}{M}RT \quad \text{or} \quad M = \frac{gRT}{PV} \quad \text{(modified ideal gas law)}$$

which allows us to calculate the molar mass, M, for any substance in the gaseous state.

This form of the ideal gas law is most useful in problems containing mass instead of moles.

ENHANCED EXAMPLE 12.14 WileyPLUS

Calculate the molar mass of butane gas if 3.69 g occupy 1.53 L at 20.°C and 1.00 atm.

SOLUTION

Read **Knowns:**
$P = 1.00$ atm
$V = 1.53$ L
$T = 20.°\text{C} = 293$ K
$m = 3.69$ g (mass)

Plan Since we know m and not the number of moles (n) of the gas, we need to use the ideal gas law, modifying it since $n = g/M$, where M is the molar mass of the gas:

$$PV = \frac{g}{M}RT$$

$$M = \frac{gRT}{PV}$$

Calculate

$$M = \frac{(3.69\ \text{g})\left(0.0821\ \frac{\cancel{\text{L}} \cdot \cancel{\text{atm}}}{\text{mol} \cdot \cancel{\text{K}}}\right)(293\ \cancel{\text{K}})}{(1.00\ \cancel{\text{atm}})(1.53\ \cancel{\text{L}})} = 58.0\ \text{g/mol}$$

Check If we look up the formula for butane, we find it is C_4H_{10} with a molar mass of 58.0 g/mol, so our calculation is correct.

ONLINE LEARNING MODULE
Using the Ideal Gas Law in a Molar Mass Calculation
➡ WileyPLUS

© Emil Schreiner/iStockphoto

Butane

PRACTICE 12.8

A 0.286-g sample of a certain gas occupies 50.0 mL at standard temperature and 76.0 cm Hg. Determine the molar mass of the gas.

The Kinetic-Molecular Theory

The data accumulated from the gas laws allowed scientists to formulate a general theory to explain the behavior and properties of gases. This theory is called the **kinetic-molecular theory (KMT)**. The KMT has since been extended to cover, in part, the behavior of liquids and solids.

CHECK YOUR UNDERSTANDING
12.13 Modified Ideal Gas Law for Mass Calculations
➡ WileyPLUS

The KMT is based on the motion of particles, particularly gas molecules. A gas that behaves exactly as outlined by the theory is known as an **ideal gas**. No ideal gases exist, but under certain conditions of temperature and pressure, real gases approach ideal behavior or at least show only small deviations from it.

The principal assumptions of the kinetic-molecular theory are as follows:

1. Gases consist of tiny particles.
2. The distance between particles is large compared with the size of the particles themselves. The volume occupied by a gas consists mostly of empty space.
3. Gas particles have no attraction for one another.
4. Gas particles move in straight lines in all directions, colliding frequently with one another and with the walls of the container.

5. No energy is lost by the collision of a gas particle with another gas particle or with the walls of the container. All collisions are perfectly elastic.
6. The average kinetic energy for particles is the same for all gases at the same temperature, and its value is directly proportional to the Kelvin temperature.

   The kinetic energy (KE) of a particle is expressed by the equation

$$KE = \frac{1}{2}mv^2$$

where m is the mass and v is the velocity of the particle.

All gases have the same kinetic energy at the same temperature. Therefore, from the kinetic energy equation we can see that, if we compare the velocities of the molecules of two gases, the lighter molecules will have a greater velocity than the heavier ones. For example, calculations show that the velocity of a hydrogen molecule is four times the velocity of an oxygen molecule.

Real Gases

All the gas laws are based on the behavior of an ideal gas—that is, a gas with a behavior that is described exactly by the gas laws for all possible values of P, V, and T. Most real gases actually do behave very nearly as predicted by the gas laws over a fairly wide range of temperatures and pressures. However, when conditions are such that the gas molecules are crowded closely together (high pressure and/or low temperature), they show marked deviations from ideal behavior. Deviations occur because molecules have finite volumes and also have intermolecular attractions, which result in less compressibility at high pressures and greater compressibility at low temperatures than predicted by the gas laws. Many gases become liquids at high pressure and low temperature.

CHECK YOUR UNDERSTANDING

12.14 Ideal Gas Conditions

➡ WileyPLUS

12.7 Dalton's Law of Partial Pressures

LEARNING OBJECTIVE Use Dalton's law of partial pressures to calculate the total pressure from a mixture of gases or the pressure of a single gas in a mixture of gases.

KEY TERMS

Dalton's law of partial pressures

partial pressure

If gases behave according to the kinetic-molecular theory, there should be no difference in the pressure–volume–temperature relationships whether the gas molecules are all the same or different. This similarity in the behavior of gases is the basis for an understanding of **Dalton's law of partial pressures**:

> The total pressure of a mixture of gases is the sum of the partial pressures exerted by each of the gases in the mixture.

Each gas in the mixture exerts a pressure that is independent of the other gases present. These pressures are called **partial pressures** (FIGURE 12.15). Thus, if we have a mixture of hydrogen and oxygen gases (H_2 and O_2), exerting partial pressures of 0.5 atm and 0.5 atm, respectively, the total pressure will be 1.0 atm:

$$P_{\text{total}} = P_{H_2} + P_{O_2}$$

$$P_{\text{total}} = 0.5 \text{ atm} + 0.5 \text{ atm} = 1.0 \text{ atm}$$

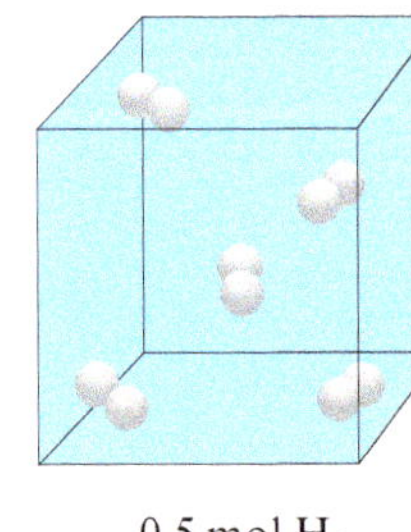

0.5 mol H_2
$P = 0.5$ atm

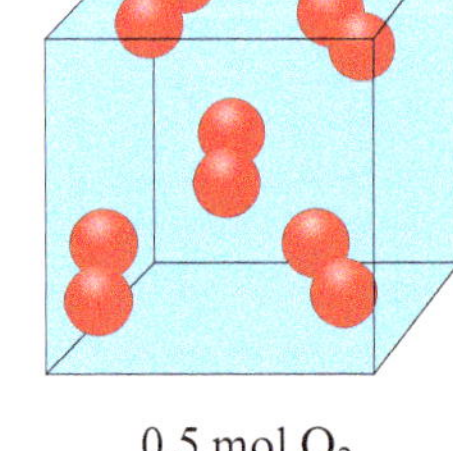

0.5 mol O_2
$P = 0.5$ atm

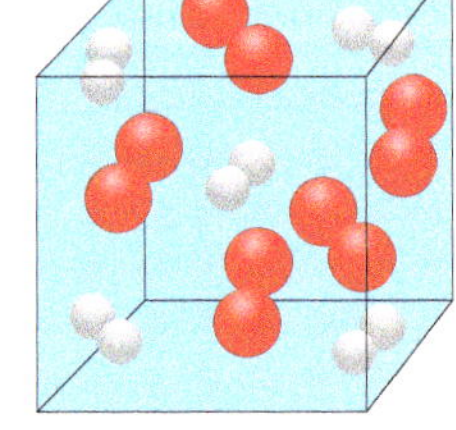

0.5 mol H_2 + 0.5 mol O_2
$P = 1$ atm

FIGURE 12.15 Each gas in a mixture of gases (H_2 and O_2) has a partial pressure, which is the same as though the gas were alone in the container.

CHEMISTRY *IN ACTION* Air Quality

Chemical reactions occur among the gases that are emitted into our atmosphere. In recent years, there has been growing concern over the effects these reactions have on our environment and our lives.

The outer portion (stratosphere) of the atmosphere plays a significant role in determining the conditions for life at the surface of the Earth. This stratosphere protects the surface from the intense radiation and particles bombarding our planet. Some of the high-energy radiation from the sun acts upon oxygen molecules, O_2, in the stratosphere, converting them into ozone, O_3.

$$O_2 \xrightarrow{\text{sunlight}} \underbrace{O + O}_{\text{oxygen atoms}}$$

$$O_2 + O \longrightarrow \underset{\text{ozone}}{O_3}$$

Ultraviolet radiation from the sun is highly damaging to living tissues of plants and animals. The ozone layer, however, shields the Earth by absorbing ultraviolet radiation and thus prevents most of this lethal radiation from reaching the Earth's surface.

In the lower atmosphere, ozone is a harmful pollutant. Ozone is formed in the atmosphere during electrical storms and by the photochemical action of ultraviolet radiation on a mixture of nitrogen dioxide and oxygen. Areas with high air pollution are subject to high atmospheric ozone concentrations. Ozone is not a desirable low-altitude constituent of the atmosphere because it is known to cause extensive plant damage, cracking of rubber, and the formation of eye-irritating substances.

In addition to ozone, the air in urban areas contains nitrogen oxides, which are components of smog. The term *smog* refers to air pollution in urban environments. Often the chemical reactions occur as part of a *photochemical process*. Nitrogen monoxide (NO) is oxidized in the air or in automobile engines to produce nitrogen dioxide (NO_2). In addition to nitrogen oxides, combustion of fossil fuels releases CO_2, CO, and sulfur oxides. Incomplete combustion releases unburned and partially burned hydrocarbons. Society is continually attempting to discover, understand, and control emissions that contribute to this sort of atmospheric chemistry.

Hong Kong skyline comparing poor and good air quality.

We can see an application of Dalton's law in the collection of insoluble gases over water. When prepared in the laboratory, oxygen is commonly collected by the downward displacement of water. Thus the oxygen is not pure but is mixed with water vapor (FIGURE 12.16). When the water levels are adjusted to the same height inside and outside the bottle, the pressure of the oxygen plus water vapor inside the bottle is equal to the atmospheric pressure:

$$P_{\text{atm}} = P_{O_2} + P_{H_2O}$$

To determine the amount of O_2 or any other gas collected over water, we subtract the pressure of the water vapor (vapor pressure) from the total pressure of the gases:

$$P_{O_2} = P_{\text{atm}} - P_{H_2O}$$

The vapor pressure of water at various temperatures is tabulated in Appendix 4.

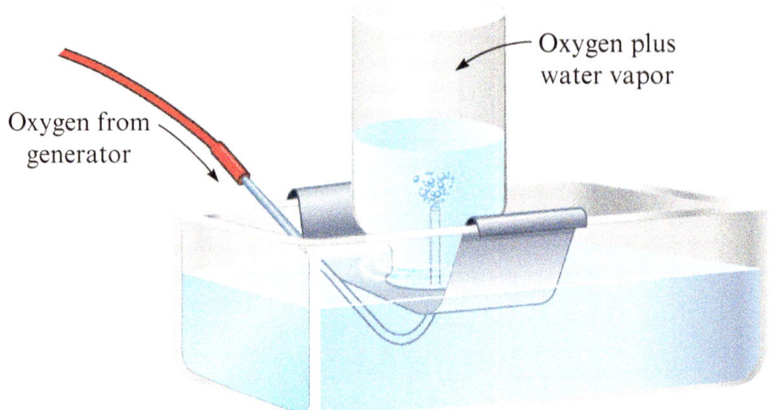

FIGURE 12.16 Oxygen collected over water.

ONLINE LEARNING MODULE

Determining the Volume of a Dry Gas

➡ WileyPLUS

ENHANCED EXAMPLE 12.15

WileyPLUS

A 500.-mL sample of oxygen was collected over water at 23°C and 760. torr. What volume will the dry O_2 occupy at 23°C and 760. torr? The vapor pressure of water at 23°C is 21.2 torr.

SOLUTION

Read • **Knowns:** $V_{\text{oxygen and water vapor}} = 500.\ \text{mL}$

$P_{\text{oxygen and water vapor}} = 760\ \text{torr}$

$P_{\text{water vapor}} = 21.2\ \text{torr}$

Temperature constant at 23°C

Solving for: $V_{\text{dry oxygen}}$ (at 760 torr)

Plan • First, we must find the pressure of the dry oxygen, and then determine the volume of the dry oxygen using Boyle's law (since the temperature is constant).

$$P_1V_1 = P_2V_2$$

$$P_{\text{dry oxygen}} = P_{\text{oxygen and water vapor}} - P_{\text{water vapor}}$$

$$= 760.\ \text{torr} - 21.2\ \text{torr} = 739\ \text{torr}$$

$$P_1 = 739\ \text{torr} \qquad P_2 = 760.\ \text{torr}$$

$$V_1 = 500.\ \text{mL}$$

Solution map: pressure increases → volume decreases

Since $P_1V_1 = P_2V_2$, $V_2 = \dfrac{P_1V_1}{P_2}$

Calculate • $V_2 = \dfrac{(739\ \cancel{\text{torr}})(500.\ \text{mL})}{760.\ \cancel{\text{torr}}} = 486\ \text{mL dry } O_2$

Check • The final volume is smaller than the original as predicted by Dalton's law.

PRACTICE 12.9

Hydrogen gas was collected by downward displacement of water. A volume of 600.0 mL of gas was collected at 25.0°C and 740.0 torr. What volume will the dry hydrogen occupy at STP?

12.8 Density of Gases

LEARNING OBJECTIVE

Calculate the density of a gas.

The density, d, of a gas is its mass per unit volume, which is generally expressed in grams per liter as follows:

$$d = \frac{\text{mass}}{\text{volume}} = \frac{\text{g}}{\text{L}}$$

Because the volume of a gas depends on temperature and pressure, both should be given when stating the density of a gas. The volume of a solid or liquid is hardly affected by changes in pressure and is changed only slightly when the temperature is varied. Increasing the temperature from 0°C to 50°C will reduce the density of a gas by about 18% if the gas is allowed to expand, whereas a 50°C rise in the temperature of water (0°C → 50°C) will change its density by less than 0.2%.

The density of a gas at any temperature and pressure can be determined by calculating the mass of gas present in 1 L. At STP, in particular, the density can be calculated by multiplying the molar mass of the gas by 1 mol/22.4 L:

$$d_{\text{STP}} = (\text{molar mass})\left(\frac{1\text{ mol}}{22.4\text{ L}}\right)$$

$$\text{molar mass} = (d_{\text{STP}})\left(\frac{22.4\text{ L}}{1\text{ mol}}\right)$$

TABLE 12.4 lists the densities of some common gases at STP.

TABLE 12.4 Density of Common Gases at STP

Gas	Molar mass (g/mol)	Density (g/L at STP)
H_2	2.016	0.0900
CH_4	16.04	0.716
NH_3	17.03	0.760
C_2H_2	26.04	1.16
HCN	27.03	1.21
CO	28.01	1.25
N_2	28.02	1.25
air	**(28.9)**	**(1.29)**
O_2	32.00	1.43
H_2S	34.09	1.52
HCl	36.46	1.63
F_2	38.00	1.70
CO_2	44.01	1.96
C_3H_8	44.09	1.97
O_3	48.00	2.14
SO_2	64.07	2.86
Cl_2	70.90	3.17

ENHANCED EXAMPLE 12.16

Calculate the density of Cl_2 at STP.

SOLUTION

Read To find the density, we need the molar mass of Cl_2. Using the periodic table, we find it to be 70.90 g/mol.

Plan The unit for density for gases is g/L.

Solution map: g/mol → g/L

1 mol = 22.4 L, so the conversion factor is 1 mol /22.4 L.

Calculate $\left(\frac{70.90\text{ g}}{1\text{ mol}}\right)\left(\frac{1\text{ mol}}{22.4\text{ L}}\right) = 3.165\text{ g/L}$

Check Compare the answer 3.165 g/L to Cl_2 density in Table 12.4.

PRACTICE 12.10

The molor mass of a gas is 20. g/mol. Calculate the density of the gas at STP.

PRACTICE 12.11

List these gases in order of increasing density (same T and P): H_2, CO, C_3H_8, NO_2, O_2, N_2O, POF_3, and B_2H_6.

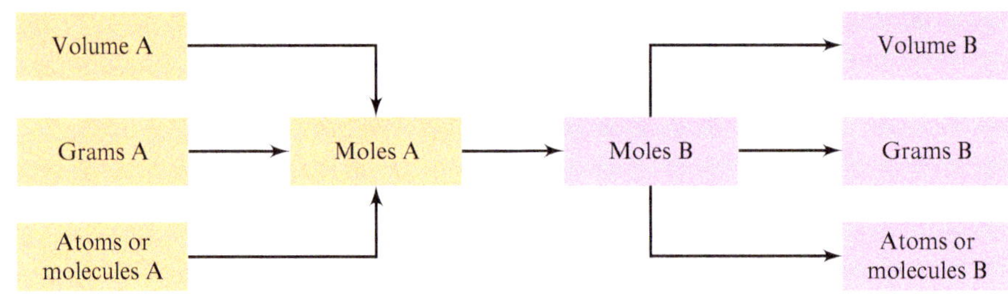

FIGURE 12.17 Summary of the primary conversions involved in stoichiometry. The conversion for volumes of gases is included.

12.9 Gas Stoichiometry

LEARNING OBJECTIVE Solve stoichiometric problems involving gases.

Mole–Volume and Mass–Volume Calculations

CHECK YOUR UNDERSTANDING

12.15 Conversion Factor Units Used for Stoichiometry Calculations

12.16 Conversions from Volume to Moles

WileyPLUS

Stoichiometric problems involving gas volumes can be solved by the general mole-ratio method outlined in Chapter 9 (summarized in FIGURE 12.17). The factors 1 mol/22.4 L and 22.4 L/1 mol are used for converting volume to moles and moles to volume, respectively. These conversion factors are used under the assumption that the gases are at STP and that they behave as ideal gases. In actual practice, gases are measured at other than STP conditions, and the volumes are converted to STP for stoichiometric calculations. The following are examples of typical problems involving gases in chemical reactions.

ONLINE LEARNING MODULE

Gas Stoichiometry-Mole–Volume

WileyPLUS

ENHANCED EXAMPLE 12.17 WileyPLUS

What volume of oxygen (at STP) can be formed from 0.500 mol of potassium chlorate?

SOLUTION

Read • **Knowns:** $P = 1.00\text{ atm}$

$T = 273\text{ K}$

$n = 0.500\text{ mol KClO}_3$

Plan • We can use the balanced equation to solve this problem since moles of a gas are directly convertible to L at STP.

See Section 9.1 if you have forgotten the Problem-Solving Strategy.

$$2\,\text{KClO}_3(s) \longrightarrow 2\,\text{KCl}(s) + 3\,\text{O}_2(g)$$

Solution map: mol $KClO_3$ → mol O_2 → L O_2

We use our Problem-Solving Strategy for Stoichiometry Problems to find the moles of oxygen produced:

$$(0.500\,\cancel{\text{mol KClO}_3})\left(\frac{3\text{ mol O}_2}{2\,\cancel{\text{mol KClO}_3}}\right) = 0.750\text{ mol O}_2$$

Calculate • Now we can convert moles of O_2 to liters of O_2.

$$(0.750\,\cancel{\text{mol}}\text{ O}_2)\left(\frac{22.4\text{ L}}{1\,\cancel{\text{mol}}}\right) = 16.8\text{ L O}_2$$

ENHANCED EXAMPLE 12.18

How many grams of aluminum must react with sulfuric acid to produce 1.25 L of hydrogen gas at STP?

SOLUTION

Read **Knowns:** $V = 1.25\ L\ H_2$
$T = 273\ K$
$P = 1.00\ atm$

Plan We can use the balanced equation to solve this problem since moles of a gas are directly convertible to liters at STP.

$$2\ Al(s) + 3\ H_2SO_4(aq) \longrightarrow Al_2(SO_4)_3(aq) + 3\ H_2(g)$$

Solution map: $L\ H_2 \rightarrow mol\ H_2 \rightarrow mol\ Al \rightarrow g\ Al$

We first use the conversion factor of 1 mol/22.4 L to convert the liters H_2 to moles. Then we use our Problem-Solving Strategy for Stoichiometry Problems to find the moles of aluminum produced:

$$(1.25\ \cancel{L}\ \cancel{H_2})\left(\frac{1\ \cancel{mol}}{22.4\ \cancel{L}}\right)\left(\frac{2\ mol\ Al}{3\ \cancel{mol}\ \cancel{H_2}}\right) = 0.0372\ mol\ Al$$

Calculate Now we can convert moles of Al to grams of Al.

$$(0.0372\ \cancel{mol\ Al})\left(\frac{26.98\ g\ Al}{\cancel{mol\ Al}}\right) = 1.00\ g\ Al$$

CHECK YOUR UNDERSTANDING

12.17 Comparing Quantity (Moles) to Properties of a Gas

➡ WileyPLUS

ENHANCED EXAMPLE 12.19

What volume of hydrogen, collected at 30.°C and 700. torr, will be formed by reacting 50.0 g of aluminum with hydrochloric acid?

SOLUTION

Read **Knowns:** $T = 30.°C = 303\ K$
$P = 700.\ torr = 0.921\ atm$
$m = 50.0\ g$

Plan We need the balanced equation to determine the number of moles of Al and then the moles of H_2.

$$2\ Al(s) + 6\ HCl(aq) \rightarrow 2\ AlCl_3(aq) + 3\ H_2(g)$$

Solution map: $g\ Al \rightarrow mol\ Al \rightarrow mol\ H_2$

We first use the molar mass for Al to convert the g Al to moles. Then we use the mole ratio from the balanced equation to find the moles of hydrogen produced:

$$(50.0\ \cancel{g\ Al})\left(\frac{1\ \cancel{mol\ Al}}{26.98\ \cancel{g\ Al}}\right)\left(\frac{3\ mol\ H_2}{2\ \cancel{mol\ Al}}\right) = 2.78\ mol\ H_2$$

Now we can use the ideal gas law to complete the calculation:

$$PV = nRT$$

Calculate $$V = \frac{(2.78\ \cancel{mol}\ H_2)\left(0.0821\dfrac{L \cdot \cancel{atm}}{\cancel{mol} \cdot \cancel{K}}\right)(303\ \cancel{K})}{(0.921\ \cancel{atm})} = 75.1\ L\ H_2$$

PRACTICE 12.12

If 10.0 g of sodium peroxide (Na_2O_2) react with water to produce sodium hydroxide and oxygen, how many liters of oxygen will be produced at 20.°C and 750. torr?

$$2\ Na_2O_2(s) + 2\ H_2O(l) \longrightarrow 4\ NaOH(aq) + O_2(g)$$

Volume–Volume Calculations

When all substances in a reaction are in the gaseous state, simplifications in the calculation can be made. These are based on Avogadro's law, which states that gases under identical conditions of temperature and pressure contain the same number of molecules and occupy the same volume. Under the standard conditions of temperature and pressure, the volumes of gases reacting are proportional to the numbers of moles of the gases in the balanced equation. Consider the reaction:

$H_2(g)$	+	$Cl_2(g)$	$\longrightarrow$	$2\ HCl(g)$
1 mol		1 mol		2 mol
1 volume		1 volume		2 volumes
Y volume		*Y* volume		2 *Y* volumes

Remember: For gases at the same *T* and *P*, equal volumes contain equal numbers of particles.

This statement is true because these volumes are equivalent to the number of reacting moles in the equation. Therefore, *Y* volume of H_2 will combine with *Y* volume of Cl_2 to give 2 *Y* volumes of HCl. For example, 100 L of H_2 react with 100 L of Cl_2 to give 200 L of HCl.

CHECK YOUR UNDERSTANDING

12.18 Volume of a Gas Used for Stoichiometry Calculations

➡ WileyPLUS

For reacting gases at constant temperature and pressure, volume–volume relationships are the same as mole–mole relationships.

ONLINE LEARNING MODULE

Gas Stoichiometry–Volume to Volume

➡ WileyPLUS

EXAMPLE 12.20

What volume of oxygen will react with 150. L of hydrogen to form water vapor? What volume of water vapor will be formed?

SOLUTION

Assume that both reactants and products are measured at standard conditions. Calculate by using reacting volumes:

$2\ H_2(g)$	+	$O_2(g)$	$\longrightarrow$	$2\ H_2O(g)$
2 mol		1 mol		2 mol
2 volumes		1 volume		2 volumes
150. L		75 L		150. L

For every two volumes of H_2 that react, one volume of O_2 reacts and two volumes of $H_2O(g)$ are produced:

$$(150.\ L\ H_2)\left(\frac{1\ \text{volume}\ O_2}{2\ \text{volumes}\ H_2}\right) = 75.0\ L\ O_2$$

$$(150.\ L\ H_2)\left(\frac{2\ \text{volumes}\ H_2O}{2\ \text{volumes}\ H_2}\right) = 150.\ L\ H_2O$$

EXAMPLE 12.21

The equation for the preparation of ammonia is

$$3\,H_2(g) + N_2(g) \xrightarrow{400°C} 2\,NH_3(g)$$

Assuming that the reaction goes to completion, determine the following:

(a) What volume of H_2 will react with 50.0 L of N_2?
(b) What volume of NH_3 will be formed from 50.0 L of N_2?
(c) What volume of N_2 will react with 100. mL of H_2?
(d) What volume of NH_3 will be produced from 100. mL of H_2?
(e) If 600. mL of H_2 and 400. mL of N_2 are sealed in a flask and allowed to react, what amounts of H_2, N_2, and NH_3 are in the flask at the end of the reaction?

SOLUTION

The answers to parts (a)–(d) are shown in the boxes and can be determined from the equation by inspection, using the principle of reacting volumes:

	$3\,H_2(g)$	+	$N_2(g)$	$\longrightarrow$	$2\,NH_3(g)$
	3 volume		1 volume		2 volumes
(a)	**[150. L]**		50.0 L		
(b)			50.0 L		**[100. L]**
(c)	100. mL		**[33.3 mL]**		
(d)	100. mL				**[66.7 mL]**

(e) Volume ratio from the equation $= \dfrac{3 \text{ volumes } H_2}{1 \text{ volume } N_2}$

$$\text{Volume ratio used} = \frac{600.\text{ mL } H_2}{400.\text{ mL } N_2} = \frac{3 \text{ volumes } H_2}{2 \text{ volumes } N_2}$$

Comparing these two ratios, we see that an excess of N_2 is present in the gas mixture. Therefore, the reactant limiting the amount of NH_3 that can be formed is H_2:

$$\underset{600.\text{ mL}}{3\,H_2(g)} + \underset{200.\text{ mL}}{N_2(g)} \longrightarrow \underset{400.\text{ mL}}{2\,NH_3(g)}$$

To have a 3:1 ratio of volumes reacting, 600. mL of H_2 will react with 200. mL of N_2 to produce 400. mL of NH_3, leaving 200. mL of N_2 unreacted. At the end of the reaction, the flask will contain 400. mL of NH_3 and 200. mL of N_2.

PRACTICE 12.13

What volume of oxygen will react with 15.0 L of propane (C_3H_8) to form carbon dioxide and water? What volume of carbon dioxide will be formed? What volume of water vapor will be formed?

$$C_3H_8(g) + 5\,O_2 \longrightarrow 3\,CO_2(g) + 4\,H_2O(g)$$

CHAPTER 12 REVIEW

12.1 Properties of Gases

KEY TERMS
pressure
atmospheric pressure
barometer
1 atmosphere

- Gases:
  - Particles are relatively far apart.
  - Particles are very mobile.
  - Gases take the shape and volume of the container.
  - Gases are easily compressible.
- Pressure is force per unit area.
- Pressure of the atmosphere is measured by using a barometer:
  - Units of pressure include:
    - Atmosphere (atm) = 760 mm Hg.
    - Pascal, 1 atm = 101,325 Pa = 101.3 kPa.
    - Torr, 1 atm = 760 torr.
- Pressure is directly related to the number of molecules in the sample.
- Pressure is directly related to the Kelvin temperature of the sample.

12.2 Boyle's Law

KEY TERM
Boyle's law

- At constant temperature, the volume of a gas is inversely proportional to the pressure of the gas:

$$V \propto \frac{1}{P} \quad \text{or} \quad P_1V_1 = P_2V_2$$

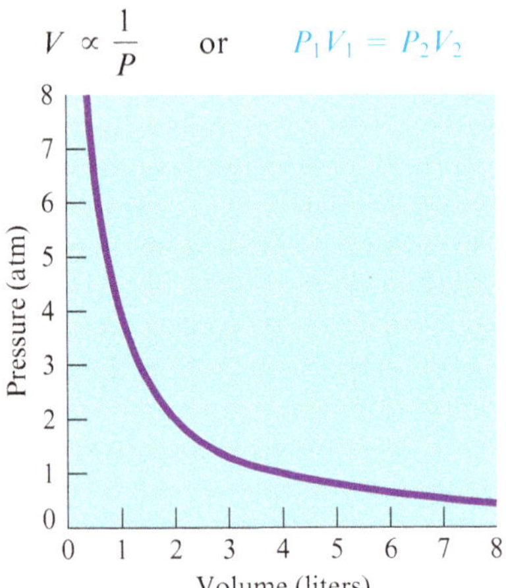

12.3 Charles' Law

KEY TERMS
absolute zero
Charles' law

- At constant pressure, the volume of a gas is directly proportional to the absolute temperature of the gas.

$$V \propto T \quad \text{or} \quad \frac{V_1}{T_1} = \frac{V_2}{T_2}$$

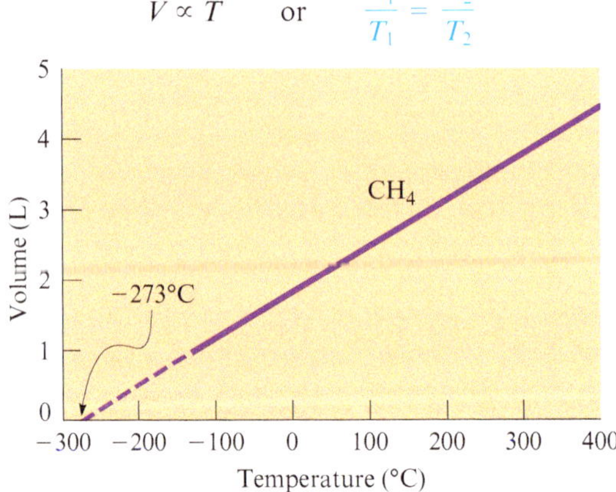

12.4 Avogadro's Law

KEY TERMS
Gay-Lussac's law of combining volumes
Avogadro's law

- Gay-Lussac's law of combining volumes states that when measured at constant T and P, the ratios of the volumes of reacting gases are small whole numbers.
- Avogadro's law states that equal volumes of different gases at the same T and P contain the same number of particles.

12.5 Combined Gas Law

KEY TERMS

standard conditions
standard temperature and pressure (STP)
molar volume

- The P, V, and T relationship for gases can be expressed in a single equation known as the combined gas law:

$$\frac{P_1V_1}{T_1} = \frac{P_2V_2}{T_2}$$

- One mole of any gas occupies 22.4 L at STP.

12.6 Ideal Gas Law

KEY TERMS

ideal gas law
kinetic-molecular theory (KMT)
ideal gas

- The ideal gas law combines all the variables involving gases into a single expression:

$$PV = nRT$$

- R is the ideal gas constant and can be expressed as

$$R = 0.0821\frac{\text{L} \cdot \text{atm}}{\text{mol} \cdot \text{K}}$$

- Kinetic-molecular theory assumptions:
  - Gases are tiny particles with no attraction for each other.
  - The distance between particles is great compared to the size of the particles.
  - Gas particles move in straight lines.
  - No energy is lost in particle collisions (perfectly elastic collisions).
  - The average kinetic energy for particles is the same for all gases at the same temperature and pressure.
- A gas that follows the KMT is an ideal gas.
- The kinetic energy of a particle is expressed as $\text{KE} = \frac{1}{2}mv^2$.
- Real gases show deviation from the ideal gas law:
  - Deviations occur at:
    - High pressure
    - Low temperature
  - These deviations occur because:
    - Molecules have finite volumes.
    - Molecules have intermolecular attractions.

12.7 Dalton's Law of Partial Pressures

KEY TERMS

Dalton's law of partial pressures
partial pressure

- The total pressure of a mixture of gases is the sum of the partial pressures of the component gases in the mixture.
- When a gas is collected over water, the pressure of the collected gas is the difference between the atmospheric pressure and the vapor pressure of water at that temperature.

12.8 Density of Gases

- The density of a gas is usually expressed in units of g/L.
- Temperature and pressure are given for a density of a gas since the volume of the gas depends on these conditions.

12.9 Gas Stoichiometry

- Stoichiometry problems involving gases are solved the same way as other stoichiometry problems. See graphic below.

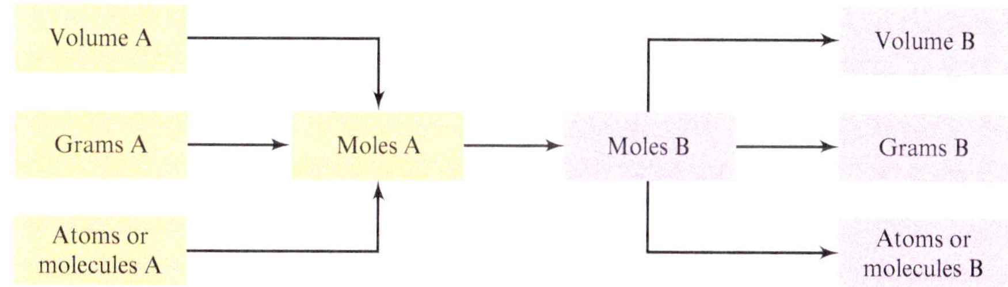

- For reacting gases (constant T and P) volume–volume relationships are the same as mole–mole relationships.

REVIEW QUESTIONS

1. What is meant by pressure (for a gas)?
2. How does the air pressure inside the balloon shown in Figure 12.3 compare with the air pressure outside the balloon? Explain.
3. According to Table 12.1, what two gases are the major constituents of dry air?
4. How does the pressure represented by 1 torr compare in magnitude to the pressure represented by 1 mm Hg? See Table 12.2.
5. In which container illustrated in Figure 12.6 are the molecules of gas moving faster? Assume both gases to be hydrogen.
6. In Figure 12.7, what gas pressure corresponds to a volume of 4 L?
7. How do the data illustrated in Figure 12.7 verify Boyle's law?
8. What effect would you observe in Figure 12.10 if T_2 were lower than T_1?
9. Explain how the reaction

$$N_2(g) + O_2(g) \xrightarrow{\Delta} 2\ NO(g)$$

proves that nitrogen and oxygen are diatomic molecules.
10. What is the reason for comparing gases to STP?
11. What are the four parameters used to describe the behavior of a gas?
12. What are the characteristics of an ideal gas?
13. How is Boyle's law related to the ideal gas law?
14. How is Charles' law related to the ideal gas law?
15. What are the basic assumptions of the kinetic-molecular theory?
16. Arrange the following gases, all at standard temperature, in order of increasing relative molecular velocities: H_2, CH_4, Rn, N_2, F_2, and He. What is your basis for determining the order?
17. List, in descending order, the average kinetic energies of the molecules in Question 16.
18. Under what condition of temperature, high or low, is a gas least likely to exhibit ideal behavior? Explain.
19. Under what condition of pressure, high or low, is a gas least likely to exhibit ideal behavior? Explain.
20. How does the kinetic-molecular theory account for the behavior of gases as described by
    (a) Boyle's law?
    (b) Charles' law?
    (c) Dalton's law of partial pressures?
21. Is the conversion of oxygen to ozone an exothermic or endothermic reaction? How do you know?
22. Write formulas for an oxygen atom, an oxygen molecule, and an ozone molecule. How many electrons are in an oxygen molecule?
23. In the diagram shown in Figure 12.16, is the pressure of the oxygen plus water vapor inside the bottle equal to, greater than, or less than the atmospheric pressure outside the bottle? Explain.
24. List five gases in Table 12.4 that are more dense than air. Explain the basis for your selection.
25. Compare, at the same temperature and pressure, equal volumes of H_2 and O_2 as to the following:
    (a) number of molecules
    (b) mass
    (c) number of moles
    (d) average kinetic energy of the molecules
    (e) density
26. When constant pressure is maintained, what effect does heating a mole of N_2 gas have on
    (a) its density?
    (b) its mass?
    (c) the average kinetic energy of its molecules?
    (d) the average velocity of its molecules?
    (e) the number of N_2 molecules in the sample?

Most of the exercises in this chapter are available for assignment via the online homework management program, WileyPLUS (www.wileyplus.com) All exercises with blue numbers have answers in Appendix 5.

PAIRED EXERCISES

1. Fill in the table below with the missing information:

	torr	in. Hg	kilopascals (kPa)
(a)		30.2	
(b)	752		
(c)			99.3

2. Fill in the table below with the missing information:

	mm Hg	lb/in.^{2}	atmospheres (atm)
(a)	789		
(b)		32	
(c)			1.4

3. Perform the following pressure conversions:
    (a) Convert 953 torr to kPa.
    (b) Convert 2.98 kPa to atm.
    (c) Convert 2.77 atm to mm Hg.
    (d) Convert 372 torr to atm.
    (e) Convert 2.81 atm to cm Hg.

4. Perform the following pressure in conversions:
    (a) Convert 649 torr to kPa.
    (b) Convert 5.07 kPa to atm.
    (c) Convert 3.64 atm to mm Hg.
    (d) Convert 803 torr to atm.
    (e) Convert 1.08 atm to cm Hg.

5. This scuba diver watch shows the air pressure in the diver's scuba tank as 1920 lb/in.^{2}. Convert this pressure to
(a) atm (b) torr (c) kPa

courtesy Aeris

6. A pressure sensor shows tire pressure as 31 lb/in.^{2}. Convert this pressure to
(a) atm (b) torr (c) kPa

© Thomas Acop/iStockphoto

7. The internal pressure on a bottle of soda is 5570 torr. The bottle cap is designed to withstand a pressure of 7.0 atm.
(a) Calculate the pressure in atm.
(b) Calculate the pressure in psi.
(c) Calculate the pressure in kPa.
(d) Is the bottle cap at risk of failing?

8. The cabin pressure has just dropped on a fighter jet. The pressure gauge says 631 torr. Anytime the cabin pressure drops below 0.850 atm, the pilot is required to put on an oxygen mask.
(a) Calculate the pressure in atm.
(b) Calculate the pressure in psi.
(c) Calculate the pressure in kPa.
(d) Does the pilot need to put on an oxygen mask?

9. A sample of a gas occupies a volume of 725 mL at 825 torr. At constant temperature, what will be the new pressure (torr) when the volume changes to the following:
(a) 283 mL
(b) 2.87 L

10. A sample of a gas occupies a volume of 486 mL at 508 torr. At constant temperature, what will be the new pressure (torr) when the volume changes to the following:
(a) 185 mL
(b) 6.17 L

11. A sample of methane gas, CH_4, occupies a volume (L) 58.2 L at a pressure of 7.25 atm. What volume will the gas occupy if the pressure is lowered to 2.03 atm?

12. A sample of nitrous oxide gas, N_2O, occupies a volume of 832 L at a pressure of 0.204 atm. What volume (L) will the gas occupy if the pressure is increased to 8.02 atm?

13. A sample of CO_2 gas occupies a volume of 125 mL at 21°C. If pressure remains constant, what will be the new volume if temperature changes to:
(a) −5°C (b) 95°F (c) 1095 K

14. A sample of CH_4 gas occupies a volume of 575 mL at −25°C. If pressure remains constant, what will be the new volume if temperature changes to:
(a) 298 K (b) 32°F (c) 45°C

15. A sample of a gas occupies a volume of 1025 mL at 75°C and 0.75 atm. What will be the new volume if temperature decreases to 35°C and pressure increases to 1.25 atm?

16. A sample of a gas occupies a volume of 25.6 L at 19°C and 678 torr. What will be the new volume if temperature increases to 35°C and pressure decreases to 595 torr?

17. An expandable balloon contains 1400. L of He at 0.950 atm pressure and 18°C. At an altitude of 22 miles (temperature 2.0°C and pressure 4.0 torr), what will be the volume of the balloon?

18. A gas occupies 22.4 L at 2.50 atm and 27°C. What will be its volume at 1.50 atm and −5.00°C?

19. A 775-mL sample of NO_2 gas is at STP. If the volume changes to 615 mL and the temperature changes to 25°C, what will be the new pressure?

20. A 2.5-L sample of SO_3 is at 19°C and 1.5 atm. What will be the new temperature in °C if the volume changes to 1.5 L and the pressure to 765 torr?

21. A sample of O_2 gas was collected over water at 23°C and 772 torr. What is the partial pressure of the O_2?
(Refer to Appendix 5 for the vapor pressure of water.)

22. A sample of CH_4 gas was collected over water at 29°C and 749 mm Hg. What is the partial pressure of the CH_4?
(Refer to Appendix 5 for the vapor pressure of water.)

23. A mixture contains H_2 at 600. torr pressure, N_2 at 200. torr pressure, and O_2 at 300. torr pressure. What is the total pressure of the gases in the system?

24. A mixture contains H_2 at 325 torr pressure, N_2 at 475 torr pressure, and O_2 at 650. torr pressure. What is the total pressure of the gases in the system?

25. A sample of methane gas, CH_4 was collected over water at 25.0°C and 720. torr. The volume of the wet gas is 2.50 L. What will be the volume of the dry methane at standard pressure?

26. A sample of propane gas, C_3H_8 was collected over water at 22.5°C and 745 torr. The volume of the wet gas is 1.25 L. What will be the volume of the dry propane at standard pressure?

27. A package of pretzels contains 0.143 L of air on the ground where the temperature is 27°C and the atmospheric pressure is 760 torr. If the temperature decreases to 18°C and the pressure decreases to 635 torr when the plane reaches its cruising altitude, what volume of air will be in the bag of pretzels?

28. Fish contain a collapsible swim bladder containing air that they use to help them maintain buoyancy so they can swim at any level and not sink or float. If a swordfish has a fish bladder with a volume of 32.7 L at sea level where the pressure is 755 torr and the temperature is 28°C, what will the volume of the swim bladder be if the swordfish dives under the water to catch some mackerel where the pressure is 2577 torr and the temperature is 12°C?

29. Explain why a sealed bag of chips expands when you take it to a higher altitude.

30. Explain why you need to add more air to your tires when the weather turns colder in the winter.

31. What volume will each of the following occupy at STP?
(a) 6.02×10^{23} molecules of CO_2
(b) 2.5 mol CH_4
(c) 12.5 g oxygen

32. What volume will each of the following occupy at STP?
(a) 1.80×10^{24} molecules of SO_3
(b) 7.5 mol C_2H_6
(c) 25.2 g chlorine

33. How many grams of NH_3 are present in 725 mL of the gas at STP?

34. How many grams of C_3H_6 are present in 945 mL of the gas at STP?

35. How many liters of CO_2 gas at STP will contain 1025 molecules?

36. How many molecules of CO_2 gas at STP are present in 10.5 L?

37. What volume would result if a balloon were filled with 10.0 grams of chlorine gas at STP?

38. How many grams of methane gas were used to fill a balloon to a volume of 3.0 L at STP?

39. At 22°C and 729 torr pressure, what will be the volume of 75 mol of NH_3 gas?

40. At 39°C and 1.5 atm, what will be the volume of 105 mol of CH_4?

41. How many moles of O_2 are contained in 5.25 L at 26°C and 1.2 atm?

42. How many moles of CO_2 are contained in 9.55 L at 45°C and 752 torr?

43. At what Kelvin temperature will 25.2 mol of Xe occupy a volume of 645 L at a pressure of 732 torr?

44. At what Kelvin temperature will 37.5 mol of Ar occupy a volume of 725 L at a pressure of 675 torr?

45. Calculate the density of each of the following gases at STP:
(a) He (b) HF (c) C_3H_6 (d) CCl_2F_2

46. Calculate the density of each of the following gases at STP:
(a) Rn (b) NO_2 (c) SO_3 (d) C_2H_4

47. Calculate the density of each of the following gases:
(a) NH_3 at 25°C and 1.2 atm
(b) Ar at 75°C and 745 torr

48. Calculate the density of each of the following gases:
(a) C_2H_4 at 32°C and 0.75 atm
(b) He at 57°C and 791 torr

49. In the lab, students decomposed a sample of calcium carbonate by heating it over a Bunsen burner and collected carbon dioxide according to the following equation.

$$CaCO_3(s) \longrightarrow CaO(s) + CO_2(g)$$

(a) How many mL of carbon dioxide gas were generated by the decomposition of 6.24 g of calcium carbonate at STP?
(b) If 52.6 L of carbon dioxide at STP were needed, how many moles of calcium carbonate would be required?

50. In the lab, students generated and collected hydrogen gas according to the following equation:

$$Mg(s) + 2HCl(aq) \longrightarrow MgCl_2(aq) + H_2(g)$$

(a) How many mL of hydrogen gas at STP were generated from 42.9 g of magnesium metal?
(b) If 825 mL of hydrogen gas at STP were needed, how many moles of HCl would be required?

51. Consider the following equation:

$$4\ NH_3(g) + 5\ O_2(g) \longrightarrow 4\ NO(g) + 6\ H_2O(g)$$

(a) How many liters of oxygen are required to react with 2.5 L NH_3? Both gases are at STP.
(b) How many grams of water vapor can be produced from 25 L NH_3 if both gases are at STP?
(c) How many liters of NO can be produced when 25 L O_2 are reacted with 25 L NH_3? All gases are at the same temperature and pressure.

52. Consider the following equation:

$$C_3H_8(g) + 5\ O_2(g) \longrightarrow 3\ CO_2(g) + 4\ H_2O(g)$$

(a) How many liters of oxygen are required to react with 7.2 L C_3H_8? Both gases are at STP.
(b) How many grams of CO_2 will be produced from 35 L C_3H_8 if both gases are at STP?
(c) How many liters of water vapor can be produced when 15 L C_3H_8 are reacted with 15 L O_2? All gases are at the same temperature and pressure.

53. Oxygen gas can be generated by the decomposition of potassium chlorate according to the following equation:

$$2\ KClO_3(s) \longrightarrow 2\ KCl(s) + 3\ O_2(g)$$

How many liters of oxygen at STP will be produced when 0.525 kg of KCl is also produced?

54. When glucose is burned in a closed container, carbon dioxide gas and water are produced according to the following equation:

$$C_6H_{12}O_6(s) + 6\ O_2(g) \longrightarrow 6\ CO_2(g) + 6\ H_2O(l)$$

How many liters of CO_2 at STP can be produced when 1.50 kg of glucose are burned?

ADDITIONAL EXERCISES

55. How is it possible that 1 mole of liquid water occupies a volume of 18 mL, but 1 mole of gaseous water occupies a volume of 22.4 L? (Figure 12.2.)

56. Why are gases more compressible than liquids?

57. Look at the three pictures below. If each of these pictures represents a gas at 85°C, which of the gases is exerting the greatest pressure?

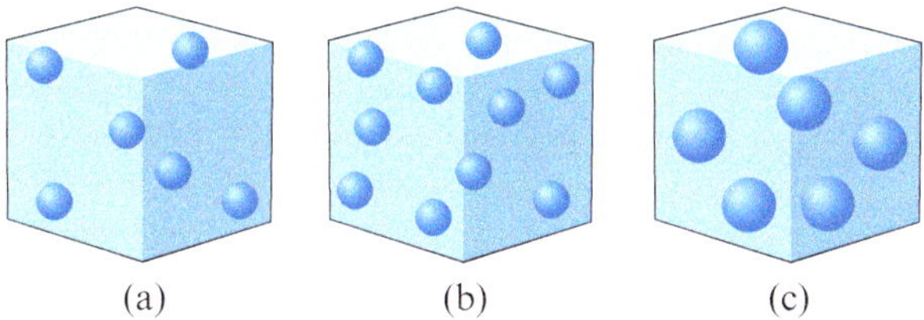

58. The balloon below is filled with helium at a temperature of 37°C. If the balloon is placed into a freezer with a temperature of −20°C, which diagram best represents the molecules inside the balloon?

59. You have a 10-L container filled with 0.5 mol of O_2 gas at a temperature of 30.°C with a pressure of 945 torr.

(a) What will happen to the pressure if the container size is doubled while keeping the temperature and number of moles constant?

(b) What will happen to the pressure when the temperature is doubled while keeping the size of the container and the number of moles constant?

(c) What will happen to the pressure when the amount of O_2 gas is cut in half while keeping the size of the container and the temperature constant?

(d) What will happen to the pressure if 1 mole of N_2 gas is added to the container while keeping the temperature and size of the container the same?

60. Sketch a graph to show each of the following relationships:

(a) P vs. V at constant temperature and number of moles

(b) T vs. V at constant pressure and number of moles

(c) n vs. V at constant temperature and pressure

61. For gases the relationship among the factors P, V, T, and n can be used to solve many different problems. Using the factors and the graphs below, complete the table.

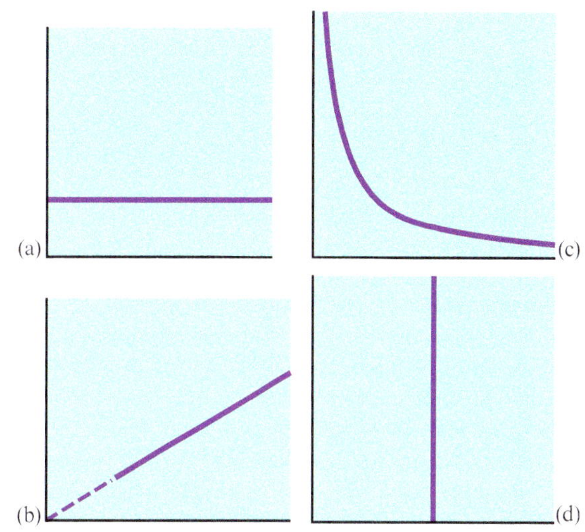

Law	Factors that are constant	Factors that are variable	Graph showing the relationship of variable factors
Boyle's law			
Charles' law			
Avogadro's law			

62. Why is it dangerous to incinerate an aerosol can?

63. What volume does 1 mol of an ideal gas occupy at standard conditions?

64. Which of these occupies the greatest volume?

(a) 0.2 mol of chlorine gas at 48°C and 80 cm Hg

(b) 4.2 g of ammonia at 0.65 atm and −11°C

(c) 21 g of sulfur trioxide at 55°C and 110 kPa

65. Which of these contains the largest number of molecules?

(a) 1.00 L of CH_4 at STP

(b) 3.29 L of N_2 at 952 torr and 235°F

(c) 5.05 L of Cl_2 at 0.624 atm and 0°C

66. Which of these has the greatest density?

(a) SF_6 at STP

(b) C_2H_6 at room conditions

(c) He at −80°C and 2.15 atm

67. For each of the following pairs, predict which sample would have the greater density.

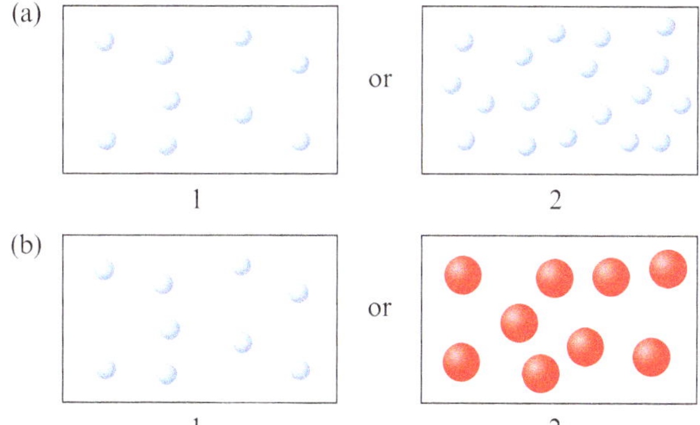

68. A chemist carried out a chemical reaction that produced a gas. It was found that the gas contained 80.0% carbon and 20.0% hydrogen. It was also noticed that 1500 mL of the gas at STP had a mass of 2.01 g.
(a) What is the empirical formula of the compound?
(b) What is the molecular formula of the compound?
(c) What Lewis structure fits this compound?

69. Three gases were added to the same 2.0-L container. The total pressure of the gases was 790 torr at room temperature (25.0°C). If the mixture contained 0.65 g of oxygen gas, 0.58 g of carbon dioxide, and an unknown amount of nitrogen gas, determine the following:
(a) the total number of moles of gas in the container
(b) the number of grams of nitrogen in the container
(c) the partial pressure of each gas in the mixture

70. When carbon monoxide and oxygen gas react, carbon dioxide results. If 500. mL of O_2 at 1.8 atm and 15°C are mixed with 500. mL of CO at 800 mm Hg and 60°C, how many milliliters of CO_2 at STP could possibly result?

71. One of the methods for estimating the temperature at the center of the sun is based on the ideal gas law. If the center is assumed to be a mixture of gases whose average molar mass is 2.0 g/mol, and if the density and pressure are 1.4 g/cm^3 and 1.3×10^9 atm, respectively, calculate the temperature.

72. A soccer ball of constant volume 2.24 L is pumped up with air to a gauge pressure of 13 lb/in.^{2} at 20.0°C. The molar mass of air is about 29 g/mol.
(a) How many moles of air are in the ball?
(b) What mass of air is in the ball?
(c) During the game, the temperature rises to 30.0°C. What mass of air must be allowed to escape to bring the gauge pressure back to its original value?

73. A balloon will burst at a volume of 2.00 L. If it is partially filled at 20.0°C and 65 cm Hg to occupy 1.75 L, at what temperature will it burst if the pressure is exactly 1 atm at the time that it bursts?

74. Given a sample of a gas at 27°C, at what temperature would the volume of the gas sample be doubled, the pressure remaining constant?

75. A gas sample at 22°C and 740 torr pressure is heated until its volume is doubled. What pressure would restore the sample to its original volume?

76. A gas occupies 250. mL at 700. torr and 22°C. When the pressure is changed to 500. torr, what temperature (°C) is needed to maintain the same volume?

77. The tires on an automobile were filled with air to 30. psi at 71.0°F. When driving at high speeds, the tires become hot. If the tires have a bursting pressure of 44 psi, at what temperature (°F) will the tires "blow out"?

78. What pressure will 800. mL of a gas at STP exert when its volume is 250. mL at 30°C?

79. How many L of SO_2 gas at STP will contain 9.14 g of sulfur dioxide gas?

80. How many gas molecules are present in 600. mL of N_2O at 40°C and 400. torr pressure? How many atoms are present? What would be the volume of the sample at STP?

81. In the future when commercial spacecraft have become safer, you decide to take a trip to a remote planet that has lots of laughing gas (nitrous oxide, N_2O). On this planet the atmosphere is composed of oxygen gas (346 atm), nitrous oxide gas (93 atm), and nitrogen gas. If the total pressure on this planet is 950 atm, what is the pressure due to nitrogen gas?

82. An automobile tire has a bursting pressure of 60 lb/in.^{2}. The normal pressure inside the tire is 32 lb/in.^{2}. When traveling at high speeds, a tire can get quite hot. Assuming that the temperature of the tire is 25°C before running it, determine whether the tire will burst when the inside temperature gets to 212°F. Show your calculations.

83. If you prepared a barometer using water instead of mercury, how high would the column of water be at one atmosphere pressure? (Neglect the vapor pressure of water.)

84. How many moles of oxygen are in a 55-L cylinder at 27°C and 2.20×10^3 lb/in.^{2}?

85. How many moles of Cl_2 are in one cubic meter (1.00 m^3) of Cl_2 gas at STP?

86. At STP, 560. mL of a gas have a mass of 1.08 g. What is the molar mass of the gas?

87. A gas has a density at STP of 1.78 g/L. What is its molar mass?

88. Using the ideal gas law, $PV = nRT$, calculate the following:
(a) the volume of 0.510 mol of H_2 at 47°C and 1.6 atm pressure
(b) the number of grams in 16.0 L of CH_4 at 27°C and 600. torr pressure
(c) the density of CO_2 at 4.00 atm pressure and 20.0°C
(d) the molar mass of a gas having a density of 2.58 g/L at 27°C and 1.00 atm pressure.

89. Acetylene (C_2H_2) and hydrogen fluoride (HF) react to give difluoroethane:

$$C_2H_2(g) + 2\,HF(g) \longrightarrow C_2H_4F_2(g)$$

When 1.0 mol of C_2H_2 and 5.0 mol of HF are reacted in a 10.0-L flask, what will be the pressure in the flask at 0°C when the reaction is complete?

90. What volume of hydrogen at STP can be produced by reacting 8.30 mol of Al with sulfuric acid? The equation is

$$2\,Al(s) + 3\,H_2SO_4(aq) \longrightarrow Al_2(SO_4)_3(aq) + 3\,H_2(g)$$

91. A gas has a percent composition by mass of 85.7% carbon and 14.3% hydrogen. At STP the density of the gas is 2.50 g/L. What is the molecular formula of the gas?

92. Assume that the reaction

$$2\,CO(g) + O_2(g) \longrightarrow 2\,CO_2(g)$$

goes to completion. When 10. mol of CO and 8.0 mol of O_2 react in a closed 10.-L vessel,
(a) how many moles of CO, O_2 and CO_2 are present at the end of the reaction?
(b) what will be the total pressure in the flask at 0°C?

93. If 250 mL of O_2 measured at STP, are obtained by the decomposition of the $KClO_3$ in a 1.20-g mixture of KCl and $KClO_3$,

$$2\,KClO_3(s) \longrightarrow 2\,KCl(s) + 3\,O_2(g)$$

what is the percent by mass of $KClO_3$ in the mixture?

CHALLENGE EXERCISES

94. A glass containing 345 mL of a soft drink (a carbonated beverage) was left sitting out on a kitchen counter. If the CO_2 released at room temperature (25°C) and pressure (1 atm) occupies 1.40 L, at a minimum, what is the concentration (in ppm) of the CO_2 in the original soft drink? (Assume the density of the original soft drink is 0.965 g/mL.)

95. (a) If you filled up three balloons with equal volumes of hydrogen, argon, and carbon dioxide gas, all at the same temperature and pressure, which balloon would weigh the most? the least? Explain.

(b) If you filled up three balloons with equal masses of nitrogen, oxygen, and neon, all to the same volume at the same temperature, which would have the lowest pressure?

96. Air has a density of 1.29 g/L at STP. Calculate the density of air on Pikes Peak, where the pressure is 450 torr and the temperature is 17°C.

97. Consider the arrangement of gases shown below. If the valve between the gases is opened and the temperature is held constant, determine the following:

(a) the pressure of each gas

(b) the total pressure in the system

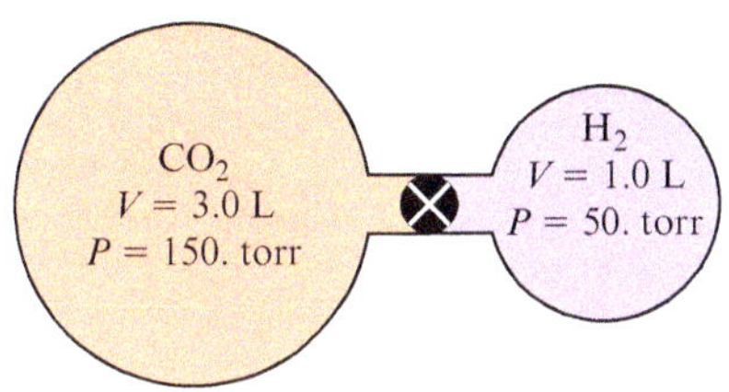

98. A steel cylinder contains 50.0 L of oxygen gas under a pressure of 40.0 atm and at a temperature of 25°C. What was the pressure in the cylinder during a storeroom fire that caused the temperature to rise 152°C? (Be careful!)

99. A balloon has a mass of 0.5 g when completely deflated. When it is filled with an unknown gas, the mass increases to 1.7 g. You notice on the canister of the unknown gas that it occupies a volume of 0.4478 L at a temperature of 50°C. You note the temperature in the room is 25°C. Identify the gas.

100. You are responsible for ensuring that the giant American eagle balloon stays inflated at the local Veterans Day parade. You inflate the balloon to a pressure of 976 torr using 5.27×10^5 mol of helium in the morning when the temperature is 12°C. At the end of the day the temperature increases to 31°C and 15.0% of the helium seeps out of the balloon.

(a) How many moles of air will be left in the balloon at the end of the day after 15.0% is lost?

(b) What will the pressure of the balloon be at the end of the day if the volume is unchanged?

101. A baker is making strawberry cupcakes and wants to ensure they are very light. To do this, he must add enough baking soda to increase the volume of the cupcakes by 55.0%. All of the carbon dioxide produced by the decomposition of the baking soda is incorporated into the cupcakes. The baking soda or sodium bicarbonate will react with citric acid in the batter according to the reaction

$$3\ NaHCO_3 + H_3C_6H_5O_7 \longrightarrow Na_3C_6H_5O_7 + 3\ H_2O + 3\ CO_2$$

which occurs with a 63.7% yield. How many grams of baking soda must be added to 1.32 L of cupcake batter at a baking temperature of 325°F and a pressure of 738 torr to achieve the correct consistency?

ANSWERS TO PRACTICE EXERCISES

12.1 (a) 851 torr, (b) 851 mm Hg, (c) 113 kPa

12.2 0.596 atm

12.3 4.92 L

12.4 84.7 L

12.5 861 K (588°C)

12.6 1.4×10^2 g/mol

12.7 0.953 mol

12.8 128 g/mol

12.9 518 mL

12.10 0.89 g/L

12.11 $H_2 < B_2H_6 < CO < O_2 < N_2O < C_3H_8 < NO_2 < POF_3$

12.12 1.55 L O_2

12.13 75.0 L O_2, 45.0 L CO_2, 60.0 L H_2O

13

Liquids

CHAPTER OUTLINE

Planet Earth, that magnificient blue sphere we enjoy viewing from space, is spectacular. Over 75% of Earth is covered with water. We are born from it, drink it, bathe in it, cook with it, enjoy its beauty in waterfalls and rainbows, and stand in awe of the majesty of icebergs. Water supports and enhances life.

In chemistry, water provides the medium for numerous reactions. The shape of the water molecule is the basis for hydrogen bonds. These bonds determine the unique properties and reactions of water. The tiny water molecule holds the answers to many of the mysteries of chemical reactions.

13.1 States of Matter: A Review

In the last chapter, we found that gases contain particles that are far apart, in rapid random motion, and essentially independent of each other. The kinetic-molecular theory, along with the ideal gas law, summarizes the behavior of most gases at relatively high temperatures and low pressures.

Solids are obviously very different from gases. Solids contain particles that are very close together; solids have a high density, compress negligibly, and maintain their shape regardless of container. These characteristics indicate large attractive forces between particles. The model for solids is very different from the one for gases.

Liquids, on the other hand, lie somewhere between the extremes of gases and solids. Liquids contain particles that are close together; liquids are essentially incompressible and have definite volume. These properties are very similar to those of solids. But liquids also take the shape of their containers; this is closer to the model of a gas.

Although liquids and solids show similar properties, they differ tremendously from gases (FIGURE 13.1). No simple mathematical relationship, like the ideal gas law, works well for liquids or solids. Instead, these models are directly related to the forces of attraction between molecules. With these general statements in mind, let's consider some specific properties of liquids.

CHECK YOUR UNDERSTANDING

13.1 Surface Tension

13.2 Capillary Action

13.3 Water Striders Can "Walk on Water"

WileyPLUS

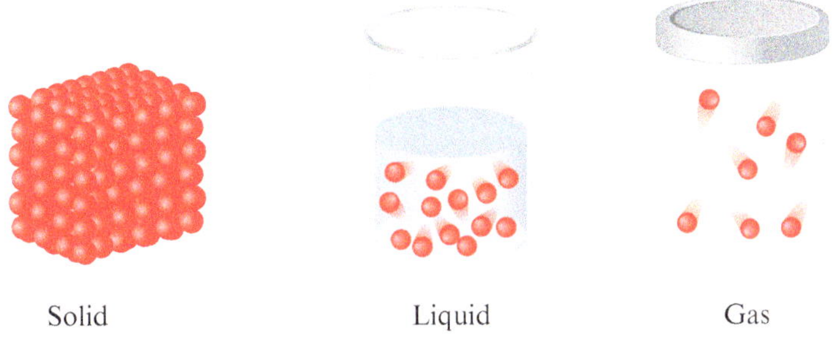

FIGURE 13.1 The three states of matter. Solid—water molecules are held together rigidly and are very close to each other. Liquid—water molecules are close together but are free to move around and slide over each other. Gas—water molecules are far apart and move freely and randomly.

13.2 Properties of Liquids

LEARNING OBJECTIVE

Explain why liquids tend to form drops and explain the process of evaporation and its relationship to vapor pressure.

KEY TERM

surface tension
capillary action
meniscus
evaporation
vaporization
sublimation
condensation
vapor pressure
volatile

Surface Tension

Have you ever observed water and mercury in the form of small drops? These liquids form drops because liquids have *surface tension.* A droplet of liquid that is not falling or under the influence of gravity (as on the space station) will form a sphere. Spheres minimize the ratio of surface area to volume. The molecules within the liquid are attracted to the surrounding liquid molecules, but at the liquid's surface, the attraction is nearly all inward. This pulls the surface into a spherical shape. The resistance of a liquid to an increase in its surface area is called the **surface tension** of the liquid. Substances with large attractive forces between molecules have high surface tensions. The effect of surface tension in water is illustrated by floating a needle on the surface of still water. Other examples include a water strider walking across a calm pond (FIGURE 13.2) and water beading on a freshly waxed car. Surface tension is temperature dependent, decreasing with increasing temperature.

Charles D. Winters/Photo Researchers

Mercury droplets.

Liquids also exhibit a phenomenon called **capillary action**, the spontaneous rising of a liquid in a narrow tube. This action results from the *cohesive forces* within the liquid and the *adhesive forces* between the liquid and the walls of the container. If the forces between the liquid and the container are greater than those within the liquid itself, the liquid will climb the walls of the container.

FIGURE 13.2 A water strider skims the surface of the water as a result of surface tension. At the molecular level, the surface tension results from the net attraction of the water molecules toward the liquid below. In the interior of the water, the forces are balanced in all directions.

For example, consider the California sequoia, a tree that reaches over 200 feet in height. Although water rises only 33 feet in a glass tube (under atmospheric pressure), capillary action causes water to rise from the sequoia's roots to all its parts.

The meniscus in liquids is further evidence of cohesive and adhesive forces. When a liquid is placed in a glass cylinder, the surface of the liquid shows a curve called the **meniscus** (FIGURE 13.3). The concave shape of water's meniscus shows that the adhesive forces between the glass and water are stronger than the cohesive forces within the water. In a nonpolar substance such as mercury, the meniscus is convex, indicating that the cohesive forces within mercury are greater than the adhesive forces between the glass wall and the mercury.

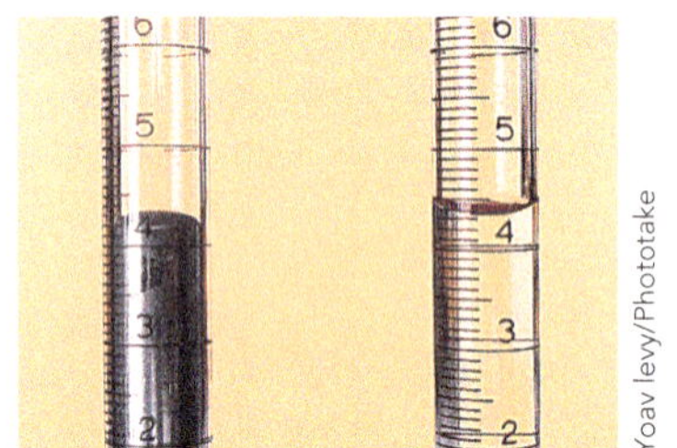

FIGURE 13.3 The meniscus of mercury (left) and water (right). The meniscus is the characteristic curve of the surface of a liquid in a narrow tube.

Evaporation

When beakers of water, ethyl ether, and ethyl alcohol (all liquids at room temperature) are allowed to stand uncovered, their volumes gradually decrease. The process by which this change takes place is called *evaporation*.

Attractive forces exist between molecules in the liquid state. Not all of these molecules, however, have the same kinetic energy. Molecules that have greater-than-average kinetic energy can overcome the attractive forces and break away from the surface of the liquid to become a gas (FIGURE 13.4). **Evaporation**, or **vaporization**, is the escape of molecules from the liquid state to the gas or vapor state.

In evaporation, molecules of greater-than-average kinetic energy escape from a liquid, leaving it cooler than it was before they escaped. For this reason, evaporation of perspiration is one way the human body cools itself and keeps its temperature constant. When volatile liquids such as ethyl chloride (C_2H_5Cl) are sprayed on the skin, they evaporate rapidly, cooling the area by removing heat. The numbing effect of the low temperature produced by evaporation of ethyl chloride allows it to be used as a local anesthetic for minor surgery.

Solids such as iodine, camphor, naphthalene (mothballs), and, to a small extent, even ice will go directly from the solid to the gaseous state, bypassing the liquid state. This change is a form of evaporation and is called **sublimation**:

$$\text{liquid} \xrightarrow{\text{evaporation}} \text{vapor}$$

$$\text{solid} \xrightarrow{\text{sublimation}} \text{vapor}$$

FIGURE 13.4 The high-energy molecules escape from the surface of a liquid in a process known as evaporation.

CHECK YOUR UNDERSTANDING

13.4 Vaporization

WileyPLUS

EXAMPLE 13.1

Which will evaporate more quickly: 60 mL of water in a beaker with a diameter of 5 cm or 60 mL of water in a shallow dish with a diameter of 15 cm? Why?

SOLUTION

The water in the shallow dish will evaporate more quickly since there is more surface area in the dish than in the beaker.

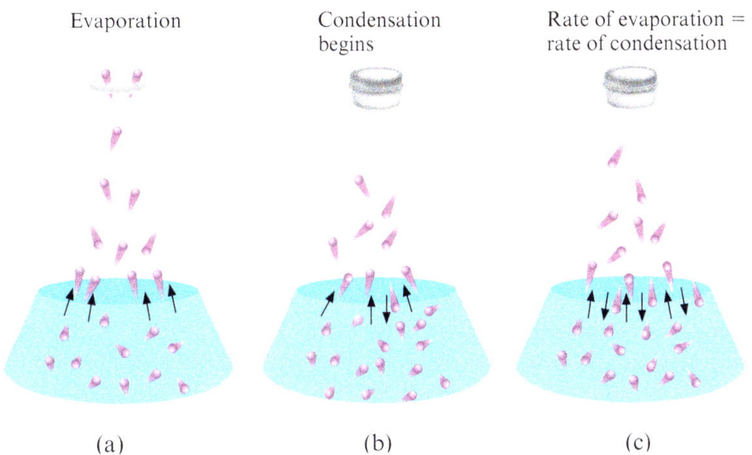

FIGURE 13.5 (a) Molecules in an open flask evaporate from the liquid and disperse into the atmosphere. (b) When the flask is stoppered, the number of gaseous molecules increases inside the flask. Some gaseous molecules collide with the surface of the liquid and stick—condensation occurs. (c) When the rate of evaporation equals the rate of condensation, an equilibrium between the liquid and vapor is established.

Vapor Pressure

When a liquid vaporizes in a closed system like that shown in FIGURE 13.5B, some of the molecules in the vapor or gaseous state strike the surface and return to the liquid state by the process of **condensation**. The rate of condensation increases until it is equal to the rate of vaporization. At this point, the space above the liquid is said to be saturated with vapor, and an equilibrium, or steady state, exists between the liquid and the vapor. The equilibrium equation is

$$\text{liquid} \underset{\text{condensation}}{\overset{\text{evaporation}}{\rightleftharpoons}} \text{vapor}$$

This equilibrium is dynamic; both processes—vaporization and condensation—are taking place, even though we cannot see or measure a change. The number of molecules leaving the liquid in a given time interval is equal to the number of molecules returning to the liquid.

At equilibrium, the molecules in the vapor exert a pressure like any other gas. The pressure exerted by a vapor in equilibrium with its liquid is known as the **vapor pressure** of the liquid. The vapor pressure may be thought of as a measure of the "escaping" tendency of molecules to go from the liquid to the vapor state. The vapor pressure of a liquid is independent of the amount of liquid and vapor present, but it increases as the temperature rises. We can measure the vapor pressure of a liquid by using a simple barometer as shown in FIGURE 13.6. Mercury is so dense that if we inject a sample of liquid at

CHECK YOUR UNDERSTANDING
13.5 Vapor Pressure
WileyPLUS

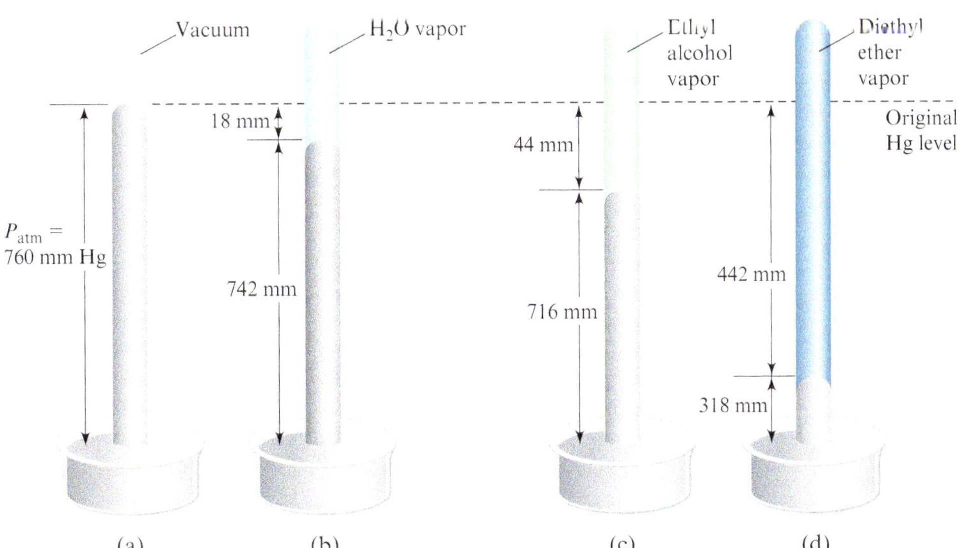

FIGURE 13.6
(a) A barometer measures atmospheric pressure (values given for 20°C).
To measure the vapor pressure of a liquid, a tiny amount is injected into the barometer.
(b) The water vapor presses the Hg down 18 mm (vapor pressure = 18 mm).
(c) Ethyl alcohol presses the Hg level down 44 mm (vapor pressure = 44 mm).
(d) Ethyl ether has a vapor pressure of 442 mm Hg.

the bottom of the Hg tube, the liquid rises to the top of the Hg column. It is then trapped in a closed space where it produces a vapor that pushes down the Hg column. When equilibrium between the liquid and vapor is reached, we can measure the vapor pressure by the change in height of the Hg column.

When equal volumes of water, ethyl ether, and ethyl alcohol are placed in beakers and allowed to evaporate at the same temperature, we observe that the ether evaporates faster than the alcohol, which evaporates faster than the water. This order of evaporation is consistent with the fact that ether has a higher vapor pressure at any particular temperature than ethyl alcohol or water. One reason for this higher vapor pressure is that the attraction is less between ether molecules than between alcohol or water molecules.

Substances that evaporate readily are said to be **volatile**. A volatile liquid has a relatively high vapor pressure at room temperature. Ethyl ether is a very volatile liquid; water is not too volatile; and mercury, which has a vapor pressure of 0.0012 torr at 20°C, is essentially a nonvolatile liquid. Most substances that are normally in a solid state are nonvolatile (solids that sublime are exceptions).

PRACTICE 13.1

How is the vapor pressure of a substance defined? How does it depend on temperature?

PRACTICE 13.2

A liquid in a closed container has a constant vapor pressure. What is the relationship between the rate of evaporation of the liquid and the rate of condensation of the vapor in the container?

13.3 Boiling Point and Melting Point

LEARNING OBJECTIVE Define boiling point and melting point and determine the boiling point of a liquid from a graph of temperature versus vapor pressure.

KEY TERM
boiling point
normal boiling point
vapor pressure curve
melting point (or freezing point)

The boiling temperature of a liquid is related to its vapor pressure. We've seen that vapor pressure increases as temperature increases. When the internal or vapor pressure of a liquid becomes equal to the external pressure, the liquid boils. (By external pressure we mean the pressure of the atmosphere above the liquid.) The boiling temperature of a pure liquid remains constant as long as the external pressure does not vary.

The boiling point (bp) of water is 100°C at 1 atm pressure. FIGURE 13.7 shows that the vapor pressure of water at 100°C is 760 torr. The significant fact here is that the boiling point is the temperature at which the vapor pressure of the water or other liquid is equal to standard, or atmospheric, pressure at sea level. **Boiling point** is the temperature at which the vapor pressure of a liquid is equal to the external pressure above the liquid.

CHECK YOUR UNDERSTANDING
13.6 Normal Boiling Point
WileyPLUS

We can readily see that a liquid has an infinite number of boiling points. When we give the boiling point of a liquid, we should also state the pressure. When we express the boiling point without stating the pressure, we mean it to be the **normal boiling point** at standard pressure (760 torr). Using Figure 13.7 again, we see that the normal boiling point of ethyl ether is between 30°C and 40°C, and for ethyl alcohol it is between 70°C and 80°C because for each compound 760 torr lies within these stated temperature ranges. At the normal boiling point, 1 g of a liquid changing to a vapor (gas) absorbs an amount of energy equal to its heat of vaporization (TABLE 13.1).

The boiling point at various pressures can be evaluated by plotting the vapor pressures of various liquids on the graph in Figure 13.7, where temperature is plotted horizontally along the x-axis and vapor pressure is plotted vertically along the y-axis. The resulting curve is known as the **vapor pressure curve**. Any point on this curve represents a vapor–liquid equilibrium at a particular temperature and pressure. We can find the boiling point at any pressure by tracing a horizontal line from the designated pressure to a point on the vapor pressure curve. From this point, we draw a vertical line to obtain the boiling point on the temperature axis. Three such points are shown in Figure 13.7; they

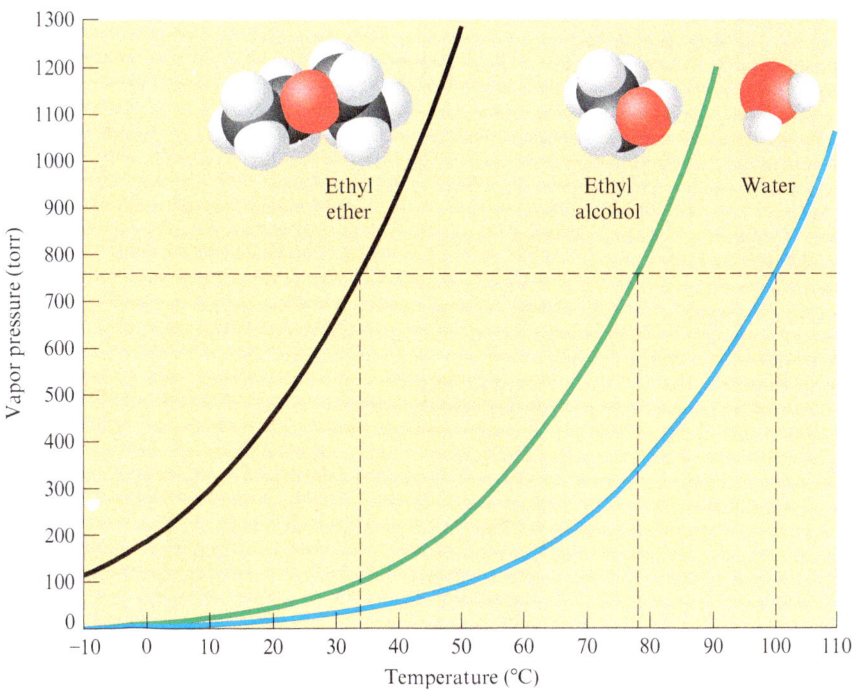

FIGURE 13.7 Vapor pressure–temperature curves for ethyl ether, ethyl alcohol, and water.

represent the normal boiling points of the three compounds at 760 torr. By reversing this process, you can ascertain at what pressure a substance will boil at a specific temperature. The boiling point is one of the most commonly used physical properties for characterizing and identifying substances.

ENHANCED EXAMPLE 13.2

WileyPLUS

Determine the boiling point of ethyl alcohol at a vapor pressure of 500 torr.

SOLUTION

Use Figure 13.7. Trace a horizontal line from 500 torr to the green ethyl alcohol curve. Then trace a vertical line down from the curve to the *x*-axis. The boiling point is 69°C.

ONLINE LEARNING MODULE
Vapor Pressure - Temperature Curves
WileyPLUS

PRACTICE 13.3

Use the graph in Figure 13.7 to determine the boiling points of ethyl ether, ethyl alcohol, and water at 600 torr.

PRACTICE 13.4

The average atmospheric pressure in Denver is 0.83 atm. What is the boiling point of water in Denver?

TABLE 13.1 Physical Properties of Ethyl Chloride, Ethyl Ether, Ethyl Alcohol, and Water

Substance	Boiling point (°C)	Melting point (°C)	Heat of vaporization J/g (cal/g)	Heat of fusion J/g (cal/g)
Ethyl chloride	12.3	−139	387 (92.5)	—
Ethyl ether	34.6	−116	351 (83.9)	—
Ethyl alcohol	78.4	−112	855 (204.3)	104 (24.9)
Water	100.0	0	2259 (539.9)	335 (80.1)

CHEMISTRY *IN ACTION* Chemical Eye Candy

The chemical reaction between Mentos and diet soda has been seen everywhere from chemistry classrooms to YouTube. The Mentos candy fizzes, the diet soda sprays out in a tall fountain, and everyone laughs. But just what is happening during the reaction? Lots of informal explanations have been given including one in *Chemical and Engineering News* and also one on "MythBusters" on the Discovery Channel. Yet none of these have been systematic experiments to discover the details of the reaction. Now Tonya Coffey and her students at Appalachian State University have explained the details of the reaction's parameters.

Coffey and her students discovered that the type of Mentos did not matter and neither did the type of diet soda (caffeinated or clear). They also found that the reaction was not a simple acid–base reaction; the acidity of the soda was the same before and after the reaction, and Mentos candy is not basic. The key component of the reaction is the ability of some of the reactants to reduce the surface tension of the soda, allowing the carbon dioxide to escape more quickly, producing the fountain. It turns out that the gum Arabic in the Mentos's outer coating is an excellent surfactant (a substance that reduces the surface tension of a liquid). In addition, two chemicals in the diet soda (aspartame and potassium benzoate) also reduce surface tension and assist in the formation of bubbles. Coffey and her students also determined (using a scanning electron microscope) that the surface of the Mentos candy is very rough, providing many "nucleation sites" for carbon dioxide bubbles. The Appalachian State University students tested other materials (sand, dishwashing detergent, salt) to determine whether they might produce fountains and found that they did but not as spectacular as the ones Mentos produced.

Several other factors contribute to the exciting nature of the reaction. Mentos candy is very dense, so it sinks to the bottom of the soda rapidly. The students also found that temperature matters: Warm soda leads to a higher fountain.

The Mentos in Diet Coke reactions

As heat is removed from a liquid, the liquid becomes colder and colder, until a temperature is reached at which it begins to solidify. A liquid changing into a solid is said to be *freezing*, or *solidifying*. When a solid is heated continuously, a temperature is reached at which the solid begins to liquefy. A solid that is changing into a liquid is said to be *melting*. The temperature at which the solid phase of a substance is in equilibrium with its liquid phase is known as the **melting point** or **freezing point** of that substance. The equilibrium equation is

$$\text{solid} \underset{\text{freezing}}{\overset{\text{melting}}{\rightleftharpoons}} \text{liquid}$$

When a solid is slowly and carefully heated so that a solid–liquid equilibrium is achieved and then maintained, the temperature will remain constant as long as both phases are present. The energy is used solely to change the solid to the liquid. The melting point is another physical property that is commonly used for characterizing substances.

The most common example of a solid–liquid equilibrium is ice and water. In a well-stirred system of ice and water, the temperature remains at 0°C as long as both phases are present. The melting point changes only slightly with pressure unless the pressure change is very large.

CHECK YOUR UNDERSTANDING

13.7 Melting

WileyPLUS

13.4 Changes of State

LEARNING OBJECTIVE

Calculate the amount of energy involved in a change of state.

KEY TERM

heat of fusion

heat of vaporization

The majority of solids undergo two changes of state upon heating. A solid changes to a liquid at its melting point, and a liquid changes to a gas at its boiling point. This warming process can be represented by a graph called a *heating curve* (FIGURE 13.8). This figure

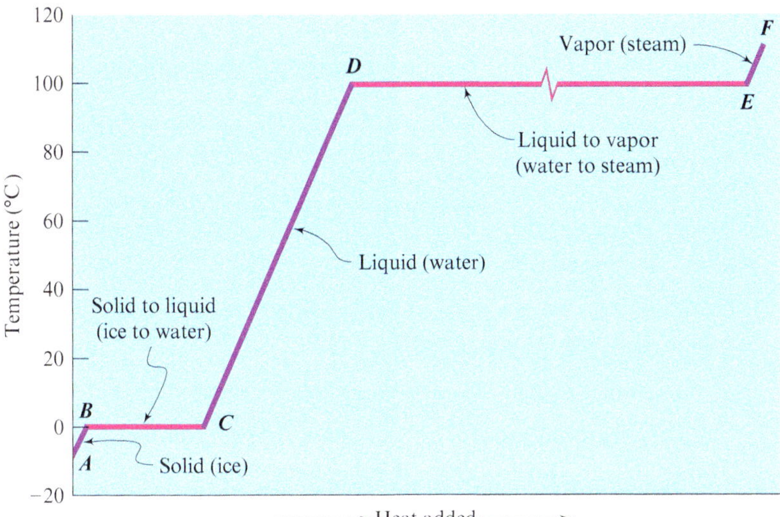

FIGURE 13.8 Heating curve for a pure substance—the absorption of heat by a substance from the solid state to the vapor state. Using water as an example, the *AB* interval represents the ice phase; *BC* interval, the melting of ice to water; *CD* interval, the elevation of the temperature of water from 0°C to 100°C; *DE* interval, the boiling of water to steam; and *EF* interval, the heating of steam.

shows ice being heated at a constant rate. As energy flows into the ice, the vibrations within the crystal increase and the temperature rises ($A \longrightarrow B$). Eventually, the molecules begin to break free from the crystal and melting occurs ($B \longrightarrow C$). During the melting process, all energy goes into breaking down the crystal structure; the temperature remains constant.

The energy required to change exactly one gram of a solid at its melting point into a liquid is called the **heat of fusion**. When the solid has completely melted, the temperature once again rises ($C \longrightarrow D$); the energy input is increasing the molecular motion within the water. At 100°C, the water reaches its boiling point; the temperature remains constant while the added energy is used to vaporize the water to steam ($D \longrightarrow E$). The **heat of vaporization** is the energy required to change exactly one gram of liquid to vapor at its normal boiling point. The attractive forces between the liquid molecules are overcome during vaporization. Beyond this temperature, all the water exists as steam and is being heated further ($E \longrightarrow F$).

CHECK YOUR UNDERSTANDING
13.8 Heat of Fusion
➡ WileyPLUS

ENHANCED EXAMPLE 13.3

WileyPLUS

How many joules of energy are needed to change 10.0 g of ice at 0.00°C to water at 20.0°C?

SOLUTION

Read **Knowns:** $m = 10.0$ g ice
$t_{\text{initial}} = 0.00°\text{C}$
$t_{\text{final}} = 20.0°\text{C}$

Solving for: energy

Plan The process involves two steps:

1. melting the ice (absorbing 335 J/g) and
2. warming the water from 0.00°C to 20.0°C, which requires 4.184 J/g°C.

Setup 1. Energy needed to melt the ice

$$(10.0\text{ g})\left(\frac{335\text{ J}}{1\text{ g}}\right) = 3350\text{ J} = 3.35 \times 10^3\text{ J}$$

2. Energy needed to warm the water (mass)(specific heat)(Δt) = energy

$$(10.0\text{ g})\left(\frac{4.184\text{ J}}{1\text{ g°C}}\right)(20.0°\text{C}) = 837\text{ J}$$

Calculate The total amount of energy required for the process is the sum of the two steps:

$$3350\text{ J} + 837\text{ J} = 4.19 \times 10^3\text{ J}$$

ONLINE LEARNING MODULE
Energy to Change State
➡ WileyPLUS

ENHANCED EXAMPLE 13.4 WileyPLUS

How many kilojoules of energy are needed to change 20.0 g of water at 20.°C to steam at 100.°C?

SOLUTION

Read • **Knowns:** m = 20.0 g water
t_{initial} = 20.°C
t_{final} = 100.°C

Solving for: energy

Plan • The process involves two steps:

1. warming the water from 20.°C to 100.°C, which requires 4.184 J/g°C, and
2. vaporizing the water to steam (absorbing 2.26 kJ/g).

Setup • 1. Energy needed to warm the water

$$(\text{mass})(\text{specific heat})(\Delta t) = \text{energy}$$

$$(20.0\ \text{g})\left(\frac{4.184\ \text{J}}{1\ \text{g}\,^\circ\text{C}}\right)(80.\,^\circ\text{C}) = 6.7 \times 10^3\ \text{J} = 6.7\ \text{kJ}$$

2. Energy needed to vaporize the water to steam

$$(20.0\ \text{g})\left(\frac{2.26\ \text{kJ}}{1\ \text{g}}\right) = 45.2\ \text{kJ}$$

Calculate • The total amount of energy required for the process is the sum of the two steps:

$$6.7\ \text{kJ} + 45.2\ \text{kJ} = 51.9\ \text{kJ}$$

PRACTICE 13.5

How many kilojoules of energy are required to change 50.0 g of ethyl alcohol from 60.0°C to vapor at 78.4°C? The specific heat of ethyl alcohol is 2.138 J/g°C.

13.5 Intermolecular Forces

LEARNING OBJECTIVE Describe the three types of intermolecular forces and explain their significance in liquids.

KEY TERM

intermolecular forces
intramolecular forces
dipole–dipole attractions
hydrogen bond
London dispersion forces

Why do molecules of a liquid stay together instead of floating away from each other like molecules of a gas? **Intermolecular forces** are the forces of attraction between molecules. These forces hold molecules close together and allow for the formation of liquids and solids. These intermolecular forces can also help us to predict some of the properties of liquids. For example, if the molecules of a liquid are strongly held together, it will be difficult to pull the molecules apart, resulting in a low vapor pressure and a high boiling point. Molecules with strong intermolecular forces are also likely to have strong cohesive forces, resulting in a high surface tension as well.

Molecules are held together by bonds, sometimes called **intramolecular forces**, that occur inside the molecules. These covalent bonding forces arise from sharing electrons between atoms. Intermolecular forces occur *between* molecules. How do these forces happen? There are several different kinds of intermolecular forces. Some of the most common are dipole–dipole attractions, hydrogen bonds, and London dispersion forces. Let's look at each of these types of intermolecular forces in more detail.

CHECK YOUR UNDERSTANDING

13.9 Intermolecular versus Intramolecular Forces

➡ WileyPLUS

Dipole–Dipole Attractions

As we learned in Chapter 11, some covalent bonds do not share electrons equally because of differences in the electronegativity of the bonded atoms. When this happens, the polar bonds that form may create polar molecules or molecules with a positive end and a negative end. Some examples of polar molecules, such as HF, CO, and ICl, are easy to identify because they have only two atoms. Others are more difficult because of their complex molecular geometry.

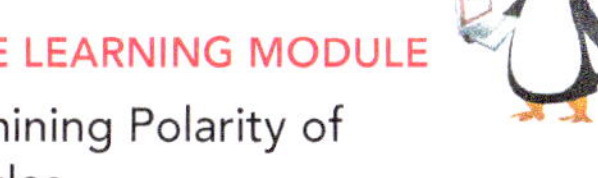

EXAMPLE 13.5

Determine which of the following molecules are polar: HCl, CO_2, and NBr_3.

SOLUTION

The Lewis structures and three-dimensional drawings for these three molecules are shown here. Arrows represent the dipole moments of the individual bonds.

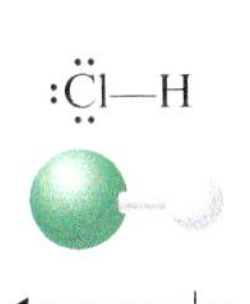

Chlorine is more electronegative than hydrogen, so both the H—Cl bond and the molecule are polar.

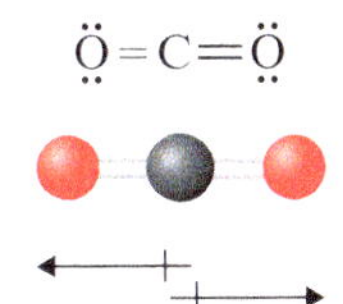

The carbon dioxide molecule is nonpolar because the carbon–oxygen dipoles cancel each other by acting in equal and opposite directions.

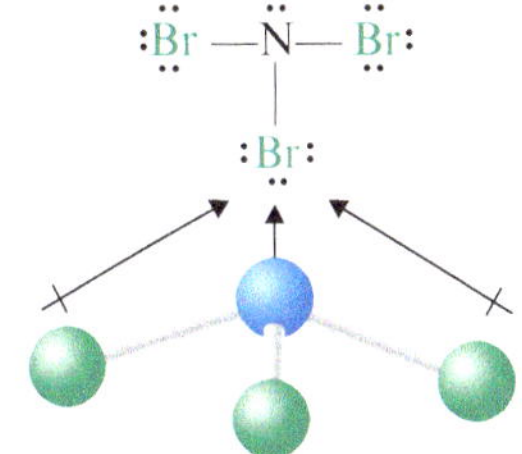

The nitrogenbrominee bonds are all polar and all three dipoles point toward the nitrogen atom, resulting in a net dipole up. This is a polar molecule.

PRACTICE 13.6

Determine which of the following molecules are polar: SO_2, Br_2, and CH_3F.

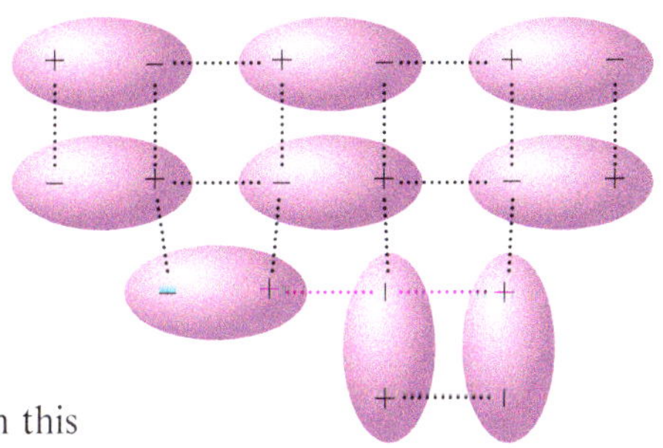

As we saw in Chapter 11 water is a polar molecule—it has a dipole moment. When molecules with dipole moments are put together, they orient themselves to take advantage of their charge distribution. The positive ends of the molecules are attracted to the negative ends of other molecules. Because of this attraction, the molecules tend to stick together creating **dipole–dipole attraction**.

The Hydrogen Bond

TABLE 13.2 compares the physical properties of H_2O, H_2S, H_2Se, and H_2Te. From this comparison, it is apparent that four physical properties of water—melting point, boiling point, heat of fusion, and heat of vaporization—are extremely high and do not fit the trend relative to the molar masses of the four compounds. For example, if the properties of water followed the progression shown by the other three compounds, we would expect the melting point of water to be below −85°C and the boiling point to be below −60°C.

Why does water exhibit these anomalies? Because liquid water molecules are held together more strongly than other molecules in the same family. The intermolecular force

TABLE 13.2 Physical Properties of Water and Other Hydrogen Compounds of Group 6A Elements

Formula	Color	Molar mass (g/mol)	Melting point (°C)	Boiling point, 1 atm (°C)	Heat of fusion J/g (cal/g)	Heat of vaporization J/g (cal/g)
H_2O	Colorless	18.02	0.00	100.0	335 (80.1)	2.26×10^3 (540)
H_2S	Colorless	34.09	−85.5	−60.3	69.9 (16.7)	548 (131)
H_2Se	Colorless	80.98	−65.7	−41.3	31 (7.4)	238 (57.0)
H_2Te	Colorless	129.6	−49	−2	—	179 (42.8)

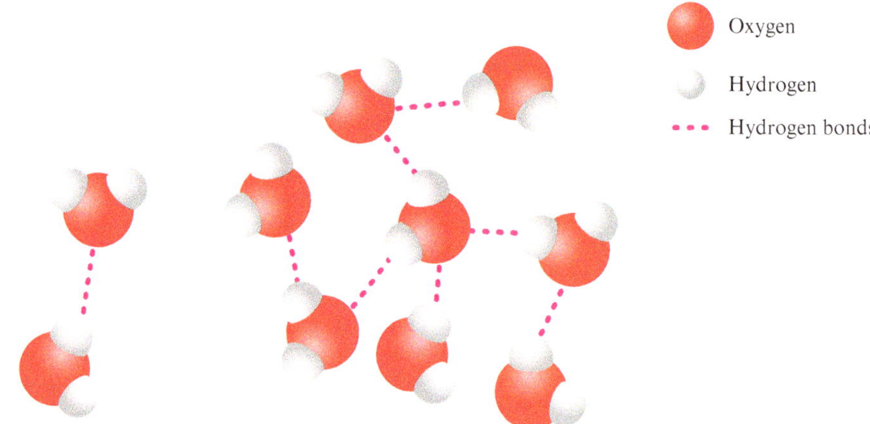

FIGURE 13.9 Hydrogen bonding. Water in the liquid and solid states exists as aggregates in which the water molecules are linked together by hydrogen bonds.

acting between water molecules is called a **hydrogen bond**, which acts like a very weak bond between two polar molecules. A hydrogen bond is formed between polar molecules that contain hydrogen covalently bonded to a small, highly electronegative atom such as fluorine, oxygen, or nitrogen (F—H, O—H, N—H). A hydrogen bond is actually the dipole–dipole attraction between polar molecules containing these three types of polar bonds.

Compounds that have significant hydrogen-bonding ability are those that contain H covalently bonded to F, O, or N.

Because a hydrogen atom has only one electron, it forms only one covalent bond. When it is bonded to a strong electronegative atom such as oxygen, a hydrogen atom will also be attracted to an oxygen atom of another molecule, forming a dipole–dipole attraction (H bond) between the two molecules. Water has two types of bonds: covalent bonds that exist between hydrogen and oxygen atoms within a molecule and hydrogen bonds that exist between hydrogen and oxygen atoms in *different* water molecules.

Hydrogen bonds are *intermolecular* bonds; that is, they are formed between atoms in different molecules. They are somewhat ionic in character because they are formed by electrostatic attraction. Hydrogen bonds are much weaker than the ionic or covalent bonds that unite atoms to form compounds. Despite their weakness, they are of great chemical importance.

The oxygen atom in water can form two hydrogen bonds—one through each of the unbonded pairs of electrons. FIGURE 13.9 shows two water molecules linked by a hydrogen bond and eight water molecules linked by hydrogen bonds. A dash (–) is used for the covalent bond and a dotted line (····) for the hydrogen bond. In water each molecule is linked to others through hydrogen bonds to form a three-dimensional aggregate of water molecules. This intermolecular hydrogen bonding effectively gives water the properties of a much larger, heavier molecule, explaining in part its relatively high melting point, boiling point, heat of fusion, and heat of vaporization. As water is heated and energy absorbed, hydrogen bonds are continually being broken until at 100°C, with the absorption of an additional 2.26 kJ/g, water separates into individual molecules, going into the gaseous state. Sulfur, selenium, and tellurium are not sufficiently electronegative for their hydrogen compounds to behave like water. The lack of hydrogen bonding is one reason H_2S is a gas, not a liquid, at room temperature.

CHECK YOUR UNDERSTANDING

13.10 Properties of Hydrogen Bonds

WileyPLUS

Fluorine, the most electronegative element, forms the strongest hydrogen bonds. This bonding is strong enough to link hydrogen fluoride molecules together as *dimers*, H_2F_2, or as larger $(HF)_n$ molecular units. The dimer structure may be represented in this way:

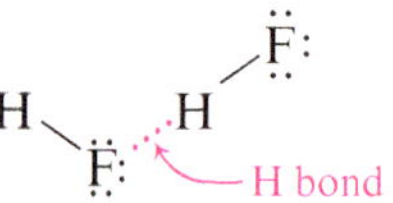

CHEMISTRY *IN ACTION* How Sweet It Is!

Did you think artificial sweeteners were a product of the post–World War II chemical industry? Not so—many of them have been around a long time, and several of the important ones were discovered quite by accident. In 1878, Ira Remsen was working late in his laboratory and realized he was about to miss a dinner with friends. In his haste to leave the lab, he forgot to wash his hands. Later at dinner he broke a piece of bread and tasted it only to discover that it was very sweet. The sweet taste had to be the chemical he had been working with in the lab. Back at the lab, he isolated saccharin—the first of the artificial sweeteners.

In 1937, Michael Sveda was smoking a cigarette in his laboratory (a very dangerous practice to say the least!). He touched the cigarette to his lips and was surprised by the exceedingly sweet taste. The chemical on his hands turned out to be cyclamate, which soon became a staple of the artificial sweetener industry.

In 1965, James Schlatter was researching antiulcer drugs for the pharmaceutical firm G. D. Searle. In the course of his work, he accidentally ingested a small amount of a preparation and found to his surprise that it had an extremely sweet taste. He had discovered aspartame, a molecule consisting of two amino acids joined together. Since only very small quantities of aspartame are necessary to produce sweetness, it proved to be an excellent low-calorie artificial sweetener. More than 50 different molecules have a sweet taste, and it is difficult to find a single binding site that could interact with all of them.

Our taste receptors are composed of proteins that can form hydrogen bonds with other molecules. The proteins contain —NH and —OH groups (with hydrogen available to bond) as well as C=O groups (providing oxygen for hydrogen bonding). "Sweet molecules" also contain H-bonding groups including —OH, —NH_2, and O or N. These molecules not only must have the proper atoms to form hydrogen bonds, but they must also contain a hydrophobic region (repels H_2O). A new model for binding to a sweetness receptor has been developed at Senomyx in La Jolla, California. The model shows four binding sites that can act independently. Small molecules bind to a pocket on a subunit as shown in the model. Large molecules (such as proteins) bind to a different site above one of the pockets.

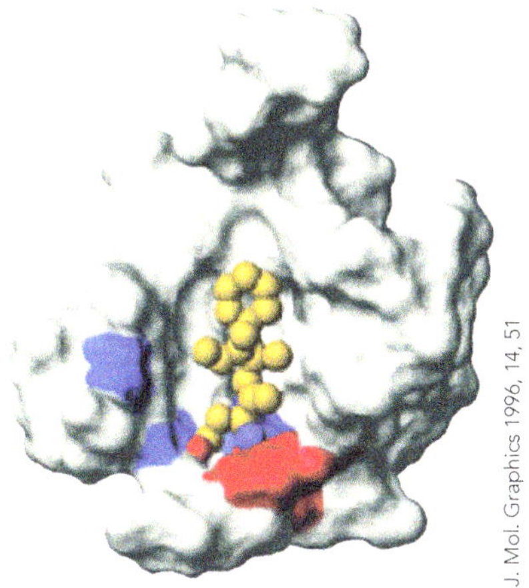

Model shows how the sweetener aspartame binds to a site on the sweetness receptor's T1R3 subunit. Red and blue are hydrogen-bond donor and acceptor residues, respectively; aspartame is in gold, except for its carboxylate (red) and ammonium (blue) groups. Model prepared with MOLMOL. Adapted from *J. Med. Chem.*

Hydrogen bonding can occur between two different atoms that are capable of forming H bonds. Thus, we may have an O····H—N or O—H····N linkage in which the hydrogen atom forming the H bond is between an oxygen and a nitrogen atom. This form of H bond exists in certain types of protein molecules and many biologically active substances.

ENHANCED EXAMPLE 13.6 WileyPLUS

ONLINE LEARNING MODULE
Hydrogen Bonding
➡ WileyPLUS

Would you expect hydrogen bonding to occur between molecules of these substances?

(a)

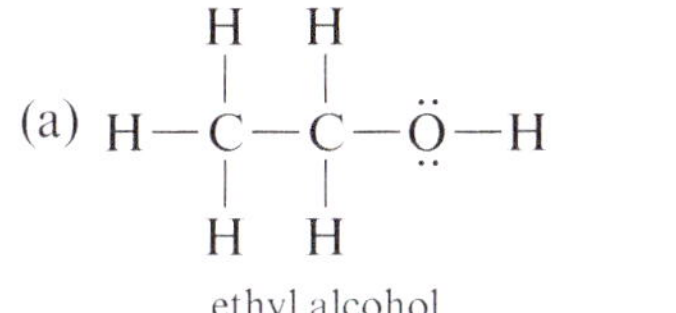

(b)

```
     H       H
     |       |
 H — C — Ö — C — H
     |   ..  |
     H       H
```

dimethyl ether

SOLUTION

(a) Hydrogen bonding should occur in ethyl alcohol because one hydrogen atom is bonded to an oxygen atom:

```
     H   H             H   H   H
     |   |             |   |   |
 H — C — C — Ö — H····:O — C — C — H
     |   |   ..     ↑  ..  |   |
     H   H        H bond   H   H
```

(b) There is no hydrogen bonding in dimethyl ether because all the hydrogen atoms are bonded only to carbon atoms.

Both ethyl alcohol and dimethyl ether have the same molar mass (46.07 g/mol). Although both compounds have the same molecular formula, C_2H_6O, ethyl alcohol has a much higher boiling point (78.4°C) than dimethyl ether (−23.7°C) because of hydrogen bonding between the alcohol molecules.

PRACTICE 13.7

Would you expect hydrogen bonding to occur between molecules of these substances?

```
         H   H   H                     H   H   H                           H
         |   |   |                     |   |   |                           |
(a)  H — C — C — N — H      (b)  H — C — N — C — H    (c)      H   H   H — C — H
         |   |   ..                    |   ..  |               |   |       |
         H   H                         H       H           H — C — C ——— N:
                                                               |   |       |
                                                               H   H   H — C — H
                                                                           |
                                                                           H
```

London Dispersion Forces

Even molecules without dipole moments must exert forces on each other. We know this because all substances, even noble gases, exist in the liquid and solid states (at very low temperatures). The forces that exist between nonpolar molecules and also between noble gas molecules are called **London dispersion forces**. Let's consider a couple of noble gas atoms. We would assume the electrons are evenly distributed around the nucleus.

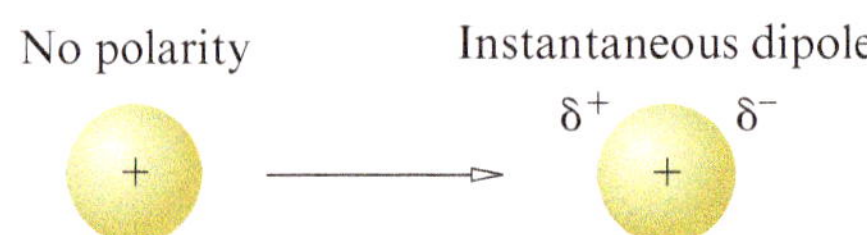

But the electrons are in motion around the nucleus, and so this even distribution is not true all of the time. Atoms can develop an instantaneous dipolar arrangement of charge. This instantaneous dipole can induce a similar dipole in a nearby atom.

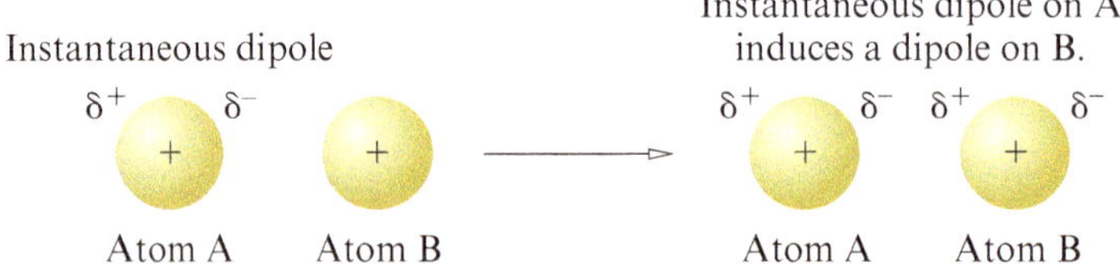

CHECK YOUR UNDERSTANDING

13.11 London Dispersion Forces

13.12 Dispersion Forces versus Molecular Mass

➡ WileyPLUS

The interaction is weak and short-lived but can be important in large atoms and molecules. London dispersion forces become more significant as the atoms or molecules get larger. Larger size means more electrons are available to form dipoles.

The strength of this attraction depends on the molecular mass of the substance. Nonpolar molecules with very high molar masses such as paraffin wax, $C_{25}H_{52}$, have strong enough intermolecular forces to form soft solids.

ONLINE LEARNING MODULE

London Dispersion Forces

➡ WileyPLUS

ENHANCED EXAMPLE 13.7 WileyPLUS

Predict which of the following compounds will have the strongest London dispersion forces: C_4H_{10}, $C_{15}H_{32}$, and C_8H_{18}.

SOLUTION

London dispersion forces depend on molar mass. The molar masses of these compounds are 58.12 amu, 114.2 amu, and 212.4 amu, so $C_{15}H_{32}$ will have the strongest London dispersion forces.

PRACTICE 13.8

Predict which of the following compounds will have the strongest London dispersion forces: F_2, Br_2, and I_2.

13.6 Hydrates

Explain what hydrates are, write formulas for hydrates, and calculate the percent water in a hydrate.

LEARNING OBJECTIVE

KEY TERM
hydrate
water of hydration
water of crystallization

When certain solutions containing ionic compounds are allowed to evaporate, some water molecules remain as part of the crystalline compound that is left after evaporation is complete. Solids that contain water molecules as part of their crystalline structure are known as **hydrates**. Water in a hydrate is known as **water of hydration** or **water of crystallization**.

Formulas for hydrates are expressed by first writing the usual anhydrous (without water) formula for the compound and then adding a dot followed by the number of water molecules present. An example is $BaCl_2 \cdot 2\,H_2O$. This formula tells us that each formula unit of this compound contains one barium ion, two chloride ions, and two water molecules. A crystal of the compound contains many of these units in its crystalline lattice.

In naming hydrates, we first name the compound exclusive of the water and then add the term *hydrate*, with the proper prefix representing the number of water molecules in the formula. For example, $BaCl_2 \cdot 2\,H_2O$ is called *barium chloride dihydrate*. Hydrates are true compounds and follow the law of definite composition. The molar mass of $BaCl_2 \cdot 2\,H_2O$ is 244.2 g/mol; it contains 56.22% barium, 29.03% chlorine, and 14.76% water.

Water molecules in hydrates are bonded by electrostatic forces between polar water molecules and the positive or negative ions of the compound. These forces are not as strong as covalent or ionic chemical bonds. As a result, water of crystallization can be removed by moderate heating of the compound. A partially dehydrated or completely anhydrous compound may result. When $BaCl_2 \cdot 2\,H_2O$ is heated, it loses its water at about 100°C:

CHECK YOUR UNDERSTANDING
13.13 Hydrates
➡ WileyPLUS

$$BaCl_2 \cdot 2\,H_2O(s) \xrightarrow{100°C} BaCl_2(s) + 2\,H_2O(g)$$

When a solution of copper(II) sulfate ($CuSO_4$) is allowed to evaporate, beautiful blue crystals containing 5 moles water per 1 mole $CuSO_4$ are formed (FIGURE 13.10A). The formula for this hydrate is $CuSO_4 \cdot 5\,H_2O$; it is called copper(II) sulfate pentahydrate.

Richard Megna/Fundamental Photographs

(a)

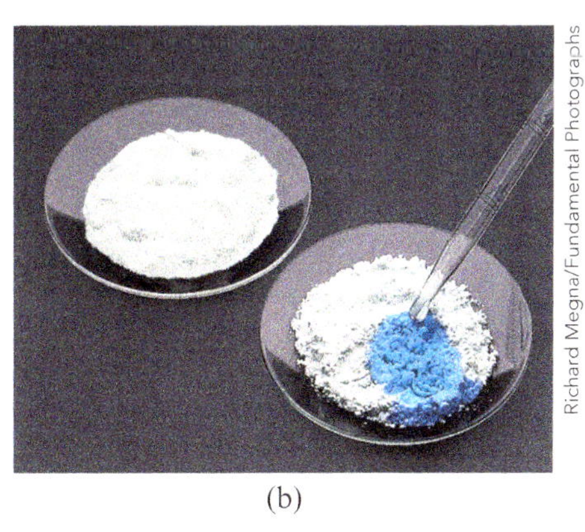

Richard Megna/Fundamental Photographs

(b)

FIGURE 13.10 (a) When these blue crystals of $CuSO_4 \cdot 5\,H_2O$ are dissolved in water, a blue solution forms. (b) The anhydrous crystals of $CuSO_4$ are pale green. When water is added, they immediately change color to blue $CuSO_4 \cdot 5\,H_2O$ crystals.

When $CuSO_4 \cdot 5\ H_2O$ is heated, water is lost, and a pale green-white powder, anhydrous $CuSO_4$, is formed:

$$CuSO_4 \cdot 5\ H_2O(s) \xrightarrow{150°C} CuSO_4(s) + 5\ H_2O(g)$$

When water is added to anhydrous copper(II) sulfate, the foregoing reaction is reversed, and the compound turns blue again (FIGURE 13.10B). Because of this outstanding color change, anhydrous copper(II) sulfate has been used as an indicator to detect small amounts of water. The formation of the hydrate is noticeably exothermic.

The formula for plaster of paris is $(CaSO_4)_2 \cdot H_2O$. When mixed with the proper quantity of water, plaster of paris forms a dihydrate and sets to a hard mass. It is therefore useful for making patterns for the production of art objects, molds, and surgical casts. The chemical reaction is

$$(CaSO_4)_2 \cdot H_2O(s) + 3\ H_2O(l) \longrightarrow 2\ CaSO_4 \cdot 2\ H_2O(s)$$

TABLE 13.3 lists a number of common hydrates.

TABLE 13.3 | Selected Hydrates

Hydrate	Name	Hydrate	Name
$CaCl_2 \cdot 2\ H_2O$	calcium chloride dihydrate	$Na_2CO_3 \cdot 10\ H_2O$	sodium carbonate decahydrate
$Ba(OH)_2 \cdot 8\ H_2O$	barium hydroxide octahydrate	$(NH_4)_2C_2O_4 \cdot H_2O$	ammonium oxalate monohydrate
$MgSO_4 \cdot 7\ H_2O$	magnesium sulfate heptahydrate	$NaC_2H_3O_2 \cdot 3\ H_2O$	sodium acetate trihydrate
$SnCl_2 \cdot 2\ H_2O$	tin(II) chloride dihydrate	$Na_2B_4O_7 \cdot 10\ H_2O$	sodium tetraborate decahydrate
$CoCl_2 \cdot 6\ H_2O$	cobalt(II) chloride hexahydrate	$Na_2S_2O_3 \cdot 5\ H_2O$	sodium thiosulfate pentahydrate

ENHANCED EXAMPLE 13.8 WileyPLUS

Write the formula for calcium chloride hexahydrate.

SOLUTION

Begin by writing the formula for calcium chloride, $CaCl_2$. Remember the Ca is +2 and the Cl is −1, so we need to have two Cl atoms in the formula for the charge to be neutral. Next determine the number of water molecules in the hydrate. The prefix *hexa-* means 6, so the formula for the hydrate would be $CaCl_2 \cdot 6\ H_2O$.

PRACTICE 13.9

Write formulas for
(a) beryllium carbonate tetrahydrate
(b) cadmium permanganate hexahydrate
(c) chromium(III) nitrate nonahydrate
(d) platinum(IV) oxide trihydrate

ENHANCED EXAMPLE 13.9 WileyPLUS

Calculate the percent water in $CuSO_4 \cdot 5\ H_2O$.

SOLUTION

Using the periodic table, find the molar mass of $CuSO_4$. It is 159.6 g/mol. The molar mass for water is 18.02 g/mol. Since there are 5 mol of water associated with each mol of $CuSO_4$ in the hydrate, the water has a mass of $5 \times 18.02 = 90.10$ g. The percent water is

$$\frac{90.10\text{ g}}{159.6\text{ g} + 90.10\text{ g}} \times 100\% = 36.08\%\text{ water}$$

PRACTICE 13.10

Calculate the percent water in Epsom salts, $MgSO_4 \cdot 7\ H_2O$.

13.7 Water, a Unique Liquid

LEARNING OBJECTIVE

Describe the characteristics of water in terms of its structure and list the sources of drinking water.

KEY TERM

semipermeable membrane

Water is our most common natural resource. It covers about 75% of Earth's surface. Not only is it found in the oceans and seas, in lakes, rivers, streams, and glacial ice deposits, but also it is always present in the atmosphere and in cloud formations.

About 97% of Earth's water is in the oceans. This *saline* water contains vast amounts of dissolved minerals. More than 70 elements have been detected in the mineral content of seawater. Only four of these—chlorine, sodium, magnesium, and bromine—are now commercially obtained from the sea. The world's *fresh* water comprises the other 3%, of which about two-thirds is locked up in polar ice caps and glaciers. The remaining fresh water is found in groundwater, lakes, rivers, and the atmosphere.

Water is an essential constituent of all living matter. It is the most abundant compound in the human body, making up about 70% of total body mass. About 92% of blood plasma is water; about 80% of muscle tissue is water; and about 60% of a red blood cell is water. Water is more important than food in the sense that we can survive much longer without food than without water.

Physical Properties of Water

Water is a colorless, odorless, tasteless liquid with a melting point of 0°C and a boiling point of 100°C at 1 atm. The heat of fusion of water is 335 J/g (80 cal/g). The heat of vaporization of water is 2.26 kJ/g (540 cal/g). The values for water for both the heat of fusion and the heat of vaporization are high compared with those for other substances; this indicates strong attractive forces between the molecules.

Ice and water exist together in equilibrium at 0°C, as shown in FIGURE 13.11. When ice at 0°C melts, it absorbs 335 J/g in changing into a liquid; the temperature remains at 0°C. To refreeze the water, 335 J/g must be removed from the liquid at 0°C.

In Figure 13.11, both boiling water and steam are shown to have a temperature of 100°C. It takes 418 J to heat 1 g of water from 0°C to 100°C, but water at its boiling point absorbs 2.26 kJ/g in changing to steam. Although boiling water and steam are both at the same temperature, steam contains considerably more heat per gram and can cause more severe burns than hot water.

The maximum density of water is 1.000 g/mL at 4°C. Water has the unusual property of contracting in volume as it is cooled to 4°C and then expanding when cooled from 4°C to 0°C. Therefore, 1 g of water occupies a volume greater than 1 mL at all temperatures except 4°C. Although most liquids contract in volume all the way down to the point at which they solidify, a large increase (about 9%) in volume occurs when water changes from a liquid at 0°C to a solid (ice) at 0°C. The density of ice at 0°C is 0.917 g/mL, which means that ice, being less dense than water, will float in water.

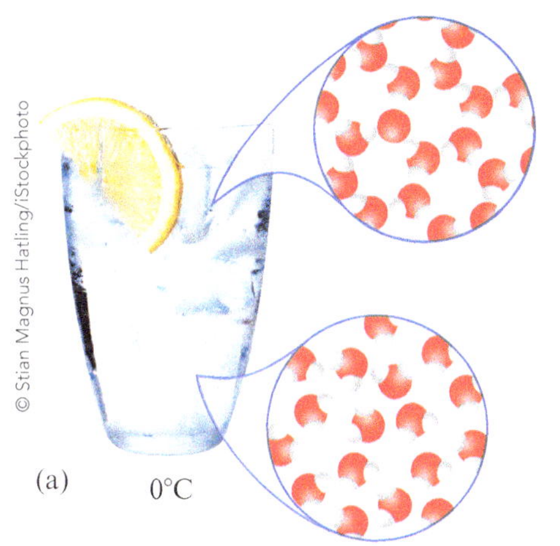

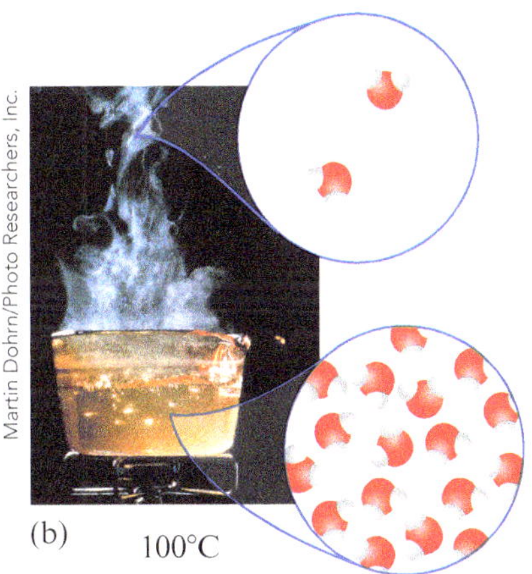

FIGURE 13.11 (a) Ice and water in equilibrium at 0°C. When ice melts, 335 J/g are needed to change it from solid to liquid. The same amount is released in freezing. (b) Boiling water and steam in equilibrium at 100°C. It takes 2.26 kJ/g to convert water to steam.

CHEMISTRY *IN ACTION* Reverse Osmosis?

If you have ever made a salad and been disappointed to find that the lettuce has become limp and the dressing has become watery, you have experienced osmosis. *Osmosis* is the process by which water flows through a membrane from a region of more pure water to a region of less pure water. (Section 14.6.) In the case of your salad, lettuce contains lots of water in its cells. When the lettuce is surrounded by salad dressing (water with lots of things dissolved in it), the water in the lettuce leaf will flow through the cell membranes, resulting in diluted dressing and wilted lettuce.

This same process can be pushed in the reverse direction as well. Suppose you were to soak a raisin in water. The water would flow into the raisin to dilute the concentrated sugar solution inside the raisin, resulting in a plumper raisin. What would happen if you put the raisin into a box and put pressure on the sides so that the raisin could not expand? The water would not be able to enter the raisin. And if you increased the pressure, some of the water would even flow out of the raisin. This is *reverse osmosis*.

This is the process used in the reverse osmosis purification of water. A sample of water containing contaminant molecules and ions (such as seawater) is put into a vessel with a semipermeable membrane. A **semipermeable membrane** allows the passage of water (solvent) molecules through it in either direction but prevents the passage of larger solute molecules or ions. As pressure is applied to the water sample, water molecules will pass through the membrane and pure water can be isolated.

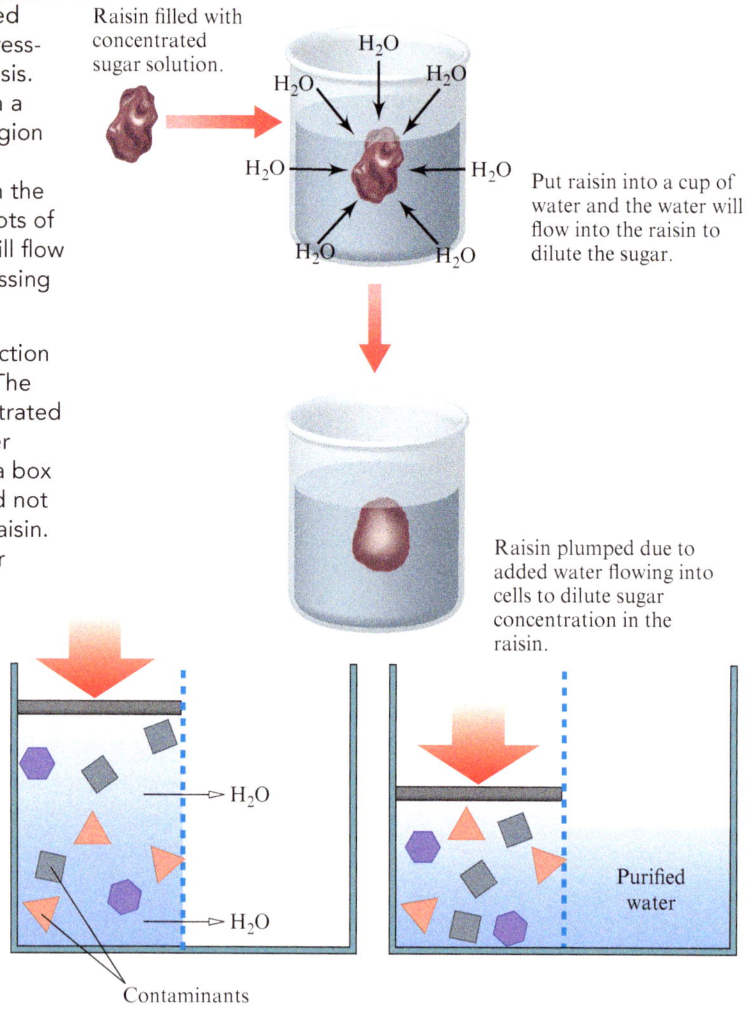

Structure of the Water Molecule

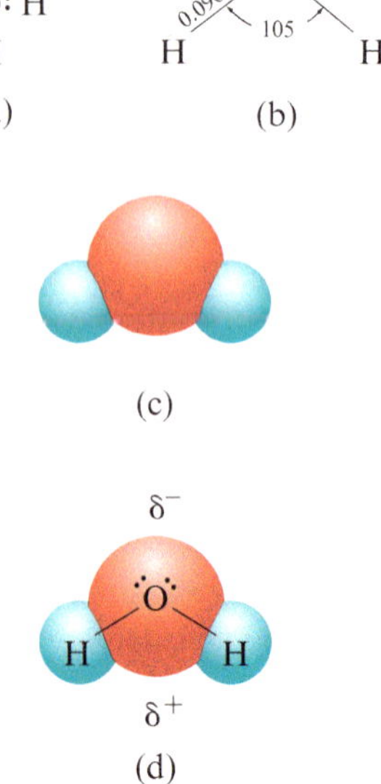

A single water molecule consists of two hydrogen atoms and one oxygen atom. Each hydrogen atom is bonded to the oxygen atom by a single covalent bond. This bond is formed by the overlap of the 1*s* orbital of hydrogen with an unpaired 2*p* orbital of oxygen. The average distance between the two nuclei is known as the *bond length*. The O—H bond length in water is 0.096 nm. The water molecule is nonlinear and has a bent structure with an angle of about 105° between the two bonds (FIGURE 13.12).

Oxygen is the second most electronegative element. As a result, the two covalent OH bonds in water are polar. If the three atoms in a water molecule were aligned in a linear structure, such as $H \leftrightarrow O \leftarrow H$, the two polar bonds would be acting in equal and opposite directions and the molecule would be nonpolar. However, water is a highly polar molecule. It therefore does not have a linear structure. When atoms are bonded in a nonlinear fashion, the angle formed by the bonds is called the *bond angle*. In water the HOH bond angle is 105°. The two polar covalent bonds and the bent structure result in a partial negative charge on the oxygen atom and a partial positive charge on each hydrogen atom. The polar nature of water is responsible for many of its properties, including its behavior as a solvent.

FIGURE 13.12 Diagrams of a water molecule: (a) electron distribution, (b) bond angle and O—H bond length, (c) molecular structure, and (d) dipole representation.

Sources of Water for a Thirsty World

One of the greatest challenges facing our planet is finding sufficient potable water for the global population that now surpasses the 7 billion mark. A U.S. citizen, on average, consumes between 300-400 liters of water a day (US Geological Survey) Thus, it is not difficult to see why water is becoming a very precious resource. To satisfy the insatiable growing demand, fresh water is being depleted from streams, aquifers, and reservoirs faster than it can be replenished. Climate change also contributes to the water problem: Changes in weather patterns are predicted to cause severe drought in many parts of the globe. Reclamation of wastewater and desalination of seawater are currently being considered to augment the supplies of fresh water.

© Btrenkel/iStockphoto

Bottled water provides a convenient source of drinking water for people on the go.

1. Reclamation of wastewater is actually an excellent source of fresh water. Although it may be very difficult to get comfortable with the idea of drinking reclaimed wastewater (the "ick factor" as termed by the reclamation industry), most of the water provided by municipalities is in fact reclaimed water from upstream cities. Many cities are using reclaimed water for irrigation and for industrial uses. Some cities are beginning to augment their water supplies using this reclaimed water. As an example, Singapore has been a leader in the challenge to reclaim wastewater for many uses. NEWater (provided by the Singapore water treatment department) is water that has been filtered, purified by reverse osmosis (see Chemistry in Action, Reverse Osmosis?), and reused for agricultural and industrial use. Some of this water is also treated with UV light to further purify it and it is bottled for drinking. Thirty percent of Singapore's water is currently reclaimed, and plans are to increase this to 50% by 2060. Orange County in California pumps reclaimed water into a local aquifer for storage and reuse. San Diego County has a plan to pump reclaimed water to upstream reservoirs to be mixed with "natural" water to be reused. These are excellent models for other cities that suffer from a shortage of available fresh water.
2. Desalination of seawater is a good source of fresh water for people living near the ocean. Unfortunately, desalination is very expensive, and it is inefficient to extract fresh water from the ocean. On average it takes about 4 liters of seawater to make 1 liter of fresh water. The wastewater, which has a very high salt concentration, may also have an adverse impact on the environment. Generally, salt water is desalinated using the process of reverse osmosis in which a sample of salt water is placed in a vessel with a semipermeable membrane and pressure is applied to push the water molecules through the membrane. (FIGURE 13.13.) This method of water purification has been championed by Israel, a nation with easy access to the sea but limited fresh water. Israel produces 40% of its water by desalination and will produce over 50% of its water in this way by 2016.

Low-temperature distillation and combustion of hydrogen are two other promising technologies being investigated as possible new sources of fresh water.

- In low-temperature distillation, water is subjected to a very low pressure environment. When the pressure is lowered, the boiling temperature is also lowered. This could possibly result in producing distilled water at low temperature and with a lower energy cost.
- To produce hydrogen gas for combustion, an electric current is run through a sample of water, breaking it down into its component elements. In regions with lots of sunshine, solar panels can create electricity to decompose water. The hydrogen and oxygen gas can then be collected and recombined later to produce electricity and pure water. This provides both a source of fresh water and a storage method for energy produced from the sun.

Though these technologies are still in the experimental stage, hopefully they will become important sources of fresh water in the future.

FIGURE 13.13 Catalina Island, California, gets most of its water from a desalination plant.

© Rarpia/iStockphoto

CHAPTER 13 REVIEW

13.1 States of Matter: A Review

- Solids
  - Particles close together
  - High density
  - Incompressible
  - Maintain shape
- Liquids
  - Particles close together, yet free to move
  - Incompressible
  - Definite volume, but take the shape of the container
- Gases
  - Particles far apart
  - Low density
  - Compressible
  - Take volume and shape of container

13.2 Properties of Liquids

KEY TERMS

surface tension
capillary action
meniscus
evaporation
vaporization
sublimation
condensation
vapor pressure
volatile

- The resistance of a liquid to an increase in its surface area is the surface tension of the liquid.
- Capillary action is caused by the cohesive forces within the liquid and adhesive forces between the liquid and the walls of the container:
  - If the forces between the liquid and the container are greater than those within the liquid, the liquid will climb the walls of the container.
- A meniscus is evidence of cohesive and adhesive forces:
  - It is concave if the adhesive forces are stronger than the cohesive forces.
  - It is convex if the cohesive forces are stronger than the adhesive forces.
- During evaporation, molecules of greater-than-average kinetic energy escape from the liquid.
- Sublimation is the evaporation of a solid directly to a gas.
- Gaseous molecules return to the liquid state through condensation.
- At equilibrium the rate of evaporation equals the rate of condensation.
- The pressure exerted by the vapor in equilibrium with its liquid in a closed container is the vapor pressure of the liquid.
- A volatile substance evaporates readily.

13.3 Boiling Point and Melting Point

KEY TERMS

boiling point
normal boiling point
vapor pressure curve
melting point (or freezing point)

- The boiling point of a liquid is the temperature at which its vapor pressure equals the atmospheric pressure:
  - At 1 atm the boiling point is called the normal boiling point for a liquid.
- The temperature at which a solid is in equilibrium with its liquid phase is the freezing point or melting point.

13.4 Changes of State

KEY TERMS

heat of fusion
heat of vaporization

- A graph of the warming of a liquid is called a heating curve:
  - Horizontal lines on the heating curve represent changes of state:
    - The energy required to change 1 g of solid to liquid at its melting point is the heat of fusion.
    - The energy required to change 1 g of liquid to gas at its normal boiling point is the heat of vaporization.
    - The energy required to change the phase of a sample (at its melting or boiling point) is energy = (mass)(heat of fusion (or vaporization))
  - The energy required to heat molecules without a phase change is determined by energy = (mass)(specific heat)(Δt)

13.5 Intermolecular forces

KEY TERMS

intermolecular forces
intramolecular forces
dipole–dipole attractions
hydrogen bond
London dispersion forces

- Dipole–dipole attractions exist among polar molecules.

- A hydrogen bond is the dipole–dipole attraction between polar molecules containing any of these types of bonds: F—H, O—H, N—H.
- London dispersion forces cause attractions among nonpolar atoms and molecules.

13.6 Hydrates

KEY TERMS
hydrate
water of hydration
water of crystallization

- Solids that contain water molecules as part of their crystalline structure are called hydrates.
- Formulas of hydrates are given by writing the formula for the anhydrous compound followed by a dot and then the number of water molecules present.
- Water molecules in hydrates are bonded by electrostatic forces, which are weaker than covalent or ionic bonds:
  - Water of hydration can be removed by heating the hydrate to form a partially dehydrated compound or the anhydrous compound.

13.7 Water, a Unique Liquid

KEY TERM
semipermeable membrane

- Water is our most common resource, covering 75% of the Earth's surface.
- Water is a colorless, odorless, tasteless liquid with a melting point of 0°C and a boiling point of 100°C at 1 atm.
- The water molecule consists of 2 H atoms and 1 O atom bonded together at a 105° bond angle, making the molecule polar.
- Water can be formed in a variety of ways including:
  - Reclamation of wastewater
  - Reverse osmosis
  - Desalination
  - Low-temperature distillation
  - Hydrogen combustion

REVIEW QUESTIONS

1. In what state (solid, liquid, or gas) would H_2S, H_2Se, and H_2Te be at 0°C? (Table 13.2)

2. What property or properties of liquids are similar to solids?

3. What property or properties of liquids are similar to gases?

4. If water were placed in the containers in Figure 13.5, would they all have the same vapor pressure at the same temperature? Explain.

5. Why doesn't the vapor pressure of a liquid depend on the amount of liquid and vapor present?

6. In Figure 13.5, in which case, (a), (b) or (c), will the atmosphere above the liquid reach a point of saturation?

7. Suppose a solution of ethyl ether and ethyl alcohol is placed in the closed bottle in Figure 13.5. (Use Figure 13.7 for information on the substances.)
(a) Are both substances present in the vapor?
(b) If the answer to part (a) is yes, which has more molecules in the vapor?

8. Explain why rubbing alcohol warmed to body temperature still feels cold when applied to your skin.

9. The vapor pressure at 20°C for the following substances is

methyl alcohol	96 torr
acetic acid	11.7 torr
benzene	74.7 torr
bromine	173 torr
water	17.5 torr
carbon tetrachloride	91 torr
mercury	0.0012 torr
toluene	23 torr

(a) Arrange these substances in order of increasing rate of evaporation.
(b) Which substance listed has the highest boiling point? the lowest?

10. On the basis of the kinetic-molecular theory, explain why vapor pressure increases with temperature.

11. The temperature of the water in the pan on the burner (Figure 13.11) reads 100°C. What is the pressure of the atmosphere?

12. If ethyl alcohol is boiling in a flask and the atmospheric pressure is 543 torr, what is the temperature of the boiling liquid? (Use Figure 13.7.)

13. At approximately what temperature would each of the substances shown in Figure 13.7 boil when the pressure is 30 torr?

14. Use the graph in Figure 13.7 to find the following:
(a) boiling point of water at 500 torr
(b) normal boiling point of ethyl alcohol
(c) boiling point of ethyl ether at 0.50 atm

15. Suggest a method whereby water could be made to boil at 50°C.

16. Explain why a higher temperature is obtained in a pressure cooker than in an ordinary cooking pot.

17. What is the relationship between vapor pressure and boiling point?

18. The boiling point of ammonia, NH_3, is −33.4°C and that of sulfur dioxide, SO_2, is −10.0°C. Which has the higher vapor pressure at −40°C?

19. Explain what is occurring physically when a substance is boiling.

20. At what specific temperature will ethyl ether have a vapor pressure of 760 torr?

21. Compare the potential energy of the three states of water shown in Figure 13.11.

22. Consider Figure 13.8.
(a) Why is line *BC* horizontal? What is happening in this interval?
(b) What phases are present in the interval *BC*?
(c) When heating is continued after point *C*, another horizontal line, *DE*, is reached at a higher temperature. What does this line represent?

23. Account for the fact that an ice–water mixture remains at 0°C until all the ice is melted, even though heat is applied to it.
24. Which contains less energy: ice at 0°C or water at 0°C? Explain.
25. Why does a boiling liquid maintain a constant temperature when heat is continuously being added?
26. Define the terms *intermolecular forces* and *intramolecular forces*.
27. What kind of covalent bond is formed when atoms do not share electrons equally?
28. What property of a molecule most affects its ability to form instantaneous dipoles?
29. If water molecules were linear instead of bent, would the heat of vaporization be higher or lower? Explain.
30. The heat of vaporization for ethyl ether is 351 J/g and that for ethyl alcohol is 855 J/g. From these data, which of these compounds has hydrogen bonding? Explain.
31. Would there be more or less H bonding if water molecules were linear instead of bent? Explain.
32. In which condition are there fewer hydrogen bonds between molecules: water at 40°C or water at 80°C? Explain.
33. Which compound

$$H_2NCH_2CH_2NH_2 \quad \text{or} \quad CH_3CH_2CH_2NH_2$$

would you expect to have the higher boiling point? Explain. (Both compounds have similar molar masses.)

34. Why does water have such a relatively high boiling point?
35. Explain why HF (bp = 19.4°C) has a higher boiling point than HCl (bp = −85°C), whereas F_2 (bp = −188°C) has a lower boiling point than Cl_2 (bp = −34°C).
36. How do we specify 1, 2, 3, 4, 5, 6, 7, and 8 molecules of water in the formulas of hydrates? (Table 13.3)
37. Diagram a water molecule and point out the negative and positive ends of the dipole.
38. If the water molecule were linear, with all three atoms in a straight line rather than in the shape of a V, as shown in Figure 13.12, what effect would this have on the physical properties of water?
39. List six physical properties of water.
40. What condition is necessary for water to have its maximum density? What is its maximum density?
41. Why does ice float in water? Would ice float in ethyl alcohol (d = 0.789 g/mL)? Explain.
42. What water temperature would you theoretically expect to find at the bottom of a very deep lake? Explain.
43. Is the formation of hydrogen and oxygen from water an exothermic or an endothermic reaction? How do you know?

PAIRED EXERCISES

Most of the exercises in this chapter are available for assignment via the online homework management program, WileyPLUS (www.wileyplus.com). All exercises with blue numbers have answers in Appendix 5.

1. Rank the following molecules in order of increasing polarity: CH_3Cl, CH_3F, and CH_3Br.

2. Rank the following molecules in order of increasing polarity: H_2S, H_2Se, and H_2O.

3. Rank the following molecules in order of increasing London dispersion forces: CCl_4, CI_4, and CBr_4.

4. Rank the following molecules in order of increasing London dispersion forces: CO_2, SO_2, and CS_2.

5. In which of the following substances would you expect to find hydrogen bonding?
(a) C_3H_7OH
(b) H_2O_2
(c) $CHCI_3$
(d) PH_3
(e) HF

6. In which of the following substances would you expect to find hydrogen bonding?
(a) HI
(b) NH_3
(c) CH_2F_2
(d) C_2H_5OH
(e) H_2O

7. What is surface tension? How does it depend on intermolecular forces?

8. What is boiling point? How does it depend on intermolecular forces?

9. For each of the compounds in Question 5 that form hydrogen bonds, draw a diagram of the two molecules using a dotted line to indicate where the hydrogen bonding will occur.

10. For each of the compounds in Question 6 that form hydrogen bonds, draw a diagram of the two molecules, using a dotted line to indicate where the hydrogen bonding will occur.

11. You leave the house wearing a cotton T-shirt and are surprised by a sudden rainstorm. You notice that the water soaks into the T-shirt whereas it just beads up on your raincoat. In this example, are the adhesive forces or cohesive forces stronger between water and your T-shirt? Between water and your raincoat? Explain.

12. Rainex is a product that causes water to "bead" instead of spread when sprayed on car windshields. When the water forms these droplets, are adhesive forces or cohesive forces stronger? Explain.

13. Ether has a higher vapor pressure than water. Which of these two liquids has stronger intermolecular forces? Which liquid would you expect to have the higher boiling point?

14. Glycerol has a higher surface tension than water. What can you say about the relative strength of the intermolecular forces in the two compounds? Which has a higher boiling point?

15. Name these hydrates:
(a) $BaBr_2 \cdot 2\,H_2O$
(b) $AlCl_3 \cdot 6\,H_2O$
(c) $FePO_4 \cdot 4\,H_2O$

16. Name these hydrates:
(a) $MgNH_4PO_4 \cdot 6\,H_2O$
(b) $FeSO_4 \cdot 7\,H_2O$
(c) $SnCl_4 \cdot 5\,H_2O$

17. How many moles of compound are in 25.0 g of $Na_2CO_3 \cdot 10\ H_2O$?

18. How many moles of compound are in 25.0 g of $Na_2B_4O_7 \cdot 10\ H_2O$?

19. Upon heating 125 g $MgSO_4 \cdot 7\ H_2O$:
(a) how many grams of water can be obtained?
(b) how many grams of anhydrous compound can be obtained?

20. Upon heating 125 g $AlCl_3 \cdot 6\ H_2O$:
(a) how many grams of water can be obtained?
(b) how many grams of anhydrous compound can be obtained?

21. Cobalt(II) chloride hexahydrate, $CoCl_2 \cdot 6\,H_2O$, is often used as a humidity indicator. This is due to the fact that the hydrate is a deep magenta color while the anhydrous form is a pale blue. As the humidity level changes, the color changes as well. What is the mass percent water in cobalt(II) hexahydrate?

22. Gypsum, $CaSO_4 \cdot 2\ H_2O$, is used to manufacture cement. What is the mass percent of water in this compound?

23. A hydrated iron chloride compound was found to contain 20.66% Fe, 39.35% Cl, and 39.99% water. Determine the empirical formula of this hydrated compound.

24. A hydrated nickel chloride compound was found to contain 24.69% Ni, 29.83% Cl, and 45.48% water. Determine the empirical formula of this hydrated compound.

25. How many joules of energy are needed to change 275 g of water from 15°C to steam at 100.°C?

26. How many joules of energy must be removed from 325 g water at 35°C to form ice at 0°C?

27. Suppose 100. g of ice at 0°C are added to 300. g of water at 25°C. Is this sufficient ice to lower the temperature of the system to 0°C and still have ice remaining? Show evidence for your answer.

28. Suppose 35.0 g of steam at 100.°C are added to 300. g of water at 25°C. Is this sufficient steam to heat all the water to 100.°C and still have steam remaining? Show evidence for your answer.

29. If 75 g of ice at 0.0°C were added to 1.5 L of water at 75°C, what would be the final temperature of the mixture?

30. If 9560 J of energy were absorbed by 500. g of ice at 0.0°C, what would be the final temperature?

31. What are intermolecular forces and how are they different from covalent bonds?

32. Give an example of a nonpolar molecule. What kind of intermolecular forces are most important in this molecule?

ADDITIONAL EXERCISES

33. You walk out to your car just after a rain shower and notice that small droplets of water are scattered on the hood. How was it possible for the water to form those droplets?

34. Which causes a more severe burn: liquid water at 100°C or steam at 100°C? Why?

35. You have a shallow dish of alcohol set into a tray of water. If you blow across the tray, the alcohol evaporates, while the water cools significantly and eventually freezes. Explain why.

36. Regardless of how warm the outside temperature may be, we always feel cool when stepping out of a swimming pool, the ocean, or a shower. Why is this so?

37. Sketch a heating curve for a substance X whose melting point is 40°C and whose boiling point is 65°C.
(a) Describe what you will observe as a 60.-g sample of X is warmed from 0°C to 100°C.
(b) If the heat of fusion of X is 80. J/g, the heat of vaporization is 190. J/g, and if 3.5 J are required to warm 1 g of X each degree, how much energy will be needed to accomplish the change in (a)?

38. For the heating curve of water (Figure 13.8), why doesn't the temperature increase when ice is melting into liquid water and liquid water is changing into steam?

39. Why does the vapor pressure of a liquid increase as the temperature is increased?

40. Mount McKinley in Alaska is the tallest peak in North America with a summit at 20,320 ft or 6194 m. It is also known as Denali, meaning "the high one" in the native Koyukon Athabaskan language. The average atmospheric pressure at the top of Denali is approximately 330 torr. Use Figure 13.7 to determine the approximate boiling temperature of water on Denali.

41. Explain how anhydrous copper(II) sulfate ($CuSO_4$) can act as an indicator for moisture.

42. Write formulas for magnesium sulfate heptahydrate and disodium hydrogen phosphate dodecahydrate.

43. Analyze the graph below and draw conclusions as appropriate. The boiling point and molar mass are graphed for the noble gases. What generalizations can you draw from these data?

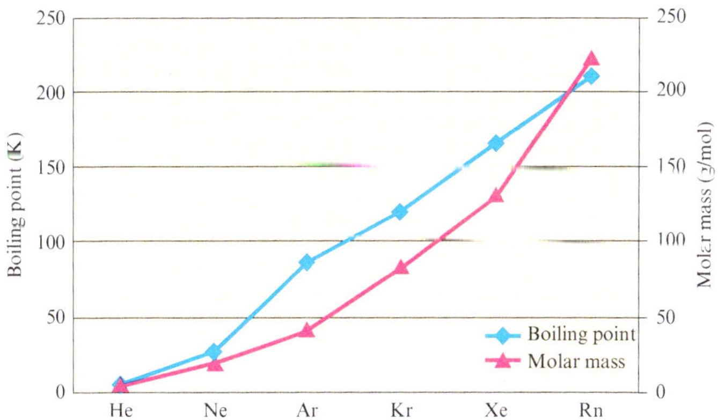

44. You decide to go to the mountains of Santa Fe, New Mexico (7200 ft elevation), for your spring holiday. One of the important tasks for you to complete is to boil eggs to color and hide for the egg hunt. You are in a hurry and know that eggs boiled for exactly 8 minutes at home at the beach in Santa Barbara, California, come out perfectly cooked. As everyone is cracking their eggs for breakfast after the egg hunt, you discover to your dismay that your eggs are not fully cooked. Explain this observation.

45. Look at the drawing on the next page of the water in the glass-graduated cylinder and the water in the plastic graduated cylinder. What differences do you observe in the shape of the meniscus? Plastics are composed of long-chain hydrocarbons, and glass is composed of silicon oxides. Based on this information, propose a reason for the differences in these two water samples.

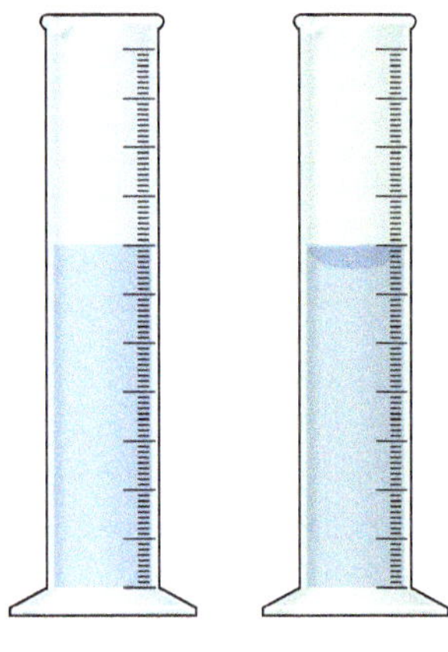

Plastic cylinder Glass cylinder

46. The flow rate of a liquid is related to the strength of the cohesive forces between molecules. The stronger the cohesive forces, the harder it is for the molecules to slide past one another. Honey is a very thick liquid that pours slowly at ordinary temperatures. A student tried an experiment. She filled five eyedroppers with honey at different temperatures and then measured the rate of dripping for each sample (see the table below). Graph the rate of dripping versus temperature for the honey and propose a relationship between temperature and flow rate. Propose an explanation for the trend that you observe.

Temperature (°C)	Number of drops/minute
10	3
20	17
30	52
50	87
70	104

47. You work in a hardware store and notice that whenever you spill water on the waxed floors it tends to bead up and stay fairly well confined. When you spill hexane, however, it spreads over the entire floor and is a big mess to clean up. Based on your knowledge of adhesive and cohesive forces, explain this difference in behavior.

48. Isobutyl propionate, the ester responsible for the flavor of rum, has a boiling point of 137°C and octyl acetate, the ester responsible for the flavor of oranges, has a boiling point of 210°C. Based on this information, which compound would you expect to have the higher vapor pressure? Explain how you came to this conclusion.

49. Define the term *boiling point*. How does the boiling point of a liquid change as intermolecular forces become stronger? Using kinetic-molecular theory (KMT), explain how differences in intermolecular forces affect boiling point.

50. What are dispersion forces? Which has greater dispersion forces, C_3H_8 or C_9H_{20}? Justify your answer.

51. A sample of pure acetic acid is also known as glacial acetic acid. This name comes from the fact that flasks of acetic acid would often freeze in the old laboratories of Europe. When acetic acid freezes, it forms many cracks resembling a glacier and thus the name. A heating curve for a sample of glacial acetic acid is pictured below.

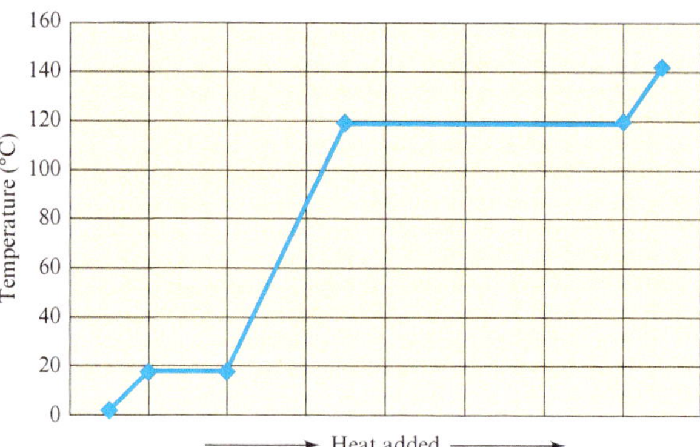

(a) Using this graph, determine the melting point and the boiling point of acetic acid.

(b) Redraw this graph and indicate regions where the solid, liquid, and gas phases will exist.

(c) Indicate regions where a transition of states is occurring and tell what phases are present.

52. How many calories are required to change 225 g of ice at 0°C to steam at 100.°C?

53. The molar heat of vaporization is the number of joules required to change 1 mol of a substance from liquid to vapor at its boiling point. What is the molar heat of vaporization of water?

54. The specific heat of zinc is 0.096 cal/g°C. Determine the energy required to raise the temperature of 250. g of zinc from room temperature (20.0°C) to 150.°C.

55. How many joules of energy would be liberated by condensing 50.0 mol of steam at 100.0°C and allowing the liquid to cool to 30.0°C?

56. How many kilojoules of energy are needed to convert 100. g of ice at −10.0°C to water at 20.0°C? (The specific heat of ice at −10.0°C is 2.01 J/g°C.)

57. What mass of water must be decomposed to produce 25.0 L of oxygen at STP?

58. Suppose 1.00 mol of water evaporates in 1.00 day. How many water molecules, on average, leave the liquid each second?

59. Compare the volume occupied by 1.00 mol of liquid water at 0°C and 1.00 mol of water vapor at STP.

60. A mixture of 80.0 mL of hydrogen and 60.0 mL of oxygen is ignited by a spark to form water.

(a) Does any gas remain unreacted? Which one, H_2 or O_2?

(b) What volume of which gas (if any) remains unreacted? (Assume the same conditions before and after the reaction.)

61. A student (with slow reflexes) puts his hand in a stream of steam at 100.°C until 1.5 g of water have condensed. If the water then cools to room temperature (20.0°C), how many joules have been absorbed by the student's hand?

62. Determine which of the following molecules would hydrogen-bond with other molecules like it. For those that do not hydrogen-bond, explain why. For those that hydrogen-bond, draw a diagram of two molecules using a dotted line to indicate where the hydrogen bonding will occur.

(a) Br_2
(b) $CH_3—O—H$
(c) $CH_3—O—CH_3$
(d) H_2O
(e) H_2S

63. The heat of fusion of a substance is given in units of J/g. The specific heat of a substance is given in units of J/g °C. Why is a temperature factor not needed in the units for heat of fusion?
64. How many joules of energy are required to change 50.0 g Cu from 25.0°C to a liquid at its melting point, 1083°C?
Specific heat of Cu = 0.385 J/g°C
Heat of fusion for Cu = 134 J/g
65. You pour a steaming hot bowl of soup. After one taste, you burn your tongue and decide to cool it down with some ice. If you add 75 g of ice at 0°C to your soup and the final temperature of the ice–soup mixture is 87°C, how many joules of energy did the ice remove from the bowl of soup?
66. Write and balance the chemical equation for the decomposition of water into its component elements.
67. Use Figure 13.7 to determine the temperature in degrees Fahrenheit required to purify water by distillation at atmospheric pressures of 500 torr, 300 torr, and 100 torr.

CHALLENGE EXERCISES

68. You buy a box of borax ($Na_2B_4O_7 \cdot 10\ H_2O$) from the corner market in Phoenix, Arizona, in the middle of the summer. You open up the box and pour the borax into a weighed beaker. After all, you do want to be sure that you were not cheated by the manufacturer. You are distracted from your task and do not get back to weigh the filled beaker for several days. Upon weighing the beaker, you get the following data from the 5.0-lb box of borax.

Empty beaker	492.5 g
Filled beaker	2467.4 g

Were you cheated by the manufacturer? Why or why not?
69. Why does a lake freeze from the top down? What significance does this have for life on Earth?
70. Suppose 150. g of ice at 0.0°C are added to 0.120 L of water at 45°C. If the mixture is stirred and allowed to cool to 0.0°C, how many grams of ice remain?
71. Three students at Jamston High—Zack, Gaye, and Lamont—each had the opportunity to travel over spring break. As part of a project, each of them measured the boiling point of pure water at their vacation spot. Zack found a boiling point of 93.9°C, Gaye measured 101.1°C, and Lamont read 100.°C. Which student most likely went snow skiing near Ely, Nevada, and which student most likely went water skiing in Honolulu? From the boiling point information, what can you surmise about the Dead Sea region, the location of the third student's vacation? Explain.
72. Why does a change in pressure of a gas significantly affect its volume, whereas a change in pressure on a solid or liquid has negligible effect on their respective volumes? If the accompanying picture represents a liquid at the molecular level, draw what you might expect a solid and a gas to look like.

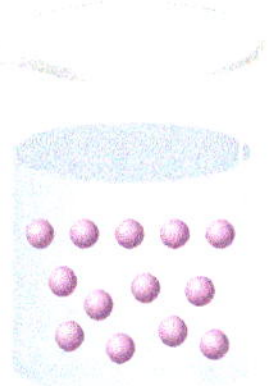

73. Ray ran a double-displacement reaction using 0.050 mol $CuCl_2$ and 0.10 mol $AgNO_3$ The resulting white precipitate was removed by filtration. The filtrate was accidentally left open on the lab bench for over a week, and when Ray returned the flask contained solid, blue crystals. Ray weighed the crystals and found they had a mass of 14.775 g. Was Ray expecting this number? If not, what did he expect?

ANSWERS TO PRACTICE EXERCISES

13.1 Vapor pressure is the pressure exerted by a vapor in equilibrium with its liquid. The vapor pressure of a liquid increases with increasing temperature.

13.2 They are the same.

13.3 28°C, 73°C, 93°C

13.4 approximately 95°C

13.5 44.8 kJ

13.6 SO_2 and CH_3F are polar.

13.7 (a) yes, (b) yes, (c) no

13.8 I_2

13.9 (a) $BeCO_3 \cdot 4\ H_2O$ (b) $Cd(MnO_4)_2 \cdot 6\ H_2O$ (c) $Cr(NO_3)_3 \cdot 9\ H_2O$ (d) $PtO_2 \cdot 3\ H_2O$

13.10 51.17% H_2O

14

Solutions

Steve Cole/Exactostock/SuperStock

CHAPTER OUTLINE

Most substances we encounter in our daily lives are mixtures. Often they are homogeneous mixtures, which are called *solutions*. Some solutions we commonly encounter are shampoo, soft drinks, and wine. Blood plasma is a complex mixture composed of compounds and ions dissolved in water and proteins suspended in the solution. These solutions all have water as a main component, but many common items, such as air, gasoline, and steel, are also solutions that do not contain water. Brass, a solid solution of zinc and copper, is used to make musical instruments like the trumpet in the photo above. What are the necessary components of a solution? Why do some substances dissolve, while others do not? What effect does a dissolved substance have on the properties of the solution? Answering these questions is the first step in understanding the solutions we encounter in our daily lives.

14.1 General Properties of Solutions

List the properties of a true solution.

LEARNING OBJECTIVE

KEY TERMS

solution
solute
solvent

The term **solution** is used in chemistry to describe a system in which one or more substances are homogeneously mixed or dissolved in another substance. A simple solution has two components: a solute and a solvent. The **solute** is the component that is dissolved or is the least abundant component in the solution. The **solvent** is the dissolving agent or the most abundant component in the solution. For example, when salt is dissolved in water to form a solution, salt is the solute and water is the solvent. Complex solutions containing more than one solute and/or more than one solvent are common.

The three states of matter—solid, liquid, and gas—give us nine different types of solutions: solid dissolved in solid, solid dissolved in liquid, solid dissolved in gas, liquid dissolved in liquid, and so on. Of these, the most common solutions are solid dissolved in liquid, liquid dissolved in liquid, gas dissolved in liquid, and gas dissolved in gas. Some common types of solutions are listed in TABLE 14.1.

TABLE 14.1 | Common Types of Solutions

Phase of solution	Solute	Solvent	Example
Gas	gas	gas	air
Liquid	gas	liquid	soft drinks
Liquid	liquid	liquid	antifreeze
Liquid	solid	liquid	salt water
Solid	gas	solid	H_2 in Pt
Solid	solid	solid	brass

A true solution is one in which the particles of dissolved solute are molecular or ionic in size, generally in the range of 0.1 to 1 nm (10^{-8} to 10^{-7} cm). The properties of a true solution are as follows:

1. A mixture of two or more components—solute and solvent—is homogeneous and has a variable composition; that is, the ratio of solute to solvent can be varied.
2. The dissolved solute is molecular or ionic in size.
3. It is either colored or colorless and is usually transparent.
4. The solute remains uniformly distributed throughout the solution and will not settle out with time.
5. The solute can generally be separated from the solvent by purely physical means (for example, by evaporation).

Let's illustrate these properties using water solutions of sugar and of potassium permanganate.

EXAMPLE 14.1

Consider two sugar solutions: solution A containing 10 g of sugar added to 100 mL of water and solution B containing 20 g of sugar added to 100 mL of water. Each is stirred until all the sugar dissolves. Explain why these solutions are considered to be true solutions.

SOLUTION

1. All the sugar dissolves in each solution, showing we can vary the composition of a sugar solution.
2. Every portion of each solution has the same sweet taste uniformly (although you should not taste solutions in the laboratory) because the sugar molecules are uniformly distributed throughout.

3. If the solution is confined so that no water is lost, the solution will look and taste the same a week or more later.
4. The solutions cannot be separated by filtering.
5. Careful evaporation will separate the solution into its components: sugar and water.

Richard Megna/Fundamental Photographs

Note the beautiful purple trails of $KMnO_4$ as the crystals dissolve.

PRACTICE 14.1

Identify the solute and the solvent in each of these solutions:

(a) air (b) seawater (c) carbonated water

To observe the dissolving of potassium permanganate ($KMnO_4$), we drop a few crystals of it in a beaker of water. Almost at once, the beautiful purple color of dissolved permanganate ions (MnO_4^-) appears and streams to the bottom of the beaker as the crystals dissolve. After a while, the purple color disperses until it's evenly distributed throughout the solution. This dispersal demonstrates that molecules and ions move about freely and spontaneously (diffuse) in a liquid or solution.

Solution permanency is explained in terms of the kinetic-molecular theory (see Section 12.6). According to this theory, both the solute and solvent particles (molecules or ions) are in constant random motion. This motion is energetic enough to prevent the solute particles from settling out under the influence of gravity.

CHECK YOUR UNDERSTANDING

14.1 Properties of True Solutions

➡ WileyPLUS

14.2 Solubility

LEARNING OBJECTIVE Define solubility and the factors that affect it.

KEY TERMS

- solubility
- miscible
- immiscible
- concentration of a solution
- saturated solution
- unsaturated solution
- supersaturated solution

The term **solubility** describes the amount of one substance (solute) that will dissolve in a specified amount of another substance (solvent) under stated conditions. For example, 36.0 g of sodium chloride will dissolve in 100 g of water at 20°C. We say then that the solubility of NaCl in water is 36.0 g/100 g H_2O at 20°C. Solubility is often used in a relative way. For instance, we say that a substance is very soluble, moderately soluble, slightly soluble, or insoluble. Although these terms do not accurately indicate how much solute will dissolve, they are frequently used to describe the solubility of a substance qualitatively.

Two other terms often used to describe solubility are *miscible* and *immiscible*. Liquids that are capable of mixing and forming a homogeneous solution are **miscible**; those that do not form solutions or are generally insoluble in each other are **immiscible**. Methyl alcohol and water are miscible in each other in all proportions. Oil and water are immiscible, forming two separate layers when they are mixed, as shown in **FIGURE 14.1**.

The general guidelines for the solubility of common ionic compounds (salts) are given in **FIGURE 14.2**. These guidelines have some exceptions, but they provide a solid foundation for the compounds considered in this course. The solubilities of over 200 compounds are given in the Solubility Table in Appendix 4. Solubility data for thousands of compounds can be found by consulting standard reference sources.*

The quantitative expression of the amount of dissolved solute in a particular quantity of solvent is known as the **concentration of a solution**. Several methods of expressing concentration are described in Section 14.4.

Predicting solubilities is complex and difficult. Many variables, such as size of ions, charge on ions, interaction between ions, interaction between solute and solvent, and temperature, complicate the problem. Because of the factors involved, the general rules of solubility given in Figure 14.2 have many exceptions. However, these rules are useful because they do apply to many of the more common compounds that we encounter in the study of chemistry. Keep in mind that these are rules, not laws, and are therefore subject to exceptions. Fortunately, the solubility of a solute is relatively easy to determine experimentally. Now let's examine the factors related to solubility.

Kip Peticolas/FundamentalPhotographs

FIGURE 14.1 An immiscible mixture of oil and water.

*Two commonly used handbooks are *Lange's Handbook of Chemistry*, 16th ed. (New York: McGraw-Hill, 2009), and the *CRC Handbook of Chemistry and Physics*, 93rd ed. (Cleveland: Chemical Rubber Co., 2012).

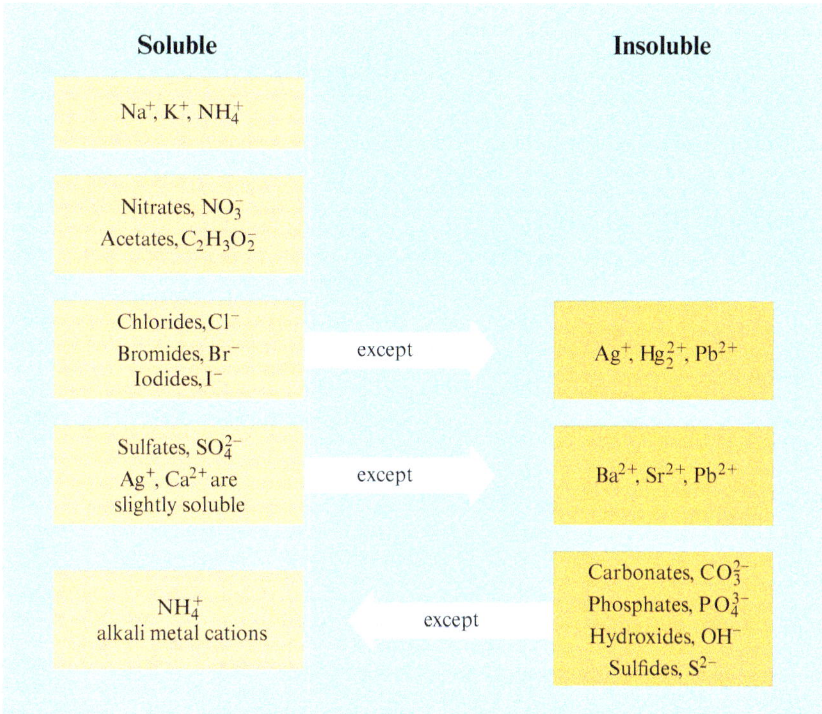

FIGURE 14.2 The solubility of various common ions. Substances containing the ions on the left are generally soluble in cold water, while those substances containing the ions on the right are insoluble in cold water. The arrows point to the exceptions.

CHECK YOUR UNDERSTANDING

The Nature of the Solute and Solvent

The old saying "like dissolves like" has merit, in a general way. Polar or ionic substances tend to be more compatible with other polar substances. Nonpolar substances tend to be compatible with other nonpolar substances and less soluble with polar substances. Thus ionic compounds, which are polar, tend to be much more soluble in water, which is polar, than in solvents such as ether, hexane, or benzene, which are essentially nonpolar. Sodium chloride, an ionic substance, is soluble in water, slightly soluble in ethyl alcohol (less polar than water), and insoluble in ether and benzene. Pentane, C_5H_{12}, a nonpolar substance, is only slightly soluble in water but is very soluble in benzene and ether.

At the molecular level the formation of a solution from two nonpolar substances, such as hexane and benzene, can be visualized as a process of simple mixing. These non-polar molecules, having little tendency to either attract or repel one another, easily intermingle to form a homogeneous solution.

Solution formation between polar substances is much more complex. See, for example, the process by which sodium chloride dissolves in water (FIGURE 14.3). Water molecules are very polar and are attracted to other polar molecules or ions. When salt crystals are put into water, polar water molecules become attracted to the sodium and chloride ions on the crystal surfaces enabling the Na^+ and Cl^- ions to separate. The positive end of the water dipole is attracted to the Cl^- ions, and the negative end of the water dipole to the Na^+ ions this permits the ions to move apart, making room for more water dipoles. Thus the surface ions are surrounded by water molecules, becoming hydrated ions, $Na^+(aq)$ and $Cl^-(aq)$, and slowly diffuse away from the crystals and dissolve in solution:

$$NaCl(crystal) \xrightarrow{H_2O} Na^+(aq) + Cl^-(aq)$$

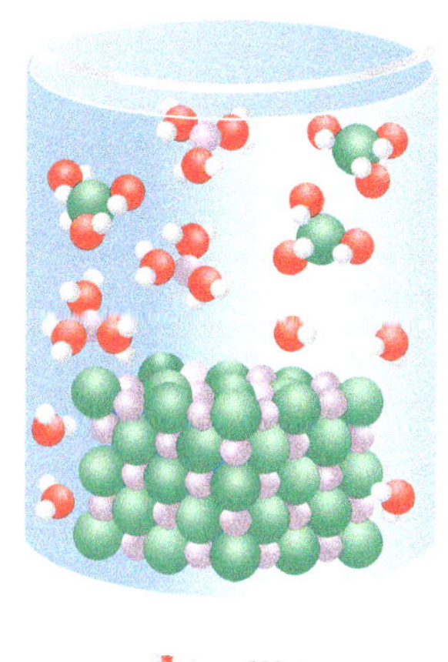

FIGURE 14.3 Dissolution of sodium chloride in water. Polar water molecules are attracted to Na^+ and Cl^- ions in the salt crystal, this enables the ions move apart and become surrounded by water dipoles. The hydrated ions slowly diffuse away from the crystal to become dissolved in solution.

Examination of the data in TABLE 14.2 raises some complex questions regarding the relationship between temperature and solubility?

TABLE 14.2 Solubility of Alkali Metal Halides in Water

Salt	Solubility (g salt/100 g H_2O) 0°C	Solubility (g salt/100 g H_2O) 100°C
LiF	0.12	0.14 (at 35°C)
LiCl	63.7	127.5
LiBr	143	266
LiI	151	481
NaF	4	5
NaCl	35.7	39.8
NaBr	79.5	121
NaI	158.7	302
KF	92.3 (at 18°C)	Very soluble
KCl	27.6	57.6
KBr	53.5	56.7
KI	127.5	102

The Effect of Temperature on Solubility

Temperature affects the solubility of most substances, as shown by the data in TABLE 14.2. Most solutes have a limited solubility in a specific solvent at a fixed temperature. For most solids dissolved in a liquid, an increase in temperature results in increased solubility (FIGURE 14.4). However, no single rule governs the solubility of solids in liquids with change in temperature. Some solids increase in solubility only slightly with increasing temperature (see NaCl in Figure 14.4); other solids decrease in solubility with increasing temperature (see Li_2SO_4 in Figure 14.4).

On the other hand, the solubility of a gas in water usually decreases with increasing temperature (see HCl and SO_2 in Figure 14.4). The tiny bubbles that form when water is heated are due to the decreased solubility of air at higher temperatures. The decreased solubility of gases at higher temperatures is explained in terms of the kinetic-molecular theory (KMT) by assuming that, in order to dissolve, the gas molecules must have attraction of some sort with the molecules of the liquid. An increase in temperature decreases the solubility of the gas because it increases the kinetic energy (speed) of the gas molecules and thereby decreases their ability to interact with the liquid molecules.

CHECK YOUR UNDERSTANDING

14.6 Relationship between Temperature and Solubility

WileyPLUS

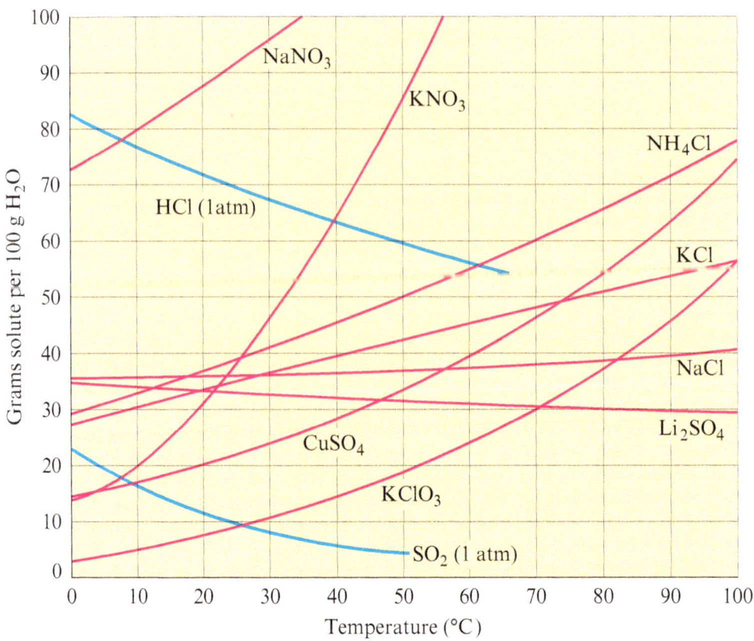

FIGURE 14.4 Solubility of various compounds in water. Solids are shown in red and gases are shown in blue.

The Effect of Pressure on Solubility

Small changes in pressure have little effect on the solubility of solids in liquids or liquids in liquids but have a marked effect on the solubility of gases in liquids. The solubility of a gas in a liquid is directly proportional to the pressure of that gas above the solution. Thus the amount of a gas dissolved in solution will double if the pressure of that gas over the solution is doubled. For example, carbonated beverages contain dissolved carbon dioxide under pressures greater than atmospheric pressure. When a can of carbonated soda is opened, the pressure is immediately reduced to the atmospheric pressure, and the excess dissolved carbon dioxide bubbles out of the solution.

Pouring root beer into a glass illustrates the effect of pressure on solubility. The escaping CO_2 produces the foam.

CHECK YOUR UNDERSTANDING

14.7 Relationship between Pressure and Solubility

➡ WileyPLUS

Saturated, Unsaturated, and Supersaturated Solutions

At a specific temperature there is a limit to the amount of solute that will dissolve in a given amount of solvent. When this limit is reached, the resulting solution is said to be *saturated*. For example, when we put 40.0 g of KCl into 100 g of H_2O at 20°C, we find that 34.0 g of KCl dissolve and 6.0 g of KCl remain undissolved. The solution formed is a saturated solution of KCl.

Two processes are occurring simultaneously in a saturated solution. The solid is dissolving into solution, and, at the same time, the dissolved solute is crystallizing out of solution. This may be expressed as

$$\text{solute (undissolved)} \rightleftharpoons \text{solute (dissolved)}$$

When these two opposing processes are occurring at the same rate, the amount of solute in solution is constant, and a condition of equilibrium is established between dissolved and undissolved solute. Therefore, a **saturated solution** contains dissolved solute in equilibrium with undissolved solute. Thus, any point on any solubility curve (Figure 14.4) represents a saturated solution of that solute. For example, a solution containing 60 g NH_4Cl per 100 g H_2O is saturated at 70°C.

It's important to state the temperature of a saturated solution because a solution that is saturated at one temperature may not be saturated at another. If the temperature of a saturated solution is changed, the equilibrium is disturbed, and the amount of dissolved solute will change to reestablish equilibrium.

A saturated solution may be either dilute or concentrated, depending on the solubility of the solute. A saturated solution can be conveniently prepared by dissolving a little more than the saturated amount of solute at a temperature somewhat higher than room temperature. Then the amount of solute in solution will be in excess of its solubility at room temperature, and, when the solution cools, the excess solute will crystallize, leaving the solution saturated. (In this case, the solute must be more soluble at higher temperatures and must not form a supersaturated solution.) Examples expressing the solubility of saturated solutions at two different temperatures are given in TABLE 14.3.

An **unsaturated solution** contains less solute per unit of volume than does its corresponding saturated solution. In other words, additional solute can be dissolved in an unsaturated solution without altering any other conditions. Consider a solution made by adding 40 g of KCl to 100 g of H_2O at 20°C (see Table 14.3). The solution formed will be saturated and will contain about 6 g of undissolved salt because the maximum amount of KCl that can dissolve in 100 g of H_2O at 20°C is 34 g. If the solution is now heated and maintained at 50°C, all the salt will dissolve and even more can be dissolved. The solution at 50°C is unsaturated.

CHECK YOUR UNDERSTANDING

14.8 Saturated Solutions

➡ WileyPLUS

TABLE 14.3 Saturated Solutions at 20°C and 50°C

	Solubility (g solute/100 g H_2O)	
Solute	**20°C**	**50°C**
NaCl	36.0	37.0
KCl	34.0	42.6
$NaNO_3$	88.0	114.0
$KClO_3$	7.4	19.3
$AgNO_3$	222.0	455.0
$C_{12}H_{22}O_{11}$	203.9	260.4

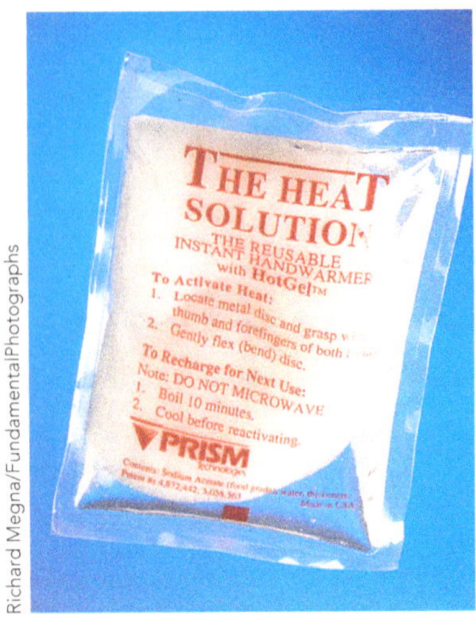

Richard Megna/FundamentalPhotographs

The heat released in this hot pack results from the crystallization of a supersaturated solution of sodium acetate.

In some circumstances, solutions can be prepared that contain more solute than needed for a saturated solution at a particular temperature. These solutions are said to be supersaturated. However, we must qualify this definition by noting that a **supersaturated solution** is unstable. Disturbances, such as jarring, stirring, scratching the walls of the container, or dropping in a "seed" crystal, cause the supersaturation to return to saturation, releasing heat. When a supersaturated solution is disturbed, the excess solute crystallizes out rapidly, returning the solution to a saturated state.

Supersaturated solutions are not easy to prepare but may be made from certain substances by dissolving, in warm solvent, an amount of solute greater than that needed for a saturated solution at room temperature. The warm solution is then allowed to cool very slowly. With the proper solute and careful cooling, a supersaturated solution will result.

EXAMPLE 14.2

Will a solution made by adding 2.5 g of $CuSO_4$ to 10 g of H_2O be saturated or unsaturated at 20°C?

SOLUTION

We first need to know the solubility of $CuSO_4$ at 20°C. From Figure 14.4, we see that the solubility of $CuSO_4$ at 20°C is about 20 g per 100 g of H_2O. This amount is equivalent to 2.1 g of $CuSO_4$ per 10 g of H_2O.

Since 2.5 g per 10 g of H_2O is greater than 2.0 g per 10 g of H_2O, the solution will be saturated and 0.5 g of $CuSO_4$ will be undissolved.

ONLINE LEARNING MODULE
Solubility
WileyPLUS

PRACTICE 14.2

Will a solution made by adding 9.0 g NH_4Cl to 20 g of H_2O be saturated or unsaturated at 50°C?

14.3 Rate of Dissolving Solids

LEARNING OBJECTIVE Describe the factors that affect the rate at which a solid dissolves.

The rate at which a solid dissolves is governed by (1) the size of the solute particles, (2) the temperature, (3) the concentration of the solution, and (4) agitation or stirring. Let's look at each of these conditions:

1. *Particle size.* A solid can dissolve only at the surface that is in contact with the solvent. Because the surface-to-volume ratio increases as size decreases, smaller crystals dissolve faster than large ones. For example, if a salt crystal 1 cm on a side (a surface area of 6 cm^2) is divided into 1000 cubes, each 0.1 cm on a side, the total surface of the smaller cubes is 60 cm^2—a 10-fold increase in surface area (FIGURE 14.5).

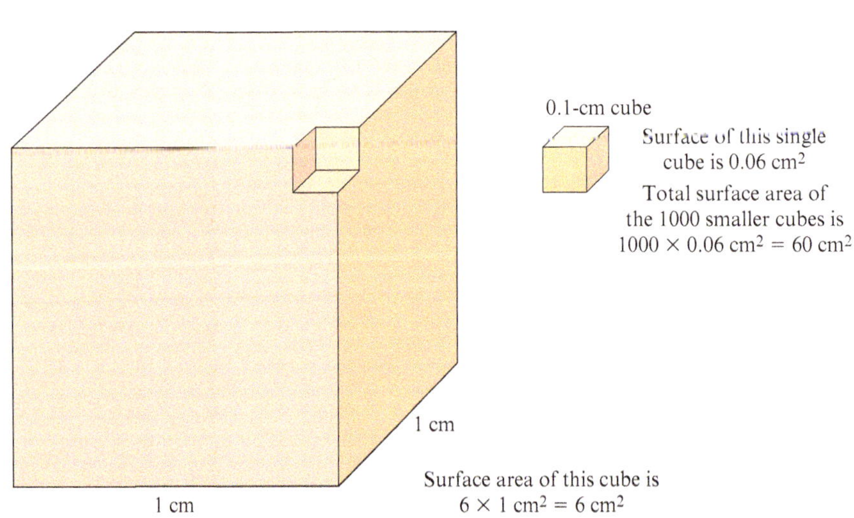

FIGURE 14.5 Surface area of crystals. A crystal 1 cm on a side has a surface area of 6 cm^2. Subdivided into 1000 smaller crystals, each 0.1 cm on a side, the total surface area is increased to 60 cm^2.

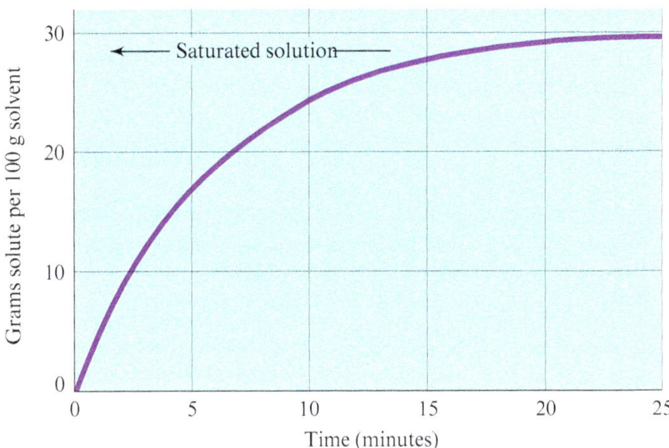

FIGURE 14.6 Rate of dissolution of a solid solute in a solvent. The rate is maximum at the beginning and decreases as the concentration approaches saturation.

2. *Temperature.* In most cases, the rate of dissolving of a solid increases with temperature. This increase is due to kinetic effects. The solvent molecules move more rapidly at higher temperatures and strike the solid surfaces more often, causing the rate of dissolving to increase.
3. *Concentration of the solution.* When the solute and solvent are first mixed, the rate of dissolving is at its maximum. As the concentration of the solution increases and the solution becomes more nearly saturated with the solute, the rate of dissolving decreases greatly. The rate of dissolving is graphed in FIGURE 14.6. Note that about 17 g dissolve in the first five-minute interval, but only about 1 g dissolves in the fourth five-minute interval. Although different solutes show different rates, the rate of dissolving always becomes very slow as the concentration approaches the saturation point.
4. *Agitation or stirring.* The effect of agitation or stirring is kinetic. When a solid is first put into water, it comes in contact only with solvent in its immediate vicinity. As the solid dissolves, the amount of dissolved solute around the solid becomes more and more concentrated, and the rate of dissolving slows down. If the mixture is not stirred, the dissolved solute diffuses very slowly through the solution; weeks may pass before the solid is entirely dissolved. Stirring distributes the dissolved solute rapidly through the solution, and more solvent is brought into contact with the solid, causing it to dissolve more rapidly.

EXAMPLE 14.3

What does it mean to increase the surface area of a solid? Explain how this change causes an increase in the rate of dissolving.

SOLUTION

Increasing the surface area means breaking up the solid to form smaller particles that have a larger total surface area. Increasing the surface area speeds up dissolving because more solid comes in contact with the solute more quickly.

PRACTICE 14.3

Use a molecular explanation to explain why increasing the temperature speeds up the rate of dissolving a solid in a liquid.

CHECK YOUR UNDERSTANDING

14.9 Factors Affecting the Rate of Dissolution

WileyPLUS

14.4 Concentration of Solutions

Solve problems involving mass percent, volume percent, molarity, and dilution.

LEARNING OBJECTIVE

KEY TERMS

dilute solution
concentrated solution
parts per million (ppm)
molarity (M)

Many solids must be put into solution to undergo appreciable chemical reaction. We can write the equation for the double-displacement reaction between sodium chloride and silver nitrate:

$$NaCl + AgNO_3 \longrightarrow AgCl + NaNO_3$$

But suppose we mix solid NaCl and solid $AgNO_3$ and look for a chemical change. If any reaction occurs, it is slow and virtually undetectable. In fact, the crystalline structures of NaCl and $AgNO_3$ are so different that we could separate them by tediously picking out each kind of crystal from the mixture. But if we dissolve the NaCl and $AgNO_3$ separately in water and mix the two solutions, we observe the immediate formation of a white, curd-like precipitate of silver chloride.

Molecules or ions must collide with one another in order to react. In the foregoing example, the two solids did not react because the ions were securely locked within their crystal structures. But when the NaCl and $AgNO_3$ are dissolved, their crystal lattices are broken down and the ions become mobile. When the two solutions are mixed, the mobile Ag^+ and Cl^- ions come into contact and react to form insoluble AgCl, which precipitates out of solution. The soluble Na^+ and NO_3^- ions remain mobile in solution but form the crystalline salt $NaNO_3$ when the water is evaporated:

$$NaCl(aq) + AgNO_3(aq) \longrightarrow AgCl(s) + NaNO_3(aq)$$

$$\underbrace{Na^+(aq) + Cl^-(aq)}_{\text{sodium chloride solution}} + \underbrace{Ag^+(aq) + NO_3^-(aq)}_{\text{silver nitrate solution}} \longrightarrow \underbrace{AgCl(s)}_{\text{silver chloride}} + \underbrace{Na^+(aq) + NO_3^-(aq)}_{\text{sodium nitrate in solution}}$$

The mixture of the two solutions provides a medium or space in which the Ag^+ and Cl^- ions can react. (See Chapter 15 for further discussion of ionic reactions.)

Solutions also function as diluting agents in reactions in which the undiluted reactants would combine with each other too violently. Moreover, a solution of known concentration provides a convenient method for delivering a specific amount of reactant.

The concentration of a solution expresses the amount of solute dissolved in a given quantity of solvent or solution. Because reactions are often conducted in solution, it's important to understand the methods of expressing concentration and to know how to prepare solutions of particular concentrations. The concentration of a solution may be expressed qualitatively or quantitatively. Let's begin with a look at the qualitative methods of expressing concentration.

Dilute and Concentrated Solutions

When we say that a solution is *dilute* or *concentrated*, we are expressing, in a relative way, the amount of solute present. One gram of a compound and 2 g of a compound in solution are both dilute solutions when compared with the same volume of a solution containing 20 g of a compound. Ordinary concentrated hydrochloric acid contains 12 mol of HCl per liter of solution. In some laboratories, the dilute acid is made by mixing equal volumes of water and the concentrated acid. In other laboratories the concentrated acid is diluted with two or three volumes of water, depending on its use. The term **dilute solution**, then, describes a solution that contains a relatively small amount of dissolved solute. Conversely, a **concentrated solution** contains a relatively large amount of dissolved solute.

Mass Percent Solution

The mass percent method expresses the concentration of the solution as the percent of solute in a given mass of solution. It says that for a given mass of solution, a certain percent of that mass is solute. Suppose we take a bottle from the reagent shelf that reads "sodium hydroxide, NaOH, 10%." This statement means that for every 100 g of this solution, 10 g will be NaOH and 90 g will be water. (Note that this amount of solution is 100 g, not 100 mL.) We could also make this same concentration of solution by dissolving 2.0 g of NaOH in 18 g of water. Mass percent concentrations are most generally used for solids dissolved in liquids:

Note that mass percent is independent of the formula for the solute.

$$\text{mass percent} = \frac{\text{g solute}}{\text{g solute} + \text{g solvent}} \times 100 = \frac{\text{g solute}}{\text{g solution}} \times 100$$

As instrumentation advances are made in chemistry, our ability to measure the concentration of dilute solutions is increasing as well. In addition to mass percent, chemists now commonly use **parts per million (ppm)**:

$$\text{parts per million} = \frac{\text{g solute}}{\text{g solute} + \text{g solvent}} \times 1{,}000{,}000$$

Currently, air and water contaminants, drugs in the human body, and pesticide residues are measured in parts per million.

ENHANCED EXAMPLE 14.4 WileyPLUS

What is the mass percent of sodium hydroxide in a solution that is made by dissolving 8.00 g NaOH in 50.0 g H_2O?

ONLINE LEARNING MODULE
Calculating Mass Percent for a Solution
➡ WileyPLUS

SOLUTION

Read **Knowns:** 8.00 g NaOH
50.0 g H_2O

Solving for: mass percent

Calculate $\left(\dfrac{8.00 \text{ g NaOH}}{8.00 \text{ g NaOH} + 50.0 \text{ g } H_2O}\right) = 13.8\%$ NaOH solution

ENHANCED EXAMPLE 14.5 WileyPLUS

What masses of potassium chloride and water are needed to make 250. g of 5.00% solution?

SOLUTION

Read **Knowns:** 250. g solution
5.00% solution

Solving for: masses KCl and water

Plan 5.00% of 250. g = (0.0500)(250. g) = 12.5 g KCl (solute)

Calculate 250. g − 12.5 g = 238 g H_2O
Dissolving 12.5 g KCl in 238 g H_2O gives a 5.00% solution.

ENHANCED EXAMPLE 14.6 WileyPLUS

A 34.0% sulfuric acid solution had a density of 1.25 g/mL. How many grams of H_2SO_4 are contained in 1.00 L of this solution?

ONLINE LEARNING MODULE
Using Concentration to Determine Mass of a Solute
➡ WileyPLUS

SOLUTION

Read **Knowns:** $d = 1.25$ g/mL
$V = 1.00$ L
34.0% H_2SO_4 solution

Solving for: mass H_2SO_4

Plan Find the mass of the solution from the density and then use the mass percent to determine the mass of H_2SO_4.

Calculate $d = \text{mass}/V$

$$\text{mass of solution} = \left(\frac{1.25 \text{ g}}{\cancel{\text{mL}}}\right)(1.00 \times 10^3 \cancel{\text{mL}}) = 1250 \text{ g (solution)}$$

$$\text{mass percent} = \left(\frac{\text{g solute}}{\text{g solution}}\right)100$$

$$\text{g solute} = \frac{(\text{mass percent})(\text{g solution})}{100}$$

$$\text{g solute} = \frac{(34.0)(1250 \text{ g})}{100} = 425 \text{ g } H_2SO_4$$

Therefore, 1.00 L of 34.0% H_2SO_4 solution contains 425 g H_2SO_4.

PRACTICE 14.4

What is the mass percent of Na_2SO_4 in a solution made by dissolving 25.0 g Na_2SO_4 in 225.0 g H_2O?

Mass/Volume Percent (m/v)

This method expresses concentration as grams of solute per 100 mL of solution. With this system, a 10.0%-m/v-glucose solution is made by dissolving 10.0 g of glucose in water, diluting to 100 mL, and mixing. The 10.0%-m/v solution could also be made by diluting 20.0 g to 200 mL, 50.0 g to 500 mL, and so on. Of course, any other appropriate dilution ratio may be used:

$$\text{mass/volume percent} = \frac{\text{g solute}}{\text{mL solution}} \times 100$$

ENHANCED EXAMPLE 14.7 **WileyPLUS**

A 3.0% (m/V) H_2O_2 solution is commonly used as a topical antiseptic to prevent infection. What volume of this solution will contain 10. g of H_2O_2?

SOLUTION

First solve the mass/volume percent equation for mL of solution:

$$\text{mL solution} = \frac{(\text{g solute})}{(\text{m/v percent})}(100)$$

$$\text{mL solution} = \frac{(10.\ \text{g solute})}{(3.0\ \text{m/v percent})}(100) = 330\ \text{mL}$$

Volume Percent

Solutions that are formulated from two liquids are often expressed as *volume percent* with respect to the solute. The volume percent is the volume of a liquid in 100 mL of solution. The label on a bottle of ordinary rubbing alcohol reads "isopropyl alcohol, 70% by volume." Such a solution could be made by mixing 70 mL of alcohol with water to make a total volume of 100 mL, but we cannot use 30 mL of water because the two volumes are not necessarily additive:

$$\text{volume percent} = \frac{\text{volume of liquid in question}}{\text{total volume of solution}} \times 100$$

Volume percent is used to express the concentration of alcohol in beverages. Wines generally contain 12% alcohol by volume. This translates into 12 mL of alcohol in each 100 mL of wine. The beverage industry also uses the concentration unit of *proof* (twice the volume percent). Pure ethyl alcohol is 100% and therefore 200 proof. Scotch whiskey is 86 proof, or 43% alcohol by volume.

Molarity

Mass percent doesn't give any information regarding the number of moles of solute in a solution. For example, 1000. g of 10.0% NaOH solution contains 100. g NaOH; 1000. g of 10.0% KOH solution contains 100. g KOH. In terms of moles of NaOH and KOH, these solutions contain

$$\text{mol NaOH} = (100.\ \text{g NaOH})\left(\frac{1\ \text{mol NaOH}}{40.00\ \text{g NaOH}}\right) = 2.50\ \text{mol NaOH}$$

$$\text{mol KOH} = (100.\ \text{g KOH})\left(\frac{1\ \text{mol KOH}}{56.11\ \text{g KOH}}\right) = 1.78\ \text{mol KOH}$$

From these figures, we see that the two 10.0% solutions do not contain the same number of moles of NaOH and KOH. As a result, we find that a 10.0% NaOH solution contains more reactive base than a 10.0% KOH solution.

FIGURE 14.7 Preparation of a 1 *M* solution.

We need a method of expressing concentration that will easily indicate how many moles of solute are present per unit volume of solution. For this purpose, the concentration unit known as molarity is used in calculations involving chemical reactions.

A 1-molar solution contains 1 mol of solute per liter of solution. For example, to make a 1-molar solution of sodium hydroxide (NaOH) we dissolve 40.00 g NaOH (1 mol) in water and dilute the solution with more water to a volume of 1 L. The solution contains 1 mol of the solute in 1 L of solution and is said to be 1 molar in concentration. FIGURE 14.7 illustrates the preparation of a 1-molar solution. Note that the volume of the solute and the solvent together is 1 L.

The concentration of a solution can, of course, be varied by using more or less solute or solvent; but in any case the **molarity (*M*)** of a solution is the number of moles of solute per liter of solution. The abbreviation for molarity is *M*. The units of molarity are moles per liter. The expression "2.0 *M* NaOH" means a 2.0-molar solution of NaOH (2.0 mol, or 80. g, of NaOH dissolved in water to make 1.0 L of solution):

$$\text{molarity} = M = \frac{\text{number of moles of solute}}{\text{liter of solution}} = \frac{\text{moles}}{\text{liter}}$$

Flasks that are calibrated to contain specific volumes at a particular temperature are used to prepare solutions of a desired concentration. These *volumetric flasks* have a calibration mark on the neck that accurately indicates the measured volume. Molarity is based on a specific volume of solution and therefore will vary slightly with temperature because volume varies with temperature (1000 mL H_2O at 20°C = 1001 mL at 25°C).

Suppose we want to make 500 mL of 1 *M* solution. This solution can be prepared by determining the mass of 0.5 mol of the solute and diluting with water in a 500-mL (0.5-L) volumetric flask. The molarity will be

$$M = \frac{0.5 \text{ mol solute}}{0.5 \text{ L solution}} = 1 \text{ molar}$$

You can see that it isn't necessary to have a liter of solution to express molarity. All we need to know is the number of moles of dissolved solute and the volume of solution. Thus, 0.001 mol NaOH in 10 mL of solution is 0.1 *M*:

$$\left(\frac{0.001 \text{ mol}}{10 \text{ \cancel{mL}}}\right)\left(\frac{1000 \text{ \cancel{mL}}}{1 \text{ L}}\right) = 0.1\ M$$

Molarities of concentrated acids commonly used in the laboratory:

HCl	12 *M*
$HC_2H_3O_2$	17 *M*
HNO_3	16 *M*
H_2SO_4	18 *M*

When we stop to think that a balance is not calibrated in moles, but in grams, we can incorporate grams into the molarity formula. We do so by using the relationship

$$\text{moles} = \frac{\text{grams of solute}}{\text{molar mass}}$$

Substituting this relationship into our expression for molarity, we get

$$M = \frac{\text{mol}}{\text{L}} = \frac{\text{g solute}}{\text{molar mass solute} \times \text{L solution}}$$

We can now determine the mass of any amount of a solute that has a known formula, dilute it to any volume, and calculate the molarity of the solution using this formula.

CHECK YOUR UNDERSTANDING

14.10 Molarity

WileyPLUS

ENHANCED EXAMPLE 14.8 WileyPLUS

What is the molarity of a solution containing 1.4 mol of acetic acid ($HC_2H_3O_2$) in 250. mL of solution?

SOLUTION

By the unit-conversion method, we note that the concentration given in the problem statement is 1.4 mol per 250. mL (mol/mL). Since molarity = mol/L, the needed conversion is

Solution map: $\frac{\text{mol}}{\text{mL}} \longrightarrow \frac{\text{mol}}{\text{L}} = M$

Calculate $\left(\frac{1.4\text{ mol}}{250.\text{ }\cancel{\text{mL}}}\right)\left(\frac{1000\text{ }\cancel{\text{mL}}}{\text{L}}\right) = \frac{5.6\text{ mol}}{\text{L}} = 5.6\ M$

ONLINE LEARNING MODULE

Calculating Molarity

WileyPLUS

ENHANCED EXAMPLE 14.9 WileyPLUS

What is the molarity of a solution made by dissolving 2.00 g of potassium chlorate in enough water to make 150. mL of solution?

SOLUTION

Knowns: mass $KClO_3$ = 2.00 g
volume = 150. mL
molar mass $KClO_3$ = 122.6 g/mol

Solution map: $\frac{\text{g KClO}_3}{\text{mL}} \longrightarrow \frac{\text{g KClO}_3}{\text{L}} \longrightarrow \frac{\text{mol KClO}_3}{\text{L}} = M$

Calculate $\left(\frac{2.00\text{ }\cancel{\text{g KClO}_3}}{150.\text{ }\cancel{\text{mL}}}\right)\left(\frac{1000\text{ }\cancel{\text{mL}}}{\text{L}}\right)\left(\frac{1\text{ mol KClO}_3}{122.6\text{ }\cancel{\text{g KClO}_3}}\right) = \frac{0.109\text{ mol}}{\text{L}}$

$= 0.109\ M\ KClO_3$

ENHANCED EXAMPLE 14.10 WileyPLUS

How many grams of potassium hydroxide are required to prepare 600. mL of 0.450 *M* KOH solution?

SOLUTION

Plan

Solution map: milliliters ⟶ liters ⟶ moles ⟶ grams

Knowns: volume = 600. mL $\quad M = \frac{0.450\text{ mol}}{\text{L}}$

molar mass KOH = $\frac{56.11\text{ g}}{\text{mol}}$

Calculate $(600.\text{ }\cancel{\text{mL}})\left(\frac{1\text{ }\cancel{\text{L}}}{1000\text{ }\cancel{\text{mL}}}\right)\left(\frac{0.450\text{ }\cancel{\text{mol}}}{\cancel{\text{L}}}\right)\left(\frac{56.11\text{ g KOH}}{\cancel{\text{mol}}}\right) = 15.1\text{ g KOH}$

PRACTICE 14.5

What is the molarity of a solution made by dissolving 7.50 g of magnesium nitrate [$Mg(NO_3)_2$] in enough water to make 25.0 mL of solution?

PRACTICE 14.6

How many grams of sodium chloride are needed to prepare 125 mL of a 0.037 *M* NaCl solution?

ENHANCED EXAMPLE 14.11 WileyPLUS

Calculate the number of moles of nitric acid in 325 mL of 16 *M* HNO_3 solution.

SOLUTION

Read **Knowns:** $V = 325\ \text{mL}$

16 *M* HNO_3 solution

Plan Use the equation moles = liters × *M*

Calculate $\text{moles} = (0.325\ \cancel{\text{L}})\left(\frac{16\ \text{mol HNO}_3}{1\ \cancel{\text{L}}}\right) = 5.2\ \text{mol HNO}_3$

ENHANCED EXAMPLE 14.12 WileyPLUS

ONLINE LEARNING MODULE
Using Molarity to Determine Volume of a Solution
➡ WileyPLUS

What volume of 0.250 *M* solution can be prepared from 16.0 g of potassium carbonate?

SOLUTION

Read **Knowns:** mass = 16.0 g K_2CO_3

$M = 0.250\ \text{mol/L}$

Molar mass K_2CO_3 = 138.2 g/mol

Plan Find the volume that can be prepared from 16.0 g K_2CO_3.

Solution map: g K_2CO_3 → mol K_2CO_3 → L solution

Calculate $(16.0\ \cancel{\text{g K}_2\text{CO}_3})\left(\frac{1\ \cancel{\text{mol K}_2\text{CO}_3}}{138.2\ \cancel{\text{g K}_2\text{CO}_3}}\right)\left(\frac{1\ \text{L}}{0.250\ \cancel{\text{mol K}_2\text{CO}_3}}\right) = 0.463\ \text{L}$

$= 463\ \text{mL}$

ENHANCED EXAMPLE 14.13 WileyPLUS

ONLINE LEARNING MODULE
Solution Stoichiometry
➡ WileyPLUS

How many milliliters of 2.00 *M* HCl will react with 28.0 g NaOH?

SOLUTION

Read **Knowns:** mass = 28.0 g NaOH

$M = 2.00\ \text{mol/L HCl}$

Plan Write a balanced equation for the reaction

$$HCl(aq) + NaOH(aq) \longrightarrow NaCl(aq) + H_2O(l)$$

Find the number of moles NaOH in 28.0 g NaOH.

Solution map: grams NaOH → moles NaOH

$$(28.0\ \cancel{\text{g NaOH}})\left(\frac{1\ \text{mole}}{40.00\ \cancel{\text{g NaOH}}}\right) = 0.700\ \text{mol NaOH}$$

Setup **Solution map:** mol NaOH → mol HCl → L HCl → mL HCl

Calculate $(0.700\ \cancel{\text{mol NaOH}})\left(\frac{1\ \cancel{\text{mol HCl}}}{1\ \cancel{\text{mol NaOH}}}\right)\left(\frac{1\ \text{L HCl}}{2.00\ \cancel{\text{mol HCl}}}\right) = 0.350\ \text{L HCl}$

$= 350.\ \text{mL HCl}$

PRACTICE 14.7

What volume of 0.035 M $AgNO_3$ can be made from 5.0 g of $AgNO_3$?

PRACTICE 14.8

How many milliliters of 0.50 M NaOH are required to react with 25.00 mL of 1.5 M HCl?

We've now examined several ways to measure concentration of solutions quantitatively. A summary of these concentration units is found in **TABLE 14.4**.

Molality is covered in Section 14.5.

TABLE 14.4 Concentration Units for Solutions

Units	Symbol	Definition
Mass percent	% m/m	$\frac{\text{mass solute}}{\text{mass solution}} \times 100$
Parts per million	ppm	$\frac{\text{mass solute}}{\text{mass solution}} \times 1{,}000{,}000$
Mass/volume percent	% m/v	$\frac{\text{mass solute}}{\text{mL solution}} \times 100$
Volume percent	% v/v	$\frac{\text{mL solute}}{\text{mL solution}} \times 100$
Molarity	M	$\frac{\text{moles solute}}{\text{L solution}}$
Molality	m	$\frac{\text{moles solute}}{\text{kg solvent}}$

Dilution Problems

Chemists often find it necessary to dilute solutions from one concentration to another by adding more solvent to the solution. If a solution is diluted by adding pure solvent, the volume of the solution increases, but the number of moles of solute in the solution remains the same. Thus the moles/liter (molarity) of the solution decreases. Always read a problem carefully to distinguish between (1) how much solvent must be added to dilute a solution to a particular concentration and (2) to what volume a solution must be diluted to prepare a solution of a particular concentration.

ONLINE LEARNING MODULE
Dilutions
WileyPLUS

ENHANCED EXAMPLE 14.14 **WileyPLUS**

Calculate the molarity of a sodium hydroxide solution that is prepared by mixing 100. mL of 0.20 M NaOH with 150. mL of water. Assume that the volumes are additive.

SOLUTION

Read • **Knowns:** $V_1 = 100.$ mL NaOH

$V_2 = 150.$ mL water

$V_{solution} = 250.$ mL

$M_{initial} = 0.20\ M$

Plan • In the dilution, the moles NaOH remain the same; the molarity and volume change.

1. Calculate the moles of NaOH in the original solution.

$$\text{mol} = (0.100\ \cancel{\text{L}})\left(\frac{0.20\ \text{mol NaOH}}{1\ \cancel{\text{L}}}\right) = 0.020\ \text{mol NaOH}$$

2. Divide the mol NaOH by the final volume of the solution to find the new molarity.

Calculate $M = \dfrac{0.020 \text{ mol NaOH}}{0.250 \text{ L}} = 0.080\ M \text{ NaOH}$

Check If we double the volume of the solution, the concentration is half. Therefore the concentration of the new solution here should be less than 0.10 M.

PRACTICE 14.9

Calculate the molarity of a solution prepared by diluting 125 mL of 0.400 M $K_2Cr_2O_7$ with 875 mL of water.

ENHANCED EXAMPLE 14.15 WileyPLUS

How many grams of silver chloride will be precipitated by adding sufficient silver nitrate to react with 1500. mL of 0.400 M barium chloride solution?

$$2\,AgNO_3(aq) + BaCl_2(aq) \longrightarrow 2\,AgCl(s) + Ba(NO_3)_2(aq)$$

SOLUTION

Read **Knowns:** $V = 1500.\text{ mL } BaCl_2$

$M = 0.400\ M\ BaCl_2$

Plan Solve as a stoichiometry problem.

Determine the mol $BaCl_2$ in 1500. mL of 0.400 M solution.

$$\text{mol} = M \times V = (1.500\ \cancel{L})\left(\frac{0.400 \text{ mol } BaCl_2}{\cancel{L}}\right) = 0.600 \text{ mol } BaCl_2$$

Solution map: moles $BaCl_2$ → moles AgCl → grams AgCl

Calculate $(0.600\ \cancel{\text{mol } BaCl_2})\left(\dfrac{2\ \cancel{\text{mol AgCl}}}{1\ \cancel{\text{mol } BaCl_2}}\right)\left(\dfrac{143.4 \text{ g AgCl}}{\cancel{\text{mol AgCl}}}\right) = 172 \text{ g AgCl}$

PRACTICE 14.10

How many grams of lead(II) iodide will be precipitated by adding sufficient $Pb(NO_3)_2$ to react with 750 mL of 0.250 M KI solution?

$$2\,KI(aq) + Pb(NO_3)_2(aq) \longrightarrow PbI_2(s) + 2\,KNO_3(aq)$$

14.5 Colligative Properties of Solutions

LEARNING OBJECTIVE Use the concept of colligative properties to calculate molality, freezing point, boiling point, freezing point depression, and boiling point elevation of various solutions.

KEY TERM
colligative properties
molality (m)

Two solutions—one containing 1 mol (60.06 g) of urea (NH_2CONH_2) and the other containing 1 mol (342.3 g) of sucrose ($C_{12}H_{22}O_{11}$) each in 1 kg of water—both have a freezing point of $-1.86°C$, not 0°C as for pure water. Urea and sucrose are distinct substances, yet they lower the freezing point of the water by the same amount. The only thing apparently common to these two solutions is that each contains 1 mol (6.022×10^{23} molecules) of solute and 1 kg of solvent. In fact, when we dissolve 1 mol of any nonionizable solute in 1 kg of water, the freezing point of the resulting solution is $-1.86°C$.

These results lead us to conclude that the freezing point depression for a solution containing 6.022×10^{23} solute molecules (particles) and 1 kg of water is a constant, namely, 1.86°C. Freezing point depression is a general property of solutions. Furthermore, the amount by which the freezing point is depressed is the same for all solutions made with a

TABLE 14.5 | Freezing Point Depression and Boiling Point Elevation Constants of Selected Solvents

Solvent	Freezing point of pure solvent (°C)	Freezing point depression constant, K_f $\left(\frac{\text{°C kg solvent}}{\text{mol solute}}\right)$	Boiling point of pure solvent (°C)	Boiling point elevation constant, K_b $\left(\frac{\text{°C kg solvent}}{\text{mol solute}}\right)$
Water	0.00	1.86	100.0	0.512
Acetic acid	16.6	3.90	118.5	3.07
Benzene	5.5	5.1	80.1	2.53
Camphor	178	40	208.2	5.95

CHECK YOUR UNDERSTANDING
14.12 Freezing Point Depression
➡ WileyPLUS

CHECK YOUR UNDERSTANDING
14.13 Colligative Properties
➡ WileyPLUS

CHECK YOUR UNDERSTANDING
14.14 Vapor Pressure
➡ WileyPLUS

given solvent; that is, each solvent shows a characteristic *freezing point depression constant*. Freezing point depression constants for several solvents are given in TABLE 14.5.

The solution formed by the addition of a nonvolatile solute to a solvent has a lower freezing point, a higher boiling point, and a lower vapor pressure than that of the pure solvent. These effects are related and are known as colligative properties. The **colligative properties** are properties that depend only on the number of solute particles in a solution, not on the nature of those particles. Freezing point depression, boiling point elevation, and vapor pressure lowering are colligative properties of solutions.

The colligative properties of a solution can be considered in terms of vapor pressure. The vapor pressure of a pure liquid depends on the tendency of molecules to escape from its surface. If 10% of the molecules in a solution are nonvolatile solute molecules, the vapor pressure of the solution is 10% lower than that of the pure solvent. The vapor pressure is lower because the surface of the solution contains 10% nonvolatile molecules and 90% of the volatile solvent molecules. A liquid boils when its vapor pressure equals the pressure of the atmosphere. We can thus see that the solution just described as having a lower vapor pressure will have a higher boiling point than the pure solvent. The solution with a lowered vapor pressure doesn't boil until it has been heated above the boiling point of the solvent (FIGURE 14.8A). Each solvent has its own characteristic boiling point elevation constant (Table 14.5). The boiling point elevation constant is based on a solution that contains 1 mol of solute particles per kilogram of solvent. For example, the boiling point elevation constant for a solution containing 1 mol of solute particles per kilogram of water is 0.512°C, which means that this water solution will boil at 100.512°C.

The freezing behavior of a solution can also be considered in terms of lowered vapor pressure. FIGURE 14.8B shows the vapor pressure relationships of ice, water, and a solution containing 1 mol of solute per kilogram of water. The freezing point of water is at the intersection of the liquid and solid vapor pressure curves (i.e., at the point where water and ice have the same vapor pressure). Because the vapor pressure of the liquid is lowered

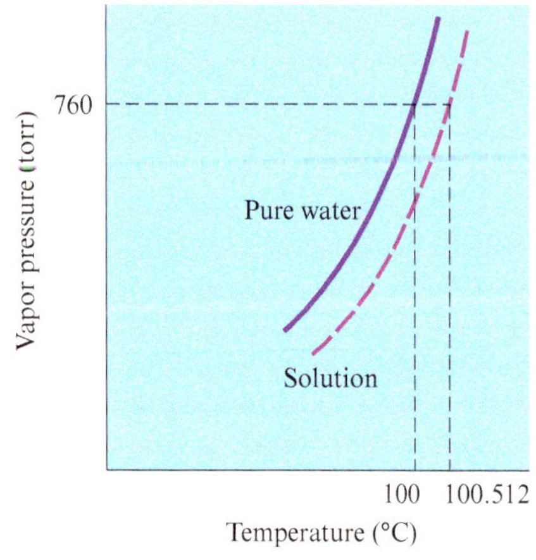

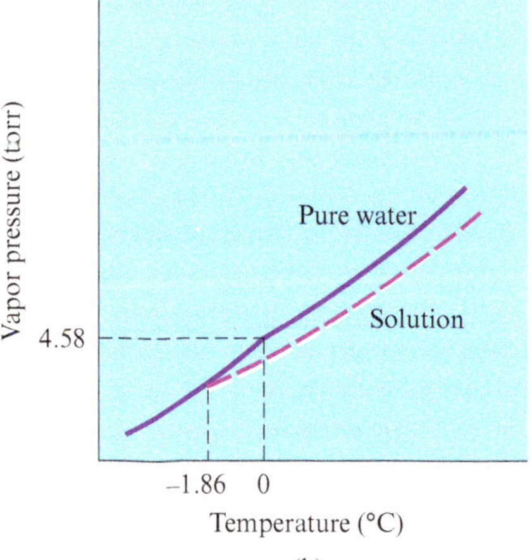

FIGURE 14.8 Vapor pressure curves of pure water and water solutions, showing (a) boiling point elevation and (b) freezing point depression effects (concentration: 1 mol solute/1 kg water).

by the solute, the vapor pressure curve of the solution does not intersect the vapor pressure curve of the solid until the solution has been cooled below the freezing point of pure water. So the solution must be cooled below 0°C in order for it to freeze.

The foregoing discussion dealing with freezing point depressions is restricted to *un-ionized* substances. The discussion of boiling point elevations is restricted to *nonvolatile* and un-ionized substances.

Some practical applications involving colligative properties are (1) use of salt–ice mixtures to provide low freezing temperatures for homemade ice cream, (2) use of sodium chloride or calcium chloride to melt ice from streets, and (3) use of ethylene glycol–water mixtures as antifreeze in automobile radiators (ethylene glycol also raises the boiling point of radiator fluid, thus allowing the engine to operate at a higher temperature).

Brian H. Thomas/Alamy

Engine coolant is one application of colligative properties. The addition of coolant to the water in a radiator raises its boiling point and lowers its freezing point.

Sodium chloride or calcium chloride is used to melt ice on snowy streets and highways.

© J Morrill Photo/Photolibrary/GettyImages, Inc.

Both the freezing point depression and the boiling point elevation are directly proportional to the number of moles of solute per kilogram of solvent. When we deal with the colligative properties of solutions, another concentration expression, *molality*, is used. The **molality (*m*)** of a solution is the number of moles of solute per kilogram of solvent:

$$m = \frac{\text{mol solute}}{\text{kg solvent}}$$

Note that a lowercase *m* is used for molality concentrations and a capital *M* for molarity. The difference between molality and molarity is that molality refers to moles of solute *per kilogram of solvent*, whereas molarity refers to moles of solute *per liter of solution*. For un-ionized substances, the colligative properties of a solution are directly proportional to its molality.

Molality is independent of volume. It is a mass-to-mass relationship of solute to solvent and allows for experiments, such as freezing point depression and boiling point elevation, to be conducted at variable temperatures.

The following equations are used in calculations involving colligative properties and molality:

$$\Delta t_f = mK_f \qquad \Delta t_b = mK_b \qquad m = \frac{\text{mol solute}}{\text{kg solvent}}$$

m = molality; mol solute/kg solvent

Δt_f = freezing point depression; °C

Δt_b = boiling point elevation; °C

K_f = freezing point depression constant; °C kg solvent/mol solute

K_b = boiling point elevation constant; °C kg solvent/mol solute

ONLINE LEARNING MODULE
Calculating Molality
➡ WileyPLUS

ENHANCED EXAMPLE 14.16

WileyPLUS

What is the molality (m) of a solution prepared by dissolving 2.70 g CH_3OH in 25.0 g H_2O?

SOLUTION

$$\text{Since } m = \frac{\text{mol solute}}{\text{kg solvent}}, \text{ the conversion is}$$

$$\frac{2.70 \text{ g } CH_3OH}{25.0 \text{ g } H_2O} \longrightarrow \frac{\text{mol } CH_3OH}{25.0 \text{ g } H_2O} \longrightarrow \frac{\text{mol } CH_3OH}{1 \text{ kg } H_2O}$$

The molar mass of CH_3OH is (12.01 + 4.032 + 16.00), or 32.04 g/mol:

$$\left(\frac{2.70 \text{ g } \cancel{CH_3OH}}{25.0 \text{ g } \cancel{H_2O}}\right)\left(\frac{1 \text{ mol } CH_3OH}{32.04 \text{ g } \cancel{CH_3OH}}\right)\left(\frac{1000 \text{ g } \cancel{H_2O}}{1 \text{ kg } H_2O}\right) = \frac{3.37 \text{ mol } CH_3OH}{1 \text{ kg } H_2O}$$

The molality is 3.37 m.

PRACTICE 14.11

What is the molality of a solution prepared by dissolving 150.0 g $C_6H_{12}O_6$ in 600.0 g H_2O?

ONLINE LEARNING MODULE
Freezing Point Depression
➡ WileyPLUS

ENHANCED EXAMPLE 14.17

WileyPLUS

A solution is made by dissolving 100. g of ethylene glycol ($C_2H_6O_2$) in 200. g of water. What is the freezing point of this solution?

SOLUTION

To calculate the freezing point of the solution, we first need to calculate Δt_f, the change in freezing point. Use the equation

$$\Delta t_f = mK_f = \frac{\text{mol solute}}{\text{kg solvent}} \times K_f$$

$$K_f \text{ (for water): } \frac{1.86°\text{C kg solvent}}{\text{mol solute}} \text{ (from Table 14.5)}$$

$$\text{mol solute: } (100. \text{ g } \cancel{C_2H_6O_2})\left(\frac{1 \text{ mol } C_2H_6O_2}{62.07 \text{ g } \cancel{C_2H_6O_2}}\right) = 1.61 \text{ mol } C_2H_6O_2$$

$$\text{kg solvent: } (200. \cancel{\text{g}} \ H_2O)\left(\frac{1 \text{ kg}}{1000 \cancel{\text{g}}}\right) = 0.200 \text{ kg } H_2O$$

$$\Delta t_f = \left(\frac{1.61 \ \cancel{\text{mol } C_2H_6O_2}}{0.200 \ \cancel{\text{kg } H_2O}}\right)\left(\frac{1.86°\text{C} \ \cancel{\text{kg } H_2O}}{1 \ \cancel{\text{mol } C_2H_6O_2}}\right) = 15.0°\text{C}$$

The freezing point depression, 15.0°C, must be subtracted from 0°C, the freezing point of the pure solvent (water):

$$\begin{aligned}\text{freezing point of solution} &= \text{freezing point of solvent} - \Delta t_f \\ &= 0.0°\text{C} - 15.0°\text{C} = -15.0°\text{C}\end{aligned}$$

Therefore, the freezing point of the solution is −15.0°C. This solution will protect an automobile radiator down to −15.0°C (5°F).

ENHANCED EXAMPLE 14.18

WileyPLUS

A solution made by dissolving 4.71 g of a compound of unknown molar mass in 100.0 g of water has a freezing point of −1.46°C. What is the molar mass of the compound?

SOLUTION

First substitute the data in $\Delta t_f = mK_f$ and solve for m:

$\Delta t_f = +1.46$ (since the solvent, water, freezes at 0°C)

$$K_f = \frac{1.86°\text{C kg H}_2\text{O}}{\text{mol solute}}$$

$$1.46°\text{C} = mK_f = m \times \frac{1.86°\text{C kg H}_2\text{O}}{\text{mol solute}}$$

$$m = \frac{1.46°\cancel{\text{C}} \times \text{mol solute}}{1.86°\cancel{\text{C}} \times \text{kg H}_2\text{O}} = \frac{0.785 \text{ mol solute}}{\text{kg H}_2\text{O}}$$

Now convert the data, 4.71 g solute/100.0 g H_2O, to g/mol:

$$\left(\frac{4.71 \text{ g } \cancel{\text{solute}}}{100.0 \text{ g } \cancel{\text{H}_2\text{O}}}\right)\left(\frac{1000 \text{ g } \cancel{\text{H}_2\text{O}}}{1 \text{ kg } \cancel{\text{H}_2\text{O}}}\right)\left(\frac{1 \text{ kg } \cancel{\text{H}_2\text{O}}}{0.785 \text{ mol } \cancel{\text{solute}}}\right) = 60.0 \text{ g/mol}$$

The molar mass of the compound is 60.0 g/mol.

PRACTICE 14.12

What is the freezing point of the solution in Practice Exercise 14.11? What is the boiling point?

CHEMISTRY *IN ACTION* The Scoop on Ice Cream

Ice cream is mainly composed of water (from milk and cream), milk solids, milk fats, and, frequently, various sweeteners (corn syrup or sugar), flavorings, emulsifiers, and stabilizers. But that smooth, creamy, rich flavor and texture are the result of the chemistry of the mixing and freezing process. The rich, smooth texture of great ice cream results from the milk fat. By law, if the carton is labeled "ice cream," it must contain a minimum of 10% milk fat. That carton of ice cream also contains 20%–50% air whipped into the ingredients during the initial mixing process.

H. Douglas Goff, a professor of food science and an ice cream expert from Ontario, Canada, says, "There are no real chemical reactions that take place when you make ice cream, but that doesn't mean that there isn't plenty of chemistry." The structure of ice cream contributes greatly to its taste. Tiny air bubbles are formed in the initial whipping process. These bubbles are distributed through a network of fat globules and liquid water. The milk fat has surface proteins on the globules in the milk to keep the fat dissolved in solution. Ice cream manufacturers destabilize these globules using emulsifiers (such as egg yolks, mono- or diglycerides) and let them come together in larger networks sort of like grape clusters.

Once the mixture is fully whipped, it is cooled to begin the freezing process. But ice cream does not freeze at 0°C even though it is 55%–64% water. The freezing point is depressed as a colligative property of the ice cream solution. Once the freezing of the water begins, the concentration of the solution increases, which continues to depress the freezing point. Goff tells us that even at a typical serving temperature (−16°C for most ice cream), only about 72% of the water in the ice cream is frozen. The unfrozen solution keeps the ice cream "scoopable."

One last factor that affects the quality of ice cream is the size of the ice crystals. For very smooth ice cream, tiny crystals are needed. To produce these, the ice cream must freeze very slowly. Large crystals give a coarse, grainy texture. Now, as you savor that premium ice cream cone, you'll know just how colligative properties and the chemistry of freezing helped make it so delicious!

14.6 Osmosis and Osmotic Pressure

LEARNING OBJECTIVE Discuss osmosis and osmotic pressure and their importance in living systems.

KEY TERM

semipermeable membrane
osmosis
osmotic pressure

When red blood cells are put into distilled water, they gradually swell and in time may burst. If red blood cells are put in a 5% urea (or a 5% salt) solution, they gradually shrink and take on a wrinkled appearance. The cells behave in this fashion because they are enclosed in semipermeable membranes. A **semipermeable membrane** allows the passage of water (solvent) molecules through it in either direction but prevents the passage of larger solute molecules or ions. When two solutions of different concentrations (or water and a water solution) are separated by a semipermeable membrane, water diffuses through the membrane from the solution of lower concentration into the solution of higher concentration. The diffusion of water, either from a dilute solution or from pure water, through a semipermeable membrane into a solution of higher concentration is called **osmosis**.

A 0.90% (0.15 *M*) sodium chloride solution is known as a *physiological saline solution* because it is *isotonic* with blood plasma; that is, it has the same concentration of NaCl as blood plasma. Because each mole of NaCl yields about 2 mol of ions when in solution, the solute particle concentration in physiological saline solution is nearly 0.30 *M*. A 5% glucose solution (0.28 *M*) is also approximately isotonic with blood plasma. Blood cells neither swell nor shrink in an isotonic solution. The cells described in the preceding paragraph swell in water because water is *hypotonic* to cell plasma. The cells shrink in 5% urea solution because the urea solution is *hypertonic* to the cell plasma. To prevent possible injury to blood cells by osmosis, fluids for intravenous use are usually made up at approximately isotonic concentration.

CHECK YOUR UNDERSTANDING

14.15 Osmosis

WileyPLUS

All solutions exhibit *osmotic pressure*, which is another colligative property. Osmotic pressure is a pressure difference between the system and atmospheric pressure. The osmotic pressure of a system can be measured by applying enough pressure to stop the flow of water due to osmosis in the system. The difference between the applied pressure and atmospheric pressure is the **osmotic pressure**. When pressure greater than the osmotic pressure is applied to a system, the flow of water can be reversed from that of osmosis. This process can be used to obtain useful drinking water from seawater and is known as *reverse osmosis.* Osmotic pressure is dependent only on the concentration of the solute particles and is independent of their nature. The osmotic pressure of a solution can be measured by determining the amount of counterpressure needed to prevent osmosis; this pressure can be very large. The osmotic pressure of a solution containing 1 mol of

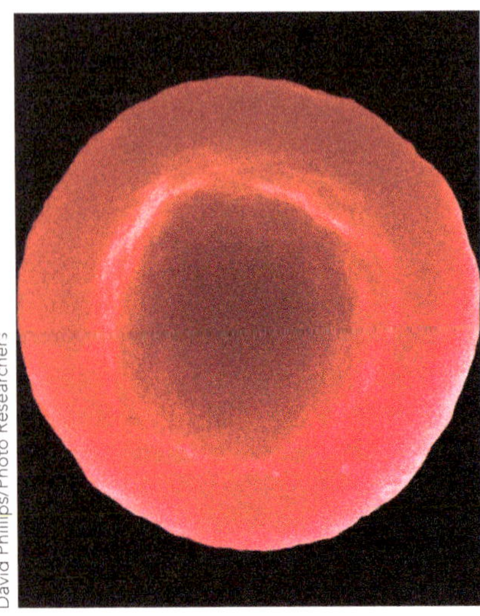
David Phillips/Photo Researchers

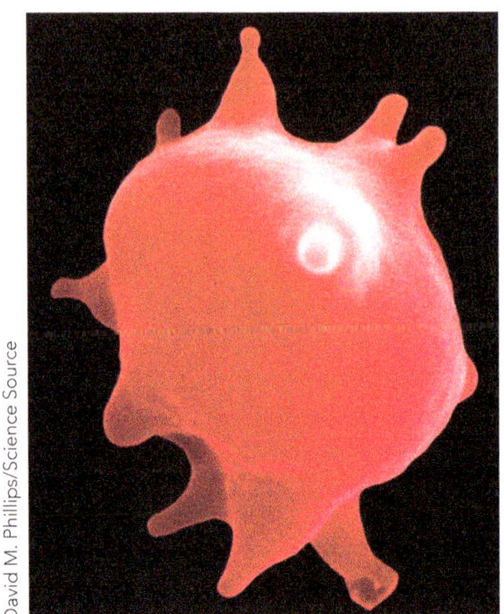
David M. Phillips/Science Source

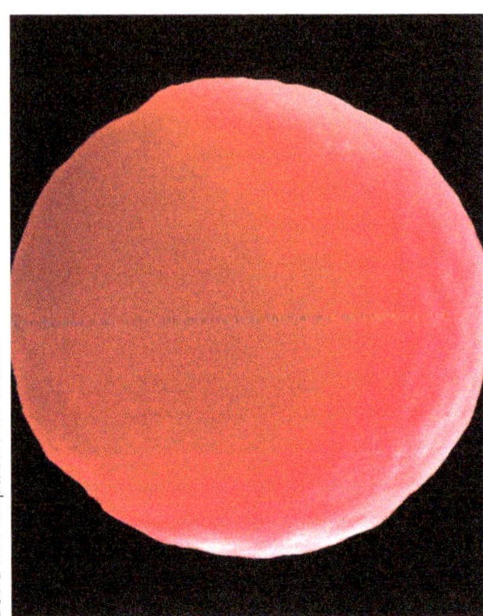
David M. Phillips/Science Source

Human red blood cells. Left: In an isotonic solution, the concentration is the same inside and outside the cell (0.9% saline). Center: In a hypertonic solution (1.6% saline), water leaves the cells, causing them to crenate (shrink). Right: In a hypotonic solution (0.2% saline), the cells swell as water moves into the cell center. Magnification is 260,000×.

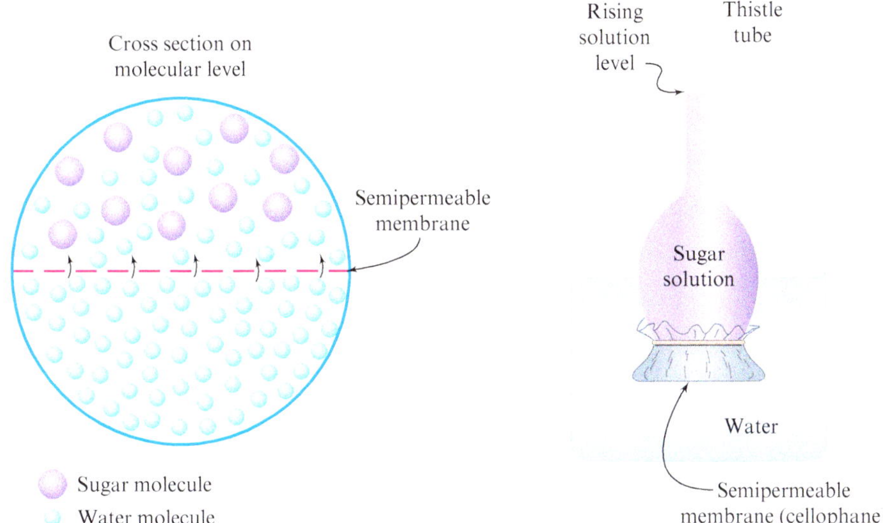

FIGURE 14.9 Laboratory demonstration of osmosis: As a result of osmosis, water passes through the membrane, causing the solution to rise in the thistle tube.

solute particles in 1 kg of water is about 22.4 atm, which is about the same as the pressure exerted by 1 mol of a gas confined in a volume of 1 L at 0°C.

Osmosis has a role in many biological processes, and semipermeable membranes occur commonly in living organisms. An example is the roots of plants, which are covered with tiny structures called root hairs; soil water enters the plant by osmosis, passing through the semipermeable membranes covering the root hairs. Artificial or synthetic membranes can also be made.

Osmosis can be demonstrated with the simple laboratory setup shown in FIGURE 14.9. As a result of osmotic pressure, water passes through the cellophane membrane into the thistle tube, causing the solution level to rise. In osmosis, the net transfer of water is always from a less concentrated to a more concentrated solution; that is, the effect is toward equalization of the concentration on both sides of the membrane. Note that the effective movement of water in osmosis is always from the region of *higher water concentration* to the region of *lower water concentration*.

Osmosis can be explained by assuming that a semipermeable membrane has passages that permit water molecules and other small molecules to pass in either direction. Both sides of the membrane are constantly being struck by water molecules in random motion. The number of water molecules crossing the membrane is proportional to the number of water molecule-to-membrane impacts per unit of time. Because the solute molecules or ions reduce the concentration of water, there are more water molecules and thus more water molecule impacts on the side with the lower solute concentration (more dilute solution). The greater number of water molecule-to-membrane impacts on the dilute side thus causes a net transfer of water to the more concentrated solution. Again, note that the overall process involves the net transfer, by diffusion through the membrane, of water molecules from a region of higher water concentration (dilute solution) to one of lower water concentration (more concentrated solution).

This is a simplified picture of osmosis. No one has ever seen the hypothetical passages that allow water molecules and other small molecules or ions to pass through them. Alternative explanations have been proposed, but our discussion has been confined to water solutions. Osmotic pressure is a general colligative property, however, and is known to occur in nonaqueous systems.

CHAPTER 14 REVIEW

14.1 General Properties of Solutions

KEY TERMS
solution
solute
solvent

- A solution is a homogeneous mixture of two or more substances:
  - Consists of solvent—the dissolving agent—and solute—the component(s) dissolved in the solvent
  - Is a homogeneous mixture
  - Contains molecular or ionic particles
  - Can be colored or colorless
  - Can be separated into solute and solvent by a physical separation process

14.2 Solubility

KEY TERMS
solubility
miscible
immiscible
concentration of a solution
saturated solution
unsaturated solution
supersaturated solution

- Solubility describes the amount of solute that will dissolve in a specified amount of solvent.
- Solubility can also be qualitative.
- General guidelines for ionic solubility are:

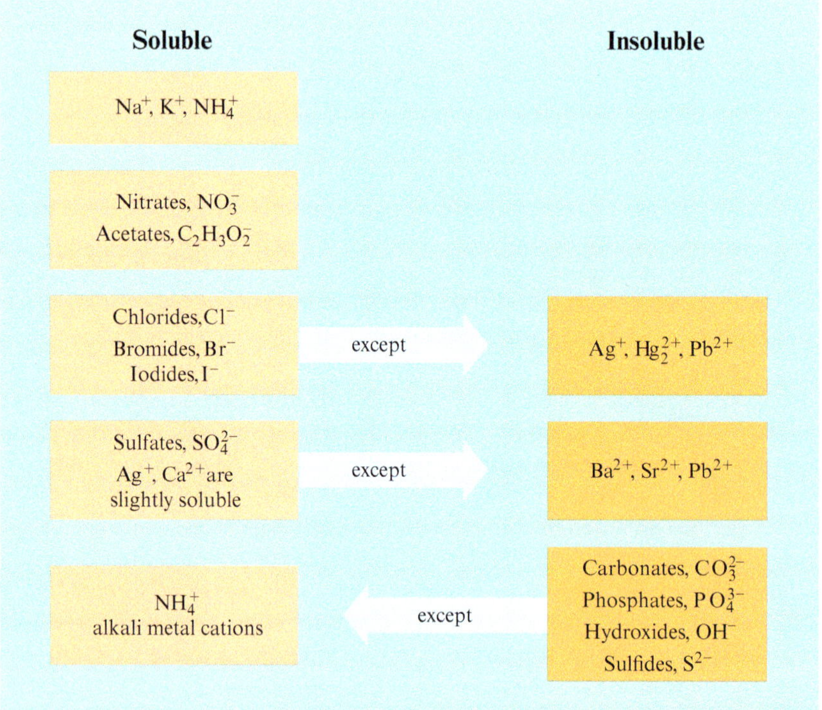

- Liquids can also be classified as miscible (soluble in each other) or immiscible (not soluble in each other).
- The concentration of a solution is the quantitative measurement of the amount of solute that is dissolved in a solution.
- Like tends to dissolve like is a general rule for solvents and solutes.
- As temperature increases:
  - Solubility of a solid in a liquid tends to increase.
  - Solubility of a gas in a liquid tends to decrease.
- As pressure increases:
  - Solubility of a solid in a liquid remains constant.
  - Solubility of a gas in a liquid tends to increase.
- At a specific temperature, the amount of solute that can dissolve in a solvent has a limit:
  - Unsaturated solutions contain less solute than the limit.
  - Saturated solutions contain dissolved solute at the limit.
  - Supersaturated solutions contain more solute than the limit and are therefore unstable:
    - If disturbed, the excess solute will precipitate out of solution.

14.3 Rate of Dissolving Solids

- The rate at which a solute dissolves is determined by these factors:
  - Particle size
  - Temperature
  - Concentration of solution
  - Agitation

14.4 Concentration of Solutions

KEY TERMS
dilute solution
concentrated solution
parts per million (ppm)
molarity (M)

- Molecules or ions must collide in order to react.
- Solutions provide a medium for the molecules or ions to collide.
- Concentrations can be measured in many ways:

Mass percent	% m/m	$\frac{\text{mass solute}}{\text{mass solution}} \times 100$
Parts per million	ppm	$\frac{\text{mass solute}}{\text{mass solution}} \times 1{,}000{,}000$
Mass/volume percent	% m/v	$\frac{\text{mass solute}}{\text{mL solution}} \times 100$
Volume percent	% v/v	$\frac{\text{mL solute}}{\text{mL solution}} \times 100$
Molarity	M	$\frac{\text{moles solute}}{\text{L solution}}$
Molality	m	$\frac{\text{moles solute}}{\text{kg solvent}}$

- Dilution of solutions requires the addition of more solvent to an existing solution:
  - The number of moles in the diluted solution is the same as that in the original solution.
  - $M_1V_1 = M_2V_2$.

14.5 Colligative Properties of Solutions

KEY TERMS
colligative properties
molality (m)

- Properties of a solution that depend only on the number of solute particles in solution are called colligative properties:
  - Freezing point depression

    $\Delta t_f = mK_f$

  - Boiling point elevation

    $\Delta t_b = mK_b$

  - Osmotic pressure
- Molality is used in working with colligative properties.

14.6 Osmosis and Osmotic Pressure

KEY TERMS
semipermeable membrane
osmosis
osmotic pressure

- Osmosis is the diffusion of water through a semipermeable membrane:
  - Occurs from dilute solution to a solution of higher concentration
- Osmosis results in osmotic pressure, which is a colligative property of a solution.

REVIEW QUESTIONS

1. What is a true solution?
2. Name and distinguish between the two components of a solution.
3. Is it always apparent in a solution which component is the solute, for example, in a solution of a liquid in a liquid?
4. Explain why the solute does not settle out of a solution.
5. Is it possible to have one solid dissolved in another? Explain.
6. An aqueous solution of KCl is colorless, $KMnO_4$ is purple, and $K_2Cr_2O_7$ is orange. What color would you expect of an aqueous solution of $Na_2Cr_2O_7$? Explain.
7. Why is air considered to be a solution?
8. Sketch the orientation of water molecules (a) about a single sodium ion and (b) about a single chloride ion in solution.
9. Refer to Table 14.2 and estimate the number of grams of potassium bromide that would dissolve in 50 g water at 0°C.
10. Refer to Figure 14.4 to determine the solubility of each of these substances at 25°C:
    (a) ammonium chloride
    (b) copper(II) sulfate
    (c) sodium nitrate

11. As you go down Group (1)A from lithium to potassium, what is the trend in solubilities of the chlorides and bromides of these metals? (Table 14.2)
12. State the solubility, in grams of solute per 100 g of H_2O, of: (Figure 14.4)
    (a) NH_4Cl at 35°C
    (b) $CuSO_4$ at 60°C
    (c) SO_2 gas at 30°C
    (d) $NaNO_3$ at 15°C
13. Which solid substance in Figure 14.4 shows an overall decrease in solubility with an increase in temperature?
14. At a temperature of 45°C, will a solution made with 6.5 g KNO_3 in 15.0 g water be saturated or unsaturated? What will be the mass percent of solute in the solution? (Figure 14.4)
15. If 40. g Li_2SO_4 are added to 75.0 g water at 40°C, will the solution be saturated or unsaturated? What will be the mass percent of solute in the solution? (Figure 14.4)
16. Explain how a supersaturated solution of $NaC_2H_3O_2$ can be prepared and proven to be supersaturated.
17. Explain why hexane will dissolve benzene but will not dissolve sodium chloride.
18. Some drinks like tea are consumed either hot or cold, whereas others like Coca-Cola are drunk only cold. Why?
19. What is the effect of pressure on the solubility of gases in liquids? of solids in liquids?
20. In a saturated solution containing undissolved solute, solute is continuously dissolving, but the concentration of the solution remains unchanged. Explain.
21. Champagne is usually cooled in a refrigerator prior to opening. It's also opened very carefully. What would happen if a warm bottle of champagne were shaken and opened quickly and forcefully?
22. What would be the total surface area if the 1-cm cube in Figure 14.5 were cut into cubes 0.01 cm on a side?
23. When a solid solute is put into a solvent, does the rate of dissolving increase or decrease as the dissolution proceeds? Explain. (Figure 14.6)
24. In which will a teaspoonful of sugar dissolve more rapidly, 200 mL of iced tea or 200 mL of hot coffee? Explain in terms of the KMT.
25. Why do smaller particles dissolve faster than large ones?
26. Explain why there is no apparent reaction when crystals of $AgNO_3$ and NaCl are mixed, but a reaction is apparent immediately when solutions of $AgNO_3$ and NaCl are mixed.
27. What do we mean when we say that concentrated nitric acid (HNO_3) is 16 molar?
28. Will 1 L of 1 *M* NaCl contain more chloride ions than 0.5 L of 1 *M* $MgCl_2$? Explain.
29. Describe how you would prepare 750 mL of 5.0 *M* NaCl solution.
30. Arrange the following bases (in descending order) according to the volume of each that will react with 1 L of 1 *M* HCl:
    (a) 1 *M* NaOH
    (b) 1.5 *M* $Ca(OH)_2$
    (c) 2 *M* KOH
    (d) 0.6 *M* $Ba(OH)_2$
31. Explain in terms of vapor pressure why the boiling point of a solution containing a nonvolatile solute is higher than that of the pure solvent.
32. Explain why the freezing point of a solution is lower than the freezing point of the pure solvent.
33. When water and ice are mixed, the temperature of the mixture is 0°C. But if methyl alcohol and ice are mixed, a temperature of −10°C is readily attained. Explain why the two mixtures show such different temperature behavior.
34. Which would be more effective in lowering the freezing point of 500. g of water?
    (a) 100. g of sucrose, $C_{12}H_{22}O_{11}$, or 100. g of ethyl alcohol, C_2H_5OH
    (b) 100. g of sucrose or 20.0 g of ethyl alcohol
    (c) 20.0 g of ethyl alcohol or 20.0 g of methyl alcohol, CH_3OH.
35. What is the difference between molarity and molality?
36. Is the molarity of a 5 *m* aqueous solution of NaCl greater or less than 5 *M*? Explain.
37. Why do salt trucks distribute salt over icy roads in the winter?
38. Assume that the thistle tube in Figure 14.9 contains 1.0 *M* sugar solution and that the water in the beaker has just been replaced by a 2.0 *M* solution of urea. Would the solution level in the thistle tube continue to rise, remain constant, or fall? Explain.
39. Explain in terms of the KMT how a semipermeable membrane functions when placed between pure water and a 10% sugar solution.
40. Which has the higher osmotic pressure, a solution containing 100 g of urea (NH_2CONH_2) in 1 kg H_2O or a solution containing 150 g of glucose, $C_6H_{12}O_6$, in 1 kg H_2O?
41. Explain why a lettuce leaf in contact with salad dressing containing salt and vinegar soon becomes wilted and limp whereas another lettuce leaf in contact with plain water remains crisp.
42. A group of shipwreck survivors floated for several days on a life raft before being rescued. Those who had drunk some seawater were found to be suffering the most from dehydration. Explain.

PAIRED EXERCISES

Most of the exercises in this chapter are available for assignment via the online homework management program, WileyPLUS (www.wileyplus.com). All exercises with blue numbers have answers in Appendix 5.

1. Of the following substances, which ones are generally soluble in water? (See Figure 14.2 or Appendix 4.)
   (a) AgCl
   (b) K_2SO_4
   (c) Na_3PO_4
   (d) NaOH
   (e) PbI_2
   (f) $SnCO_3$
2. Of the following substances, which ones are generally soluble in water? (See Figure 14.2 or Appendix 4.)
   (a) $Ba_3(PO_4)_2$
   (b) $Cu(NO_3)_2$
   (c) $Fe(OH)_3$
   (d) $NH_4C_2H_3O_2$
   (e) MgO
   (f) $AgNO_3$

3. Predict whether each of the following is soluble in water, hexane, or both. Justify your answer.
   (a) $CH_3CH_2{-}OH$
   (b)
```
        H  H
    H    \/    H
     \   C    /
  H—C  /   \  C—H
      |       |
  H—C  \   /  C—H
     /   C    \
    H   / \    H
       H   H
```
   (c) KBr

5. Determine whether the following solutions will be saturated at a temperature of 20°C. Use the data in Table 14.3 to answer this question.
   (a) 24.9 g of potassium chloride dissolved in 100. g of water
   (b) 44.0 g of sodium nitrate dissolved in 50.0 g of water
   (c) 29.6 g of potassium chlorate dissolved in 400. g of water
   (d) 115 g of sucrose, $C_6H_{22}O_{11}$, dissolved in 115 g of water

7. Calculate the mass percent of the following solutions:
   (a) 15.0 g KCl + 100.0 g H_2O
   (b) 2.50 g Na_3PO_4 + 10.0 g H_2O
   (c) 0.20 mol $NH_4C_2H_3O_2$ + 125 g H_2O
   (d) 1.50 mol NaOH in 33.0 mol H_2O

9. A bleaching solution requires 5.23 g of sodium hypochlorite. How many grams of a 21.5% by mass solution of sodium hypochorite should be used?

11. In 25 g of a 7.5% by mass solution of $CaSO_4$
   (a) how many grams of solute are present?
   (b) how many grams of solvent are present?

13. Determine the mass/volume percent of a solution made by dissolving:
   (a) 15.0 g of C_2H_5OH (ethanol) in water to make 150.0 mL of solution
   (b) 25.2 g of NaCl in water to make 125.5 mL of solution

15. Determine the volume percent of a solution made by dissolving:
   (a) 50.0 mL of hexanol in enough ethanol to make 125 mL of solution
   (b) 2.0 mL of ethanol in enough methanol to make 15.0 mL of solution
   (c) 15.0 mL of acetone in enough hexane to make 325 mL of solution

17. Calculate the molarity of the following solutions:
   (a) 0.25 mol of solute in 75.0 mL of solution
   (b) 1.75 mol of KBr in 0.75 L of solution
   (c) 35.0 g of $NaC_2H_3O_2$ in 1.25 L of solution
   (d) 75 g of $CuSO_4 \cdot 5\,H_2O$ in 1.0 L of solution

19. Calculate the number of moles of solute in each of the following solutions:
   (a) 1.5 L of 1.20 *M* H_2SO_4
   (b) 25.0 mL of 0.0015 *M* $BaCl_2$
   (c) 125 mL of 0.35 *M* K_3PO_4

21. Calculate the grams of solute in each of the following solutions:
   (a) 2.5 L of 0.75 *M* K_2CrO_4
   (b) 75.2 mL of 0.050 *M* $HC_2H_3O_2$
   (c) 250 mL of 16 *M* HNO_3

23. How many milliliters of 0.750 *M* H_3PO_4 will contain the following?
   (a) 0.15 mol H_3PO_4
   (b) 35.5 g H_3PO_4
   (c) 7.34×10^{22} molecules of H_3PO_4

4. Predict whether each of the following is soluble in water, hexane, or both. Justify your answer.
   (a) $CH_3{-}CH_2{-}CH_2{-}CH_2{-}CH_2{-}CH_2{-}CH_3$
   (b) Li_3PO_4
   (c)
```
  OH              OH
  |               |
  CH2-CH2-CH2-CH2
```

6. Determine whether the following solutions will be saturated at a temperature of 50°C. Use the data in Table 14.3 to answer this question.
   (a) 38.6 g of potassium chlorate dissolved in 200. g of water
   (b) 300. g of silver nitrate dissolved in 300. g of water
   (c) 9.25 g of sodium chloride dissolved in 25.0 g of water
   (d) 492 g of sodium nitrate dissolved in 500. g of water

8. Calculate the mass percent of the following solutions:
   (a) 25.0 g $NaNO_3$ in 125.0 g H_2O
   (b) 1.25 g $CaCl_2$ in 35.0 g H_2O
   (c) 0.75 mol K_2CrO_4 in 225 g H_2O
   (d) 1.20 mol H_2SO_4 in 72.5 mol H_2O

10. A paint requires 42.8 g of iron(III) oxide to give it the right yellow tint. How many grams of a 30.0% by mass solution of iron(III) oxide should be used?

12. In 75 g of a 12.0% by mass solution of $BaCl_2$
   (a) how many grams of solute are present?
   (b) how many grams of solvent are present?

14. Determine the mass/volume percent of a solution made by dissolving:
   (a) 175.2 g of table sugar, $C_{12}H_{22}O_{11}$, in water to make 275.5 mL of solution
   (b) 35.5 g of CH_3OH (methanol) in water to make 75.0 mL of solution

16. Determine the volume percent of a solution made by dissolving:
   (a) 37.5 mL of butanol in enough ethanol to make 275 mL of solution
   (b) 4.0 mL of methanol in enough water to make 25.0 mL of solution
   (c) 45.0 mL of isoamyl alcohol in enough acetone to make 750. mL of solution

18. Calculate the molarity of the following solutions:
   (a) 0.50 mol of solute in 125 mL of solution
   (b) 2.25 mol of $CaCl_2$ in 1.50 L of solution
   (c) 275 g $C_6H_{12}O_6$ in 775 mL of solution
   (d) 125 g $MgSO_4 \cdot 7\,H_2O$ in 2.50 L of solution

20. Calculate the number of moles of solute in each of the following solutions:
   (a) 0.75 L of 1.50 *M* HNO_3
   (b) 10.0 mL of 0.75 *M* $NaClO_3$
   (c) 175 mL of 0.50 *M* LiBr

22. Calculate the grams of solute in each of the following solutions:
   (a) 1.20 L of 18 *M* H_2SO_4
   (b) 27.5 mL of 1.50 *M* $KMnO_4$
   (c) 120 mL of 0.025 *M* $Fe_2(SO_4)_3$

24. How many milliliters of 0.250 *M* NH_4Cl will contain the following?
   (a) 0.85 mol NH_4Cl
   (b) 25.2 g NH_4Cl
   (c) 2.06×10^{20} formula units of NH_4Cl

25. Some people like to make their own hand lotion because it can be less expensive than buying it and they have the ability to add only the fragrances that they like. A recipe for hand lotion calls for mixing 327 g of water (H_2O) with 311 g of glycerin ($C_3H_8O_3$). This recipe makes a total of 550. mL of hand lotion.
(a) What is the mass percent of glycerin in the hand lotion?
(b) What is the molarity of glycerin in the hand lotion?

26. Some people have begun to formulate their own window-cleaning solutions using commonly available household chemicals. One formulation calls for mixing vinegar (dilute acetic acid, $HC_2H_3O_2$) with water and a small amount of detergent. A commonly used recipe contains 14.3 g of acetic acid mixed with 525.0 g of water and 1.3 g of detergent. This recipe makes a total of 540.0 mL of solution.
(a) What is the mass percent of acetic acid in the window cleaner?
(b) What is the molarity of acetic acid in the window cleaner?

27. What will be the molarity of the resulting solutions made by mixing the following? Assume that volumes are additive.
(a) 125 mL of 5.0 M H_3PO_4 with 775 mL of H_2O
(b) 250 mL of 0.25 M Na_2SO_4 with 750 mL of H_2O
(c) 75 mL of 0.50 M HNO_3 with 75 mL of 1.5 M HNO_3

28. What will be the molarity of the resulting solutions made by mixing the following? Assume that volumes are additive.
(a) 175 mL of 3.0 M H_2SO_4 with 275 mL of H_2O
(b) 350 mL of 0.10 M $CuSO_4$ with 150 mL of H_2O
(c) 50.0 mL of 0.250 M HCl with 25.0 mL of 0.500 M HCl

29. Calculate the volume of concentrated reagent required to prepare the diluted solutions indicated:
(a) 15 M H_3PO_4 to prepare 750 mL of 3.0 M H_3PO_4
(b) 16 M HNO_3 to prepare 250 mL of 0.50 M HNO_3

30. Calculate the volume of concentrated reagent required to prepare the diluted solutions indicated:
(a) 18 M H_2SO_4 to prepare 225 mL of 2.0 M H_2SO_4
(b) 15 M NH_3 to prepare 75 mL of 1.0 M NH_3

31. Gingerone, a molecule found in ginger, is converted to zingerone by cooking. This is the molecule that contributes the flavor to foods cooked with ginger. You are perfecting the recipe for your punch and have discovered that to achieve the perfect flavor you need to have a final concentration of zingerone of 6.05×10^{-5} M. How much of a 1.15×10^{-3} M solution of zingerone do you need to make 10.0 liters of punch?

32. Anethole is the molecule found in anise that contributes the flavor to foods cooked with anise or fennel. You are making pho, a Vietnamese soup, and have discovered that a concentration of 7.15×10^{-6} M anethole will give the soup just the right flavor. How much of a 3.85×10^{-3} M solution of anethole do you need to make 2.50 liters of pho?

33. Calculate the molarity of the solutions made by mixing 125 mL of 6.0 M $HC_2H_3O_2$ with the following:
(a) 525 mL of H_2O
(b) 175 mL of 1.5 M $HC_2H_3O_2$

34. Calculate the molarity of the solutions made by mixing 175 mL of 3.0 M HCl with the following:
(a) 250 mL of H_2O
(b) 115 mL of 6.0 M HCl

35. Use the equation to calculate the following:

$$3\,Ca(NO_3)_2(aq) + 2\,Na_3PO_4(aq) \longrightarrow Ca_3(PO_4)_2(s) + 6\,NaNO_3(aq)$$

(a) the moles of $Ca_3(PO_4)_2$ produced from 2.7 mol Na_3PO_4
(b) the moles of $NaNO_3$ produced from 0.75 mol $Ca(NO_3)_2$
(c) the moles of Na_3PO_4 required to react with 1.45 L of 0.225 M $Ca(NO_3)_2$
(d) the grams of $Ca_3(PO_4)_2$ that can be obtained from 125 mL of 0.500 M $Ca(NO_3)_2$
(e) the volume of 0.25 M Na_3PO_4 needed to react with 15.0 mL of 0.50 M $Ca(NO_3)_2$
(f) the molarity (M) of the $Ca(NO_3)_2$ solution when 50.0 mL react with 50.0 mL of 2.0 M Na_3PO_4

36. Use the equation to calculate the following:

$$2\,NaOH(aq) + H_2SO_4(aq) \longrightarrow Na_2SO_4(aq) + 2\,H_2O(l)$$

(a) the moles of Na_2SO_4 produced from 3.6 mol H_2SO_4
(b) the moles of H_2O produced from 0.025 mol NaOH
(c) the moles of NaOH required to react with 2.50 L of 0.125 M H_2SO_4
(d) the grams of Na_2SO_4 that can be obtained from 25 mL of 0.050 M NaOH
(e) the volume of 0.250 M H_2SO_4 needed to react with 25.5 mL of 0.750 M NaOH
(f) the molarity (M) of the NaOH solution when 48.20 mL react with 35.72 mL of 0.125 M H_2SO_4

37. Use the equation to calculate the following:

$$2\,KMnO_4(aq) + 16\,HCl(aq) \longrightarrow 2\,MnCl_2(aq) + 5\,Cl_2(g) + 8\,H_2O(l) + 2\,KCl(aq)$$

(a) the moles of H_2O that can be obtained from 15.0 mL of 0.250 M HCl
(b) the volume of 0.150 M $KMnO_4$ needed to produce 1.85 mol $MnCl_2$
(c) the volume of 2.50 M HCl needed to produce 125 mL of 0.525 M KCl
(d) the molarity (M) of the HCl solution when 22.20 mL react with 15.60 mL of 0.250 M $KMnO_4$
(e) the liters of Cl_2 gas at STP produced by the reaction of 125 mL of 2.5 M HCl
(f) the liters of Cl_2 gas at STP produced by the reaction of 15.0 mL of 0.750 M HCl and 12.0 mL of 0.550 M $KMnO_4$

38. Use the equation to calculate the following:

$$K_2CO_3(aq) + 2\,HC_2H_3O_2(aq) \longrightarrow 2\,KC_2H_3O_2(aq) + H_2O(l) + CO_2(g)$$

(a) the moles of H_2O that can be obtained from 25.0 mL of 0.150 M $HC_2H_3O_2$
(b) the volume of 0.210 M K_2CO_3 needed to produce 17.5 mol $KC_2H_3O_2$
(c) the volume of 1.25 M $HC_2H_3O_2$ needed to react with 75.2 mL 0.750 M K_2CO_3
(d) the molarity (M) of the $HC_2H_3O_2$ solution when 10.15 mL react with 18.50 mL of 0.250 M K_2CO_3
(e) the liters of CO_2 gas at STP produced by the reaction of 105 mL of 1.5 M $HC_2H_3O_2$
(f) the liters of CO_2 gas at STP produced by the reaction of 25.0 mL of 0.350 M K_2CO_3 and 25.0 mL of 0.250 M $HC_2H_3O_2$

39. Calculate the molality of each of the following solutions:
(a) 2.0 mol HCl in 175 g water
(b) 14.5 g $C_{12}H_{22}O_{11}$ in 550.0 g water
(c) 25.2 mL methanol, CH_3OH (d = 0.791 g/mL) in 595 g ethanol, CH_3CH_2OH

40. Calculate the molality of each of the following solutions:
(a) 1.25 mol $CaCl_2$ in 750.0 g water
(b) 2.5 g $C_6H_{12}O_6$ in 525 g water
(c) 17.5 mL 2-propanol, $(CH_3)_2CHOH$ (d = 0.785 g/mL) in 35.5 mL H_2O (d = 1.00 g/mL)

41. What is the (a) molality, (b) freezing point, and (c) boiling point of a solution containing 2.68 g of naphthalene ($C_{10}H_8$) in 38.4 g of benzene (C_6H_6)?

42. What is the (a) molality, (b) freezing point, and (c) boiling point of a solution containing 100.0 g of ethylene glycol ($C_2H_6O_2$) in 150.0 g of water?

43. The freezing point of a solution of 8.00 g of an unknown compound dissolved in 60.0 g of acetic acid is 13.2°C. Calculate the molar mass of the compound.

44. What is the molar mass of a compound if 4.80 g of the compound dissolved in 22.0 g of H_2O given a solution that freezes at −2.50°C?

45. Identify which of the following substances are examples of true solutions.
(a) jasmine tea
(b) chromium metal
(c) muddy water
(d) gasoline

46. Identify which of the following substances are examples of true solutions.
(a) red paint
(b) Concord grape juice
(c) oil and vinegar salad dressing
(d) stainless steel

47. For those substances in Exercise 45 that are not true solutions, explain why.

48. For those substances in Exercise 46 that are not true solutions, explain why.

49. For each pairing below, predict which will dissolve faster and explain why.
(a) a teaspoon of sugar or a large crystal of sugar dissolving in hot coffee
(b) 15.0 grams of copper(II) sulfate in 100 mL of water or in 100 mL of a 15% copper(II) sulfate solution
(c) a packet of artificial sweetener dissolving in a cup of iced tea or a cup of hot tea
(d) 20 grams of silver nitrate in 1.5 L of water sitting on a table or in 1.5 L of water sloshing around in a speeding car

50. For each pairing below, predict which will dissolve faster and explain why.
(a) 0.424 g of the amino acid phenylalanine in 50.0 g of isopropyl alcohol at 25°C or in 50.0 g of isopropyl alcohol at 75°C
(b) a 1.42-g crystal of sodium acetate in 300 g of 37°C water or 1.42 g of powdered sodium acetate in 300 g of 37°C water
(c) a 500-g carton of table salt (NaCl) in 5.0 L of water at 35°C or a 500-g salt lick (NaCl) for livestock in 5.0 L of water at 21°C
(d) 25 mg of acetaminophen in 100 mL of infant pain medication containing 160 mg of acetaminophen per 15 mL or in 100 mL of adult pain medication containing 320 mg of acetaminophen per 15 mL

51. Law enforcement uses a quick and easy test for the presence of the illicit drug PCP, reacting it with potassium iodide. The PCP will form a crystalline solid with a long branching needle-like structure with KI. What is the molarity of KI in a stock solution prepared by dissolving 396.1 g of KI to a total volume of 750.0 mL?

52. Drug enforcement agents have a variety of chemical tests they can use to identify crystalline substances they may find at a crime scene. Heroin can be detected by treating a sample with a solution of mercury(II) chloride. The heroin will form a rosette of needle-shaped crystals. What is the molarity of $HgCl_2$ in a stock solution prepared by dissolving 74.15 g of $HgCl_2$ to a final volume of 250.0 mL?

ADDITIONAL EXERCISES

53. What happens to salt (NaCl) crystals when they are dissolved in water?

54. What happens to sugar molecules ($C_{12}H_{22}O_{11}$) when they are dissolved in water?

55. Why do sugar and salt behave differently when dissolved in water?

56. How does solubility vary with temperature? Compare the effect of temperature change on gases, liquids, and solids.

57. A hospital saline solution is analyzed to verify its concentration. A 50.0-mL sample of the solution with a mass of 50.320 g is evaporated to dryness. If the solid sodium chloride residue has a mass of 3.84 g, what was the mass percent of sodium chloride in the original solution?

58. Why don't blood cells shrink or swell in an isotonic sodium chloride solution (0.9% saline)?

59. In the picture of dissolving $KMnO_4$ found in Section 14.1, the compound is forming purple streaks as it dissolves. Why?

60. In Figure 14.4, observe the line for KNO_3. Explain why it slopes up from left to right. How does the slope compare to the slopes of the other substances? What does this mean?

61. An IV bag contains 9.0 g of sodium chloride per liter of solution. What is the molarity of sodium chloride in this solution?

62. How many grams of solution, 10.0% NaOH by mass, are required to neutralize 250.0 g of a 1.0 m solution of HCl?

63. A sugar syrup solution contains 15.0% sugar, $C_{12}H_{22}O_{11}$, by mass and has a density of 1.06 g/mL.
(a) How many grams of sugar are in 1.0 L of this syrup?
(b) What is the molarity of this solution?
(c) What is the molality of this solution?

64. A solution of 3.84 g C_4H_2N (empirical formula) in 250.0 g of benzene depresses the freezing point of benzene 0.614°C. What is the molecular formula for the compound?

65. Hydrochloric acid (HCl) is sold as a concentrated aqueous solution (12.0 mol/L). If the density of the solution is 1.18 g/mL, determine the molality of the solution.

66. How many grams of KNO_3 are needed to make 450 mL of a solution that is to contain 5.5 mg/mL of potassium ion? Calculate the molarity of the solution.

67. Witch hazel solution, an astringent for skin, contains 14% ethyl alcohol, C_2H_5OH, by volume. How many mL of ethyl alcohol are contained in a 16 fluid ounce bottle of witch hazel?

68. Given a solution containing 16.10 g $C_2H_6O_2$ in 82.0 g H_2O that has a boiling point of 101.62°C, verify that the boiling point elevation constant K_f for water is 0.512°C kg H_2O/mole solute.

69. Physiological saline (NaCl) solutions used in intravenous injections have a concentration of 0.90% NaCl (mass/volume).
   (a) How many grams of NaCl are needed to prepare 500.0 mL of this solution?
   (b) How much water must evaporate from this solution to give a solution that is 9.0% NaCl (mass/volume)?

70. As you are studying for finals, you calculate the concentration of caffeine in an energy drink and discover that it is 0.02080 *M* in caffeine. If a cup of your favorite coffee contains 254 mg of caffeine, how many mL of the energy drink will provide the same amount of caffeine? The molecular formula of caffeine is $C_8H_{10}N_4O_2$.

71. You love the smell of jasmine and have bought a solution of methyl jasmonate, one of the compounds responsible for the characteristic odor of jasmine. If the solution is composed of a 0.5298-molar solution of methyl jasmonate in acetone, how many molecules of methyl jasmonate will you spread around the room if you spritz 0.414 mL of the solution into the air?

72. A solution is made from 50.0 g KNO_3 and 175 g H_2O. How many grams of water must evaporate to give a saturated solution of KNO_3 in water at 20°C? (Figure 14.4.)

73. How many liters of a 0.25% (v/v) oil of wintergreen solution can you prepare if you have only 7.35 mL of oil of wintergreen on hand?

74. At 20°C, an aqueous solution of HNO_3 that is 35.0% HNO_3 by mass has a density of 1.21 g/mL.
   (a) How many grams of HNO_3 are present in 1.00 L of this solution?
   (b) What volume of this solution will contain 500. g HNO_3?

75. What is the molarity of a phosphoric acid solution if the solution is 85% by mass H_3PO_4 and has a density of 1.7 g/mL?

76. To what volume must a solution of 80.0 g H_2SO_4 in 500.0 mL of solution be diluted to give a 0.10 *M* solution?

77. How many pounds of glycerol, $C_3H_8O_3$, are present in 30.0 gallons of a 4.28 *M* solution?

78. (a) How many moles of hydrogen will be liberated from 200.0 mL of 3.00 *M* HCl reacting with an excess of magnesium? The equation is

$$Mg(s) + 2\,HCl(aq) \longrightarrow MgCl_2(aq) + H_2(g)$$

   (b) How many liters of hydrogen gas (H_2) measured at 27°C and 720 torr will be obtained?
   (*Hint*: Use the ideal gas law.)

79. Which will be more effective in neutralizing stomach acid, HCl: a tablet containing 1.20 g $Mg(OH)_2$ or a tablet containing 1.00 g $Al(OH)_3$? Show evidence for your answer.

80. Which would be more effective as an antifreeze in an automobile radiator? A solution containing
   (a) 10 kg of methyl alcohol (CH_3OH) or 10 kg of ethyl alcohol (C_2H_5OH)?
   (b) 10 *m* solution of methyl alcohol or 10 *m* solution of ethyl alcohol?

81. Automobile battery acid is 38% H_2SO_4 and has a density of 1.29 g/mL. Calculate the molality and the molarity of this solution.

82. What is the (a) molality and (b) boiling point of an aqueous sugar, $C_{12}H_{22}O_{11}$, solution that freezes at −5.4°C?

83. A solution of 6.20 g $C_2H_6O_2$ in water has a freezing point of −0.372°C. How many grams of H_2O are in the solution?

84. What (a) mass and (b) volume of ethylene glycol ($C_2H_6O_2$, density = 1.11 g/mL) should be added to 12.0 L of water in an automobile radiator to protect it from freezing at −20°C? (c) To what temperature Fahrenheit will the radiator be protected?

85. If 150 mL of 0.055 *M* HNO_3 are needed to completely neutralize 1.48 g of an *impure* sample of sodium hydrogen carbonate (baking soda), what percent of the sample is baking soda?

86. A pot of chai tea has a volume of 600. mL, a mass of 618. g, and contains 866 mg of cinnamaldehyde (C_9H_8O), the principal component of cinnamon.
   (a) What is the molarity of cinnamaldehyde in the tea?
   (b) What is the mass percent of cinnamaldehyde in the tea?
   (c) What is the mass/volume percent of cinnamaldehyde in the tea?

87. If 6.04 mL of a 2.75 *M* solution of calcium acetate is added to enough orange juice to make 1.50 L of orange juice, what will the molarity of calcium acetate be in the calcium-fortified orange juice?

88. (a) How much water must be added to concentrated sulfuric acid (H_2SO_4) (17.8 *M*) to prepare 8.4 L of 1.5 *M* sulfuric acid solution?
   (b) How many moles of H_2SO_4 are in each milliliter of the original concentrate?
   (c) How many moles are in each milliliter of the diluted solution?

89. How would you prepare a 6.00 *M* HNO_3 solution if only 3.00 *M* and 12.0 *M* solutions of the acid are available for mixing?

90. A 20.0-mL portion of an HBr solution of unknown strength is diluted to exactly 240 mL. If 100.0 mL of this diluted solution requires 88.4 mL of 0.37 *M* NaOH to achieve complete neutralization, what was the strength of the original HBr solution?

91. When 80.5 mL of 0.642 *M* $Ba(NO_3)_2$ are mixed with 44.5 mL of 0.743 *M* KOH, a precipitate of $Ba(OH)_2$ forms. How many grams of $Ba(OH)_2$ do you expect?

92. A 0.25 *M* solution of lithium carbonate (Li_2CO_3), a drug used to treat manic depression, is prepared.
   (a) How many moles of Li_2CO_3 are present in 45.8 mL of the solution?
   (b) How many grams of Li_2CO_3 are in 750 mL of the same solution?
   (c) How many milliliters of the solution would be needed to supply 6.0 g of the solute?
   (d) If the solution has a density of 1.22 g/mL, what is its mass percent?

93. Forensic chemists use Super Glue® to help them see fingerprints left behind at crime scenes. If cyanoacrylate, the main component of Super Glue®, is vaporized, it will be attracted to the oils left behind after someone touches an object. Forensic investigators will then rinse the print with a solution of europium chloride hexahydrate, $EuCl_3 \cdot 6\,H_2O$, which fluoresces when exposed to ultraviolet light. If 0.625 g of europium chloride hexahydrate is dissolved in enough water to make 500.0 mL of solution, what is the molarity of the resulting solution?

94. The pictures below represent solutions of sodium phosphate, potassium chloride, and sucrose dissolved in water. Identify each representation below:

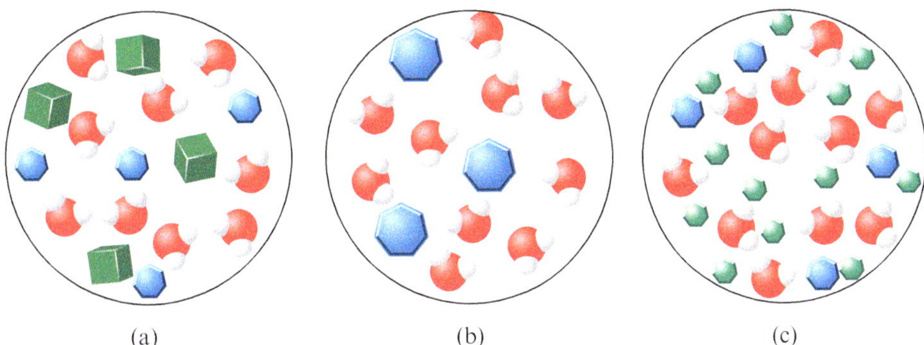

Represents H_2O molecules

95. Dark streaks have been observed on the surface of Mars. These streaks sometimes seem to run or flow, and other times they appear static. It has been hypothesized that these streaks are salt (sodium chloride) water rivers that are sometimes melted, allowing them to flow, and at other times these rivers are frozen. If the salt water mixtures melt when the temperature rises to $-12.7°C$, what is the molality of salt in the rivers? (*Hint*: The freezing point is determined by the molality of ions. How will the molality of sodium chloride be related to the molality of the ions?)

96. Snails are very sensitive to the concentration of salt in their environment. Because their "skin" is a semipermeable membrane, water can easily flow through this membrane. If you spill salt onto the "skin" of a snail, water from the inside of the body will flow out very quickly. Explain this phenomenon in terms of the chemistry described in this chapter.

97. Eggplant is often considered to be very bitter. One of the secrets chefs use to remove the bitterness is to coat the surface of the eggplant with salt. As the eggplant sits, it becomes wet and when the wetness is rinsed or wiped away the bitterness is gone. Explain what is happening to improve the flavor of the eggplant.

98. To make taffy a sugar water solution is heated until the boiling point is 127°C. What is the concentration (m) of sugar in taffy?

CHALLENGE EXERCISES

99. When solutions of hydrochloric acid and sodium sulfite react, a salt, water, and sulfur dioxide gas are produced. How many liters of sulfur dioxide gas at 775 torr and 22°C can be produced when 125 mL of 2.50 M hydrochloric acid react with 75.0 mL of 1.75 M sodium sulfite?

100. Consider a saturated solution at 20°C made from 5.549 moles of water and an unknown solute. You determine the mass of the container containing the solution to be 563 g. The mass of the empty container is 375 g. Identify the solute.

101. Dina and Murphy were trying to react 100. mL of a 0.10 M HCl solution with KOH. The procedure called for a 10% KOH solution. Dina made a 10% mass/volume solution, while Murphy made a 10% by mass solution. (Assume there is no volume change upon dissolving KOH.) Which solution required less volume to fully react with 100. mL of the HCl solution?

102. A flask containing 825 mL of a solution containing 0.355 mol of $CuSO_4$ was left open overnight. The next morning the flask only contained 755 mL of solution. What is the concentration (molarity) of the $CuSO_4$ solution remaining in the flask?

103. Which of the pathways shown below best represents the evaporation of water, and why are the others wrong?

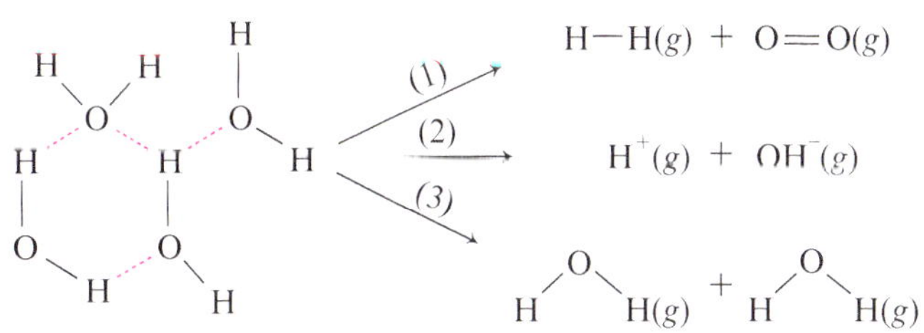

104. Why is it often advantageous or even necessary to run reactions in solution rather than mixing two solids? Would you expect reactions run in the gas phase to be more similar to solutions or to solids? Why?

105. A solution of 5.36 g of a molecular compound dissolved in 76.8 g benzene (C_6H_6) has a boiling point of 81.48°C. What is the molar mass of the compound?

Boiling point for benzene = 80.1°C

K_b for benzene = 2.53°C kg solvent/mol solute

ANSWERS TO PRACTICE EXERCISES

14.1 (a) solute, oxygen, CO_2 other gases; solvent, nitrogen (b) solute, salt; solvent, water (c) solute CO_2; solvent, water

14.2 unsaturated

14.3 The rate of dissolving increases as the temperature is increased because the molecules of solvent are moving faster and strike the surfaces of the solute more frequently.

14.4 10.0% Na_2SO_4 solution

14.5 2.02 *M*

14.6 0.27 g NaCl

14.7 0.84 L (840 mL)

14.8 75 mL NaOH solution

14.9 5.00×10^{-2} *M*

14.10 43 g

14.11 1.387 *m*

14.12 freezing point = −2.58°C,
boiling point = 100.71°C

15

Acids, Bases, and Salts

Jan Rihak/iStockphoto

CHAPTER OUTLINE

Acids are important chemicals. They are used in cooking to produce the surprise of tartness (from lemons and limes) and to release CO_2 bubbles from leavening agents in baking. Vitamin C is an acid that is an essential nutrient in our diet. Our stomachs release acid to aid in digestion. Excess stomach acid can produce heartburn and indigestion. Bacteria in our mouths produce acids that can dissolve tooth enamel to form cavities. In our recreational activities we are concerned about acidity levels in swimming pools and spas. Acids are essential in the manufacture of detergents, plastics, and storage batteries. The acid–base properties of substances are found in all areas of our lives. In this chapter we consider the properties of acids, bases, and salts.

15.1 Acids and Bases

LEARNING OBJECTIVE Compare the definitions of acids and bases, including Arrhenius, Brønsted–Lowry, and Lewis acids and bases.

KEY TERM
hydronium ion

Blend Images/Getty Images, Inc.

A characteristic of acids is sour taste, such as when sucking a lemon.

The word *acid* is derived from the Latin *acidus*, meaning "sour" or "tart," and is also related to the Latin word *acetum*, meaning "vinegar." Vinegar has been around since antiquity as a product of the fermentation of wine and apple cider. The sour constituent of vinegar is acetic acid ($HC_2H_3O_2$). Characteristic properties commonly associated with acids include the following:

1. sour taste
2. the ability to change the color of litmus, a vegetable dye, from blue to red
3. the ability to react with
   - metals such as zinc and magnesium to produce hydrogen gas
   - hydroxide bases to produce water and an ionic compound (salt)
   - carbonates to produce carbon dioxide

These properties are due to the hydrogen ions, H^+, released by acids in a water solution.

Classically, a *base* is a substance capable of liberating hydroxide ions, OH^-, in water solution. Hydroxides of the alkali metals (Group 1A(1)) and alkaline earth metals (Group 2A (2)), such as LiOH, NaOH, KOH, $Mg(OH)_2$, $Ca(OH)_2$, and $Ba(OH)_2$, are the most common inorganic bases. Water solutions of bases are called *alkaline solutions* or *basic solutions*. Some of the characteristic properties commonly associated with bases include the following:

1. bitter or caustic taste
2. a slippery, soapy feeling
3. the ability to change litmus from red to blue
4. the ability to interact with acids

Several theories have been proposed to answer the question "What is an acid and what is a base?" One of the earliest, most significant of these theories was advanced in 1884 by Svante Arrhenius (1859–1927), a Swedish scientist, who stated that "an acid is a hydrogen-containing substance that dissociates to produce hydrogen ions, and a base is a hydroxide-containing substance that dissociates to produce hydroxide ions in aqueous solutions." Arrhenius postulated that the hydrogen ions are produced by the dissociation of acids in water and that the hydroxide ions are produced by the dissociation of bases in water:

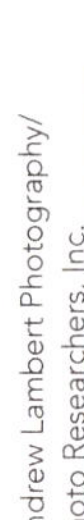
Andrew Lambert Photography/Photo Researchers, Inc.

Acids change litmus paper from blue to pink.

$$\underset{\text{acid}}{HA} \xrightarrow{H_2O} H^+(aq) + A^-(aq)$$

$$\underset{\text{base}}{MOH} \xrightarrow{H_2O} M^+(aq) + OH^-(aq)$$

An Arrhenius acid solution contains an excess of H^+ ions.
An Arrhenius base solution contains an excess of OH^- ions.

In 1923, the Brønsted–Lowry proton transfer theory was introduced by J. N. Brønsted (1897–1947), a Danish chemist, and T. M. Lowry (1847–1936), an English chemist. This theory states that an acid is a proton donor and a base is a proton acceptor.

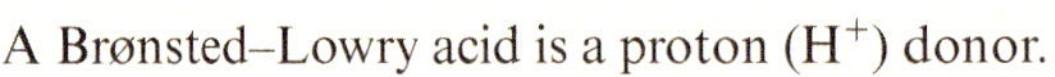

A Brønsted–Lowry acid is a proton (H^+) donor.
A Brønsted–Lowry base is a proton (H^+) acceptor.

Charles D. Winters/Photo Researchers, Inc.

Bases change litmus paper from red to blue.

Consider the reaction of hydrogen chloride gas with water to form hydrochloric acid:

$$HCl(g) + H_2O(l) \longrightarrow H_3O^+(aq) + Cl^-(aq) \quad (1)$$

In the course of the reaction, HCl donates, or gives up, a proton to form a Cl^- ion, and H_2O accepts a proton to form the H_3O^+ ion. Thus, HCl is an acid and H_2O is a base, according to the Brønsted–Lowry theory.

A hydrogen ion (H^+) is nothing more than a bare proton and does not exist by itself in an aqueous solution. In water H^+ combines with a polar water molecule to form a hydrated hydrogen ion (H_3O^+) commonly called a **hydronium ion**. The H^+ is attracted to a polar water molecule, forming a bond with one of the two pairs of unshared electrons:

$$H^+ + H\!:\!\ddot{\underset{\cdot\cdot}{O}}\!:\underset{H}{} \longrightarrow \left[H\!:\!\ddot{\underset{\cdot\cdot}{O}}\!:\!H\right]^+$$

hydronium ion

Note the electron structure of the hydronium ion. For simplicity we often use H^+ instead of H_3O^+ in equations, with the explicit understanding that H^+ is always hydrated in aqueous solution.

When a Brønsted–Lowry acid donates a proton, as illustrated in the equation below, it forms the conjugate base of that acid. When a base accepts a proton, it forms the conjugate acid of that base. A conjugate acid and base are produced as products. The formulas of a conjugate acid–base pair differ by one proton (H^+). Consider what happens when $HCl(g)$ is bubbled through water, as shown by this equation:

conjugate acid–base pair

$$\underset{\text{acid}}{HCl(g)} + \underset{\text{base}}{H_2O(l)} \longrightarrow \underset{\text{base}}{Cl^-(aq)} + \underset{\text{acid}}{H_3O^+(aq)}$$

conjugate acid–base pair

The conjugate acid–base pairs are $HCl—Cl^-$ and $H_3O^+—H_2O$. The conjugate base of HCl is Cl^-, and the conjugate acid of Cl^- is HCl. The conjugate base of H_3O^+ is H_2O, and the conjugate acid of H_2O is H_3O^+.

Another example of conjugate acid–base pairs can be seen in this equation:

$$\underset{\text{acid}}{NH_4^+} + \underset{\text{base}}{H_2O} \longrightarrow \underset{\text{acid}}{H_3O^+} + \underset{\text{base}}{NH_3}$$

Here the conjugate acid–base pairs are $NH_4^+—NH_3$ and $H_3O^+—H_2O$.

NH_3 $\xrightarrow[H^+]{\text{Add}}$ NH_4^+

CHECK YOUR UNDERSTANDING

15.1 Properties of Acids

15.2 Properties of Bases

15.3 Acids and Bases

WileyPLUS

EXAMPLE 15.1

Write the formula for (a) the conjugate base of H_2O and of HNO_3, and (b) the conjugate acid of SO_4^{2-} and of $C_2H_3O_2^-$.

SOLUTION

(a) To write the conjugate base of an acid, remove one proton from the acid formula:

$$H_2O \xrightarrow{-H^+} OH^- \quad \text{(conjugate base)}$$

$$HNO_3 \xrightarrow{-H^+} NO_3^- \quad \text{(conjugate base)}$$

Note that, by removing an H^+, the conjugate base becomes more negative than the acid by one minus charge.

(b) To write the conjugate acid of a base, add one proton to the formula of the base:

$$SO_4^{2-} \xrightarrow{+H^+} HSO_4^- \quad \text{(conjugate acid)}$$

$$C_2H_3O_2^- \xrightarrow{+H^+} HC_2H_3O_2 \quad \text{(conjugate acid)}$$

In each case the conjugate acid becomes more positive than the base by a +1 charge due to the addition of H^+.

ONLINE LEARNING MODULE

Conjugate Acid-Base Pairs

WileyPLUS

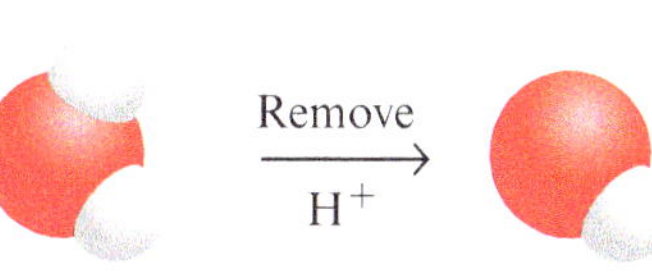

Remember: The difference between an acid or a base and its conjugate is one proton, H^+.

CHECK YOUR UNDERSTANDING

15.4 Conjugate Bases

15.5 Conjugate Acids

15.6 Identifying Conjugate Acids and Bases in a Reaction

➥ WileyPLUS

PRACTICE 15.1

Write formulas for the conjugate base for these acids:

(a) H_2CO_3 (b) HNO_2 (c) $HC_2H_3O_2$

PRACTICE 15.2

Write formulas for the conjugate acid for these bases:

(a) HSO_4^- (b) NH_3 (c) OH^-

A more general concept of acids and bases was introduced by Gilbert N. Lewis. The Lewis theory deals with the way in which a substance with an unshared pair of electrons reacts in an acid–base type of reaction. According to this theory, a base is any substance that has an unshared pair of electrons (electron pair donor), and an acid is any substance that will attach itself to or accept a pair of electrons.

A Lewis acid is an electron pair acceptor.
A Lewis base is an electron pair donor.

Consider the following reaction:

$$\underset{\text{acid}}{H^+} + \underset{\text{base}}{\begin{array}{c} H \\ :\ddot{N}:H \\ H \end{array}} \longrightarrow \left[\begin{array}{c} H \\ H:\ddot{N}:H \\ \ddot{H} \end{array}\right]^+$$

The H^+ is a Lewis acid, and the $:NH_3$ is a Lewis base. According to the Lewis theory, substances other than proton donors (e.g., BF_3) behave as acids:

$$\underset{\text{acid}}{\begin{array}{c} F \\ F:\ddot{B} \\ F \end{array}} + \underset{\text{base}}{\begin{array}{c} H \\ :\ddot{N}:H \\ H \end{array}} \longrightarrow \begin{array}{c} F\ H \\ F:\ddot{B}:\ddot{N}:H \\ F\ H \end{array}$$

These three theories, which explain how acid–base reactions occur, are summarized in **TABLE 15.1**. We will generally use the theory that best explains the reaction under consideration. Most of our examples will refer to aqueous solutions. Note that in an aqueous acidic solution the H^+ ion concentration is always greater than the OH^- ion concentration. And vice versa—in an aqueous basic solution the OH^- ion concentration is always greater than the H^+ ion concentration. When the H^+ and OH^- ion concentrations in a solution are equal, the solution is *neutral*; that is, it is neither acidic nor basic.

CHECK YOUR UNDERSTANDING

15.7 Acid/Conjugate Base Differences

➥ WileyPLUS

TABLE 15.1 | Summary of Acid–Base Definitions

Theory	Acid	Base
Arrhenius	A hydrogen-containing substance that produces hydrogen ions in aqueous solution	A hydroxide-containing substance that produces hydroxide ions in aqueous solution
Brønsted–Lowry	A proton (H^+) donor	A proton (H^+) acceptor
Lewis	Any species that will bond to an unshared pair of electrons (electron pair acceptor)	Any species that has an unshared pair of electrons (electron pair donor)

CHEMISTRY *IN ACTION* Drug Delivery: An Acid–Base Problem

Have you ever looked at the package insert for some of the injections you have received at the doctor's office? You may have been given lidocaine · HCl while receiving a filling at the dentist, proparacaine · HCl drops during an eye exam, or been prescribed pseudoephedrine · HCl for congestion. Why do all of these drugs have HCl in them? Are we taking acid? The answer to that question is a qualified yes. It is not like the acid we use in the laboratory, but it is indeed a salt according to the rules of chemistry. Many pharmaceuticals on the market are molecules with a structure very similar to ammonia. They are called amines, and they are synthesized by replacing one or more of the hydrogen atoms in the ammonia molecule with an organic or carbon-based group. Ammonia and several common amines are pictured below:

NH_3

Ammonia

$H{-}\ddot{N}H{-}CH_3 \qquad H{-}\ddot{N}(CH_3){-}CH_3 \qquad H{-}\ddot{N}H{-}CH_2{-}CH_3$

Amines

Most pharmaceuticals are dissolved in water to be administered to patients. For amine-based drugs this can be a problem because many of the pharmacologically active amines are not soluble in water. These amines are generally fairly volatile, meaning they evaporate easily and thus could be vaporized and inhaled, but this is an inconvenient mode of drug delivery. Pharmaceutical chemists discovered a better method of delivering these drugs by taking advantage of one of the chemical properties of ammonia and the amines. These molecules are all Lewis bases and will react with acids to form a salt as shown in the following reaction:

$$H{-}\ddot{N}H{-}CH_3 + H{-}\ddot{Cl}{:} \longrightarrow \left[H{-}NH_2{-}CH_3\right]^+ \ {:}\ddot{Cl}{:}^-$$

These salts are much more soluble in water than the original amine, which makes it much easier to administer the medication. It can be dissolved in water to be injected into the bloodstream, be dropped into the aqueous environment of the eye or other mucous membranes, or dissolved into an aqueous medium to be consumed orally. So, yes, the hydrochloride salts of many amine-based drugs are very useful. Check the package insert next time you receive an injection or are given a prescription.

Some common pharmacological drugs are shown here.

Lidocaine

Proparacaine

Amines used as local anesthetic

Pseudoephedrine

Phenylephrine

Amines used as decongestants

Acid Reactions

In aqueous solutions, the H^+ or H_3O^+ ions are responsible for the characteristic reactions of acids. The following reactions are in an aqueous medium:

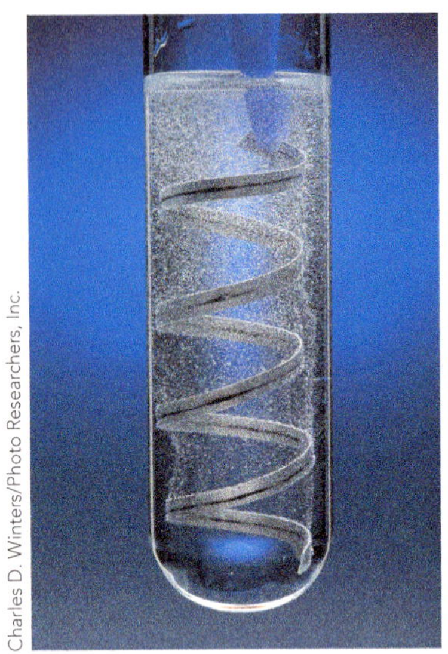

Charles D. Winters/Photo Researchers, Inc.

Magnesium ribbon reacts in HCl solution to produce hydrogen bubbles.

Reaction with metals Acids react with metals that lie above hydrogen in the activity series of elements to produce hydrogen and an ionic compound (salt) (see Section 17.5):

$$\text{acid} + \text{metal} \longrightarrow \text{hydrogen} + \text{ionic compound}$$
$$2\,HCl(aq) + Mg(s) \longrightarrow H_2(g) + MgCl_2(aq)$$
$$H_2SO_4(aq) + Mg(s) \longrightarrow H_2(g) + MgSO_4(aq)$$
$$6\,HC_2H_3O_2(aq) + 2\,Al(s) \longrightarrow 3\,H_2(g) + 2\,Al(C_2H_3O_2)_3(aq)$$

Acids such as nitric acid (HNO_3) are oxidizing substances (see Chapter 17) and react with metals to produce water instead of hydrogen. For example,

$$3\,Zn(s) + 8\,HNO_3(\text{dilute}) \longrightarrow 3\,Zn(NO_3)_2(aq) + 2\,NO(g) + 4\,H_2O(l)$$

Reaction with bases The interaction of an acid and a base is called a *neutralization reaction.* In aqueous solutions, the products of this reaction are a salt and water:

$$\text{acid} + \text{base} \longrightarrow \text{salt} + \text{water}$$
$$HBr(aq) + KOH(aq) \longrightarrow KBr(aq) + H_2O(l)$$
$$2\,HNO_3(aq) + Ca(OH)_2(aq) \longrightarrow Ca(NO_3)_2(aq) + 2\,H_2O(l)$$
$$2\,H_3PO_4(aq) + 3\,Ba(OH)_2(aq) \longrightarrow Ba_3(PO_4)_2(s) + 6\,H_2O(l)$$

EXAMPLE 15.2

Complete and balance these equations.

(a) $Mg(s) + HBr(aq) \longrightarrow$

(b) $HCl(aq) + Ba(OH)_2(aq) \longrightarrow$

(c) $H_3PO_4(aq) + NaOH(aq) \longrightarrow$

SOLUTION

(a) $Mg + 2\,HBr \longrightarrow MgBr_2(aq) + H_2(g)$

(b) $2\,HCl(aq) + Ba(OH)_2(aq) \longrightarrow BaCl_2(aq) + 2\,H_2O(l)$

(c) $2\,H_3PO_4(aq) + 6\,NaOH(aq) \longrightarrow 2\,Na_3PO_4(aq) + 6\,H_2O(l)$

Base Reactions

The OH^- ions are responsible for the characteristic reactions of bases. Bases react with acids to produce a salt and water. See reaction of acids with bases earlier in this section.

15.2 Salts

LEARNING OBJECTIVE Explain how a salt is formed and predict the formula of a salt given an acid and a base precursor.

Chemists use the terms *ionic compound* and *salt* interchangeably.

Salts are very abundant in nature. Most of the rocks and minerals of Earth's mantle are salts of one kind or another. Huge quantities of dissolved salts also exist in the oceans. Salts can be considered compounds derived from acids and bases. They consist of positive metal or ammonium ions combined with negative nonmetal ions (OH^- and O^{2-} excluded). The positive ion is the base counterpart, and the nonmetal ion is the acid counterpart:

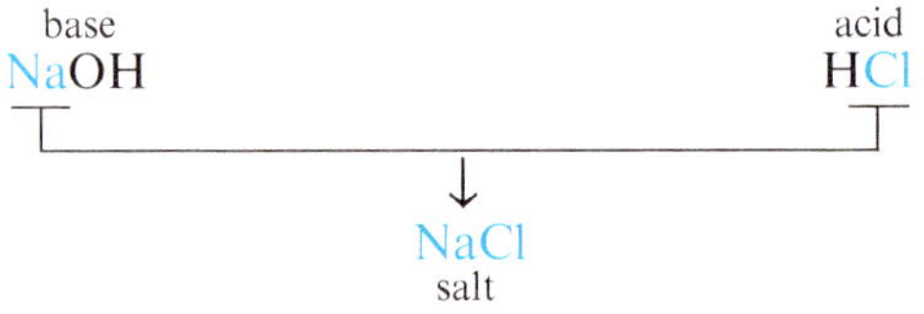

Salts are usually crystalline and have high melting and boiling points.

From a single acid such as hydrochloric acid (HCl), we can produce many chloride compounds by replacing the hydrogen with metal ions (e.g., NaCl, KCl, RbCl, $CaCl_2$, $NiCl_2$). Hence, the number of known salts greatly exceeds the number of known acids and bases. If the hydrogen atoms of a binary acid are replaced by a nonmetal, the resulting compound has covalent bonding and is therefore not considered to be ionic (e.g., PCl_3, S_2Cl_2, Cl_2O, NCl_3, ICl).

You may want to review Chapter 6 for nomenclature of acids, bases, and salts.

PRACTICE 15.3

Write the formulas and names of the acid and base from which these salts are formed.

(a) K_3PO_4 (b) $MgBr_2$ (c) LiCl (d) $FeCO_3$

EXAMPLE 15.3

Why are there many more salts than there are acids? Write the formulas for five salts derived from sulfuric acid.

SOLUTION

For any particular acid you can write formulas for many salts by replacing the hydrogen(s) of the acid with a cation: Na_2SO_4, $MgSO_4$, $Fe_2(SO_4)_3$, $KHSO_4$, $Sn(SO_4)_2$, and many more.

PRACTICE 15.4

Given the formulas for three acids and three bases, write the formulas for all the salts that can be formed from the following: HCl, H_2SO_4, H_3PO_4, NaOH, $Mg(OH)_2$, and $Fe(OH)_3$.

CHEMISTRY *IN ACTION* A Cool Fizz

Try this experiment with your friends: Chill some soft drinks well. Next, pour the soft drink into a glass. Stick your tongue into the liquid and time how long you can keep it there.

What causes the tingling we feel in our mouth (or on our tongue)? Many people believe it is the bubbles of carbon dioxide, but scientists have found that is not the answer. The tingling is caused by "chemisthesis." Bruce Bryant at Monell Chemical Senses Center in Philadelphia says that chemisthesis is a chemically induced sensation that does not involve taste or odor receptors. The tongue-tingling response to the soft drink is caused by production of protons (H^+ ions) released when an enzyme (carbonic anhydrase) acts on CO_2. The H^+ ions acidify nerve endings, producing the sensation of tingling.

Carbon dioxide also stimulates other neurons when the drink is cold. This means that at a constant pressure of CO_2, a cold drink will produce more tingling than a room-temperature drink. If the drink is at room temperature, CO_2 will increase the cool feeling. Chilling a soda increases the effect of the protons on the nerve endings. At the same time, the high concentration of CO_2 in a freshly opened soda makes the cold soda feel even colder. Together these chemical effects make it really painful to keep your tongue in the chilled soda.

Chemisthesis is important for survival, as well as for having fun with friends.

Bryant says, "It tells you that something is chemically impinging on your body, that tissue is in imminent danger. Burrowing animals can sense toxic levels of carbon dioxide and even feel a sting when exposed to those levels."

Tongue papillae contain nerve endings that send temperature and tactile information to the brain.

15.3 Electrolytes and Nonelectrolytes

LEARNING OBJECTIVE Describe properties, ionization, dissociation, and strength of electrolytes and compare them to nonelectrolytes.

KEY TERMS
electrolyte
nonelectrolyte
dissociation
ionization
strong electrolyte
weak electrolyte

We can show that solutions of certain substances are conductors of electricity with a simple conductivity apparatus, which consists of a pair of electrodes connected to a voltage source through a light bulb and switch (FIGURE 15.1). If the medium between the electrodes is a conductor of electricity, the light bulb will glow when the switch is closed. When chemically pure water is placed in the beaker and the switch is closed, the light does not glow, indicating that water is a virtual nonconductor. When we dissolve a small amount of sugar in the water and test the solution, the light still does not glow, showing that a sugar solution is also a nonconductor. But when a small amount of salt, NaCl, is dissolved in water and this solution is tested, the light glows brightly. Thus the salt solution conducts electricity. A fundamental difference exists between the chemical bonding in sugar and that in salt. Sugar is a covalently bonded (molecular) substance; common salt is a substance with ionic bonds.

CHECK YOUR UNDERSTANDING
15.8 Electrolytes
WileyPLUS

Substances whose aqueous solutions are conductors of electricity are called **electrolytes**. Substances whose solutions are nonconductors are known as **nonelectrolytes**. The major classes of compounds that are electrolytes are acids, bases, and other ionic compounds (salts). Solutions of certain oxides also are conductors because the oxides form an acid or a base when dissolved in water. One major difference between electrolytes and nonelectrolytes is that electrolytes are capable of producing ions in solution, whereas nonelectrolytes do not have this property. Solutions that contain a sufficient number of ions will conduct an electric current. Although pure water is essentially a nonconductor, many city water supplies contain enough dissolved ionic matter to cause the light to glow dimly when the water is tested in a conductivity apparatus. TABLE 15.2 lists some common electrolytes and nonelectrolytes.

TABLE 15.2 Representative Electrolytes and Nonelectrolytes

Electrolytes		Nonelectrolytes	
H_2SO_4	$HC_2H_3O_2$	$C_{12}H_{22}O_{11}$ (sugar)	CH_3OH (methyl alcohol)
HCl	NH_3	C_2H_5OH (ethyl alcohol)	$CO(NH_2)_2$ (urea)
HNO_3	K_2SO_4	$C_2H_4(OH)_2$ (ethylene glycol)	O_2
NaOH	$NaNO_3$	$C_3H_5(OH)_3$ (glycerol)	H_2O

Dissociation and Ionization of Electrolytes

Arrhenius received the 1903 Nobel Prize in chemistry for his work on electrolytes. He found that a solution conducts electricity because the solute dissociates immediately upon dissolving into electrically charged particles (ions). The movement of these ions toward oppositely charged electrodes causes the solution to be a conductor. According to Arrhenius's theory, solutions that are relatively poor conductors contain electrolytes that

(a) Pure water

(b) Nonelectrolyte

(c) Strong electrolyte

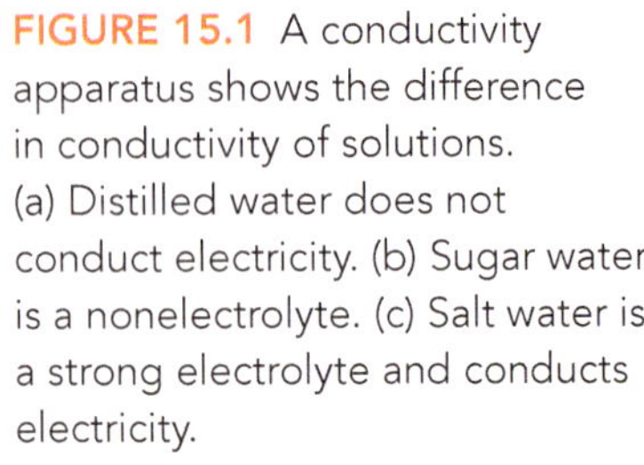
FIGURE 15.1 A conductivity apparatus shows the difference in conductivity of solutions. (a) Distilled water does not conduct electricity. (b) Sugar water is a nonelectrolyte. (c) Salt water is a strong electrolyte and conducts electricity.

are only partly dissociated. Arrhenius also believed that ions exist in solution whether or not an electric current is present. In other words, the electric current does not cause the formation of ions. Remember that positive ions are cations; negative ions are anions.

Acids, bases, and salts are electrolytes.

EXAMPLE 15.4

What is an electrolyte?

SOLUTION

An electrolyte is a substance that forms ions in solution and makes the solution a conductor of electricity.

We have seen that sodium chloride crystals consist of sodium and chloride ions held together by ionic bonds. **Dissociation** is the process by which the ions of a salt separate as the salt dissolves. When placed in water, the sodium and chloride ions are attracted by the polar water molecules, which surround each ion as it dissolves. In water, the salt dissociates, forming hydrated sodium and chloride ions (FIGURE 15.2). The sodium and chloride ions in solution are surrounded by a specific number of water molecules and have less attraction for each other than they had in the crystalline state. The equation representing this dissociation is

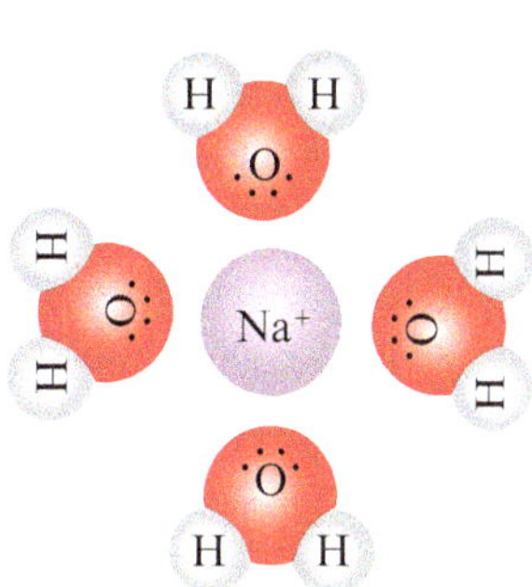

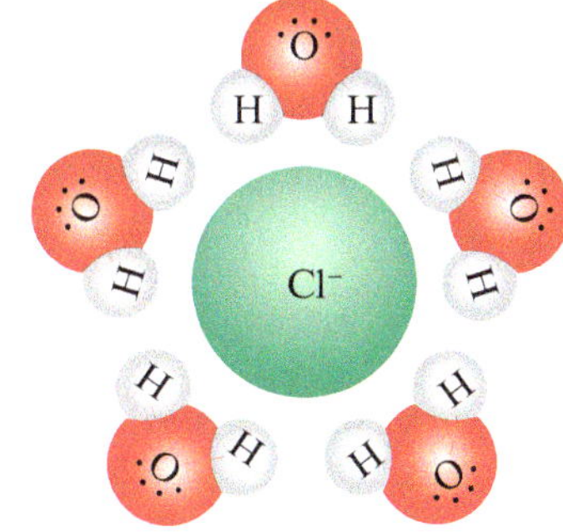

FIGURE 15.2 Hydrated sodium and chloride ions. The negative end of the water dipole is attracted to the Na^+ ion, and the positive end is attracted to the Cl^- ion.

$$NaCl(s) + (x + y)H_2O \longrightarrow Na^+(H_2O)_x + Cl^-(H_2O)_y$$

A simplified dissociation equation in which the water is omitted but understood to be present is

$$NaCl(s) \longrightarrow Na^+(aq) + Cl^-(aq)$$

Remember that sodium chloride exists in an aqueous solution as hydrated ions, not as NaCl units, even though the formula NaCl (or $Na^+ + Cl^-$) is often used in equations.

The chemical reactions of salts in solution are the reactions of their ions. For example, when sodium chloride and silver nitrate react and form a precipitate of silver chloride, only the Ag^+ and Cl^- ions participate in the reaction. The Na^+ and NO_3^- remain as ions in solution:

$$Ag^+(aq) + Cl^-(aq) \longrightarrow AgCl(s)$$

Ionization is the formation of ions; it occurs as a result of a chemical reaction of certain substances with water. Glacial acetic acid (100% $HC_2H_3O_2$) is a liquid that behaves as a nonelectrolyte when tested by the method described earlier. But a water solution of acetic acid conducts an electric current (as indicated by the dull-glowing light of the conductivity apparatus). The equation for the reaction with water, which forms hydronium and acetate ions, is

$$\underset{\text{acid}}{HC_2H_3O_2} + \underset{\text{base}}{H_2O} \rightleftharpoons \underset{\text{acid}}{H_3O^+} + \underset{\text{base}}{C_2H_3O_2^-}$$

or, in the simplified equation,

$$HC_2H_3O_2 \rightleftharpoons H^+ + C_2H_3O_2^-$$

In this ionization reaction, water serves not only as a solvent but also as a base according to the Brønsted–Lowry theory.

Hydrogen chloride is predominantly covalently bonded, but when dissolved in water, it reacts to form hydronium and chloride ions:

$$HCl(g) + H_2O(l) \longrightarrow H_3O^+(aq) + Cl^-(aq)$$

When a hydrogen chloride solution is tested for conductivity, the light glows brilliantly, indicating many ions in the solution.

Ionization occurs in each of the preceding two reactions with water, producing ions in solution. The necessity for water in the ionization process can be demonstrated by dissolving hydrogen chloride in a nonpolar solvent such as hexane and testing the solution for conductivity. The solution fails to conduct electricity, indicating that no ions are produced.

The terms *dissociation* and *ionization* are often used interchangeably to describe processes taking place in water. But, strictly speaking, the two are different. In the dissociation of a salt, the salt already exists as ions; when it dissolves in water, the ions separate, or dissociate, and increase in mobility. In the ionization process, ions are produced by the reaction of a compound with water.

Strong and Weak Electrolytes

Electrolytes are classified as strong or weak depending on the degree, or extent, of dissociation or ionization. **Strong electrolytes** are essentially 100% ionized in solution; **weak electrolytes** are much less ionized (based on comparing 0.1 *M* solutions). Most electrolytes are either strong or weak, with a few classified as moderately strong or weak. Most salts are strong electrolytes. Acids and bases that are strong electrolytes (highly ionized) are called *strong acids* and *strong bases.* Acids and bases that are weak electrolytes (slightly ionized) are called *weak acids* and *weak bases.*

For equivalent concentrations, solutions of strong electrolytes contain many more ions than do solutions of weak electrolytes. As a result, solutions of strong electrolytes are better conductors of electricity. Consider two solutions, 1 *M* HCl and 1 *M* $HC_2H_3O_2$. Hydrochloric acid is almost 100% ionized; acetic acid is about 1% ionized. (**FIGURE 15.3**.) Thus, HCl is a strong acid and $HC_2H_3O_2$ is a weak acid. Hydrochloric acid has about 100 times as many hydronium ions in solution as acetic acid, making the HCl solution much more acidic.

We can distinguish between strong and weak electrolytes experimentally using the apparatus described earlier. A 1 *M* HCl solution causes the light to glow brilliantly, but a 1 *M* $HC_2H_3O_2$ solution causes only a dim glow. The strong base sodium hydroxide (NaOH) can be distinguished in a similar fashion from the weak base ammonia (NH_3). The ionization of a weak electrolyte in water is represented by an equilibrium equation showing that both the un-ionized and ionized forms are present in solution. In the equilibrium equation of $HC_2H_3O_2$ and its ions, we say that the equilibrium lies "far to the left" because relatively few hydrogen and acetate ions are present in solution:

$$HC_2H_3O_2(aq) \rightleftharpoons H^+(aq) + C_2H_3O_2^-(aq)$$

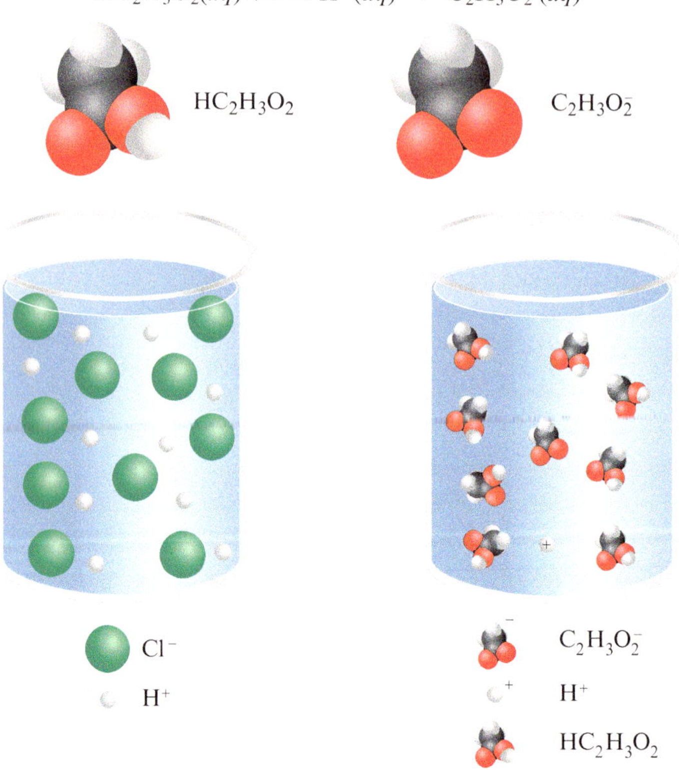

FIGURE 15.3 HCl solution (left) is 100% ionized, while in $HC_2H_3O_2$ solution (bottom) almost all of the solute is in molecular form. HCl is a strong acid, while $HC_2H_3O_2$ is a weak acid.
Note: The water molecules in the solution are not shown in this figure.

We have previously used a double arrow in an equation to represent reversible processes in the equilibrium between dissolved and undissolved solute in a saturated solution. A double arrow ($\rightleftharpoons$) is also used in the ionization equation of soluble weak electrolytes to indicate that the solution contains a considerable amount of the un-ionized compound in equilibrium with its ions in solution. (See Section 16.2 for a discussion of reversible reactions.) A single arrow is used to indicate that the electrolyte is essentially all in the ionic form in the solution. For example, nitric acid is a strong acid; nitrous acid is a weak acid. Their ionization equations in water may be indicated as

$$HNO_3(aq) \xrightarrow{H_2O} H^+(aq) + NO_3^-(aq)$$

$$HNO_3(aq) \xrightleftharpoons{H_2O} H^+(aq) + NO_2^-(aq)$$

Practically all soluble salts, acids (such as sulfuric, nitric, and hydrochloric acids), and bases (such as sodium, potassium, calcium, and barium hydroxides) are strong electrolytes. Weak electrolytes include numerous other acids and bases such as acetic acid, nitrous acid, carbonic acid, and ammonia. The terms *strong acid, strong base, weak acid,* and *weak base* refer to whether an acid or a base is a strong or weak electrolyte. A brief list of strong and weak electrolytes is given in **TABLE 15.3**.

Electrolytes yield two or more ions per formula unit upon dissociation—the actual number being dependent on the compound. Dissociation is complete or nearly complete for nearly all soluble ionic compounds and for certain other strong electrolytes, such as those given in Table 15.3. The following are dissociation equations for several strong electrolytes. In all cases, the ions are actually hydrated:

$$NaOH \xrightarrow{H_2O} Na^+(aq) + OH^-(aq) \qquad \text{2 ions in solution per formula unit}$$

$$Na_2SO_4 \xrightarrow{H_2O} 2\ Na^+(aq) + SO_4^{2-}(aq) \qquad \text{3 ions in solution per formula unit}$$

$$Fe_2(SO_4)_3 \xrightarrow{H_2O} 2\ Fe^{3+}(aq) + 3\ SO_4^{2-}(aq) \qquad \text{5 ions in solution per formula unit}$$

TABLE 15.3 | **Strong and Weak Electrolytes**

Strong electrolytes		Weak electrolytes	
Most soluble salts	$HClO_4$	$HC_2H_3O_2$	$H_2C_2O_4$
H_2SO_4	$NaOH$	H_2CO_3	H_3BO_3
HNO_3	KOH	HNO_2	$HClO$
HCl	$Ca(OH)_2$	H_2SO_3	NH_3
HBr	$Ba(OH)_2$	H_2S	HF

One mole of NaCl will give 1 mol of Na^+ ions and 1 mol of Cl^- ions in solution, assuming complete dissociation of the salt. One mole of $CaCl_2$ will give 1 mol of Ca^{2+} ions and 2 mol of Cl^- ions in solution:

$$\underset{1\ \text{mol}}{NaCl} \xrightarrow{H_2O} \underset{1\ \text{mol}}{Na^+(aq)} + \underset{1\ \text{mol}}{Cl^-(aq)}$$

$$\underset{1\ \text{mol}}{CaCl_2} \xrightarrow{H_2O} \underset{1\ \text{mol}}{Ca^{2+}(aq)} + \underset{2\ \text{mol}}{2\ Cl^-(aq)}$$

ONLINE LEARNING MODULE
Ion Concentrations in Solutions
➡ WileyPLUS

ENHANCED EXAMPLE 15.5 **WileyPLUS**

What is the molarity of each ion in a solution of (a) 2.0 *M* NaCl and (b) 0.40 *M* K_2SO_4? Assume complete dissociation.

SOLUTION

(a) According to the dissociation equation,

$$\underset{1\ \text{mol}}{\text{NaCl}} \xrightarrow{H_2O} \underset{1\ \text{mol}}{Na^+(aq)} + \underset{1\ \text{mol}}{Cl^-(aq)}$$

the concentration of Na^+ is equal to that of NaCl:
1 mol NaCl $\longrightarrow$ 1 mol Na^+ and the concentration of Cl^- is also equal to that of NaCl. Therefore, the concentrations of the ions in 2.0 *M* NaCl are 2.0 *M* Na^+ and 2.0 *M* Cl^-.

(b) According to the dissociation equation,

$$\underset{1\ \text{mol}}{K_2SO_4} \xrightarrow{H_2O} \underset{2\ \text{mol}}{2\,K^+(aq)} + \underset{1\ \text{mol}}{SO_4^{2-}(aq)}$$

the concentration of K^+ is twice that of K_2SO_4 and the concentration of SO_4^{2-} is equal to that of K_2SO_4. Therefore, the concentrations of the ions in 0.40 *M* K_2SO_4 are 0.80 *M* K^+ and 0.40 *M* SO_4^{2-}.

PRACTICE 15.5

What is the molarity of each ion in a solution of (a) 0.050 *M* $MgCl_2$ and (b) 0.070 *M* $AlCl_3$?

Ionization of Water

Pure water is a *very* weak electrolyte, but it does ionize slightly. Two equations commonly used to show how water ionizes are

$$\underset{\text{acid}}{H_2O} + \underset{\text{base}}{H_2O} \rightleftharpoons \underset{\text{acid}}{H_3O^+} + \underset{\text{base}}{OH^-}$$

and

$$H_2O \rightleftharpoons H^+ + OH^-$$

The first equation represents the Brønsted–Lowry concept, with water reacting as both an acid and a base, forming a hydronium ion and a hydroxide ion. The second equation is a simplified version, indicating that water ionizes to give a hydrogen and a hydroxide ion. Actually, the proton (H^+) is hydrated and exists as a hydronium ion. In either case equal molar amounts of acid and base are produced so that water is neutral, having neither H^+ nor OH^- ions in excess. The ionization of water at 25°C produces an H^+ ion concentration of 1.0×10^{-7} mol/L and an OH^- ion concentration of 1.0×10^{-7} mol/L. Square brackets, [ ], are used to indicate that the concentration is in moles per liter. Thus $[H^+]$ means the concentration of H^+ is in moles per liter. These concentrations are usually expressed as

$$[H^+] \text{ or } [H_3O^+] = 1.0 \times 10^{-7}\ \text{mol/L}$$
$$[OH^-] = 1.0 \times 10^{-7}\ \text{mol/L}$$

CHECK YOUR UNDERSTANDING
15.9 Dissociation of Pure Water
➡ WileyPLUS

These figures mean that about two out of every billion water molecules are ionized. This amount of ionization, small as it is, is a significant factor in the behavior of water in many chemical reactions.

15.4 Introduction to pH

Calculate the pH of a solution from the hydrogen ion concentration.

LEARNING OBJECTIVE

KEY TERM
pH
logarithm (log)

The acidity of an aqueous solution depends on the concentration of hydrogen or hydronium ions. The pH scale of acidity provides a simple, convenient, numerical way to state the acidity of a solution. Values on the pH scale are obtained by mathematical conversion of H^+ ion concentrations to pH by the expression

$$\text{pH} = -\log[H^+]$$

where $[H^+] = H^+$ or H_3O^+ ion concentration in moles per liter. The **pH** is defined as the *negative* logarithm of the H^+ or H_3O^+ concentration in moles per liter:

$$\text{pH} = -\log[H^+] = -\log(1 \times 10^{-7}) = -(-7) = 7$$

For example, the pH of pure water at 25°C is 7 and is said to be neutral; that is, it is neither acidic nor basic, because the concentrations of H^+ and OH^- are equal. Solutions that contain more H^+ ions than OH^- ions have pH values less than 7, and solutions that contain fewer H^+ ions than OH^- ions have pH values greater than 7.

$\text{pH} < 7.00$ is an acidic solution
$\text{pH} = 7.00$ is a neutral solution
$\text{pH} > 7.00$ is a basic solution

$$\text{When } [H^+] = 1 \times 10^{-5}\ \text{mol/L, pH} = 5\ \text{(acidic)}$$
$$\text{When } [H^+] = 1 \times 10^{-9}\ \text{mol/L, pH} = 9\ \text{(basic)}$$

Instead of saying that the hydrogen ion concentration in the solution is 1×10^{-5} mol/L, it's customary to say that the pH of the solution is 5. The smaller the pH value, the more acidic the solution (**FIGURE 15.4**).

CHECK YOUR UNDERSTANDING
15.10 Definition of pH
WileyPLUS

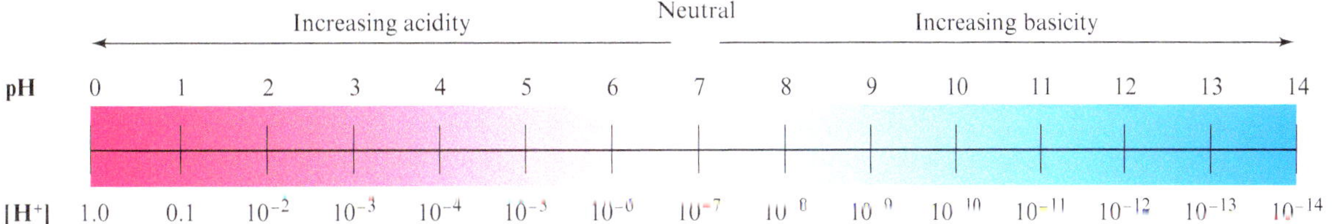

FIGURE 15.4 The pH scale of acidity and basicity.

The pH scale, along with its interpretation, is given in **TABLE 15.4**, and **TABLE 15.5** lists the pH of some common solutions. Note that a change of only one pH unit means a 10-fold increase or decrease in H^+ ion concentration. For example, a solution with a pH of 3.0 is 10 times more acidic than a solution with a pH of 4.0. A simplified method of determining pH from $[H^+]$ follows:

CHECK YOUR UNDERSTANDING
15.11 Acidic Substances
WileyPLUS

$[H^+] = 1 \times 10^{-5}$ ← pH = this number (5); pH = 5.0
(when this number is exactly 1)

$[H^+] = 2 \times 10^{-5}$ ← pH is between this number and next lower number (4 and 5); pH = 4.7
(when this number is between 1 and 10)

Calculating the pH value for H^+ ion concentrations requires the use of logarithms, which are exponents. The **logarithm (log)** of a number is simply the power to which 10 must be raised to give that number. Thus the log of 100 is 2 ($100 = 10^2$), and the log of 1000 is 3 ($1000 = 10^3$). The log of 500 is 2.70, but you can't determine this value easily without a scientific calculator.

Help on using calculators is found in Appendix II.

Let's determine the pH of a solution with $[H^+] = 2 \times 10^{-5}$ using a calculator. The number $-4.69\ldots$ will be displayed. The pH is then

$$\text{pH} = -\log[H^+] = -(-4.69\ldots) = 4.7$$

Remember: Change the sign on your calculator since $\text{pH} = -\log[H^+]$.

TABLE 15.4 pH Scale for Expressing Acidity

$[H^+]$ (mol/L)	pH	
1×10^{-14}	14	↑
1×10^{-13}	13	
1×10^{-12}	12	Increasing basicity
1×10^{-11}	11	
1×10^{-10}	10	
1×10^{-9}	9	
1×10^{-8}	8	
1×10^{-7}	7	Neutral
1×10^{-6}	6	
1×10^{-5}	5	
1×10^{-4}	4	
1×10^{-3}	3	Increasing acidity
1×10^{-2}	2	
1×10^{-1}	1	
1×10^{0}	0	↓

TABLE 15.5 The pH of Common Solutions

Solution	pH
Gastric juice	1.0
0.1 *M* HCl	1.0
Lemon juice	2.3
Vinegar	2.8
0.1 *M* $HC_2H_3O_2$	2.9
Orange juice	3.7
Tomato juice	4.1
Coffee, black	5.0
Urine	6.0
Milk	6.6
Pure water (25°C)	7.0
Blood	7.4
Household ammonia	11.0
1 *M* NaOH	14.0

Next we must determine the correct number of significant figures in the logarithm. The rules for logs are different from those we use in other math operations. The number of decimal places for a log must equal the number of significant figures in the original number. Since 2×10^{-5} has one significant figure, we should round the log to one decimal place (4.69...) = 4.7.

ONLINE LEARNING MODULE
Calculating pH

ENHANCED EXAMPLE 15.6 WileyPLUS

What is the pH of a solution with an $[H^+]$ of (a) 1.0×10^{-11}, (b) 6.0×10^{-4}, and (c) 5.47×10^{-8}?

SOLUTION

(a) $[H^+] = 1.0 \times 10^{-11}$
(2 significant figures)
$\text{pH} = -\log(1.0 \times 10^{-11})$
$\text{pH} = 11.00$
(2 decimal places)

(b) $[H^+] = 6.0 \times 10^{-4}$
(2 significant figures)
$\log(6.0 \times 10^{-4}) = -3.22$
$\text{pH} = -\log[H^+]$
$\text{pH} = -(-3.22) = 3.22$
(2 decimal places)

(c) $[H^+] = 5.47 \times 10^{-8}$
(3 significant figures)
$\log(5.47 \times 10^{-8}) = -7.262$
$\text{pH} = -\log[H^+]$
$\text{pH} = -(-7.262) = 7.262$
(3 decimal places)

PRACTICE 15.6

What is the pH of a solution with $[H^+]$ of (a) 3.9×10^{-12} *M*, (b) 1.3×10^{-3} *M*, and (c) 3.72×10^{-6} *M*?

CHECK YOUR UNDERSTANDING
15.12 Classifying Acids and Bases Using pH
➡ WileyPLUS

The measurement and control of pH is extremely important in many fields. Proper soil pH is necessary to grow certain types of plants successfully. The pH of certain foods is too acidic for some diets. Many biological processes are delicately controlled pH systems. The pH of human blood is regulated to very close tolerances through the uptake or release of H^+ by mineral ions, such as HCO_3^-, HPO_4^{2-}, and $H_2PO_4^-$. Changes in the pH of the blood by as little as 0.4 pH unit result in death.

Compounds with colors that change at particular pH values are used as indicators in acid–base reactions. Commercially available pH test paper contains chemical indicators. The indicator in the paper takes on different colors when wetted with solutions of different pH. Thus the pH of a solution can be estimated by placing a drop on the test paper and comparing the color of the test paper with a color chart calibrated at different pH values. Common applications of pH test indicators are the kits used to measure and adjust the pH of swimming pools, hot tubs, and saltwater aquariums. Electronic pH meters are used for making rapid and precise pH determinations (**FIGURE 15.5**).

© Leslie Garland Picture Library / Alamy Stock Photo

FIGURE 15.5 The pH of a substance (in this case garden soil) can be measured by using a pH meter.

15.5 Neutralization

Describe a neutralization reaction and do calculations involving titrations.

LEARNING OBJECTIVE

KEY TERMS

neutralization
spectator ion
titration

The reaction of an acid and a base to form a salt and water is known as **neutralization**. We've seen this reaction before, but now with our knowledge about ions and ionization, let's reexamine the process of neutralization.

Consider the reaction that occurs when solutions of sodium hydroxide and hydrochloric acid are mixed. The ions present initially are Na^+ and OH^- from the base and H^+ and Cl^- from the acid. The products, sodium chloride and water, exist as Na^+ and Cl^- ions and H_2O molecules. A chemical equation representing this reaction is

$$HCl(aq) + NaOH(aq) \longrightarrow NaCl(aq) + H_2O(l)$$

This equation, however, does not show that HCl, NaOH, and NaCl exist as ions in solution. The following total ionic equation gives a better representation of the reaction:

$$(H^+ + Cl^-) + (Na^+ + OH^-) \longrightarrow Na^+ + Cl^- + H_2O(l)$$

This equation shows that the Na^+ and Cl^- ions did not react. These ions are called **spectator ions** because they were present but did not take part in the reaction. The only reaction that occurred was that between the H^+ and OH^- ions. Therefore the equation for the neutralization can be written as this net ionic equation:

$$\underset{\text{acid}}{H^+(aq)} + \underset{\text{base}}{OH^-(aq)} \longrightarrow \underset{\text{water}}{H_2O(l)}$$

CHECK YOUR UNDERSTANDING

15.13 Acid and Base Combinations That Form Salts

➥ **WileyPLUS**

This simple net ionic equation represents not only the reaction of sodium hydroxide and hydrochloric acid but also the reaction of any strong acid with any water-soluble hydroxide base in an aqueous solution. The driving force of a neutralization reaction is the ability of an H^+ ion and an OH^- ion to react and form a molecule of water.

The amount of acid, base, or other species in a sample can be determined by **titration**, which measures the volume of one reagent required to react with a measured mass or volume of another reagent. Consider the titration of an acid with a base shown in **FIGURE 15.6**. A measured volume of acid of unknown concentration is placed in a flask, and a few drops of an indicator solution are added. Base solution of known concentration is slowly added from a buret to the acid until the indicator changes color. The indicator selected is one that changes color when the stoichiometric quantity (according to the equation) of base has been added to the acid. At this point, known as the *end point of the titration*, the titration is complete, and the volume of base used

CHECK YOUR UNDERSTANDING
15.14 Titration
➡ WileyPLUS

to neutralize the acid is read from the buret. The concentration or amount of acid in solution can be calculated from the titration data and the chemical equation for the reaction. Let's look at some examples.

Before
colorless and clear

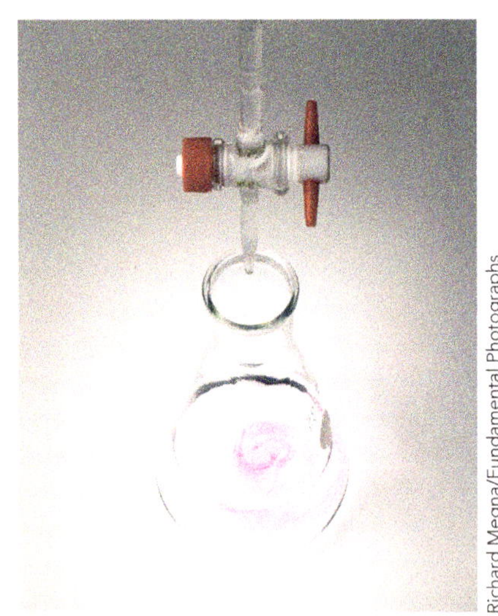

End point
pale pink persists

After end point
bright pink (excess base)

FIGURE 15.6 The titration process with phenolphthalein indicator. Phenolphthalein, an organic compound, is colorless in acid solution and changes to pink at a pH of 8.3. When a solution of sodium hydroxide is added to a hydrochloric acid solution containing phenolphthalein, the change in color (from colorless to pink) indicates that all the acid is neutralized.

ONLINE LEARNING MODULE
Titration
➡ WileyPLUS

ENHANCED EXAMPLE 15.7 WileyPLUS

Suppose that 42.00 mL of 0.150 M NaOH solution are required to neutralize 50.00 mL of hydrochloric acid solution. What is the molarity of the acid solution?

SOLUTION

Read • **Knowns:** 42.00 mL 0.150 M NaOH solution
50.00 mL acid solution

Solving for: acid molarity

Plan • The equation for the reaction is

$$NaOH(aq) + HCl(aq) \longrightarrow NaCl(aq) + H_2O(l)$$

Determine the moles NaOH in the solution.

Set Up • $$(0.04200\ \cancel{L})\left(\frac{0.150\text{ mol NaOH}}{1\ \cancel{L}}\right) = 0.00630\text{ mol NaOH}$$

Since the acid and base react in a 1 : 1 ratio (from the equation for the reaction), the mol base = mol acid.

Calculate • $$M = \frac{\text{mol}}{\text{L}} = \frac{0.00630\text{ mol HCl}}{0.05000\text{ L}} = 0.126\ M\text{ HCl}$$

ENHANCED EXAMPLE 15.8

WileyPLUS

Suppose that 42.00 mL of 0.150 *M* NaOH solution are required to neutralize 50.00 mL of H_2SO_4 solution. What is the molarity of the acid solution?

SOLUTION

Read **Knowns:** 42.00 mL 0.150 *M* NaOH solution
50.00 mL H_2SO_4 solution

Solving for: acid molarity

Plan The equation for the reaction is

$$2\,\text{NaOH}(aq) + \text{H}_2\text{SO}_4(aq) \longrightarrow \text{Na}_2\text{SO}_4(aq) + 2\,\text{H}_2\text{O}(l)$$

Determine the moles NaOH in the solution.

Set Up

$$(0.04200\,\cancel{\text{L}})\left(\frac{0.150\text{ mol NaOH}}{1\,\cancel{\text{L}}}\right) = 0.00630\text{ mol NaOH}$$

Since the acid and base react in a 1 : 2 ratio (from the equation for the reaction) 2 mol base = 1 mol acid.

$$(0.00630\,\cancel{\text{mol NaOH}})\left(\frac{1\text{ mol H}_2\text{SO}_4}{2\,\cancel{\text{mol NaOH}}}\right) = 0.00315\text{ mol H}_2\text{SO}_4$$

Calculate

$$M = \frac{\text{mol}}{\text{L}} = \frac{0.00315\text{ mol H}_2\text{SO}_4}{0.05000\text{ L}} = 0.0630\,M\text{ H}_2\text{SO}_4$$

ONLINE LEARNING MODULE
Calculating Molarity

WileyPLUS

ENHANCED EXAMPLE 15.9

WileyPLUS

A 25.00-mL sample of H_2SO_4 solution required 14.26 mL of 0.2240 *M* NaOH solution for complete neutralization. What is the molarity of the sulfuric acid?

SOLUTION

Read **Knowns:** 14.26 mL 0.2240 *M* NaOH solution
25.00 mL H_2SO_4 solution

Solving for: sulfuric acid molarity

Plan The equation for the reaction is

$$2\,\text{NaOH}(aq) + \text{H}_2\text{SO}_4(aq) \longrightarrow \text{Na}_2\text{SO}_4(aq) + 2\,\text{H}_2\text{O}(l)$$

Determine the moles NaOH in solution.

Set Up

$$(0.01426\,\cancel{\text{L}})\left(\frac{0.2240\text{ mol NaOH}}{1\,\cancel{\text{L}}}\right) = 0.003194\text{ mol NaOH}$$

Since the mole ratio of acid to base is $\frac{1\text{ mol H}_2\text{SO}_4}{2\text{ mol NaOH}}$,

$$(0.0031940\text{ mol NaOH})\left(\frac{1\text{ mol H}_2\text{SO}_4}{2\text{ mol NaOH}}\right) = 0.001597\text{ mol H}_2\text{SO}_4$$

Calculate

$$M = \frac{\text{mol}}{\text{L}} = \frac{0.001597\text{ mol H}_2\text{SO}_4}{0.02500\text{ L}} = 0.06388\,M\text{ H}_2\text{SO}_4$$

PRACTICE 15.7

A 50.0-mL sample of HCl required 24.81 mL of 0.1250 *M* NaOH for neutralization. What is the molarity of the acid?

15.6 Writing Net Ionic Equations

LEARNING OBJECTIVE Write net ionic equations using the stated rules.

KEY TERMS
formula equation
total ionic equation
net ionic equation

In Section 15.5, we wrote the reaction of hydrochloric acid and sodium hydroxide in three different equations:

1. $HCl(aq) + NaOH(aq) \longrightarrow NaCl(aq) + H_2O(l)$
2. $(H^+ + Cl^-) + (Na^+ + OH^-) \longrightarrow Na^+ + Cl^- + H_2O(l)$
3. $H^+ + OH^- \longrightarrow H_2O$

In the **formula equation** (1), compounds are written in their molecular, or formula, expressions. In the **total ionic equation** (2), compounds are written to show the form in which they are predominantly present: strong electrolytes as ions in solution and nonelectrolytes, weak electrolytes, precipitates, and gases in their molecular forms. In the **net ionic equation** (3), only those molecules or ions that have reacted are included in the equation; ions or molecules that do not react (the spectators) are omitted.

When balancing equations thus far, we've been concerned only with the atoms of the individual elements. Because ions are electrically charged, ionic equations often end up with a net electrical charge. A balanced equation must have the same net charge on each side, whether that charge is positive, negative, or zero. Therefore, when balancing ionic equations, we must make sure that both the same number of each kind of atom and the same net electrical charge are present on each side.

RULES FOR WRITING IONIC EQUATIONS

1. Strong electrolytes in solution are written in their ionic form.
2. Weak electrolytes are written in their molecular form.
3. Nonelectrolytes are written in their molecular form.
4. Insoluble substances, precipitates, and gases are written in their molecular forms.
5. The net ionic equation should include only substances that have undergone a chemical change. Spectator ions are omitted from the net ionic equation.
6. Equations must be balanced, both in atoms and in electrical charge.

Study the following examples. In each one, the formula equation is given. Write the total ionic equation and the net ionic equation for each.

ONLINE LEARNING MODULE
Writing a Net Ionic Equation
WileyPLUS

EXAMPLE 15.10

$$HNO_3(aq) + KOH(aq) \longrightarrow KNO_3(aq) + H_2O(l)$$

formula equation

SOLUTION

$$(H^+ + NO_3^-) + (K^+ + OH^-) \longrightarrow K^+ + NO_3^- + H_2O$$

total ionic equation

$$H^+ + OH^- \longrightarrow H_2O$$

net ionic equation

The HNO_3, KOH, and KNO_3 are soluble, strong electrolytes. The K^+ and NO_3^- ions are spectator ions, have not changed, and are not included in the net ionic equation. Water is a nonelectrolyte and is written in the molecular form.

EXAMPLE 15.11

$$2\,AgNO_3(aq) + BaCl_2(aq) \longrightarrow 2\,AgCl(s) + Ba(NO_3)_2(aq)$$
formula equation

SOLUTION

$$(2\,Ag^+ + 2\,NO_3^-) + (Ba^{2+} + 2\,Cl^-) \longrightarrow 2\,AgCl(s) + Ba^{2+} + 2\,NO_3^-$$
total ionic equation

$$Ag^+ + Cl^- \longrightarrow AgCl(s)$$
net ionic equation

Although AgCl is an ionic compound, it is written in the un-ionized form on the right side of the ionic equations because most of the Ag^+ and Cl^- ions are no longer in solution but have formed a precipitate of AgCl. The Ba^{2+} and NO_3^- ions are spectator ions.

EXAMPLE 15.12

ONLINE LEARNING MODULE
Ionic Equations Involving Gas Production
WileyPLUS

$$Na_2CO_3(aq) + H_2SO_4(aq) \longrightarrow Na_2SO_4(aq) + H_2O(l) + CO_2(g)$$
formula equation

SOLUTION

$$(2\,Na^+ + CO_3^{2-}) + (2\,H^+ + SO_4^{2-}) \longrightarrow 2\,Na^+ + SO_4^{2-} + H_2O(l) + CO_2(g)$$
total ionic equation

$$CO_3^{2-} + 2\,H^+ \longrightarrow H_2O(l) + CO_2(g)$$
net ionic equation

Carbon dioxide (CO_2) is a gas and evolves from the solution; Na^+ and SO_4^{2-} are spectator ions.

EXAMPLE 15.13

$$HC_2H_3O_2(aq) + NaOH(aq) \longrightarrow NaC_2H_3O_2(aq) + H_2O(l)$$
formula equation

SOLUTION

$$HC_2H_3O_2 + (Na^+ + OH^-) \longrightarrow Na^+ + C_2H_3O_2^- + H_2O$$
total ionic equation

$$HC_2H_3O_2 + OH^- \longrightarrow C_2H_3O_2^- + H_2O$$
net ionic equation

Acetic acid, $HC_2H_3O_2$, a weak acid, is written in the molecular form, but sodium acetate, $NaC_2H_3O_2$, a soluble salt, is written in the ionic form. The Na^+ ion is the only spectator ion in this reaction. Both sides of the net ionic equation have a -1 electrical charge.

EXAMPLE 15.14

$$Mg(s) + 2\,HCl(aq) \longrightarrow MgCl_2(aq) + H_2(g)$$
formula equation

SOLUTION

$$Mg + (2\,H^+ + 2\,Cl^-) \longrightarrow Mg^{2+} + 2\,Cl^+ + H_2(g)$$
total ionic equation

$$Mg + 2\,H^+ \longrightarrow Mg^{2+} + H_2(g)$$
net ionic equation

The net electrical charge on both sides of the equation is +2.

EXAMPLE 15.15

$$H_2SO_4(aq) + Ba(OH)_2(aq) \longrightarrow BaSO_4(s) + 2\,H_2O(l)$$
formula equation

SOLUTION

$$(2\,H^+ + SO_4^{2-}) + (Ba^{2+} + 2\,OH^-) \longrightarrow BaSO_4(s) + 2\,H_2O(l)$$
total ionic equation

$$2\,H^+ + SO_4^{2-} + Ba^{2+} + 2\,OH^- \longrightarrow BaSO_4(s) + 2\,H_2O(l)$$
net ionic equation

Barium sulfate ($BaSO_4$) is a highly insoluble salt. If we conduct this reaction using the conductivity apparatus described in Section 15.4, the light glows brightly at first but goes out when the reaction is complete because almost no ions are left in solution. The $BaSO_4$ precipitates out of solution, and water is a nonconductor of electricity.

PRACTICE 15.8

Write the net ionic equation for

$$3\,H_2S(aq) + 2\,Bi(NO_3)_3(aq) \longrightarrow Bi_2S_3(s) + 6\,HNO_3(aq)$$

CHEMISTRY *IN ACTION* Ocean Corals Threatened by Increasing Atmospheric CO_2 Levels

As we continue to burn fossil fuels, we are increasing the atmospheric CO_2 levels beyond anything seen in recent history. Some of this CO_2 is removed from the atmosphere by dissolving in the water of the oceans, as shown in the following reaction:

$$CO_2(g) + H_2O(l) \longrightarrow H_2CO_3(aq) \longrightarrow H^+(aq) + HCO_3^-(aq)$$

This at first seems like an excellent way to remove excess atmospheric CO_2 until we look at the effect it has on the world's oceans. The pH of seawater has dropped from about 8.2 to between 8.05 and 8.1 since the beginning of the Industrial Revolution. Although this may seem like a small change, it has had dramatic effects on the corals living in the oceans. In a study published in the November 8, 2010, issue of *PNAS*, Rebecca Albright of the University of Miami reported a significant decrease in the number of elkhorn coral eggs fertilized as the pH of seawater drops. Additionally, the ability of fertilized eggs to attach and begin growing has also been affected. Albright suggests that there may be as much as a 73% decrease in the number of new corals able to establish themselves by the year 2100 if acidification of the oceans continues at the current rate. The CO_2 in the atmosphere obviously has significant effects on the chemistry of the oceans.

Darryl Torckler/Taxi/Getty Images, Inc.

Healthy coral.

Daryl Leniuk/Digital Vision/Getty Images, Inc.

Unhealthy coral.

CHAPTER 15 REVIEW

15.1 Acids and Bases

KEY TERM
hydronium ion

- Characteristic properties of acids include:
  - They taste sour.
  - They change litmus from blue to red.
  - They react with:
    - Metals to form hydrogen gas
    - Hydroxide bases to form water and a salt
- Characteristic properties of bases include:
  - They taste bitter or caustic.
  - They have a slippery feeling.
  - They change litmus from red to blue.
  - They interact with acids to form water and a salt.
- Arrhenius definition of acids and bases:
  - Acids contain excess H^+ ions in aqueous solutions.
  - Bases contain excess OH^- ions in aqueous solutions.
- Brønsted–Lowry definition of acids and bases:
  - Acids are proton donors.
  - Bases are proton acceptors.
- Lewis definition of acids and bases:
  - Acids are electron pair acceptors.
  - Bases are electron pair donors.
- Acids react with metals above hydrogen in the activity series to form hydrogen gas and a salt.
- Acids react with bases to form water and a salt (neutralization reaction).
- Bases react with acids to form water and a salt (neutralization reaction).

15.2 Salts

- Salts can be considered to be derived from the reaction of an acid and a base.
- Salts are crystalline and have high melting and boiling points.

15.3 Electrolytes and Nonelectrolytes

KEY TERMS
electrolyte
nonelectrolyte
dissociation
ionization
strong electrolyte
weak electrolyte

- A substance whose aqueous solution conducts electricity is called an electrolyte:
  - Aqueous solution contains ions.
- A substance whose aqueous solution does not conduct electricity is called a nonelectrolyte:
  - Aqueous solution contains molecules, not ions.
- Dissociation is the process by which the ions of a salt separate as the salt dissolves.
- Ionization is the formation of ions that occurs as a result of a chemical reaction with water.
- Strong electrolytes are 100% ionized in solution:
  - Strong acids and strong bases
- Weak electrolytes are much less ionized in solution:
  - Weak acids and bases
- Water can self-ionize to form H^+ and OH^- ions.
- Concentrations of ions in water at 25°C:
  - $[H^+] = 1.0 \times 10^{-7}$
  - $[OH^-] = 1.0 \times 10^{-7}$

15.4 Introduction to pH

KEY TERMS
pH
logarithm (log)

- $pH = -\log[H^+]$:
  - $pH < 7$ for acidic solution
  - $pH = 7$ in neutral solution
  - $pH > 7$ for basic solution
- The number of decimal places in a logarithm equals the number of significant figures in the original number.

15.5 Neutralization

KEY TERMS

neutralization
spectator ion
titration

- The reaction of an acid and a base to form water and a salt is called neutralization:
  - The general equation for a neutralization reaction is

$$H^+(aq) + OH^-(aq) \longrightarrow H_2O(l)$$

- The quantitative study of a neutralization reaction is called a titration.

15.6 Writing Net Ionic Equations

KEY TERMS

formula equation
total ionic equation
net ionic equation

- In a formula equation the compounds are written in their molecular, or formula, expressions.
- In the total ionic equation compounds are written as ions if they are strong electrolytes in solution and as molecules if they are precipitates, nonelectrolytes, or weak electrolytes in solution.
- The net ionic equation shows only the molecules or ions that have changed:
  - The spectators (nonreactors) are omitted.

REVIEW QUESTIONS

1. Since a hydrogen ion and a proton are identical, what differences exist between the Arrhenius and Brønsted–Lowry definitions of an acid? (Table 15.1)
2. Use the three acid–base theories (Arrhenius, Brønsted–Lowry, and Lewis) to define an acid and a base.
3. For each acid–base theory referred to in Question 2, write an equation illustrating the neutralization of an acid with a base.
4. Write the Lewis structure for the (a) bromide ion, (b) hydroxide ion, and (c) cyanide ion. Why are these ions considered to be bases according to the Brønsted–Lowry and Lewis acid–base theories?
5. What kinds of substances will form hydrogen gas when they react with an acid?
6. Write the formulas and names of three salts that can be formed from nitric acid.
7. Write the formulas and names of three salts that can be formed from the base lithium hydroxide.
8. According to Figure 15.1, what type of substance must be in solution for the bulb to light?
9. Which of the following classes of compounds are electrolytes: acids, alcohols, bases, salts? (Table 15.2)
10. What two differences are apparent in the arrangement of water molecules about the hydrated ions as depicted in Figure 15.2?
11. Into what three classes of compounds do electrolytes generally fall?
12. Name each compound listed in Table 15.3.
13. A solution of HCl in water conducts an electric current, but a solution of HCl in hexane does not. Explain this behavior in terms of ionization and chemical bonding.
14. How do ionic compounds exist in their crystalline structure? What occurs when they are dissolved in water?
15. An aqueous methyl alcohol, CH_3OH, solution does not conduct an electric current, but a solution of sodium hydroxide, NaOH, does. What does this information tell us about the OH group in the alcohol?
16. Why does molten NaCl conduct electricity?
17. Explain the difference between dissociation of ionic compounds and ionization of molecular compounds.
18. Distinguish between strong and weak electrolytes.
19. Explain why ions are hydrated in aqueous solutions.
20. What is the main distinction between water solutions of strong and weak electrolytes?
21. The solubility of HCl gas in water, a polar solvent, is much greater than its solubility in hexane, a nonpolar solvent. How can you account for this difference?
22. The pH of a solution with a hydrogen ion concentration of 0.003 M is between what two whole numbers? (Table 15.4)
23. Which is more acidic, tomato juice or blood? (Table 15.5)
24. What are the relative concentrations of $H^+(aq)$ and $OH^-(aq)$ in (a) a neutral solution, (b) an acid solution, and (c) a basic solution?
25. Pure water, containing equal concentrations of both acid and base ions, is neutral. Why?
26. A solution with a pH of 7 is neutral. A solution with a pH less than 7 is acidic. A solution with a pH greater than 7 is basic. What do these statements mean?
27. Explain the purpose of a titration.
28. Write the net ionic equation for the reaction of a strong acid with a water-soluble hydroxide base in an aqueous solution.

PAIRED EXERCISES

Most of the exercises in this chapter are available for assignment via the online homework management program, WileyPLUS (www.wileyplus.com) All exercises with blue numbers have answers in Appendix 5.

1. Identify the conjugate acid–base pairs in each of the following equations:
   (a) $NH_3 + H_2O \rightleftharpoons NH_4^+ + OH^-$
   (b) $HC_2H_3O_2 + H_2O \rightleftharpoons C_2H_3O_2^- + H_3O^+$
   (c) $H_2PO_4^- + OH^- \rightleftharpoons HPO_4^{2-} + H_2O$
   (d) $HCl + H_2O \longrightarrow Cl^- + H_3O^+$

2. Identify the conjugate acid–base pairs in each of the following equations:
   (a) $H_2S + NH_3 \rightleftharpoons NH_4^+ + HS^-$
   (b) $HSO_4^- + NH_3 \rightleftharpoons SO_4^{2-} + NH_4^+$
   (c) $HBr + CH_3O^- \rightleftharpoons Br^- + CH_3OH$
   (d) $HNO_3 + H_2O \longrightarrow NO_3^- + H_3O^+$

3. Complete and balance these equations:
   (a) $Zn(s) + HCl(aq) \longrightarrow$
   (b) $Al(OH)_3(s) + H_2SO_4(aq) \longrightarrow$
   (c) $Ca(HCO_3)_2(s) + HBr(aq) \longrightarrow$
   (d) $KOH(aq) + H_3PO_4(aq) \longrightarrow$

4. Complete and balance these equations:
   (a) $Mg(s) + HClO_4(aq) \longrightarrow$
   (b) $Al(s) + H_2SO_4(aq) \longrightarrow$
   (c) $NaOH(aq) + H_2CO_3(aq) \longrightarrow$
   (d) $Ba(OH)_2(s) + HClO_4(aq) \longrightarrow$

5. For each of the formula equations in Question 3, write total and net ionic equations.

6. For each of the formula equations in Question 4, write total and net ionic equations.

7. Write an equation to illustrate the acid–base reaction that will take place between propanoic acid, $HC_3H_5O_2$, and ammonia, NH_3. Identify the acids, bases, and their conjugates.

8. Write an equation to illustrate the acid–base reaction that will take place between perchloric acid, $HClO_4$, and a hydroxide ion, OH^-. Identify the acids, bases, and their conjugates.

9. Write the equation for the reaction that will occur when each pair of substances is allowed to react.
   (a) HNO_3 and NaOH
   (b) $HC_2H_3O_2$ and $Ba(OH)_2$
   (c) $HClO_4$ and NH_4OH

10. Write the equation for the reaction that will occur when each pair of substances is allowed to react.
   (a) HBr and $Mg(OH)_2$
   (b) H_3PO_4 and KOH
   (c) H_2SO_4 and NH_4OH

11. The bisulfate ion, HSO_4^-, has the ability to act as both an acid and a base.
   (a) What is the conjugate acid of HSO_4^-?
   (b) What is the conjugate base of HSO_4^-?

12. The bicarbonate ion, HCO_3^-, has the ability to act as both an acid and a base.
   (a) What is the conjugate acid of HCO_3^-?
   (b) What is the conjugate base of HCO_3^-?

13. Predict the acid and base that must be reacted together to make the salt Li_2S. Write the equation for the reaction to form this salt.

14. Predict the acid and base that must be reacted together to make the salt $CaCO_3$. Write the equation for the reaction to form this salt.

15. Determine whether each of the following substances is an electrolyte or a nonelectrolyte. All are mixed with water.
   (a) SO_3
   (b) K_2CO_3
   (c) CH_3OH (methyl alcohol)
   (d) O_2 (insoluble)
   (e) $CuBr_2$
   (f) HI

16. Determine whether each of the following substances is an electrolyte or a nonelectrolyte. All are mixed with water.
   (a) $C_6H_{12}O_6$ (glucose)
   (b) P_2O_5
   (c) NaClO
   (d) LiOH
   (e) C_2H_5OH (ethyl alcohol)
   (f) $KMnO_4$

17. Determine the molarity of each of the ions present in the following aqueous salt solutions: (assume 100% ionization)
   (a) 1.25 M $CuBr_2$
   (b) 0.75 M $NaHCO_3$
   (c) 3.50 M K_3AsO_4
   (d) 0.65 M $(NH_4)_2SO_4$

18. Determine the molarity of each of the ions present in the following aqueous salt solutions: (assume 100% ionization)
   (a) 2.25 M $FeCl_3$
   (b) 1.20 M $MgSO_4$
   (c) 0.75 M NaH_2PO_4
   (d) 0.35 M $Ca(ClO_3)_2$

19. In Question 17, how many grams of each ion would be present in 100. mL of each solution?

20. In Question 18, how many grams of each ion would be present in 100. mL of each solution?

21. Calculate the $[H^+]$ for
   (a) black coffee, with a pH of 5.0
   (b) black tea with a pH of 6.8
   (c) green tea with a pH of 7.9
   (d) laundry detergent with a pH of 9.7

22. Calculate the $[H^+]$ for
   (a) pure water
   (b) lemon lime soda with a pH of 4.3
   (c) root beer with a pH of 5.8
   (d) window cleaner with a pH of 10.3

23. Determine the molar concentration of each ion present in the solutions that result from each of the following mixtures: (Disregard the concentration of H^+ and OH^- from water and assume that volumes are additive.)
   (a) 55.5 mL of 0.50 M HCl and 75.0 mL of 1.25 M HCl
   (b) 125 mL of 0.75 M $CaCl_2$ and 125 mL of 0.25 M $CaCl_2$
   (c) 35.0 mL of 0.333 M NaOH and 22.5 mL of 0.250 M HCl
   (d) 12.5 mL of 0.500 M H_2SO_4 and 23.5 mL of 0.175 M NaOH

24. Determine the molar concentration of each ion present in the solutions that result from each of the following mixtures: (Disregard the concentration of H^+ and OH^- from water and assume that volumes are additive.)
   (a) 45.5 mL of 0.10 M NaCl and 60.5 mL of 0.35 M NaCl
   (b) 95.5 mL of 1.25 M HCl and 125.5 mL of 2.50 M HCl
   (c) 15.5 mL of 0.10 M $Ba(NO_3)_2$ and 10.5 mL of 0.20 M $AgNO_3$
   (d) 25.5 mL of 0.25 M NaCl and 15.5 mL of 0.15 M $Ca(C_2H_3O_2)_2$

25. Write an equation showing how nitric acid ionizes in water. Would this be a strong electrolyte? A strong acid?

26. Write an equation showing how hydrocyanic acid ionizes in water. How is this ionization different from the ionization of nitric acid? Would this be a strong electrolyte? A strong acid?

27. Given the data for the following separate titrations, calculate the molarity of the HCl:

	mL HCl	Molarity HCl	mL NaOH	Molarity NaOH
(a)	40.13	M	37.70	0.728
(b)	19.00	M	33.66	0.306
(c)	27.25	M	18.00	0.555

28. Given the data for the following separate titrations, calculate the molarity of the NaOH:

	mL HCl	Molarity HCl	mL NaOH	Molarity NaOH
(a)	37.19	0.126	31.91	M
(b)	48.04	0.482	24.02	M
(c)	13.13	1.425	39.39	M

29. Balance each of the following equations; then change them into balanced net ionic equations:
(a) $K_3PO_4(aq) + Ca(NO_3)_2(aq) \longrightarrow Ca_3(PO_4)_2(s) + KNO_3(aq)$
(b) $Al(s) + H_2SO_4(aq) \longrightarrow H_2(g) + Al_2(SO_4)_3(aq)$
(c) $Na_2CO_3(aq) + HCl(aq) \longrightarrow NaCl(aq) + H_2O(l) + CO_2(g)$

30. Balance each of the following equations; then change them into balanced net ionic equations:
(a) $Mg(s) + Cu(NO_3)_2(aq) \longrightarrow Cu(s) + Mg(NO_3)_2(aq)$
(b) $H_2SO_4(aq) + NaOH(aq) \longrightarrow Na_2SO_4(aq) + H_2O(l)$
(c) $K_2SO_3(aq) + HClO_4(aq) \longrightarrow KClO_4(aq) + H_2O(l) + SO_2(g)$

31. For each of the given pairs, determine which solution is more acidic. All are water solutions. Explain your answer.
(a) 1 M HCl or 1 M H_2SO_4?
(b) 1 M HCl or 1 M $HC_2H_3O_2$?

32. For each of the given pairs, determine which solution is more acidic. All are water solutions. Explain your answer.
(a) 1 M HCl or 2 M HCl?
(b) 1 M HNO_3 or 1 M H_2SO_4?

33. Determine how many milliliters of 0.525 M $HClO_4$ will be required to neutralize 50.25 g $Ca(OH)_2$ according to the reaction:

$$2\ HClO_4(aq) + Ca(OH)_2(s) \longrightarrow Ca(ClO_4)_2(aq) + 2\ H_2O(l)$$

34. Determine how many grams of $Al(OH)_3$ will be required to neutralize 275 mL of 0.125 M HCl according to the reaction:

$$3\ HCl(aq) + Al(OH)_3(s) \longrightarrow AlCl_3(aq) + 3\ H_2O(l)$$

35. A 0.200-g sample of impure NaOH requires 18.25 mL of 0.2406 M HCl for neutralization. What is the percent of NaOH in the sample?

36. A batch of sodium hydroxide was found to contain sodium chloride as an impurity. To determine the amount of impurity, a 1.00-g sample was analyzed and found to require 49.90 mL of 0.466 M HCl for neutralization. What is the percent of NaCl in the sample?

37. What volume of H_2 gas, measured at 27°C and 700. torr, can be obtained by reacting 5.00 g of zinc metal with 100. mL of 0.350 M HCl? The equation is

$$Zn(s) + 2\ HCl(aq) \longrightarrow ZnCl_2(aq) + H_2(g)$$

38. What volume of H_2 gas, measured at 27°C and 700. torr, can be obtained by reacting 5.00 g of zinc metal with 200. mL of 0.350 M HCl? The equation is

$$Zn(s) + 2\ HCl(aq) \longrightarrow ZnCl_2(aq) + H_2(g)$$

39. Calculate the pH of solutions having the following $[H^+]$:
(a) 0.35 M (b) 1.75 M (c) 2.0×10^{-5} M

40. Calculate the pH of solutions having the following $[H^+]$:
(a) 0.0020 M (b) 7.0×10^{-8} M (c) 3.0 M

41. Calculate the pH of
(a) orange juice, 3.7×10^{-4} M H^+
(b) vinegar, 2.8×10^{-3} M H^+
(c) shampoo, 2.4×10^{-6} M H^+
(d) dishwashing detergent, 3.6×10^{-8} M H^+

42. Calculate the pH of
(a) black coffee, 5.0×10^{-5} M H^+
(b) limewater, 3.4×10^{-11} M H^+
(c) fruit punch, 2.1×10^{-4} M H^+
(d) cranberry apple drink, 1.3×10^{-3} M H^+

43. Determine whether each of the following is a strong acid, weak acid, strong base, or weak base. Then write an equation describing the process that occurs when the substance is dissolved in water.
(a) NH_3 (b) HCl (c) KOH (d) $HC_2H_3O_2$

44. Determine whether each of the following is a strong acid, weak acid, strong base, or weak base. Then write an equation describing the process that occurs when the substance is dissolved in water.
(a) $H_2C_2O_4$ (b) $Ba(OH)_2$ (c) $HClO_4$ (d) HBr

ADDITIONAL EXERCISES

45. Write the formula for the conjugate acid of the following bases:
(a) NH_2CH_3 (b) HS^-

46. Write the formula for the conjugate base of the following acids:
(a) $HBrO_3$ (b) NH_4^+ (c) $H_2PO_4^-$

47. Write the equation for the reaction between the acid $HC_2H_3O_2$ and the base NH_3. Identify the conjugate acid–base pairs in the reaction.

48. Write a reaction between the sulfide ion (S^{2-}) and water showing how it acts as a Brønsted–Lowry base.

49. Write the reaction of hydrochloric acid with the active metal magnesium.

50. Write an equation showing how sodium sulfate dissociates in water.

51. Determine whether each of the following describes a substance that is acidic, basic, or neutral:
(a) Red litmus paper turns blue
(b) pH = 6
(c) $[H^+] = 1 \times 10^{-7}$
(d) $[OH^-] = 1 \times 10^{-8}$
(e) Blue litmus paper turns red
(f) pH = 12

52. Draw pictures similar to those found in Figure 15.2 to show what happens to the following compounds when each is dissolved in water.
(a) $CaCl_2$ (b) KF (c) $AlBr_3$

53. What is the concentration of Al^{3+} ions in a solution of $AlBr_3$ having a Br^- ion concentration of 0.142 M?

54. In an acid–base titration, 25.22 mL of H_2SO_4 were used to neutralize 35.22 mL of 0.313 M NaOH. What was the molarity of the H_2SO_4 solution?

$$H_2SO_4 + 2\ NaOH \longrightarrow Na_2SO_4 + 2\ H_2O$$

55. How does a hydronium ion differ from a hydrogen ion?

56. At 100°C the H^+ concentration in water is about 1×10^{-6} mol/L, about 10 times that of water at 25°C. At which of these temperatures is
(a) the pH of water the greater?
(b) the hydrogen ion (hydronium ion) concentration the higher?
(c) the water neutral?

57. What is the relative difference in H^+ concentration in solutions that differ by one pH unit?

58. What volume (mL) of 0.4233 M H_2SO_4 is needed to neutralize 6.38 g KOH?

59. How many grams of KOH are required to neutralize 50.00 mL of 0.240 M HNO_3?

60. Two drops (0.1 mL) of 1.0 *M* HCl are added to water to make 1.0 L of solution. What is the pH of this solution if the HCl is 100% ionized?

61. When 250.0 mL of water are added to 10.0 mL of 12 *M* HCl, what will be the new molarity?

62. If 3.0 g NaOH are added to 500. mL of 0.10 *M* HCl, will the resulting solution be acidic or basic? Show evidence for your answer.

63. If 380 mL of 0.35 *M* $Ba(OH)_2$ are added to 500. mL of 0.65 *M* HCl, will the mixture be acidic or basic? Find the pH of the resulting solution.

64. If 50.00 mL of 0.2000 *M* HCl are titrated with 0.2000 *M* NaOH, find the pH of the solution after the following amounts of base have been added:
(a) 0.000 mL (d) 49.00 mL (f) 49.99 mL
(b) 10.00 mL (e) 49.90 mL (g) 50.00 mL
(c) 25.00 mL
Plot your answers on a graph with pH on the *y*-axis and mL NaOH on the *x*-axis.

65. Sulfuric acid reacts with NaOH:
(a) Write a balanced equation for the reaction producing Na_2SO_4.
(b) How many milliliters of 0.10 *M* NaOH are needed to react with 0.0050 mol H_2SO_4?
(c) How many grams of Na_2SO_4 will also form?

66. A 10.0-mL sample of HNO_3 was diluted to a volume of 100.00 mL. Then 25 mL of that diluted solution were needed to neutralize 50.0 mL of 0.60 *M* KOH. What was the concentration of the original nitric acid?

67. The pH of a solution of a strong acid was determined to be 3. If water is then added to dilute this solution, would the pH change? Why or why not? Could enough water ever be added to raise the pH of an acid solution above 7?

68. If 425 mL of 0.94 *M* H_2SO_4 are added to 750. mL of 0.83 *M* NaOH, will the mixture be acidic or basic? Calculate the pH of the resulting solution.

69. For each of the following substances determine the concentration of each ion/molecule in solution.
(a) 1 *M* $CuSO_4$ (d) $CaS(s)$
(b) 1 *M* HNO_3 (e) 1 *M* $HC_2H_3O_2$
(c) 1 *M* H_2SO_4

70. Write the formula, total ionic, and net ionic equations for the following reactions:
(a) HNO_3 + LiOH (c) HF + NaOH
(b) HBr + $Ba(OH)_2$

71. Sodium rhodizonate is often used in forensic investigations to detect the presence of any residue from shooting a firearm. Burned and unburned particles from the gunpowder will be ejected with the projectile. The residue is then stained red using sodium rhodizonate at a pH of 2.8. How many mL of 0.60 *M* hydrochloric acid need to be added to a neutral solution of sodium rhodizonate to make 500.0 mL of solution with a pH of 2.800?

72. One of the tests used to detect the presence of LSD or some of the other psilocybin-like alkaloids often found in psychedelic mushrooms is the Ehrlich test. For this test a solution of *p*-dimethylaminobenzaldehyde, $C_9H_{11}NO$, in 3.25 *M* HCl is used to detect the presence of any of these compounds. How many mL of concentrated HCl (12.1 *M*) are required to make 3.00 L of 3.25 *M* HCl for preparing Ehrlich reagent?

CHALLENGE EXERCISES

73. One of the tests used by geologists to identify rocks containing calcium carbonate, such as limestone and marble, is to drop dilute hydrochloric acid onto the rock. Fizzing or bubbling results if there are carbonates in the rocks. Write the balanced chemical reaction between calcium carbonate and aqueous hydrochloric acid.

74. An HCl solution has a pH of 0.300. What volume of water must be added to 200 mL of this solution to change the pH to 0.150?

75. Lactic acid (found in sour milk) has an empirical formula of $HC_3H_5O_3$. A 1.0-g sample of lactic acid required 17.0 mL of 0.65 *M* NaOH to reach the end point of a titration. What is the molecular formula for lactic acid?

76. 50. mL of 0.10 *M* HCl solution are poured equally into two flasks.
(a) What is the pH of the HCl solution?
(b) Next, 0.050 mol Zn is added to Flask A and 0.050 mol Cu is added to Flask B. Determine the pH of each solution after approximately 20 minutes.

ANSWERS TO PRACTICE EXERCISES

15.1 (a) HCO_3^-, (b) NO_2^-, (c) $C_2H_3O_2^-$

15.2 (a) H_2SO_4, (b) NH_4^+, (c) H_2O

15.3 (a) KOH, potassium hydroxide; H_3PO_4, phosphoric acid
(b) $Mg(OH)_2$, magnesium hydroxide; HBr, hydrobromic acid
(c) LiOH, lithium hydroxide; HCl, hydrochloric acid
(d) $Fe(OH)_2$, iron(II) hydroxide; H_2CO_3, carbonic acid

15.4 The cation of a salt comes from the cation of the base, and the anion of a salt comes from the anion of the acid.

NaCl, Na_2SO_4, Na_3PO_4,
$MgCl_2$, $MgSO_4$, $Mg_3(PO_4)_2$,
$FeCl_3$, $Fe_2(SO_4)_3$, $FePO_4$.

15.5 (a) 0.050 *M* Mg^{2+}, 0.10 *M* Cl^-,
(b) 0.070 *M* Al^{3+}, 0.21 *M* Cl^-

15.6 (a) 11.41, (b) 2.89, (c) 5.429

15.7 0.0620 *M* HCl

15.8 $3\ H_2S(aq) + 2\ Bi^{3+}(aq) \longrightarrow Bi_2S_3(s) + 6\ H^+(aq)$

16

Chemical Equilibrium

CHAPTER OUTLINE

Thus far, we've considered chemical change as proceeding from reactants to products. Does that mean that the change then stops? No, but often it appears to be the case at the macroscopic level. A solute dissolves until the solution becomes saturated. Once a solid remains undissolved in a container, the system appears to be at rest. The human body is a marvelous chemical factory, yet from day to day it appears to be quite the same. For example, the blood remains at a constant pH, even though all sorts of chemical reactions are taking place. Another example is keeping fish in an aquarium which requires maintaining an equilibrium among the living organisms and the water. Or an antacid, which neutralizes excess stomach acid. In all of these cases, reactions are proceeding, even though visible signs of chemical change are absent. Similarly, when a system is at equilibrium, chemical reactions are dynamic at the molecular level. In this chapter, we will consider chemical systems as they approach equilibrium conditions.

16.1 Rates of Reaction

List the factors that affect the rate of a chemical reaction.

LEARNING OBJECTIVE

KEY TERM

chemical kinetics

Every reaction has a rate, or speed, at which it proceeds. Some reactions are fast; some are extremely slow. The study of reaction rates and reaction mechanisms is known as **chemical kinetics**.

The rate of a reaction is variable and depends on the concentration of the reacting species, the temperature, the presence of catalysts, and the nature of the reactants. Consider the hypothetical reaction

$$A + B \longrightarrow C + D \quad \text{(forward reaction)}$$

$$C + D \longrightarrow A + B \quad \text{(reverse reaction)}$$

in which a collision between A and B is necessary for a reaction to occur. The rate at which A and B react depends on the concentration, or the number, of A and B molecules present; it will be fastest, for a fixed set of conditions, when they are first mixed (as shown by the height of the red line in **FIGURE 16.1**). As the reaction proceeds, the number of A and B molecules available for reaction decreases, and the rate of reaction slows down (seen as the red line flattens in Figure 16.1). If the reaction is reversible, the speed of the reverse reaction is zero at first (blue line in Figure 16.1) and gradually increases as the concentrations of C and D increase. As the number of A and B molecules decreases, the forward rate slows down because A and B cannot find one another as often in order to accomplish a reaction. To counteract this diminishing rate of reaction, an excess of one reagent is often used to keep the reaction from becoming impractically slow. Collisions between molecules may be compared to video games. When many objects are on the screen, collisions occur frequently; but if only a few objects are present, collisions can usually be avoided.

CHECK YOUR UNDERSTANDING

16.1 Kinetics

WileyPLUS

EXAMPLE 16.1

When two reactants are mixed, what happens to the forward and the reverse reactions as the reaction proceeds to equal rates of reaction?

SOLUTION

The rate of the forward reaction slows down and the reverse reaction speeds up until equilibrium is reached in which both reactions are proceeding at the same rate.

PRACTICE 16.1

When two reactants are mixed, why does the rate of the forward reaction slow down?

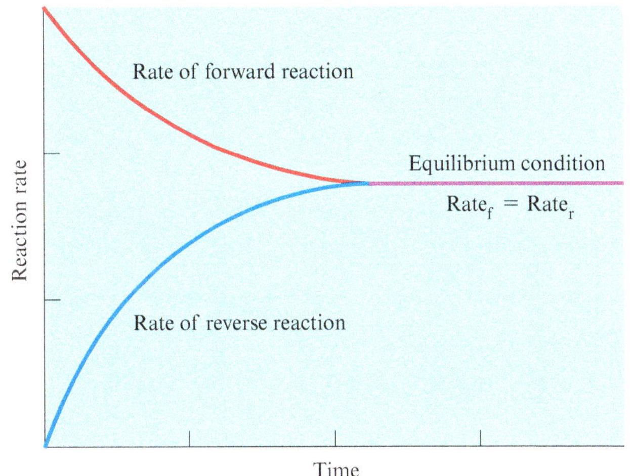

FIGURE 16.1 The rates of the forward and reverse reactions become equal at some point in time. The forward reaction rate (red) decreases as a result of decreasing amounts of reactants. The reverse reaction rate (blue) starts at zero and increases as the amount of product increases. When the two rates become equal (purple), a state of chemical equilibrium has been reached.

16.2 Chemical Equilibrium

LEARNING OBJECTIVE Define a reversible chemical reaction and explain what is occurring in a chemical system at equilibrium.

KEY TERMS

reversible chemical reaction
equilibrium
chemical equilibrium

Reversible Reactions

In the preceding chapters, we treated chemical reactions mainly as reactants changing to products. However, many reactions do not go to completion. Some reactions do not go to completion because they are reversible; that is, when the products are formed, they react to produce the starting reactants.

We've encountered reversible systems before. One is the vaporization of a liquid by heating and its subsequent condensation by cooling:

$$\text{liquid} + \text{heat} \longrightarrow \text{vapor}$$
$$\text{vapor} + \text{cooling} \longrightarrow \text{liquid}$$

The conversion between nitrogen dioxide, NO_2, and dinitrogen tetroxide, N_2O_4, shows us visible evidence of the reversibility of a reaction. The NO_2 is a reddish-brown gas that changes with cooling to N_2O_4, a colorless gas. The reaction is reversible by heating N_2O_4:

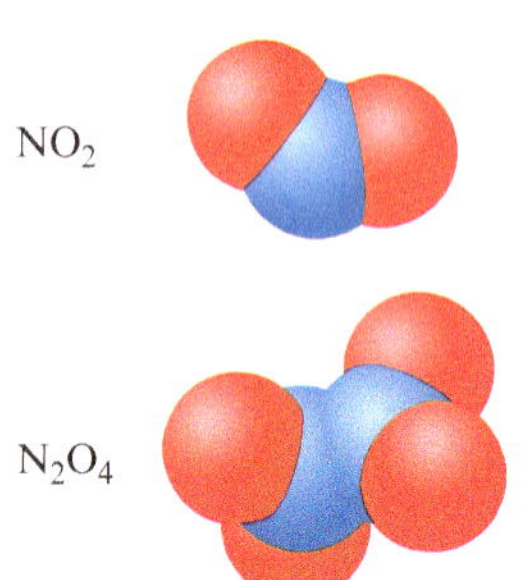

$$2\ NO_2(g) \xrightarrow{\text{cooling}} N_2O_4(g)$$

$$N_2O_4(g) \xrightarrow{\text{heating}} 2\ NO_2(g)$$

These two reactions may be represented by a single equation with a double arrow, $\rightleftharpoons$, to indicate that the reactions are taking place in both directions at the same time:

$$2\ NO_2(g) \rightleftharpoons N_2O_4(g)$$

This reversible reaction can be demonstrated by sealing samples of NO_2 in two tubes and placing one tube in warm water and the other in ice water (**FIGURE 16.2**).

A **reversible chemical reaction** is one in which the products formed react to produce the original reactants. Both the forward and reverse reactions occur simultaneously. The forward reaction is also called *the reaction to the right*, and the reverse reaction is also called *the reaction to the left*. A double arrow is used in the equation to indicate that the reaction is reversible.

Any system at **equilibrium** represents a dynamic state in which two or more opposing processes are taking place at the same time and at the same rate. A chemical equilibrium is a dynamic system in which two or more opposing chemical reactions are going on at the

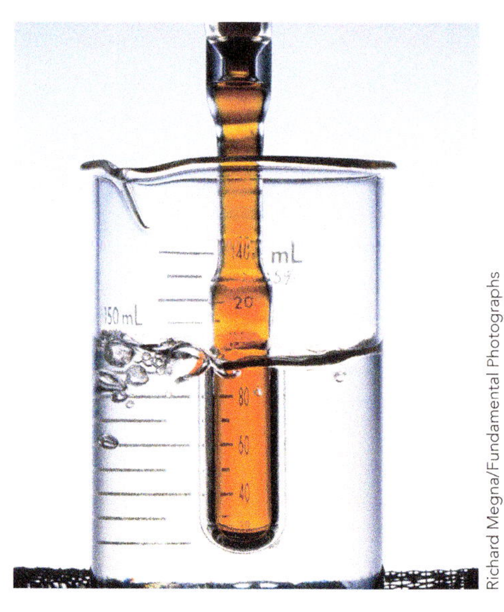

FIGURE 16.2 Reversible reaction of NO_2 and N_2O_4. More of the dark brown molecules are visible in the heated container on the right than in the room-temperature tube on the left.

same time and at the same rate. When the rate of the forward reaction is exactly equal to the rate of the reverse reaction, a condition of **chemical equilibrium** exists (see purple line in Figure 16.1). The concentrations of the products and the reactants are not changing, and the system appears to be at a standstill because the products are reacting at the same rate at which they are being formed.

Chemical equilibrium:
rate of forward reaction = rate of reverse reaction

A saturated salt solution is in a condition of equilibrium:

$$NaCl(s) \rightleftharpoons Na^+(aq) + Cl^-(aq)$$

At equilibrium, salt crystals are continuously dissolving, and Na^+ and Cl^- ions are continuously crystallizing. Both processes are occurring at the same rate.

The ionization of weak electrolytes is another chemical equilibrium system:

$$HC_2H_3O_2(aq) + H_2O(l) \rightleftharpoons H_3O^+(aq) + C_2H_3O_2^-(aq)$$

In this reaction, the equilibrium is established in a 1 M solution when the forward reaction has gone about 1%—that is, when only 1% of the acetic acid molecules in solution have ionized. Therefore, only a relatively few ions are present, and the acid behaves as a weak electrolyte.

The reaction represented by

$$H_2(g) + I_2(g) \xrightleftharpoons{700\ K} 2\ HI(g)$$

provides another example of chemical equilibrium. Theoretically, 1.00 mol of hydrogen should react with 1.00 mol of iodine to yield 2.00 mol of hydrogen iodide. Actually, when 1.00 mol H_2 and 1.00 mol I_2 are reacted at 700 K, only 1.58 mol HI are present when equilibrium is attained. Since 1.58 is 79% of the theoretical yield of 2.00 mol HI, the forward reaction is only 79% complete at equilibrium. The equilibrium mixture will also contain 0.21 mol each of unreacted H_2 and I_2 (1.00 mol − 0.79 mol = 0.21 mol):

$$H_2(g) + I_2(g) \xrightarrow{700\ K} 2\ HI(g)$$

This equation represents the condition if the reaction were 100% complete; 2.00 mol HI would be formed, and no H_2 and I_2 would be left unreacted.

$$\underset{0.21\ mol}{H_2(g)} + \underset{0.21\ mol}{I_2(g)} \xrightleftharpoons{700\ K} \underset{1.58\ mol}{2\ HI(g)}$$

This equation represents the actual equilibrium attained starting with 1.00 mol each of H_2 and I_2. It shows that the forward reaction is only 79% complete.

CHECK YOUR UNDERSTANDING
16.2 Properties of Equilibrium
➡ WileyPLUS

EXAMPLE 16.2

Is a reversible chemical reaction at equilibrium a static or a dynamic system? Explain.

SOLUTION

A reversible chemical reaction is a dynamic system in which two opposing reactions are taking place at the same time and at the same rate of reaction.

PRACTICE 16.2

What symbolism is used in a chemical equation to indicate that a chemical reaction is reversible?

16.3 Le Châtelier's Principle

LEARNING OBJECTIVE Use Le Châtelier's principle to predict the changes that occur when concentration, temperature, or volume is changed in a system at equilibrium.

KEY TERMS
Le Châtelier's principle
catalyst
activation energy

In 1888, the French chemist Henri Le Châtelier (1850–1936) set forth a simple, far-reaching generalization on the behavior of equilibrium systems. This generalization, known as **Le Châtelier's principle**, states:

> If a stress is applied to a system in equilibrium, the system will respond in such a way as to relieve that stress and restore equilibrium under a new set of conditions.

The application of Le Châtelier's principle helps us predict the effect of changing conditions in chemical reactions. We will examine the effect of changes in concentration, temperature, and volume.

Effect of Concentration on Equilibrium

The manner in which the rate of a chemical reaction depends on the concentration of the reactants must be determined experimentally. Many simple, one-step reactions result from a collision between two molecules or ions. The rate of such one-step reactions can be altered by changing the concentration of the reactants or products. An increase in concentration of the reactants provides more individual reacting species for collisions and results in an increase in the rate of reaction.

An equilibrium is disturbed when the concentration of one or more of its components is changed. As a result, the concentration of all species will change, and a new equilibrium mixture will be established. Consider the hypothetical equilibrium represented by the equation

$$A + B \rightleftharpoons C + D$$

where A and B react in one step to form C and D. When the concentration of B is increased, the following results occur:

1. The rate of the reaction to the right (forward) increases. This rate is proportional to the concentration of A times the concentration of B.
2. The rate to the right becomes greater than the rate to the left.
3. Reactants A and B are used faster than they are produced; C and D are produced faster than they are used.
4. After a period of time, rates to the right and left become equal, and the system is again in equilibrium.
5. In the new equilibrium the concentration of A is less, and the concentrations of B, C, and D are greater than in the original equilibrium.
   Conclusion: The equilibrium has shifted to the right.

Concentration	Change
$[H_2]$	?
$[I_2]$	increase
$[HI]$	?

Applying this change in concentration to the equilibrium mixture of 1.00 mol of hydrogen and 1.00 mol of iodine from Section 16.2, we find that, when an additional 0.20 mol I_2 is added, the yield of HI (based on H_2) is 85% (1.70 mol) instead of 79%. Here is how the two systems compare after the new equilibrium mixture is reached:

Original equilibrium	$H_2 + I_2 \rightleftharpoons 2\,HI$	**New equilibrium**
1.00 mol H_2 + 1.00 mol I_2		1.00 mol H_2 + 1.20 mol I_2
Yield: 79% HI		Yield: 85% HI (based on H_2)
Equilibrium mixture contains:		Equilibrium mixture contains:
1.58 mol HI		1.70 mol HI
0.21 mol H_2		0.15 mol H_2
0.21 mol I_2		0.35 mol I_2

Analyzing this new system, we see that, when 0.20 mol I_2 is added, the equilibrium shifts to the right to counteract the increase in I_2 concentration. Some of the H_2 reacts with added I_2 and produces more HI, until an equilibrium mixture is established again. When I_2 is added, the concentration of I_2 increases, the concentration of H_2 decreases, and the concentration of HI increases.

Concentration	Change
$[H_2]$	decrease
$[I_2]$	increase
[HI]	increase

PRACTICE 16.3

Use a chart like the one in the margin to show what would happen to the concentrations of each substance in the system

$$H_2(g) + I_2(g) \rightleftharpoons 2\,HI(g)$$

upon adding (a) more H_2 and (b) more HI.

The equation

$$\underset{\text{pale yellow}}{Fe^{3+}(aq)} + \underset{\text{colorless}}{SCN^-(aq)} \rightleftharpoons \underset{\text{red}}{Fe(SCN)^{2+}(aq)}$$

represents an equilibrium that is used in certain analytical procedures as an indicator because of the readily visible, intense red color of the complex $Fe(SCN)^{2+}$ ion. A very dilute solution of iron(III) (Fe^{3+}) and thiocyanate (SCN^-) is light red. When the concentration of either Fe^{3+} or SCN^- is increased, the equilibrium shift to the right is observed by an increase in the intensity of the red color, resulting from the formation of additional $Fe(SCN)^{2+}$.

If either Fe^{3+} or SCN^- is removed from solution, the equilibrium will shift to the left, and the solution will become lighter in color. When Ag^+ is added to the solution, a white precipitate of silver thiocyanate (AgSCN) is formed, thus removing SCN^- ion from the equilibrium:

$$Ag^+(aq) + SCN^-(aq) \rightleftharpoons AgSCN(s)$$

The system accordingly responds to counteract the change in SCN^- concentration by shifting the equilibrium to the left. This shift is evident by a decrease in the intensity of the red color due to a decreased concentration of $Fe(SCN)^{2+}$.

Now consider the effect of changing the concentrations in the equilibrium mixture of chlorine water. The equilibrium equation is

$$Cl_2(aq) + 2\,H_2O(l) \rightleftharpoons HOCl(aq) + H_3O^+(aq) + Cl^-(aq)$$

The variation in concentrations and the equilibrium shifts are tabulated in the following table. An X in the second or third column indicates that the reagent is increased or decreased. The fourth column indicates the direction of the equilibrium shift.

	Concentration		
Reagent	Increase	Decrease	Equilibrium shift
Cl_2	—	X	Left
H_2O	X	—	Right
HOCl	X	—	Left
H_3O^+	—	X	Right
Cl^-	X	—	Left

CHEMISTRY *IN ACTION* New Ways in Fighting Cavities and Avoiding the Drill

Dentists have understood for more than 20 years what causes cavities, but, until now, there have been only a limited number of over-the-counter products to help us avoid our dates with the drill. Bacteria in the mouth break down sugars remaining in the mouth after eating. Acids produced during this process slip through tooth enamel, dissolving minerals below the surface in a process called demineralization. Saliva works to rebuild teeth by adding calcium and phosphate back in a process called remineralization. Under ideal conditions (assuming that you brush after eating), these two processes form an equilibrium.

Unfortunately, bacteria in plaque (resulting from not brushing) shift the equilibrium toward demineralization (shown in the figure), and a cavity can begin to form. Scientists realized that fluoride encourages remineralization in teeth by replacing hydroxyl ions in nature's calcium phosphate (hydroxyapatite). The substitution changes the hydroxyapatite to fluorapatite, which is more acid resistant.

The good news is that scientists have now figured out how to shift the equilibrium between demineralization and remineralization toward remineralization. This new understanding has resulted in products that actually stop cavities and prevent the need for a filling. Scientists now know that adding calcium and phosphate to the mouth regularly speeds up remineralization of the teeth. The largest problem in doing it this way is to figure out how to dispense the calcium and phosphate through the surface enamel gradually. If a calcium ion finds a phosphate ion before this happens, a precipitate forms that can't penetrate the enamel. Researchers have developed new products such as chewing gum and sticky substances (see photo) that can be applied between the teeth. These products work better than brushing or rinsing because they last longer—a brushing or rinsing lasts only a minute.

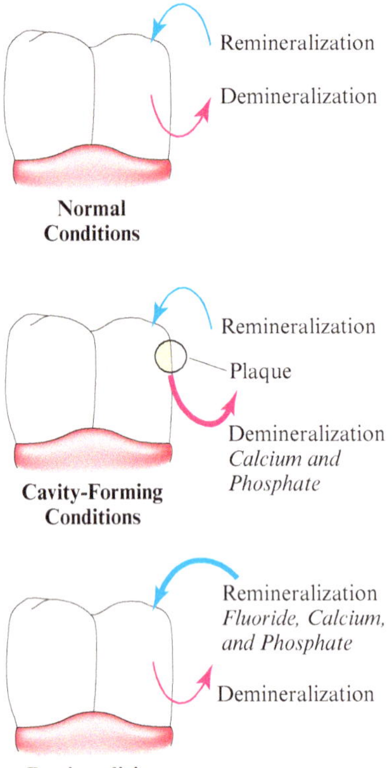

Several types of remineralizing chewing gums are currently available, including Trident Advantage and Trident for Kids. The calcium and phosphate ions are stablilized by using a system that mimics nature. Researchers looked at how newborns get their large supply of calcium and phosphate. Eric Reynolds of the University of Melbourne in Australia says, "Nature's evolved this system in milk to carry very high levels of calcium and phosphate in a highly bioavailable form." They used casein (a milk protein) to produce microscopic particles of stabilized calcium phosphate that diffuse through the surface of teeth. When these particles were placed into the chewing gum and chewed, the results were better than hoped. In Japan, researchers at Tokyo's FAP Dental Institute have designed a toothpaste that repairs cavities smaller than 50 μm. The toothpaste grows nanocrystals of hydroxyapatite, which is treated with fluoride right on the cavity. The synthetic enamel not only repairs the tiny cavity as you brush it but also strengthens the natural enamel. Brushing can now repair cavities and prevent new ones at the same time. New research suggests that a weak electrical current can be applied to a region of a tooth that will draw the charged calcium and phosphate ions toward this area to speed up remineralization in a process known as Electrically Accelerated and Enhanced Remineralization or EAER. Researchers at King's College in London are currently looking for a way to bring this procedure to your local dentist's office.

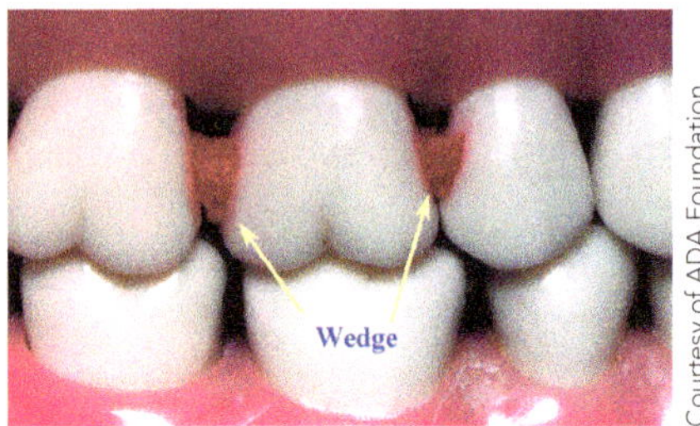

These sticky wedges dissolve slowly to release fluoride, calcium, and phosphate to teeth.

Consider the equilibrium in a 0.100 *M* acetic acid solution:

$$HC_2H_3O_2(aq) + H_2O(l) \rightleftharpoons H_3O^+(aq) + C_2H_3O_2^-(aq)$$

In this solution, the concentration of the hydronium ion (H_3O^+), which is a measure of the acidity, is 1.34×10^{-3} mol/L, corresponding to a pH of 2.87. What will happen to the acidity when 0.100 mol of sodium acetate ($NaC_2H_3O_2$) is added to 1 L of 0.100 *M* acetic acid ($HC_2H_3O_2$)? When $NaC_2H_3O_2$ dissolves, it dissociates into sodium ions (Na^+) and acetate ions ($C_2H_3O_2^-$). The acetate ion from the salt is a common ion to the acetic acid equilibrium system and increases the total acetate ion concentration in the solution. As a result, the equilibrium shifts to the left, decreasing the hydronium ion concentration and lowering the acidity of the solution. Evidence of this decrease in acidity is shown by the fact that the pH of a solution that is 0.100 *M* in $HC_2H_3O_2$ and 0.100 *M* in $NaC_2H_3O_2$ is 4.74. The pH of several different solutions of $HC_2H_3O_2$ and $NaC_2H_3O_2$ is shown in the table that follows. Each time the acetate ion is increased, the pH increases, indicating a further shift in the equilibrium toward un-ionized acetic acid.

Concentration	Change
$[HC_2H_3O_2]$	increase
$[H_3O^+]$	decrease
$[C_2H_3O_2^-]$	increase

Solution	pH
1 L 0.100 *M* $HC_2H_3O_2$	2.87
1 L 0.100 *M* $HC_2H_3O_2$ + 0.100 mol $NaC_2H_3O_2$	4.74
1 L 0.100 *M* $HC_2H_3O_2$ + 0.200 mol $NaC_2H_3O_2$	5.05
1 L 0.100 *M* $HC_2H_3O_2$ + 0.300 mol $NaC_2H_3O_2$	5.23

In summary, we can say that when the concentration of a reagent on the left side of an equation is increased, the equilibrium shifts to the right. When the concentration of a reagent on the right side of an equation is increased, the equilibrium shifts to the left. In accordance with Le Châtelier's principle, the equilibrium always shifts in the direction that tends to reduce the concentration of the added reactant.

PRACTICE 16.4

Aqueous chromate ion, CrO_4^{2-}, exists in equilibrium with aqueous dichromate ion, $Cr_2O_7^{2-}$, in an acidic solution. What effect will (a) increasing the dichromate ion and (b) adding HCl have on the equilibrium?

$$2\,CrO_4^{2-}(aq) + 2\,H^+(aq) \rightleftharpoons Cr_2O_7^{2-}(aq) + H_2O(l)$$

Effect of Volume on Equilibrium

Changes in volume significantly affect the reaction rate only when one or more of the reactants or products is a gas and the reaction is run in a closed container. In these cases the effect of decreasing the volume of the reacting gases is equivalent to increasing their concentrations. In the reaction

$$CaCO_3(s) \overset{\Delta}{\rightleftharpoons} CaO(s) + CO_2(g)$$

calcium carbonate decomposes into calcium oxide and carbon dioxide when heated about 825°C. Decreasing the volume of the container speeds up the reverse reaction and causes the equilibrium to shift to the left. Decreasing the volume increases the concentration of CO_2, the only gaseous substance in the reaction.

If the volume of the container is decreased, the pressure of the gas will increase. In a system composed entirely of gases, this decrease in the volume of the container will cause the reaction and the equilibrium to shift to the side that contains the smaller number of molecules. To clarify what's happening: When the container volume is decreased, the pressure in the container is increased. The system tries to lower this pressure by reducing the number of molecules. Let's consider an example that shows these effects.

Prior to World War I, Fritz Haber (1868–1934) invented the first major process for the fixation of nitrogen. In this process, nitrogen and hydrogen are reacted together in the presence of a catalyst at moderately high temperature and pressure to produce ammonia:

Haber received the Nobel Prize in chemistry for this process in 1918.

$$N_2(g) + 3\,H_2(g) \rightleftharpoons 2\,NH_3(g) + 92.5\text{ kJ}$$

$N_2(g)$	$3\,H_2(g)$	$2\,NH_3(g)$
1 mol	3 mol	2 mol
1 volume	3 volumes	2 volumes

© Charles Brutlag/iStockphoto

Gaseous ammonia is often used to add nitrogen to the fields before planting and during early growth.

The left side of the equation in the Haber process represents 4 mol of gas combining to give 2 mol of gas on the right side of the equation. A decrease in the volume of the container shifts the equilibrium to the right. This decrease in volume results in a higher concentration of both reactants and products. The equilibrium shifts to the right toward fewer molecules.

When the total number of gaseous molecules on both sides of an equation is the same, a change in volume does not cause an equilibrium shift. The following reaction is an example:

$N_2(g)$	+	$O_2(g)$	$\rightleftharpoons$	$2\,NO(g)$
1 mol		1 mol		2 mol
1 volume		1 volume		2 volumes
6.022×10^{23} molecules		6.022×10^{23} molecules		$2(6.022 \times 10^{23})$ molecules

When the volume of the container is decreased, the rate of both the forward and the reverse reactions will increase because of the higher concentrations of N_2,O_2, and NO. But the equilibrium will not shift because the number of molecules is the same on both sides of the equation and the effects on concentration are the same on both forward and reverse rates.

ONLINE LEARNING MODULE
Le Châtelier's Principle
WileyPLUS

ENHANCED EXAMPLE 16.3 WileyPLUS

What effect would a decrease in volume of the container have on the position of equilibrium in these reactions?

(a) $2\,SO_2(g) + O_2(g) \rightleftharpoons 2\,SO_3(g)$
(b) $H_2(g) + Cl_2(g) \rightleftharpoons 2\,HCl(g)$
(c) $N_2O_4(g) \rightleftharpoons 2\,NO_2(g)$

SOLUTION

(a) The equilibrium will shift to the right because the substance on the right has a smaller number of moles than the substances on the left.
(b) The equilibrium position will be unaffected because the moles of gases on both sides of the equation are the same.
(c) The equilibrium will shift to the left because $N_2O_4(g)$ represents the smaller number of moles.

PRACTICE 16.5

What effect would a decrease in the container's volume have on the position of the equilibrium in these reactions?

(a) $2\,NO(g) + Cl_2(g) \rightleftharpoons 2\,NOCl(g)$
(b) $COBr_2(g) \rightleftharpoons CO(g) + Br_2(g)$

Effect of Temperature on Equilibrium

When the temperature of a system is raised, the rate of reaction increases because of increased kinetic energy and more frequent collisions of the reacting species. In a reversible reaction, the rate of both the forward and the reverse reactions is increased by an increase in temperature; however, the reaction that absorbs heat increases to a greater extent, and the equilibrium shifts to favor that reaction.

High temperatures can also cause the destruction or decomposition of the reactants or products.

An increase in temperature generally increases the rate of reaction. Molecules at elevated temperatures have more kinetic energy; their collisions are thus more likely to result in a reaction.

When heat is applied to a system in equilibrium, the reaction that absorbs heat is favored. When the process, as written, is endothermic, the forward reaction is increased. When the reaction is exothermic, the reverse reaction is favored. In this sense heat may be treated as a reactant in endothermic reactions or as a product in exothermic reactions. Therefore, temperature is analogous to concentration when applying Le Châtelier's principle to heat effects on a chemical reaction.

Hot coke (C) is a very reactive element. In the reaction

$$C(s) + CO_2(g) + \text{heat} \rightleftharpoons 2\,CO(g)$$

very little if any CO is formed at room temperature. At 1000°C, the equilibrium mixture contains about an equal number of moles of CO and CO_2. Since the reaction is endothermic, the equilibrium is shifted to the right at higher temperatures.

When phosphorus trichloride reacts with dry chlorine gas to form phosphorus pentachloride, the reaction is exothermic:

$$PCl_3(l) + Cl_2(g) \rightleftharpoons PCl_5(s) + 88\text{ kJ}$$

Heat must continuously be removed during the reaction to obtain a good yield of the product. According to Le Châtelier's principle, heat will cause the product, PCl_5, to decompose, re-forming PCl_3 and Cl_2. The equilibrium mixture at 200°C contains 52% PCl_5, and at 300°C it contains 3% PCl_5, verifying that heat causes the equilibrium to shift to the left.

ENHANCED EXAMPLE 16.4 **WileyPLUS**

What effect would an increase in temperature have on the position of the equilibrium in these reactions?

$$4\,HCl(g) + O_2(g) \rightleftharpoons 2\,H_2O(g) + 2\,Cl_2(g) + 95.4\text{ kJ} \quad (1)$$

$$H_2(g) + Cl_2(g) \rightleftharpoons 2\,HCl(g) + 185\text{ kJ} \quad (2)$$

$$CH_4(g) + 2\,O_2(g) \rightleftharpoons CO_2(g) + 2\,H_2O(g) + 890\text{ kJ} \quad (3)$$

$$N_2O_4(g) + 58.6\text{ kJ} \rightleftharpoons 2\,NO_2(g) \quad (4)$$

$$2\,CO_2(g) + 566\text{ kJ} \rightleftharpoons 2\,CO(g) + O_2(g) \quad (5)$$

$$H_2(g) + I_2(g) + 51.9\text{ kJ} \rightleftharpoons 2\,HI(g) \quad (6)$$

SOLUTION

Reactions 1, 2, and 3 are exothermic; an increase in temperature will cause the equilibrium to shift to the left. Reactions 4, 5, and 6 are endothermic; an increase in temperature will cause the equilibrium to shift to the right.

PRACTICE 16.6

What effect would an increase in temperature have on the position of the equilibrium in these reactions?

(a) $2\,SO_2(g) + O_2(g) \rightleftharpoons 2\,SO_3(g) + 198\text{ kJ}$
(b) $H_2(g) + CO_2(g) + 41\text{ kJ} \rightleftharpoons H_2O(g) + CO(g)$

Effect of Catalysts on Equilibrium

A **catalyst** is a substance that influences the rate of a chemical reaction and can be recovered essentially unchanged at the end of the reaction. A catalyst does not shift the equilibrium of a reaction; it affects only the speed at which the equilibrium is reached. It does this by lowering the activation energy for the reaction (FIGURE 16.3). **Activation energy** is the minimum energy required for the reaction to occur. A catalyst speeds up a reaction by lowering the activation energy while not changing the energies of reactants or products. If a catalyst does not affect the equilibrium, then it follows that it must affect the rate of both the forward and the reverse reactions equally.

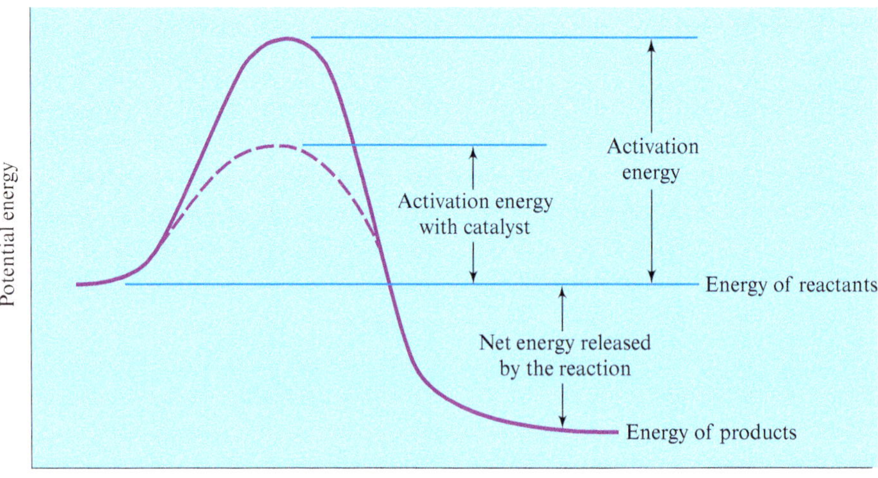

FIGURE 16.3 Energy diagram for an exothermic reaction. Energy is put into the reaction (activation energy) to initiate the process. In the reaction shown, all of the activation energy and the net energy are released as the reaction proceeds to products. Note that the presence of a catalyst lowers the activation energy but does not change the energies of the reactants or the products.

The reaction between phosphorus trichloride and sulfur is highly exothermic, but it's so slow that very little product, thiophosphoryl chloride, is obtained, even after prolonged heating. When a catalyst such as aluminum chloride is added, the reaction is complete in a few seconds:

$$PCl_3(l) + S(s) \xrightarrow{AlCl_3} PSCl_3(l)$$

The lab preparation of oxygen uses manganese dioxide as a catalyst to increase the rates of decomposition of both potassium chlorate and hydrogen peroxide:

$$2\,KClO_3(s) \xrightarrow[\Delta]{MnO_2} 2\,KCl(s) + 3\,O_2(g)$$

$$2\,H_2O_2(aq) \xrightarrow{MnO_2} 2\,H_2Ol) + O_2(g)$$

Catalysts are extremely important to industrial chemistry. Hundreds of chemical reactions that are otherwise too slow to be of practical value have been put to commercial use once a suitable catalyst was found. And in the area of biochemistry, catalysts are of supreme importance because nearly all chemical reactions in all forms of life are completely dependent on biochemical catalysts known as *enzymes*.

CHECK YOUR UNDERSTANDING
16.3 Catalysts
➡ WileyPLUS

16.4 Equilibrium Constants

LEARNING OBJECTIVE Write the general expression for the equilibrium constant and calculate equilibrium constants.

KEY TERM
equilibrium constant, K_{eq}

In a reversible chemical reaction at equilibrium, the concentrations of the reactants and products are constant. At equilibrium, the rates of the forward and reverse reactions are equal, and an equilibrium constant expression can be written relating the products to the reactants. For the general reaction

$$aA + bB \rightleftharpoons cC + dD$$

at a given temperature, the equilibrium constant expression can be written as

$$K_{eq} = \frac{[C]^c[D]^d}{[A]^a[B]^b}$$

Note: The exponents (*a*, *b*, *c*, and *d*) are the same as the coefficients in the balanced equation.

where K_{eq}, the **equilibrium constant**, is constant at a particular temperature. The quantities in brackets are the concentrations of each substance in moles per liter. The superscript letters *a*, *b*, *c*, and *d* are the coefficients of the substances in the balanced equation. According to convention, we place the concentrations of the products (the substances on the right side of the equation as written) in the numerator and the concentrations of the reactants in the denominator.

ENHANCED EXAMPLE 16.5 **WileyPLUS**

ONLINE LEARNING MODULE
Equilibrium Constants
WileyPLUS

Write equilibrium constant expressions for

(a) $3\ H_2(g) + N_2(g) \rightleftharpoons 2\ NH_3(g)$

(b) $CO(g) + 2\ H_2(g) \rightleftharpoons CH_3OH(g)$

SOLUTION

(a) The only product, NH_3, has a coefficient of 2. Therefore, the numerator will be $[NH_3]^2$. Two reactants are present: H_2, with a coefficient of 3, and N_2, with a coefficient of 1. The denominator will thus be $[H_2]^3[N_2]$. The equilibrium constant expression is

$$K_{eq} = \frac{[NH_3]^2}{[H_2]^3[N_2]}$$

(b) For this equation, the numerator is $[CH_3OH]$ and the denominator is $[CO][H_2]^2$. The equilibrium constant expression is

$$K_{eq} = \frac{[CH_3OH]}{[CO][H_2]^2}$$

PRACTICE 16.7

Write equilibrium constant expressions for

(a) $2\ N_2O_5(g) \rightleftharpoons 4\ NO_2(g) + O_2(g)$

(b) $4\ NH_3(g) + 3\ O_2(g) \rightleftharpoons 2\ N_2(g) + 6\ H_2O(g)$

The magnitude of an equilibrium constant indicates the extent to which the forward and reverse reactions take place. When K_{eq} is greater than 1, the amount of products at equilibrium is greater than the amount of reactants. When K_{eq} is less than 1, the amount of reactants at equilibrium is greater than the amount of products. A very large value for K_{eq} indicates that the forward reaction goes essentially to completion. A very small K_{eq} means that the reverse reaction goes nearly to completion and that the equilibrium is far to the left (toward the reactants). Consider the following two examples:

$$H_2(g) + I_2(g) \rightleftharpoons 2\ HI(g) \qquad K_{eq} = 54.8 \text{ at } 425°C$$

$$COCl_2(g) \rightleftharpoons CO(g) + Cl_2(g) \qquad K_{eq} = 7.6 \times 10^{-4} \text{ at } 400°C$$

In the first example, K_{eq} indicates that more product than reactant exists at equilibrium. In the second equation, K_{eq} indicates that $COCl_2$ is stable and that very little decomposition to CO and Cl_2 occurs at 400°C. The equilibrium is far to the left.

When the molar concentrations of all species in an equilibrium reaction are known, the K_{eq} can be calculated by substituting the concentrations into the equilibrium constant expression.

CHECK YOUR UNDERSTANDING

16.4 Equilibrium Constant for a Reaction

16.5 Properties of an Equilibrium Constant

WileyPLUS

Units are generally not included in values of K_{eq} for reasons beyond the scope of this book.

ONLINE LEARNING MODULE
Equilibrium Calculations
WileyPLUS

EXAMPLE 16.6

Calculate K_{eq} for the following reaction based on concentrations of PCl_5 = 0.030 mol/L, PCl_3 = 0.97 mol/L, and Cl_2 = 0.97 mol/L at 300°C.

$$PCl_5(g) \rightleftharpoons PCl_3(g) + Cl_2(g)$$

SOLUTION

First write the K_{eq} expression; then substitute the respective concentrations into this equation and solve:

$$K_{eq} = \frac{[PCl_3][Cl_2]}{[PCl_5]} = \frac{(0.97)(0.97)}{(0.030)} = 31$$

Remember: Units are not included for K_{eq}.

This K_{eq} is considered to be a fairly large value, indicating that at 300°C the decomposition of PCl_5 proceeds far to the right.

PRACTICE 16.8

Calculate the K_{eq} for this reaction. Is the forward or the reverse reaction favored?

$$2\,NO(g) + O_2(g) \rightleftharpoons 2\,NO_2(g)$$

when [NO] = 0.050 M, $[O_2]$ = 0.75 M, and $[NO_2]$ = 0.25 M.

16.5 Ion Product Constant for Water

LEARNING OBJECTIVE Calculate the concentrations of H^+, OH^-, pH, and pOH in a solution using the ion product constant for water.

KEY TERM
ion product constant for water, K_w

We've seen that water ionizes to a slight degree. This ionization is represented by these equilibrium equations:

$$H_2O + H_2O \rightleftharpoons H_3O^+ + OH^- \quad (1)$$

$$H_2O \rightleftharpoons H^+ + OH^- \quad (2)$$

Equation 1 is the more accurate representation of the equilibrium because free protons (H^+) do not exist in water. Equation 2 is a simplified and often-used representation of the water equilibrium. The actual concentration of H^+ produced in pure water is minute and amounts to only 1.00×10^{-7} mol/L at 25°C. In pure water,

$$[H^+] = [OH^-] = 1.00 \times 10^{-7}\ \text{mol/L}$$

since both ions are produced in equal molar amounts, as shown in equation 2.

The $H_2O \rightleftharpoons H^+ + OH^-$ equilibrium exists in water and in all water solutions. A special equilibrium constant called the **ion product constant for water, K_w**, applies to this equilibrium. The constant K_w is defined as the product of the H^+ ion concentration and the OH^- ion concentration, each in moles per liter:

$$K_w = [H^+][OH^-]$$

The numerical value of K_w is 1.00×10^{-14} for pure water at 25°C,

$$K_w = [H^+][OH^-] = (1.00 \times 10^{-7})(1.00 \times 10^{-7}) = 1.00 \times 10^{-14}$$

The value of K_w for all water solutions at 25°C is the constant 1.00×10^{-14}. It is important to realize that as the concentration of one of these ions, H^+ or OH^-, increases, the other decreases. However, the product of $[H^+]$ and $[OH^-]$ always equals 1.00×10^{-14}. This relationship can be seen in the examples shown in **TABLE 16.1**. If the concentration of one ion is known, the concentration of the other can be calculated from the K_w expression.

TABLE 16.1 Relationship of H^+ and OH^- Concentrations in Water Solutions

$[H^+]$	$[OH^-]$	K_w	pH	pOH
1.00×10^{-2}	1.00×10^{-12}	1.00×10^{-14}	2.00	12.00
1.00×10^{-4}	1.00×10^{-10}	1.00×10^{-14}	4.00	10.00
2.00×10^{-6}	5.00×10^{-9}	1.00×10^{-14}	5.70	8.30
1.00×10^{-7}	1.00×10^{-7}	1.00×10^{-14}	7.00	7.00
1.00×10^{-9}	1.00×10^{-5}	1.00×10^{-14}	9.00	5.00

$$K_w = [H^+][OH^-] \qquad [H^+] = \frac{K_w}{[OH^-]} \qquad [OH^-] = \frac{K_w}{[H^+]}$$

CHECK YOUR UNDERSTANDING

16.6 Using K_w

➡ WileyPLUS

ENHANCED EXAMPLE 16.7

WileyPLUS

What is the concentration of (a) H^+ and (b) OH^- in a 0.001 M HCl solution? Remember that HCl is 100% ionized.

SOLUTION

(a) Since all the HCl is ionized, $H^+ = 0.001$ mol/L, or 1×10^{-3} mol/L:

$$\underset{}{HCl} \longrightarrow \underset{0.001\,M}{H^+} + \underset{0.001\,M}{Cl^-}$$

$$[H^+] = 1 \times 10^{-3}\ \text{mol/L}$$

(b) To calculate the $[OH^-]$ in this solution, use the following equation and substitute the values for K_w and $[H^+]$:

$$[OH^-] = \frac{K_w}{[H^+]}$$

$$[OH^-] = \frac{1.00 \times 10^{-14}}{1 \times 10^{-3}} = 1 \times 10^{-11}\ \text{mol/L}$$

ONLINE LEARNING MODULE

Ion Product Constant - Water

➡ WileyPLUS

PRACTICE 16.9

Determine the $[H^+]$ and $[OH^-]$ in

(a) 5.0×10^{-5} M HNO_3 (b) 2.0×10^{-6} M KOH

ENHANCED EXAMPLE 16.8

WileyPLUS

What is the pH of a 0.010 M NaOH solution? Assume that NaOH is 100% ionized.

SOLUTION

Since all the NaOH is ionized, $[OH^-] = 0.010$ mol/L or 1.0×10^{-2} mol/L.

$$NaOH \longrightarrow \underset{0.010\,M}{Na^+} + \underset{0.010\,M}{OH^-}$$

To find the pH of the solution, we first calculate the H^+ ion concentration. Use the following equation and substitute the values for K_w and $[OH^-]$:

$$[H^+] = \frac{K_w}{[OH^-]} = \frac{1.00 \times 10^{-14}}{1.0 \times 10^{-2}} = 1.0 \times 10^{-12} \text{ mol/L}$$

$$pH = -\log[H^+] = -\log(1.0 \times 10^{-12}) = 12.00$$

PRACTICE 16.10

Determine the pH for the following solutions:

(a) 5.0×10^{-5} *M* HNO_3 (b) 2.0×10^{-6} *M* KOH

Just as pH is used to express the acidity of a solution, pOH is used to express the basicity of an aqueous solution. The pOH is related to the OH^- ion concentration in the same way that the pH is related to the H^+ ion concentration:

$$pOH = -\log[OH^-]$$

Thus a solution in which $[OH^-] = 1.0 \times 10^{-2}$, as in Example 16.8, will have pOH = 2.00.

In pure water, where $[H^+] = 1.00 \times 10^{-7}$ and $[OH^-] = 1.00 \times 10^{-7}$, the pH is 7.0 and the pOH is 7.0. The sum of the pH and pOH is always 14.00:

$$pH + pOH = 14.00$$

In Example 16.8, the pH can also be found by first calculating the pOH from the OH^- ion concentration and then subtracting from 14.00.

$$pH = 14.00 - pOH = 14.00 - 2.00 = 12.00$$

Table 16.1 summarizes the relationship between $[H^+]$ and $[OH^-]$ in water solutions.

16.6 Ionization Constants

LEARNING OBJECTIVE Use the ionization constant of a reactant in an equilibrium expression to find the percent ionization of a substance in solution and to find the pH of a weak acid.

KEY TERM

acid ionization constant, K_a

In addition to K_w, several other equilibrium constants are commonly used. Strong acids are essentially 100% ionized. Weak acids are only slightly ionized. Let's consider the equilibrium constant for acetic acid in solution. Because it is a weak acid, an equilibrium is established between molecular $HC_2H_3O_2$ and its ions in solution:

$$HC_2H_3O_2(aq) \rightleftharpoons H^+(aq) + C_2H_3O_2^-(aq)$$

The ionization constant expression is the concentration of the products divided by the concentration of the reactants:

$$K_a = \frac{[H^+][C_2H_3O_2^-]}{[HC_2H_3O_2]}$$

The constant is called the **acid ionization constant, K_a**, a special type of equilibrium constant. The concentration of water in the solution is large compared to other concentrations and does not change appreciably. It is therefore part of the constant K_{eq}.

At 25°C, a 0.100 *M* $HC_2H_3O_2$ solution is 1.34% ionized and has an $[H^+]$ of 1.34×10^{-3} mol/L. From this information we can calculate the ionization constant for acetic acid.

A 0.100 *M* solution initially contains 0.100 mol of acetic acid per liter. Of this 0.100 mol, only 1.34%, or 1.34×10^{-3} mol, is ionized, which gives an $[H^+] = 1.34 \times 10^{-3}$ mol/L. Because each molecule of acid that ionizes yields one H^+ and one $C_2H_3O_2^-$, the concentration of $C_2H_3O_2^-$ ions is also 1.34×10^{-3} mol/L. This ionization leaves $0.100 - 0.00134 = 0.099$ mol/L of un-ionized acetic acid.

Substituting these concentrations in the equilibrium expression, we obtain the value for K_a:

Acid	Initial concentration (mol/L)	Equilibrium concentration (mol/L)
$[HC_2H_3O_2]$	0.100	0.099
$[H^+]$	0	0.00134
$[C_2H_3O_2^-]$	0	0.00134

$$K_a = \frac{[H^+][C_2H_3O_2^-]}{[HC_2H_3O_2]} = \frac{(1.34 \times 10^{-3})(1.34 \times 10^{-3})}{(0.099)} = 1.8 \times 10^{-5}$$

The K_a for acetic acid, 1.8×10^{-5}, is small and indicates that the position of the equilibrium is far toward the un-ionized acetic acid. In fact, a 0.100 M acetic acid solution is 99% un-ionized.

Once the K_a for acetic acid is established, it can be used to describe other systems containing H^+, $C_2H_3O_2^-$, and $HC_2H_3O_2$ in equilibrium at 25°C. The ionization constants for several other weak acids are listed in TABLE 16.2.

TABLE 16.2 Ionization Constants (K_a) of Weak Acids at 25°C

Acid	Formula	K_a	Acid	Formula	K_a
Acetic	$HC_2H_3O_2$	1.8×10^{-5}	Hydrocyanic	HCN	4.0×10^{-10}
Benzoic	$HC_7H_5O_2$	6.3×10^{-5}	Hypochlorous	HClO	3.5×10^{-8}
Carbolic (phenol)	HC_6H_5O	1.3×10^{-10}	Nitrous	HNO_2	4.5×10^{-4}
Cyanic	HCNO	2.0×10^{-4}	Hydrofluoric	HF	6.5×10^{-4}
Formic	$HCHO_2$	1.8×10^{-4}			

CHECK YOUR UNDERSTANDING

16.7 Writing K_a for a Reaction

16.8 Understanding the Meaning of an Equilibrium Constant

ENHANCED EXAMPLE 16.9

WileyPLUS

What is the $[H^+]$ in a 0.50 M $HC_2H_3O_2$ solution? The ionization constant, K_a, for $HC_2H_3O_2$ is 1.8×10^{-5}.

SOLUTION

To solve this problem, first write the equilibrium equation and the K_a expression:

$$HC_2H_3O_2 \rightleftharpoons H^+ + C_2H_3O_2^- \quad K_a = \frac{[H^+][C_2H_3O_2^-]}{[HC_2H_3O_2]} = 1.8 \times 10^{-5}$$

We know that the initial concentration of $HC_2H_3O_2$ is 0.50 M. We also know from the ionization equation that one $C_2H_3O_2^-$ is produced for every H^+ produced; that is, the $[H^+]$ and the $[C_2H_3O_2^-]$ are equal. To solve, let $Y = [H^+]$, which also equals the $[C_2H_3O_2^-]$. The un-ionized $[HC_2H_3O_2]$ remaining will then be $0.50 - Y$, the starting concentration minus the amount that ionized:

	Initial	Equilibrium
$[H^+]$	0	Y
$[C_2H_3O_2^-]$	0	Y
$[HC_2H_3O_2]$	0.50	0.50−Y

$$[H^+] = [C_2H_3O_2^-] = Y \qquad [HC_2H_3O_2] = 0.50 - Y$$

Substituting these values into the K_a expression, we obtain

$$K_a = \frac{(Y)(Y)}{0.50 - Y} = \frac{Y^2}{0.50 - Y} = 1.8 \times 10^{-5}$$

An exact solution of this equation for Y requires the use of a mathematical equation known as the *quadratic equation*. However, an approximate solution is obtained if we assume that Y is small and can be neglected compared with 0.50. Then $0.50 - Y$ will be equal to approximately 0.50. The equation now becomes

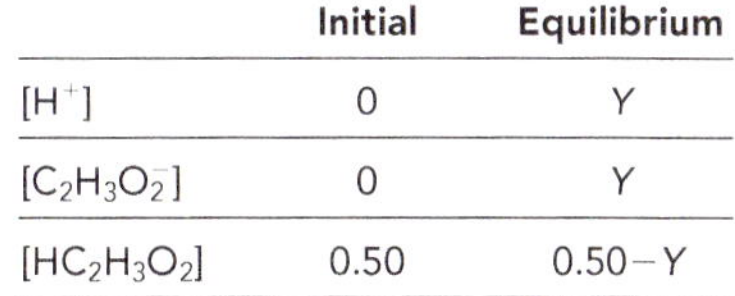

$$\frac{Y^2}{0.50} = 1.8 \times 10^{-5}$$

$$Y^2 = 0.50 \times 1.8 \times 10^{-5} = 0.90 \times 10^{-5} = 9.0 \times 10^{-6}$$

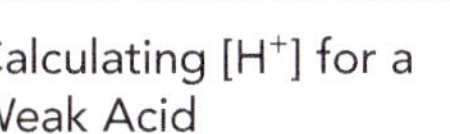

WileyPLUS

Taking the square root of both sides of the equation, we obtain

$$Y = \sqrt{9.0 \times 10^{-6}} = 3.0 \times 10^{-3}\ \text{mol/L}$$

Thus, the $[H^+]$ is approximately 3.0×10^{-3} mol/L in a 0.50 M $HC_2H_3O_2$ solution. The exact solution to this problem, using the quadratic equation, gives a value of 2.99×10^{-3} mol/L for $[H^+]$, showing that we were justified in neglecting Y compared with 0.50.

PRACTICE 16.11

Calculate the hydrogen ion concentration in (a) 0.100 M hydrocyanic acid (HCN) solution and (b) 0.0250 M carbolic acid (HC_6H_5O) solution.

EXAMPLE 16.10

Calculate the percent ionization in a 0.50 M $HC_2H_3O_2$ solution.

SOLUTION

The percent ionization of a weak acid, $HA(aq) \rightleftharpoons H^+(aq) + A^-(aq)$, is found by dividing the concentration of the H^+ or A^- ions at equilibrium by the initial concentration of HA and multiplying by 100. For acetic acid,

$$\frac{\text{concentration of } [H^+] \text{ or } [C_2H_3O_2^-]}{\text{initial concentration of } [HC_2H_3O_2]} \times 100 = \text{percent ionized}$$

To solve this problem, we first need to calculate the $[H^+]$. This calculation has already been done for a 0.50 M $HC_2H_3O_2$ solution:

$$[H^+] = 3.0 \times 10^{-3}\ \text{mol/L in a 0.50 } M \text{ solution} \quad \text{(from Example 16.9)}$$

This $[H^+]$ represents a fractional amount of the initial 0.50 M $HC_2H_3O_2$. Therefore,

$$\frac{3.0 \times 10^{-3}\ \text{mol/L}}{0.50\ \text{mol/L}} \times 100 = 0.60\%\ \text{ionized}$$

A 0.50 M $HC_2H_3O_2$ solution is 0.60% ionized.

PRACTICE 16.12

Calculate the percent ionization for

(a) 0.100 M hydrocyanic acid (HCN) (b) 0.0250 M carbolic acid (HC_6H_5O)

16.7 Solubility Product Constant

LEARNING OBJECTIVE Use the solubility product constant to calculate the solubility of a slightly soluble salt and to determine whether a precipitate will form in a solution.

KEY TERMS

solubility product constant, K_{sp}

common ion effect

The **solubility product constant, K_{sp}**, is the equilibrium constant of a slightly soluble salt. To evaluate K_{sp}, consider this example. The solubility of AgCl in water is 1.3×10^{-5} mol/L at 25°C. The equation for the equilibrium between AgCl and its ions in solution is

$$AgCl(s) \rightleftharpoons Ag^+(aq) + Cl^-(aq)$$

The equilibrium constant expression is

$$K_{eq} = \frac{[Ag^+][Cl^-]}{[AgCl(s)]}$$

The amount of solid AgCl does not affect the equilibrium system, provided that some is present. In other words, the concentration of solid AgCl is constant whether 1 mg or 10 g of the salt are present. Therefore, the product obtained by multiplying the two constants K_{eq} and $[AgCl(s)]$ is also a constant. This is the solubility product constant, K_{sp}:

$$K_{eq} \times [AgCl(s)] = [Ag^+][Cl^-] = K_{sp}$$

$$K_{sp} = [Ag^+][Cl^-]$$

The K_{sp} is equal to the product of the $[Ag^+]$ and the $[Cl^-]$, each in moles per liter. When 1.3×10^{-5} mol/L of AgCl dissolves, it produces 1.3×10^{-5} mol/L each of Ag^+ and Cl^-. From these concentrations, the K_{sp} can be calculated:

$$[Ag^+] = 1.3 \times 10^{-5}\ \text{mol/L} \qquad [Cl^-] = 1.3 \times 10^{-5}\ \text{mol/L}$$

$$K_{sp} = [Ag^+][Cl^-] = (1.3 \times 10^{-5})(1.3 \times 10^{-5}) = 1.7 \times 10^{-10}$$

Once the K_{sp} value for AgCl is established, it can be used to describe other systems containing Ag^+ and Cl^-.

The K_{sp} expression does not have a denominator. It consists only of the concentrations (mol/L) of the ions in solution. As in other equilibrium expressions, each of these concentrations is raised to a power that is the same number as its coefficient in the balanced equation. Here are equilibrium equations and the K_{sp} expressions for several other substances:

$$AgBr(s) \rightleftharpoons Ag^+(aq) + Br^-(aq) \qquad K_{sp} = [Ag^+][Br^-]$$

$$BaSO_4(s) \rightleftharpoons Ba^{2+}(aq) + SO_4^{2-}(aq) \qquad K_{sp} = [Ba^{2+}][SO_4^{2-}]$$

$$Ag_2CrO_4(s) \rightleftharpoons 2Ag^+(aq) + CrO_4^{2-}(aq) \qquad K_{sp} = [Ag^+]^2[CrO_4^{2-}]$$

$$CuS(s) \rightleftharpoons Cu^{2+}(aq) + S^{2-}(aq) \qquad K_{sp} = [Cu^{2+}][S^{2-}]$$

$$Mn(OH)_2(s) \rightleftharpoons Mn^{2+}(aq) + 2\,OH^-(aq) \qquad K_{sp} = [Mn^{2+}][OH^-]^2$$

$$Fe(OH)_3(s) \rightleftharpoons Fe^{3+}(aq) + 3\,OH^-(aq) \qquad K_{sp} = [Fe^{3+}][OH^-]^3$$

TABLE 16.3 lists K_{sp} values for these and several other substances.

When the product of the molar concentration of the ions in solution (each raised to its proper power) is greater than the K_{sp} for that substance, precipitation will occur. If the ion product is less than the K_{sp} value, no precipitation will occur.

TABLE 16.3 Solubility Product Constants (K_{sp}) at 25°C

Compound	K_{sp}	Compound	K_{sp}
AgCl	1.7×10^{-10}	CaF_2	3.9×10^{-11}
AgBr	5.2×10^{-13}	CuS	8.5×10^{-45}
AgI	8.5×10^{-17}	$Fe(OH)_3$	6.1×10^{-38}
$AgC_2H_3O_2$	2.1×10^{-3}	PbS	3.4×10^{-28}
Ag_2CrO_4	1.9×10^{-12}	$PbSO_4$	1.3×10^{-8}
$BaCrO_4$	8.5×10^{-11}	$Mn(OH)_2$	2.0×10^{-13}
$BaSO_4$	1.5×10^{-9}		

ONLINE LEARNING MODULE
Writing Solubility Constant Expressions
➡ WileyPLUS

EXAMPLE 16.11

Write K_{sp} expressions for AgI and PbI_2, both of which are slightly soluble salts.

SOLUTION

First write the equilibrium equations:

$$AgI(s) \rightleftharpoons Ag^+(aq) + I^-(aq)$$
$$PbI_2(s) \rightleftharpoons Pb^{2+}(aq) + 2\,I^-(aq)$$

Since the concentration of the solid crystals is constant, the K_{sp} equals the product of the molar concentrations of the ions in solution. In the case of PbI_2, the $[I^-]$ must be squared:

$$K_{sp} = [Ag^+][I^-] \qquad K_{sp} = [Pb^{2+}][I^-]^2$$

ONLINE LEARNING MODULE
K_{sp} Calculations
➡ WileyPLUS

EXAMPLE 16.12

The K_{sp} value for lead sulfate is 1.3×10^{-8}. Calculate the solubility of $PbSO_4$ in grams per liter.

SOLUTION

First write the equilibrium equation and the K_{sp} expression:

$$PbSO_4(s) \rightleftharpoons Pb^{2+}(aq) + SO_4^{2-}(aq)$$
$$K_{sp} = [Pb^{2+}][SO_4^{2-}] = 1.3 \times 10^{-8}$$

Since the lead sulfate that is in solution is completely dissociated, the $[Pb^{2+}]$ or $[SO_4^{2-}]$ is equal to the solubility of $PbSO_4$ in moles per liter. Let

$$Y = [Pb^{2+}] = [SO_4^{2-}]$$

Substitute Y into the K_{sp} equation and solve:

$$[Pb^{2+}][SO_4^{2-}] = (Y)(Y) = 1.3 \times 10^{-8}$$
$$Y^2 = 1.3 \times 10^{-8}$$
$$Y = 1.1 \times 10^{-4}\ \text{mol/L}$$

The solubility of $PbSO_4$, therefore, is 1.1×10^{-4} mol/L.
Now convert mol/L to g/L:

1 mol of $PbSO_4$ has a mass of

$$(207.2\ g + 32.07\ g + 64.00\ g) = 303.3\ g$$

$$\left(\frac{1.1 \times 10^{-4}\ \cancel{\text{mol}}}{\text{L}}\right)\left(\frac{303.3\ \text{g}}{\cancel{\text{mol}}}\right) = 3.3 \times 10^{-2}\ \text{g/L}$$

The solubility of $PbSO_4$ is 3.3×10^{-2} g/L.

PRACTICE 16.13

Write the K_{sp} expression for

(a) $Cr(OH)_3$ (b) $Cu_3(PO_4)_2$

PRACTICE 16.14

The K_{sp} value for CuS is 8.5×10^{-45}. Calculate the solubility of CuS in grams per liter.

An ion added to a solution already containing that ion is called a *common ion*. When a common ion is added to an equilibrium solution of a weak electrolyte or a slightly soluble salt, the equilibrium shifts according to Le Châtelier's principle. For example, when silver nitrate ($AgNO_3$) is added to a saturated solution of AgCl,

$$AgCl(s) \rightleftharpoons Ag^+(aq) + Cl^-(aq)$$

the equilibrium shifts to the left due to the increase in the $[Ag^+]$. As a result, the $[Cl^-]$ and the solubility of AgCl decrease. The AgCl and $AgNO_3$ have the common ion Ag^+. A shift in the equilibrium position upon addition of an ion already contained in the solution is known as the **common ion effect**.

EXAMPLE 16.13

Silver nitrate is added to a saturated AgCl solution until the $[Ag^+]$ is 0.10 *M*. What will be the $[Cl^-]$ remaining in solution?

SOLUTION

This is an example of the common ion effect. The addition of $AgNO_3$ puts more Ag^+ in solution; the Ag^+ combines with Cl^- and causes the equilibrium to shift to the left, reducing the $[Cl^-]$ in solution. After the addition of Ag^+ to the mixture, the $[Ag^+]$ and $[Cl^-]$ in solution are no longer equal.

We use the K_{sp} to calculate the $[Cl^-]$ remaining in solution. The K_{sp} is constant at a particular temperature and remains the same no matter how we change the concentration of the species involved:

$$K_{sp} = [Ag^+][Cl^-] = 1.7 \times 10^{-10} \qquad [Ag^+] = 0.10 \text{ mol/L}$$

We then substitute the $[Ag^+]$ into the K_{sp} expression and calculate the $[Cl^-]$:

$$[0.10][Cl^-] = 1.7 \times 10^{-10}$$

$$[Cl^-] = \frac{1.7 \times 10^{-10}}{0.10} = 1.7 \times 10^{-9} \text{ mol/L}$$

This calculation shows a 10,000-fold reduction of Cl^- ions in solution. It illustrates that Cl^- ions may be quantitatively removed from solution with an excess of Ag^+ ions.

PRACTICE 16.15

Sodium sulfate (Na_2SO_4) is added to a saturated solution of $BaSO_4$ until the concentration of the sulfate ion is 2.0×10^{-7} *M*. What will be the concentration of the Ba^{2+} ions remaining in solution?

16.8 Buffer Solutions: The Control of pH

LEARNING OBJECTIVE

Describe the function of a buffer solution.

KEY TERM

buffer solution

The control of pH within narrow limits is critically important in many chemical applications and vitally important in many biological systems. For example, human blood must be maintained between pH 7.35 and 7.45 for the efficient transport of oxygen from the lungs to the cells. This narrow pH range is maintained by buffer systems in the blood.

A **buffer solution** resists changes in pH when diluted or when small amounts of acid or base are added. Two common types of buffer solutions are (1) a weak acid mixed with a salt of its conjugate base and (2) a weak base mixed with a salt of its conjugate acid.

The action of a buffer system can be understood by considering a solution of acetic acid and sodium acetate. The weak acid ($HC_2H_3O_2$) is mostly un-ionized and is in equilibrium with its ions in solution. The sodium acetate is completely ionized:

$$HC_2H_3O_2(aq) \rightleftharpoons H^+(aq) + C_2H_3O_2^-(aq)$$

$$NaC_2H_3O_2(aq) \longrightarrow Na^+(aq) + C_2H_3O_2^-(aq)$$

Mark E. Gibson/Corbis

The aquariums at the MGM Grand Hotel are a buffer system.

Because the sodium acetate is completely ionized, the solution contains a much higher concentration of acetate ions than would be present if only acetic acid were in solution. The acetate ion represses the ionization of acetic acid and also reacts with water, causing the solution to have a higher pH (be more basic) than an acetic acid solution (Section 16.3). Thus, a 0.1 *M* acetic acid solution has a pH of 2.87, but a solution that is 0.1 *M* in acetic acid and 0.1 *M* in sodium acetate has a pH of 4.74. This difference in pH is the result of the common ion effect.

A buffer solution has a built-in mechanism that counteracts the effect of adding acid or base. Consider the effect of adding HCl or NaOH to an acetic acid–sodium acetate buffer. When a small amount of HCl is added, the acetate ions of the buffer combine with the H^+ ions from HCl to form un-ionized acetic acid, thus neutralizing the added acid and maintaining the approximate pH of the solution. When NaOH is added, the OH^- ions react

CHEMISTRY *IN ACTION* Exchange of Oxygen and Carbon Dioxide in the Blood

The transport of oxygen and carbon dioxide between the lungs and tissues is a complex process that involves several reversible reactions, each of which behaves in accordance with Le Châtelier's principle.

The binding of oxygen to hemoglobin is a reversible reaction. The oxygen molecule must attach to the hemoglobin (Hb) and then later detach. The equilibrium equation for this reaction can be written:

$$Hb + O_2 \rightleftharpoons HbO_2$$

In the lungs the concentration of oxygen is high and favors the forward reaction. Oxygen quickly binds to the hemoglobin until it is saturated with oxygen.

In the tissues the concentration of oxygen is lower and in accordance with Le Châtelier's principle: The equilibrium position shifts to the left, and the hemoglobin releases oxygen to the tissues. Approximately 45% of the oxygen diffuses out of the capillaries into the tissues, where it may be picked up by *myoglobin*, another carrier molecule.

Myoglobin functions as an oxygen-storage molecule, holding the oxygen until it is required in the energy-producing portions of the cell. The reaction between myoglobin (Mb) and oxygen can be written as an equilibrium reaction:

$$Mb + O_2 \rightleftharpoons MbO_2$$

The hemoglobin and myoglobin equations are very similar, so what accounts for the transfer of the oxygen from the hemoglobin to the myoglobin? Although both equilibria involve similar interactions, the affinity between oxygen and hemoglobin is different from the affinity between myoglobin and oxygen. In the tissues the position of the hemoglobin equilibrium is such that it is 55% saturated with oxygen, whereas the myoglobin is at 90% oxygen saturation. Under these conditions hemoglobin will release oxygen, while myoglobin will bind oxygen. Thus oxygen is loaded onto hemoglobin in the lungs and unloaded in the tissue's cells.

Carbon dioxide produced in the cells must be removed from the tissues. Oxygen-depleted hemoglobin molecules accomplish this by becoming carriers of carbon dioxide. The carbon dioxide does not bind at the heme site as the oxygen does, but rather at one end of the protein chain. When carbon dioxide dissolves in water, some of the CO_2 reacts to release hydrogen ions:

$$CO_2 + H_2O \rightleftharpoons HCO_3^- + H^+$$

To facilitate the removal of CO_2 from the tissues, this equilibrium needs to be moved toward the right. This shift is accomplished by the removal of H^+ from the tissues by the hemoglobin molecule. The deoxygenated hemoglobin molecule can bind H^+ ions as well as CO_2. In the lungs this whole process is reversed so the CO_2 is removed from the hemoglobin and exhaled.

Molecules that are similar in structure to the oxygen molecule can become involved in competing equilibria. Hemoglobin is capable of binding with carbon monoxide (CO), nitrogen monoxide (NO), and cyanide (CN^-). The extent of the competition depends on the affinity. Since these molecules have a greater affinity for hemoglobin than oxygen, they will effectively displace oxygen from hemoglobin. For example,

$$HbO_2 + CO \rightleftharpoons HbCO + O_2$$

Since the affinity of hemoglobin for CO is 150 times stronger than its affinity for oxygen, the equilibrium position lies far to the right. This explains why CO is a poisonous substance and why oxygen is administered to victims of CO poisoning. The hemoglobin molecules can only transport oxygen if the CO is released and the oxygen shifts the equilibrium toward the left.

Charles D. Winters/Photo Researchers

Oxygen and carbon dioxide are exchanged in the red blood cells when they are in capillaries.

with acetic acid to neutralize the added base and thus maintain the approximate pH. The equations for these reactions are

$$H^+(aq) + C_2H_3O_2^-(aq) \rightleftharpoons HC_2H_3O_2(aq)$$

$$OH^-(aq) + HC_2H_3O_2(aq) \rightleftharpoons H_2O(l) + C_2H_3O_2^-(aq)$$

Data comparing the changes in pH caused by adding HCl and NaOH to pure water and to an acetic acid–sodium acetate buffer solution are shown in **TABLE 16.4**. Notice that adding 0.01 mol HCl or 0.01 mol NaOH to 100 mL water changes the pH of the solution by 5 units. But adding the same amount of HCl or NaOH to a buffer solution causes a change of only 0.08 or 0.09 pH units. So buffers really protect the pH of a solution.

TABLE 16.4 Changes in pH Caused by the Addition of HCl and NaOH

Solution	pH	Change in pH
H_2O (1000 mL)	7	—
H_2O + 0.010 mol HCl	2	5
H_2O + 0.010 mol NaOH	12	5
Buffer solution (1000 mL)		
0.10 M $HC_2H_3O_2$ + 0.10 M $NaC_2H_3O_2$	4.74	—
Buffer + 0.010 mol HCl	4.66	0.08
Buffer + 0.010 mol NaOH	4.83	0.09

Buffers cannot withstand the addition of large amounts of acids or bases. They overpower the capacity of the buffer to absorb acid or base. The maximum buffering effect occurs when the weak acid or base and its conjugate are of equal molar concentrations.

The human body has a number of buffer systems. One of these, the hydrogen carbonate–carbonic acid buffer, HCO_3^-—H_2CO_3, maintains the blood plasma at a pH of 7.4. The phosphate system, HPO_4^{2-}—$H_2PO_4^-$, is an important buffer in the red blood cells, as well as in other places in the body.

EXAMPLE 16.14

What is the purpose of a buffer solution?

SOLUTION

A buffer solution resists change in pH when small amounts of acid or base are added to it.

PRACTICE 16.16

Describe how a buffer solution made from HF and NaF behaves when a small amount of a strong acid is added to it: $HF \rightleftharpoons H^+ + F^-$.

ONLINE LEARNING MODULE

Buffers

➡ WileyPLUS

CHAPTER 16 REVIEW

16.1 Rates of Reaction

KEY TERM

chemical kinetics

- The rate of a reaction is variable and depends on:
  - Concentration of reacting species
  - Temperature
  - Presence of catalysts
  - Nature of the reactants

16.2 Chemical Equilibrium

KEY TERMS

reversible chemical reaction
equilibrium
chemical equilibrium

- In a reversible chemical reaction, the products formed react to produce the original reaction mixture.
- The forward reaction is called the reaction to the right.
- The reverse reaction is called the reaction to the left.
- A system at equilibrium is dynamic.
- At chemical equilibrium, the rate of the forward reaction equals the rate of the reverse reaction.

16.3 Le Châtelier's Principle

KEY TERMS

Le Châtelier's principle
catalyst
activation energy

- If stress is applied to a system at equilibrium, the system will respond in such a way as to relieve the stress and restore equilibrium under a new set of conditions.
- When the concentration of a reactant is increased, the equilibrium shifts toward the right.
- When the concentration of a product on the right side of the equation is increased, the equilibrium shifts to the left.
- If the concentration of a reactant is decreased, the equilibrium shifts toward the side where the reactant is decreased.
- In gaseous reactions:
  - A decrease in volume causes the equilibrium position to shift toward the side of the reaction with the fewest molecules.
  - If the number of molecules is the same on both sides of a reaction, a decrease in volume has no effect on the equilibrium position.
- When heat is added to a reaction, the side of the equation that absorbs heat is favored:
  - If the reaction is endothermic, the forward reaction increases.
  - If the reaction is exothermic, the reverse reaction increases.
- A catalyst does not shift the equilibrium of a reaction; it only affects the speed at which equilibrium is reached.
- The activation energy for a reaction is the minimum energy required for the reaction to occur:
  - A catalyst lowers the activation energy for a reaction by providing a different pathway for the reaction.

16.4 Equilibrium Constants

KEY TERM

equilibrium constant, K_{eq}

- For the general reaction $a\text{A} + b\text{B} \rightleftharpoons c\text{C} + d\text{D}$

$$K_{eq} = \frac{[\text{C}]^c[\text{D}]^d}{[\text{A}]^a[\text{B}]^b}$$

- The magnitude of K_{eq} indicates the extent of the reaction:
  - If $K_{eq} > 1$: Amount of products is greater than reactants at equilibrium.
  - If $K_{eq} < 1$: Amount of reactants is less than products at equilibrium.

16.5 Ion Product Constant for Water

KEY TERM

ion product constant for water, K_w

- $K_w = [\text{H}^+][\text{OH}^-] = 1 \times 10^{-14}$
- $[\text{H}^+] = \dfrac{K_w}{[\text{OH}^-]}$
- $[\text{OH}^-] = \dfrac{K_w}{[\text{H}^+]}$

16.6 Ionization Constants

KEY TERM

acid ionization constant, K_a

- The ionization constant can be used in the equilibrium expression to calculate the concentration of a reactant, the percent ionization of a substance, or the pH of a weak acid.

16.7 Solubility Product Constant

KEY TERMS

solubility product constant, K_{sp}
common ion effect

- The solubility product constant is used to calculate the solubility of a slightly soluble salt in water.
- The solubility product constant can also be used to determine whether or not precipitation will occur in a solution:
  - If the product of the molar concentrations of the ions in solution is greater than the K_{sp}, precipitation will occur.
- A shift in the equilibrium position upon addition of an ion already contained in the solution is known as the common ion effect.

16.8 Buffer Solutions: The Control of pH

KEY TERM

buffer solution

- A buffer solution could contain:
  - A weak acid and a salt of its conjugate base
  - A weak base and a salt of its conjugate acid
- A buffer solution resists a change in pH by neutralizing small amounts of acid or base added to it.

REVIEW QUESTIONS

1. At equilibrium how do the forward and reverse reaction rates compare? (Figure 16.1)

2. Why does the rate of a reaction usually increase when the concentration of one of the reactants is increased?

3. Why does an increase in temperature cause the rate of reaction to increase?

4. How would you expect the two tubes in Figure 16.2 to appear if both are at 25°C?

5. Is the reaction $N_2O_4 \rightleftharpoons 2\ NO_2$ exothermic or endothermic? (Figure 16.2)

6. If pure hydrogen iodide, HI, is placed in a vessel at 700 K, will it decompose? Explain.

7. Of the acids listed in Table 16.2, which ones are stronger than acetic acid and which are weaker?

8. Why is heat treated as a reactant in an endothermic process and as a product in an exothermic process?

9. What does a catalyst do?

10. Is it important to specify the temperature when reporting an equilibrium constant? Why or why not?

11. If the value of an equilibrium constant is very large, will the equilibrium lie far to the right or far to the left?

12. If the value of an equilibrium constant is very small, will the equilibrium lie far to the right or far to the left?

13. Why don't free protons (H^+) exist in water?

14. For each solution in Table 16.1, what is the sum of the pH plus the pOH? What would be the pOH of a solution whose pH is -1?

15. What would cause two separate samples of pure water to have slightly different pH values?

16. Why are the pH and pOH equal in pure water?

17. With dilution, aqueous solutions of acetic acid ($HC_2H_3O_2$) show increased ionization. For example, a 1.0 *M* solution of acetic acid is 0.42% ionized, whereas a 0.10 *M* solution is 1.34% ionized. Explain the behavior using the ionization equation and equilibrium principles.

18. A 1.0 *M* solution of acetic acid ionizes less and has a higher concentration of H^+ ions than a 0.10 *M* acetic acid solution. Explain this behavior. (See Question 17 for data.)

19. Explain why silver acetate is more soluble in nitric acid than in water. (*Hint*: Write the equilibrium equation first and then consider the effect of the acid on the acetate ion.) What would happen if hydrochloric acid were used in place of nitric acid?

20. Dissolution of sodium acetate ($NaC_2H_3O_2$) in pure water gives a basic solution. Why?
(*Hint*: A small amount of $HC_2H_3O_2$ is formed.)

21. Tabulate the relative order of molar solubilities of AgCl, AgBr, AgI, $AgC_2H_3O_2$, $PbSO_4$, $BaSO_4$, $BaCrO_4$, and PbS. List the most soluble first. (Table 16.3)

22. Which compound in the following pairs has the greater molar solubility? (Table 16.3)
(a) $Mn(OH)_2$ or Ag_2CrO_4
(b) $BaCrO_4$ or Ag_2CrO_4

23. Explain why a precipitate of NaCl forms when HCl gas is passed into a saturated aqueous solution of NaCl.

24. Describe the similarities and differences between K_a, K_b, K_w, and K_{sp}.

25. Explain how the acetic acid–sodium acetate buffer system maintains its pH when 0.010 mol of HCl is added to 1 L of the buffer solution. (Table 16.4)

26. Describe why the pH of a buffer solution remains almost constant when a small amount of acid or base is added to it.

27. Describe how equilibrium is reached when the substances A and B are first mixed and react as

$$A + B \rightleftharpoons C + D$$

PAIRED EXERCISES

Most of the exercises in this chapter are available for assignment via the online homework management program, WileyPLUS (www.wileyplus.com). All exercises with blue numbers have answers in Appendix 5.

1. Write equations for the following reversible systems:
(a) solid $KMnO_4$ in a saturated aqueous solution of $KMnO_4$
(b) a mixture of solid and gaseous forms of CO_2 in a closed container

2. Write equations for the following reversible systems:
(a) a closed container of solid and gaseous forms of I_2
(b) solid $NaNO_3$ in a saturated aqueous solution of $NaNO_3$

3. Consider the following system at equilibrium:

$$SiF_4(g) + 2\,H_2O(g) + 103.8\text{ kJ} \rightleftharpoons SiO_2(g) + 4\,HF(g)$$

(a) Is the reaction endothermic or exothermic?
(b) If HF is added, in which direction will the reaction shift in order to reestablish equilibrium? After the new equilibrium has been established, will the final molar concentrations of SiF_4, H_2O, SiO_2, and HF increase, decrease, or remain the same?
(c) If heat is added, in which direction will the equilibrium shift?

4. Consider the following system at equilibrium:

$$4\,HCl(g) + O_2(g) \rightleftharpoons 2\,H_2O(g) + 2\,Cl_2(g) + 114.4\text{ kJ}$$

(a) Is the reaction endothermic or exothermic?
(b) If O_2 is added, in which direction will the reaction shift in order to reestablish equilibrium? After the new equilibrium has been established, will the final molar concentrations of HCl, O_2, H_2O, and Cl_2 increase or decrease?
(c) If heat is added, in which direction will the equilibrium shift?

5. Consider the following system at equilibrium:

$$N_2(g) + 3\,H_2(g) \rightleftharpoons 2\,NH_3(g) + 92.5\text{ kJ}$$

Complete the table that follows. Indicate changes in moles by entering I, D, N, or ? in the table. (I = increase, D = decrease, N = no change, ? = insufficient information to determine.)

Change of stress imposed on the system at equilibrium	Direction of reaction, left or right, to reestablish equilibrium	Change in number of moles		
		N_2	H_2	NH_3
(a) Add N_2				
(b) Remove H_2				
(c) Decrease volume of reaction vessel				
(d) Increase temperature				

6. Consider the following system at equilibrium:

$$N_2(g) + 3\,H_2(g) \rightleftharpoons 2\,NH_3(g) + 92.5\text{ kJ}$$

Complete the table that follows. Indicate changes in moles by entering I, D, N, or ? in the table. (I = increase, D = decrease, N = no change, ? = insufficient information to determine.)

Change of stress imposed on the system at equilibrium	Direction of reaction, left or right, to reestablish equilibrium	Change in number of moles		
		N_2	H_2	NH_3
(a) Add NH_3				
(b) Increase volume of reaction vessel				
(c) Add catalyst				
(d) Add both H_2 and NH_3				

7. For the following equations, tell in which direction, left or right, the equilibrium will shift when these changes are made: The temperature is increased, the pressure is increased by decreasing the volume of the reaction vessel, and a catalyst is added.
(a) $3\,O_2(g) + 271\text{ kJ} \rightleftharpoons 2\,O_3(g)$
(b) $CH_4(g) + Cl_2(g) \rightleftharpoons CH_3Cl(g) + HCl(g) + 110\text{ kJ}$
(c) $2\,NO(g) + 2\,H_2(g) \rightleftharpoons N_2(g) + 2\,H_2O(g) + 665\text{ kJ}$

8. For the following equations, tell in which direction, left or right, the equilibrium will shift when these changes are made: The temperature is increased, the pressure is increased by decreasing the volume of the reaction vessel, and a catalyst is added.
(a) $2\,SO_3(g) + 197\text{ kJ} \rightleftharpoons 2\,SO_2(g) + O_2(g)$
(b) $4\,NH_3(g) + 3\,O_2(g) \rightleftharpoons 2\,N_2(g) + 6\,H_2O(g) + 1531\text{ kJ}$
(c) $OF_2(g) + H_2O(g) \rightleftharpoons O_2(g) + 2\,HF(g) + 318\text{ kJ}$

9. Applying Le Châtelier's principle, in which direction will the equilibrium shift (if at all)?

$$CH_4(g) + 2\,O_2(g) \rightleftharpoons CO_2(g) + 2\,H_2O(g) + 802.3\text{ kJ}$$

(a) if the temperature is increased
(b) if a catalyst is added
(c) if CH_4 is added
(d) if the volume of the reaction vessel is decreased

10. Applying Le Châtelier's principle, in which direction will the equilibrium shift (if at all)?

$$2\,CO_2(g) + N_2(g) + 1095.9\text{ kJ} \rightleftharpoons C_2N_2(g) + 2\,O_2(g)$$

(a) if the volume of the reaction vessel is increased
(b) if O_2 is added
(c) if the temperature is increased
(d) if the concentration of N_2 is increased

11. Write the equilibrium constant expression for these reactions:
(a) $2\,NO_2(g) + 7\,H_2(g) \rightleftharpoons 2\,NH_3(g) + 4\,H_2O(g)$
(b) $H_2CO_3(aq) \rightleftharpoons H^+(aq) + HCO_3^-(aq)$
(c) $2\,COF_2(g) \rightleftharpoons CO_2(g) + CF_4(g)$

12. Write the equilibrium constant expression for these reactions:
(a) $H_3PO_4(aq) \rightleftharpoons H^+(aq) + H_2PO_4^-(aq)$
(b) $CS_2(g) + 4\,H_2(g) \rightleftharpoons CH_4(g) + 2\,H_2S(g)$
(c) $4\,NO_2(g) + O_2(g) \rightleftharpoons 2\,N_2O_5(g)$

13. Write the solubility product expression, K_{sp}, for these substances:
(a) AgCl
(b) $PbCrO_4$
(c) $Zn(OH)_2$
(d) $Ca_3(PO_4)_2$

14. Write the solubility product expression, K_{sp}, for these substances:
(a) $MgCO_3$
(b) CaC_2O_4
(c) $Tl(OH)_3$
(d) $Pb_3(AsO_4)_2$

15. What effect will decreasing the $[H^+]$ of a solution have on (a) pH, (b) pOH, (c) $[OH^-]$, and (d) K_w?

16. What effect will increasing the $[H^+]$ of a solution have on (a) pH, (b) pOH, (c) $[OH^-]$, and (d) K_w?

17. One of the important pH-regulating systems in the blood consists of a carbonic acid–sodium hydrogen carbonate buffer:

$$H_2CO_3(aq) \rightleftharpoons H^+(aq) + HCO_3^-(aq)$$
$$NaHCO_3(aq) \longrightarrow Na^+(aq) + HCO_3^-(aq)$$

Explain how this buffer resists changes in pH when excess acid, H^+, gets into the bloodstream.

18. One of the important pH-regulating systems in the blood consists of a carbonic acid–sodium hydrogen carbonate buffer:

$$H_2CO_3(aq) \rightleftharpoons H^+(aq) + HCO_3^-(aq)$$
$$NaHCO_3(aq) \longrightarrow Na^+(aq) + HCO_3^-(aq)$$

Explain how this buffer resists changes in pH when excess base, OH^-, gets into the bloodstream.

19. For a solution of 1.2 *M* H_2CO_3 ($K_a = 4.4 \times 10^{-7}$), calculate:
(a) $[H^+]$
(b) pH
(c) percent ionization

20. For a solution of 0.025 *M* lactic acid, $HC_3H_5O_2$ ($K_a = 8.4 \times 10^{-4}$), calculate:
(a) $[H^+]$
(b) pH
(c) percent ionization

21. A 0.025 *M* solution of a weak acid, HA, is 0.45% ionized. Calculate the ionization constant, K_a, for the acid.

22. A 0.500 *M* solution of a weak acid, HA, is 0.68% ionized. Calculate the ionization constant, K_a, for the acid.

23. Calculate the percent ionization and the pH of each of the following solutions of phenol, HC_6H_5O ($K_a = 1.3 \times 10^{-10}$):
(a) 1.0 *M*
(b) 0.10 *M*
(c) 0.010 *M*

24. Calculate the percent ionization and the pH of each of the following solutions of benzoic acid, $HC_7H_5O_2$ ($K_a = 6.3 \times 10^{-5}$):
(a) 1.0 *M*
(b) 0.10 *M*
(c) 0.010 *M*

25. A 0.37 *M* solution of a weak acid (HA) has a pH of 3.7. What is the K_a for this acid?

26. A 0.23 *M* solution of a weak acid (HA) has a pH of 2.89. What is the K_a for this acid?

27. A student needs a sample of 1.0 *M* NaOH for a laboratory experiment. Calculate the $[H^+]$, $[OH^-]$, pH, and pOH of this solution.

28. A laboratory cabinet contains a stock solution of 3.0 *M* HNO_3. Calculate the $[H^+]$, $[OH^-]$, pH, and pOH of this solution.

29. Calculate the pH and the pOH of these solutions:
(a) 0.250 *M* HBr
(b) 0.333 *M* KOH
(c) 0.895 *M* $HC_2H_3O_2$ ($K_a = 1.8 \times 10^{-5}$)

30. Calculate the pH and the pOH of these solutions:
(a) 0.0010 *M* NaOH
(b) 0.125 *M* HCl
(c) 0.0250 *M* HC_6H_5O ($K_a = 1.3 \times 10^{-10}$)

31. Calculate the $[OH]^-$ in each of the following solutions:
(a) $[H^+] = 1.0 \times 10^{-2}$
(b) $[H^+] = 3.2 \times 10^{-7}$
(c) 1.25 *M* KOH
(d) 0.75 *M* $HC_2H_3O_2$

32. Calculate the $[OH^-]$ in each of the following solutions:
(a) $[H^+] = 4.0 \times 10^{-9}$
(b) $[H^+] = 1.2 \times 10^{-5}$
(c) 1.25 *M* HCN
(d) 0.333 *M* NaOH

33. Calculate the $[H^+]$ in each of the following solutions:
(a) $[OH^-] = 1.0 \times 10^{-8}$
(b) $[OH^-] = 2.0 \times 10^{-4}$

34. Calculate the $[H^+]$ in each of the following solutions:
(a) $[OH^-] = 4.5 \times 10^{-2}$
(b) $[OH^-] = 5.2 \times 10^{-9}$

35. Given the following solubility data, calculate the solubility product constant for each substance:
(a) $BaSO_4$, 3.9×10^{-5} mol/L
(b) Ag_2CrO_4, 7.8×10^{-5} mol/L
(c) $CaSO_4$, 0.67 g/L
(d) AgCl, 0.0019 g/L

36. Given the following solubility data, calculate the solubility product constant for each substance:
(a) ZnS, 3.5×10^{-12} mol/L
(b) $Pb(IO_3)_2$, 4.0×10^{-5} mol/L
(c) Ag_3PO_4, 6.73×10^{-3} g/L
(d) $Zn(OH)_2$, 2.33×10^{-4} g/L

37. Calculate the molar solubility for these substances:
(a) CaF_2, $K_{sp} = 3.9 \times 10^{-11}$
(b) $Fe(OH)_3$, $K_{sp} = 6.1 \times 10^{-38}$

38. Calculate the molar solubility for these substances:
(a) $PbSO_4$, $K_{sp} = 1.3 \times 10^{-8}$
(b) $BaCrO_4$, $K_{sp} = 8.5 \times 10^{-11}$

39. For each substance in Question 37, calculate the solubility in grams per 100. mL of solution.

40. For each substance in Question 38, calculate the solubility in grams per 100. mL of solution.

41. Solutions containing 100. mL of 0.010 *M* Na_2SO_4 and 100. mL of 0.001 *M* $Pb(NO_3)_2$ are mixed. Show by calculation whether or not a precipitate will form. Assume that the volumes are additive. (K_{sp} for $PbSO_4 = 1.3 \times 10^{-8}$)

42. Solutions containing 50.0 mL of 1.0×10^{-4} *M* $AgNO_3$ and 100. mL of 1.0×10^{-4} *M* NaCl are mixed. Show by calculation whether or not a precipitate will form. Assume the volumes are additive. (K_{sp} for AgCl $= 1.7 \times 10^{-10}$)

43. How many moles of AgBr will dissolve in 1.0 L of 0.10 *M* NaBr? ($K_{sp} = 5.2 \times 10^{-13}$ for AgBr)

44. How many moles of AgBr will dissolve in 1.0 L of 0.10 *M* $MgBr_2$? ($K_{sp} = 5.2 \times 10^{-13}$ for AgBr)

45. Calculate the $[H^+]$ and pH of a buffer solution that is 0.20 *M* in $HC_2H_3O_2$ and contains sufficient sodium acetate to make the $[C_2H_3O_2^-]$ equal to 0.10 *M*. (K_a for $HC_2H_3O_2 = 1.8 \times 10^{-5}$)

46. Calculate the $[H^+]$ and pH of a buffer solution that is 0.20 *M* in $HC_2H_3O_2$ and contains sufficient sodium acetate to make the $[C_2H_3O_2^-]$ equal to 0.20 *M*. (K_a for $HC_2H_3O_2 = 1.8 \times 10^{-5}$)

47. When 1.0 mL of 1.0 *M* HCl is added to 50. mL of 1.0 *M* NaCl, the $[H^+]$ changes from 1×10^{-7} *M* to 2.0×10^{-2} *M*. Calculate the initial pH and the pH change in the solution.

48. When 1.0 mL of 1.0 *M* HCl is added to 50. mL of a buffer solution that is 1.0 *M* in $HC_2H_3O_2$ and 1.0 *M* in $NaC_2H_3O_2$, the $[H^+]$ changes from 1.8×10^{-5} *M* to 1.9×10^{-5} *M*. Calculate the initial pH and the pH change in the solution.

ADDITIONAL EXERCISES

49. In a K_{sp} expression, the concentration of solid salt is not included. Why?

50. The energy diagram in Figure 16.3 is for an exothermic reaction. How can you tell?

51. Sketch the energy diagram for an endothermic reaction using a solid line. Sketch what would happen to the energy profile if a catalyst is added to the reaction using a dotted line.

52. What is the maximum number of moles of HI that can be obtained from a reaction mixture containing 2.30 mol I_2 and 2.10 mol H_2?

53. (a) How many moles of HI are produced when 2.00 mol H_2 and 2.00 mol I_2 are reacted at 700 K? (Reaction is 79% complete.)
(b) Addition of 0.27 mol I_2 to the system increases the yield of HI to 85%. How many moles of H_2, I_2, and HI are now present?
(c) From the data in part (a), calculate K_{eq} for the reaction at 700 K.

54. After equilibrium is reached in the reaction of 6.00 g H_2 with 200. g I_2 at 500. K, analysis shows that the flask contains 64.0 g of HI. How many moles of H_2, I_2, and HI are present in this equilibrium mixture?

55. What is the equilibrium constant of the reaction

$$PCl_3(g) + Cl_2(g) \rightleftharpoons PCl_5(g)$$

if a 20.-L flask contains 0.10 mol PCl_3, 1.50 mol Cl_2, and 0.22 mol PCl_5?

56. If the rate of a reaction doubles for every 10°C rise in temperature, how much faster will the reaction go at 100°C than at 30°C?

57. Would you believe that an NH_4Cl solution is acidic? Calculate the pH of a 0.30 *M* NH_4Cl solution.

$$(NH_4^+(aq) \rightleftharpoons NH_3(g) + H^+(aq),\ K_{eq} = 5.6 \times 10^{-10})$$

58. Calculate the pH of an acetic acid buffer composed of $HC_2H_3O_2$ (0.30 *M*) and $C_2H_3O_2^-$ (0.20 *M*).

59. Solutions of 50.0 mL of 0.10 *M* $BaCl_2$ and 50.0 mL of 0.15 *M* Na_2CrO_4 are mixed, forming a precipitate of $BaCrO_4$. Calculate the concentration of Ba^{2+} that remains in solution.

60. Calculate the ionization constant for the given acids. Each acid ionizes as follows: $HA \rightleftharpoons H^+ + A^-$.

Acid	Acid concentration	$[H^+]$
Hypochlorous, HOCl	0.10 *M*	5.9×10^{-5} mol/L
Propanoic, $HC_3H_5O_2$	0.15 *M*	1.4×10^{-3} mol/L
Hydrocyanic, HCN	0.20 *M*	8.9×10^{-6} mol/L

61. The K_{sp} of CaF_2 is 3.9×10^{-11}. Calculate (a) the molar concentrations of Ca^{2+} and F^- in a saturated solution and (b) the grams of CaF_2 that will dissolve in 500. mL of water.

62. Calculate whether or not a precipitate will form in the following mixed solutions:
(a) 100 mL of 0.010 *M* Na_2SO_4 and 100 mL of 0.001 *M* $Pb(NO_3)_2$
(b) 50.0 mL of 1.0×10^{-4} *M* $AgNO_3$ and 100. mL of 1.0×10^{-4} *M* NaCl
(c) 1.0 g $Ca(NO_3)_2$ in 150 mL H_2O and 250 mL of 0.01 *M* NaOH

$$K_{sp}PbSO_4 = 1.3 \times 10^{-8}$$
$$K_{sp}AgCl = 1.7 \times 10^{-10}$$
$$K_{sp}Ca(OH)_2 = 1.3 \times 10^{-6}$$

63. If $BaCl_2$ is added to a saturated $BaSO_4$ solution until the $[Ba^{2+}]$ is 0.050 *M*,
(a) what concentration of SO_4^{2-} remains in solution?
(b) how much $BaSO_4$ remains dissolved in 100. mL of the solution? ($K_{sp} = 1.5 \times 10^{-9}$ for $BaSO_4$)

64. The K_{sp} for $PbCl_2$ is 2.0×10^{-5}. Will a precipitate form when 0.050 mol $Pb(NO_3)_2$ and 0.010 mol NaCl are in 1.0 L solution? Show evidence for your answer.

65. Suppose the concentration of a solution is 0.10 *M* Ba^{2+} and 0.10 *M* Sr^{2+}. Which sulfate, $BaSO_4$ or $SrSO_4$, will precipitate first when a dilute solution of H_2SO_4 is added dropwise to the solution? Show evidence for your answer. ($K_{sp} = 1.5 \times 10^{-9}$ for $BaSO_4$ and $K_{sp} = 3.5 \times 10^{-7}$ for $SrSO_4$)

66. Calculate the K_{eq} for the reaction

$$SO_2(g) + O_2(g) \rightleftharpoons SO_3(g)$$

when the equilibrium concentrations of the gases at 530°C are $[SO_3] = 11.0$ *M*, $[SO_2] = 4.20$ *M*, and $[O_2] = 0.60 \times 10^{-3}$ *M*.

67. If it takes 0.048 g BaF_2 to saturate 15.0 mL of water, what is the K_{sp} of BaF_2?

68. The K_{eq} for the formation of ammonia gas from its elements is 4.0. If the equilibrium concentrations of nitrogen gas and hydrogen gas are both 2.0 *M*, what is the equilibrium concentration of the ammonia gas?

69. The K_{sp} of $SrSO_4$ is 7.6×10^{-7}. Should precipitation occur when 25.0 mL of 1.0×10^{-3} *M* $SrCl_2$ solution are mixed with 15.0 mL of 2.0×10^{-3} *M* Na_2SO_4? Show proof.

70. The solubility of Hg_2I_2 in H_2O is 3.04×10^{-7} g/L. The reaction $Hg_2I_2(s) \rightleftharpoons Hg_2^{2+}(aq) + 2I^-(aq)$ represents the equilibrium. Calculate the K_{sp}.

71. Under certain circumstances, when oxygen gas is heated, it can be converted into ozone according to the following reaction equation:

$$3\,O_2(g) + \text{heat} \rightleftharpoons 2\,O_3(g)$$

Name three different ways that you could increase the production of the ozone.

72. One day in a laboratory, some water spilled on a table. In just a few minutes the water had evaporated. Some days later, a similar amount of water spilled again. This time, the water remained on the table after 7 or 8 hours. Name three conditions that could have changed in the lab to cause this difference.

73. For the reaction $CO(g) + H_2O(g) \rightleftharpoons CO_2(g) + H_2(g)$ at a certain temperature, K_{eq} is 1. At equilibrium which of the following would you expect to find?
(a) only CO and H_2
(b) mostly CO_2 and H_2
(c) about equal concentrations of CO and H_2O, compared to CO_2 and H_2
(d) mostly CO and H_2O
(e) only CO and H_2O
Explain your answer briefly.

74. Write the equilibrium constant expressions for these reactions:
(a) $3\,O_2(g) \rightleftharpoons 2\,O_3(g)$
(b) $H_2O(g) \rightleftharpoons H_2O(l)$
(c) $MgCO_3(s) \rightleftharpoons MgO(s) + CO_2(g)$
(d) $2\,Bi^{3+}(aq) + 3\,H_2S(aq) \rightleftharpoons Bi_2S_3(s) + 6\,H^+(aq)$

75. Reactants A and B are mixed, each initially at a concentration of 1.0 *M*. They react to produce C according to this equation:

$$2\,A + B \rightleftharpoons C$$

When equilibrium is established, the concentration of C is found to be 0.30 *M*. Calculate the value of K_{eq}.

76. At a certain temperature, K_{eq} is 2.2×10^{-3} for the reaction

$$2\,ICl(g) \rightleftharpoons I_2(g) + Cl_2(g)$$

Now calculate the K_{eq} value for the reaction

$$I_2(g) + Cl_2(g) \rightleftharpoons 2\,ICl(g)$$

77. One drop of 1 *M* OH^- ion is added to a 1 *M* solution of HNO_2. What will be the effect of this addition on the equilibrium concentration of the following?

$$HNO_2(aq) \rightleftharpoons H^+(aq) + NO_2^-(aq)$$

(a) $[OH^-]$
(b) $[H^+]$
(c) $[NO_2^-]$
(d) $[HNO_2]$

78. The picture below represents the equilibrium condition of $2\,A_3X \rightleftharpoons 2\,A_2X + A_2$.

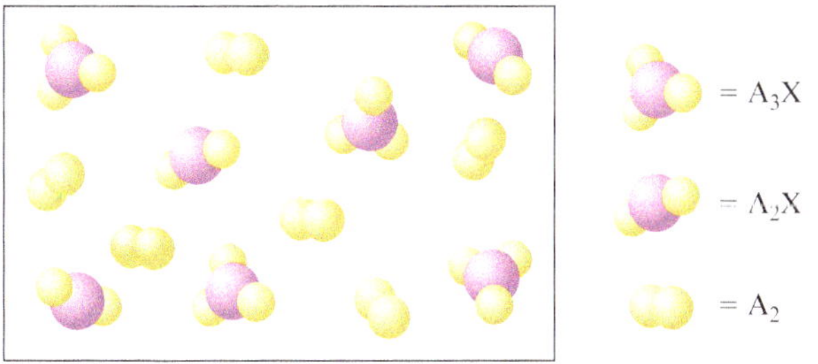

(a) What is the equilibrium constant?
(b) Does the equilibrium lie to the left or to the right?

79. The picture below represents the equilibrium condition of the reaction $X_2 + 2G \rightleftharpoons X_2G_2$.

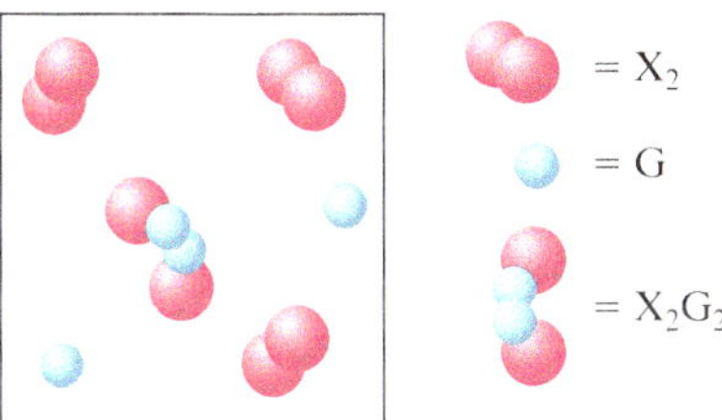

(a) What is the equilibrium constant?
(b) If the ratio of reactants to products increased when the temperature was raised, was the reaction exothermic or endothermic?
(c) Provide a logical explanation for why the equilibrium shifts to the right when the pressure is increased at constant temperature.

80. How many grams of $CaSO_4$ will dissolve in 600. mL of water? ($K_{sp} = 2.0 \times 10^{-4}$ for $CaSO_4$)

81. A student found that 0.098 g of PbF_2 was dissolved in 400. mL of saturated PbF_2. What is the K_{sp} for the lead(II) fluoride?

82. Humidity indicators are available that use the equilibrium reaction between the cobalt(II) hexahydrate ion (pink) and chloride ions, which forms the cobalt(II) tetrachloride ion (blue) and water. The equilibrium reaction used by these indicators is

$$Co(H_2O)_6^{2+} + 4\,Cl^- \rightleftharpoons [CoCl_4]^{2-} + 6\,H_2O$$

If you started with a pink indicator, what would you expect to happen if you
(a) blew a hair dryer on it?
(b) put it over a steam vent?
(c) left it in Houston before a rainstorm?
(d) left it in Phoenix in the middle of the summer?

83. Using Le Châtelier's principle, predict how the following reaction will respond to the specified changes.

$$HCHO_2(g) \rightleftharpoons CO(g) + H_2O(g)$$

(a) increase the concentration of $HCHO_2(g)$
(b) increase the concentration of $CO(g)$
(c) decrease the concentration of $H_2O(g)$
(d) decrease the volume of the reaction vessel

84. Aspirin is synthesized by reacting salicyclic acid with acetic anhydride, as shown in the following reaction:

$$\text{Salicylic acid } (C_7H_6O_3) + CH_3{-}C(=O){-}O{-}C(=O){-}CH_3 \; (C_4H_6O_3, \text{ Acetic anhydride}) \rightleftharpoons \text{Aspirin } (C_9H_8O_4) + CH_3{-}C(=O){-}OH \; (C_2H_4O_2, \text{ Acetic acid})$$

$C_7H_6O_3$ Salicylic acid $C_4H_6O_3$ Acetic anhydride $C_9H_8O_4$ Aspirin $C_2H_4O_2$ Acetic acid

The cost of salicylic acid is \$58.90 per 500.0 g and acetic anhydride costs \$53.50 for 1.00 kg. Given these data, which is the more cost-effective reagent to use in excess to shift the equilibrium far to the right and get the maximum possible yield?

85. Maintaining the proper ratio of carbonic acid and bicarbonate ion in the blood is important to keeping the pH of the blood in the correct range.
(a) Write the reaction of the bicarbonate ion (HCO_3^-) with an acid (H^+).
(b) Write the reaction of the bicarbonate ion with a base (OH^-).

CHALLENGE EXERCISES

86. All a snowbound skier had to eat were walnuts! He was carrying a bag holding 12 dozen nuts. With his mittened hands, he cracked open the shells. Each nut that was opened resulted in one kernel and two shell halves. When he tired of the cracking and got ready to do some eating, he discovered he had 194 total pieces (whole nuts, shell halves, and kernels). What is the K_{eq} for this reaction?

87. At 500°C, the reaction

$$SO_2(g) + NO_2(g) \rightleftharpoons NO(g) + SO_3(g)$$

has $K_{eq} = 81$. What will be the equilibrium concentrations of the four gases if the two reactants begin with equal concentrations of 0.50 *M*?

88. (a) Write a balanced acid–base reaction that produces Na_2S.
(b) If Na_2S is added to an aqueous solution of H_2S ($K_a = 9.1—10^{-8}$), will the pH of the solution rise or fall? Explain.

89. (a) Would you expect a reaction to take place between HCN (*aq*) and $AgNO_3$ (*aq*)

$$(K_{sp} \text{ for AgCN} = 5.97 \times 10^{-17})?$$

Explain, and if a reaction occurs, write the net ionic equation.
(b) If NaCN is added to distilled water, would you expect the solution to be acidic, basic, or neutral? Explain using any chemical equations that may be appropriate.

ANSWERS TO PRACTICE EXERCISES

16.1 The reactants must come into contact to produce the product(s). When the reactants are first mixed, the rate of the reaction is at its peak. As the reaction proceeds, the amount of the reactants available for reaction becomes smaller and the number of contacts between reactants becomes smaller so the rate of the reaction slows down.

16.2 A double arrow between reactants and products. $\rightleftharpoons$

16.3 (a)

Concentration	Change
$[H_2]$	increase
$[I_2]$	decrease
[HI]	increase

(b)

Concentration	Change
$[H_2]$	increase
$[I_2]$	increase
[HI]	increase

16.4 (a) equilibrium shifts to the left;
(b) equilibrium shifts to the right

16.5 (a) equilibrium shifts to the right;
(b) equilibrium shifts to the left

16.6 (a) equilibrium shifts to the left;
(b) equilibrium shifts to the right

16.7 (a) $K_{eq} = \dfrac{[NO_2]^4[O_2]}{[N_2O_5]^2}$; (b) $K_{eq} = \dfrac{[N_2]^2[H_2O]^6}{[NH_3]^4[O_2]^3}$

16.8 $K_{eq} = 33$; the forward reaction is favored

16.9 (a) $[H^+] = 5.0 \times 10^{-5}$ $[OH^-] = 2.0 \times 10^{-10}$
(b) $[H^+] = 5.0 \times 10^{-9}$ $[OH^-] = 2.0 \times 10^{-6}$

16.10 (a) 4.30 (b) 8.30

16.11 (a) 6.3×10^{-6} (b) 1.8×10^{-6}

16.12 (a) 6.3×10^{-3} % ionized
(b) 7.2×10^{-3} % ionized

16.13 (a) $K_{sp} = [Cr^{3+}][OH^-]^3$
(b) $K_{sp} = [Cu^{2+}]^3[PO_4^{3-}]^2$

16.14 8.8×10^{-21} g/L

16.15 7.5×10^{-8} mol/L

16.16 Addition of an acid (H^+) increases the acidity of the buffer solution. The increased acid will combine with F^- in the solution to form un-ionized HF, neutralizing the H^+ added. Thus, the pH of the buffer solution will remain about the same as in the original solution.

17

Oxidation–Reduction

Igor Karasi/Shutterstock

CHAPTER OUTLINE

The variety of oxidation–reduction reactions that affect us every day is amazing. Our society runs on batteries—in our calculators, laptop computers, cars, toys, radios, televisions, and more. The electric car in the photo is powered by an electric motor and lithium-ion batteries. The batteries are recharged by plugging the car into a household circuit. Oxidation–reduction reactions provide the energy in the batteries to drive the electric motor. We paint iron railings and galvanize nails to combat corrosion. We electroplate jewelry and computer chips with very thin coatings of gold or silver. We bleach our clothes using chemical reactions that involve electron transfer. We test for glucose in urine or alcohol in the breath with reactions that show vivid color changes. Plants turn energy into chemical compounds through a series of reactions called photosynthesis. These reactions all involve the transfer of electrons between substances in a chemical process called *oxidation–reduction*.

17.1 Oxidation Number

LEARNING OBJECTIVE Assign oxidation numbers to the atoms in a compound.

KEY TERMS
- oxidation number
- oxidation state
- oxidation–reduction
- redox
- oxidation
- reduction
- oxidizing agent
- reducing agent

The oxidation number of an atom (sometimes called its *oxidation state*) represents the number of electrons lost, gained, or unequally shared by an atom. Oxidation numbers can be zero, positive, or negative. An oxidation number of zero means the atom has the same number of electrons assigned to it as there are in the free neutral atom. A positive oxidation number means the atom has fewer electrons assigned to it than in the neutral atom, and a negative oxidation number means the atom has more electrons assigned to it than in the neutral atom.

The oxidation number of an atom that has lost or gained electrons to form an ion is the same as the positive or negative charge of the ion. (TABLE 17.1.) In the ionic compound NaCl, the oxidation numbers are clearly +1 for the Na^+ ion and −1 for the Cl^- ion. The Na^+ ion has one less electron than the neutral Na atom, and the Cl^- ion has one more electron than the neutral Cl atom. In $MgCl_2$, two electrons have transferred from the Mg atom to the two Cl atoms; the oxidation number of Mg is +2.

TABLE 17.1 Oxidation Numbers for Common Ions

Ion	Oxidation number
H^+	+1
Na^+	+1
K^+	+1
Li^+	+1
Ag^+	+1
Cu^{2+}	+2
Ca^{2+}	+2
Ba^{2+}	+2
Fe^{2+}	+2
Mg^{2+}	+2
Zn^{2+}	+2
Al^{3+}	+3
Fe^{3+}	+3
Cl^-	−1
Br^-	−1
F^-	−1
I^-	−1
S^{2-}	−2
O^{2-}	−2

In covalently bonded substances, where electrons are shared between two atoms, oxidation numbers are assigned by an arbitrary system based on relative electronegativities. For symmetrical covalent molecules such as H_2 and Cl_2, each atom is assigned an oxidation number of zero because the bonding pair of electrons is shared equally between two like atoms, neither of which is more electronegative than the other:

H:H   :Cl:Cl:

When the covalent bond is between two unlike atoms, the bonding electrons are shared unequally because the more electronegative element has a greater attraction for them. In this case the oxidation numbers are determined by assigning both electrons to the more electronegative element.

Thus in compounds with covalent bonds such as NH_3 and H_2O,

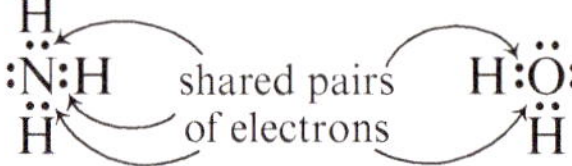

the pairs of electrons are unequally shared between the atoms and are attracted toward the more electronegative elements, N and O. This causes the N and O atoms to be relatively negative with respect to the H atoms. At the same time, it causes the H atoms to be relatively positive with respect to the N and O atoms. In H_2O, both pairs of shared electrons are assigned to the O atom, giving it two electrons more than the neutral O atom, and each H atom is assigned one electron less than the neutral H atom. Therefore, the oxidation number of the O atom is −2, and the oxidation number of each H atom is +1. In NH_3, the three pairs of shared electrons are assigned to the N atom, giving it three electrons more than the neutral N atom, and each H atom has one electron less than the neutral atom. Therefore, the oxidation number of the N atom is −3, and the oxidation number of each H atom is +1.

Assigning correct oxidation numbers to elements is essential for balancing oxidation–reduction equations. The **oxidation number** or **oxidation state** of an element is an integer value assigned to each element in a compound or ion that allows us to keep track of electrons associated with each atom. Oxidation numbers have a variety of uses in chemistry—from writing formulas to predicting properties of compounds and assisting in the balancing of oxidation–reduction reactions in which electrons are transferred.

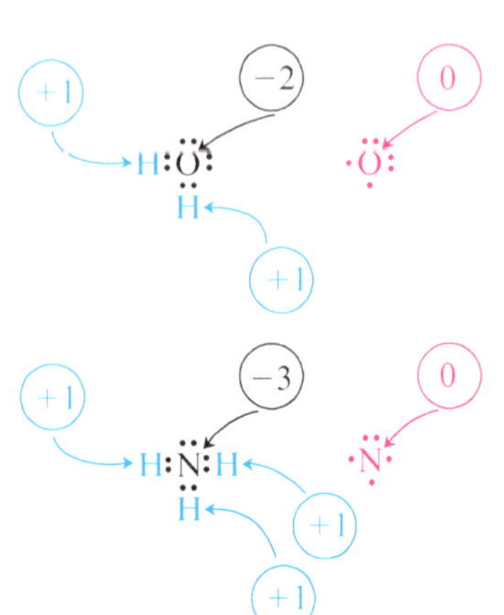

As a starting point, the oxidation number of an uncombined element, regardless of whether it is monatomic or diatomic, is zero. Rules for assigning oxidation numbers are summarized below.

RULES FOR ASSIGNING OXIDATION NUMBERS

1. All elements in their free state (uncombined with other elements) have an oxidation number of zero (e.g., Na, Cu, Mg, H_2, O_2, Cl_2, N_2).
2. H is +1, except in metal hydrides, where it is −1 (e.g., NaH, CaH_2).
3. O is −2, except in peroxides, where it is −1 and in OF_2, where it is +2.

4. The metallic element in an ionic compound has a positive oxidation number.
5. In covalent compounds the negative oxidation number is assigned to the most electronegative atom.
6. The algebraic sum of the oxidation numbers of the elements in a compound is zero.
7. The algebraic sum of the oxidation numbers of the elements in a polyatomic ion is equal to the charge of the ion.

CHECK YOUR UNDERSTANDING

17.1 Oxidation Numbers

17.2 Oxidation Numbers of Elements

➡ WileyPLUS

Many elements have multiple oxidation numbers; for example, nitrogen:

	N_2	N_2O	NO	N_2O_3	NO_2	N_2O_5	NO_3^-
Oxidation number	0	+1	+2	+3	+4	+5	+5

Tom Pantages

The oxidation number for Cu metal is 0 while the oxidation number for Cu^{2+} ions in the crystal is +2.

PROBLEM-SOLVING STRATEGY

Finding the Oxidation Number of an Element in a Compound

1. Write the oxidation number of each known atom below the atom in the formula.
2. Multiply each oxidation number by the number of atoms of that element in the compound.
3. Write an expression indicating the sum of all the oxidation numbers in the compound. *Remember:* The sum of the oxidation numbers in a compound must equal zero.

ENHANCED EXAMPLE 17.1

WileyPLUS

Determine the oxidation number for carbon in carbon dioxide:

SOLUTION CO_2

1. -2
2. $(-2)2$
3. $C + (-4) = 0$

$C = +4$ (oxidation number for carbon)

ONLINE LEARNING MODULE

Oxidation Numbers 1

➡ WileyPLUS

ENHANCED EXAMPLE 17.2

WileyPLUS

Determine the oxidation number for sulfur in sulfuric acid:

SOLUTION H_2SO_4

1. $+1$ $\quad -2$
2. $2(+1) = +2 \qquad 4(-2) = -8$
3. $+2 + S + (-8) = 0$

$S = +6$ (oxidation number for sulfur)

ONLINE LEARNING MODULE

Oxidation Numbers 2

➡ WileyPLUS

PRACTICE 17.1

Determine the oxidation number of (a) S in Na_2SO_4, (b) As in K_3AsO_4, and (c) C in $CaCO_3$.

Oxidation numbers in a polyatomic ion (ions containing more than one atom) are determined in a similar fashion, except that in a polyatomic ion the sum of the oxidation numbers must equal the charge on the ion instead of zero.

ONLINE LEARNING MODULE
Oxidation Numbers 3
➡ WileyPLUS

ENHANCED EXAMPLE 17.3 WileyPLUS

Determine the oxidation number for manganese in the permanganate ion MnO_4^-:

SOLUTION MnO_4^-

1. -2
2. $(-2)4$
3. $Mn + (-8) = -1$ (the charge on the ion)

$Mn = +7$ (oxidation number for manganese)

ONLINE LEARNING MODULE
Oxidation Numbers 4
➡ WileyPLUS

ENHANCED EXAMPLE 17.4 WileyPLUS

Determine the oxidation number for carbon in the oxalate ion $C_2O_4^{2-}$:

SOLUTION $C_2O_4^{2-}$

-2

1. $(-2)4$
2. $2C + (-8) = -2$ (the charge on the ion)
3. $2C = +6$

$C = +3$ (oxidation number for C)

PRACTICE 17.2

Determine the oxidation numbers of (a) N in NH_4^+, (b) Cr in $Cr_2O_7^{2-}$, and (c) P in PO_4^{3-}.

EXAMPLE 17.5

Determine the oxidation number of each element in (a) KNO_3 and (b) SO_4^{2-}.

SOLUTION

(a) Potassium is a Group 1A (1) metal; therefore, it has an oxidation number of +1. The oxidation number of each O atom is −2 (see Rules for Assigning Oxidation Numbers, Rule 3). Using these values and the fact that the sum of the oxidation numbers of all the atoms in a compound is zero, we can determine the oxidation number of N:

$$KNO_3$$
$$+1 + N + 3(-2) = 0$$
$$N = +6 - 1 = +5$$

The oxidation numbers are K, +1; N, +5; O, −2.

(b) Because SO_4^{2-} is an ion, the sum of oxidation numbers of the S and the O atoms must be −2, the charge of the ion. The oxidation number of each O atom is −2 (Rule 3). Then

$$SO_4^{2-}$$
$$S + 4(-2) = -2, \quad S - 8 = -2$$
$$S = -2 + 8 = +6$$

The oxidation numbers are S, +6; O, −2.

PRACTICE 17.3

Determine the oxidation number of each element in these species:
(a) $BeCl_2$ (b) HClO (c) H_2O_2 (d) NH_4^+ (e) BrO_3^-

Oxidation–Reduction

Oxidation–reduction, also known as **redox**, is a chemical process in which the oxidation number of an element is changed. The process may involve the complete transfer of electrons to form ionic bonds or only a partial transfer or shift of electrons to form covalent bonds.

Oxidation occurs whenever the oxidation number of an element increases as a result of losing electrons. Conversely, **reduction** occurs whenever the oxidation number of an element decreases as a result of gaining electrons. For example, a change in oxidation number from +2 to +3 or from −1 to 0 is oxidation (FIGURE 17.1); a change from +5 to +2 or from −2 to −4 is reduction (FIGURE 17.2). Oxidation and reduction occur simultaneously in a chemical reaction; one cannot take place without the other.

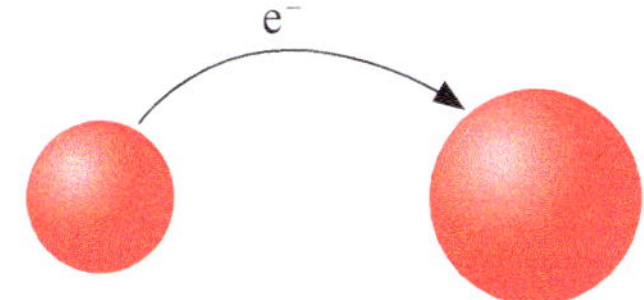

FIGURE 17.1 In an oxidation–reduction reaction, the element losing an electron is oxidized, while the element gaining an electron is reduced.

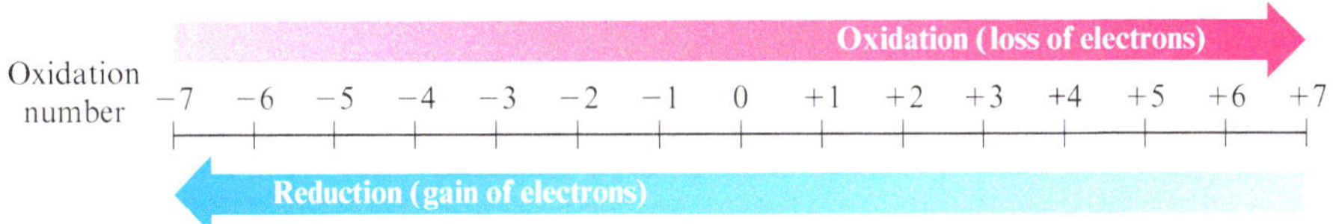

FIGURE 17.2 Oxidation and reduction. Oxidation results in an increase in the oxidation number, and reduction results in a decrease in the oxidation number.

Many combination, decomposition, and single-displacement reactions involve oxidation–reduction. Let's examine the combustion of hydrogen and oxygen from this point of view:

$$2\,H_2 + O_2 \longrightarrow 2\,H_2O$$

Both reactants, hydrogen and oxygen, are elements in the free state and have an oxidation number of zero. In the product (water), hydrogen has been oxidized to +1 and oxygen reduced to −2. The substance that causes an increase in the oxidation state of another substance is called an **oxidizing agent**. The substance that causes a decrease in the oxidation state of another substance is called a **reducing agent**. In this reaction the oxidizing agent is free oxygen, and the reducing agent is free hydrogen. In the reaction

$$Zn(s) + H_2SO_4(aq) \longrightarrow ZnSO_4(aq) + H_2(g)$$

metallic zinc is oxidized, and hydrogen ions are reduced. Zinc is the reducing agent, and hydrogen ions are the oxidizing agent. Electrons are transferred from the zinc metal to the hydrogen ions. The electron transfer is more clearly expressed as

$$Zn^0 + 2H^+ + SO_4^{2-} \longrightarrow Zn^{2+} + SO_4^{2-} + H_2^0$$

Oxidation:	Increase in oxidation number Loss of electrons
Reduction:	Decrease in oxidation number Gain of electrons

The oxidizing agent is reduced and gains electrons. The reducing agent is oxidized and loses electrons. The transfer of electrons is characteristic of all redox reactions.

CHECK YOUR UNDERSTANDING

17.3 Redox Reactions

WileyPLUS

In the reaction between Zn and H_2SO_4, the Zn is oxidized while hydrogen is reduced.

CHECK YOUR UNDERSTANDING

17.4 Oxidized Elements

17.5 Reducing Agents

WileyPLUS

ONLINE LEARNING MODULE

Redox Reactions

WileyPLUS

EXAMPLE 17.6

Analyze the following balanced redox reactions and identify the element oxidized, the element reduced, the oxidizing agent, and the reducing agent.

(a) $Mg(s) + CuSO_4(aq) \longrightarrow Cu(s) + MgSO_4(aq)$

(b) $C_6H_{12}(l) + 9O_2(g) \longrightarrow 6CO_2(g) + 6H_2O(g)$

SOLUTION

(a)

1. Assign oxidation numbers to each element.

$Mg(s) + CuSO_4(aq) \longrightarrow Cu(s) + MgSO_4(aq)$

0 +2 +6 −2 0 +2 +6 −2

(b)

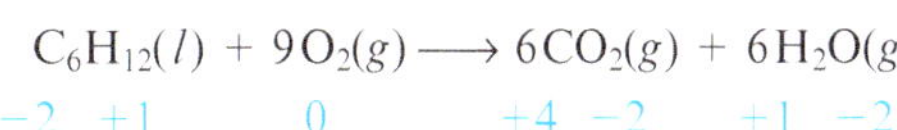

$C_6H_{12}(l) + 9O_2(g) \longrightarrow 6CO_2(g) + 6H_2O(g)$

−2 +1 0 +4 −2 +1 −2

2. Determine which element loses electrons. This is the element oxidized.

Mg^0 loses 2 electrons to become Mg^{2+}. Magnesium is oxidized.

C^{2-} loses 6 electrons to become C^{4+}. Carbon is oxidized.

3. Determine which element gains electrons. This is the element reduced.

Cu^{2+} gains 2 electrons to become Cu^0. Copper is reduced.

O^0 gains 2 electrons to become O^{2-}. Oxygen is reduced.

4. The substance that is oxidized is the reducing agent. (Remember the reducing agent must be a substance, not just an atom.)

Mg is oxidized, so it is the reducing agent.

C is oxidized, so C_6H_{12} is the reducing agent.

5. The substance that is reduced is the oxidizing agent. (Remember the oxidizing agent must be a substance, not just an atom.)

Cu is reduced, so $CuSO_4$ is the oxidizing agent.

O is reduced, so O_2 is the oxidizing agent.

PRACTICE 17.4

Analyze the following balanced redox reactions and identify the element oxidized, the element reduced, the oxidizing agent, and the reducing agent.

(a) $2\,Al(s) + 6\,HNO_3(aq) \longrightarrow 2\,Al(NO_3)_3(aq) + 3\,H_2(g)$

(b) $Ni(s) + CuSO_4(aq) \longrightarrow Cu(s) + NiSO_4(aq)$

(c) $2\,C_2H_6(s) + 7\,O_2(g) \longrightarrow 4\,CO_2(g) + 6\,H_2O(g)$

17.2 Balancing Oxidation–Reduction Equations

LEARNING OBJECTIVE Balance equations for oxidation–reduction reactions.

Many simple redox equations can be balanced readily by inspection, or by trial and error:

$$Na + Cl_2 \longrightarrow NaCl \qquad \text{(unbalanced)}$$

$$2\,Na + Cl_2 \longrightarrow 2\,NaCl \qquad \text{(balanced)}$$

Balancing this equation is certainly not complicated. But as we study more complex reactions and equations such as

$$P + HNO_3 + H_2O \longrightarrow NO + H_3PO_4 \qquad \text{(unbalanced)}$$

$$3\,P + 5\,HNO_3 + 2\,H_2O \longrightarrow 5\,NO + 3\,H_3PO_4 \qquad \text{(balanced)}$$

the trial-and-error method of balancing equations takes an unnecessarily long time.

One systematic method for balancing oxidation–reduction equations is based on the transfer of electrons between the oxidizing and reducing agents. Consider the first equation again:

The superscript 0 shows that the oxidation number is 0 for elements in their uncombined state.

$$Na^0 + Cl_2^0 \longrightarrow Na^+Cl^- \qquad \text{(unbalanced)}$$

In this reaction, sodium metal loses one electron per atom when it changes to a sodium ion. At the same time chlorine gains one electron per atom. Because chlorine is diatomic, two electrons per molecule are needed to form a chloride ion from each atom. These electrons are furnished by two sodium atoms. Stepwise, the reaction may be written as two half-reactions, the oxidation half-reaction and the reduction half-reaction:

$$2\,Na^0 \longrightarrow 2\,Na^+ + \cancel{2e^-} \qquad \text{oxidation half-reaction}$$

$$Cl_2^0 + \cancel{2e^-} \longrightarrow 2\,Cl^- \qquad \text{reduction half-reaction}$$

$$Cl_2^0 + 2\,Na^0 \longrightarrow 2\,Na^+Cl^-$$

When the two half-reactions, each containing the same number of electrons, are added together algebraically, the electrons cancel out. In this reaction there are no excess electrons; the two electrons lost by the two sodium atoms are utilized by chlorine. In all redox reactions, the loss of electrons by the reducing agent must equal the gain of electrons by the oxidizing agent. Here sodium is oxidized and chlorine is reduced. Chlorine is the oxidizing agent; sodium is the reducing agent.

In the following examples, we use the change-in-oxidation-number method, a system for balancing more complicated redox equations.

PROBLEM-SOLVING STRATEGY

Using Change in Oxidation Numbers to Balance Oxidation–Reduction Reactions

1. Assign oxidation numbers for each element to identify the elements being oxidized and reduced. Look for those elements that have changed oxidation number.
2. Write two half-reactions using only the elements that have changed oxidation numbers. One half-reaction must produce electrons, and the other must use electrons.
3. Multiply the half-reactions by the smallest whole numbers that will make the electrons lost by oxidation equal the electrons gained by reduction.
4. Transfer the coefficient in front of each substance in the balanced half-reactions to the corresponding substance in the original equation.
5. Balance the remaining elements that are not oxidized or reduced to give the final balanced equation.
6. Check to make sure both sides of the equation have the same number of atoms of each element.

ONLINE LEARNING MODULE

Balancing Redox Reactions Using Half Reactions

➡ WileyPLUS

EXAMPLE 17.7

Balance the equation

$$Sn + HNO_3 \longrightarrow SnO_2 + NO_2 + H_2O \quad \text{(unbalanced)}$$

SOLUTION

Use the Change in Oxidation Number Problem-Solving Strategy.

1. Assign oxidation numbers to each element to identify the elements being oxidized and those being reduced. Write the oxidation numbers below each element to avoid confusing them with ionic charge:

$$Sn + HNO_3 \longrightarrow SnO_2 + NO_2 + H_2O$$

Sn	H	N	O	Sn	O	N	O	H	O
0	+1	+5	−2	+4	−2	+4	−2	+1	−2

Note that the oxidation numbers of Sn and N have changed.

2. Now write two new equations (half-reactions), using only the elements that change in oxidation number. Then add electrons to bring the equations into electrical balance. One equation represents the oxidation step; the other represents the reduction step. *Remember:* Oxidation produces electrons; reduction uses electrons.

$$Sn^0 \longrightarrow Sn^{4+} + 4e^- \quad \text{(oxidation)}$$
Sn^0 loses 4 electrons

$$N^{5+} + 1e^- \longrightarrow N^{4+} \quad \text{(reduction)}$$
N^{5+} gains 1 electron

3. Multiply the two equations by the smallest whole numbers that will make the electrons lost by oxidation equal to the number of electrons gained by reduction. In this reaction the oxidation step is multiplied by 1 and the reduction step by 4. The equations become

$$Sn^0 \longrightarrow Sn^{4+} + 4e^- \quad \text{(oxidation)}$$
Sn^0 loses 4 electrons

$$4N^{5+} + 4e^- \longrightarrow 4N^{4+} \quad \text{(reduction)}$$
$4N^{5+}$ gains 4 electrons

We have now established the ratio of the oxidizing to the reducing agent as being four atoms of N to one atom of Sn.

4. Transfer the coefficient in front of each substance in the balanced oxidation–reduction equations to the corresponding substance in the original equation. We need to use 1 Sn, 1 SnO_2, 4 HNO_3, and 4 NO_2:

$$Sn + 4HNO_3 \longrightarrow SnO_2 + 4NO_2 + H_2O \quad \text{(unbalanced)}$$

5. In the usual manner, balance the remaining elements that are not oxidized or reduced to give the final balanced equation:

$$Sn + 4HNO_3 \longrightarrow SnO_2 + 4NO_2 + 2H_2O \quad \text{(balanced)}$$

In balancing the final elements, we must not change the ratio of the elements that were oxidized and reduced.

6. Finally, check to ensure that both sides of the equation have the same number of atoms of each element. The final balanced equation contains 1 atom of Sn, 4 atoms of N, 4 atoms of H, and 12 atoms of O on each side.

Because each new equation presents a slightly different problem and because proficiency in balancing equations requires practice, let's work through two more examples.

EXAMPLE 17.8

Balance the equation

$$I_2 + Cl_2 + H_2O \longrightarrow HIO_3 + HCl \quad \text{(unbalanced)}$$

SOLUTION

Use the Problem-Solving Strategy: Using Change in Oxidation Numbers to Balance Oxidation–Reduction Reactions.

1. Assign oxidation numbers:

$$\underset{0}{I_2} + \underset{0}{Cl_2} + \underset{+1\ -2}{H_2O} \longrightarrow \underset{+1\ +5\ -2}{HIO_3} + \underset{+1\ -1}{HCl}$$

The oxidation numbers of I_2 and Cl_2 have changed, I_2 from 0 to +5, and Cl_2 from 0 to −1.

2. Write the oxidation and reduction steps. Balance the number of atoms and then balance the electrical charge using electrons:

$$I_2 \longrightarrow 2I^{5+} + 10e^- \quad \text{(oxidation)} \quad (10e^- \text{ are needed to balance the } +10 \text{ charge})$$

I_2 loses 10 electrons

$$Cl_2 + 2e^- \longrightarrow 2Cl^- \quad \text{(reduction)} \quad (2e^- \text{ are needed to balance the } -2 \text{ charge})$$

Cl_2 gains 2 electrons

3. Adjust loss and gain of electrons so that they are equal. Multiply the oxidation step by 1 and the reduction step by 5:

$$I_2 \longrightarrow 2I^{5+} + 10e^- \quad \text{(oxidation)}$$

I_2 loses 10 electrons

$$5Cl_2 + 10e^- \longrightarrow 10Cl^- \quad \text{(reduction)}$$

$5Cl_2$ gain 10 electrons

4. Transfer the coefficients from the balanced redox equations into the original equation. We need to use 1 I_2, 2 HIO_3, 5 Cl_2, and 10 HCl:

$$I_2 + 5\,Cl_2 + H_2O \longrightarrow 2\,HIO_3 + 10\,HCl \quad \text{(unbalanced)}$$

5. Balance the remaining elements, H and O:

$$I_2 + 5Cl_2 + 6H_2O \longrightarrow 2HIO_3 + 10HCl \quad \text{(balanced)}$$

6. *Check:* The final balanced equation contains 2 atoms of I, 10 atoms of Cl, 12 atoms of H, and 6 atoms of O on each side.

EXAMPLE 17.9

Balance the equation

$$K_2Cr_2O_7 + FeCl_2 + HCl \longrightarrow CrCl_3 + KCl + FeCl_3 + H_2O \quad \text{(unbalanced)}$$

SOLUTION

Use the Problem-Solving Strategy: Using the Change in Oxidation Numbers to Balance Oxidation–Reduction Reactions.

1. Assign oxidation numbers (Cr and Fe have changed):

$$\underset{+1\ +6\ -2}{K_2Cr_2O_7} + \underset{+2\ -1}{FeCl_2} + \underset{+1\ -1}{HCl} \longrightarrow \underset{+3\ -1}{CrCl_3} + \underset{+1\ -1}{KCl} + \underset{+3\ -1}{FeCl_3} + \underset{+1\ -2}{H_2O}$$

2. Write the oxidation and reduction steps. Balance the number of atoms and then balance the electrical charge using electrons:

$$Fe^{2+} \longrightarrow Fe^{3+} + 1e^- \quad \text{(oxidation)}$$

Fe^{2+} loses 1 electron

$$2\,Cr^{6+} + 6\,e^- \longrightarrow 2\,Cr^{3+} \quad \text{(reduction)}$$

$2Cr^{6+}$ gain 6 electrons

3. Balance the loss and gain of electrons. Multiply the oxidation step by 6 and the reduction step by 1 to equalize the transfer of electrons.

$$6Fe^{2+} \longrightarrow 6Fe^{3+} + 6e^- \quad \text{(oxidation)}$$

$6Fe^{2+}$ lose 6 electrons

$$2Cr^{6+} + 6e^- \longrightarrow 2Cr^{3+} \quad \text{(reduction)}$$

$2Cr^{6+}$ gain 6 electrons

4. Transfer the coefficients from the balanced redox equations into the original equation. (Note that one formula unit of $K_2Cr_2O_7$ contains two Cr atoms.) We need to use 1 $K_2Cr_2O_7$, 2 $CrCl_3$, 6 $FeCl_2$, and 6 $FeCl_3$:

$$K_2Cr_2O_7 + 6\,FeCl_2 + HCl \longrightarrow 2CrCl_3 + KCl + 6FeCl_3 + H_2O \quad \text{(unbalanced)}$$

5. Balance the remaining elements in the order K, Cl, H, O:

$$K_2Cr_2O_7 + 6FeCl_2 + 14HCl \longrightarrow 2CrCl_3 + 2KCl + 6FeCl_3 + 7H_2O \quad \text{(balanced)}$$

6. *Check:* The final balanced equation contains 2 K atoms, 2 Cr atoms, 7 O atoms, 6 Fe atoms, 26 Cl atoms, and 14 H atoms on each side.

PRACTICE 17.5

Balance these equations using the change-in-oxidation-number method:

(a) $HNO_3 + S \longrightarrow NO_2 + H_2SO_4 + H_2O$

(b) $CrCl_3 + MnO_2 + H_2O \longrightarrow MnCl_2 + H_2CrO_4$

(c) $KMnO_4 + HCl + H_2S \longrightarrow KCl + MnCl_2 + S + H_2O$

17.3 Balancing Ionic Redox Equations

LEARNING OBJECTIVE Balance equations for ionic oxidation–reduction reactions.

The main difference between balancing ionic redox equations and molecular redox equations is in how we handle the ions. In the ionic redox equations, besides having the same number of atoms of each element on both sides of the final equation, we must also have equal net charges. In assigning oxidation numbers, we must therefore remember to consider the ionic charge.

Several methods are used to balance ionic redox equations, including, with slight modification, the oxidation-number method just shown for molecular equations. But the most popular method is probably the ion–electron method.

The ion–electron method uses ionic charges and electrons to balance ionic redox equations. Oxidation numbers are not formally used, but it is necessary to determine what is being oxidized and what is being reduced.

PROBLEM-SOLVING STRATEGY

Ion–Electron Strategy for Balancing Oxidation–Reduction Reactions

1. Write the two half-reactions that contain the elements being oxidized and reduced using the entire formula of the ion or molecule.
2. Balance the elements other than oxygen and hydrogen.
3. Balance oxygen and hydrogen.

   Acidic solution: For reactions in acidic solution, use H^+ and H_2O to balance oxygen and hydrogen. For each oxygen needed, use one H_2O. Then add H^+ as needed to balance the hydrogen atoms.

   Basic solution: For reactions in alkaline solutions, first balance as though the reaction were in an acidic solution, using Steps 1–3. Then add as many OH^- ions to each side of the equation as there are H^+ ions in the equation. Now combine the H^+ and OH^- ions into water (for example, $4H^+$ and $4OH^-$ give $4H_2O$). Rewrite the equation, canceling equal numbers of water molecules that appear on opposite sides of the equation.
4. Add electrons (e^-) to each half-reaction to bring them into electrical balance.
5. Since the loss and gain of electrons must be equal, multiply each half-reaction by the appropriate number to make the number of electrons the same in each half-reaction.
6. Add the two half-reactions together, canceling electrons and any other identical substances that appear on opposite sides of the equation.

ONLINE LEARNING MODULE
Balancing Redox Reactions Using the Ion-Electron Method
WileyPLUS

EXAMPLE 17.10

Balance this equation using the ion–electron method:

$$MnO_4^- + S^{2-} \longrightarrow Mn^{2+} + S^0 \quad \text{(acidic solution)}$$

SOLUTION

1. Write two half-reactions, one containing the element being oxidized and the other the element being reduced (use the entire molecule or ion):

$$S^{2-} \longrightarrow S^0 \qquad \text{(oxidation)}$$
$$MnO_4^- \longrightarrow Mn^{2+} \qquad \text{(reduction)}$$

2. Balance elements other than oxygen and hydrogen (accomplished in Step 1: 1 S and 1 Mn on each side).

3. Balance O and H. Remember the solution is acidic. The oxidation requires neither O nor H, but the reduction equation needs $4H_2O$ on the right and $8H^+$ on the left:

$$S^{2-} \longrightarrow S^0$$
$$8H^+ + MnO_4^- \longrightarrow Mn^{2+} + 4H_2O$$

4. Balance each half-reaction electrically with electrons:

$$S^{2-} \longrightarrow S^0 + 2e^-$$

net charge = −2 on each side

$$5e^- + 8H^+ + MnO_4^- \longrightarrow Mn^{2+} + 4H_2O$$

net charge = +2 on each side

5. Equalize loss and gain of electrons. In this case, multiply the oxidation equation by 5 and the reduction equation by 2:

$$5S^{2-} \longrightarrow 5S^0 + 10e^-$$

$$10e^- + 16H^+ + 2MnO_4^- \longrightarrow 2Mn^{2+} + 8H_2O$$

6. Add the two half-reactions together, canceling the $10e^-$ from each side, to obtain the balanced equation:

$$5S^{2-} \longrightarrow 5S^0 + \cancel{10e^-}$$

$$\cancel{10e^-} + 16H^+ + 2MnO_4^- \longrightarrow 2Mn^{2+} + 8H_2O$$

$$16H^+ + 2MnO_4^- + 5S^{2-} \longrightarrow 2Mn^{2+} + 5S^0 + 8H_2O \quad \text{(balanced)}$$

Check: Both sides of the equation have a charge of +4 and contain the same number of atoms of each element.

EXAMPLE 17.11

Balance this equation:

$$CrO_4^{2-} + Fe(OH)_2 \longrightarrow Cr(OH)_3 + Fe(OH)_3 \quad \text{(basic solution)}$$

SOLUTION

Use the Ion–Electron Problem-Solving Strategy.

1. Write the two half-reactions:

$$Fe(OH)_2 \longrightarrow Fe(OH)_3 \quad \text{(oxidation)}$$

$$CrO_4^{2-} \longrightarrow Cr(OH)_3 \quad \text{(reduction)}$$

2. Balance elements other than H and O (accomplished in Step 1).
3. Remember the solution is basic. Balance O and H as though the solution were acidic. Use H_2O and H^+. To balance O and H in the oxidation equation, add $1\,H_2O$ on the left and $1\,H^+$ on the right side:

$$Fe(OH)_2 + H_2O \longrightarrow Fe(OH)_3 + H^+$$

Add $1\,OH^-$ to each side:

$$Fe(OH)_2 + H_2O + OH^- \longrightarrow Fe(OH)_3 + H^+ + OH^-$$

Combine H^+ and OH^- as H_2O and rewrite, canceling H_2O on each side:

$$Fe(OH)_2 + \cancel{H_2O} + OH^- \longrightarrow Fe(OH)_3 + \cancel{H_2O}$$

$$\boxed{Fe(OH)_2 + OH^- \longrightarrow Fe(OH)_3} \quad \text{(oxidation)}$$

To balance O and H in the reduction equation, add $1\,H_2O$ on the right and $5\,H^+$ on the left:

$$CrO_4^{2-} + 5H^+ \longrightarrow Cr(OH)_3 + H_2O$$

Add $5\,OH^-$ to each side:

$$CrO_4^{2-} + 5H^+ + 5OH^- \longrightarrow Cr(OH)_3 + H_2O + 5OH^-$$

Combine $5H^+ + 5OH^- \longrightarrow 5H_2O$:

$$CrO_4^{2-} + 5H_2O \longrightarrow Cr(OH)_3 + H_2O + 5OH^-$$

Rewrite, canceling $1\,H_2O$ from each side:

$$CrO_4^{2-} + 4H_2O \longrightarrow Cr(OH)_3 + 5OH^- \quad \text{(reduction)}$$

4. Balance each half-reaction electrically with electrons:

$$Fe(OH)_2 + OH^- \longrightarrow Fe(OH)_3 + e^- \quad \text{(balanced oxidation equation)}$$

$$CrO_4^{2-} + 4H_2O + 3e^- \longrightarrow Cr(OH)_3 + 5OH^- \quad \text{(balanced reduction equation)}$$

5. Equalize the loss and gain of electrons. Multiply the oxidation reaction by 3:

$$3Fe(OH)_2 + 3OH^- \longrightarrow 3Fe(OH)_3 + 3e^-$$

$$CrO_4^{2-} + 4H_2O + 3e^- \longrightarrow Cr(OH)_3 + 5OH^-$$

6. Add the two half-reactions together, canceling the $3e^-$ and $3OH^-$ from each side of the equation:

$$3Fe(OH)_2 + 3OH^- \longrightarrow 3Fe(OH)_3 + \cancel{3e^-}$$

$$CrO_4^{2-} + 4H_2O + \cancel{3e^-} \longrightarrow Cr(OH)_3 + 5OH^-$$

$$CrO_4^{2-} + 3Fe(OH)_2 + 4H_2O \longrightarrow Cr(OH)_3 + 3Fe(OH)_3 + 2OH^-$$

(balanced)

Check: Each side of the equation has a charge of -2 and contains the same number of atoms of each element.

PRACTICE 17.6

Balance these equations using the ion–electron method:

(a) $I^- + NO_2^- \longrightarrow I_2 + NO$ (acidic solution)

(b) $Cl_2 + IO_3^- \longrightarrow IO_4^- + Cl^-$ (basic solution)

(c) $AuCl_4^- + Sn^{2+} \longrightarrow Sn^{4+} + AuCl + Cl^-$ (acidic solution)

Ionic equations can also be balanced using the change-in-oxidation-number method shown in Example 17.7. To illustrate this method, let's use the equation from Example 17.11.

ONLINE LEARNING MODULE

Balancing Redox Reactions Using the Oxidation Number Method

WileyPLUS

EXAMPLE 17.12

Balance this equation using the change-in-oxidation-number method:

$$CrO_4^{2-} + Fe(OH)_2 \longrightarrow Cr(OH)_3 + Fe(OH)_3 \quad \text{(basic solution)}$$

SOLUTION

1. and 2. Assign oxidation numbers and balance the charges with electrons:

$$Cr^{6+} + 3e^- \longrightarrow Cr^{3+} \quad \text{(reduction)}$$

Cr^{6+} gains $3e^-$

$$Fe^{2+} \longrightarrow Fe^{3+} + e^- \quad \text{(oxidation)}$$

Fe^{2+} loses $1e^-$

3. Equalize the loss and gain of electrons, by multiplying the oxidation step by 3:

$$Cr^{6+} + 3e^- \longrightarrow Cr^{3+} \quad \text{(reduction)}$$

Cr^{6+} gains $3e^-$

$$3Fe^{2+} \longrightarrow 3Fe^{3+} + 3e^- \quad \text{(oxidation)}$$

$3Fe^{2+}$ loses $3e^-$

4. Transfer coefficients back to the original equation:

$$CrO_4^{2-} + 3\,Fe(OH)_2 \longrightarrow Cr(OH)_3 + 3\,Fe(OH)_3$$

5. Balance electrically. Because the solution is basic, use OH^- to balance charges. The charge on the left side is -2 and on the right side is 0. Add $2OH^-$ ions to the right side of the equation:

$$CrO_4^{2-} + 3\,Fe(OH)_2 \longrightarrow Cr(OH)_3 + 3\,Fe(OH)_3 + 2OH^-$$

Adding $4\,H_2O$ to the left side balances the equation:

$$CrO_4^{2-} + 3\,Fe(OH)_2 + 4\,H_2O \longrightarrow Cr(OH)_3 + 3\,Fe(OH)_3 + 2OH^- \quad \text{(balanced)}$$

Check: Each side of the equation has a charge of -2 and contains the same number of atoms of each element.

PRACTICE 17.7

Balance these equations using the change-in-oxidation-number method:

(a) $Zn \longrightarrow Zn(OH)_4^{2-} + H_2$ (basic solution)
(b) $H_2O_2 + Sn^{2+} \longrightarrow Sn^{4+}$ (acidic solution)
(c) $Cu + Cu^{2+} \longrightarrow Cu_2O$ (basic solution)

17.4 Activity Series of Metals

Use the activity series of metals to predict whether a reaction will occur.

LEARNING OBJECTIVE

KEY TERM

activity series of metals

Knowledge of the relative chemical reactivities of the elements helps us predict the course of many chemical reactions. For example, calcium reacts with cold water to produce hydrogen, and magnesium reacts with steam to produce hydrogen. Therefore, calcium is considered a more reactive metal than magnesium:

$$Ca(s) + 2H_2O(l) \longrightarrow Ca(OH)_2(aq) + H_2(g)$$

$$Mg(s) + \underset{\text{steam}}{H_2O(g)} \longrightarrow MgO(s) + H_2(g)$$

The difference in their activity is attributed to the fact that calcium loses its two valence electrons more easily than magnesium and is therefore more reactive and/or more readily oxidized than magnesium.

When a coil of copper is placed in a solution of silver nitrate ($AgNO_3$), free silver begins to plate out on the copper. (FIGURE 17.3.) After the reaction has continued for some time, we can observe a blue color in the solution, indicating the presence of copper(II) ions. The equations are

$$Cu^0(s) + 2AgNO_3(aq) \longrightarrow 2Ag^0(s) + Cu(NO_3)_2(aq)$$

$$Cu^0(s) + 2Ag^+(aq) \longrightarrow 2Ag^0(s) + Cu^{2+}(aq) \quad \text{(net ionic equation)}$$

$$Cu^0(s) \longrightarrow Cu^{2+}(aq) + 2e^- \quad \text{(oxidation of } Cu^0\text{)}$$

$$Ag^+(aq) + e^- \longrightarrow Ag^0(s) \quad \text{(reduction of } Ag^+\text{)}$$

If a coil of silver is placed in a solution of copper(II) nitrate, $Cu(NO_3)_2$, no reaction is visible.

$$Ag^0(s) + Cu(NO_3)_2(aq) \longrightarrow \text{no reaction}$$

In the reaction between Cu and $AgNO_3$, electrons are transferred from Cu^0 atoms to Ag^+ ions in solution. Copper has a greater tendency than silver to lose electrons, so an electrochemical force is exerted upon silver ions to accept electrons from copper atoms. When an Ag^+ ion accepts an electron, it is reduced to an Ag^0 atom and is no longer soluble in solution.

Peticolas/Megna/Fundamental Photographs

FIGURE 17.3 A coil of copper placed in a silver nitrate solution forms silver crystals on the wire. The pale blue of the solution indicates the presence of copper ions.

CHECK YOUR UNDERSTANDING

17.6 Activity Series of Metals 1

17.7 Activity Series of Metals 2

WileyPLUS

At the same time, Cu^0 is oxidized and goes into solution as Cu^{2+} ions. From this reaction, we can conclude that copper is more reactive than silver.

Metals such as sodium, magnesium, zinc, and iron that react with solutions of acids to liberate hydrogen are more reactive than hydrogen. Metals such as copper, silver, and mercury that do not react with solutions of acids to liberate hydrogen are less reactive than hydrogen. By studying a series of reactions such as these, we can list metals according to their chemical activity, placing the most active at the top and the least active at the bottom. This list is called the **activity series of metals**. TABLE 17.2 lists some of the common metals in the series. The arrangement corresponds to the ease with which the elements are oxidized or lose electrons, with the most easily oxidizable element listed first. More extensive tables are available in chemistry reference books.

TABLE 17.2 Activity Series of Metals

Ease of oxidation (↑)	Reaction
	$K \longrightarrow K^+ + e^-$
	$Ba \longrightarrow Ba^{2+} + 2e^-$
	$Ca \longrightarrow Ca^{2+} + 2e^-$
	$Na \longrightarrow Na^+ + e^-$
	$Mg \longrightarrow Mg^{2+} + 2e^-$
	$Al \longrightarrow Al^{3+} + 3e^-$
	$Zn \longrightarrow Zn^{2+} + 2e^-$
	$Cr \longrightarrow Cr^{3+} + 3e^-$
	$Fe \longrightarrow Fe^{2+} + 2e^-$
	$Ni \longrightarrow Ni^{2+} + 2e^-$
	$Sn \longrightarrow Sn^{2+} + 2e^-$
	$Pb \longrightarrow Pb^{2+} + 2e^-$
	$\mathbf{H_2 \longrightarrow 2H^+ + 2e^-}$
	$Cu \longrightarrow Cu^{2+} + 2e^-$
	$As \longrightarrow As^{3+} + 3e^-$
	$Ag \longrightarrow Ag^+ + e^-$
	$Hg \longrightarrow Hg^{2+} + 2e^-$
	$Au \longrightarrow Au^{3+} + 3e^-$

PROBLEM-SOLVING STRATEGY

Using the Activity Series to Predict Reactions

1. The reactivity of the metals listed decreases from top to bottom.
2. A free metal can displace the ion of a second metal from solution, provided that the free metal is above the second metal in the activity series.
3. Free metals above hydrogen react with nonoxidizing acids in solution to liberate hydrogen gas.
4. Free metals below hydrogen do not liberate hydrogen from acids.
5. Conditions such as temperature and concentration may affect the relative position of some of these elements.

Here are two examples using the activity series of metals.

ENHANCED EXAMPLE 17.13 WileyPLUS

Will zinc metal react with dilute sulfuric acid?

SOLUTION

From Table 17.2, we see that zinc is above hydrogen; therefore, zinc atoms will lose electrons more readily than hydrogen atoms. Hence, zinc atoms will reduce hydrogen ions from the acid to form hydrogen gas and zinc ions. In fact, these reagents are commonly used for the laboratory preparation of hydrogen. The equation is

$$Zn(s) + H_2SO_4(aq) \longrightarrow ZnSO_4(aq) + H_2(g)$$

$$Zn(s) + 2H^+(aq) \longrightarrow Zn^{2+}(aq) + H_2(g) \quad \text{(net ionic equation)}$$

ENHANCED EXAMPLE 17.14 WileyPLUS

Will a reaction occur when copper metal is placed in an iron(II) sulfate solution?

SOLUTION

No, copper lies below iron in the series, loses electrons less easily than iron, and therefore will not displace iron(II) ions from solution. In fact, the reverse is true. When an iron nail is dipped into a copper(II) sulfate solution, it becomes coated with free copper. The equations are

$$Cu(s) + FeSO_4(aq) \longrightarrow \text{no reaction}$$

$$Fe(s) + CuSO_4(aq) \longrightarrow FeSO_4(aq) + Cu(s)$$

From Table 17.2, we may extract the following pair in their relative position to each other:

$$Fe \longrightarrow Fe^{2+} + 2e^-$$
$$Cu \longrightarrow Cu^{2+} + 2e^-$$

According to Step 2 in Problem-Solving Strategy: Using the Activity Series, we can predict that free iron will react with copper(II) ions in solution to form free copper metal and iron(II) ions in solution:

$$Fe(s) + Cu^{2+}(aq) \longrightarrow Fe^{2+}(aq) + Cu(s) \quad \text{(net ionic equation)}$$

PRACTICE 17.8

Indicate whether these reactions will occur:

(a) Sodium metal is placed in dilute hydrochloric acid.
(b) A piece of lead is placed in magnesium nitrate solution.
(c) Mercury is placed in a solution of silver nitrate.

17.5 Electrolytic and Voltaic Cells

Compare the reactions and functions of electrolytic and voltaic cells.

LEARNING OBJECTIVE

KEY TERMS
electrolysis
electrolytic cell
cathode
anode
voltaic cell

The process in which electrical energy is used to bring about chemical change is known as **electrolysis**. An **electrolytic cell** uses electrical energy to produce a chemical reaction. The use of electrical energy has many applications in industry—for example, in the production of sodium, sodium hydroxide, chlorine, fluorine, magnesium, aluminum, and pure hydrogen and oxygen, and in the purification and electroplating of metals.

What happens when an electric current is passed through a solution? Let's consider a hydrochloric acid solution in a simple electrolytic cell, as shown in FIGURE 17.4. The cell consists of a source of direct current (a battery) connected to two electrodes that are immersed in a solution of hydrochloric acid. The negative electrode is called the **cathode** because cations are attracted to it. The positive electrode is called the **anode** because anions are attracted to it. The cathode is attached to the negative pole and the anode to the positive pole of the battery. The battery supplies electrons to the cathode.

When the electric circuit is completed, positive hydronium ions (H_3O^+) migrate to the cathode, where they pick up electrons and evolve as hydrogen gas. At the same time the negative chloride ions (Cl^-) migrate to the anode, where they lose electrons and evolve as chlorine gas.

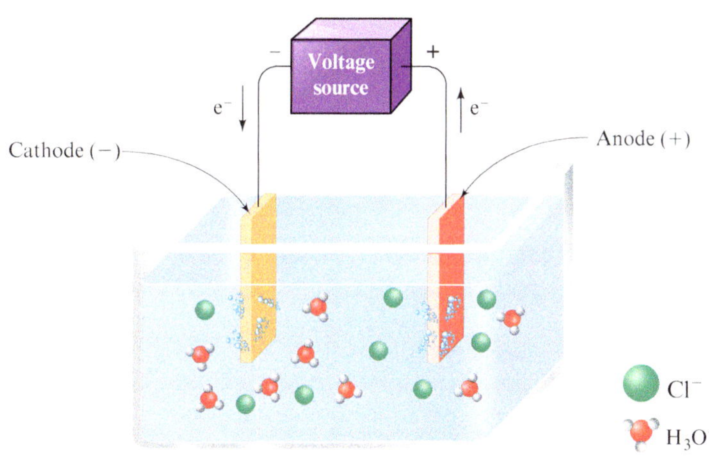

FIGURE 17.4 During the electrolysis of a hydrochloric acid solution, positive hydronium ions are attracted to the cathode, where they gain electrons and form hydrogen gas. Chloride ions migrate to the anode, where they lose electrons and form chlorine gas. The equation for this process is $2\,HCl(aq) \longrightarrow H_2(g) + Cl_2(g)$.

Reaction at the cathode:

$$H_3O^+ + 1\,e^- \longrightarrow H^0 + H_2O \qquad \text{(reduction)}$$
$$H^0 + H^0 \longrightarrow H_2$$

Reaction at the anode:

$$Cl^- \longrightarrow Cl^0 + 1\,e^- \qquad \text{(oxidation)}$$
$$Cl^0 + Cl^0 \longrightarrow Cl_2$$
$$2\,HCl(aq) \xrightarrow{\text{electrolysis}} H_2(g) + Cl_2(g) \qquad \text{(net reaction)}$$

Note that oxidation–reduction has taken place. Chloride ions lost electrons (were oxidized) at the anode, and hydronium ions gained electrons (were reduced) at the cathode.

Oxidation always occurs at the anode and reduction at the cathode.

When concentrated sodium chloride solutions (brines) are electrolyzed, the products are sodium hydroxide, hydrogen, and chlorine. The overall reaction is

$$2\,Na^+(aq) + 2\,Cl^-(aq) + 2\,H_2O(l) \xrightarrow{\text{electrolysis}} 2\,Na^+(aq) + 2\,OH^-(aq) + H_2(g) + Cl_2(g)$$

The net ionic equation is

$$2Cl^-(aq) + 2H_2O(l) \longrightarrow 2OH^-(aq) + H_2(g) + Cl_2(g)$$

During electrolysis, Na^+ ions move toward the cathode and Cl^- ions move toward the anode. The anode reaction is similar to that of hydrochloric acid; the chlorine is liberated:

$$2Cl^-(aq) \longrightarrow Cl_2(g) + 2e^-$$

Even though Na^+ ions are attracted by the cathode, the facts show that hydrogen is liberated there. No evidence of metallic sodium is found, but the area around the cathode tests alkaline from the accumulated OH^- ions. The reaction at the cathode is

$$2H_2O(l) + 2e^- \longrightarrow H_2(g) + 2OH^-(aq)$$

If electrolysis is allowed to continue until all the chloride is reacted, the solution remaining will contain only sodium hydroxide, which on evaporation yields solid NaOH. Large amounts of sodium hydroxide and chlorine are made by this process.

When molten sodium chloride (without water) is subjected to electrolysis, metallic sodium and chlorine gas are formed:

$$2\,Na^+(l) + 2\,Cl^-(l) \xrightarrow{\text{electrolysis}} 2\,Na(s) + Cl_2(g)$$

James L. Amos/Corbis Images

Jewelry and eyeglasses are electroplated with rhodium to prevent tarnishing.

An important electrochemical application is the electroplating of metals. Electroplating is the art of covering a surface or an object with a thin adherent electrodeposited metal coating. Electroplating is done for protection of the surface of the base metal or for a purely decorative effect. The layer deposited is surprisingly thin, varying from as little as 5×10^{-5}cm to 2×10^{-3}cm, depending on the metal and the intended use. The object to be plated is set up as the cathode and is immersed in a solution containing ions of the plating metal. When an electric current passes through the solution, metal ions that migrate to the cathode are reduced, depositing on the object as the free metal. In most cases, the metal deposited on the object is replaced in the solution by using an anode of the same metal. The following equations show the chemical changes in the electroplating of nickel:

Reaction at the cathode: $Ni^{2+}(aq) + 2e^- \longrightarrow Ni(s)$ (Ni plated out on an object)

Reaction at the anode: $Ni(s) \longrightarrow Ni^{2+}(aq) + 2e^-$ (Ni replenished in solution)

Metals commonly used in commercial electroplating are copper, nickel, zinc, lead, cadmium, chromium, tin, gold, and silver.

In the electrolytic cell shown in Figure 17.4, electrical energy from the voltage source is used to bring about nonspontaneous redox reactions. The hydrogen and chlorine produced have more potential energy than was present in the hydrochloric acid before electrolysis.

Conversely, some spontaneous redox reactions can be made to supply useful amounts of electrical energy. When a piece of zinc is put in a copper(II) sulfate solution, the zinc quickly becomes coated with metallic copper. We expect this coating to happen because zinc is above copper in the activity series; copper(II) ions are therefore reduced by zinc atoms:

$$Zn^0(s) + Cu^{2+}(aq) \longrightarrow Zn^{2+}(aq) + Cu^0(s)$$

This reaction is clearly a spontaneous redox reaction, but simply dipping a zinc rod into a copper(II) sulfate solution will not produce useful electric current. However, when we carry out this reaction in the cell shown in FIGURE 17.5, an electric current is produced. The cell consists of a piece of zinc immersed in a zinc sulfate solution and connected by a wire through a voltmeter to a piece of copper immersed in copper(II) sulfate solution. The two solutions are connected by a salt bridge. Such a cell produces an electric current and a potential of about 1.1 volts when both solutions are 1.0 *M* in concentration. A cell that produces electric current from a spontaneous chemical reaction is called a **voltaic cell**. A voltaic cell is also known as a *galvanic cell.*

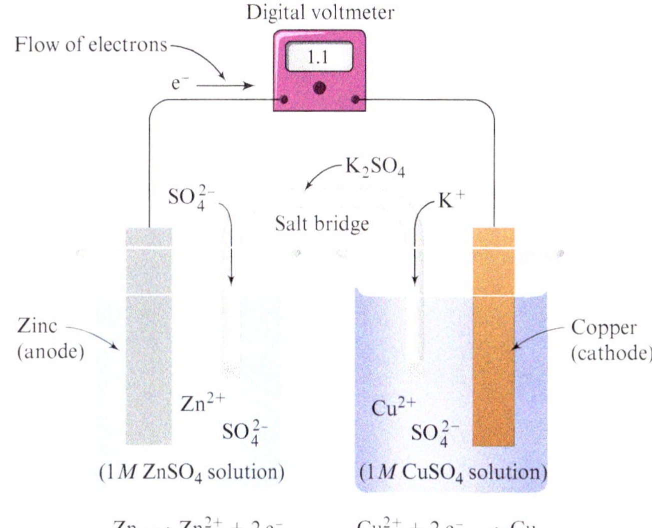

FIGURE 17.5 Zinc–copper voltaic cell. The cell has a potential of 1.1 volts when $ZnSO_4$ and $CuSO_4$ solutions are 1.0 *M*. The salt bridge provides electrical contact between the two half-cells.

The driving force responsible for the electric current in the zinc–copper cell originates in the great tendency of zinc atoms to lose electrons relative to the tendency of copper(II) ions to gain electrons. In the cell shown in Figure 17.5, zinc atoms lose electrons and are converted to zinc ions at the zinc electrode surface; the electrons flow through the wire (external circuit) to the copper electrode. Here copper(II) ions pick up electrons and are reduced to copper atoms, which plate out on the copper electrode. Sulfate ions flow from the $CuSO_4$ solution via the salt bridge into the $ZnSO_4$ solution (internal circuit) to complete the circuit. The equations for the reactions of this cell are

anode	$Zn^0(s) \longrightarrow Zn^{2+}(aq) + 2e^-$	(oxidation)
cathode	$Cu^{2+}(aq) + 2e^- \longrightarrow Cu^0(s)$	(reduction)
net ionic	$Zn^0(s) + Cu^{2+}(aq) \longrightarrow Zn^{2+}(aq) + Cu^0(s)$	
overall	$Zn(s) + CuSO_4(aq) \longrightarrow ZnSO_4(aq) + Cu(s)$	

The redox reaction, the movement of electrons in the metallic or external part of the circuit, and the movement of ions in the solution or internal part of the circuit of the zinc–copper cell are very similar to the actions that occur in the electrolytic cell of Figure 17.4. The only important difference is that the reactions of the zinc–copper cell are spontaneous. This spontaneity is the crucial difference between all voltaic and electrolytic cells.

Voltaic cells use chemical reactions to produce electrical energy.
Electrolytic cells use electrical energy to produce chemical reactions.

Calculators, watches, radios, cell phones, and portable CD players are powered by small efficient voltaic cells called *dry cell batteries*. These batteries are called dry cells because they do not contain a liquid electrolyte (like the voltaic cells discussed earlier). Dry cell batteries are found in several different versions.

The *acid*-type dry cell battery contains a zinc inner case that functions as the anode. A carbon (graphite) rod runs through the center and is in contact with the zinc case at one end and a moist paste of solid MnO_2, NH_4Cl, and carbon that functions as the cathode. (FIGURE 17.6A.) The cell produces about 1.5 volts. The *alkaline*-type dry cell battery is the same as the acid type except the NH_4Cl is replaced by either KOH or NaOH. These dry cells typically last longer because the zinc anode corrodes more slowly in basic conditions. A third type of dry cell is the *zinc–mercury* cell shown in FIGURE 17.6B. The reactions occurring in this cell are

anode	$Zn^0 + 2\,OH^- \longrightarrow ZnO + H_2O + 2\,e^-$	(oxidation)
cathode	$HgO + H_2O + 2\,e^- \longrightarrow Hg^0 + 2\,OH^-$	(reduction)
net ionic	$Zn^0 + Hg^{2+} \longrightarrow Zn^{2+} + Hg^0$	
overall	$Zn^0 + HgO \longrightarrow ZnO + Hg^0$	

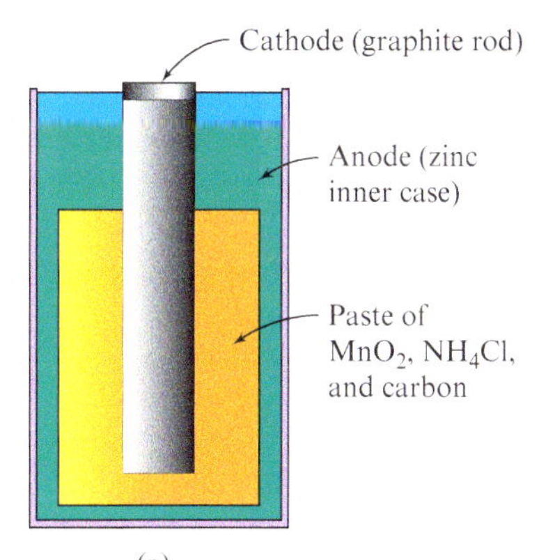

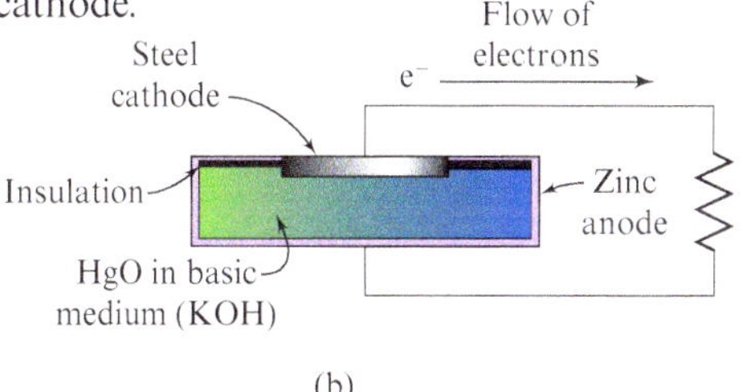

FIGURE 17.6 (a) A common acid-type dry cell. (b) Diagram of an alkaline zinc–mercury cell.

Charles D. Winters/Photo Researchers

To offset the relatively high initial cost, this cell (1) provides current at a very steady potential of about 1.5 volts; (2) has an exceptionally long service life—that is, high energy output to weight ratio; (3) is completely self-contained; and (4) can be stored for relatively long periods of time when not in use.

An automobile storage battery is an energy reservoir. The charged battery acts as a voltaic cell and through chemical reactions furnishes electrical energy to operate the starter, lights, radio, and so on. When the engine is running, a generator or alternator produces and forces an electric current through the battery and, by electrolytic chemical action, restores it to the charged condition.

The cell unit consists of a lead plate filled with spongy lead and a lead(IV) oxide plate, both immersed in dilute sulfuric acid solution, which serves as the electrolyte (FIGURE 17.7). When the cell is discharging, or acting as a voltaic cell, these reactions occur:

Pb plate (anode):	$Pb^0 \longrightarrow Pb^{2+} + 2e^-$	(oxidation)
PbO_2 plate (cathode):	$PbO_2 + 4H^+ + 2e^- \longrightarrow Pb^{2+} + 2H_2O$	(reduction)
Net ionic redox reaction:	$Pb^0 + PbO_2 + 4H^+ \longrightarrow 2Pb^{2+} + 2H_2O$	
Precipitation reaction on plates:	$Pb^{2+}(aq) + SO_4^{2-}(aq) \longrightarrow PbSO_4(s)$	

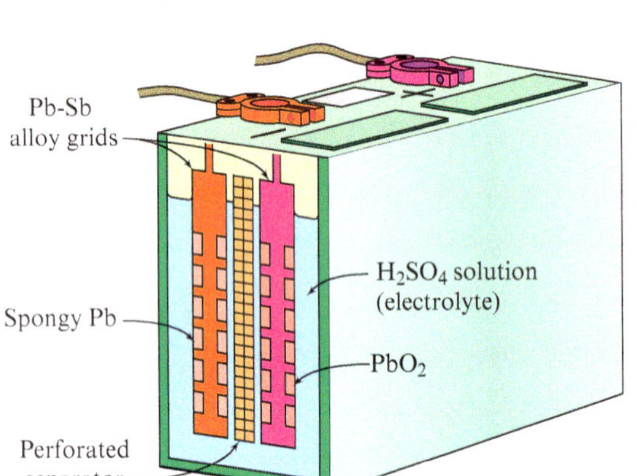

FIGURE 17.7 Cross-sectional diagram of a lead storage battery cell.

Because lead(II) sulfate is insoluble, the Pb^{2+} ions combine with SO_4^{2-} ions to form a coating of $PbSO_4$ on each plate. The overall chemical reaction of the cell is

$$Pb(s) + PbO_2(s) + 2\,H_2SO_4(aq) \xrightarrow[\text{cycle}]{\text{discharge}} 2\,PbSO_4(s) + 2\,H_2O(l)$$

The cell can be recharged by reversing the chemical reaction. This reversal is accomplished by forcing an electric current through the cell in the opposite direction. Lead sulfate and water are reconverted to lead, lead(IV) oxide, and sulfuric acid:

$$2\,PbSO_4(s) + 2\,H_2O(l) \xrightarrow[\text{cycle}]{\text{discharge}} Pb(s) + PbO_2(s) + 2\,H_2SO_4(aq)$$

The electrolyte in a lead storage battery is a 38% by mass sulfuric acid solution having a density of 1. 29 g/mL. As the battery is discharged, sulfuric acid is removed, thereby decreasing the density of the electrolyte solution. The state of charge or discharge of the battery can be estimated by measuring the density (or specific gravity) of the electrolyte solution with a hydrometer. When the density has dropped to about 1. 05 g/mL, the battery needs recharging.

In a commerical battery, each cell consists of a series of cell units of alternating lead–lead(IV) oxide plates separated and supported by wood, glass wool, or fiberglass. The energy storage capacity of a single cell is limited, and its electrical potential is only about 2 volts. Therefore a bank of six cells is connected in series to provide the 12-volt output of the usual automobile battery.

EXAMPLE 17.15

(a) What is the main difference between a voltaic cell and an electrolytic cell?
(b) In Figure 17.5 what is the purpose of the salt bridge?

SOLUTION

(a) A voltaic cell uses a chemical reaction to produce electrical energy. An electrolytic cell uses electrical energy to produce a chemical reaction.
(b) The salt bridge is used to complete the electrical circuit between the cathode and the anode half-cells.

PRACTICE 17.9

Consider the reaction

$$Sn^{2+}(aq) + Cu(s) \longrightarrow Sn(s) + Cu^{2+}(aq)$$

(a) Which metal is oxidized, and which is reduced?
(b) Write the reaction occurring at the anode and at the cathode.

CHAPTER 17 REVIEW

KEY TERMS

oxidation number
oxidation state
oxidation–reduction
redox
oxidation
reduction
oxidizing agent
reducing agent

17.1 Oxidation Number

- To assign an oxidation number:

RULES FOR ASSIGNING OXIDATION NUMBERS

1. All elements in their free state (uncombined with other elements) have an oxidation number of zero (e.g., Na, Cu, Mg, H_2, O_2, Cl_2, N_2).
2. H is $+1$, except in metal hydrides, where it is -1 (e.g., NaH, CaH_2).
3. O is -2, except in peroxides, where it is -1, and in OF_2, where it is $+2$.
4. The metallic element in an ionic compound has a positive oxidation number.
5. In covalent compounds the negative oxidation number is assigned to the most electronegative atom.
6. The algebraic sum of the oxidation numbers of the elements in a compound is zero.
7. The algebraic sum of the oxidation numbers of the elements in a polyatomic ion is equal to the charge of the ion.

- Oxidation–reduction is a chemical process in which electrons are transferred from one atom to another to change the oxidation number of the atom.
- When the oxidation number increases, oxidation occurs, resulting in the loss of electrons.
- When the oxidation number decreases, reduction occurs, resulting in the gain of electrons.

17.2 Balancing Oxidation–Reduction Equations

- Trial and error or inspection methods.
- Can be balanced by writing and balancing the half-reactions for the overall reaction.
- Can be balanced by the change-in-oxidation-number method.

PROBLEM-SOLVING STRATEGY

Using Change in Oxidation Numbers to Balance Oxidation–Reduction Reactions

1. Assign oxidation numbers for each element to identify the elements being oxidized and reduced. Look for those elements that have changed oxidation number.
2. Write two half-reactions using only the elements that have changed oxidation numbers. One half-reaction must produce electrons, and the other must use electrons.
3. Multiply the half-reactions by the smallest whole numbers that will make the electrons lost by oxidation equal the electrons gained by reduction.
4. Transfer the coefficient in front of each substance in the balanced half-reactions to the corresponding substance in the original equation.
5. Balance the remaining elements that are not oxidized or reduced to give the final balanced equation.
6. Check to make sure both sides of the equation have the same number of atoms of each element.

17.3 Balancing Ionic Redox Equations

- To balance equations that are ionic, charge must also be balanced (in addition to atoms and ions).
- The ion–electron method for balancing equations is:

PROBLEM-SOLVING STRATEGY

Ion–Electron Strategy for Balancing Oxidation–Reduction Reactions

1. Write the two half-reactions that contain the elements being oxidized and reduced using the entire formula of the ion or molecule.
2. Balance the elements other than oxygen and hydrogen.
3. Balance oxygen and hydrogen.

   Acidic solution: For reactions in acidic solution, use H^+ and H_2O to balance oxygen and hydrogen. For each oxygen needed, use one H_2O. Then add H^+ as needed to balance the hydrogen atoms.

   Basic solution: For reactions in alkaline solutions, first balance as though the reaction were in an acidic solution, using Steps 1–3. Then add as many OH^- ions to each side of the equation as there are H^+ ions in the equation. Now combine the H^+ and OH^- ions into water (for example, $4H^+$ and $4OH^-$ give $4H_2O$). Rewrite the equation, canceling equal numbers of water molecules that appear on opposite sides of the equation.
4. Add electrons (e^-) to each half-reaction to bring them into electrical balance.
5. Since the loss and gain of electrons must be equal, multiply each half-reaction by the appropriate number to make the number of electrons the same in each half-reaction.
6. Add the two half-reactions together, canceling electrons and any other identical substances that appear on opposite sides of the equation.

KEY TERM

activity series of metals

17.4 Activity Series of Metals

- The activity series lists metals from most to least reactive.
- A free metal can displace anything lower on the activity series.

PROBLEM-SOLVING STRATEGY

Using the Activity Series to Predict Reactions

1. The reactivity of the metals listed decreases from top to bottom.
2. A free metal can displace the ion of a second metal from solution, provided that the free metal is above the second metal in the activity series.
3. Free metals above hydrogen react with nonoxidizing acids in solution to liberate hydrogen gas.
4. Free metals below hydrogen do not liberate hydrogen from acids.
5. Conditions such as temperature and concentration may affect the relative position of some of these elements.

KEY TERMS

electrolysis
electrolytic cell
cathode
anode
voltaic cell

17.5 Electrolytic and Voltaic Cells

- Electrolysis is the process of using electricity to bring about chemical change.
- Oxidation always occurs at the anode and reduction at the cathode.
- A cell that produces electric current from a spontaneous chemical reaction is a voltaic or galvanic cell.
- A typical electrolytic cell is shown here:

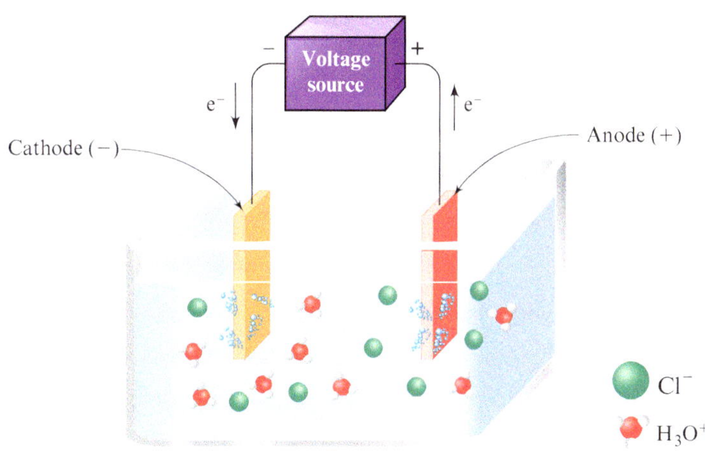

REVIEW QUESTIONS

1. In the equation
$$I_2 + 5Cl_2 + 6H_2O \longrightarrow 2HIO_3 + 10HCl$$
(a) has iodine been oxidized, or has it been reduced?
(b) has chlorine been oxidized, or has it been reduced? (Figure 17.2)
2. What is the difference between an oxidation number for an atom in an ionic compound and an oxidation number for an atom in a covalently bonded compound?
3. Why are oxidation and reduction said to be complementary processes?
4. Why are oxidation–reduction equations not usually balanced by inspection?
5. How many electrons are needed to balance the following half-reaction? Should the electrons be placed on the right or left side of the equation?
$$Mn^{2+} \longrightarrow Mn^{+6}$$
6. How many electrons are needed to balance the following half-reaction? Should the electrons be placed on the right or left side of the equation?
$$O^0 \longrightarrow O^{2-}$$
7. What is the important first step in balancing oxidation–reduction reactions by the ion–electron strategy?
8. Describe how you would balance the following balanced half-reaction in a basic solution. Write the equation in basic solution.
$$2\,IO_3^- + 12H^+ + 10e^- \longrightarrow I_2 + 6H_2O$$
9. Why do we say that the more active a metal is, the more easily it will be oxidized?
10. Which element in each pair is more chemically reactive? (Table 17.2)
(a) Mg or Ca (b) Fe or Ag (c) Zn or H
11. Complete and balance the equation for all of the reactions that will occur. If the reaction will not occur, explain why. (Table 17.2)
(a) $Al(s) + ZnCl_2(aq) \rightarrow$
(b) $Sn(s) + HCl(aq) \rightarrow$
(c) $Ag(s) + H_2SO_4(aq) \rightarrow$
(d) $Fe(s) + AgNO_3(aq) \rightarrow$
(e) $Cr(s) + Ni^{2+}(aq) \rightarrow$
(f) $Mg(s) + Ca^{2+}(aq) \rightarrow$
(g) $Cu(s) + H^+(aq) \rightarrow$
(h) $Ag(s) + Al^{3+}(aq) \rightarrow$
12. If a copper wire is placed into a solution of lead (II) nitrate, will a reaction occur? Explain.
13. The reaction between powdered aluminum and iron(III) oxide (in the thermite process) producing molten iron is very exothermic.
(a) Write the equation for the chemical reaction that occurs.
(b) Explain in terms of Table 17.2 why a reaction occurs.
(c) Would you expect a reaction between powdered iron and aluminum oxide?
(d) Would you expect a reaction between powdered aluminum and chromium(III) oxide?
14. Write equations for the chemical reaction of aluminum, chromium, gold, iron, copper, magnesium, mercury, and zinc with dilute solutions of (a) hydrochloric acid and (b) sulfuric acid. If a reaction will not occur, write "no reaction" as the product. (Table 17.2)
15. State the charge and purpose of the anode and the cathode in an electrolytic or volatic cell.
16. A $NiCl_2$ solution is placed in the apparatus shown in Figure 17.4, instead of the HCl solution shown. Write equations for the following:
(a) the anode reaction
(b) the cathode reaction
(c) the net electrochemical reaction
17. What is the major distinction between the reactions occurring in Figures 17.4 and 17.5?
18. In the cell shown in Figure 17.5,
(a) what would be the effect of removing the voltmeter and connecting the wires shown coming to the voltmeter?
(b) what would be the effect of removing the salt bridge?

19. When molten $CaBr_2$ is electrolyzed, calcium metal and bromine are produced. Write equations for the two half-reactions that occur at the electrodes. Label the anode half-reaction and the cathode half-reaction.
20. Why is direct current used instead of alternating current in the electroplating of metals?
21. What property of lead(IV) oxide and lead(II) sulfate makes it unnecessary to have salt bridges in the cells of a lead storage battery?
22. Explain why the density of the electrolyte in a lead storage battery decreases during the discharge cycle.
23. In one type of alkaline cell used to power devices such as portable radios, Hg^{2+} ions are reduced to metallic mercury when the cell is being discharged. Does this reduction occur at the anode or the cathode? Explain.
24. Differentiate between an electrolytic cell and a voltaic cell.
25. Why is a porous barrier or a salt bridge necessary in some voltaic cells?

PAIRED EXERCISES

Most of the exercises in this chapter are available for assignment via the online homework management program, WileyPLUS (www.wileyplus.com). All exercises with blue numbers have answers in Appendix 5.

1. Determine the oxidation number of each element in the compound:
   (a) $CuCO_3$
   (b) CH_4
   (c) IF
   (d) CH_2Cl_2
   (e) SO_2
   (f) $(NH_4)_2CrO_4$

3. Determine the oxidation number for each of the elements in blue:
   (a) PO_4^{3-}
   (b) $CaSO_4$
   (c) $NaHCO_3$
   (d) BrO_4^-

5. Many ionic compounds containing metal atoms are brightly colored. Several of these compounds are pictured below. For each compound determine the oxidation number of the specified element(s).

(a) Sodium chromate (Na_2CrO_4). Determine the oxidation state of chromium.

Courtesy Ondřej Mangl

(b) Potassium ferricyanide ($K_3[Fe(CN)_6]$). Determine the oxidation state of iron.

© Eric Nelson/Custom Medical Stock Photo—All rights reserved.

(c) Cobalt chloride $(CoCl)_2$. Determine the oxidation state of cobalt and chlorine.

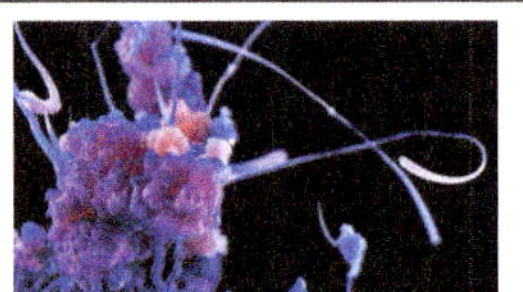

W. Oelen/http://woelen.homescience.net/science/index.html

(d) Nickel chloride hexahydrate ($NiCl_2 \cdot 6\,H_2O$). Determine the oxidation state of nickel.

Courtesy Benjah-bmm27 via Wikimedia

7. Determine whether each of the following half-reactions represents an oxidation or a reduction. Supply the correct number of electrons to the appropriate side to balance the equation.
   (a) $Na \longrightarrow Na^+$
   (b) $C_2O_4^{2-} \longrightarrow 2\,CO_2$
   (c) $2\,I^- \longrightarrow I_2$
   (d) $Cr_2O_7^{2-} + 14\,H^+ \longrightarrow 2\,Cr^{3+} + 7\,H_2O$

2. Determine the oxidation number of each element in the compound:
   (a) CHF_3
   (b) P_2O_5
   (c) SF_6
   (d) $SnSO_4$
   (e) CH_3OH
   (f) H_3PO_4

4. Determine the oxidation number for each of the elements in blue:
   (a) CO_3^{2-}
   (b) H_2SO_4
   (c) NaH_2PO_4
   (d) $Cr_2O_7^{2-}$

6. Many ionic compounds containing metal atoms are brightly colored. Several of these compounds are pictured below. For each compound determine the oxidation number of the specified element(s).

(a) Cupric sulfate pentahydrate ($CuSO_4 \cdot 5\,H_2O$). Determine the oxidation state of copper and sulfur.

© 1995 Richard Megna/Fundamental Photographs

(b) Potassium permanganate ($KMnO_4$). Determine the oxidation state of manganese.

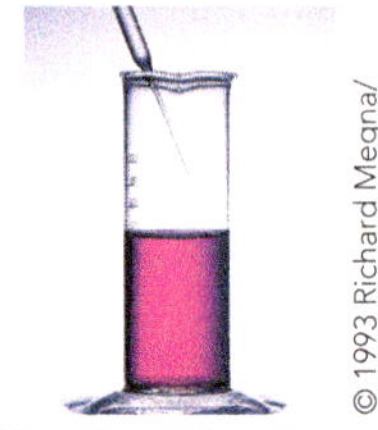

© 1993 Richard Megna/Fundamental Photographs

(c) Manganese dioxide (MnO_2). Determine the oxidation state of manganese.

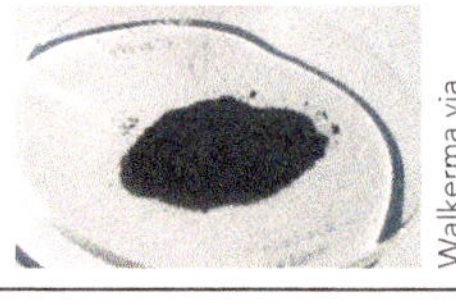

Walkerma via Wikimedia

(d) Potassium dichromate ($K_2Cr_2O_7$). Determine the oxidation state of chromium.

Courtesy Benjah-bmm27 via Wikimedia

8. Determine whether each of the following half-reactions represents an oxidation or a reduction. Supply the correct number of electrons to the appropriate side to balance the equation.
   (a) $Cu^{2+} \longrightarrow Cu^{1+}$
   (b) $F_2 \longrightarrow 2\,F^-$
   (c) $2\,IO_4^- + 16\,H^+ \longrightarrow I_2 + 8\,H_2O$
   (d) $Mn \longrightarrow Mn^{2+}$

9. In each of the following unbalanced equations, identify the oxidized element and the reduced element. Then identify the oxidizing agent and the reducing agent.
(a) $Cu + AgNO_3 \longrightarrow Ag + Cu(NO_3)_2$
(b) $Zn + HCl \longrightarrow H_2 + ZnCl_2$

10. In each of the following unbalanced equations, identify the oxidized element and the reduced element. Then identify the oxidizing agent and the reducing agent.
(a) $CH_4 + O_2 \longrightarrow CO_2 + H_2O$
(b) $Mg + FeCl_3 \longrightarrow Fe + MgCl_2$

11. Determine whether the following oxidation–reduction reactions are balanced correctly. If they are not, provide the correct balanced reaction.
(a) unbalanced:
$CH_4(g) + O_2(g) \longrightarrow CO_2(g) + H_2O(g)$
balanced:
$CH_4(g) + 2O_2(g) \longrightarrow CO_2(g) + 2H_2O(g)$
(b) unbalanced:
$NO^{2-}(aq) + Al(s) \longrightarrow NH_3(g) + AlO_2^-(aq)$ (basic solution)
balanced:
$2H_2O(l) + Al(s) + NO^{2-}(aq) \longrightarrow AlO_2^-(aq) + NH_3(g) + OH^-(aq)$
(c) unbalanced:
$Mg(s) + HCl(aq) \longrightarrow Mg^{2+}(aq) + Cl^-(aq) + H_2(g)$
balanced:
$Mg(s) + 2HCl(aq) \longrightarrow Mg^{2+}(aq) + Cl^-(aq) + H_2(g)$
(d) unbalanced:
$CH_3OH(aq) + Cr_2O_7^{2-}(aq) \longrightarrow CH_2O(aq) + Cr^{3+}(aq)$ (acidic solution)
balanced:
$3CH_3OH(aq) + 14H^+(aq) + Cr_2O_7^{2-}(aq) \longrightarrow 2Cr^{3+}(aq) + 3CH_2O(aq) + 7H_2O(l) + 6H^+(aq)$

12. Determine whether the following oxidation–reduction reactions are balanced correctly. If they are not, provide the correct balanced reaction.
(a) unbalanced:
$MnO_2(s) + Al(s) \longrightarrow Mn(s) + Al_2O_3(s)$
balanced:
$MnO_2(s) + 2Al(s) \longrightarrow Mn(s) + Al_2O_3(s)$
(b) unbalanced:
$Cu(s) + Ag^+(aq) \longrightarrow Cu^{2+}(aq) + Ag(s)$
balanced:
$Cu(s) + 2Ag^+(aq) \longrightarrow Cu^{2+}(aq) + 2Ag(s)$
(c) unbalanced:
$Br^-(aq) + MnO_4^-(aq) \longrightarrow Br_2(l) + Mn^{2+}(aq)$ (acidic solution)
balanced:
$16H^+(aq) + 10Br^-(aq) + 2MnO_4^-(aq) \longrightarrow 5Br_2(l) + 2Mn^{2+}(aq) + 8H_2O(l)$
(d) unbalanced:
$MnO_4^-(aq) + S^{2-}(aq) \longrightarrow MnS(s) + S(s)$ (basic solution)
balanced:
$8H^+(aq) + MnO_4^-(aq) + S^{2-}(aq) \longrightarrow S(s) + MnS(s) + 4H_2O(l)$

13. Balance each of the following equations using the change-in-oxidation-number method:
(a) $Cu + O_2 \longrightarrow CuO$
(b) $KClO_3 \longrightarrow KCl + O_2$
(c) $Ca + H_2O \longrightarrow Ca(OH)_2 + H_2$
(d) $PbS + H_2O_2 \longrightarrow PbSO_4 + H_2O$
(e) $CH_4 + NO_2 \longrightarrow N_2 + CO_2 + H_2O$

14. Balance each of the following equations using the change-in-oxidation-number method:
(a) $Cu + AgNO_3 \longrightarrow Ag + Cu(NO_3)_2$
(b) $MnO_2 + HCl \longrightarrow MnCl_2 + Cl_2 + H_2O$
(c) $HCl + O_2 \longrightarrow Cl_2 + H_2O$
(d) $Ag + H_2S + O_2 \longrightarrow Ag_2S + H_2O$
(e) $KMnO_4 + CaC_2O_4 + H_2SO_4 \longrightarrow K_2SO_4 + MnSO_4 + CaSO_4 + CO_2 + H_2O$

15. Balance these ionic redox equations using the ion electron method. These reactions occur in acidic solution.
(a) $Zn + NO_3^- \longrightarrow Zn^{2+} + NH_4^+$
(b) $NO_3^- + S \longrightarrow NO_2 + SO_4^{2-}$
(c) $PH_3 + I_2 \longrightarrow H_3PO_2 + I^-$
(d) $Cu + NO_3^- \longrightarrow Cu^{2+} + NO$
(e) $ClO_3^- + Cl^- \longrightarrow Cl_2$

16. Balance these ionic redox equations using the ion–electron method. These reactions occur in acidic solution.
(a) $ClO_3^- + I^- \longrightarrow I_2 + Cl^-$
(b) $Cr_2O_7^{2-} + Fe^{2+} \longrightarrow Cr^{3+} + Fe^{3+}$
(c) $MnO_4^- + SO_2 \longrightarrow Mn^{2+} + SO_4^{2-}$
(d) $H_3AsO_3 + MnO_4^- \longrightarrow H_3AsO_4 + Mn^{2+}$
(e) $Cr_2O_7^{2+} + H_3AsO_3 \longrightarrow Cr^{3+} + H_3AsO_4$

17. Balance these ionic redox equations using the ion–electron method. These reactions occur in basic solution.
(a) $Cl_2 + IO_3^- \longrightarrow Cl^- + IO_4^-$
(b) $MnO_4^- + ClO_2^- \longrightarrow MnO_2 + ClO_4^-$
(c) $Se \longrightarrow Se^{2-} + SeO_3^{2-}$
(d) $Fe_3O_4 + MnO_4^- \longrightarrow Fe_2O_3 + MnO_2$
(e) $BrO^- + Cr(OH)_4^- \longrightarrow Br^- + CrO_4^{2-}$

18. Balance these ionic redox equations using the ion–electron method. These reactions occur in basic solution.
(a) $MnO_4^- + SO_3^{2-} \longrightarrow MnO_2 + SO_4^{2-}$
(b) $ClO_2 + SbO_2^- \longrightarrow ClO_2^- + Sb(OH)_6^-$
(c) $Al + NO_3^- \longrightarrow NH_3 + Al(OH)_4^-$
(d) $P_4 \longrightarrow HPO_3^{2-} + PH_3$
(e) $Al + OH^- \longrightarrow Al(OH)_4^- + H_2$

19. Balance these reactions:
(a) $IO_3^-(aq) + I^-(aq) \longrightarrow I_2(aq)$ (acidic solution)
(b) $Mn^{2+}(aq) + S_2O_8^{2-}(aq) \longrightarrow MnO_4^-(aq) + SO_4^{2-}(aq)$ (acidic solution)
(c) $Co(NO_2)_6^{3-}(aq) + MnO_4^-(aq) \longrightarrow Co^{2+}(aq) + Mn^{2+}(aq) + NO_3^-(aq)$ (acidic solution)

20. Balance these reactions:
(a) $Mo_2O_3(s) + MnO_4^-(aq) \longrightarrow MoO_3(s) + Mn^{2+}(aq)$ (acidic solution)
(b) $BrO^-(aq) + Cr(OH)_4^-(aq) \longrightarrow Br^-(aq) + CrO_4^{2-}(aq)$ (basic solution)
(c) $S_2O_3^{2-}(aq) + MnO_4^-(aq) \longrightarrow SO_4^{2-}(aq) + MnO_2(s)$ (basic solution)

ADDITIONAL EXERCISES

Questions 21–23 pertain to the activity series below.

K Ca Mg Al Zn Fe H Cu Ag

21. Which of these pairs will not react in water solution?
    (a) Zn, $CuSO_4$
    (b) Cu, $Al_2(SO_4)_3$
    (c) Fe, $AgNO_3$
    (d) Zn, $Ca(NO_3)_2$
    (e) Ag, HCl
    (f) Mg, $Al_2(SO_4)_3$
22. Which element is the most easily oxidized? Which element is the least easily oxidized?
23. Which elements will reduce Cu^{2+} to Cu but will not reduce Zn^{2+} to Zn?
24. Vitamin C (ascorbic acid) is often touted as a good nutritional supplement because of its antioxidant properties. Vitamin C is very easily oxidized and, therefore, will be preferentially oxidized in the presence of an oxidizing agent. The hope is that vitamin C will coexist in cellular solutions and thus prevent the oxidation of other important biomolecules. Write and balance the equation for the oxidation of vitamin C ($C_6H_8O_6$) by oxygen gas. The products of this reaction are dehydroascorbic acid ($C_6H_6O_6$) and water.
25. Most explosive reactions are complex redox reactions with multiple oxidations and reductions. The reaction of gunpowder is shown below. Determine the element(s) oxidized and reduced.

$$4\,KNO_3(s) + 7C(s) + S(s) \longrightarrow 3\,CO_2(g) + 3\,CO(g) + 2\,N_2(g) + K_2CO_3(s) + K_2S(s)$$

26. Draw a picture of an electrolytic cell made from an aqueous HBr solution.
27. The chemical reactions taking place during discharge in a lead storage battery are

$$Pb + SO_4^{2-} \longrightarrow PbSO_4$$
$$PbO_2 + SO_4^{2-} + 4H^+ \longrightarrow PbSO_4 + 2H_2O$$

    (a) Complete each half-reaction by supplying electrons.
    (b) Which reaction is oxidation, and which is reduction?
    (c) Which reaction occurs at the anode of the battery?
28. Use this unbalanced redox equation

$$KMnO_4 + HCl \longrightarrow KCl + MnCl_2 + H_2 + Cl_2$$

    to indicate
    (a) the oxidizing agent
    (b) the reducing agent
    (c) the number of electrons that are transferred per mole of oxidizing agent
29. Brass is an alloy of zinc and copper. When brass is in contact with salt water, it corrodes as the zinc dissolves from the alloy, leaving almost pure copper. Explain why the zinc is preferentially dissolved.
30. How many liters of NO_2 gas at STP will be formed by the reaction of 75.5 g of copper with 55.0 g of nitric acid? Which substance is oxidized, and which one is reduced?

$$Cu + HNO_3 \longrightarrow Cu(NO_3)_2 + NO_2 + H_2O \quad \text{(acidic solution)}$$

31. How many liters of CO_2 gas at STP will be formed by the reaction of 25.5 g of $K_2S_2O_8$ and 35.5 g of $H_2C_2O_4$? Which substance is oxidized, and which one is reduced?

$$K_2S_2O_8 + H_2C_2O_4 \longrightarrow K_2SO_4 + H_2SO_4 + CO_2$$

32. The amount of vitamin C can be determined experimentally by titration. The balanced chemical reaction between vitamin C or ascorbic acid and iodine is shown below. If the vitamin C is extracted from 100.0 grams of green peppers and reacted with 0.03741 *M* potassium triiodide, it requires 32.61 mL of the solution to react with all of the vitamin C. Calculate the number of milligrams of vitamin C per gram of green pepper.

$$C_6H_8O_6(aq) + KI_3(aq) \longrightarrow C_6H_6O_6(aq) + KI(aq) + 2\,HI(aq)$$

33. Alcohols can be oxidized very easily by potassium dichromate. This fact formed the basis for many of the early breathalyzers used by law enforcement officers to determine the approximate blood alcohol levels of suspected drunk drivers. The unbalanced equation for the reaction is

$$CH_3CH_2OH + Cr_2O_7^{2-} \longrightarrow CH_3COOH + Cr^{3+}$$

    In this reaction, the potassium dichromate is orange and the chromium(III) ions are blue. The bluer a solution of potassium dichromate becomes, the greater the concentration of alcohol in the breath. Write a balanced ionic equation for this reaction in an acidic environment.
34. What mass of $KMnO_4$ is needed to react with 100. mL H_2O_2 solution? (d = 1.031 g/mL, 9.0% H_2O_2 by mass)

$$H_2O_2 + KMnO_4 + H_2SO_4 \longrightarrow O_2 + MnSO_4 + K_2SO_4 + H_2O \quad \text{(acidic solution)}$$

35. How many grams of zinc are required to reduce Fe^{3+} when 25.0 mL of 1.2 *M* $FeCl_3$ are reacted?

$$Zn + Fe^{3+} \longrightarrow Zn^{2+} + Fe$$

36. What volume of 0.200 *M* $K_2Cr_2O_7$ will be required to oxidize the Fe^{2+} ion in 60.0 mL of 0.200 *M* $FeSO_4$ solution?

$$Cr_2O_7^{2-} + Fe^{2+} \longrightarrow Cr^{3+} + Fe^{3+} \quad \text{(acidic solution)}$$

37. How many moles of H_2 can be produced from 100.0 g Al according to this reaction?

$$Al + OH^- \longrightarrow Al(OH)_4^- + H_2 \quad \text{(basic solution)}$$

38. There is something incorrect about these half-reactions:
    (a) $Cu^+ + e^- \longrightarrow Cu^{2+}$ (b) $Pb^{2+} + e^{2-} \longrightarrow Pb$
    Identify what is wrong and correct it.
39. Why can oxidation *never* occur without reduction?
40. The following observations were made concerning metals A, B, C, and D.
    (a) When a strip of metal A is placed in a solution of B^{2+} ions, no reaction is observed.
    (b) Similarly, A in a solution containing C^+ ions produces no reaction.
    (c) When a strip of metal D is placed in a solution of C^+ ions, black metallic C deposits on the surface of D, and the solution tests positively for D^{2+} ions.
    (d) When a piece of metallic B is placed in a solution of D^{2+} ions, metallic D appears on the surface of B and B^{2+} ions are found in the solution.
    Arrange the ions—A^+, B^{2+}, C^+, and D^{2+}—in order of their ability to attract electrons. List them in order of increasing ability.
41. Tin normally has oxidation numbers of 0, +2, and +4. Which of these species can be an oxidizing agent, which can be a reducing agent, and which can be both? In each case, what product would you expect as the tin reacts?
42. Manganese is an element that can exist in numerous oxidation states. In each of these compounds, identify the oxidation number of the manganese. Which compound would you expect to be the best oxidizing agent and why?
    (a) $Mn(OH)_2$
    (b) MnF_3
    (c) MnO_2
    (d) K_2MnO_4
    (e) $KMnO_4$

43. Which equations represent oxidations?
(a) $Mg \longrightarrow Mg^{2+}$
(b) $SO_2 \longrightarrow SO_3$
(c) $KMnO_4 \longrightarrow MnO_2$
(d) $Cl_2O_3 \longrightarrow Cl^-$

44. In the following equation, note the reaction between manganese(IV) oxide and bromide ions:

$$MnO_2 + Br^- \longrightarrow Br_2 + Mn^{2+}$$

(a) Balance this redox reaction in acidic solution.
(b) How many grams of MnO_2 would be needed to produce 100.0 mL of 0.05 M Mn^{2+}?
(c) How many liters of bromine vapor at 50°C and 1.4 atm would also result?

45. Use the table shown to complete the following reactions. If no reaction occurs, write NR:
(a) $F_2 + Cl^- \longrightarrow$
(b) $Br_2 + Cl^- \longrightarrow$
(c) $I_2 + Cl^- \longrightarrow$
(d) $Br_2 + I^- \longrightarrow$

Activity
F_2
Cl_2
Br_2
I_2

(Ease of reduction ↑)

46. Manganese metal reacts with HCl to give hydrogen gas and the Mn^{2+} ion in solution. Write a balanced equation for the reaction.

47. If zinc is allowed to react with dilute nitric acid, zinc is oxidized to the 2+ ion, while the nitrate ion can be reduced to ammonium, NH_4^+. Write a balanced equation for the reaction in acidic solution.

48. In the following equations, identify the
(a) atom or ion oxidized
(b) atom or ion reduced
(c) oxidizing agent
(d) reducing agent
(e) change in oxidation number associated with each oxidizing process
(f) change in oxidation number associated with each reducing process

1. $C_3H_8 + O_2 \longrightarrow CO_2 + H_2O$
2. $HNO_3 + H_2S \longrightarrow NO + S + H_2O$
3. $CuO + NH_3 \longrightarrow N_2 + H_2O + Cu$
4. $H_2O_2 + Na_2SO_3 \longrightarrow Na_2SO_4 + H_2O$
5. $H_2O_2 \longrightarrow H_2O + O_2$

49. In the galvanic cell shown in the diagram, a strip of silver is placed in a solution of silver nitrate, and a strip of lead is placed in a solution of lead(II) nitrate. The two beakers are connected with a salt bridge. Determine
(a) the anode
(b) the cathode
(c) where oxidation occurs
(d) where reduction occurs
(e) in which direction electrons flow through the wire
(f) in which direction ions flow through the solution

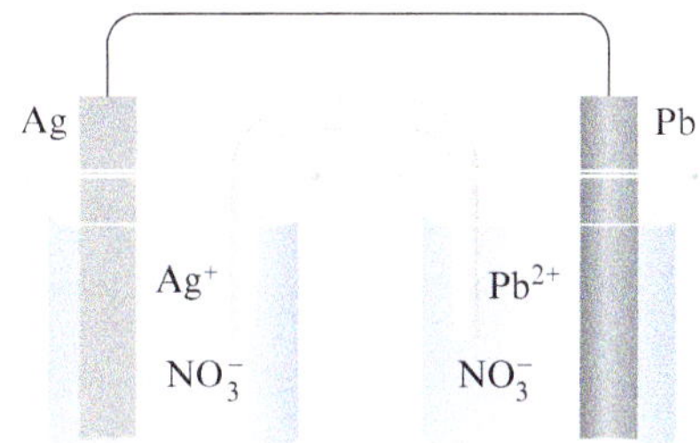

CHALLENGE EXERCISES

50. For each set of beakers below, draw a picture you might expect to see when the contents of the beakers are mixed together and allowed to react.
(a)

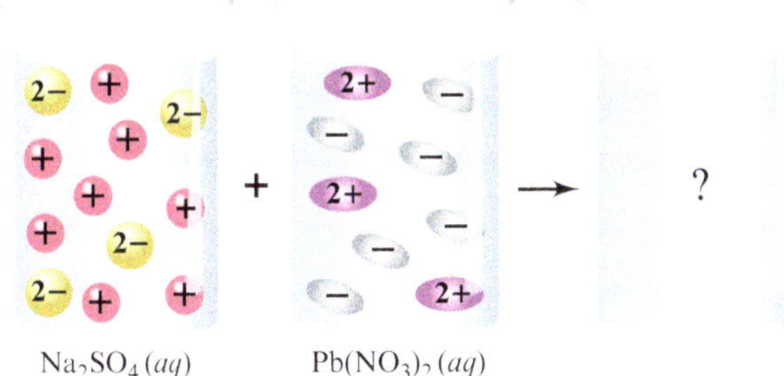

(b)

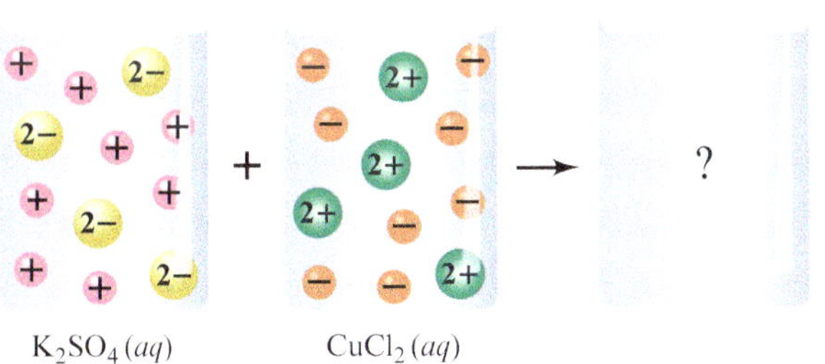

51. The hydroxide ion concentration of a solution is 3.4×10^{-10} M. Balance the following equation:

$$Fe^{2+}(aq) + MnO_4^-(aq) \longrightarrow Fe^{3+}(aq) + Mn^{2+}(aq)$$

52. A sample of crude potassium iodide was analyzed using this reaction (not balanced):

$$I^- + SO_4^{2-} \longrightarrow I_2 + H_2S \quad \text{(acidic solution)}$$

If a 4.00-g sample of crude KI produced 2.79 g of iodine, what is the percent purity of the KI?

53. What volume of NO gas, measured at 28°C and 744 torr, will be formed by the reaction of 0.500 mol Ag reacting with excess nitric acid?

$$Ag + HNO_3 \longrightarrow AgNO_3 + NO + H_2O \quad \text{(acidic solution)}$$

ANSWERS TO PRACTICE EXERCISES

17.1 (a) $S = +6$, (b) $As = +5$, (c) $C = +4$

17.2 (*Note:* $H = +1$ even though it comes second in the formula: N is a nonmetal.)

(a) $N = -3$, (b) $Cr = +6$, (c) $P = +5$

17.3 (a) $Be = +2$; $Cl = -1$,

(b) $H = +1$; $Cl = +1$; $O = -2$,

(c) $H = +1$; $O = -1$,

(d) $N = -3$; $H = +1$,

(e) $Br = +5$; $O = -2$

17.4 (a) Al is oxidized and loses 3 electrons. H is reduced and gains 1 electron. Al is the reducing agent and HNO_3 is the oxidizing agent.

(b) Ni is oxidized and loses 2 electrons. Cu is reduced and gains 2 electrons. Ni is the reducing agent and $CuSO_4$ is the oxidizing agent.

(c) C is oxidized and loses 1 electron. O is reduced and gains 2 electrons. C_2H_6 is the reducing agent and O_2 is the oxidizing agent.

17.5 (a) $6HNO_3 + S \longrightarrow 6NO_2 + H_2SO_4 + 2H_2O$

(b) $2CrCl_3 + 3MnO_2 + 2H_2O \longrightarrow 3MnCl_2 + 2H_2CrO_4$

(c) $2KMnO_4 + 6HCl + 5H_2S \longrightarrow 2KCl + 2MnCl_2 + 5S + 8H_2O$

17.6 (a) $4H^+ + 2I^- + 2NO_2^- \longrightarrow I_2 + 2NO + 2H_2O$

(b) $2OH^- + Cl_2 + IO_3^- \longrightarrow IO_4^- + H_2O + 2Cl^-$

(c) $AuCl_4^- + Sn^{2+} \longrightarrow Sn^{4+} + AuCl + 3Cl^-$

17.7 (a) $Zn + 2H_2O + 2OH^- \longrightarrow Zn(OH)_4^{2-} + H_2$

(b) $H_2O_2 + Sn^{2+} + 2H^+ \longrightarrow Sn^{4+} + 2H_2O$

(c) $Cu + Cu^{2+} + 2OH^- \longrightarrow Cu_2O + H_2O$

17.8 (a) yes, (b) no, (c) no

17.9 (a) Cu is oxidized; Sn is reduced

(b) anode: $Cu(s) \longrightarrow Cu^{2+}(aq) + 2e^-$

cathode: $Sn^{2+}(aq) + 2e^- \longrightarrow Sn(s)$

18

Nuclear Chemistry

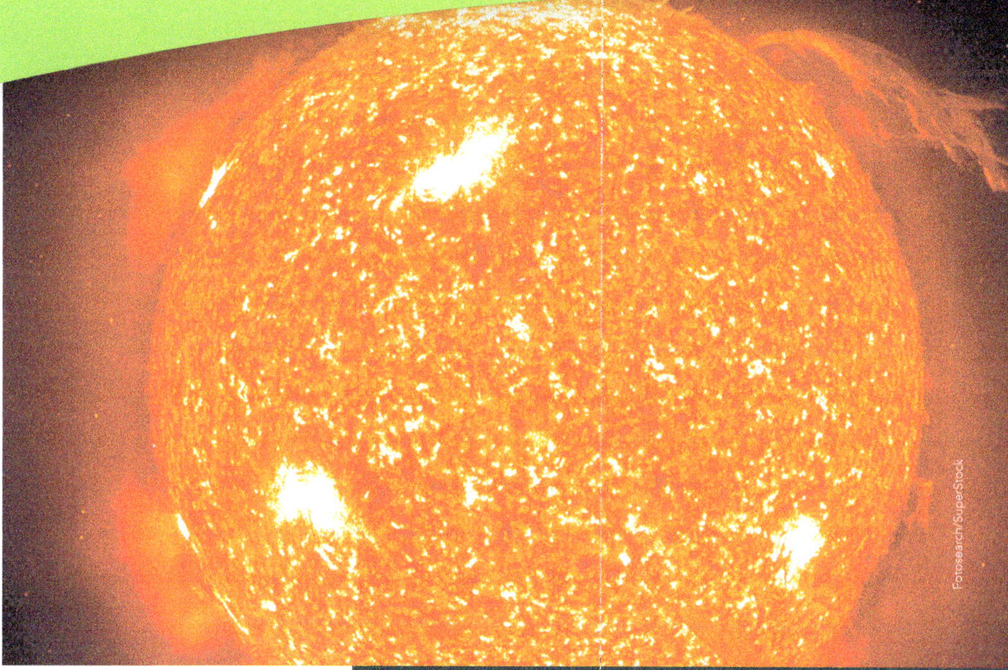

CHAPTER OUTLINE

The nucleus of the atom is a source of tremendous energy. The energy of the Sun comes from nuclear reactions. Solar flares are an indication of fusion reactions occurring at a temperature of millions of degrees. Harnessing this energy has enabled us to fuel power stations, treat cancer, and preserve food. We use isotopes in medicine to diagnose illness and to detect minute quantities of drugs or hormones. Researchers use radioactive tracers to sequence the human genome. We also use nuclear processes to detect explosives in luggage and to establish the age of objects such as human artifacts and rocks. In this chapter we'll consider properties of atomic nuclei and their applications in our lives.

18.1 Discovery of Radioactivity

LEARNING OBJECTIVE Describe the particles associated with nuclear chemistry and radioactivity and use half-life to calculate the age of an object.

KEY TERMS
radioactivity
nucleon
nuclide
radioactive decay
half-life ($t_{1/2}$)

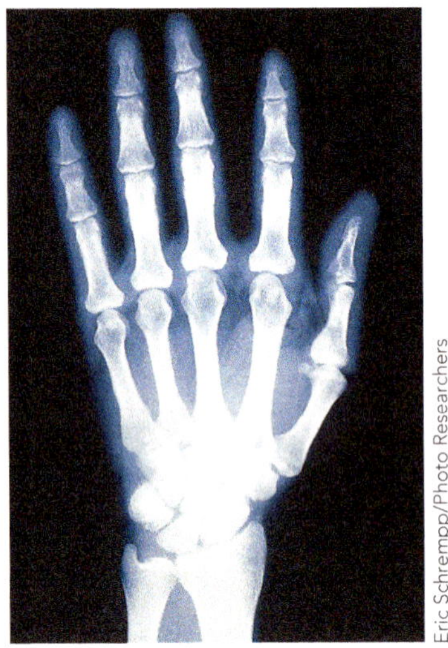

X-ray technology opened the door to a new world of diagnosis and treatment.

In 1895, Wilhelm Conrad Roentgen (1845–1923) made an important breakthrough that eventually led to the discovery of radioactivity. Roentgen discovered X-rays when he observed that a vacuum discharge tube enclosed in a thin, black cardboard box caused a nearby piece of paper coated with the salt barium platinocyanide to glow with a brilliant phosphorescence. From this and other experiments, he concluded that certain rays, which he called X-rays, were emitted from the discharge tube, penetrated the box, and caused the salt to glow. Roentgen's observations that X-rays could penetrate other bodies and affect photographic plates led to the development of X-ray photography.

Shortly after this discovery, Antoine Henri Becquerel (1852–1908) attempted to show a relationship between X-rays and the phosphorescence of uranium salts. In one experiment he wrapped a photographic plate in black paper, placed a sample of uranium salt on it, and exposed it to sunlight. The developed photographic plate showed that rays emitted from the salt had penetrated the paper. When Becquerel attempted to repeat the experiment, the sunlight was intermittent, so he placed the entire setup in a drawer. Several days later he developed the photographic plate, expecting to find it only slightly affected. To his amazement, he found an intense image on the plate. He repeated the experiment in total darkness and obtained the same results, proving that the uranium salt emitted rays that affected the photographic plate without its being exposed to sunlight. Thus did the discovery of radioactivity come about, a combination of numerous experiments by the finest minds of the day—and serendipity. Becquerel later showed that the rays coming from uranium are able to ionize air and are also capable of penetrating thin sheets of metal.

The name *radioactivity* was coined two years later (in 1898) by Marie Curie. **Radioactivity** is the spontaneous emission of particles and/or rays from the nucleus of an atom. Elements having this property are said to be radioactive.

In 1898, Marie Sklodowska Curie (1867–1934) and her husband Pierre Curie (1859–1906) turned their research interests to radioactivity. In a short time, the Curies discovered two new elements, polonium and radium, both of which are radioactive. To confirm their work on radium, they processed 1 ton of pitchblende residue ore to obtain 0.1 g of pure radium chloride, which they used to make further studies on the properties of radium and to determine its atomic mass.

In 1899, Ernest Rutherford began to investigate the nature of the rays emitted from uranium. He found two particles, which he called *alpha* and *beta particles.* Soon he realized that uranium, while emitting these particles, was changing into another element. By 1912, over 30 radioactive isotopes were known, and many more are known today. The *gamma ray*, a third type of emission from radioactive materials similar to an X-ray, was discovered by Paul Villard (1860–1934) in 1900. Rutherford's description of the nuclear atom led scientists to attribute the phenomenon of radioactivity to reactions taking place in the nuclei of atoms.

The symbolism and notation we described for isotopes in Chapter 5 are also very useful in nuclear chemistry:

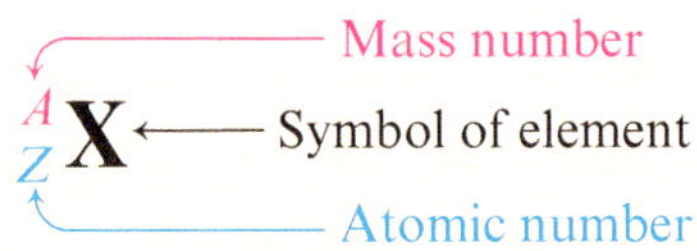

For example, $^{238}_{92}U$ represents a uranium isotope with an atomic number of 92 and a mass number of 238. This isotope is also designated as U-238, or uranium-238, and contains 92 protons and 146 neutrons. The protons and neutrons collectively are

TABLE 18.1 Isotopic Notation for Several Particles (and Small Isotopes) Associated with Nuclear Chemistry

Particle	Symbol	Z Atomic number	A Mass number
Neutron	${}^{1}_{0}n$	0	1
Proton	${}^{1}_{1}H$	1	1
Beta particle (electron)	${}^{0}_{-1}e$	−1	0
Positron (positive electron)	${}^{0}_{+1}e$	1	0
Alpha particle (helium nucleus)	${}^{4}_{2}He$	2	4
Deuteron (heavy hydrogen nucleus)	${}^{2}_{1}H$	1	2

known as **nucleons**. The mass number is the total number of nucleons in the nucleus. TABLE 18.1 shows the isotopic notations for several particles associated with nuclear chemistry.

When we speak of isotopes, we mean atoms of the same element with different masses, such as ${}^{16}_{8}O$, ${}^{17}_{8}O$, ${}^{18}_{8}O$. In nuclear chemistry we use the term **nuclide** to mean any isotope of any atom. Thus, ${}^{16}_{8}O$ and ${}^{235}_{92}U$ are referred to as nuclides. Nuclides that spontaneously emit radiation are referred to as *radionuclides*.

CHECK YOUR UNDERSTANDING
18.1 Isotopic Notation
➥ WileyPLUS

Natural Radioactivity

Radioactive elements continuously undergo **radioactive decay**, or disintegration, to form different elements. The chemical properties of an element are associated with its electronic structure, but radioactivity is a property of the nucleus. Therefore, neither ordinary changes of temperature and pressure nor the chemical or physical state of an element has any effect on its radioactivity.

The principal emissions from the nuclei of radionuclides are known as alpha particles, beta particles, and gamma rays. Upon losing an alpha or beta particle, the radioactive element changes into a different element. We will explain this process in detail later.

Each radioactive nuclide disintegrates at a specific and constant rate, which is expressed in units of half-life. The **half-life ($t_{1/2}$)** is the time required for one-half of a specific amount of a radioactive nuclide to disintegrate. The half-lives of the elements range from a fraction of a second to billions of years. To illustrate, suppose we start with 1.0 g of ${}^{226}_{88}Ra$ ($t_{1/2}$ = 1620 years):

$$1.0\,g\,{}^{226}_{88}Ra \xrightarrow[1620\ years]{t_{1/2}} 0.50\,g\,{}^{226}_{88}Ra \xrightarrow[1620\ years]{t_{1/2}} 0.25\,g\,{}^{226}_{88}Ra$$

Element	$t_{1/2}$
${}^{238}_{92}U$	4.5×10^{9} years
${}^{226}_{88}Ra$	1620 years
${}^{15}_{6}C$	2.4 seconds

The half-lives of the various radioisotopes of the same element differ dramatically. Half-lives for certain isotopes of radium, carbon, and uranium are listed in TABLE 18.2.

TABLE 18.2 Half-Lives for Radium, Carbon, and Uranium Isotopes

Isotope	Half-life	Isotope	Half-life
Ra-223	11.7 days	C-14	5730 years
Ra-224	3.64 days	C-15	2.4 seconds
Ra-225	14.8 days	U-235	7.1×10^{8} years
Ra-226	1620 years	U-238	4.5×10^{9} years
Ra-228	6.7 years		

CHECK YOUR UNDERSTANDING
18.2 Half-Life
➥ WileyPLUS

ENHANCED EXAMPLE 18.1

WileyPLUS

The half-life of $^{131}_{53}I$ is 8 days. How much $^{131}_{53}I$ from a 32-g sample remains after five half-lives?

SOLUTION

Using the following graph, we can find the number of grams of ^{131}I remaining after one half-life:

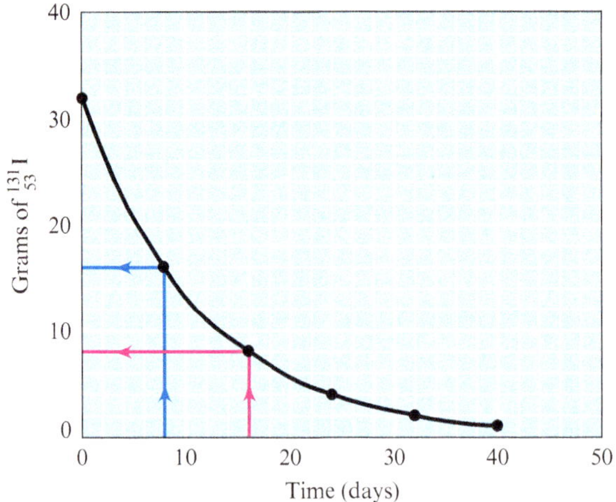

Trace a perpendicular line from 8 days on the x-axis to the line on the graph. Now trace a horizontal line from this point on the plotted line to the y-axis and read the corresponding grams of ^{131}I. Continue this process for each half-life, adding 8 days to the previous value on the x-axis:

Half-lives	0	1	2	3	4	5
Number of days		8	16	24	32	40
Amount remaining	32 g	16 g	8 g	4 g	2 g	1 g

Starting with 32 g, one gram of ^{131}I remains after five half-lives (40 days).

ONLINE LEARNING MODULE

Radioactive Decay

WileyPLUS

ENHANCED EXAMPLE 18.2

WileyPLUS

In how many half-lives will 10.0 g of a radioactive nuclide decay to less than 10% of its original value?

SOLUTION

We know that 10% of the original amount is 1.0 g. After the first half-life, half the original material remains and half has decayed (5.00 g). After the second half-life, one-fourth of the original material remains (i.e., one-half of 5.00 g). This progression continues, reducing the quantity remaining by half for each half-life that passes.

Half-lives	0	1	2	3	4
Percent remaining	100%	50%	25%	12.5%	6.25%
Amount remaining	10.0 g	5.00 g	2.50 g	1.25 g	0.625 g

Therefore, the amount remaining will be less than 10% sometime between the third and the fourth half-lives.

PRACTICE 18.1

The half-life of $^{14}_{6}C$ is 5730 years. How much $^{14}_{6}C$ will remain after six half-lives in a sample that initially contains 25.0 g?

Nuclides are said to be either *stable* (nonradioactive) or *unstable* (radioactive). Elements that have atomic numbers greater than 83 (bismuth) are naturally radioactive, although some of the nuclides have extremely long half-lives. Some of the naturally occurring nuclides of elements 81, 82, and 83 are radioactive, and some are stable. Only a few naturally occurring elements that have atomic numbers less than 81 are radioactive. However, no stable isotopes of element 43 (technetium) or of element 61 (promethium) are known.

Radioactivity is believed to be a result of an unstable ratio of neutrons to protons in the nucleus. Stable nuclides of elements up to about atomic number 20 generally have about a 1:1 neutron-to-proton ratio. In elements above number 20, the neutron-to-proton ratio in the stable nuclides gradually increases to about 1.5:1 in element number 83 (bismuth). When the neutron-to-proton ratio is too high or too low, alpha, beta, or other particles are emitted to achieve a more stable nucleus.

18.2 Alpha Particles, Beta Particles, and Gamma Rays

LEARNING OBJECTIVE

Describe the change in atomic number or atomic mass associated with emission of an alpha particle, beta particle, and gamma ray.

KEY TERMS

alpha particle (α)
beta particle (β)
gamma ray (γ)

The classical experiment proving that alpha and beta particles are oppositely charged was performed by Marie Curie (FIGURE 18.1). She placed a radioactive source in a hole in a lead block and positioned two poles of a strong electromagnet so that the radiations that were given off passed between them. The paths of three different kinds of radiation were detected by means of a photographic plate placed some distance beyond the electromagnet. The lighter beta particles were strongly deflected toward the positive pole of the electromagnet; the heavier alpha particles were less strongly deflected and in the opposite direction. The uncharged gamma rays were not affected by the electromagnet and struck the photographic plates after traveling along a path straight out of the lead block.

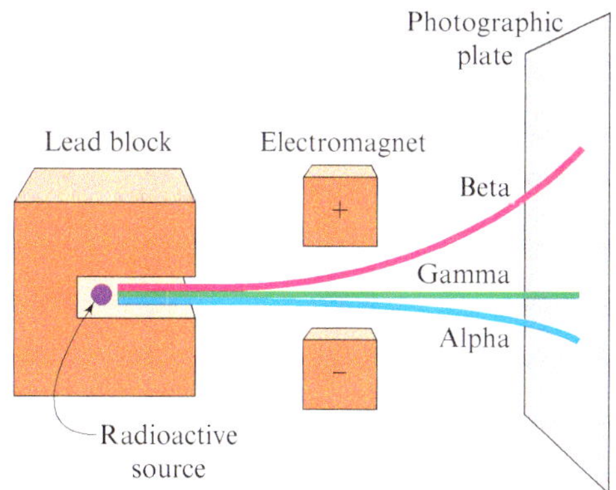

FIGURE 18.1 The effect of an electromagnetic field on alpha particles, beta particles, and gamma rays. Lighter beta particles are deflected considerably more than alpha particles. Alpha and beta particles are deflected in opposite directions. Gamma radiation is not affected by the electromagnetic field.

Alpha Particles

An **alpha particle (α)** consists of two protons and two neutrons, has a mass of about 4 amu, and has a charge of +2. It is a helium nucleus that is usually given one of the following symbols: α or ${}^{4}_{2}\text{He}$. When an alpha particle is emitted from the nucleus, a different element is formed. The atomic number of the new element is 2 less, and the mass is 4 amu less, than that of the starting element.

Loss of an alpha particle from the nucleus results in
loss of 4 in the mass number (A)
loss of 2 in the atomic number (Z)

For example, when $^{238}_{92}U$ loses an alpha particle, $^{234}_{90}Th$ is formed, because two neutrons and two protons are lost from the uranium nucleus. This disintegration may be written as a nuclear equation:

$$^{238}_{92}U \longrightarrow \, ^{234}_{90}Th + \alpha \quad \text{or} \quad ^{238}_{92}U \longrightarrow \, ^{234}_{90}Th + \, ^{4}_{2}He$$

CHECK YOUR UNDERSTANDING
18.3 Alpha Particles
WileyPLUS

For the loss of an alpha particle from $^{226}_{88}Ra$, the equation is

$$^{226}_{88}Ra \longrightarrow \, ^{222}_{86}Rn + \, ^{4}_{2}He \quad \text{or} \quad ^{226}_{88}Ra \longrightarrow \, ^{222}_{86}Rn + \alpha$$

A nuclear equation, like a chemical equation, consists of reactants and products and must be balanced. To have a balanced nuclear equation, the sum of the mass numbers (superscripts) on both sides of the equation must be equal and the sum of the atomic numbers (subscripts) on both sides of the equation must be equal:

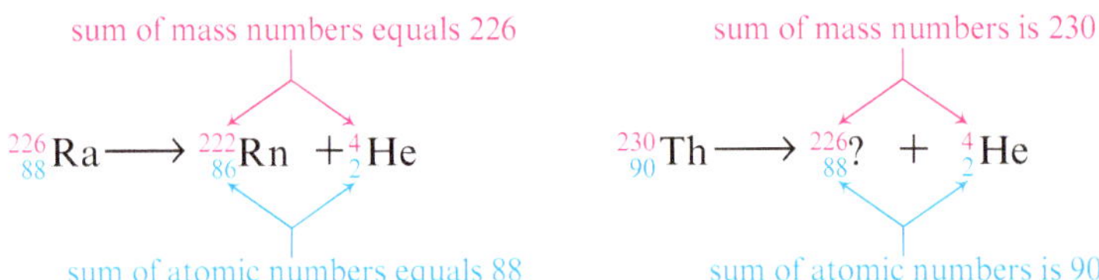

What new nuclide will be formed when $^{230}_{90}Th$ loses an alpha particle? The new nuclide will have a mass of 226 amu and will contain 88 protons, so its atomic number is 88. Locate the corresponding element on the periodic table—in this case, $^{226}_{88}Ra$ or radium-226.

Beta Particles

The **beta particle (β)** is identical in mass and charge to an electron; its charge is −1. Both a beta particle and a proton are produced by the decomposition of a neutron:

$$^{1}_{0}n \longrightarrow \, ^{1}_{1}p + \, ^{0}_{-1}e$$

The beta particle leaves, and the proton remains in the nucleus. When an atom loses a beta particle from its nucleus, a different element is formed that has essentially the same mass but an atomic number that is 1 greater than that of the starting element. The beta particle is written as β or $^{0}_{-1}e$.

Loss of a beta particle from the nucleus results in
 no change in the mass number (A)
 increase of 1 in the atomic number (Z)

CHECK YOUR UNDERSTANDING
18.4 Beta Particles
WileyPLUS

Examples of equations in which a beta particle is lost are

$$^{234}_{90}Th \longrightarrow \, ^{234}_{91}Pa + \beta$$

$$^{234}_{91}Pa \longrightarrow \, ^{234}_{92}U + \, ^{0}_{-1}e$$

$$^{210}_{82}Pb \longrightarrow \, ^{210}_{83}Bi + \beta$$

Gamma Rays

Gamma rays (γ) are photons of energy. A gamma ray is similar to an X-ray but is more energetic. It has no electrical charge and no measurable mass. Gamma rays are released from the nucleus in many radioactive changes along with either alpha or beta particles. Gamma radiation does not result in a change of atomic number or the mass of an element.

Loss of a gamma ray from the nucleus results in
 no change in mass number (A) or atomic number (Z)

ENHANCED EXAMPLE 18.3

WileyPLUS

(a) Write an equation for the loss of an alpha particle from the nuclide $^{194}_{78}Pt$.

(b) What nuclide is formed when $^{228}_{88}Ra$ loses a beta particle from its nucleus?

SOLUTION

(a) Loss of an alpha particle, $^{4}_{2}He$, results in a decrease of 4 in the mass number and a decrease of 2 in the atomic number:

Mass of new nuclide:	$A - 4$	or	$194 - 4 = 190$
Atomic number of new nuclide:	$Z - 2$	or	$78 - 2 = 76$

Looking up element number 76 on the periodic table, we find it to be osmium, Os. The equation then is

$$^{194}_{78}Pt \longrightarrow {}^{190}_{76}Os + {}^{4}_{2}He$$

(b) The loss of a beta particle from a $^{228}_{88}Ra$ nucleus means a gain of 1 in the atomic number with no essential change in mass. The new nuclide will have an atomic number of $(Z + 1)$, or 89, which is actinium, Ac:

$$^{228}_{88}Ra \longrightarrow {}^{228}_{89}Ac + {}^{0}_{-1}e$$

ONLINE LEARNING MODULE
Alpha Emission Nuclear Equations
➡ WileyPLUS

EXAMPLE 18.4

What nuclide will be formed when $^{214}_{82}Pb$ successively emits two beta particles, then one alpha particle from its nucleus? Write successive equations showing these changes.

ONLINE LEARNING MODULE
Nuclear Decay
➡ WileyPLUS

SOLUTION

The changes brought about in the three steps outlined are as follows:

β loss:	Increase of 1 in the atomic number; no change in mass
β loss:	Increase of 1 in the atomic number; no change in mass
α loss:	Decrease of 2 in the atomic number; decrease of 4 in the mass

The equations are

$$^{214}_{82}Pb \longrightarrow {}^{214}_{83}X + \beta \longrightarrow {}^{214}_{84}X + \beta \longrightarrow {}^{210}_{82}X + \alpha$$

where X stands for the new nuclide formed. Looking up each of these elements by their atomic numbers, we rewrite the equations

$$^{214}_{82}Pb \xrightarrow[\beta]{} {}^{214}_{83}Bi \xrightarrow[\beta]{} {}^{214}_{84}Po \xrightarrow[\alpha]{} {}^{210}_{82}Pb$$

PRACTICE 18.2

What nuclide will be formed when $^{222}_{86}Rn$ emits an alpha particle?

The ability of radioactive rays to pass through various objects is in proportion to the speed at which they leave the nucleus. Gamma rays travel at the velocity of light (186,000 miles per second) and are capable of penetrating several inches of lead. The velocities of beta particles are variable, the fastest being about nine-tenths the velocity of light. Alpha particles have velocities less than one-tenth the velocity of light. FIGURE 18.2 illustrates the relative penetrating power of these rays. A few sheets of paper will stop alpha particles; a thin sheet of aluminum will stop both alpha and beta particles; and a 5-cm block of lead will reduce, but not completely stop, gamma radiation. In fact, it is difficult to stop all gamma radiation. TABLE 18.3 summarizes the properties of alpha, beta, and gamma radiation.

CHECK YOUR UNDERSTANDING
18.5 Symbols for Nuclear Particles
➡ WileyPLUS

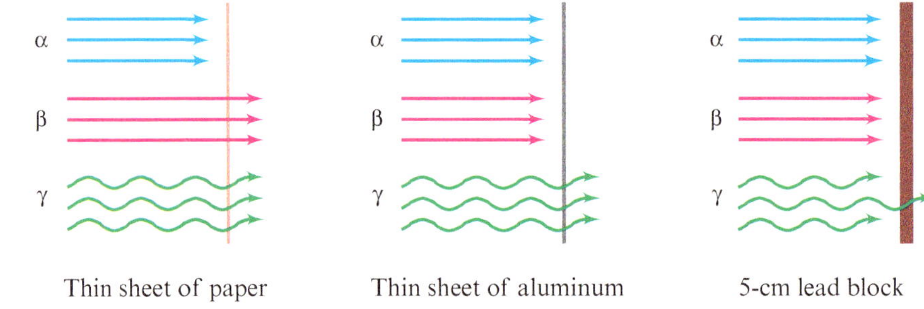

FIGURE 18.2 Relative penetrating ability of alpha, beta, and gamma radiation.

TABLE 18.3 Characteristics of Nuclear Radiation

Radiation	Symbol	Mass (amu)	Electrical charge	Velocity	Composition	Ionizing power
Alpha	α, ${}^{4}_{2}He$	4	+2	Variable, less than 10% the speed of light	He nucleus	High
Beta	β, ${}^{0}_{-1}e$	$\frac{1}{1837}$	−1	Variable, up to 90% the speed of light	Identical to an electron	Moderate
Gamma	γ	0	0	Speed of light	Photons or electromagnetic waves of energy	Almost none

18.3 Radioactive Disintegration Series

LEARNING OBJECTIVE Discuss the transmutation of elements and both natural and artificial disintegration series.

KEY TERMS
transmutation
artificial radioactivity
induced radioactivity
transuranium elements

These three series begin with the elements uranium, thorium, and actinium.

The naturally occurring radioactive elements with a higher atomic number than lead (Pb) fall into three orderly disintegration series. Each series proceeds from one element to the next by the loss of either an alpha or a beta particle, finally ending in a nonradioactive nuclide. The uranium series starts with ${}^{238}_{92}U$ and ends with ${}^{206}_{82}Pb$. The thorium series starts with ${}^{232}_{90}Th$ and ends with ${}^{208}_{82}Pb$. The actinium series starts with ${}^{235}_{92}U$ and ends with ${}^{207}_{82}Pb$. A fourth series, the neptunium series, starts with the synthetic element ${}^{241}_{94}Pu$ and ends with the stable bismuth nuclide ${}^{209}_{83}Bi$. The uranium series is shown in FIGURE 18.3. Gamma radiation, which accompanies alpha and beta radiation, is not shown.

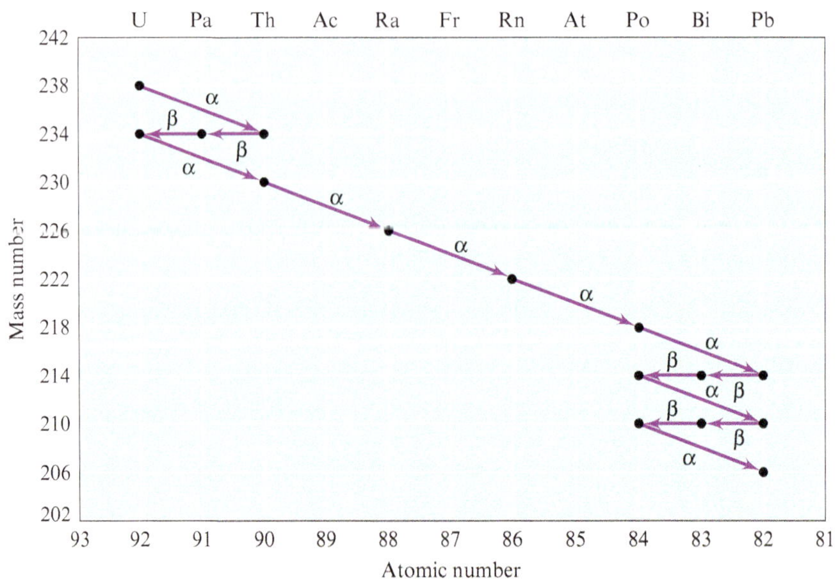

FIGURE 18.3 The uranium disintegration series. ${}^{238}_{92}U$ decays by a series of alpha (α) and beta (β) emissions to the stable nuclide ${}^{206}_{82}Pb$.

By using these series and the half-lives of their members, scientists have been able to approximate the age of certain geologic deposits. This approximation is done by comparing the amount of $^{238}_{92}U$ with the amount of $^{206}_{82}Pb$ and other nuclides in the series that are present in a particular geologic formation. Rocks found in Canada and Finland have been calculated to be about 3.0×10^9 (3 billion) years old. Some meteorites have been determined to be 4.5×10^9 years old.

EXAMPLE 18.5

In Figure 18.3 identify the last five nuclides in the disintegration series of $^{381}_{92}U \longrightarrow ^{206}_{82}Pb$. Write the symbol, the mass number, and the atomic number of each nuclide.

SOLUTION

The five nuclides are $^{214}_{84}Po \longrightarrow ^{210}_{82}Pb \longrightarrow ^{210}_{83}Bi \longrightarrow ^{210}_{84}Po \longrightarrow ^{206}_{82}Pb$.

PRACTICE 18.3

What nuclides are formed when $^{238}_{92}U$ undergoes the following decays?

(a) loss of an alpha particle and a beta particle
(b) loss of three alpha particles and two beta particles

Transmutation of Elements

Transmutation is the conversion of one element into another by either natural or artificial means. Transmutation occurs spontaneously in natural radioactive disintegrations. Alchemists tried for centuries to convert lead and mercury into gold by artificial means, but transmutation by artificial means was not achieved until 1919, when Ernest Rutherford succeeded in bombarding the nuclei of nitrogen atoms with alpha particles and produced oxygen nuclides and protons. The nuclear equation for this transmutation can be written as

$$^{14}_{7}N + \alpha \longrightarrow ^{17}_{8}O + ^{1}_{1}H \quad \text{or} \quad ^{14}_{7}N + ^{4}_{2}He \longrightarrow ^{17}_{8}O + ^{1}_{1}H$$

It is believed that the alpha particle enters the nitrogen nucleus, forming $^{18}_{9}F$ as an intermediate, which then decomposes into the products.

CHECK YOUR UNDERSTANDING
18.6 Transmutation
WileyPLUS

Rutherford's experiments opened the door to nuclear transmutations of all kinds. Atoms were bombarded by alpha particles, neutrons, protons, deuterons ($^{2}_{1}H$), electrons, and so forth. Massive instruments were developed for accelerating these particles to very high speeds and energies to aid their penetration of the nucleus. The famous cyclotron was developed by E. O. Lawrence (1901–1958) at the University of California; later instruments include the Van de Graaf electrostatic generator, the betatron, and the electron and proton synchrotrons. With these instruments many nuclear transmutations became possible. Equations for a few of these are as follows:

$$^{7}_{3}Li + ^{1}_{1}H \longrightarrow 2\,^{4}_{2}He$$

$$^{40}_{18}Ar + ^{1}_{1}H \longrightarrow ^{40}_{19}K + ^{1}_{0}n$$

$$^{23}_{11}Na + ^{1}_{1}H \longrightarrow ^{23}_{12}Mg + ^{1}_{0}n$$

$$^{209}_{83}Bi + ^{2}_{1}H \longrightarrow ^{210}_{84}Po + ^{1}_{0}n$$

$$^{16}_{8}O + ^{1}_{0}n \longrightarrow ^{13}_{6}C + ^{4}_{2}He$$

$$^{238}_{92}U + ^{12}_{6}C \longrightarrow ^{244}_{98}Cf + 6\,^{1}_{0}n$$

SLAC National Accelerator Laboratory

Electrons and positrons are accelerated along the tunnels at the SLAC National Accelerator Laboratory.

On November 1, 2004, the IUPAC officially approved the name roentgenium, symbol Rg, for the element of atomic number 111. Roentgenium was named for Wilhelm Conrad Roentgen, who discovered X-rays in 1895. The nuclear equation for its formation is

$$^{209}_{83}Bi + ^{64}_{28}Ni \longrightarrow ^{272}_{111}Rg + ^{1}_{0}n$$

The IUPAC has recently named elements 112, 114, and 116, as copernicium (Cn), flerovium (Fl), and livermorium (Lv), respectively.

Artificial Radioactivity

Irene Joliot-Curie (daughter of Pierre and Marie Curie) and her husband Frederic Joliot-Curie observed that when aluminum-27 is bombarded with alpha particles, neutrons and positrons (positive electrons) are emitted as part of the products. When the source of alpha particles is removed, neutrons cease to be produced, but positrons continue to be emitted. This observation suggested that the neutrons and positrons come from two separate reactions. It also indicated that a product of the first reaction is radioactive. After further investigation, they discovered that, when aluminum-27 is bombarded with alpha particles, phosphorus-30 and neutrons are produced. Phosphorus-30 is radioactive, has a half-life of 2.5 minutes, and decays to silicon-30 with the emission of a positron. The equations for these reactions are

$$^{27}_{13}\text{Al} + ^{4}_{2}\text{He} \longrightarrow ^{30}_{15}\text{P} + ^{1}_{0}\text{n}$$

$$^{30}_{15}\text{P} \longrightarrow ^{30}_{14}\text{Si} + ^{0}_{+1}\text{e}$$

The radioactivity of nuclides produced in this manner is known as **artificial radioactivity** or **induced radioactivity**. Artificial radionuclides behave like natural radioactive elements in two ways: They disintegrate in a definite fashion, and they have a specific half-life. The Joliot-Curies received the Nobel Prize in chemistry in 1935 for the discovery of artificial, or induced, radioactivity.

Transuranium Elements

The elements that follow uranium on the periodic table and that have atomic numbers greater than 92 are known as the **transuranium elements**. They are synthetic radioactive elements; none of them occur naturally.

The first transuranium element, number 93, was discovered in 1939 by Edwin M. McMillan (1907–1991) at the University of California while he was investigating the fission of uranium. He named it neptunium for the planet Neptune. In 1941, element 94, plutonium, was identified as a beta-decay product of neptunium:

$$^{238}_{93}\text{Np} \longrightarrow ^{238}_{94}\text{Pu} + ^{0}_{-1}\text{e}$$

$$^{239}_{93}\text{Np} \longrightarrow ^{239}_{94}\text{Pu} + ^{0}_{-1}\text{e}$$

Plutonium is one of the most important fissionable elements known today.

Since 1964, the discoveries of 15 new transuranium elements, numbers 104–118, have been announced. These elements have been produced in minute quantities by high-energy particle accelerators.

18.4 Measurement of Radioactivity

LEARNING OBJECTIVE Describe the units and instruments used in the measurement of radioactivity.

KEY TERMS
curie (Ci)
roentgen (R)
rad (radiation absorbed dose)
rem (roentgen equivalent to man)

Radiation from radioactive sources is so energetic that it is called *ionizing radiation*. When it strikes an atom or a molecule, one or more electrons are knocked off, and an ion is created. The Geiger counter, an instrument commonly used to detect and measure radioactivity, depends on this property. The instrument consists of a Geiger–Müller detecting tube and a counting device. The detector tube is a pair of oppositely charged electrodes in an argon gas–filled chamber fitted with a thin window. When radiation, such as a beta particle, passes through the window into the tube, some argon is ionized, and a momentary pulse of current (discharge) flows between the electrodes. These current pulses are electronically amplified in the counter and appear as signals in the form of audible clicks, flashing lights, meter deflections, or numerical readouts (FIGURE 18.4).

The amount of radiation that an individual encounters can be measured by a film badge. This badge contains a piece of photographic film in a light-proof holder and is worn in areas where radiation might be encountered. The silver grains in the film will darken when exposed to radiation. The badges are processed after a predetermined time interval to determine the amount of radiation the wearer has been exposed to.

A scintillation counter is used to measure radioactivity for biomedical applications. A scintillator contains molecules that emit light when exposed to ionizing radiation. A light-sensitive detector counts the flashes and converts them to a numerical readout.

Rennie Van Munchow/Phototake

FIGURE 18.4 Geiger–Müller survey meter.

EXAMPLE 18.6

What three types of radiation detectors are commonly used to detect and measure radioactivity?

SOLUTION

Three types of radiation detectors are a Geiger–Müller detecting tube and counter, a radiation film badge, and a scintillation counter.

PRACTICE 18.4

How does a radiation film badge detect radiation?

The *curie* is the unit used to express the amount of radioactivity produced by an element. One **curie (Ci)** is defined as the quantity of radioactive material giving 3.7×10^{10} disintegrations per second. The basis for this figure is pure radium, which has an activity of 1 Ci/g. Because the curie is such a large quantity, the millicurie and microcurie, representing one-thousandth and one-millionth of a curie, respectively, are more practical and more commonly used.

The curie only measures radioactivity emitted by a radionuclide. Different units are required to measure exposure to radiation. The **roentgen (R)** quantifies exposure to gamma or X-rays; 1 roentgen is defined as the amount of radiation required to produce 2.1×10^9 ions/cm^3 of dry air. The **rad (radiation absorbed dose)** is defined as the amount of radiation that provides 0.01 J of energy per kilogram of matter. The amount of radiation absorbed will change depending on the type of matter. The roentgen and the rad are numerically similar; 1 roentgen of gamma radiation provides 0.92 rad in bone tissue.

Neither the rad nor the roentgen indicates the biological damage caused by radiation. One rad of alpha particles has the ability to cause 10 times more damage than 1 rad of gamma rays or beta particles. Another unit, **rem (roentgen equivalent to man)**, takes into account the degree of biological effect caused by the type of radiation exposure; 1 rem is equal to the dose in rads multiplied by a factor specific to the form of radiation. The factor is 10 for alpha particles and 1 for both beta particles and gamma rays. Units of radiation are summarized in TABLE 18.4.

TABLE 18.4 Radiation Units

Unit	Measure	Equivalent
curie (Ci)	rate of decay of a radioactive substance	1 Ci = 3.7×10^{10} disintegrations/sec
roentgen (R)	exposure based on the quantity of ionization produced in air	1 R = 2.1×10^9 ions/cm^3
rad	absorbed dose of radiation	1 rad = 0.01 J/kg matter
rem	roentgen equivalent in man	1 rem = 1 rad × factor
gray (Gy) (SI unit)	energy absorbed by tissue	1 Gy = 1 J/kg tissue (1 Gy = 100 rad)

18.5 Nuclear energy

LEARNING OBJECTIVE Describe the characteristics of a chain reaction and compare nuclear fission and nuclear fusion.

KEY TERMS
nuclear fission
chain reaction
critical mass
nuclear fusion

Nuclear Fission

In **nuclear fission** a heavy nuclide splits into two or more intermediate-sized fragments when struck in a particular way by a neutron. The fragments are called *fission products.* As the atom splits, it releases energy and two or three neutrons, each of which can cause another nuclear fission. The first instance of nuclear fission was reported in January 1939 by the German scientists Otto Hahn (1879–1968) and Fritz Strassmann (1902–1980). Detecting isotopes of barium, krypton, cerium, and lanthanum after bombarding uranium with neutrons led scientists to believe that the uranium nucleus had been split.

Characteristics of nuclear fission are as follows:

1. Upon absorption of a neutron, a heavy nuclide splits into two or more smaller nuclides (fission products).
2. The mass of the nuclides formed ranges from about 70 to 160 amu.
3. Two or more neutrons are produced from the fission of each atom.
4. Large quantities of energy are produced as a result of the conversion of a small amount of mass into energy.
5. Most nuclides produced are radioactive and continue to decay until they reach a stable nucleus.

CHECK YOUR UNDERSTANDING
18.7 Nuclear Fission
WileyPLUS

CHEMISTRY IN ACTION Does Your Food Glow in the Dark?

The Food and Drug Administration (FDA) has approved irradiation to delay ripening or kill microbes and insects in wheat, potatoes, fresh fruits, and poultry as well as in spices. But the only regular use of irradiation on food in the United States has been on spices.

Now the FDA has approved irradiation of red meat, and new legislation allows the labels about irradiated food to be in much smaller type. The general public has not yet accepted the use of radiation to reduce microbes and help preserve food. Why? Many people believe that irradiating food makes it radioactive. This is not true; in fact, the most that irradiation of food does is to produce compounds similar to those created by cooking and also to reduce the vitamin content of some food.

How does irradiation of food work? The radioactive source currently used is cobalt-60. It is contained in a concrete cell (see diagram) with 6-foot-thick walls. Inside the cell is a pool of water with racks of thin Co-60 rods suspended above. When the rods are not being used, they are submerged in the water, which absorbs the gamma radiation. When food to be irradiated moves into the cells, the Co-60 rods are lifted from the water, and the boxes of food move among them on a conveyor being irradiated from all sides.

Scientists who have investigated the process say that food irradiation is safe. The Centers for Disease Control and Prevention in Atlanta estimates that food-borne illness causes as many as 9000 deaths per year. Irradiation provides one way to reduce microbial contamination in food. It does not solve the problem of careless handling of food by processors, and long-term studies on humans have not yet been concluded regarding irradiated food supplies. Once again, we face the issue of balancing benefit and risk.

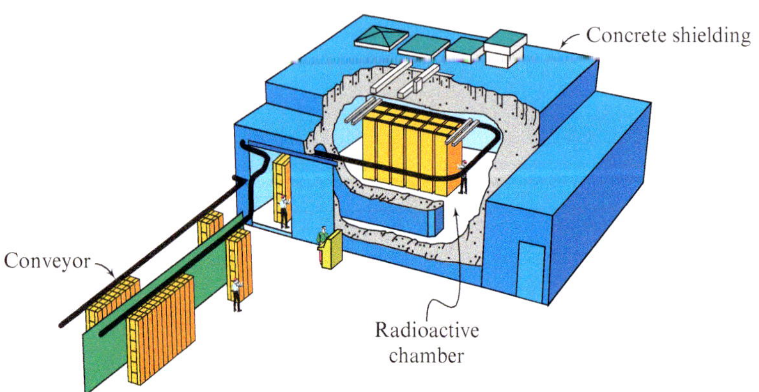

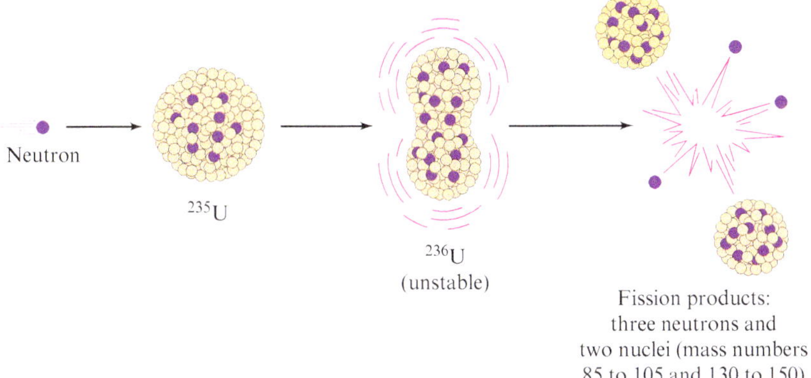

FIGURE 18.5 The fission process. When a neutron is captured by a heavy nucleus, the nucleus becomes more unstable. The more energetic nucleus begins to deform, resulting in fission. Two nuclear fragments and three neutrons are produced by this fission process.

One process by which this fission takes place is illustrated in FIGURE 18.5. When a heavy nucleus captures a neutron, the energy increase may be sufficient to cause deformation of the nucleus until the mass finally splits into two fragments, releasing energy and usually two or more neutrons.

In a typical fission reaction, a $^{235}_{92}U$ nucleus captures a neutron and forms unstable $^{236}_{92}U$. This $^{236}_{92}U$ nucleus undergoes fission, quickly disintegrating into two fragments, such as $^{139}_{56}Ba$ and $^{94}_{36}Kr$, and three neutrons. The three neutrons in turn may be captured by three other $^{235}_{92}U$ atoms, each of which undergoes fission, producing nine neutrons, and so on. A reaction of this kind, in which the products cause the reaction to continue or magnify, is known as a **chain reaction**. For a chain reaction to continue, enough fissionable material must be present so that each atomic fission causes, on average, at least one additional fission. The minimum quantity of an element needed to support a self-sustaining chain reaction is called the **critical mass**. Since energy is released in each atomic fission, chain reactions provide a steady supply of energy. (A chain reaction is illustrated in FIGURE 18.6.)

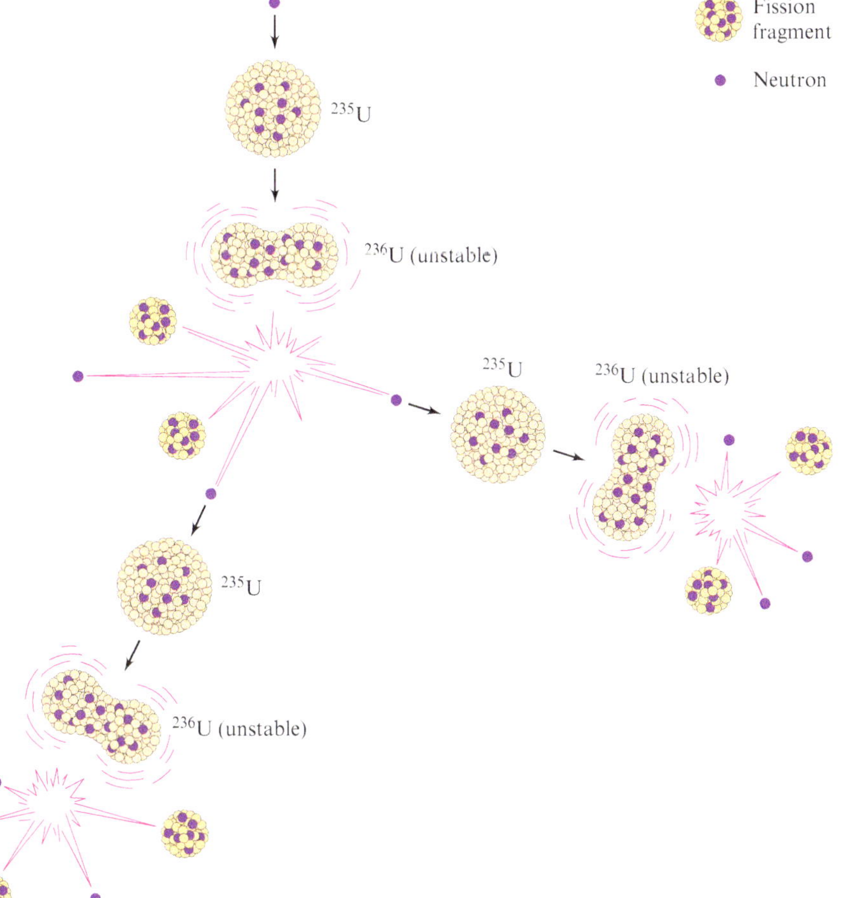

FIGURE 18.6 Fission and chain reaction of $^{235}_{92}U$. Each fission produces two major fission fragments and three neutrons, which may be captured by other $^{235}_{92}U$ nuclei, continuing the chain reaction.

Two of the many possible ways in which $^{235}_{92}U$ may fission are shown by these nuclear equations:

$$^{235}_{92}U + ^{1}_{0}n \longrightarrow ^{139}_{56}Ba + ^{94}_{36}Kr + 3\,^{1}_{0}n$$

$$^{235}_{92}U + ^{1}_{0}n \longrightarrow ^{144}_{54}Xe + ^{90}_{38}Sr + 2\,^{1}_{0}n$$

Nuclear Power

Nearly all electricity for commercial use is produced by machines consisting of a turbine linked by a driveshaft to an electrical generator. The energy required to run the turbine can be supplied by falling water, as in hydroelectric power plants, or by steam, as in thermal power plants.

The world's demand for energy, largely from fossil fuels, is heavy. At the present rates of consumption, the estimated world supply of fossil fuels is sufficient for only a few centuries. Although the United States has large coal and oil shale deposits, it currently imports over 40% of its oil supply. We clearly need to develop alternative energy sources. At present, uranium is the most productive alternative energy source, and about 20% of the electrical energy used in the United States is generated from power plants using uranium fuel. The first new permits to build nuclear plants in the United States were approved in 2012.

A nuclear power plant is a thermal power plant in which heat is produced by a nuclear reactor instead of by combustion of fossil fuel. The major components of a nuclear reactor are

1. an arrangement of nuclear fuel, called the reactor core
2. a control system, which regulates the rate of fission and thereby the rate of heat generation
3. a cooling system, which removes the heat from the reactor and also keeps the core at the proper temperature

One type of reactor uses metal slugs containing uranium enriched from the normal 0.7% U-235 to about 3% U-235. The self-sustaining fission reaction is moderated, or controlled, by adjustable control rods containing substances that slow down and capture some of the neutrons produced. Ordinary water, heavy water, and molten sodium are typical coolants used. Energy obtained from nuclear reactions in the form of heat is used in the production of steam to drive turbines for generating electricity. (FIGURE 18.7.)

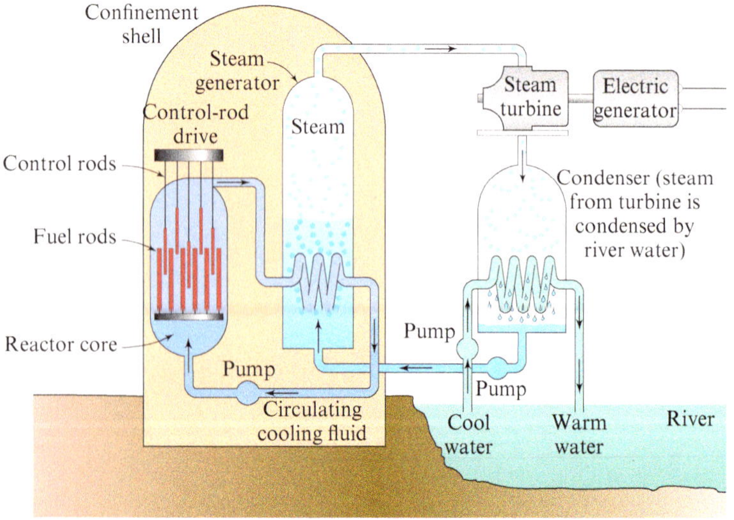

FIGURE 18.7 Diagram of a nuclear power plant. Heat produced by the reactor core is carried by a cooling fluid to a steam generator, which turns a turbine that produces electricity.

The potential dangers of nuclear power were tragically demonstrated by the accidents at Three Mile Island, Pennsylvania (1979), Chernobyl in the former Soviet Union (1986), and the earthquake and tsunami in Japan in 2011. These crises resulted from the loss of coolant to the reactor core. The reactors at Three Mile Island were covered by concrete containment buildings and, therefore, released a relatively small amount of radioactive material into the atmosphere. But because the Soviet Union did not require containment structures on nuclear power plants, the Chernobyl accident resulted in 31 deaths and the resettlement of 135,000 people. The release of large quantities of I-131, Cs-134, and Cs-137 appears to be causing long-term health problems in that exposed population. Japan is currently dealing with radiation contamination in areas around its damaged plants and estimates it could take 40 years to clean up the reactors completely.

Another problem with nuclear power plants is their radioactive waste production. The United

States has debated where and how to permanently store nuclear wastes. In July 2002, an underground site was authorized at Yucca Mountain, Nevada, about 90 miles north of Las Vegas. The Yucca Mountain facility was closed in 2009. A permanent site for nuclear waste storage has not yet been decided upon.

In the United States, reactors designed for commercial power production use uranium oxide, U_3O_8, that is enriched with the relatively scarce fissionable U-235 isotope. Because the supply of U-235 is limited, a new type of reactor known as the *breeder reactor* has been developed. Breeder reactors produce additional fissionable material at the same time that the fission reaction is occurring. In a breeder reactor, excess neutrons convert nonfissionable isotopes, such as U-238 or Th-232, to fissionable isotopes, Pu-239 or U-233, as shown below:

$$^{238}_{92}U + ^{1}_{0}n \longrightarrow ^{239}_{92}U \xrightarrow{\beta} ^{239}_{93}Np \xrightarrow{\beta} ^{239}_{94}Pu$$

$$^{232}_{90}Th + ^{1}_{0}n \longrightarrow ^{233}_{90}Th \xrightarrow{\beta} ^{233}_{91}Pa \xrightarrow{\beta} ^{233}_{92}U$$

These transmutations make it possible to greatly extend the supply of fuel for nuclear reactors. No breeder reactors are presently in commercial operation in the United States, but a number of them are being operated in Europe and Great Britain.

EXAMPLE 18.7

What are the functions of the component parts of a nuclear power plant?

SOLUTION

A nuclear power plant consists of a nuclear reactor core that generates heat energy to change water into steam. The steam drives a turbine that produces electrical energy. Surplus steam is condensed back to water and recycled to the nuclear reactor.

The Atomic Bomb

The atomic bomb is a fission bomb; it operates on the principle of a very fast chain reaction that releases a tremendous amount of energy. An atomic bomb and a nuclear reactor both depend on self-sustaining nuclear fission chain reactions. The essential difference is that in a bomb the fission is "wild," or uncontrolled, whereas in a nuclear reactor the fission is moderated and carefully controlled. A minimum critical mass of fissionable material is needed for a bomb; otherwise a major explosion will not occur. When a quantity smaller than the critical mass is used, too many neutrons formed in the fission step escape without combining with another nucleus, and a chain reaction does not occur. Therefore, the fissionable material of an atomic bomb must be stored as two or more subcritical masses and brought together to form the critical mass at the desired time of explosion. The temperature developed in an atomic bomb is believed to be about 10 million degrees Celsius.

The nuclides used in atomic bombs are U-235 and Pu-239. Uranium deposits contain about 0.7% of the U-235 isotope, the remainder being U-238. Uranium-238 does not undergo fission except with very high-energy neutrons. It was discovered, however, that U-238 captures a low-energy neutron without undergoing fission and that the product, U-239, changes to Pu-239 (plutonium) by a beta-decay process. Plutonium-239 readily undergoes fission upon capture of a neutron and is therefore useful for nuclear weapons. The equations for the nuclear transformations are

$$^{238}_{92}U + ^{1}_{0}n \longrightarrow ^{239}_{92}U \xrightarrow{\beta} ^{239}_{93}Np \xrightarrow{\beta} ^{239}_{94}Pu$$

The hazards of an atomic bomb explosion include not only shock waves from the explosive pressure and tremendous heat but also intense radiation in the form of alpha particles, beta particles, gamma rays, and ultraviolet rays. Gamma rays and X-rays can penetrate deeply into the body, causing burns, sterilization, and gene mutation, which can adversely affect future generations. Both radioactive fission products and unfissioned material are present after the explosion. If the bomb explodes near the ground, many tons of dust are lifted into the air. Radioactive material adhering to this dust, known as *fallout*, is spread by air currents over wide areas of the land and constitutes a lingering source of radiation hazard.

Today nuclear war is probably the most awesome threat facing civilization. Only two rather primitive fission-type atom bombs were used to destroy the Japanese cities of Hiroshima and Nagasaki and bring World War II to an early end. The threat of nuclear war is increased by the fact that the number of nations possessing nuclear weapons is steadily increasing.

Nuclear Fusion

The process of uniting the nuclei of two light elements to form one heavier nucleus is known as **nuclear fusion**. Such reactions can be used for producing energy, because the masses of the two nuclei that fuse into a single nucleus are greater than the mass of the nucleus formed by their fusion. The mass differential is liberated in the form of energy. Fusion reactions are responsible for the tremendous energy output of the Sun. Thus, aside from relatively small amounts from nuclear fission and radioactivity, fusion reactions are the ultimate source of our energy, even the energy from fossil fuels. They are also responsible for the devastating power of the thermonuclear, or hydrogen, bomb.

Fusion reactions require temperatures on the order of tens of millions of degrees for initiation. Such temperatures are present in the Sun but have been produced only momentarily on Earth. For example, the hydrogen, or fusion, bomb is triggered by the temperature of an exploding fission bomb. Two typical fusion reactions are

$$\underset{\text{tritium}}{{}^{3}_{1}\text{H}} + \underset{\text{deuterium}}{{}^{2}_{1}\text{H}} \longrightarrow {}^{4}_{2}\text{He} + {}^{1}_{0}\text{n} + \text{energy}$$

$$\underset{\substack{3.0150\\ \text{amu}}}{{}^{3}_{1}\text{H}} + \underset{\substack{1.0079\\ \text{amu}}}{{}^{1}_{1}\text{H}} \longrightarrow \underset{\substack{4.0026\\ \text{amu}}}{{}^{4}_{2}\text{He}} + \text{energy}$$

The total mass of the reactants in the second equation is 4.0229 amu, which is 0.0203 amu greater than the mass of the product. This difference in mass is manifested in the great amount of energy liberated.

Evidence to date leads us to believe that we can develop a practical fusion power reactor. Fusion power, if we can develop it, will be far superior to fission power for the following reasons:

1. Virtually infinite amounts of energy are possible from fusion. Uranium supplies for fission power are limited, but heavy hydrogen, or deuterium (the most likely fusion fuel), is abundant. It is estimated that the deuterium in a cubic mile of seawater used as fusion fuel could provide more energy than the petroleum reserves of the entire world.
2. From an environmental viewpoint, fusion power is much "cleaner" than fission power because fusion reactions (in contrast to uranium and plutonium fission reactions) do not produce large amounts of long-lived and dangerously radioactive isotopes.

NASA/Science Source/Photo Researchers, Inc.

Remnants of a supernova show temperatures of about 10 million degrees Celsius created when the star exploded.

PRACTICE 18.5

Briefly distinguish between nuclear fission and nuclear fusion.

18.6 Mass–Energy Relationship in Nuclear Reactions

LEARNING OBJECTIVE

Explain mass defect and binding energy.

KEY TERMS

mass defect
nuclear binding energy

Large amounts of energy are released in nuclear reactions; thus, significant amounts of mass are converted to energy. We stated earlier that the amount of mass converted to energy in chemical changes is insignificant compared to the amount of energy released in a nuclear reaction. In fission reactions, about 0.1% of the mass is converted into energy. In fusion reactions as much as 0.5% of the mass may be changed into energy. The Einstein equation, $E = mc^2$, can be used to calculate the energy liberated, or available, when the mass loss is known. For example, in the fusion reaction

$$\underset{7.016\text{ g}}{^{7}_{3}\text{Li}} + \underset{1.008\text{ g}}{^{1}_{1}\text{H}} \longrightarrow \underset{4.003\text{ g}}{^{4}_{2}\text{He}} + \underset{4.003\text{ g}}{^{4}_{2}\text{He}} + \text{energy}$$

the mass difference between the reactants and products (8.024 g − 8.006 g) is 0.018 g. The energy equivalent to this amount of mass is 1.62×10^{12} J. By comparison, this is more than 4 million times greater than the 3.9×10^{5} J of energy obtained from the complete combustion of 12.01 g (1 mol) of carbon.

The mass of a nucleus is actually less than the sum of the masses of the protons and neutrons that make up that nucleus. The difference between the mass of the protons and the neutrons in a nucleus and the mass of the nucleus is known as the **mass defect**. The energy equivalent to this difference in mass is known as the **nuclear binding energy**. This energy is the amount that would be required to break a nucleus into its individual protons and neutrons. The higher the binding energy, the more stable the nucleus. Elements of intermediate atomic masses have high binding energies. For example, iron (element number 26) has a very high binding energy and therefore a very stable nucleus. Just as electrons attain less energetic and more stable arrangements through ordinary chemical reactions, neutrons and protons attain less energetic and more stable arrangements through nuclear fission or fusion reactions. Thus, when uranium undergoes fission, the products have less mass (and greater binding energy) than the original uranium. In like manner, when hydrogen and lithium fuse to form helium, the helium has less mass (and greater binding energy) than the hydrogen and lithium. It is this conversion of mass to energy that accounts for the very large amounts of energy associated with both fission and fusion reactions.

Mass Defect

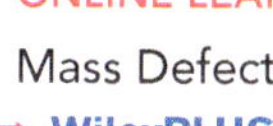

ENHANCED EXAMPLE 18.8

Calculate the mass defect and the nuclear binding energy for an α particle (helium nucleus).

SOLUTION

Knowns: proton mass = 1.0073 g/mol
neutron mass = 1.0087 g/mol
α mass = 4.0015 g/mol
$1.0\text{ g} = 9.0 \times 10^{13}$ J

Plan First calculate the sum of the individual parts of an α particle (2 protons + 2 neutrons).

2 protons: 2 × 1.0073 g/mol = 2.0146 g/mol
2 neutrons: 2 × 1.0087 g/mol = 2.0174 g/mol
4.0320 g/mol

The mass defect is the difference between the mass of the α particle and its component parts:

4.0320 g/mol
4.0015 g/mol
0.0305 g/mol (mass defect)

Calculate • To find the nuclear binding energy, convert the mass defect to its energy equivalent:

$$(0.0305 \text{ g/mol})(9.0 \times 10^{13} \text{ J/g}) = 2.7 \times 10^{12} \text{ J/mol } (6.5 \times 10^{11} \text{ cal})$$

PRACTICE 18.6

Why is the mass of an atomic nucleus less than the sum of the masses of the protons and neutrons that are in that nucleus?

18.7 Biological Effects of Radiation

LEARNING OBJECTIVE Describe the effects of ionizing radiation on living organisms.

KEY TERM
ionizing radiation

Radiation with energy to dislocate bonding electrons and create ions when passing through matter is classified as **ionizing radiation**. Alpha particles, beta particles, gamma rays, and X-rays fall into this classification. Ionizing radiation can damage or kill living cells and can be particularly devastating when it strikes the cell nuclei and affects molecules involved in cell reproduction. The effects of radiation on living organisms fall into these general categories: (1) acute or short-term effects, (2) long-term effects, and (3) genetic effects.

Acute Radiation Damage

High levels of radiation, especially from gamma rays or X-rays, produce nausea, vomiting, and diarrhea. The effect has been likened to a sunburn throughout the body. If the dosage is high enough, death will occur in a few days. The damaging effects of radiation appear to be centered in the nuclei of the cells, and cells that are undergoing rapid cell division are most susceptible to damage. It is for this reason that cancers are often treated with gamma radiation from a Co-60 source. Cancerous cells multiply rapidly and are destroyed by a level of radiation that does not seriously damage normal cells.

Long-Term Radiation Damage

Protracted exposure to low levels of any form of ionizing radiation can weaken an organism and lead to the onset of malignant tumors, even after fairly long time delays. The largest exposure to synthetic sources of radiation is from X-rays. Evidence suggests that the lives of early workers in radioactivity and X-ray technology may have been shortened by long-term radiation damage.

Strontium-90 isotopes are present in the fallout from atmospheric testing of nuclear weapons. Strontium is in the same periodic-table group as calcium, and its chemical behavior is similar to that of calcium. Hence, when foods contaminated with Sr-90 are eaten, Sr-90 ions are laid down in the bone tissue along with ordinary calcium ions. Strontium-90 is a beta emitter with a half-life of 28 years. Blood cells manufactured in bone marrow are affected by the radiation from Sr-90. Hence, there is concern that Sr-90 accumulation in the environment may cause an increase in the incidence of leukemia and bone cancers.

Genetic Effects

The information needed to create an individual of a particular species, be it a bacterial cell or a human being, is contained within the nucleus of a cell. This genetic information is encoded in the structure of DNA (deoxyribonucleic acid) molecules, which make up genes. The DNA molecules form precise duplicates of themselves when cells divide, thus passing genetic information from one generation to the next. Radiation can damage DNA molecules. If the damage is not severe enough to prevent the individual from reproducing, a mutation may result. Most mutation-induced traits are undesirable. Unfortunately, if

CHEMISTRY IN ACTION A Window into Living Organisms

Imagine viewing a living process as it is occurring. This was the dream of many scientists in the past as they tried to extract this knowledge from dead tissue. Today, because of innovations in nuclear chemistry, this dream is a common, everyday occurrence.

Compounds containing a radionuclide are described as being *labeled*, or *tagged*. These compounds undergo their normal chemical reactions, but their location can be detected because of their radioactivity. When such compounds are given to a plant or an animal, the movement of the nuclide can be traced through the organism by the use of a Geiger counter or other detecting device.

In an early use of the tracer technique, the pathway by which CO_2 becomes fixed into carbohydrate ($C_6H_{12}O_6$) during photosynthesis was determined. The net equation for photosynthesis is

$$6\,CO_2 + 6\,H_2O \longrightarrow C_6H_{12}O_6 + 6\,O_2$$

Radioactive $^{14}CO_2$ was injected into a colony of green algae, and the algae were then placed in the dark and killed at selected time intervals. When the radioactive compounds were separated by paper chromatography and analyzed, the results elucidated a series of light-independent photosynthetic reactions.

Biological research using tracer techniques have determined

1. the rate of phosphate uptake by plants, using radiophosphorus
2. the flow of nutrients in the digestive tract using radioactive barium compounds
3. the accumulation of iodine in the thyroid gland, using radioactive iodine
4. the absorption of iron by the hemoglobin of the blood, using radioactive iron

In chemistry, uses for tracers are unlimited. The study of reaction mechanisms, the measurement of the rates of chemical reactions, and the determination of physical constants are just a few of the areas of application.

Radioactive tracers are commonly used in medical diagnosis. The radionuclide must be effective at a low concentration and have a short half-life to reduce the possibility of damage to the patient.

Radioactive iodine (I-131) is used to determine the function of the thyroid, where the body concentrates iodine. In this process a small amount of radioactive potassium or sodium iodide is ingested. A detector is focused on the thyroid gland and measures the amount of iodine in the gland. This picture is then compared to that of a normal thyroid to detect any differences.

Doctors examine the heart's pumping performance and check for evidence of obstruction in coronary arteries by *nuclear scanning*. The radionuclide Tl-201, when injected into the bloodstream, lodges in healthy heart muscle. Thallium-201 emits gamma radiation, which is detected by a special imaging device called a *scintillation camera*. The data obtained are simultaneously translated into pictures by a computer. With this technique doctors can observe whether heart tissue has died after a heart attack and whether blood is flowing freely through the coronary passages.

One of the most recent applications of nuclear chemistry is the use of positron emission tomography (PET) in the measurement of dynamic processes in the body, such as oxygen use or blood flow. In this application, a compound is made that contains a positron-emitting nuclide such as C-11, O-15, or N-13. The compound is injected into the body, and the patient is placed in an instrument that detects the positron emission. A computer produces a three-dimensional image of the area.

PET scans have been used to locate the areas of the brain involved in epileptic seizures. Glucose tagged with C-11 is injected, and an image of the brain is produced. Since the brain uses glucose almost exclusively for energy, diseased areas that use glucose at a rate different than normal tissue can then be identified.

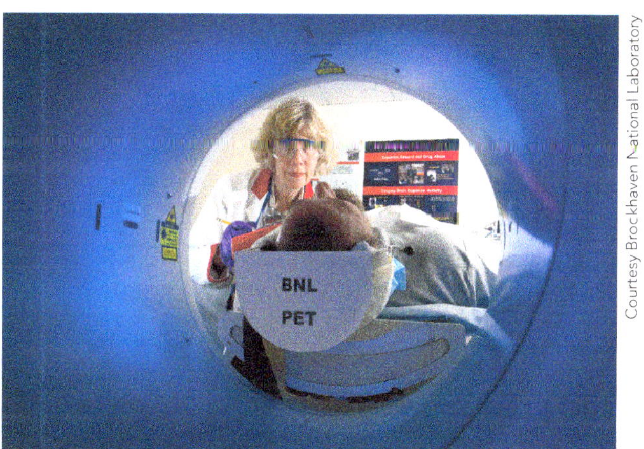

Courtesy Brockhaven National Laboratory

A radioactive tracer is injected into this patient and absorbed by the brain. The PET scanner detects photons emitted by the tracer and produces an image used in medical diagnosis and research.

the bearer of the altered genes survives to reproduce, these traits are passed along to succeeding generations. In other words, the genetic effects of increased radiation exposure are found in future generations, not only in the present generation.

Because radioactive rays are hazardous to health and living tissue, special precautions must be taken in designing laboratories and nuclear reactors, in disposing of waste materials, and in monitoring the radiation exposure of people working in this field.

CHAPTER 18 REVIEW

18.1 Discovery of Radioactivity

KEY TERMS

radioactivity
nucleon
nuclide
radioactive decay
half-life ($t_{1/2}$)

- Radioactivity is the spontaneous emission of particles and energy from the nucleus of an atom.
- Protons and neutrons are known as nucleons.
- In nuclear chemistry isotopes are also known as nuclides.
- Radioactive elements undergo radioactive decay to form different elements.
- Radioactivity is not affected by changes in temperature, pressure, or the state of the element.
- Principal emissions:
  - Alpha particles
  - Beta particles
  - Gamma rays
- The half-life of a nuclide is the time required for one-half of a specific amount of the nuclide to disintegrate.

18.2 Alpha Particles, Beta Particles, and Gamma Rays

KEY TERMS

alpha particle (α)
beta particle (β)
gamma ray (γ)

- Alpha particles:
  - Consist of 2 protons and 2 neutrons with mass of 4 amu and charge of +2
  - Loss of an alpha particle from the nucleus results in
    - Loss of 4 in mass number
    - Loss of 2 in atomic number
  - High ionizing power, low penetration
- Beta particles:
  - Same as an electron
  - Loss of a beta particle from the nucleus results in
    - No change in mass number
    - Increase of 1 in atomic number
  - Moderate ionizing power and penetration
- Gamma rays:
  - Photons of energy
  - Loss of a gamma ray from the nucleus results in
    - No change in mass number (A)
    - No change in atomic number (Z)
  - Almost no ionizing power, high penetration

18.3 Radioactive Disintegration Series

KEY TERMS

transmutation
artificial radioactivity
induced radioactivity
transuranium elements

- As elements undergo disintegration, they eventually become stable after the loss of a series of particles and energy.
- The disintegration series can be used to determine the age of geologic deposits.
- Transmutation of an element can occur spontaneously or artificially.
- Artificial radionuclides behave like natural radioactive elements in two ways:
  - They disintegrate in a definite fashion.
  - They have a specific half-life.
- Transuranium elements are the elements that follow uranium on the periodic table.

18.4 Measurement of Radioactivity

KEY TERMS

curie (Ci)
roentgen (R)
rad (radiation absorbed dose)
rem (roentgen equivalent to man)

- Radioactivity can be measured in several ways:
  - Geiger–Müller counter
  - Film badge
  - Scintillation counter
- Units for measuring radiation:
  - Gray
  - Curie
  - Roentgen
  - Rad
  - Rem

18.5 Nuclear Energy

KEY TERMS

nuclear fission
chain reaction
critical mass
nuclear fusion

- Characteristics of fission:
  - A heavy nuclide splits into two or more smaller nuclides.
  - The mass of nuclides formed is 70 to 160 amu.
  - Two or more neutrons are produced per fission.
  - Large quantities of energy are released.
  - Most nuclides produced are radioactive and continue to decay until they reach stability.
- A nuclear power plant is a thermal plant with a nuclear fission source of energy.
- Major components of a nuclear reactor:
  - Reactor core
  - Control system
  - Cooling system
- Out-of-control fission reaction results in a bomb.
- Nuclear fusion is the process of uniting 2 light nuclei to form 1 heavier nucleus.
- Nuclear fusion requires extreme temperatures for ignition.

18.6 Mass–Energy Relationship in Nuclear Reactions

KEY TERMS

mass defect
nuclear binding energy

- The mass defect is the difference between the actual mass of an atom and the calculated mass of the protons and neutrons in the nucleus of that atom.
- The nuclear binding energy is the energy equivalent to the mass defect.
- $E = mc^2$ is used to determine the energy liberated in a nuclear reaction.

18.7 Biological Effects of Radiation

KEY TERM

ionizing radiation

- Effects of radiation fall into three general categories:
  - Acute or short term
  - Long term
  - Genetic

REVIEW QUESTIONS

1. Identify these people and their associations with the early history of radioactivity:
   (a) Antoine Henri Becquerel
   (b) Marie and Pierre Curie
   (c) Wilhelm Roentgen
   (d) Ernest Rutherford
   (e) Otto Hahn and Fritz Strassmann
2. Why is the radioactivity of an element unaffected by the usual factors that affect the rate of chemical reactions, such as ordinary changes of temperature and concentration?
3. Distinguish between the terms *isotope* and *nuclide.*
4. The half-life of Pu-244 is 76 million years. If Earth's age is about 5 billion years, discuss the feasibility of finding this nuclide as a naturally occurring nuclide.
5. Describe the radiocarbon method for dating archaeological artifacts.
6. Anthropologists have found bones whose age suggests that the human line may have emerged in Africa as much as 4 million years ago. If wood or charcoal were found with such bones, would C-14 dating be useful in dating the bones? Explain.
7. To afford protection from radiation injury, which kind of radiation requires (a) the most shielding? (b) the least shielding?
8. Why is an alpha particle deflected less than a beta particle in passing through an electromagnetic field?
9. Tell how types of alpha, beta, and gamma radiation are distinguished from the standpoint of
   (a) charge
   (b) relative mass
   (c) nature of particle or ray
   (d) relative penetrating power
10. Bismuth-211 decays by alpha emission to give a nuclide that in turn decays by beta emission to yield a stable nuclide. Show these two steps with nuclear equations.
11. Name three pairs of nuclides that might be obtained by fissioning U-235 atoms.
12. Distinguish between natural and artificial radioactivity.
13. What is a radioactive disintegration series?
14. Briefly discuss the transmutation of elements.
15. Stable Pb-208 is formed from Th-232 in the thorium disintegration series by successive α, β, β, α, α, α, α, β, β, α particle emissions. Write the symbol (including mass and atomic number) for each nuclide formed in this series.
16. The nuclide Np-237 loses a total of seven alpha particles and four beta particles. What nuclide remains after these losses?
17. How might radioactivity be used to locate a leak in an underground pipe?
18. What is a scintillation counter and how does it work?
19. What unit is used to describe the amount of radioactivity produced by an element?
20. What does the unit **rem** stand for and what does it measure?
21. What were Otto Hahn and Fritz Strassmann's contribution to nuclear physics?
22. What is a breeder reactor? Explain how it accomplishes the "breeding."
23. What is the essential difference between the nuclear reactions in a nuclear reactor and those in an atomic bomb?
24. Why must a certain minimum amount of fissionable material be present before a self-supporting chain reaction can occur?

25. What are the disadvantages of nuclear power?

26. What are the hazards associated with an atomic bomb explosion?

27. What types of elements undergo fission? How about fusion?

28. Using Figure 18.7, describe how nuclear power is generated.

29. What are mass defect and nuclear binding energy?

30. Explain why radioactive rays are classified as ionizing radiation.

31. Give a brief description of the biological hazards associated with radioactivity.

32. Strontium-90 has been found to occur in radioactive fallout. Why is there so much concern about this radionuclide being found in cow's milk? (Half-life of Sr-90 is 28 years.)

33. What is a radioactive tracer? How is it used?

PAIRED EXERCISES

Most of the exercises in this chapter are available for assignment via the online homework management program, WileyPLUS (www.wileyplus.com). All exercises with blue numbers have answers in Appendix 5.

1. Indicate the number of protons, neutrons, and nucleons in these nuclei:
(a) $^{207}_{82}Pb$ (b) $^{70}_{31}Ga$

2. Indicate the number of protons, neutrons, and nucleons in these nuclei:
(a) $^{128}_{52}Te$ (b) $^{32}_{16}S$

3. How are the mass and the atomic number of a nucleus affected by the loss of an alpha particle?

4. How are the mass and the atomic number of a nucleus affected by the loss of a beta particle?

5. Write nuclear equations for the alpha decay of
(a) $^{210}_{83}Bi$ (b) $^{238}_{92}U$

6. Write nuclear equations for the alpha decay of
(a) $^{238}_{90}Th$ (b) $^{239}_{94}Pu$

7. Write nuclear equations for the beta decay of
(a) $^{13}_{7}N$ (b) $^{234}_{90}Th$

8. Write nuclear equations for the beta decay of
(a) $^{28}_{13}Al$ (b) $^{239}_{93}Np$

9. Determine the type of emission or emissions (alpha, beta, or gamma) that occurred in the following transitions:
(a) $^{226}_{88}Ra$ to $^{222}_{86}Rn$
(b) $^{222}_{88}Ra$ to $^{222}_{87}Fr$ to $^{222}_{87}Fr$
(c) $^{238}_{92}U$ to $^{238}_{93}Np$

10. Determine the type of emission or emissions (alpha, beta, or gamma) that occurred in the following transitions:
(a) $^{210}_{82}Pb$ to $^{210}_{82}Pb$
(b) $^{234}_{91}Pa$ to $^{230}_{89}Ac$ to $^{230}_{90}Th$
(c) $^{234}_{90}Th$ to $^{230}_{88}Ra$ to $^{230}_{88}Ra$

11. Write a nuclear equation for the conversion of $^{26}_{13}Al$ to $^{26}_{12}Mg$.

12. Write a nuclear equation for the conversion of $^{32}_{15}P$ to $^{32}_{16}S$.

13. Complete and balance these nuclear equations by supplying the missing particles:
(a) $^{66}_{29}Cu \longrightarrow ^{66}_{30}Zn +$ ______
(b) $^{0}_{-1}e +$ ______ $\longrightarrow ^{7}_{3}Li$
(c) $^{27}_{13}Al + ^{4}_{2}He \longrightarrow ^{30}_{14}Si +$ ______
(d) $^{85}_{37}Rb +$ ______ $\longrightarrow ^{82}_{35}Br + ^{4}_{2}He$

14. Complete and balance these nuclear equations by supplying the missing particles:
(a) $^{27}_{13}Al + ^{4}_{2}He \longrightarrow ^{30}_{15}P +$ ______
(b) $^{27}_{14}Si \longrightarrow ^{0}_{+1}e +$ ______
(c) ______ $+ ^{2}_{1}H \longrightarrow ^{13}_{7}N + ^{1}_{0}n$
(d) ______ $\longrightarrow ^{82}_{36}Kr + ^{0}_{-1}e$

15. Strontium-90 has a half-life of 28 years. If a 1.00-mg sample was stored for 112 years, what mass of Sr-90 would remain?

16. Strontium-90 has a half-life of 28 years. If a sample was tested in 1980 and found to be emitting 240 counts/min, in what year would the same sample be found to be emitting 30 counts/min? How much of the original Sr-90 would be left?

17. By what series of emissions does $^{233}_{91}Pa$ disintegrate to $^{225}_{89}Ac$?

18. By what series of emissions does $^{238}_{90}Th$ disintegrate to $^{212}_{82}Pb$?

19. Consider the fission reaction

$$^{235}_{92}U + ^{1}_{0}n \longrightarrow ^{94}_{38}Sr + ^{139}_{54}Xe + 3\,^{1}_{0}n + \text{energy}$$

Calculate the following using these mass data (1.0 g is equivalent to 9.0×10^{13} J):

U-235 = 235.0439 amu   Sr-94 = 93.9154 amu
Xe-139 = 138.9179 amu   n = 1.0087 amu

(a) the energy released in joules for a single event (one uranium atom splitting)
(b) the energy released in joules per mole of uranium splitting
(c) the percentage of mass lost in the reaction

20. Consider the fusion reaction

$$^{1}_{1}H + ^{2}_{1}H \longrightarrow ^{3}_{2}He + \text{energy}$$

Calculate the following using these mass data (1.0 g is equivalent to 9.0×10^{13} J):

$^{1}_{1}H$ = 1.00794 amu
$^{2}_{1}H$ = 2.01410 amu
$^{3}_{2}H$ = 3.01603 amu

(a) the energy released in joules per mole of He-3 formed
(b) the percentage of mass lost in the reaction

ADDITIONAL EXERCISES

21. Many radioactive isotopes are used in nuclear medicine. Some of the useful isotopes are listed below. For each of these isotopes, give the number of protons, neutrons, and electrons.
    (a) Chromium-51—this isotope is used to label red blood cells and quantify gastrointestinal protein loss.
    (b) Holmium-166—this isotope is being tested as a diagnostic and treatment tool for liver tumors.
    (c) Palladium-103—this isotope is used for treatment of early stage prostate cancer.
    (d) Strontium-89—this isotope is used to treat the pain of bone cancer
22. The Th-232 disintegration series starts with $^{232}_{90}Th$ and emits the following rays successively: α, β, β, α, α, α, β, α, β, α. The series ends with the stable $^{208}_{82}Pb$. Write the formula for each nuclide in the series.
23. The nuclide $^{223}_{87}Fr$ emits three particles losing 8 mass units and 3 atomic number units. Propose a radioactive decay series and write the symbol for the resulting nuclide.
24. Potassium-40 has a half-life of 1.25×10^9 years. How many months will it take for one-half of a 25.0-g sample to disappear?
25. When the nuclide $^{249}_{98}Cf$ was bombarded with $^{15}_{7}N$, four neutrons and a new transuranium element were formed. Write the nuclear equation for this transmutation.
26. What percent of the mass of the nuclide $^{226}_{88}Ra$ is neutrons? electrons?
27. If radium costs \$90,000 a gram, how much will 0.0100 g of $^{226}RaCl_2$ cost if the price is based only on the radium content?
28. An archaeological specimen was analyzed and found to be emitting only 25% as much C-14 radiation per gram of carbon as newly cut wood. How old is this specimen?
29. Barium-141 is a beta emitter. What is the half-life if a 16.0-g sample of the nuclide decays to 0.500 g in 90 minutes?
30. Calculate (a) the mass defect and (b) the binding energy of $^{7}_{3}Li$ using the mass data:

    $^{7}_{3}Li = 7.0160$ g   n = 1.0087 g
    p = 1.0073 g $e^- = 0.00055$ g
    $1.0 \text{ g} \equiv 9.0 \times 10^{13}$ J (from $E = mc^2$)

31. In the disintegration series $^{235}_{92}U \longrightarrow {}^{207}_{82}Pb$, how many alpha and beta particles are emitted?
32. List three devices used for radiation detection and explain their operation.
33. The half-life of I-123 is 13 hours. If 10 mg of I-123 are administered to a patient, how much I-123 remains after 3 days and 6 hours?
34. Iodine-131 is often used to treat Graves' disease or hyperthyroidism. In Graves' disease the thyroid begins to overproduce thyroid hormone. If a small amount of iodine-131 is consumed, it will concentrate in the thyroid gland and destroy its ability to function. If a tablet containing 15.0 μg of ^{131}I is ingested by a patient, how many days will it take for the amount of iodine to decay to less than 1 μg? (Iodine-131 has a half-life of 8 days.)
35. One of the ways to determine the extent of termite infestation in a wooden structure is to feed the termites a wood substitute containing a radioactive element. The extent of the infestation can then be determined by monitoring the spread of the radioactivity. If you were given the task to prepare a wood substitute, would you use bismuth-213 (half-life 46 minutes), rhenium-186 (half-life 4 days), or cobalt-60 (half-life 5 years)? Explain your reasoning.
36. Many home smoke detectors use americunium-241 as a radiation source. In the smoke detector, ^{241}Am emits alpha particles. These alpha particles are detected by the detector unless smoke particles block their passage. Thus, as long as there are no smoke particles, the detector gets a continuous stream of alpha particles hitting it; smoke causes a disruption in this stream and activates the alarm. Write out the nuclear equation for the decay of ^{241}Am by alpha emission.
37. How many half-lives are required to change 96 g of a sample of a radioactive isotope to 1.5 g over approximately 24 days? What is the approximate half-life of this isotope?
38. Clearly distinguish between fission and fusion. Give an example of each.
39. Starting with 1 g of a radioactive isotope whose half-life is 10 days, sketch a graph showing the pattern of decay for that material. On the x-axis, plot time (you may want to simply show multiples of the half-life), and on the y-axis, plot mass of material remaining. Then after completing the graph, explain why a sample never really gets to the point where *all* of its radioactivity is considered to be gone.
40. Identify each missing product (name the element and give its atomic number and mass number) by balancing the following nuclear equations:
    (a) $^{235}U + {}^{1}_{0}n \longrightarrow {}^{143}Xe + 3\,{}^{1}_{0}n +$ ________
    (b) $^{235}U + {}^{1}_{0}n \longrightarrow {}^{102}Y + 3\,{}^{1}_{0}n +$ ________
    (c) $^{14}N + {}^{1}_{0}n \longrightarrow {}^{1}H +$ ________
41. Consider these reactions:
    (a) $H_2O(l) \longrightarrow H_2O(g)$
    (b) $2\,H_2(g) + O_2(g) \longrightarrow 2\,H_2O(g)$
    (c) $^{2}_{1}H + {}^{2}_{1}H \longrightarrow {}^{3}_{1}H + {}^{1}_{1}H$

    The following energy values belong to one of these equations:

    energy$_1$ 115.6 kcal released
    energy$_2$ 10.5 kcal absorbed
    energy$_3$ 7.5×10^7 kcal released

    Match the equation to the energy value and briefly explain your choices.
42. When $^{235}_{92}U$ is struck by a neutron, the unstable isotope $^{236}_{92}U$ results. When that daughter isotope undergoes fission, there are numerous possible products. If strontium-90 and three neutrons are the results of one such fission, what is the other product?
43. Write balanced nuclear equations for
    (a) beta emission by $^{29}_{12}Mg$
    (b) alpha emission by $^{150}_{60}Nd$
    (c) positron emission by $^{72}_{33}As$
44. How much of a sample of cesium-137 ($t_{1/2}$ = 30 years) must have been present originally if, after 270 years, 15.0 g remain?
45. The curie is equal to 3.7×10^{10} disintegration/sec, and the becquerel is equivalent to just 1 disintegration/sec. Suppose a hospital has a 150-g radioactive source with an activity of 1.24 Ci. What is its activity in becquerels?
46. Cobalt-60 has a half-life of 5.26 years. If 1.00 g of ^{60}Co were allowed to decay, how many grams would be left after
    (a) one half-life?
    (b) two half-lives?
    (c) four half-lives?
    (d) ten half-lives?

47. Write balanced equations to show these changes:
    (a) alpha emission by boron-11
    (b) beta emission by strontium-88
    (c) neutron absorption by silver-107
    (d) proton emission by potassium-41
    (e) electron absorption by antimony-116
48. The $^{14}_{6}C$ content of an ancient piece of wood was found to be one-sixteenth of that in living trees. How many years old is this piece of wood if the half-life of carbon-14 is 5730 years?
49. Do you think we have radioactive isotopes in our bodies? Explain.
50. How might human exposure to ionizing radiation today affect future generations?
51. What are some of the effects of long-term exposure to low-level ionizing radiation?
52. Explain how scientists can measure the rate of phosphate uptake by plants using radiophosphorus.
53. (a) What type of process is $^{6}_{3}Li + ^{1}_{0}n \longrightarrow ^{3}_{1}H + ^{4}_{2}He$?
    (b) How is this process different from radioactive decay?
    (c) Could nuclear fission be classified as the same type process? Explain.

CHALLENGE EXERCISES

54. Rubidium-87, a beta emitter, is the product of positron emission. Identify
    (a) the product of rubidium-87 decay
    (b) the precursor of rubidium-87
55. Potassium-42 is used to locate brain tumors. Its half-life is 12.5 hours. Starting with 15.4 mg, what fraction will remain after 100 hours? If it was necessary to have at least 1 μg for a particular procedure, could you hold the original sample for 200 hours before using it?

ANSWERS TO PRACTICE EXERCISES

18.1 0.391 g

18.2 $^{218}_{84}Po$

18.3 (a) $^{234}_{91}Pa$ (b) $^{226}_{88}Ra$

18.4 The silver grains in the photographic film badge darken when exposed to ionizing radiation. Badges worn by workers around radiation sources are processed periodically to measure the amount of radiation to which workers have been exposed.

18.5 In nuclear fission a heavy nuclide is bombarded by and absorbs a neutron, which causes the nuclide to become unstable and to break (fission) into two or more lighter nuclear fragments and two or three neutrons plus energy. In a fusion reaction two light nuclides are combined (fused) with a release of energy.

18.6 The mass difference is the nuclear binding energy, which is the energy equivalent needed to hold the nucleus together.

Appendix 1 MATHEMATICAL REVIEW

Multiplication

Multiplication is a process of adding any given number or quantity to itself a certain number of times. Thus, 4 times 2 means 4 added two times, or 2 added together four times, to give the product 8. Various ways of expressing multiplication are

$$ab \qquad a \times b \qquad a \cdot b \qquad a(b) \qquad (a)(b)$$

Each of these expressions means a times b, or a multiplied by b, or b times a.

$$\text{When} \quad a = 16 \quad \text{and} \quad b = 24, \quad \text{we have} \quad 16 \times 24 = 384.$$

The expression °F = (1.8 × °C) + 32 means that we are to multiply 1.8 times the Celsius degrees and add 32 to the product. When °C equals 50,

$$°\text{F} = (1.8 \times 50) + 32 = 90 + 32 = 122°\text{F}$$

The result of multiplying two or more numbers together is known as the *product.*

Division

The word *division* has several meanings. As a mathematical expression, it is the process of finding how many times one number or quantity is contained in another. Various ways of expressing division are

$$a \div b \qquad \frac{a}{b} \qquad a/b$$

Each of these expressions means a divided by b.

$$\text{When} \quad a = 15 \quad \text{and} \quad b = 3, \quad \frac{15}{3} = 5.$$

The number above the line is called the *numerator;* the number below the line is the *denominator.* Both the horizontal and the slanted (/) division signs also mean "per." For example, in the expression for density, we determine the mass per unit volume:

$$\text{density} = \text{mass/volume} = \frac{\text{mass}}{\text{volume}} = \text{g/mL}$$

The diagonal line still refers to a division of grams by the number of milliliters occupied by that mass. The result of dividing one number into another is called the *quotient.*

Fractions and Decimals

A fraction is an expression of division, showing that the numerator is divided by the denominator. A *proper fraction* is one in which the numerator is smaller than the denominator. In an *improper fraction,* the numerator is the larger number. A decimal or a decimal fraction is a proper fraction in which the denominator is some power of 10. The decimal fraction is determined by carrying out the division of the proper fraction. Examples of proper fractions and their decimal fraction equivalents are shown in the accompanying table.

Proper fraction		**Decimal fraction**		**Proper fraction**
$\frac{1}{8}$	=	0.125	=	$\frac{125}{1000}$
$\frac{1}{10}$	=	0.1	=	$\frac{1}{10}$
$\frac{3}{4}$	=	0.75	=	$\frac{75}{100}$
$\frac{1}{100}$	=	0.01	=	$\frac{1}{100}$
$\frac{1}{4}$	=	0.25	=	$\frac{25}{100}$

Addition of Numbers with Decimals

To add numbers with decimals, we use the same procedure as that used when adding whole numbers, but we always line up the decimal points in the same column. For example, add 8.21 + 143.1 + 0.325:

$$\begin{array}{r} 8.21 \\ +143.1 \\ +0.325 \\ \hline 151.635 \end{array}$$

When adding numbers that express units of measurement, we must be certain that the numbers added together all have the same units. For example, what is the total length of three pieces of glass tubing: 10.0 cm, 125 mm, and 8.4 cm? If we simply add the numbers, we obtain a value of 143.4, but we are not certain what the unit of measurement is. To add these lengths correctly, first change 125 mm to 12.5 cm. Now all the lengths are expressed in the same units and can be added:

$$\begin{array}{r} 10.0\text{ cm} \\ 12.5\text{ cm} \\ 8.4\text{ cm} \\ \hline 30.9\text{ cm} \end{array}$$

Subtraction of Numbers with Decimals

To subtract numbers containing decimals, we use the same procedure as for subtracting whole numbers, but we always line up the decimal points in the same column. For example, subtract 20.60 from 182.49:

$$\begin{array}{r} 182.49 \\ -\ 20.60 \\ \hline 161.89 \end{array}$$

When subtracting numbers that are measurements, be certain that the measurements are in the same units. For example, subtract 22 cm from 0.62 m. First change m to cm, then do the subtraction.

$$(0.62\text{ m})\left(\frac{100\text{ cm}}{\text{m}}\right) = 62\text{ cm} \qquad \begin{array}{r} 62\text{ cm} \\ -22\text{ cm} \\ \hline 40.\text{ cm} \end{array}$$

Multiplication of Numbers with Decimals

To multiply two or more numbers together that contain decimals, we first multiply as if they were whole numbers. Then, to locate the decimal point in the product, we add together the number of digits to the right of the decimal in all the numbers multiplied together. The product should have this same number of digits to the right of the decimal point.

Multiply $2.05 \times 2.05 = 4.2025$ (total of four digits to the right of the decimal)

If a number is a measurement, the answer must be adjusted to the correct number of significant figures.

Here are more examples:

$14.25 \times 6.01 \times 0.75 = 64.231875$ (six digits to the right of the decimal)

$39.26 \times 60 = 2355.60$ (two digits to the right of the decimal)

Division of Numbers with Decimals

To divide numbers containing decimals, we first relocate the decimal points of the numerator and denominator by moving them to the right as many places as needed to make the denominator a whole number. (Move the decimal of both the numerator and the denominator the same amount and in the same direction.) For example,

$$\frac{136.94}{4.1} = \frac{1369.4}{41}$$

The decimal point adjustment in this example is equivalent to multiplying both numerator and denominator by 10. Now we carry out the division normally, locating the decimal point immediately above its position in the dividend:

$$41\overset{\displaystyle 33.4}{\overline{)1269.4}} \qquad \frac{0.441}{26.25} = \frac{44.1}{2625} = 2625\overset{\displaystyle 0.0168}{\overline{)44.1000}}$$

These examples are guides to the principles used in performing the various mathematical operations illustrated. The use of a calculator will save many hours of doing tedious calculations. After solving a problem, the student should check for errors and evaluate the answer to see if it is logical and consistent with the data given.

Algebraic Equations

Many mathematical problems that are encountered in chemistry fall into the following algebraic forms. Solutions to these problems are simplified by first isolating the desired term on one side of the equation. This rearrangement is accomplished by treating both sides of the equation in an identical manner until the desired term is isolated.

(a) $a = \dfrac{b}{c}$

To solve for b, multiply both sides of the equation by c:

$$a \times c = \frac{b}{\cancel{c}} \times \cancel{c}$$

$$b = a \times c$$

To solve for c, multiply both sides of the equation by $\dfrac{c}{a}$:

$$\cancel{a} \times \frac{c}{\cancel{a}} = \frac{b}{\cancel{c}} \times \frac{\cancel{c}}{a}$$

$$c = \frac{b}{a}$$

(b) $\dfrac{a}{b} = \dfrac{c}{d}$

To solve for a, multiply both sides of the equation by b:

$$\frac{a}{\cancel{b}} \times \cancel{b} = \frac{c}{d} \times b$$

$$a = \frac{c \times b}{d}$$

To solve for b, multiply both sides of the equation by $\dfrac{b \times d}{c}$:

$$\frac{a}{\cancel{b}} \times \frac{\cancel{b} \times d}{c} = \frac{\cancel{c}}{\cancel{d}} \times \frac{b \times \cancel{d}}{\cancel{c}}$$

$$\frac{a \times d}{c} = b$$

(c) $a \times b = c \times d$

To solve for a, divide both sides of the equation by b:

$$\frac{a \times \cancel{b}}{\cancel{b}} = \frac{c \times d}{b}$$

$$a = \frac{c \times d}{b}$$

(d) $\dfrac{b - c}{a} = d$

To solve for b, first multiply both sides of the equation by a:

$$\frac{\cancel{a}(b - c)}{\cancel{a}} = d \times a$$

$$b - c = d \times a$$

Then add c to both sides of the equation:

$$b - \cancel{c} + \cancel{c} = d \times a + c$$

$$b = (d \times a) + c$$

When $a = 1.8$, $c = 32$, and $d = 35$,

$$b = (35 \times 1.8) + 32 = 63 + 32 = 95$$

Expression of Large and Small Numbers

In scientific measurement and calculations, we often encounter very large and very small numbers—for example, 0.00000384 and 602,000,000,000,000,000,000,000. These numbers are troublesome to write and awkward to work with, especially in calculations. A convenient method of expressing these large and small numbers in a simplified form is by means of exponents, or powers, of 10. This method of expressing numbers is known as **scientific, or exponential, notation.**

An **exponent** is a number written as a superscript following another number. Exponents are often called *powers* of numbers. The term *power* indicates how many times the number is used as a factor. In the number 10^2, 2 is the exponent, and the number means 10 squared, or 10 to the second power, or $10 \times 10 = 100$. Three other examples are

$$3^2 = 3 \times 3 = 9$$

$$3^4 = 3 \times 3 \times 3 \times 3 = 81$$

$$10^3 = 10 \times 10 \times 10 = 1000$$

For ease of handling, large and small numbers are expressed in powers of 10. Powers of 10 are used because multiplying or dividing by 10 coincides with moving the decimal point in a number by one place. Thus, a number multiplied by 10^1 would move the decimal point one place to the right; 10^2, two places to the right; 10^{-2}, two places to the left. To express a number in powers of 10, we move the decimal point in the original number to a new position, placing it so that the number is a value between 1 and 10. This new decimal number is multiplied by 10 raised to the proper power. For example, to write the number 42,389 in exponential form, the decimal point is placed between the 4 and the 2 (4.2389), and the number is multiplied by 10^4; thus, the number is 4.2389×10^4:

$$\underbrace{42{,}389}_{4321} = 4.2389 \times 10^4$$

The exponent of 10 (4) tells us the number of places that the decimal point has been moved from its original position. If the decimal point is moved to the left, the exponent is a positive number; if it is moved to the right, the exponent is a negative number. To express the number 0.00248 in exponential notation (as a power of 10), the decimal point is moved three places to the right; the exponent of 10 is -3, and the number is 2.48×10^{-3}.

$$0.\underbrace{002}_{123}48 = 2.48 \times 10^{-3}$$

Study the following examples of changing a number to scientific notation.

$$1237 = 1.237 \times 10^3$$

$$988 = 9.88 \times 10^2$$

$$147.2 = 1.472 \times 10^2$$

$$2{,}200{,}000 = 2.2 \times 10^6$$

$$0.0123 = 1.23 \times 10^{-2}$$

$$0.00005 = 5 \times 10^{-5}$$

$$0.000368 = 3.68 \times 10^{-4}$$

Exponents in multiplication and division The use of powers of 10 in multiplication and division greatly simplifies locating the decimal point in the answer. In multiplication, first change all numbers to powers of 10, then multiply the numerical portion in the usual manner, and finally add the exponents of 10 algebraically, expressing them as a power of 10 in the product. In multiplication, the exponents (powers of 10) are added algebraically.

$$10^2 \times 10^3 = 10^{(2+3)} = 10^5$$

$$10^2 \times 10^2 \times 10^{-1} = 10^{(2+2-1)} = 10^3$$

Multiply:	$(40{,}000)(4200)$
Change to powers of 10:	$(4 \times 10^4)(4.2 \times 10^3)$
Rearrange:	$(4 \times 4.2)(10^4 \times 10^3)$
	$16.8 \times 10^{(4+3)}$
	16.8×10^7 or 1.68×10^8 (Answer)
Multiply:	$(380)(0.00020)$
	$(3.80 \times 10^2)(2.0 \times 10^{-4})$
	$(3.80 \times 2.0)(10^2 \times 10^{-4})$
	$7.6 \times 10^{(2-4)}$
	7.6×10^{-2} or 0.076 (Answer)
Multiply:	$(125)(284)(0.150)$
	$(1.25 \times 10^2)(2.84 \times 10^2)(1.50 \times 10^{-1})$
	$(1.25)(2.84)(1.50)(10^2 \times 10^2 \times 10^{-1})$
	$5.325 \times 10^{(2+2-1)}$
	5.33×10^3 (Answer)

In division, after changing the numbers to powers of 10, move the 10 and its exponent from the denominator to the numerator, changing the sign of the exponent. Carry out the division in the usual manner, and evaluate the power of 10. Change the sign(s) of the exponent(s) of 10 in the denominator, and move the 10 and its exponent(s) to the numerator. Then add all the exponents of 10 together. For example,

$$\frac{10^5}{10^3} = 10^5 \times 10^{-3} = 10^{(5-3)} = 10^2$$

$$\frac{10^3 \times 10^4}{10^{-2}} = 10^3 \times 10^4 \times 10^2 = 10^{(3+4+2)} = 10^9$$

Significant Figures in Calculations

The result of a calculation based on experimental measurements cannot be more precise than the measurement that has the greatest uncertainty. (See Section 2.4 for additional discussion.)

Addition and subtraction The result of an addition or subtraction should contain no more digits to the right of the decimal point than are contained in the quantity that has the least number of digits to the right of the decimal point.

Perform the operation indicated and then round off the number to the proper number of significant figures:

(a)
142.8 g
18.843 g
36.42 g
198.063 g
198.1 g (Answer)

(b)
93.45 mL
−18.0 mL
75.45 mL
75.5 mL (Answer)

(a) The answer contains only one digit after the decimal point since 142.8 contains only one digit after the decimal point.

(b) The answer contains only one digit after the decimal point since 18.0 contains one digit after the decimal point.

Multiplication and division In calculations involving multiplication or division, the answer should contain the same number of significant figures as the measurement that has the least number of significant figures. In multiplication or division, the position of the decimal point has nothing to do with the number of significant figures in the answer. Study the following examples:

	Round off to
$(2.05)(2.05) = 4.2025$	4.20
$(18.48)(5.2) = 96.096$	96
$(0.0126)(0.020) = 0.000252$ or	
$(1.26 \times 10^{-2})(2.0 \times 10^{-2}) = 2.520 \times 10^{-4}$	2.5×10^{-4}
$\frac{1369.4}{41} = 33.4$	33
$\frac{2268}{4.20} = 540.$	540.

Dimensional Analysis

Many problems of chemistry can be readily solved by dimensional analysis using the factor-label or conversion-factor method. Dimensional analysis involves the use of proper units of dimensions for all factors that are multiplied, divided, added, or subtracted in setting up and solving a problem. Dimensions are physical quantities such as length, mass, and time, which are expressed in such units as centimeters, grams, and seconds, respectively. In solving a problem, we treat these units mathematically just as though they were numbers, which gives us an answer that contains the correct dimensional units.

A measurement or quantity given in one kind of unit can be converted to any other kind of unit having the same dimension. To convert from one kind of unit to another, the original quantity or measurement is multiplied or divided by a conversion factor. The key to success lies in choosing the correct conversion factor. This general method of calculation is illustrated in the following examples.

Suppose we want to change 24 ft to inches. We need to multiply 24 ft by a conversion factor containing feet and inches. Two such conversion factors can be written relating inches to feet:

$$\frac{12\text{ in.}}{1\text{ ft}} \quad \text{or} \quad \frac{1\text{ ft}}{12\text{ in.}}$$

We choose the factor that will mathematically cancel feet and leave the answer in inches. Note that the units are treated in the same way we treat numbers, multiplying or dividing as required. Two possibilities then arise to change 24 ft to inches:

$$(24\ \cancel{\text{ft}})\left(\frac{12\text{ in.}}{1\ \cancel{\text{ft}}}\right) \quad \text{or} \quad (24\text{ ft})\left(\frac{1\text{ ft}}{12\text{ in.}}\right)$$

In the first case (the correct method), feet in the numerator and the denominator cancel, giving us an answer of 288 in. In the second case, the units of the answer are $\text{ft}^2/\text{in.}$, the answer being $2.0\ \text{ft}^2/\text{in}$. In the first case, the answer is reasonable because it is expressed in units having the proper dimensions. That is, the dimension of length expressed in feet has been converted to length in inches according to the mathematical expression

$$\cancel{\text{ft}} \times \frac{\text{in.}}{\cancel{\text{ft}}} = \text{in.}$$

In the second case, the answer is not reasonable because the units ($\text{ft}^2/\text{in.}$) do not correspond to units of length. The answer is therefore incorrect. The units are the guiding factor for the proper conversion.

The reason we can multiply 24 ft times 12 in./ft and not change the value of the measurement is that the conversion factor is derived from two equivalent quantities. Therefore, the conversion factor 12 in./ft is equal to unity. When you multiply any factor by 1, it does not change the value:

$$12\text{ in.} = 1\text{ ft} \quad \text{and} \quad \frac{12\text{ in.}}{1\text{ ft}} = 1$$

Convert 16 kg to milligrams. In this problem it is best to proceed in this fashion:

$$\text{kg} \longrightarrow \text{g} \longrightarrow \text{mg}$$

The possible conversion factors are

$$\frac{1000\text{ g}}{1\text{ kg}} \text{ or } \frac{1\text{ kg}}{1000\text{ g}} \qquad \frac{1000\text{ mg}}{1\text{ g}} \text{ or } \frac{1\text{ g}}{1000\text{ mg}}$$

We use the conversion factor that leaves the proper unit at each step for the next conversion. The calculation is

$$(16\ \cancel{\text{kg}})\left(\frac{1000\ \cancel{\text{g}}}{1\ \cancel{\text{kg}}}\right)\left(\frac{1000\text{ mg}}{1\ \cancel{\text{g}}}\right) = 1.6 \times 10^7\text{ mg}$$

Regardless of application, the basis of dimensional analysis is the use of conversion factors to organize a series of steps in the quest for a specific quantity with a specific unit.

Graphical Representation of Data

A graph is often the most convenient way to present or display a set of data. Various kinds of graphs have been devised, but the most common type uses a set of horizontal and vertical coordinates to show the relationship of two variables. It is called an x–y graph because the data of one variable are represented on the horizontal or x-axis (abscissa) and the data of the other variable are represented on the vertical or y-axis (ordinate). FIGURE I.1.

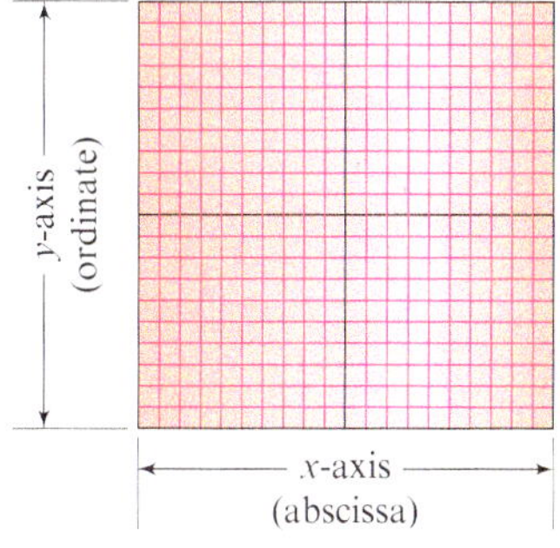

FIGURE I.1

As a specific example of a simple graph, let us graph the relationship between Celsius and Fahrenheit temperature scales. Assume that initially we have only the information in the table next to FIGURE 1.2.

On a set of horizontal and vertical coordinates (graph paper), scale off at least 100 Celsius degrees on the x-axis and at least 212 Fahrenheit degrees on the y-axis. Locate and mark the three points corresponding to the three temperatures given and draw a line connecting these points (see Figure I.2).

Here is how a point is located on the graph: Using the (50°C, 122°F) data, trace a vertical line up from 50°C on the x-axis and a horizontal line across from 122°F on the y-axis and mark the point where the two lines intersect. This process is called *plotting.* The

°C	°F
0	32
50	122
100	212

FIGURE I.2

other two points are plotted on the graph in the same way. (*Note:* The number of degrees per scale division was chosen to give a graph of convenient size. In this case, there are 5 Fahrenheit degrees per scale division and 2 Celsius degrees per scale division.)

The graph in Figure I.2 shows that the relationship between Celsius and Fahrenheit temperature is that of a straight line. The Fahrenheit temperature corresponding to any given Celsius temperature between 0 and 100° can be determined from the graph. For example, to find the Fahrenheit temperature corresponding to 40°C, trace a perpendicular line from 40°C on the x-axis to the line plotted on the graph. Now trace a horizontal line from this point on the plotted line to the y-axis and read the corresponding Fahrenheit temperature (104°F). See the dashed lines in Figure I.2. In turn, the Celsius temperature corresponding to any Fahrenheit temperature between 32° and 212° can be determined from the graph by tracing a horizontal line from the Fahrenheit temperature to the plotted line and reading the corresponding temperature on the Celsius scale directly below the point of intersection.

The mathematical relationship of Fahrenheit and Celsius temperatures is expressed by the equation $°F = (1.8 \times °C) + 32$. Figure I.2 is a graph of this equation. Because the graph is a straight line, it can be extended indefinitely at either end. Any desired Celsius temperature can be plotted against the corresponding Fahrenheit temperature by extending the scales along both axes as necessary.

FIGURE I.3 is a graph showing the solubility of potassium chlorate in water at various temperatures. The solubility curve on this graph was plotted from the data in the table next to the graph.

In contrast to the Celsius–Fahrenheit temperature relationship, there is no simple mathematical equation that describes the exact relationship between temperature and the solubility of potassium chlorate. The graph in Figure I.3 was constructed from experimentally determined solubilities at the six temperatures shown. These experimentally determined solubilities are all located on the smooth curve traced by the unbroken-line

Temperature (°C)	Solubility (g $KClO_3$/ 100 g water)
10	5.0
20	7.4
30	10.5
50	19.3
60	24.5
80	38.5

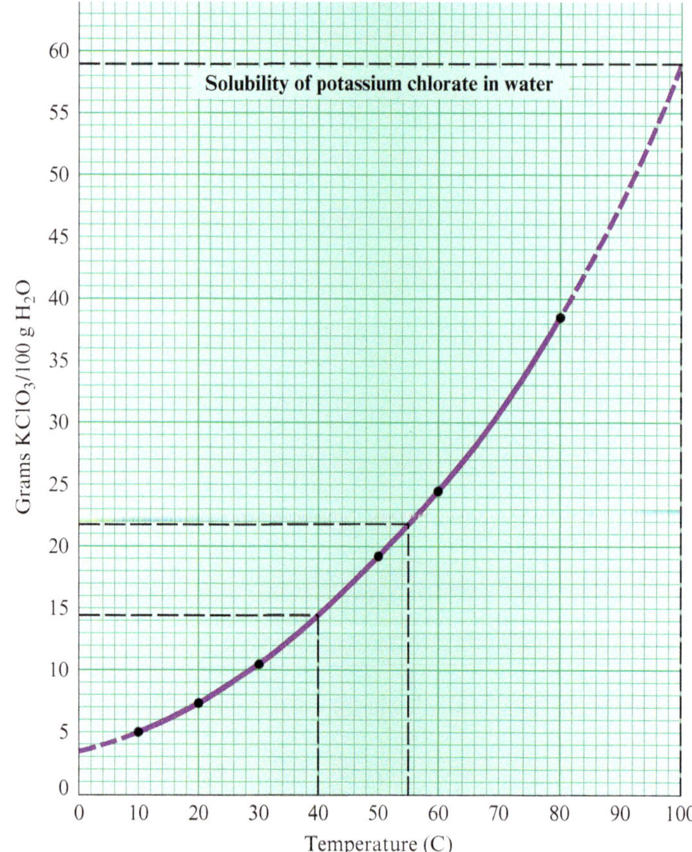

FIGURE I.3

portion of the graph. We are therefore confident that the unbroken line represents a very good approximation of the solubility data for potassium chlorate over the temperature range from 10 to 80°C. All points on the plotted curve represent the composition of saturated solutions. Any point below the curve represents an unsaturated solution.

The dashed-line portions of the curve are *extrapolations;* that is, they extend the curve above and below the temperature range actually covered by the plotted solubility data. Curves such as this one are often extrapolated a short distance beyond the range of the known data, although the extrapolated portions may not be highly accurate. Extrapolation is justified only in the absence of more reliable information.

The graph in Figure I.3 can be used with confidence to obtain the solubility of $KClO_3$ at any temperature between 10° and 80°C, but the solubilities between 0° and 10°C and between 80° and 100°C are less reliable. For example, what is the solubility of $KClO_3$ at 55°C, at 40°C, and at 100°C?

First draw a perpendicular line from each temperature to the plotted solubility curve. Now trace a horizontal line to the solubility axis from each point on the curve and read the corresponding solubilities. The values that we read from the graph are

40°C	14.2 g $KClO_3$/100 g water
55°C	22.0 g $KClO_3$/100 g water
100°C	59 g $KClO_3$/100 g water

Of these solubilities, the one at 55°C is probably the most reliable because experimental points are plotted at 50° and at 60°C. The 40°C solubility value is a bit less reliable because the nearest plotted points are at 30° and 50°C. The 100°C solubility is the least reliable of the three values because it was taken from the extrapolated part of the curve, and the nearest plotted point is 80°C. Actual handbook solubility values are 14.0 and 57.0 g of $KClO_3$/100 g water at 40°C and 100°C, respectively.

The graph in Figure I.3 can also be used to determine whether a solution is saturated or unsaturated. For example, a solution contains 15 g of $KClO_3$/100 g of water and is at a temperature of 55°C. Is the solution saturated or unsaturated? *Answer:* The solution is unsaturated because the point corresponding to 15 g and 55°C on the graph is below the solubility curve; all points below the curve represent unsaturated solutions.

Appendix 2

UNITS OF MEASUREMENT

Physical Constants

Constant	Symbol	Value
Atomic mass unit	amu	1.6606×10^{-27} kg
Avogadro's number	N	6.022×10^{23} particles/mol
Gas constant	R (at STP)	0.08205 L atm/K mol
Mass of an electron	m_e	9.109×10^{-28} g
		5.486×10^{-4} amu
Mass of a neutron	m_n	1.675×10^{-27} kg
		1.00866 amu
Mass of a proton	m_p	1.673×10^{-27} kg
		1.00728 amu
Speed of light	c	2.997925×10^{8} m/s

SI Units and Conversion Factors

Length

SI unit: meter (m)

1 meter	= 1.0936 yards
	= 100 centimeters
	= 1000 millimeters
1 centimeter	= 0.3937 inch
1 inch	= 2.54 centimeters (exactly)
1 kilometer	= 0.62137 mile
1 mile	= 5280 feet
	= 1.609 kilometers
1 angstrom	= 10^{-10} meter

Mass

SI unit: kilogram (kg)

1 kilogram	= 1000 grams
	= 2.20 pounds
1 gram	= 1000 milligrams
1 pound	= 453.59 grams
	= 0.45359 kilogram
	= 16 ounces
1 ton	= 2000 pounds
	= 907.185 kilograms
1 ounce	= 28.3 grams
1 atomic mass unit	= 1.6606×10^{-27} kilogram

Volume

SI unit: cubic meter (m^3)

1 liter	= 10^{-3} m^3
	= 1 dm^3
	= 1.0567 quarts
	= 1000 milliliters
1 gallon	= 4 quarts
	= 8 pints
	= 3.785 liters
1 quart	= 32 fluid ounces
	= 0.946 liter
1 fluid ounce	= 29.6 milliliters

Temperature

SI unit: kelvin (K)

$0\ K = -273.15°C$

$= -459.67°F$

$K = °C + 273.15$

$°C = \frac{°F - 32}{1.8}$

$°F = 1.8(°C) + 32$

$°C = \frac{5}{9}(°F - 32)$

Energy

SI unit: joule (J)

1 joule	= 1 kg m^2/s^2
	= 0.23901 calorie
1 calorie	= 4.184 joules

Pressure

SI unit: pascal (Pa)

1 pascal	= 1 kg/m s^2
1 atmosphere	= 101.325 kilopascals
	= 760 torr (mm Hg)
	= 14.70 pounds per square inch (psi)

Appendix 3

VAPOR PRESSURE OF WATER AT VARIOUS TEMPERATURES

Temperature (°C)	Vapor pressure (torr)	Temperature (°C)	Vapor pressure (torr)
0	4.6	26	25.2
5	6.5	27	26.7
10	9.2	28	28.3
15	12.8	29	30.0
16	13.6	30	31.8
17	14.5	40	55.3
18	15.5	50	92.5
19	16.5	60	149.4
20	17.5	70	233.7
21	18.6	80	355.1
22	19.8	90	525.8
23	21.2	100	760.0
24	22.4	110	1074.6
25	23.8		

Appendix 4

SOLUBILITY TABLE

	F^-	Cl^-	Br^-	I^-	O^{2-}	S^{2-}	OH^-	NO_3^-	CO_3^{2-}	SO_4^{2-}	$C_2H_3O_2^-$
H^+	*aq*	*aq*	*aq*	*aq*	*aq*	*sl.aq*	*aq*	*aq*	*sl.aq*	*aq*	*aq*
Na^+	*aq*	*aq*	*aq*	*aq*	*aq*	*aq*	*aq*	*aq*	*aq*	*aq*	*aq*
K^+	*aq*	*aq*	*aq*	*aq*	*aq*	*aq*	*aq*	*aq*	*aq*	*aq*	*aq*
NH_4^+	*aq*	*aq*	*aq*	*aq*	—	*aq*	*aq*	*aq*	*aq*	*aq*	*aq*
Ag^+	*aq*	*I*	*I*	*I*	*I*	*I*	—	*aq*	*I*	*I*	*I*
Mg^{2+}	*I*	*aq*	*aq*	*aq*	*I*	*d*	*I*	*aq*	*I*	*aq*	*aq*
Ca^{2+}	*I*	*aq*	*aq*	*aq*	*I*	*d*	*I*	*aq*	*I*	*I*	*aq*
Ba^{2+}	*I*	*aq*	*aq*	*aq*	*sl.aq*	*d*	*sl.aq*	*aq*	*I*	*I*	*aq*
Fe^{2+}	*sl.aq*	*aq*	*aq*	*aq*	*I*	*I*	*I*	*aq*	*sl.aq*	*aq*	*aq*
Fe^{3+}	*I*	*aq*	*aq*	—	*I*	*I*	*I*	*aq*	*I*	*aq*	*I*
Co^{2+}	*aq*	*aq*	*aq*	*aq*	*I*	*I*	*I*	*aq*	*I*	*aq*	*aq*
Ni^{2+}	*sl.aq*	*aq*	*aq*	*aq*	*I*	*I*	*I*	*aq*	*I*	*aq*	*aq*
Cu^{2+}	*sl.aq*	*aq*	*aq*	—	*I*	*I*	*I*	*aq*	*I*	*aq*	*aq*
Zn^{2+}	*sl.aq*	*aq*	*aq*	*aq*	*I*	*I*	*I*	*aq*	*I*	*aq*	*aq*
Hg^{2+}	*d*	*aq*	*I*	*I*	*I*	*I*	*I*	*aq*	*I*	*d*	*aq*
Cd^{2+}	*sl.aq*	*aq*	*aq*	*aq*	*I*	*I*	*I*	*aq*	*I*	*aq*	*aq*
Sn^{2+}	*aq*	*aq*	*aq*	*sl.aq*	*I*	*I*	*I*	*aq*	*I*	*aq*	*aq*
Pb^{2+}	*I*	*I*	*I*	*I*	*I*	*I*	*I*	*aq*	*I*	*I*	*aq*
Mn^{2+}	*sl.aq*	*aq*	*aq*	*aq*	*I*	*I*	*I*	*aq*	*I*	*aq*	*aq*
Al^{3+}	*I*	*aq*	*aq*	*aq*	*I*	*d*	*I*	*aq*	—	*aq*	*aq*

Key:
aq = soluble in water
sl.aq = slightly soluble in water
I = insoluble in water (less than 1 g/100 g H_2O)
d = decomposes in water

Appendix 5 ANSWERS TO SELECTED EXERCISES

CHAPTER 1

Exercises

2. Two states are present; solid and liquid.

4. The maple leaf represents a heterogeneous mixture.

6. (a) homogeneous
(b) homogeneous
(c) heterogeneous
(d) heterogeneous

9. (a) water, a pure substance
(c) salt, a pure substance

12. (a) Picture 2.
(b) Pictures (1) and (3) are heterogeneous mixtures. Picture (4) is a pure substance.
(c) Picture (1) because it is composed of multiple types of particles. Picture (2) may contain a diatomic element.

13. (a) two phases, solid and gas
(c) two phases, solid and liquid

CHAPTER 2

Exercises

2. (a) 1000 meters = 1 kilometer
(c) 0.000001 liter = 1 microliter
(e) 0.001 liter = 1 milliliter

4. (a) mg (e) Å
(c) m

6. (a) not significant (e) significant
(c) significant

8. (a) 40.0 (3 sig fig)
(b) 0.081 (2 sig fig)
(c) 129,042 (6 sig fig)
(d) 4.090×10^{-3} (4 sig fig)

10. (a) 8.87 (c) 130. (1.30×10^2)
(b) 21.3 (d) 2.00×10^6

12. (a) 4.56×10^{-2} (c) 4.030×10^1
(b) 4.0822×10^3 (d) 1.2×10^7

14. (a) 28.1 (e) 2.49×10^{-4}
(c) 4.0×10^1

16. (a) $\frac{1}{4}$ (c) $1\frac{2}{3}$ or $\frac{5}{3}$
(b) $\frac{5}{8}$ (d) $\frac{8}{9}$

18. (a) 1.0×10^2 (c) 22
(b) 4.6 mL

20. (a) 4.5×10^8 Å (e) 6.5×10^5 mg
(c) 8.0×10^6 mm (g) 468 mL

22. (a) 117 ft (c) 7.4×10^4 mm^3
(e) 75.7 L

24. 12 mi/hr

26. 8.33 gr

28. 3.54 days

30. 5.8 eggs so 6 eggs should be added

32. 2.54×10^{-3} lb

34. 5.94×10^3 cm/s

36. (a) 50 ft/min (b) 3×10^{-4} km/s

38. 5.59×10^3 cm/s

40. 4.03×10^6 tilapia

42. 160 L

44. 4×10^5 m^2

46. 6 gal

48. 5.52 g hydrogen

50. (a) 11.3%, (b) 39.6 g $NaNO_3$, (c) 442 g solution

52. 113°F Summer!

54. (a) 90.°F (c) 546 K
(b) −22.6°C (d) −300°F

56. 119°C, 392 K

58. −11.4°C = 11.4°F

60. −153°C

62. 0.810 g/mL

64. 7.2 g/mL

66. 63 mL

68. 70. lb nutty candy

70. (a) 95 g Ag/100 g sterling and 7.50 g Cu/100 g sterling,
(b) 701 sterling hearts

73. A graduated cylinder would be the best choice for adding 100 mL of solvent to a reaction. While the volumetric flask is also labeled 100 mL, volumetric flasks are typically used for doing dilutions. The other three pieces of glassware could also be used, but they hold smaller volumes, so it would take a longer time to measure out 100 mL. Also, because you would have to repeat the measurement many times using the other glassware, there is a greater chance for error.

74. 14.5 mL

75. 26 mL

77. 3.43 vials potion

78. 10.1 lb sequins

79. Yes, 120 L additional solution

82. (a) −252.88°C (b) −423.18°F

83. 97°F − 100°F, similar to mammals

88. 5.1×10^3 L

92. 0.506 km^2

94. 0.965 g/mL

96. ethyl alcohol, because it has the lower density

98. 54.3 mL

99. 18.6 mg NHP

104. 76.9 g

CHAPTER 3

Exercises

2. (d) HI (f) Cl_2 (g) CO

4. (a) magnesium, bromine
(c) hydrogen, nitrogen, oxygen
(e) aluminum, phosphorus, oxygen

6. (a) $AlBr_3$ (c) $PbCrO_4$

8. (a) $C_{55}H_{60}O_{24}$ (b) $Cl_{15}H_{14}O_6$ (c) $Cl_{30}H_{48}O_3$

10. (a) hydrogen, 4 atoms; carbon, 2 atoms; oxygen, 2 atoms
(c) magnesium, 1 atom; hydrogen, 2 atoms; sulfur 2 atoms; oxygen, 6 atoms
(e) nickel, 1 atom; carbon, 1 atom; oxygen, 3 atoms
(g) carbon, 4 atoms; hydrogen, 10 atoms

12. (a) 9 atoms (b) 12 atoms (c) 12 atoms (d) 12 atoms

14. (a) 6 atoms O (b) 4 atoms O (c) 6 atoms O (d) 21 atoms O

16. (a) mixture (c) pure substance (e) mixture

18. (c) element (d) compound

20. (a) compound (b) compound (c) mixture

22. No. The only common liquid elements (at room temperature) are mercury and bromine.

24. 72% solids

26. The formula for hydrogen peroxide is H_2O_2. There are two atoms of oxygen for every two atoms of hydrogen. The molar mass of oxygen is 16.00 g, and the molar mass of hydrogen is 1.008 g. For hydrogen peroxide, the total mass of hydrogen is 2.016 g and the total mass of oxygen is 32.00 g, for a ratio of hydrogen to oxygen of approximately 2:32. or 1:16. Therefore, there is 1 gram of hydrogen for every 16 grams of oxygen.

29. (a) 1 carbon atom and 1 oxygen atom; total number of atoms = 2
(c) 1 hydrogen atom, 1 nitrogen atom, and 3 oxygen atoms; total number of atoms = 5
(e) 1 calcium atom, 2 nitrogen atoms, and 6 oxygen atoms; total number of atoms = 9

32. 40 atoms H

34. (a) magnesium, manganese, molybdenum, mendelevium, mercury, meitnerium
(c) sodium, potassium, iron, silver, tin, antimony

36. 420 atoms

39. (a) $NaCl$ (c) K_2O (e) K_3PO_4 (g) $C_6H_{12}O_6$ (i) $Cr(NO_3)_3$

41. (a) 12C, 22H, 11O (c) 14C, 18H, 5O, 2N (e) 12C, 19H, 8O, 3Cl

CHAPTER 4

Exercises

2. (a) physical (c) chemical (e) physical (g) chemical

4. The copper wire, like the platinum wire, changed to a glowing red color when heated. Upon cooling, a new substance, black copper(II) oxide, had appeared.

6. Reactant: water
Products: hydrogen, oxygen

8. (a) physical (c) chemical (e) chemical

10. (a) kinetic (c) potential (e) kinetic

12. The transformation of kinetic energy to thermal energy

14. (a) +

16. 6.4×10^2 J

18. 5.58×10^3 J

20. 0.128 J/g°C

22. 31.7°C

27. 0.525 lb gold

30. 1.0×10^2 g Cu

32. 654°C

39. 8 g fat

40. (a) 1.12 kJ
(b) glove made of wool felt

44. A chemical change has occurred. Hydrogen molecules and oxygen molecules have combined to form water molecules.

CHAPTER 5

Exercises

2. Mg^{2+}, Cr^{3+}, Ba^{2+}, Ca^{2+}, and Y^{3+}.

4. (a) The nucleus of the atom contains most of the mass.
(c) The atom is mostly empty space.

6. The nucleus of an atom contains nearly all of its mass.

8. Electrons:
Dalton—Electrons are not part of his model.
Thomson—Electrons are scattered throughout the positive mass of matter in the atom.
Rutherford—Electrons are located out in space away from the central positive mass.
Positive matter:
Dalton—No positive matter in his model.
Thomson—Positive matter is distributed throughout the atom.
Rutherford—Positive matter is concentrated in a small central nucleus.

10. Yes. The mass of the isotope $^{12}_{6}C$, 12, is an exact number. The masses of other isotopes are not exact numbers.

12. Three isotopes of hydrogen have the same number of protons and electrons but differ in the number of neutrons.

14. All five isotopes have nuclei that contain 30 protons and 30 electrons. The numbers of neutrons are:

Isotope mass number	Neutrons
64	34
66	36
67	37
68	38
70	40

16. boron-10

18.

Proton
Neutron
Electron cloud

20. (a) $^{109}_{47}Ag$ (c) $^{57}_{26}Fe$

22. (a) 26 protons, 26 electrons, and 28 neutrons
(c) 35 protons, 35 electrons, and 44 neutrons

24. (a) 56 (c) 79 (e) $^{135}_{56}Ba$

26. (a) +2, (c) 0.001097 u

28. 50. u

30. 35.46 u; chlorine (Cl)

32. $1.5 \times 10^5 : 1.0$

34. (a) Ir−191, 77 protons, 114 neutrons; Ir−193, 77 protons, 116 neutrons (c) Ir−191, 40.28% protons, Ir−193, 39.86% protons

36. (a) and (c) are isotopes of phosphorus

37. 9.2×10^8 atoms silicon

39. ^{160}Gd, ^{156}Dy, ^{163}Ho, ^{162}Er. This order differs from that on the periodic table.

42. (a) 197 u (b) gold (Au)

45.

C^+	6 protons, 5 electrons
O^+	8 protons, 7 electrons
O^{2+}	8 protons, 6 electrons

54. (a) 0.02644% electrons (c) 0.02356% electrons

56. The electron region is the area around the nucleus where electrons are most likely to be located.

CHAPTER 6

Exercises

2. (a) Al_2S_3 (c) K_3N (d) SrO (f) $AlCl_3$

4. (a) F^- (c) I^- (e) S^{2-} (g) P^{3-} (i) $Cr_2O_7^{2-}$ (k) PO_4^{3-} (m) N^{3-}

6. (a) sodium thiosulfate (b) dinitrogen monoxide (c) aluminum oxide (d) sodium chloride (e) magnesium hydroxide (f) lead (II) sulfide

8.

Ion	SO_4^{2-}	OH^-	AsO_4^{3-}	$C_2H_3O_2^-$	CrO_4^{2-}
NH_4^+	$(NH_4)_2SO_4$	NH_4OH	$(NH_4)_3AsO_4$	$NH_4C_2H_3O_2$	$(NH_4)_2CrO_4$
Ca^{2+}	$CaSO_4$	$Ca(OH)_2$	$Ca_3(AsO_4)_2$	$Ca(C_2H_3O_2)_2$	$CaCrO_4$
Fe^{3+}	$Fe_2(SO_4)_3$	$Fe(OH)_3$	$FeAsO_4$	$Fe(C_2H_3O_2)_3$	$Fe_2(CrO_4)_3$
Ag^+	Ag_2SO_4	$AgOH$	Ag_3AsO_4	$AgC_2H_3O_2$	Ag_2CrO_4
Cu^{2+}	$CuSO_4$	$Cu(OH)_2$	$Cu_3(AsO_4)_2$	$Cu(C_2H_3O_2)_2$	$CuCrO_4$

10. NH_4^+ compounds: ammonium sulfate, ammonium hydroxide, ammonium arsenate, ammonium acetate, ammonium chromate.

Ca^{2+} compounds: calcium sulfate, calcium hydroxide, calcium arsenate, calcium acetate, calcium chromate.

Fe^{3+} compounds: iron(III) sulfate, iron(III) hydroxide, iron(III) arsenate, iron(III) acetate, iron(III) chromate.

Ag^+ compounds: silver sulfate, silver hydroxide, silver arsenate, silver acetate, silver chromate.

Cu^{2+} compounds: copper(II) sulfate, copper(II) hydroxide, copper(II) arsenate, copper(II) acetate, copper(II) chromate.

12. (a) K_3N (c) FeO (e) Ca_3N_2 (g) MnI_3

14. (a) phosphorus pentabromide
(c) dinitrogen pentasulfide
(e) silicon tetrachloride
(g) tetraphosphorus heptasulfide

16. (a) $SnBr_4$ (c) $Ni(BO_3)_2$ (d) $Hg(NO_2)_2$ (f) $Fe(C_2H_3O_2)_2$

18. (a) $HC_2H_3O_2$ (c) H_2S (d) H_3BO_3 (f) HClO

20. (a) phosphoric acid (c) iodic acid (d) hydrochloric acid (f) nitric acid (g) hydroiodic acid (i) sulfurous acid

22. (a) Na_2CrO_4 (c) $Ni(C_2H_3O_2)_2$ (e) $Mg(BrO_3)_2$ (g) $Mn(OH)_2$ (i) NaClO (k) $Cr_2(SO_3)_3$ (m) $Na_2C_2O_4$

24. (a) calcium hydrogen sulfate
(c) tin(II) nitrite
(e) potassium hydrogen carbonate
(f) bismuth(III) arsenate
(h) ammonium monohydrogen phosphate
(j) potassium permanganate

26. (a) C_2H_5OH (c) $CaSO_4 \cdot 2H_2O$ (e) HCl (g) NaOH

28. (a) $PbCO_3$, $Pb(OH)_2$
(b) PbO, $Pb(C_2H_3O_2)_2$

30. (a) sulfate (c) nitrate (d) chlorate (f) carbonate

32. (a) K_2SO_4 (c) KNO_3 (d) $KClO_3$ (f) K_2CO_3

34. Formula: KCl Name: potassium chloride

36. (a) $CaCO_3$, no concern (b) $HC_2H_3O_2$, no concern (c) H_2O, no concern

39. $Li_3Fe(CN)_6$
$AlFe(CN)_6$
$Zn_3[Fe(CN)_6]_2$

CHAPTER 7

Exercises

2. Molar masses

(a) NaOH	40.00
(c) Cr_2O_3	152.0
(e) $Mg(HCO_3)_2$	146.3
(g) $C_6H_{12}O_6$	180.2
(i) $BaCl_2 \cdot 2H_2O$	244.2

4. (a) 0.625 mol NaOH
(c) 7.18×10^{-3} mol $MgCl_2$
(e) 2.03×10^{-2} mol Na_2SO_4

6. (a) 0.0417 g H_2SO_4 (c) 0.122 g Ti
(b) 11 g CCl_4 (d) 8.0×10^{-7} g S

8. (a) 5.8×10^{24} molecules C_2H_4
(c) 4.36×10^{23} molecules CH_3OH

10. 3.209×10^{-21} g CO

12. (a) 16 atoms
(c) 4.5×10^{25} atoms
(e) 1.3×10^{24} atoms

14. 3.94×10^{16} molecules $C_{18}H_{27}NO_3$

16. (a) 2.180×10^{-22} g Xe
(c) 8.975×10^{-22} g CH_3COOH

18. (a) 4.57×10^{-22} mol W
(c) 1.277×10^{-21} g SO_2

20. (a) 0.492 mol S_2F_{10}
(c) 3.56×10^{24} total atoms
(e) 2.96×10^{24} atoms F

22. (a) 138 atoms H
(b) 1.3×10^{25} atoms H
(c) 2.2×10^{24} atoms H

24. (a) 184.3 g/mol
(b) 1.50×10^{3} g $C_{11}H_{20}O_2$
(c) 3.47×10^{12} molecules $C_{11}H_{20}O_2$
(d) 714 mol H
(e) 61.0 g C
(f) 1.45×10^{-2} g $C_{11}H_{20}O_2$
(g) 3.06×10^{-22} g $C_{11}H_{20}O_2$
(h) 1.62×10^{23} atoms C

26. (a) 1.27 g Cl (c) 0.274 g H
(b) 9.07 g H

28. (a) 47.98% Zn, 52.02% Cl (e) 23.09% Fe, 17.37% N, 59.53% O
(c) 12.26% Mg, 31.24% P, 56.48% O

30. (a) KCl 47.55% Cl (c) $SiCl_4$ 83.46% Cl
(b) $BaCl_2$ 34.05% Cl (d) LiCl 83.63% Cl
highest % Cl is LiCl
lowest % Cl is $BaCl_2$

32. 41.1% S, 6.02% H, 6.82% O, 46.0% C

34. (a) $KClO_3$ lower % Cl
(b) $KHSO_4$ higher % S
(c) Na_2CrO_4 lower % Cr

36. (a) CuCl (e) $BaCr_2O_7$
(c) Cr_2S_3

38. (a) C_6H_5Br
(c) V_2S_3

40. Ti_2O_3

42. $HgCl_2$

44. Empirical formula is $C_3H_5OS_2$, molecular formula is $C_6H_{10}O_2S_4$.

46. $C_{15}H_{26}O$.

48. The molecular formula is $C_4H_8O_2$.

50. 40.0% C, 6.73% H, 53.3% O. Empirical formula is CH_2O, molecular formula is $C_6H_{12}O_6$.

52. $Al_2(SiO_3)_3$

53. 8.43×10^{23} atoms P

55. 197 g/mol

57. 5.54×10^{19} m

58. 8.6×10^{13} dollars/person

60. (a) 10.3 cm^3 (b) 2.18 cm

62. 41.58 g Fe_2S_3

64. 1.0 ton of iron ore contains 2×10^4 g Fe

66. (a) 76.98% Hg (c) 17.27% N

70. 42.10% C
6.480% H
51.42% O

72. 4.77 g O in 8.50 g $Al_2(SO_4)_3$

74. The empirical formula is $C_4H_4CaO_6$

76. 1.910×10^{16} years

78. 1.529×10^{-7} g $C_3H_8O_3$

80. $Co_3Mo_2Cl_{11}$

82. The empirical formula is C_5H_5N.

84. (a) CH_2O (e) $C_6H_2Cl_2O$
(c) CH_2O

86. 9.0×10^{17} molecules O_2

CHAPTER 8

Exercises

2. (a) endothermic (e) exothermic
(c) endothermic

4. (a) $H_2 + Br_2 \longrightarrow 2\,HBr$ combination
(c) $Ba(ClO_3)_2 \longrightarrow BaCl_2 + 3\,O_2$ decomposition
(e) $2\,H_2O_2 \longrightarrow 2\,H_2O + O_2$ decomposition

6. (a) $2\,FeCl_3 + 3\,Li_2CO_3 \longrightarrow 6\,LiCl + Fe_2(CO_3)_3$
Double-displacement reaction
(b) $3\,Cd + 2\,TiCl_3 \longrightarrow 2\,Ti + 3\,CdCl_2$
Single-displacement reaction

(c) 

Decomposition reaction

(d) 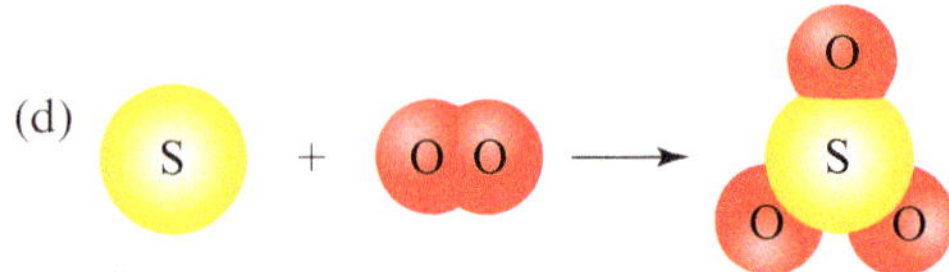

Synthesis reaction

(e) 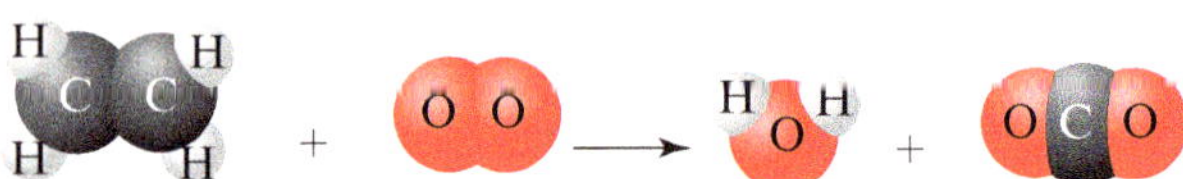

Combustion reaction

8. (a) $2\,C_4H_{10}(g) + 13\,O_2(g) \longrightarrow 8\,CO_2(g) + 10\,H_2O(g)$
(b) $C_{34}H_{64}(l) + 50\,O_2(g) \longrightarrow 34\,CO_2(g) + 32\,H_2O(g)$
(c) $2\,C_2H_2(l) + 5\,O_2(g) \longrightarrow 4\,CO_2(g) + 2\,H_2O(g)$
(d) $2\,CH_3OH(l) + 3\,O_2(g) \longrightarrow 2\,CO_2(g) + 4\,H_2O(g)$

10. (a) $2\ SO_2 + O_2 \longrightarrow 2\ SO_3$
(c) $2\ Na + 2\ H_2O \longrightarrow 2\ NaOH + H_2$
(e) $Bi_2S_3 + 6\ HCl \longrightarrow 2\ BiCl_3 + 3\ H_2S$
(g) $Hg_2(C_2H_3O_2)_2(aq) + 2\ KCl(aq) \longrightarrow Hg_2Cl_2(s) + 2\ KC_2H_3O_2(aq)$
(i) $2\ K_3PO_4 + 3\ BaCl_2 \longrightarrow 6\ KCl + Ba_3(PO_4)_2$

12. (a) $MgCO_3(s) \xrightarrow{\Delta} MgO(s) + CO_2(g)$
(c) $Fe_2(SO_4)_3(aq) + 6\ NaOH(aq) \longrightarrow 2\ Fe(OH)_3(s) + 3\ Na_2SO_4(aq)$
(e) $SO_3(g) + H_2O(l) \longrightarrow H_2SO_4(aq)$

14. (a) $CuSO_4(aq) + 2\ KOH(aq) \longrightarrow Cu(OH)_2(s) + K_2SO_4(aq)$
(c) $3\ NaHCO_3(s) + H_3PO_4(aq) \longrightarrow Na_3PO_4(aq) + 3\ H_2O(l) + 3\ CO_2(g)$

16. (a) $Cu(s) + NiCl_2\ (aq) \longrightarrow$ no reaction
(c) $I_2(s) + CaCl_2(aq) \longrightarrow$ no reaction

18. (a) $3\ NiCl_2(aq) + 2\ Al(s) \longrightarrow 2\ AlCl_3(aq) + 3\ Ni(s)$
(b) Al
(c) Ni
(d) Yes, because Al > Ni > Hg
$Ni(s) + HgCl_2(aq) \longrightarrow NiCl_2(aq) + Hg(l)$

20. (a) $Li_2O(s) + H_2O(l) \longrightarrow 2\ LiOH(aq)$
(c) $Zn(s) + CuSO_4(aq) \longrightarrow Cu(s) + ZnSO_4(aq)$

22. (a) $C + O_2 \longrightarrow CO_2$
(c) $CuBr_2 + Cl_2 \longrightarrow CuCl_2 + Br_2$
(e) $2\ NaNO_3 \xrightarrow{\Delta} 2\ NaNO_2 + O_2$

24. (a) 2 mol of Na react with 1 mol of Cl_2 to produce 2 mol of NaCl and release 822 kJ of energy. Exothermic
(b) 1 mol of PCl_5 absorbs 92.9 kJ of energy to produce 1 mol of PCl_3 and 1 mol of Cl_2. Endothermic

26. (a) $Ca(s) + 2\ H_2O(l) \longrightarrow Ca(OH)_2(aq) + H_2(g) + 635.1$ kJ
(b) $2\ BrF_3 + 601.6\ kJ \longrightarrow Br_2 + 3\ F_2$

28. (a) single displacement,
$Ni(s) + Pb(NO_3)_2(aq) \longrightarrow Pb(s) + Ni(NO_3)_2(aq)$
(c) decomposition, $2\ HgO(s) \longrightarrow 2\ Hg(l) + O_2(g)$

29. Combinations that form precipitates
$Ca(NO_3)_2(aq) + (NH_4)_2SO_4(aq) \longrightarrow CaSO_4(s) + 2\ NH_4NO_3(aq)$
$Ca(NO_3)_2(aq) + (NH_4)_2CO_3 \longrightarrow CaCO_3(s) + 2\ NH_4NO_3(aq)$
$AgNO_3(aq) + NH_4Cl(aq) \longrightarrow AgCl(s) + NH_4NO_3(aq)$
$2\ AgNO_3(aq) + (NH_4)_2SO_4 \longrightarrow Ag_2SO_4(s) + 2\ NH_4NO_3(aq)$
$2\ AgNO_3(aq) + (NH_4)_2CO_3(aq) \longrightarrow Ag_2CO_3(s) + 2\ NH_4NO_3(aq)$

30. $P_4O_{10} + 12\ HClO_4 \longrightarrow 6\ Cl_2O_7 + 4\ H_3PO_4$
58 atoms O on each side

38. (a) $4\ Cs + O_2 \longrightarrow 2\ Cs_2O$
(c) $SO_3 + H_2O \longrightarrow H_2SO_4$

41. (a) $2\ ZnO \xrightarrow{\Delta} 2\ Zn + O_2$
(c) $Na_2CO_3 \xrightarrow{\Delta} Na_2O + CO_2$

42. (a) $Mg(s) + 2\ HCl(aq) \longrightarrow H_2(g) + MgCl_2(aq)$
(c) $3\ Zn(s) + 2\ Fe(NO_3)_3(aq) \longrightarrow 2\ Fe(s) + 3\ Zn(NO_3)_2(aq)$

43. (a) $2\ (NH_4)_3PO_4(aq) + 3\ Ba(NO_3)_2(aq) \longrightarrow Ba_3(PO_4)_2(s) + 6\ NH_4NO_3(aq)$
(c) $CuSO_4(aq) + Ca(ClO_3)_2(aq) \longrightarrow CaSO_4(s) + Cu(ClO_3)_2(aq)$
(e) $H_3PO_4(aq) + 3\ KOH(aq) \longrightarrow K_3PO_4(aq) + 3\ H_2O(l)$

46. (a) $CH_4 + 2\ O_2 \longrightarrow CO_2 + 2\ H_2O$
(b) $2\ C_3H_6 + 9\ O_2 \longrightarrow 6\ CO_2 + 6\ H_2O$
(c) $C_6H_5CH_3 + 9\ O_2 \longrightarrow 7\ CO_2 + 4\ H_2O$

CHAPTER 9

Exercises

2. (a) 25.0 mol $NaHCO_3$ (c) 16 mol CO_2

4. (a) 1.31 g $NiSO_4$ (e) 18 g K_2CrO_4
(c) 373 g Bi_2S_3

6. HCl

8. (a) $\dfrac{6\ \text{mol}\ O_2}{1\ \text{mol}\ C_4H_9OH}$ (d) $\dfrac{1\ \text{mol}\ C_4H_9OH}{4\ \text{mol}\ CO_2}$
(c) $\dfrac{4\ \text{mol}\ CO_2}{5\ \text{mol}\ H_2O}$ (f) $\dfrac{4\ \text{mol}\ CO_2}{6\ \text{mol}\ O_2}$

10. (a) 34 mol NaOH (b) 11 mol Na_2SO_4

12. (a) 8.33 mol H_2O (b) 0.800 mol $Al(OH)_3$

14. $2\ Al + 6\ HBr \longrightarrow 2\ AlBr_3 + 3\ H_2$
249 g $AlBr_3$

16. 117 g H_2O, 271 g Fe

18. (a) 1.9 mol CaS (e) $2.50—10^3$ g CaS
(c) 0.4179 mol CaO

20. (a)

Li
I
neither is limiting

(b)

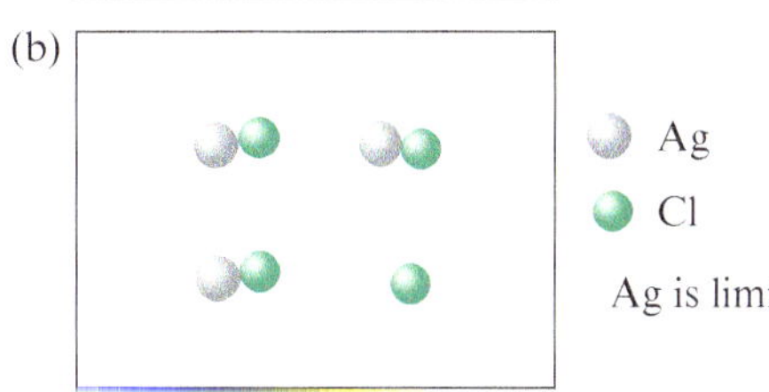

Ag
Cl
Ag is limiting

22. (a) Oxygen is the limiting reactant.

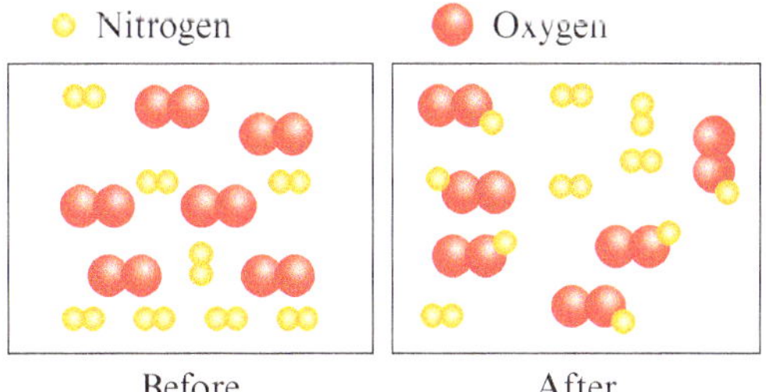

(b) Water is the limiting reactant.

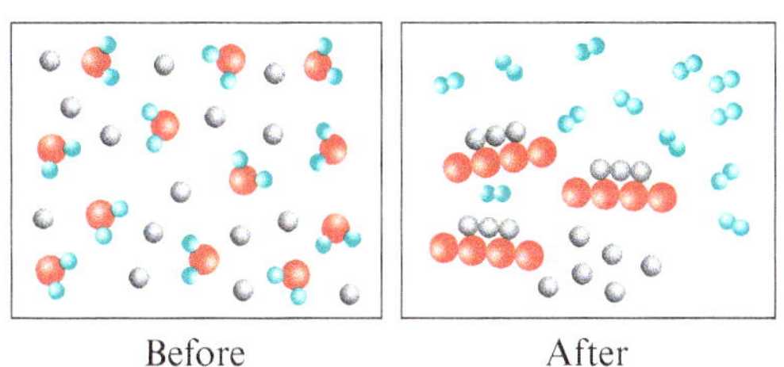

24. (a) H_2S is the limiting reactant and $Bi(NO_3)_3$ is in excess.
(b) H_2O is the limiting reactant and Fe is in excess.

26. (a) 0.193 mol PCl_3
(b) 71 g PCl_3; P_4 is in excess
(c) 23 mol PCl_3 will be produced; 9 mol P_4 unreacted

28. Zinc

30. 89.02%

32. (a) 0.32 mol $FeCl_3$ and 0.48 mol H_2O
(b) 9 g Fe_2O_3 in excess
(c) 48 g $FeCl_3$ actual yield

36. (a) 0.19 mol KO_2 (b) 28 g O_2/hr

39. 70.7 g O_2

41. (a) 0.16 mol $MnCl_2$ (e) 11 g $KMnO_4$ unreacted
(c) 82% yield

44. 46.97 g $Ca(OH)_2$

46. (a) 5.0 mol CO_2, 7.5 mol H_2O
(b) 430. g CO_2 , 264 g H_2

48. (a) 2.0 mol Cu, 2.0 mol $FeSO_4$, 1 mol $CuSO_4$
(b) 15.9 g Cu, 38.1 g $FeSO_4$, 6.0 g Fe

50. (a) 3.2×10^2 g C_2H_5OH
(b) 1.10×10^3 g $C_6H_{12}O_6$

52. 3.7×10^2 kg Li_2O

55. 223.0 g H_2SO_4; 72.90% yield

61. 67.2% $KClO_3$

CHAPTER 10

Exercises

2. (a) Na 11 total electrons; 1 valence electron
(c) P 15 total electrons; 5 valence electrons

4. (a) Mn $1s^22s^22p^63s^23p^64s^23d^5$
(c) Ga $1s^22s^22p^63s^23p^64s^23d^{10}4p^1$

6. Bohr said that a number of orbits were available for electrons, each corresponding to an energy level. When an electron falls from a higher energy orbit to a lower orbit, energy is given off as a specific wavelength of light. Only those energies in the visible range are seen in the hydrogen spectrum. Each line corresponds to a change from one orbit to another.

8. 18 electrons in third energy level; 2 in *s*, 6 in *p*, 10 in *d*

10. (a) $^{28}_{14}Si$ [↑↓] [↑↓] [↑↓|↑↓|↑↓] [↑↓] [↑|↑|]
(c) $^{40}_{18}Ar$ [↑↓] [↑↓] [↑↓|↑↓|↑↓] [↑↓] [↓↑|↓↑|↓↑]
(e) $^{31}_{15}P$ [↑↓] [↑↓] [↑↓|↑↓|↑↓] [↑↓] [↑|↑|↑]

12. (a) Li $1s^22s^1$
(c) Zn $1s^22s^22p^63s^23p^64s^23d^{10}$
(e) K $1s^22s^22p^63s^23p^64s^1$

14. (a) correct
(c) incorrect [↑↓] [↑↓] [↑↓|↑↓|↑↓] [↑↓] [↑↓|↑↓|↑↓] [↑↓]

16. (a) N (c) Ca

18. [↑↓|↑↓|↑|↑|↑]
d sublevel

20. (a) boron, B (c) lead, Pb

22. (a) Phosphorus (P)
[↑↓] [↑↓] [↑↓|↑↓|↑↓] [↑↓] [↑|↑|↑]
(c) Calcium (Ca)
[↑↓] [↑↓] [↑↓|↑↓|↑↓] [↑↓] [↑↓|↑↓|↑↓] [↑↓]

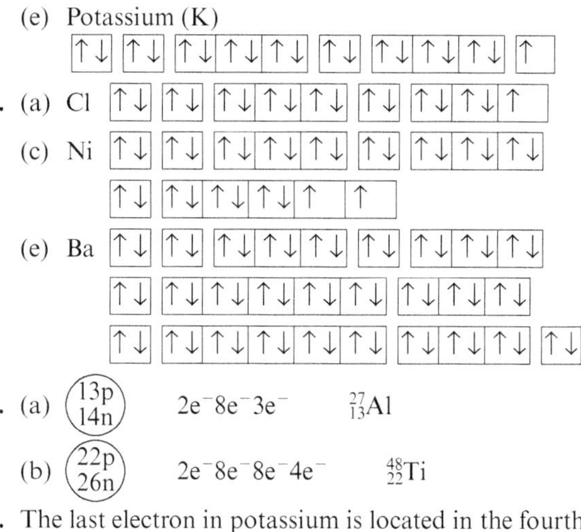

28. The last electron in potassium is located in the fourth energy level because the 4*s* orbital is at a lower energy level than the 3*d* orbital. Also, the properties of potassium are similar to the other elements in Group 1A(1).

30. Noble gases each have filled *s* and *p* orbitals in the outermost energy level.

32. The elements in a group have the same number of outer energy level electrons. They are located vertically on the periodic table.

34. Valence shell electrons
(a) 5 (c) 6 (e) 3

36. All of these elements have an s^2d^{10} electron configuration in their outermost energy levels.

38. (a) and (f)
(e) and (h)

40. 7, 33 since they are in the same periodic group.

42. (a) nonmetal, I
(c) metal, Mo

44. (a) $1s^2\ 2s^2\ 2p^6\ 3s^2\ 3p^6\ 4s^2\ 3d^{10}\ 4p^4$ (c) 6

46. Period 4, Group 3B(3)

48. Group 3A(13) contains 3 valence electrons. Group 3B(3) contains 2 electrons in the outermost level and 1 electron in an inner *d* orbital. Group A elements are representative, while Group B elements are transition elements.

50. (a) Pb (b) Sm (c) Ga (d) Ir

52. (a) Mg $1s^22s^22p^6(3s^2)$
(c) K $1s^22s^22p^63s^23p^6(4s^1)$
(d) F $1s^2(2s^22p^5)$
(f) N $1s^2(2s^22p^3)$

54. (a) 7A(17), Halogens (e) 8A(18), Noble Gases
(c) 1A(1), Alkali Metals

60. (a) Any orbital can hold a maximum of two electrons.
(c) The third principal energy level can hold two electrons in 3*s*, six electrons in 3*p*, and ten electrons in 3*d* for a total of eighteen electrons.
(e) An *f* sublevel can hold a maximum of fourteen electrons.

63. (a) Ne (c) F
(b) Ge (d) N

66. When an electron drops from a higher energy level (excited state) to a lower energy level, it releases energy in the form of light. The line spectra represent these light emissions.

68. Transition elements are found in Groups 1B–8B (3–12), lanthanides and actinides are the inner transition metals.

70. Elements 7, 15, 33, 51, 83 all have 5 electrons in their valence shell.

72. (a) sublevel *p*
(c) sublevel *f*
(b) sublevel *d*

74. If 36 is a noble gas, 35 would be in periodic Group 7A(17) and 37 would be in periodic Group 1A(1).

CHAPTER 11

Exercises

2. Drawing 1 Cl atom, drawing 2 Cl^- ion

4. (a) Fe^{2+} (e) Rubidium ion
(c) Chloride ion

6. Anions are larger because the number of electrons increases with no increase in protons.

8. Electronegativity is the attraction of an atom for bonding electrons. It increases up and to the right.

10. (a) $\overset{+}{H}\ \overset{-}{Cl}$ (e) $\overset{+}{Cs}\ \overset{-}{I}$
(c) C Cl

12. (a) ionic (c) covalent

14. (a) $K \longrightarrow K^+ + 1\ e^-$
(b) $S + 2\ e^- \longrightarrow S^{2-}$

16. (a) 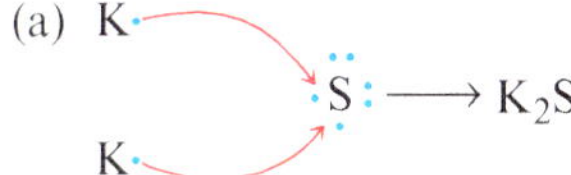

(b) 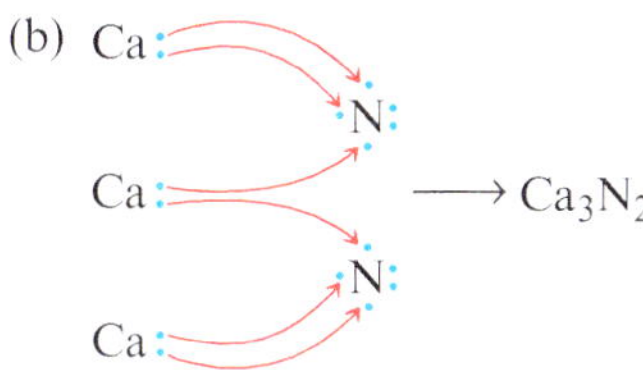

18. (a) Pb (4) (d) Cs (1)
(c) O (6) (f) Ar (8)

20. (a) sulfur atom, gain 2 e^- (c) nitrogen atom, gain 3 e^-
(b) calcium atom, lose 2 e^- (d) iodide ion, none

22. (a) $1s^22s^22p^63s^23p^64s^23d^5$
(c) Those in 4*s* are lost.

24. (a) Nonpolar covalent compound; O_2.
(c) Polar covalent compound; H_2O.

26. (a) SbH_3, Sb_2O_3 (c) HCl, Cl_2O_7

28. $BeBr_2$, beryllium bromide
$MgBr_2$, magnesium bromide
$SrBr_2$, strontium bromide
$BaBr_2$, barium bromide
$RaBr_2$, radium bromide

30. (a) Ga: (b) $[Ga]^{3+}$ (c) $[Ca]^{2+}$

32. (a) covalent (c) covalent
(b) ionic (d) covalent

34. (a) covalent (c) covalent
(b) covalent

36. (a) :Ö::Ö: (b) :B̈r:B̈r: (c) :Ï:Ï:

38. (a) :S̈:H with H below S (c) H:N̈:H with H below N

40. (a) $[:\ddot{I}:]^-$ (e) $[:\ddot{O}:N::\ddot{O}: \text{ with } :\ddot{O}: \text{ below N}]^-$
(c) $[:\ddot{O}:C::\ddot{O} \text{ with } :\ddot{O}: \text{ below C}]^{2-}$

42. (a) $[:\ddot{O}-N=\ddot{O}:]^- \longleftrightarrow [:\ddot{O}=N-\ddot{O}:]^-$ (bent, N at the vertex)

(b) $[H-C(=\ddot{O}:)-\ddot{O}:]^- \longleftrightarrow [H-C(-\ddot{O}:)=\ddot{O}:]^-$

44. (a) H_2, nonpolar (c) CH_3OH, polar
(b) NI_3, polar (d) CS_2, nonpolar

46. (a) 3 electron pairs, trigonal planar
(b) 4 electron pairs, tetrahedral
(c) 4 electron pairs, tetrahedral

48. (a) tetrahedral (c) tetrahedral
(b) trigonal pyramidal

50.

	Electron pair arrangement	Molecular structure
S (a)	trigonal planar	bent
C (b)	linear	linear
N (c)	tetrahedral	tetrahedral
O (d)	tetrahedral	bent

52. (a) tetrahedral (b) bent (c) bent

54. potassium

60. N_2H_4 14 e^- H₂N̈—N̈H₂ (each N bonded to two H) HN_3 16 e^- H—N̈=N=N̈

63. (a) Be (d) F
(b) He (e) Fr
(c) K (f) Ne

64. (a) Cl (c) Ca

68. $SnBr_2$, $GeBr_2$

71. This structure shown is incorrect since the bond is ionic. It should be represented as:

$[Na]^+[:\ddot{O}:]^{2-}[Na]^+$

72. Answer: Fluorine, oxygen, nitrogen, chlorine

77. (a) 109.5° (actual angle is closer to 105) (c) 109.5°

84. Empirical formula is SO_3.

:Ö: above S; :Ö—S=Ö:

86. (a) ionic (c) covalent
(b) both (d) covalent

CHAPTER 12

Exercises

2.

	mm Hg	lb/in.^{2}	atmospheres
a.	789	15.3	1.04
b.	1700	32	2.2
c.	1100	21	1.4

4. (a) 86.5 kPa (b) 0.0500 atm (c) 2770 mm Hg (d) 1.06 atm (e) 82.1 cm Hg

6. (a) 2.1 atm (b) 1600 torr (c) 210 kPa

8. (a) 0.830 atm (c) 84.1 kPa

10. (a) 1330 torr (b) 40.0 torr

12. 21.2 L

14. (a) 691 mL (b) 633 mL (c) 737 mL

16. 30.8 L

18. 33.4 L

20. −150°C

22. 719 mm Hg

24. 1.450×10^3 torr

26. 1.19 L C_3H_8

28. $PV = nRT$
9.07 L

30. As the temperature decreases, the air pressure in the tires also decreases. In order to bring the pressure back up, you need to add more air to the tires.

32. (a) 67.0 L SO_3 (b) 170 L C_2H_6 (c) 7.96 L Cl_2

34. 1.78 g C_3H_6

36. 2.82×10^{23} molecules CO_2

38. 2.1 g CH_4

40. 1.8×10^3 L CH_4

42. 0.362 mol CO_2

44. 209 K

46. (a) 9.91 g/L Rn (b) 2.05 g/L NO_2 (c) 3.58 g/L SO_3 (d) 1.25 g/L C_2H_4

48. (a) 0.85 g/L C_2H_4 (b) 0.154 g/L He

50. (a) 3.95×10^4 mL H_2 (b) 0.0737 mol HCl

52. (a) 36 L O_2 (b) 210 g CO_2 (c) 12 L H_2O vapor

54. 1.12×10^3 L CO_2

58. image (a)

59. (a) the pressure will be cut in half
(c) the pressure will be cut in half

63. 22.4 L

66. SF_6 has the greatest density.

68. (a) CH_3
(b) C_2H_6
(c)

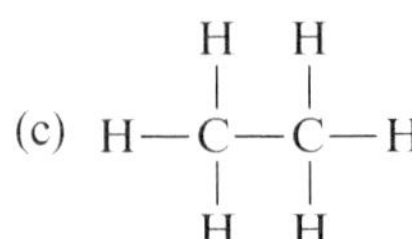

71. 2.3×10^7 K

74. 327°C

76. 211 K (−62°C)

78. 3.55 atm

83. 1.03×10^4 mm (33.8 ft)

84. 330 mol O_2

86. 43.2 g/mol (molar mass)

88. (a) 8.4 L H_2 (c) 8.48 g/L CO_2

90. 279 L H_2 at STP

91. C_4H_8

CHAPTER 13

Exercises

2. $H_2Se < H_2S < H_2O$

4. $CO_2 < SO_2 < CS_2$

6. (a) HI will not hydrogen-bond; hydrogen is not bonded to fluorine, oxygen, or nitrogen.
(b) NH_3 will hydrogen-bond; hydrogen is bonded to nitrogen.
(c) CH_2F_2 will not hydrogen-bond; hydrogen is not bonded to fluorine, oxygen, or nitrogen.
(d) C_2H_5OH will hydrogen-bond; one of the hydrogens is bonded to oxygen.
(e) H_2O will hydrogen-bond; hydrogen is bonded to oxygen.

8. The boiling point of a substance is the temperature where the vapor pressure of a liquid is equal to the external pressure above the liquid. Boiling point increases with increasing intermolecular forces.

10. (b)

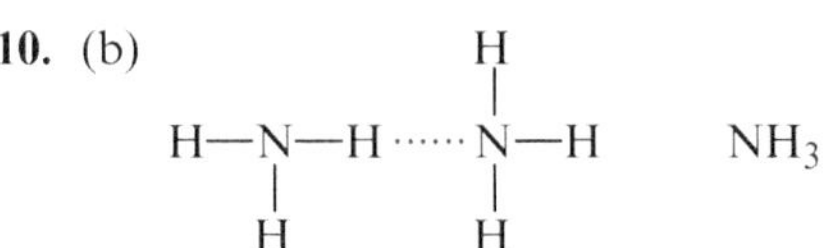

NH_3

(d)

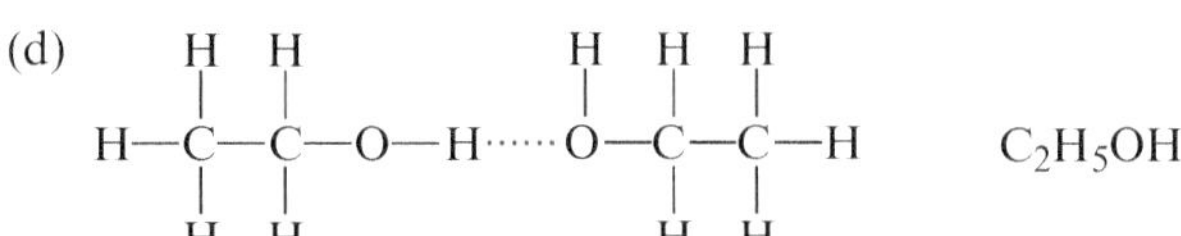

C_2H_5OH

(e)

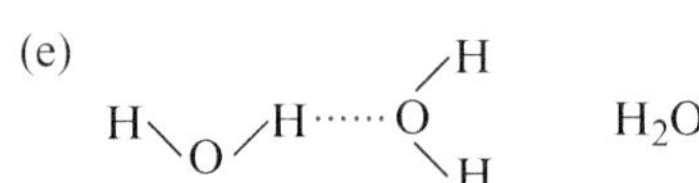

H_2O

12. Water forming beaded droplets is an example of cohesive forces. The water molecules have stronger attractive forces for other water molecules than they do for the surface.

14. Glycerol has stronger intermolecular forces because it has a higher surface tension. Glycerol should also have the higher boiling point.

16. (a) magnesium ammonium phosphate hexahydrate
(b) iron(II) sulfate heptahydrate
(c) tin(IV) chloride pentahydrate

18. 0.0655 mol $Na_2B_4O_7 \cdot 10\ H_2O$

20. (a) 56.0 g H_2O
(b) 69.0 g $AlCl_3$

22. 20.93% H_2O

24. The formula is $NiCl_2 \cdot 6\ H_2O$.

26. 1.57×10^5 J

28. 42 g of steam needed; not enough steam (35 g)

30. The system will be at 0°C. It will be a mixture of ice and water.

32. Cl_2 is nonpolar. Nonpolar molecules such as chlorine gas have dispersion forces only.

34. Steam molecules will cause a more severe burn. Steam molecules contain more energy at 100°C than water molecules at 100°C due to the energy absorbed during the vaporization state (heat of vaporization).

36. When one leaves the swimming pool, water starts to evaporate from the skin of the body. Part of the energy needed for evaporation is absorbed from the skin, resulting in the cool feeling.

42. $MgSO_4 \cdot 7\ H_2O$ $Na_2HPO_4 \cdot 12\ H_2O$

48. Isobutyl propionate, the compound with the lower boiling point will have weaker intermolecular forces and thus a higher vapor pressure.

52. 6.78×10^5 J or 1.62×10^5 cal

55. 2.30×10^6 J

57. 40.2 g H_2O

59. 1.00 mole of water vapor at STP has a volume of 22.4 L (gas); 1.00 mole of liquid water has a volume of 18.0 mL.

61. 3.9×10^3 J

64. 2.71×10^4 J

66. $2\ H_2O(l) \longrightarrow 2\ H_2(g) + O_2(g)$

CHAPTER 14

Exercises

2. Generally soluble:
(b) $Cu(NO_3)_2$
(d) $NH_4C_2H_3O_2$
(f) $AgNO_3$

4. (a) Hexane because the molecule is nonpolar and will dissolve only in a nonpolar solvent
(c) Both since the molecule contains both nonpolar and polar regions

6. (a) saturated
(c) saturated

8. (a) 16.7% $NaNO_3$
(c) 39% K_2CrO_4

10. 143 g iron(III) oxide solution

12. (a) 9.0 g $BaCl_2$
(b) 66 g solvent

14. (a) 63.59% $C_{12}H_{22}O_{11}$
(b) 47.3% CH_3OH

16. (a) 13.6% butanol
(b) 16% methanol
(c) 6.00% isoamyl alcohol

18. (a) 4.0 *M*
(c) 1.97 *M* $C_6H_{12}O_6$

20. (a) 1.1 mol HNO_3
(b) 7.5×10^{-3} mol $NaClO_3$
(c) 0.088 mol LiBr

22. (a) 2.1×10^3 g H_2SO_4
(b) 6.52 g $KMnO_4$
(c) 1.2 g $Fe_2(SO_4)_3$

24. (a) 3.4×10^3 mL
(b) 1.88×10^3 mL
(c) 1.37 mL

26. (a) 2.65% $HC_2H_3O_2$

28. (a) 1.2 *M*
(b) 0.070 *M*
(c) 0.333 *M*

30. (a) 25 mL 18 *M* H_2SO_4
(b) 5.0 mL 15 *M* NH_3

32. 4.64-mL anethole solution

34. (a) 1.2 *M* HCl
(b) 4.21 *M* HCl

36. (a) 3.6 mol Na_2SO_4
(c) 0.625 mol NaOH
(e) 38.25 mL H_2SO_4

38. (a) 1.88×10^{-3} mol H_2O
(c) 90.2 mL $HC_2H_3O_2$
(e) 1.8 L CO_2

40. (a) 1.67 *m* $CaCl_2$
(b) 0.026 *m* $C_6H_{12}O_6$
(c) 6.44 *m* $(CH_3)_2CHOH$

42. (a) 10.74 *m*
(b) 105.50°C
(c) −20.0°C

44. 163 g/mol

46. Concord grape juice and stainless steel

48. Red paint and oil and vinegar salad dressing are not a true solution because they will separate out if they are not shaken or stirred.

50. (a) The amino acid will dissolve faster in the solvent at 75°C because the rate of dissolving increases with increasing temperature.
(b) The powdered sodium acetate will dissolve faster because it has a smaller particle size.
(c) The carton of table salt will dissolve faster because it has a smaller particle size and the temperature is higher.
(d) The acetaminophen will dissolve faster in the infant pain medication because it has less acetaminophen already dissolved in it.

52. 1.092 *M* $HgCl_2$

57. 7.63% NaCl

61. 0.15 *M* NaCl

63. (a) 160 g sugar
(b) 0.47 *M* $C_{12}H_{22}O_{11}$
(c) 0.516 *m* $C_{12}H_{22}O_{11}$

65. 16.2 *m* HCl

67. 66 mL C_2H_5OH [ethyl alcohol]

69. (a) 4.5 g NaCl
(b) 450. mL H_2O must evaporate.

73. 2.94 L solution

75. 15 *M* H_3PO_4

79. 1.20 g $Mg(OH)_2$

81. 6.2 *m* H_2SO_4
5.0 *M* H_2SO_4

82. (a) 2.9 *m*
(b) 101.5°C

84. (a) 8.04×10^3 g $C_2H_6O_2$
(b) 7.24×10^3 mL $C_2H_6O_2$
(c) −4.0°F

85. 47% $NaHCO_3$

88. (a) 7.7 L H_2O must be added
(b) 0.0178 mol H_2SO_4
(c) 0.0015 mol H_2SO_4 in each mL

91. 2.84 g $Ba(OH)_2$ is formed.

92. (a) 0.011 mol Li_2CO_3
(c) 3.2×10^2 mL solution

98. 53 *m* sugar

CHAPTER 15

Exercises

2. Conjugate acid–base pairs:
(a) $H_2S - HS^-$; $NH_3 - NH_4^+$
(c) $HBr - Br^-$; $CH_3O^- - CH_3OH$

4. (a) $Mg(s) + 2\ HClO_4(aq) \longrightarrow Mg(ClO_4)_2(aq) + H_2(g)$
(c) $2\ NaOH(aq) + H_2CO_3(aq) \longrightarrow Na_2CO_3(aq) + 2\ H_2O(l)$

6. (a) $Mg + (2\ H^+ + 2\ ClO_4^-) \longrightarrow (Mg^{2+} + 2\ ClO_4^-) + H_2$
$Mg + 2\ H^+ \longrightarrow Mg^{2+} + H_2$
(c) $(2\ Na^+ + 2\ OH^-) + H_2CO_3 \longrightarrow$
$(2\ Na^+ + CO_3^{2-}) + 2\ H_2O$
$2\ OH^- + H_2CO_3 \longrightarrow CO_3^{2-} + 2\ H_2O$

8. $\underbrace{HClO_4}_{\text{acid}} + \underbrace{OH^-}_{\text{base}} \longrightarrow \underbrace{ClO_4^-}_{\text{conjugate base}} + \underbrace{H_2O}_{\text{conjugate acid}}$

10. (a) $2\ HBr + Mg(OH)_2 \longrightarrow 2\ H_2O + MgBr_2$
(b) $H_3PO_4 + 3\ KOH \longrightarrow 3\ H_2O + K_3PO_4$
(c) $H_2SO_4 + 2\ NH_4OH \longrightarrow 2\ H_2O + (NH_4)_2SO_4$

12. (a) H_2CO_3
(b) $CO_3{}^{2-}$

14. $Ca(OH)_2$ and H_2CO_3;
$Ca(OH)_2 + H_2CO_3 \longrightarrow 2\ H_2O + CaCO_3$

16. (a) $C_6H_{12}O_6$—nonelectrolyte
(c) NaClO — electrolyte
(e) C_2H_5OH — nonelectrolyte

18. (a) 2.25 *M* Fe^{3+}, 6.75 *M* Cl^-
(c) 0.75 *M* Na^+, 0.75 *M* $H_2PO_4^-$

20. (a) 12.6 g Fe^{3+}, 23.9 g Cl^-
(c) 1.7 g Na^+, 7.3 g $H_2PO_4^-$

22. (a) $[H^+] = 1 \times 10^{-7}\ M$
(b) $[H^+] = 5 \times 10^{-5}\ M$
(c) $[H^+] = 2 \times 10^{-6}\ M$
(d) $[H^+] = 5 \times 10^{-11}\ M$

24. (a) $[Na^+] = 0.25\ M$, $[Cl^-] = 0.25\ M$
(c) $[Ba^{2+}] = 0.062\ M$, $[Ag^+] = 0.081\ M$, $[NO_3^-] = 0.20\ M$

26. $HCN + H_2O \rightleftharpoons H_3O^+ + CN^-$, weak electrolyte and weak acid

28. (a) 0.147 *M* NaOH
(b) 0.964 *M* NaOH
(c) 0.4750 *M* NaOH

30. (a) $Mg(s) + Cu^{2+}(aq) \longrightarrow Cu(s) + Hg^{2+}(aq)$
(b) $H^+(aq) + OH^-(aq) \longrightarrow H_2O(l)$
(c) $SO_3^{2-}(aq) + 2\ H^+(aq) \longrightarrow H_2O(l) + SO_2(g)$

32. (a) 2 *M* HCl
(b) 1 *M* H_2SO_4

34. 0.894 g of $Al(OH)_3$

36. 7.0% NaCl in the sample

38. 0.936 L H_2

40. (a) pH = 2.70
(b) pH = 7.15
(c) pH = −0.48

42. (a) pH = 4.30
(b) pH = 10.47

44. (a) weak acid
(b) strong base
(c) strong acid
(d) strong acid

46. (a) BrO_3^-
(b) NH_3
(c) HPO_4^{2-}

48. $S^{2-}(aq) + H_2O(l) \longrightarrow HS^+(aq) + OH-(aq)$

51. (a) basic
(c) neutral
(e) acidic

54. 0.218 *M* H_2SO_4

57. As the pH changes by 1 unit, the concentration of H^+ in solution changes by a factor of 10. For example, the pH of 0.10 *M* HCl is 1.00, while the pH of 0.0100 *M* HCl is 2.00.

59. 0.673 g KOH

61. 0.462 *M* HCl

63. pH = 1.1, acidic

65. (a) $2\ NaOH(aq) + H_2SO_4(aq) \longrightarrow Na_2SO_4(aq) + 2\ H_2O(l)$
(b) 1.0×10^2 mL NaOH
(c) 0.71 g Na_2SO_4

68. acidic pH = 0.82

72. 806 mL

CHAPTER 16

Exercises

2. (a) $I_2(s) \rightleftharpoons I_2(g)$
(b) $NaNO_3(s) \rightleftharpoons Na^+(aq) + NO_3^-(aq)$

4. (a) Reaction is exothermic.
(b) Reaction shifts to the right. [HCl] will be decreased; $[O_2]$, $[H_2O]$, and $[Cl_2]$ will be increased.
(c) Reaction will shift left.

6.

(a)	left	I	I	I (add NH_3)
(b)	left	I	I	D (increase volume)
(c)	no change	N	N	N (add catalyst)
(d)	?	?	I	I (add H_2 and N_2)

8.

Reaction	Increase temperature	Increase pressure	Add catalyst
(a)	right	left	no change
(b)	left	left	no change
(c)	left	left	no change

10. Equilibrium shift
(a) no change (b) left (c) right (d) right

12. (a) $K_{eq} = \frac{[H^+][H_2PO_4^-]}{[H_3PO_4]}$ (c) $K_{eq} = \frac{[N_2O_5]^2}{[NO_2]^4[O_2]}$

(b) $K_{eq} = \frac{[CH_4][H_2S]^2}{[CS_2][H_2]^4}$

14. (a) $K_{sp} = [Mg^{2+}][CO_3^{2-}]$ (c) $K_{sp} = [Tl^{3+}][OH^-]^3$

(b) $K_{sp} = [Ca^{2+}][C_2O_4^{2-}]$ (d) $K_{sp} = [Pb^{2+}]^3[AsO_4^{3-}]^2$

16. If H^+ is increased,

(a) pH is decreased (c) OH^- is decreased

(b) pOH is increased (d) K_W remains unchanged

18. When excess base gets into the bloodstream, it reacts with H^+ to form water. Then H_2CO_3 ionizes to replace H^+, thus maintaining the approximate pH of the blood.

20. (a) $[H^+] = 4.6 \times 10^{-3}\,M$ (c) 18% ionized

(b) pH = 2.34

22. $K_a = 2.3 \times 10^{-5}$

24. (a) 0.79% ionized; pH = 2.10

(b) 2.5% ionized; pH = 2.60

(c) 7.9% ionized; pH = 3.10

26. $K_a = 7.3 \times 10^{-6}$

28. $[H^+] = 3.0\,M$; pH = −0.48; pOH = 14.48; $[OH^-] = 3.3 \times 10^{-15}$

30. (a) pOH = 3.00; pH = 11.00

(b) pH = 0.903; pOH = 13.10

(c) pH = 5.74; pOH = 8.26

32. (a) $[OH^-] = 2.5 \times 10^{-6}$

(c) $[OH^-] = 4.5 \times 10^{-10}$

34. (a) $[H^+] = 2.2 \times 10^{-13}$

(b) $[H^+] = 1.9 \times 10^{-6}$

36. (a) 1.2×10^{-23}

(c) 1.81×10^{-18}

38. (a) $1.1 \times 10^{-4}\,M$

(b) $9.2 \times 10^{-6}\,M$

40. (a) 3.3×10^{-3} g $PbSO_4$

(b) 2.3×10^{-4} g $BaCrO_4$

42. Precipitation occurs.

44. 2.6×10^{-12} mol AgBr will dissolve.

46. pH = 4.74

48. Change in pH = 4.74 − 4.72 = 0.02 units in the buffered solution. Initial pH = 4.74.

52. 4.20 mol HI

54. 0.500 mol HI, 2.73 mol H_2, 0.538 mol I_2

56. 128 times faster

58. pH = 4.57

60. Hypochlorous acid: $K_a = 3.5 \times 10^{-8}$

Propanoic acid: $K_a = 1.3 \times 10^{-5}$

Hydrocyanic acid: $K_a = 4.0 \times 10^{-10}$

62. (a) Precipitation occurs.

(b) Precipitation occurs.

(c) No precipitation occurs.

64. No precipitate of $PbCl_2$

66. $K_{eq} = 1.1 \times 10^4$

68. 8.0 M NH_3

70. $K_{sp} = 4.00 \times 10^{-28}$

72. (a) The temperature could have been cooler.

(b) The humidity in the air could have been higher.

(c) The air pressure could have been greater.

74. (a) $K_{eq} = \frac{[O_3]^2}{[O_2]^3}$ (c) $K_{sp} = [CO_2]$

(b) $K_{eq} = \frac{[H_2O(l)]}{[H_2O(g)]}$ (d) $K_{sp} = \frac{[H^+]^6}{[Bi^{3+}]^2[H_2S]^3}$

76. $K_{eq} = 450$

80. 1.1 g $CaSO_4$

83. (a) shift right (b) shift left (c) shift right (d) shift left

85. (a) $HCO_3^-(aq) + H^+(aq) \rightleftharpoons H_2CO_3(aq)$

(b) $HCO_3^-(aq) + OH^-(aq) \rightleftharpoons H_2O(l) + CO_3^{2-}(aq)$

CHAPTER 17

Exercises

2. (a) CHF_3 C = +2; H = +1; F = −1

(c) SF_6 S = +6; F = −1

(e) CH_3OH C = −2; H = +1; O = −2

4. (a) CO_3^{2-} +4

(c) NaH_2PO_4 +5

6. (a) Cu +2; S +6 (b) +7 (c) +4 (d) +6

8. (a) $Cu^{2+} + e^- \longrightarrow Cu^{1+}$ reduction

(c) $2\,IO_4^- + 16\,H^+ + 14\,e^- \longrightarrow I_2 + 8\,H_2O$ reduction

10. (a) C is oxidized, O is reduced.

CH_4 is the reducing agent, O_2 is the oxidizing agent.

(b) Mg is oxidized, Fe is reduced.

Mg is the reducing agent, $FeCl_3$ is the oxidizing agent.

12. (a) incorrectly balanced

$3\,MnO_2(s) + 4\,Al(s) \longrightarrow 3\,Mn(s) + 2\,Al_2O_3(s)$

(c) correctly balanced

14. (a) $Cu + 2\,AgNO_3 \longrightarrow 2\,Ag + Cu(NO_3)_2$

(c) $4\,HCl + O_2 \longrightarrow 2\,Cl_2 + 2\,H_2O$

(e) $2\,KMnO_4 + 5\,CaC_2O_4 + 8\,H_2SO_4 \longrightarrow K_2SO_4 + 2\,MnSO_4 + 5\,CaSO_4 + 10\,CO_2 + 8\,H_2O$

16. (a) $6\,H^+ + ClO_3^- + 6\,I^- \longrightarrow 3\,I_2 + Cl^- + 3\,H_2O$

(c) $2\,H_2O + 2\,MnO_4^- + 5\,SO_2 \longrightarrow 4\,H^+ + 2\,Mn^{2+} + 5\,SO_4^{2-}$

(e) $8\,H^+ + Cr_2O_7^{2-} + 3\,H_3AsO_3 \longrightarrow 2\,Cr^{3+} + 3\,H_3AsO_4 + 4\,H_2O$

18. (a) $H_2O + 2\,MnO_4^- + 3\,SO_3^{2-} \longrightarrow 2\,MnO_2 + 3\,SO_4^{2-} + 2\,OH^-$

(c) $8\,Al + 3\,NO_3^- + 18\,H_2O + 5\,OH^- \longrightarrow 3\,NH_3 + 8\,Al(OH)_4^-$

(e) $2\,Al + 6\,H_2O + 2\,OH^- \longrightarrow 2\,Al(OH)_4^- + 3\,H_2$

20. (a) $5\,Mo_2O_3 + 6\,MnO_4^- + 18\,H^+ \longrightarrow 10\,MoO_3 + 6\,Mn^{2+} + 9\,H_2O$

(b) $3\,BrO^- + 2\,Cr(OH)_4^- + 2\,OH^- \longrightarrow 3\,Br^- + 2\,CrO_4^{2-} + 5\,H_2O$

(c) $3\,S_2O_3^{2-} + 8\,MnO_4^- + H_2O \longrightarrow 6\,SO_4^{2-} + 8\,MnO_2 + 2\,OH^-$

21. (b), (d), (e)

23. hydrogen, iron

27. (a) $Pb + SO_4^{2-} \longrightarrow PbSO_4 + 2\,e^-$

$PbO_2 + SO_4^{2-} + 4\,H^+ + 2\,e^- \longrightarrow PbSO_4 + 2\,H_2O$

(b) The first reaction is oxidation (Pb^0 is oxidized to Pb^{2+}). The second reaction is reduction (Pb^{4+} is reduced to Pb^{2+}).

(c) The first reaction (oxidation) occurs at the anode of the battery.

29. Zinc is a more reactive metal than copper, so when corrosion occurs, the zinc preferentially reacts. Zinc is above hydrogen in the Activity Series of Metals; copper is below hydrogen.

31. 4.23 L CO_2; $H_2C_2O_4$ oxidized, $K_2S_2O_8$ reduced

35. 2.94 g Zn

37. 5.560 mol H_2

39. The electrons lost by the species undergoing oxidation must be gained (or attracted) by another species, which then undergoes reduction.

41. Sn^{4+} can only be an oxidizing agent.

Sn^0 can only be a reducing agent.

Sn^{2+} can be both oxidizing and reducing agents.

43. Equations (a) and (b) represent oxidations.

45. (a) $F_2 + 2\,Cl^- \longrightarrow 2\,F^- + Cl_2$

(b) $Br_2 + Cl^- \longrightarrow$ NR

(c) $I_2 + Cl^- \longrightarrow$ NR

(d) $Br_2 + 2\,I^- \longrightarrow 2\,Br^- + I_2$

47. $4\,Zn + NO_3^- + 10\,H^+ \longrightarrow 4\,Zn^{2+} + NH_4^+ + 3\,H_2O$

49. (a) Pb is the anode.

(c) Oxidation occurs at Pb (anode).

(e) Electrons flow from the lead through the wire to the silver.

CHAPTER 18

Exercises

2.

	Protons	Neutrons	Nucleons
(a) $^{128}_{52}Te$	52	76	128
(b) $^{32}_{16}S$	16	16	32

4. Its atomic number increases by one, and its mass number remains unchanged.

6. Equations for alpha decay:

(a) $^{238}_{90}Th \longrightarrow {}^4_2He + {}^{234}_{88}Ra$

(b) $^{239}_{94}Pu \longrightarrow {}^4_2He + {}^{235}_{92}U$

8. Equations for beta decay:

(a) $^{28}_{13}Al \longrightarrow {}^{0}_{-1}e + {}^{28}_{14}Si$

(b) $^{239}_{93}Np \longrightarrow {}^{0}_{-1}e + {}^{239}_{94}Pu$

10. (a) gamma emission

(b) alpha emission then beta emission

(c) alpha emission then gamma emission

12. $^{32}_{15}P \longrightarrow {}^{0}_{-1}e + {}^{32}_{16}S$

14. (a) $^{27}_{13}Al + {}^4_2He \longrightarrow {}^{30}_{15}P + {}^1_0n$

(c) $^{12}_{6}C + {}^2_1H \longrightarrow {}^{13}_{7}N + {}^1_0n$

16. The year 2064; 1/8 of Sr-90 remaining

18. $^{228}_{90}Th \xrightarrow{-\alpha} {}^{224}_{88}Ra \xrightarrow{-\alpha} {}^{220}_{86}Rn \xrightarrow{-\alpha} {}^{216}_{84}Po \xrightarrow{-\alpha} {}^{212}_{82}Pb$

20. (a) 5.4×10^{11} J/mol

(b) 0.199% mass loss

21. (a) 24 protons; 27 neutrons; 24 electrons

(c) 46 protons; 57 neutrons; 46 electrons

24. 1.50×10^{10} months

26. 61.59% neutrons by mass; 0.021% electrons by mass

28. 11,460 years old

30. (a) 0.0424 g/mol (mass defect)

(b) 3.8×10^{12} J/mol (binding energy)

33. 0.16 mg remaining

34. 112 days

40. (a) $^{235}_{92}U + {}^1_0n \longrightarrow {}^{143}_{54}Xe + 3\,{}^1_0n + {}^{90}_{38}Sr$

(b) $^{235}_{92}U + {}^1_0n \longrightarrow {}^{102}_{39}Y + 3\,{}^1_0n + {}^{131}_{53}I$

(c) $^{14}_{7}N + {}^1_0n \longrightarrow {}^1_1H + {}^{14}_{6}C$

42. $^{236}_{92}U \longrightarrow {}^{90}_{38}Sr + 3\,{}^1_0n + {}^{143}_{54}Xe$

44. 7680 g

46. (a) 0.500 g left

(c) 0.0625 g left

48. (a) 2.29×10^4 years

GLOSSARY

A

1 atmosphere The standard atmospheric pressure, that is, the pressure exerted by a column of mercury 760 mm high at a temperature of 0°C. *See* also atmospheric pressure. [12.1]

absolute zero −273°C, the zero point on the Kelvin (absolute) temperature scale. *See also* Kelvin temperature scale. [12.3]

acid ionization constant (K_a) The equilibrium constant for the ionization of a weak acid in water. [16.6]

activation energy The amount of energy needed to start a chemical reaction. [8.5, 16.3]

Activity series of metals A list of metals according to their relative chemical reactivity. The most active elements are placed at the top of the list and the least active are placed at the bottom. A partial activity series is shown in figure 17.2. [17.4]

actual yield The amount of product actually produced in a chemical reaction (compared with the theoretical yield). [9.5]

addition reaction In organic chemistry, a reaction in which two substances join together to produce one substance. [19.4]

alcohol An organic compound consisting of an —OH group bonded to a carbon atom in a nonaromatic hydrocarbon group; alcohols are classified as primary (1°), secondary (2°), or tertiary (3°), depending on whether the carbon atom to which the —OH group is attached is bonded to one, two, or three other carbon atoms, respectively. [19.7]

aldehyde An organic compound that contains the —CHO group. The general formula is RCHO. [19.9]

alkali metals An element (except H) from Group 1A of the periodic table. [3.2]

alkaline earth metals An element from Group 2A of the periodic table. [3.2]

alkane A compound composed of carbon and hydrogen, having only single bonds between the atoms; also known as saturated hydrocarbon. *See also* alkene and alkyne. [19.3]

alkene An unsaturated hydrocarbon whose molecules have at least one carbon–carbon double bond. [19.4]

alkyl group An organic group derived from an alkane by removal of one H atom. The general formula is C_nH_{2n+1} (e.g., CH3, methyl). Alkyl groups are generally indicated by the letter R. [19.3]

alkyne An unsaturated hydrocarbon whose molecules have at least one carbon–carbon triple bond. [19.4]

alpha particle (α) A particle emitted from a nucleus of an atom during radioactive decay; it consists of two protons and two neutrons with a mass of about 4 amu and a charge of +2; it is considered to be a doubly charged helium atom. [18.2]

amino acid An organic compound containing two functional groups—an amino group (NH_2) and a carboxyl group (COOH). Amino acids are the building blocks for proteins. [20.4]

amorphous A solid without shape or form. [1.3]

anion A negatively charged ion. *See also* ion. [3.3]

anode The electrode where oxidation occurs in an electrochemical reaction. [17.5]

aromatic compound An organic compound whose molecules contain a benzene ring or that has properties resembling benzene. [19.5]

artificial radioactivity Radioactivity produced in nuclides during some types of transmutations. Artificial radioactive nuclides behave like natural radioactive elements in two ways: They disintegrate in a definite fashion, and they have a specific half-life. Sometimes called *induced radioactivity*. [18.3]

atmospheric pressure The pressure experienced by objects on Earth as a result of the layer of air surrounding our planet. A pressure of 1 atmosphere (1 atm) is the pressure that will support a column of mercury 760 mm high at 0°C. [12.1]

atom The smallest particle of an element that can enter into a chemical reaction. [3.1]

atomic mass The average relative mass of the isotopes of an element referred to the atomic mass of carbon-12. [5.6]

atomic mass unit (amu) A unit of mass equal to one-twelfth the mass of a carbon-12 atom. [5.6]

atomic number (*Z*) The number of protons in the nucleus of an atom of a given element. *See also* isotopic notation. [5.4]

Avogadro's law Equal volumes of different gases at the same temperature and pressure contain equal numbers of molecules. [12.4]

Avogadro's number 6.022×10^{23}; the number of formula units in 1 mole. [7.1]

B

balanced equation A chemical equation having the same number of each kind of atom and the same electrical charge on each side of the equation. [8.2]

barometer A device used to measure atmospheric pressure. [12.1]

bent structure Molecular structure in which the molecule looks like a V. The electron structure is tetrahedral but there are 2 pairs of nonbonding electrons. [11.10]

beta particle (β) A particle identical in charge (−1) and mass to an electron. [18.2]

binary compounds Compounds which are composed of two different elements. [6.4]

biochemistry The branch of chemistry concerned with chemical reactions occurring in living organisms. [20.1]

boiling point The temperature at which the vapor pressure of a liquid is equal to the pressure above the liquid. It is called the normal boiling when the pressure is 1 atmosphere. [13.3]

Boyle's law The relationship between pressure and volume of a gas. At constant temperature, the volume of a fixed mass of gas is inversely proportional to the pressure (PV = constant). [12.2]

buffer solution A solution that resists changes in pH when diluted or when small amounts of a strong acid or strong base are added. [16.8]

C

calorie (cal) A commonly used unit of heat energy; 1 calorie is a quantity of heat energy that will raise the temperature of 1 g of water 1°C (e.g., from 14.5 to 15.5°C). Also, 4.184 joules = 1 calorie exactly. *See also* joule. [4.5]

capillary action The spontaneous rising of a liquid in a narrow tube, which results from the cohesive forces within the liquid and the adhesive forces between the liquid and the walls of the container. [13.2]

carbohydrate A polyhydroxy aldehyde or polyhydroxy ketone, or a compound that upon hydrolysis yields a polyhydroxy aldehyde or ketone; sugars, starch, and cellulose are examples. [20.2]

carbonyl group The structure $\rangle C{=}O$. [19.9]

carboxyl group The functional group of carboxylic acids: $-\overset{\overset{\Large O}{\|}}{C}-OH$. [19.10]

carboxylic acid An organic compound having a carboxyl group. [19.10]

catalyst A substance that influences the rate of a reaction and can be recovered essentially unchanged at the end of the reaction. [16.3]

cathode The electrode where reduction occurs in an electrochemical reaction. [17.5]

cation A positively charged ion. *See also* ion. [3.3]

chain reaction A self-sustaining nuclear or chemical reaction in which the products cause the reaction to continue or to increase in magnitude. [18.5]

Charles' law At constant pressure, the volume of a fixed mass of any gas is directly proportional to the absolute temperature (V/T = constant). [12.3]

chemical change A change producing products that differ in composition from the original substances. [4.2]

chemical equation A shorthand expression showing the reactants and the products of a chemical change (e.g., $2\ H_2O = 2\ H_2 + O_2$). [4.2, 8.1]

chemical equilibrium The state in which the rate of the forward reaction equals the rate of the reverse reaction in a chemical change. [16.2]

chemical formula A shorthand method for showing the composition of a compound, using symbols of the elements. [3.3]

chemical kinetics The study of reaction rates and reaction mechanisms. [16.1]

chemical properties The ability of a substance to form new substances either by reaction with other substances or by decomposition. [4.1]

chemistry The science of the composition, structure, properties, and reactions of matter, especially of atomic and molecular systems. [1.1]

colligative properties Properties of a solution that depend on the number of solute particles in solution and not on the nature of the solute (examples: vapor pressure lowering, freezing point depression, boiling point elevation). [14.5]

combustion reaction A reaction in which a substance reacts with oxygen to form oxides. [8.4]

common ion effect The shift of a chemical equilibrium caused by the addition of an ion common to the ions in the equilibrium. [16.7]

compound A distinct substance composed of two or more elements combined in a definite proportion by mass. [3.3]

concentrated solution A solution containing a relatively large amount of dissolved solute. [14.4]

concentration of a solution A quantitative expression of the amount of dissolved solute in a certain quantity of solvent. [14.2]

condensation The process by which molecules in the gaseous state return to the liquid state. [13.2]

conversion factor A ratio of equivalent quantities. [2.5]

copolymer A polymer containing two different kinds of monomer units. [19.12]

covalent bond A chemical bond formed between two atoms by sharing a pair of electrons. [11.5]

critical mass The minimum quantity of mass required to support a self-sustaining chain reaction. [18.5]

curie (Ci) A unit of radioactivity indicating the rate of decay of a radioactive substance: $1\ Ci = 3.7 \times 10^{10}$ disintegrations per second. [18.4]

D

Dalton's atomic model The first modern atomic theory to state that elements are composed of minute individual particles called *atoms*. [5.1]

Dalton's law of partial pressures The total pressure of a mixture of gases is the sum of the partial pressures exerted by each of the gases in the mixture. [12.7]

decomposition reaction A breaking down, or decomposition, of one substance into two or more different substances. [8.4]

density The mass of an object divided by its volume. [2.8]

diatomic molecules The molecules of elements that always contain two atoms. Seven elements occur as diatomic molecules: H_2, N_2, O_2, F_2, Cl_2, Br_2, and I_2. [3.2]

dilute solution A solution containing a relatively small amount of dissolved solute. [14.4]

dipole A molecule that is electrically asymmetrical, causing it to be oppositely charged at two points. [11.6]

dipole–dipole attractions Forces of attraction between polar molecules as a result of the dipole moment within each molecule. [13.5]

disaccharide A carbohydrate that yields two monosaccharide units when hydrolyzed. [20.2]

dissociation The process by which a salt separates into individual ions when dissolved in water. [15.3]

DNA Deoxyribonucleic acid; a high-molar-mass polymer of nucleotides, present in all living matter, that contains the genetic code that transmits hereditary characteristics. [20.6]

double-displacement reaction A reaction of two compounds to produce two different compounds by exchanging the components of the reacting compounds. [8.4]

E

electrolysis The process whereby electrical energy is used to bring about a chemical change. [17.5]

electrolyte A substance whose aqueous solution conducts electricity. [15.3]

electrolytic cell An electrolysis apparatus in which electrical energy from an outside source is used to produce a chemical change. [17.5]

electron A subatomic particle that exists outside the nucleus and carries a negative electrical charge. [5.3]

electron configuration The orbital arrangement of electrons in an atom. [10.4]

electronegativity The relative attraction that an atom has for a pair of shared electrons in a covalent bond. [11.6]

element A basic building block of matter that cannot be broken down into simpler substances by ordinary chemical changes; in 1994, there were 111 known elements. [3.1]

empirical formula A chemical formula that gives the smallest whole-number ratio of atoms in a compound—that is, the relative number of atoms of each element in the compound; also known as the simplest formula. [7.4]

endothermic reactions A chemical reaction that absorbs heat. [8.5]

energy The capacity of matter to do work. [4.4]

enzyme A protein that catalyzes a biochemical reaction. [20.5]

equilibrium A dynamic state in which two or more opposing processes are taking place at the same time and at the same rate. [16.2]

equilibrium constant (K_{eq}) A constant representing the concentrations of the reactants and the products in a chemical reaction at equilibrium. [16.4]

ester An organic compound derived from a carboxylic acid and an alcohol. The general formula is $\mathrm{R{-}\underset{\underset{\displaystyle O}{\|}}{C}{-}OR'}$. [19.11]

ether An organic compound having two hydrocarbon groups attached to an oxygen atom. The general formula is R—O—R′. [19.8]

evaporation The escape of molecules from the liquid state to the gas or vapor state. [13.2]

exothermic reactions A chemical reaction in which heat is released as a product. [8.5]

F

fats Esters of fatty acids and glycerol. *See also* triacylglycerol. [20.3]

formula equation A chemical equation in which all the reactants and products are written in their molecular, or normal, formula expression; also called a molecular equation. [15.6]

freezing point See melting point [13.3]

frequency A measurement of the number of waves that pass a particular point per second. [10.1]

functional group An atom or group of atoms that characterizes a class of organic compounds. For example, —COOH is the functional group of carboxylic acids. [19.2]

G

gamma ray (γ) High-energy photons emitted by radioactive nuclei; they have no electrical charge and no measurable mass. [18.2]

gas A state of matter that has no shape or definite volume so that the substance completely fills its container. [1.3]

Gay-Lussac's law of combining volumes When measured at the same temperature and pressure, the ratios of the volumes of reacting gases are small whole numbers. [12.4]

genes Basic units of heredity that consist primarily of DNA and proteins and occur in the chromosomes. [20.6]

ground state The lowest available energy level within an atom. [10.2]

groups or families (of elements) Vertical groups of elements in the periodic table (1A, 2A, and so on). Families of elements that have similar outer-orbital electron structures. [3.2, 10.5]

H

half-life ($t_{1/2}$) The time required for one-half of a specific amount of a radioactive nuclide to disintegrate; half-lives of the elements range from a fraction of a second to billions of years. [18.1]

halogens Group 7A of the periodic table; consists of the elements fluorine, chlorine, bromine, iodine, and astatine. [3.2]

heat Flow of energy due to a temperature difference. [2.7]

heat of fusion The energy required to change 1 gram of a solid into a liquid at its melting point. [13.4]

heat of reaction The quantity of heat produced by a chemical reaction. [8.5]

heat of vaporization The amount of heat required to change 1 gram of a liquid to a vapor at its normal boiling point. [13.4]

heterogeneous Matter without a uniform composition—having two or more components or phases. [1.4]

homogeneous Matter that has uniform properties throughout. [1.4]

homologous series A series of compounds in which the members differ from one another by a regular increment. For example, each successive member of the alkane series of hydrocarbons differs by a CH_2 group. [19.3]

hydrate A solid that contains water molecules as a part of its crystalline structure. [13.6]

hydrocarbons Compounds composed entirely of carbon and hydrogen. [8.5, 19.2]

hydrogen bond The intermolecular force acting between molecules that contain hydrogen covalently bonded to the highly electronegative elements F, O, and N. [13.5]

hydrogenation The process of adding hydrogen across a double or triple bond resulting in a more saturated compound. [19.4]

hydronium ion The result of a proton combining with a polar water molecule to form a hydrated hydrogen ion, H_3O^+. [15.1]

hypothesis A tentative explanation of certain facts to provide a basis for further experimentation. [1.2]

I

ideal gas A gas that behaves precisely according to the kinetic-molecular theory; also called a perfect gas. [12.6]

ideal gas law $PV = nRT$; that is, the volume of a gas varies *directly* with the number of gas molecules and the absolute temperature and *inversely* with the pressure. [12.6]

immiscible Incapable of mixing; immiscible liquids do not form a solution with one another. [14.2]

induced radioactivity See artificial radioactivity. [18.3]

intermolecular forces Forces of attraction between molecules. [13.5]

intramolecular forces Forces of attractions within a molecule, for example, a covalent bond. [13.5]

ion A positively or negatively charged atom or group of atoms. *See also* cation, anion. [3.3]

ion product constant for water (K_w) An equilibrium constant defined as the product of the H^+ ion concentration and the OH^+ ion concentration, each in moles per liter. $K_w = [H^+][OH^-] = 1 \times 10^{-14}$ at 25°C. [16.5]

ionic bond The chemical bond between a positively charged ion and a negatively charged ion. [11.3]

ionization energy The energy required to remove an electron from an atom, an ion, or a molecule. [11.1]

ionization The formation of ions, which occurs as the result of a chemical reaction of certain substances with water. [15.3]

ionizing radiation Radiation with enough energy to dislocate bonding electrons and create ions when passing through matter. [18.7]

isomerism The phenomenon of two or more compounds having the same molecular formula but different molecular structures. [19.3]

isomers Compounds having identical molecular formulas but different structural formulas. [19.3]

isotope An atom of an element that has the same atomic number but a different atomic mass. Since their atomic numbers are identical, isotopes vary only in the number of neutrons in the nucleus. [5.5]

J

joule (J) The SI unit of energy. *See also* calorie. [4.5]

K

ketone An organic compound that contains a carbonyl group between two other carbon atoms. The general formula is $R_2C = O$. [19.9]

kilogram (kg) The standard unit of mass in the metric system; 1 kilogram equals 2.205 pounds. [2.5]

kinetic energy (KE) The energy that matter possesses due to its motion; $KE = 1/2\ mv^2$. [4.4]

kinetic-molecular theory (KMT) A group of assumptions used to explain the behavior and properties of gases. [12.6]

L

law of conservation of energy Energy can be neither created nor destroyed, though it can be transferred from one form to another. [4.4]

law of conservation of mass No change is observed in the total mass of the substances involved in a chemical reaction; that is, the mass of the products equals the mass of the reactants. [8.1]

law of definite composition A compound always contains two or more elements in a definite proportion by mass. [3.3]

law of multiple proportions Atoms of two or more elements may combine in different ratios to produce more than one compound. [3.3]

Le Châtelier's principle If a stress is applied to a system in equilibrium, the system will respond in such a way as to relieve that stress and restore equilibrium under a new set of conditions. [16.3]

Lewis structure A method of indicating the covalent bonds between atoms in a molecule or an ion such that a pair of electrons (:) represents the valence electrons forming the covalent bond. [11.2]

limiting reactant A reactant that limits the amount of product formed because it is present in insufficient amount compared with the other reactants. [9.5]

line spectrum Colored lines generated when light emitted by a gas is passed through a spectroscope. Each element possesses a unique set of line spectra. [10.2]

linear structure In the VSEPR model, an arrangement where the pairs of electrons are arranged 180° apart for maximum separation. [11.10]

lipids Organic compounds found in living organisms that are water insoluble but soluble in such fat solvents as diethyl ether, benzene, and carbon tetrachloride; examples are fats, oils, and steroids. [20.3]

liquid A state of matter in which the particles move about freely while the substance retains a definite volume; thus, liquids flow and take the shape of their containers. [1.3]

liter (L) A unit of volume commonly used in chemistry; 1 L = 1000 mL; the volume of a kilogram of water at 4°C. [2.5]

logarithm (log) The power to which 10 must be raised to give a certain number. The log of 100 is 2.0. [15.4]

London dispersion forces Attractive forces between nonpolar molecules which result from instantaneous formation of dipoles in the electron cloud. [13.5]

M

mass defect The difference between the actual mass of an atom of an isotope and the calculated mass of the protons and neutrons in the nucleus of that atom. [18.6]

mass number (*A*) The sum of the protons and neutrons in the nucleus of a given isotope of an atom. [5.5]

mass The quantity or amount of matter that an object possesses. [2.5]

matter Anything that has mass and occupies space. [1.3]

measurement A quantitative observation which requires both a number and a unit. [2.1]

meiosis The process of cell division to form a sperm cell and an egg cell in which each cell formed contains half of the chromosomes found in the normal single cell. [20.6]

melting point (or freezing point) The temperature at which the solid and liquid states of a substance are in equilibrium. [13.3]

meniscus The shape of the surface of a liquid when placed in a glass cylinder. It can be concave or convex. [13.2]

metal An element that is solid at room temperature and whose properties include luster, ductility, malleability, and good conductivity of heat and electricity; metals tend to lose their valence electrons and become positive ions. [3.2]

metalloid An element having properties that are intermediate between those of metals and nonmetals (e.g., silicon); these elements are useful in electronics. [3.2]

meter (m) The standard unit of length in the SI and metric systems; 1 meter equals 39.37 inches. [2.5]

metric system or International System (SI) A decimal system of measurements. *See also* SI. [2.5]

miscible Capable of mixing and forming a solution. [14.2]

mitosis Ordinary cell division in which a DNA molecule is duplicated by uncoiling to single strands and then reassembling with complementary nucleotides. Each new cell contains the normal number of chromosomes. [20.6]

mixture Matter containing two or more substances, which can be present in variable amounts; mixtures can be homogeneous (sugar water) or heterogeneous (sand and water). [1.4]

molality (*m*) An expression of the number of moles of solute dissolved in 1000 g of solvent. [14.5]

molar mass The mass of Avogadro's number of atoms or molecules. The sum of the atomic masses of all the atoms in an element, compound, or ion. The mass of a mole of any formula unit. It is also known as the molecular weight. [7.1, 9.1]

molar volume The volume of 1 mol of a gas at STP equals 22.4 L/mol. [12.5]

molarity (*M*) The number of moles of solute per liter of solution. [14.4]

mole ratio A ratio between the number of moles of any two species involved in a chemical reaction; the mole ratio is used as a conversion factor in stoichiometric calculations. [9.1]

mole The amount of a substance containing the same number of formula units (6.022×10^{23}) as there are in exactly 12 g of ^{12}C. One mole is equal to the molar mass in grams of any substance. [7.1]

molecular formula The total number of atoms of each element present in one molecule of a compound; also known as the true formula. See also empirical formula. [7.5]

molecule The smallest uncharged individual unit of a compound formed by the union of two or more atoms. [3.3]

monomer The small unit or units that undergo polymerization to form a polymer. [19.12]

monosaccharide A carbohydrate that cannot be hydrolyzed to simpler carbohydrate units (e.g., simple sugars like glucose or fructose). [20.2]

N

natural law A statement summarizing general observations regarding nature. [3.3]

net ionic equation A chemical equation that includes only those molecules and ions that have changed in the chemical reaction. [15.6]

neutralization The reaction of an acid and a base to form a salt plus water. [15.5]

neutron A subatomic particle that is electrically neutral and is found in the nucleus of an atom. [5.3]

noble gases A family of elements in the periodic table—helium, neon, argon, krypton, xenon, and radon—that contains a particularly stable electron structure. [3.2]

nonelectrolyte A substance whose aqueous solutions do not conduct electricity. [15.3]

nonmetal An element that has properties the opposite of metals: lack of luster, relatively low melting point and density, and generally poor conduction of heat and electricity. Nonmetals may or may not be solid at room temperature (examples: carbon, bromine, nitrogen); many are gases. They are located mainly in the upper right-hand corner of the periodic table. [3.2]

nonpolar covalent bond A covalent bond between two atoms with the same electronegativity value; thus, the electrons are shared equally between the two atoms. [11.6]

normal boiling point The temperature at which the vapor pressure of a liquid equals 1 atm or 760 torr pressure. [13.3]

nuclear binding energy The energy equivalent to the mass defect; that is, the amount of energy required to break a nucleus into its individual protons and neutrons. [18.6]

nuclear fission A nuclear reaction in which a heavy nucleus splits into two smaller nuclei and two or more neutrons. [18.5]

nuclear fusion The uniting of two light elements to form one heavier nucleus, which is accompanied by the release of energy. [18.5]

nucleon A collective term for the neutrons and protons in the nucleus of an atom. [18.1]

nucleotide The building-block unit for nucleic acids. A phosphate group, a sugar residue, and a nitrogenous organic base are bonded together to form a nucleotide. [20.6]

nucleus The central part of an atom that contains all its protons and neutrons. The nucleus is very dense and has a positive electrical charge. [5.4]

nuclide A general term for any isotope of any atom. [18.1]

O

oils Liquid esters of fatty acids and glycerol. [20.3]

orbital A cloudlike region around the nucleus where electrons are located. Orbitals are considered to be energy sublevels (*s, p, d, f*) within the principal energy levels. *See also* principal energy levels. [10.2]

orbital diagram A way of showing the arrangement of electrons in an atom, where boxes with small arrows indicating the electrons represent orbitals. [10.4]

organic chemistry The branch of chemistry that deals with carbon compounds but does not imply that these compounds must originate from some form of life. *See also* vital-force theory. [19.1]

osmosis The diffusion of water, either pure or from a dilute solution, through a semipermeable membrane into a solution of higher concentration. [14.6]

osmotic pressure The difference between atmospheric pressure and an applied pressure that is required to prevent osmosis from occurring. [14.6]

oxidation An increase in the oxidation number of an atom as a result of losing electrons. [17.1]

oxidation number A small number representing the state of oxidation of an atom. For an ion, it is the positive or negative charge on the ion; for covalently bonded atoms, it is a positive or negative number assigned to the more electronegative atom; in free elements, it is zero. [17.1]

oxidation state *See* oxidation number. [17.1]

oxidation–reduction A chemical reaction wherein electrons are transferred from one element to another; also known as redox. [17.1]

oxidizing agent A substance that causes an increase in the oxidation state of another substance. The oxidizing agent is reduced during the course of the reaction. [17.1]

P

partial pressure The pressure exerted independently by each gas in a mixture of gases. [12.7]

parts per million (ppm) A measurement of the concentration of dilute solutions now commonly used by chemists in place of mass percent. [14.4]

Pauli exclusion principle An atomic orbital can hold a maximum of two electrons, which must have opposite spins. [10.3]

peptide linkage The amide bond in a protein molecule; bonds one amino acid to another. [20.4]

percent composition of a compound The mass percent represented by each element in a compound. [7.3]

percent A unit identifying the amount of a species in a mixture per 100 units of the mixture. A common unit is mass percent which is defined as $\left(\frac{\text{mass part}}{\text{mass whole}}\right) \times 100\%$. [Section 2.7]

percent yield The ratio of the actual yield to the theoretical yield multiplied by 100. [9.5]

period The horizontal groupings (rows) of elements in the periodic table. [10.5]

pH A method of expressing the H concentration (acidity) of a solution; pH = $-\log[H^+]$, pH = 7 is a neutral solution, pH < 7 is acidic, and pH > 7 is basic. [15.4]

phase A homogeneous part of a system separated from other parts by a physical boundary. [1.4]

photons Theoretically, a tiny packet of energy that streams with others of its kind to produce a beam of light. [10.1]

physical change A change in form (such as size, shape, or physical state) without a change in composition. [4.2]

physical properties Inherent physical characteristics of a substance that can be determined without altering its composition: color, taste, odor, state of matter, density, melting point, boiling point. [4.1]

polar covalent bond A covalent bond between two atoms with differing electronegativity values, resulting in unequal sharing of bonding electrons. [11.5]

polyatomic ion An ion composed of more than one atom. [6.5]

polyhydroxyl alcohol An alcohol that has more than one —OH group. [19.7]

polymer A natural or synthetic giant molecule (macromolecule) formed from smaller molecules (monomers). [19.12]

polymerization The process of forming large, high-molar-mass molecules from smaller units. [19.12]

polypeptide A peptide chain containing up to 50 amino acid units. [20.4]

polysaccharide A carbohydrate that can be hydrolyzed to many monosaccharide units; cellulose, starch, and glycogen are examples. [20.2]

potential energy (PE) Stored energy, or the energy of an object due to its relative position. [4.4]

pressure Force per unit area; expressed in many units, such as mm Hg, atm, lb/in.^{2}, torr, and pascal. [12.1]

primary alcohol An alcohol in which the carbon atom bonded to the —OH group is bonded to only one other carbon atom. [19.7]

principal energy levels Existing within the atom, these energy levels contain orbitals within which electrons are found. *See also* electron, orbital. [10.3]

products A chemical substance produced from reactants by a chemical change. [4.2, 8.1]

properties The characteristics, or traits, of substances that give them their unique identities. Properties are classified as physical or chemical. [4.1]

protein A polymer consisting mainly of α-amino acids linked together; occurs in all animal and vegetable matter. [20.4]

proton A subatomic particle found in the nucleus of the atom that carries a positive electrical charge and a mass of about 1 amu. An H^+ ion is a proton. [5.3]

Q

quanta Small discrete increments of energy. From the theory proposed by physicist Max Planck that energy is emitted in energy *quanta* rather than a continuous stream. [10.2]

R

rad (radiation absorbed dose) A unit of absorbed radiation indicating the energy absorbed from any ionizing radiation; 1 rad = 0.01 J of energy absorbed per kilogram of matter. [18.4]

radioactive decay The process by which a radioactive element emits particles or rays and is transformed into another element. [18.1]

radioactivity The spontaneous emission of radiation from the nucleus of an atom. [18.1]

reactants A chemical substance entering into a reaction. [4.2, 8.1]

redox *See* oxidation–reduction. [17.1]

reducing agent A substance that causes a decrease in the oxidation state of another substance; the reducing agent is oxidized during the course of a reaction. [17.1]

reduction A decrease in the oxidation number of an element as a result of gaining electrons. [17.1]

rem (*r*oentgen *e*quivalent to *m*an) A unit of radiation-dose equivalent taking into account that the energy absorbed from different sources does not produce the same degree of biological effect. [18.4]

representative elements An elements in one of the A groups in the periodic table. [3.2, 10.5]

resonance structure A molecule or ion that has multiple Lewis structures. *See also* Lewis structure. [11.8]

reversible chemical reaction A chemical reaction in which the products formed react to produce the original reactants. A double arrow is used to indicate that a reaction is reversible. [16.2]

RNA Ribonucleic acid; a high-molar-mass polymer of nucleotides present in all living matter. Its main function is to direct the synthesis of proteins. [20.6]

roentgen (R) A unit of exposure of gamma radiation based on the quantity of ionization produced in air. [18.4]

rounding off numbers The process by which the value of the last digit retained is determined after dropping nonsignificant digits. [2.3]

S

saturated hydrocarbon A hydrocarbon that has only single bonds between carbon atoms; classified as *alkanes*. [19.2]

saturated solution A solution containing dissolved solute in equilibrium with undissolved solute. [14.2]

scientific laws Simple statements of natural phenomena to which no exceptions are known under the given conditions. [1.2]

scientific method A method of solving problems by observation; recording and evaluating data of an experiment; formulating hypotheses and theories to explain the behavior of nature; and devising additional experiments to test the hypotheses and theories to *see* if they are correct. [1.2]

scientific notation A convenient way of expressing large and small numbers using powers of 10. To write a number as a power of 10 move the decimal point in the original number so that it is located after the first nonzero digit, and follow the new number by a multiplication sign and 10 with an exponent (called its *power*) that is the number of places the decimal point was moved. Example: $2468 = 2.468 \times 10^3$. [2.1]

secondary alcohol An alcohol in which the carbon atom bonded to the —OH group is bonded to two other carbon atoms. [19.7]

semipermeable membrane A membrane that allows the passage of water (solvent) molecules through it in either direction but prevents the passage of larger solute molecules or ions. [13.7, 14.6]

significant figures The number of digits that are known plus one estimated digit are considered significant in a measured quantity; also called significant digits. [2.2, Appendix I]

single-displacement reaction A reaction of an element and a compound that yields a different element and a different compound. [8.4]

solid A state of matter having a definite shape and a definite volume, whose particles cohere rigidly to one another, so that a solid can be independent of its container. [1.3]

solubility An amount of solute that will dissolve in a specific amount of solvent under stated conditions. [14.2]

solubility product constant (K_{sp}) The equilibrium constant for the solubility of a slightly soluble salt. [16.7]

solute The substance that is dissolved—or the least abundant component—in a solution. [14.1]

solution A system in which one or more substances are homogeneously mixed or dissolved in another substance. [14.1]

solution map An outline for the path of a unit conversion. [2.5]

solvent The dissolving agent or the most abundant component in a solution. [14.1]

specific heat The quantity of heat required to change the temperature of 1 g of any substance by 1°C. [4.5]

spectator ion An ion in solution that does not undergo chemical change during a chemical reaction. [15.5]

speed A measurement of how fast a wave travels through space. [10.1]

spin A property of an electron that describes its appearance of spinning on an axis like a globe; the electron can spin in only two directions, and, to occupy the same orbital, two electrons must spin in opposite directions. *See also* orbital. [10.3]

standard conditions *See* standard temperature and pressure. [12.5]

standard temperature and pressure (STP) 0°C (273 K) and 1 atm (760 torr); also known as standard conditions. [12.5]

Stock System A system that uses Roman numerals to name elements that form more than one type of cation. (For example: Fe^{2+}, iron(II); Fe^{3+}, iron(III).) [6.4]

stoichiometry The area of chemistry that deals with the quantitative relationships among reactants and products in a chemical reaction. [9.1]

strong electrolyte An electrolyte that is essentially 100% ionized in aqueous solution. [15.3]

subatomic particles Particles found within the atom, mainly protons, neutrons, and electrons. [5.3]

sublevels Each principal energy level is divided into sublevels. [10.3]

sublimation The process of going directly from the solid state to the vapor state without becoming a liquid. [13.2]

subscript Number that appears partially below the line and to the right of a symbol of an element (example: H_2SO_4). [3.3]

substance Matter that is homogeneous and has a definite, fixed composition; substances occur in two forms—as elements and as compounds. [1.4]

substrate In biochemical reactions, the substrate is the unit acted upon by an enzyme. [20.5]

supersaturated solution A solution containing more solute than needed for a saturated solution at a particular temperature. Supersaturated solutions tend to be unstable; jarring the container or dropping in a "seed" crystal will cause crystallization of the excess solute. [14.2]

surface tension The resistance of a liquid to an increase in its surface area. [13.2]

symbol In chemistry, an abbreviation for the name of an element. [3.1]

synthesis reaction A direct union or combination of two substances to produce one new substance. [8.4]

system A body of matter under consideration. [1.4]

T

temperature A measure of the intensity of heat, or of how hot or cold a system is; the SI unit is the kelvin (K). [2.7]

tertiary alcohol An alcohol in which the carbon atom bonded to the —OH group is bonded to three other carbon atoms. [19.7]

tetrahedral structure An arrangement of the VSEPR model where four pairs of electrons are placed 109.5° degrees apart to form a tetrahedron. [11.10]

theoretical yield The maximum amount of product that can be produced according to a balanced equation. [9.5]

theory An explanation of the general principles of certain phenomena with considerable evidence to support it; a well-established hypothesis. [1.2]

thermal energy A form of energy associated with the motion of small particles [2.7]

Thomson model of the atom Thomson asserted that atoms are not indivisible but are composed of smaller parts; they contain both positively and negatively charged particles— protons as well as electrons. [5.3]

titration The process of measuring the volume of one reagent required to react with a measured mass or volume of another reagent. [15.5]

total ionic equation An equation that shows compounds in the form in which they actually exist. Strong electrolytes are written as ions in solution, whereas nonelectrolytes, weak electrolytes, precipitates, and gases are written in the un-ionized form. [15.6]

transcription The process of forming RNA from DNA. [20.6]

transition elements The metallic elements characterized by increasing numbers of *d* and *f* electrons. These elements are located in Groups 1B–8B of the periodic table. [3.2, 10.5]

transmutation The conversion of one element into another element. [18.3]

transuranium elements An element that has an atomic number higher than that of uranium (>92). [18.3]

trigonal planar structure An arrangement of atoms in the VSEPR model where the three pairs of electrons are placed 120° apart on a flat plane. [11.10]

triacylglycerols (triglycerides) Another name for fats and oils derived from one molecule of glycerol and three molecules of fatty acids [20.3]

U

unsaturated solution A solution containing less solute per unit volume than its corresponding saturated solution. [14.2]

unsaturated hydrocarbon A hydrocarbon whose molecules contain one or more double or triple bonds between two carbon atoms; classified as *alkenes*, *alkynes*, and *aromatic compounds*. [19.2]

valence electrons An electron in the outermost energy level of an atom; these electrons are the ones involved in bonding atoms together to form compounds. [10.4]

vapor pressure curve A graph generated by plotting the temperature of a liquid on the x-axis and its vapor pressure on the y-axis. Any point on the curve represents an equilibrium between the vapor and liquid. [13.3]

vapor pressure The pressure exerted by a vapor in equilibrium with its liquid. [13.2]

vaporization *See* evaporation. [13.2]

vital-force theory A theory that held that organic substances could originate only from some form of living material. The theory was proved false early in the nineteenth century. [19.1]

volatile A substance that evaporates readily; a liquid with a high vapor pressure and a low boiling point. [13.2]

voltaic cell A cell that produces electric current from a spontaneous chemical reaction. [17.5]

volume The amount of space occupied by matter; measured in SI units by cubic meters (m^3) but also commonly in liters and milliliters. [2.5]

W

water of crystallization Water molecules that are part of a crystalline structure, as in a hydrate; also called water of hydration. [13.6]

water of hydration *See* water of crystallization. [13.6]

wavelength The distance between consecutive peaks and troughs in a wave; symbolized by the Greek letter lambda. [10.1]

weak electrolyte A substance that is ionized to a small extent in aqueous solution. [15.3]

weight A measure of Earth's gravitational attraction for a body (object). [2.5]

INDEX

N

Periodic Table of the Elements

Group: 1 / 1A — Current ACS and IUPAC / Preferred U.S.

Atomic number → 11; Symbol → **Na**; Name → Sodium; Atomic mass → 22.99

Atomic masses are based on carbon-12. Elements marked with † have no stable isotopes. The atomic mass given is that of the isotope with the longest known half-life.

For practical use, atomic masses have been rounded to four figures.

Period	1 1A	2 2A	3 3B	4 4B	5 5B	6 6B	7 7B	8	9 8B	10	11 1B	12 2B	13 3A	14 4A	15 5A	16 6A	17 7A	Noble Gases 18 8A
1	1 **H** Hydrogen 1.008																	2 **He** Helium 4.003
2	3 **Li** Lithium 6.941	4 **Be** Beryllium 9.012											5 **B** Boron 10.81	6 **C** Carbon 12.01	7 **N** Nitrogen 14.01	8 **O** Oxygen 16.00	9 **F** Fluorine 19.00	10 **Ne** Neon 20.18
3	11 **Na** Sodium 22.99	12 **Mg** Magnesium 24.31											13 **Al** Aluminum 26.98	14 **Si** Silicon 28.09	15 **P** Phosphorus 30.97	16 **S** Sulfur 32.07	17 **Cl** Chlorine 35.45	18 **Ar** Argon 39.95
4	19 **K** Potassium 39.10	20 **Ca** Calcium 40.08	21 **Sc** Scandium 44.96	22 **Ti** Titanium 47.87	23 **V** Vanadium 50.94	24 **Cr** Chromium 52.00	25 **Mn** Manganese 54.94	26 **Fe** Iron 55.85	27 **Co** Cobalt 58.93	28 **Ni** Nickel 58.69	29 **Cu** Copper 63.55	30 **Zn** Zinc 65.39	31 **Ga** Gallium 69.72	32 **Ge** Germanium 72.61	33 **As** Arsenic 74.92	34 **Se** Selenium 78.96	35 **Br** Bromine 79.90	36 **Kr** Krypton 83.80
5	37 **Rb** Rubidium 85.47	38 **Sr** Strontium 87.62	39 **Y** Yttrium 88.91	40 **Zr** Zirconium 91.22	41 **Nb** Niobium 92.91	42 **Mo** Molybdenum 95.94	43 **Tc** Technetium 98.00†	44 **Ru** Ruthenium 101.1	45 **Rh** Rhodium 102.9	46 **Pd** Palladium 106.4	47 **Ag** Silver 107.9	48 **Cd** Cadmium 112.4	49 **In** Indium 114.8	50 **Sn** Tin 118.7	51 **Sb** Antimony 121.8	52 **Te** Tellurium 127.6	53 **I** Iodine 126.9	54 **Xe** Xenon 131.3
6	55 **Cs** Cesium 132.9	56 **Ba** Barium 137.3	57 **La** * Lanthanum 138.9	72 **Hf** Hafnium 178.5	73 **Ta** Tantalum 180.9	74 **W** Tungsten 183.8	75 **Re** Rhenium 186.2	76 **Os** Osmium 190.2	77 **Ir** Iridium 192.2	78 **Pt** Platinum 195.1	79 **Au** Gold 197.0	80 **Hg** Mercury 200.6	81 **Tl** Thallium 204.4	82 **Pb** Lead 207.2	83 **Bi** Bismuth 209.0	84 **Po** Polonium 209†	85 **At** Astatine 210†	86 **Rn** Radon 222†
7	87 **Fr** Francium 223†	88 **Ra** Radium 226†	89 **Ac** ** Actinium 227†	104 **Rf** Rutherfordium 261†	105 **Db** Dubnium 262†	106 **Sg** Seaborgium 266†	107 **Bh** Bohrium 264†	108 **Hs** Hassium 277†	109 **Mt** Meitnerium 268†	110 **Ds** Darmstadtium 271†	111 **Rg** Roentgenium 272†	112 **Cn** Copernicium 285†	113 **Uut**	114 **Fl** Flerovium 289†	115 **Uup**	116 **Lv** Livermorium 293†	117 **Uus**	118 **Uuo**

Transition Elements: groups 3 (3B) through 12 (2B)

Inner Transition Elements

Series														
* Lanthanide Series 6	58 **Ce** Cerium 140.1	59 **Pr** Praseodymium 140.9	60 **Nd** Neodymium 144.2	61 **Pm** Promethium 145†	62 **Sm** Samarium 150.4	63 **Eu** Europium 152.0	64 **Gd** Gadolinium 157.3	65 **Tb** Terbium 158.9	66 **Dy** Dysprosium 162.5	67 **Ho** Holmium 164.9	68 **Er** Erbium 167.3	69 **Tm** Thulium 168.9	70 **Yb** Ytterbium 173.0	71 **Lu** Lutetium 175.0
** Actinide Series 7	90 **Th** Thorium 232.0	91 **Pa** Protactinium 231.0	92 **U** Uranium 238.0	93 **Np** Neptunium 237	94 **Pu** Plutonium 244†	95 **Am** Americium 243†	96 **Cm** Curium 247†	97 **Bk** Berkelium 247†	98 **Cf** Californium 251†	99 **Es** Einsteinium 252†	100 **Fm** Fermium 257†	101 **Md** Mendelevium 258†	102 **No** Nobelium 259†	103 **Lr** Lawrencium 262†

colors indicate placement of the outermost electrons: *s* orbitals, *d* orbitals, *p* orbitals, *f* orbitals

†This element has no stable isotopes. The atomic mass given is for the longest lived isotope of the element.

Names, Formulas, and Charges of Common Ions

Positive Ions (Cations)

Charge	Name	Formula
1+	Ammonium	NH_4^+
	Copper(I) (Cuprous)	Cu^+
	Hydrogen	H^+
	Potassium	K^+
	Silver	Ag^+
	Sodium	Na^+
2+	Barium	Ba^{2+}
	Cadmium	Cd^{2+}
	Calcium	Ca^{2+}
	Cobalt(II)	Co^{2+}
	Copper(II) (Cupric)	Cu^{2+}
	Iron(II) (Ferrous)	Fe^{2+}
	Lead(II)	Pb^{2+}
	Magnesium	Mg^{2+}
	Manganese(II)	Mn^{2+}
	Mercury(II) (Mercuric)	Hg^{2+}
	Nickel(II)	Ni^{2+}
	Tin(II) (Stannous)	Sn^{2+}
	Zinc	Zn^{2+}
3+	Aluminum	Al^{3+}
	Antimony(III)	Sb^{3+}
	Arsenic(III)	As^{3+}
	Bismuth(III)	Bi^{3+}
	Chromium(III)	Cr^{3+}
	Iron(III) (Ferric)	Fe^{3+}
	Titanium(III) (Titanous)	Ti^{3+}
4+	Manganese(IV)	Mn^{4+}
	Tin(IV) (Stannic)	Sn^{4+}
	Titanium(IV) (Titanic)	Ti^{4+}
5+	Antimony(V)	Sb^{5+}
	Arsenic(V)	As^{5+}

Negative Ions (Anions)

Charge	Name	Formula
1−	Acetate	$C_2H_3O_2^-$
	Bromate	BrO_3^-
	Bromide	Br^-
	Chlorate	ClO_3^-
	Chloride	Cl^-
	Chlorite	ClO_2^-
	Cyanide	CN^-
	Fluoride	F^-
	Hydride	H^-
	Hydrogen carbonate (Bicarbonate)	HCO_3^-
	Hydrogen sulfate (Bisulfate)	HSO_4^-
	Hydrogen sulfite (Bisulfite)	HSO_3^-
	Hydroxide	OH^-
	Hypochlorite	ClO^-
	Iodate	IO_3^-
	Iodide	I^-
	Nitrate	NO_3^-
	Nitrite	NO_2^-
	Perchlorate	ClO_4^-
	Permanganate	MnO_4^-
	Thiocyanate	SCN^-
2−	Carbonate	CO_3^{2-}
	Chromate	CrO_4^{2-}
	Dichromate	$Cr_2O_7^{2-}$
	Oxalate	$C_2O_4^{2-}$
	Oxide	O^{2-}
	Peroxide	O_2^{2-}
	Silicate	SiO_3^{2-}
	Sulfate	SO_4^{2-}
	Sulfide	S^{2-}
	Sulfite	SO_3^{2-}
3−	Arsenate	AsO_4^{3-}
	Borate	BO_3^{3-}
	Phosphate	PO_4^{3-}
	Phosphide	P^{3-}
	Phosphite	PO_3^{3-}

Prefixes and Numerical Values for SI Units

Prefix	Symbol	Numerical value	Power of 10 equivalent
exa	E	1,000,000,000,000,000,000	10^{18}
peta	P	1,000,000,000,000,000	10^{15}
tera	T	1,000,000,000,000	10^{12}
giga	G	1,000,000,000	10^{9}
mega	M	1,000,000	10^{6}
kilo	k	1,000	10^{3}
hecto	h	100	10^{2}
deka	da	10	10^{1}
—	—	1	10^{0}
deci	d	0.1	10^{-1}
centi	c	0.01	10^{-2}
milli	m	0.001	10^{-3}
micro	μ	0.000001	10^{-6}
nano	n	0.000000001	10^{-9}
pico	p	0.000000000001	10^{-12}
femto	f	0.000000000000001	10^{-15}
atto	a	0.000000000000000001	10^{-18}

SI Units and Conversion Factors

Length

SI unit: meter (m)

1 meter	=	1000 millimeters
	=	1.0936 yards
1 centimeter	=	0.3937 inch
1 inch	=	2.54 centimeters (exactly)
1 kilometer	=	0.62137 mile
1 mile	=	5280 feet
	=	1.609 kilometers
1 angstrom	=	10^{-10} meter

Mass

SI unit: kilogram (kg)

1 kilogram	=	1000 grams
	=	2.20 pounds
1 gram	=	1000 milligrams
1 pound	=	453.59 grams
	=	0.45359 kilogram
	=	16 ounces
1 ton	=	2000 pounds
	=	907.185 kilograms
1 ounce	=	28.3 grams
1 atomic mass unit	=	1.6606×10^{-27} kilograms

Volume

SI unit: cubic meter (m^3)

1 liter	=	1000 milliliters
	=	10^{-3} m^3
	=	1 dm^3
	=	1.0567 quarts
1 gallon	=	4 quarts
	=	8 pints
	=	3.785 liters
1 quart	=	32 fluid ounces
	=	0.946 liter
	=	4 cups
1 fluid ounce	=	29.6 mL

Temperature

SI unit: kelvin (K)

0 K	=	−273.15°C
	=	−459.67°F
K	=	°C + 273.15
°C	=	$\frac{(°F - 32)}{1.8}$
°C	=	$\frac{5}{9}(°F - 32)$
°F	=	1.8(°C) + 32

Energy

SI unit: joule (J)

1 joule	=	1 kg m^2/s^2
	=	0.23901 calorie
1 calorie	=	4.184 joules

Pressure

SI unit: pascal (Pa)

1 pascal	=	1 kg/(ms^2)
1 atmosphere	=	101.325 kilopascals
	=	760 torr
	=	760 mm Hg
	=	14.70 pounds per square inch (psi)

Atomic Masses of the Elements

Based on the 2005 IUPAC Table of Atomic Masses

Name	Symbol	Atomic Number	Atomic Mass
Actinium*	**Ac**	89	227
Aluminum	**Al**	13	26.981538
Americium*	**Am**	95	243
Antimony	**Sb**	51	121.760
Argon	**Ar**	18	39.948
Arsenic	**As**	33	74.92160
Astatine*	**At**	85	210
Barium	**Ba**	56	137.327
Berkelium*	**Bk**	97	247
Beryllium	**Be**	4	9.012182
Bismuth	**Bi**	83	208.98038
Bohrium*	**Bh**	107	264
Boron	**B**	5	10.811
Bromine	**Br**	35	79.904
Cadmium	**Cd**	48	112.411
Calcium	**Ca**	20	40.078
Californium*	**Cf**	98	251
Carbon	**C**	6	12.0107
Cerium	**Ce**	58	140.116
Cesium	**Cs**	55	132.90545
Chlorine	**Cl**	17	35.4527
Chromium	**Cr**	24	51.9961
Cobalt	**Co**	27	58.933200
Copernicium	Cn	112	
Copper	**Cu**	29	63.546
Curium*	**Cm**	96	247
Darmstadtium*	**Ds**	110	271
Dubnium*	**Db**	105	262
Dysprosium	**Dy**	66	162.500
Einsteinium*	**Es**	99	252
Erbium	**Er**	68	167.26
Europium	**Eu**	63	151.964
Fermium*	**Fm**	100	257
Flerovium	Fl	114	
Fluorine	**F**	9	18.9984032
Francium*	**Fr**	87	233
Gadolinium	**Gd**	64	157.25
Gallium	**Ga**	31	69.723
Germanium	**Ge**	32	72.61
Gold	**Au**	79	196.96655
Hafnium	**Hf**	72	178.49
Hassium*	**Hs**	108	277
Helium	**He**	2	4.002602
Holmium	**Ho**	67	164.93032
Hydrogen	**H**	1	1.00794
Indium	**In**	49	114.818
Iodine	**I**	53	126.90447
Iridium	**Ir**	77	192.217
Iron	**Fe**	26	55.845
Krypton	**Kr**	36	83.80
Lanthanum	**La**	57	138.9055
Lawrencium*	**Lr**	103	262
Lead	**Pb**	82	207.2
Lithium	**Li**	3	6.941
Livermoriuum	Lv	116	
Lutetium	**Lu**	71	174.967
Magnesium	**Mg**	12	24.3050
Manganese	**Mn**	25	54.938049
Meitnerium*	**Mt**	109	268
Mendelevium*	**Md**	101	258
Mercury	**Hg**	80	200.59
Molybdenum	**Mo**	42	95.94
Neodymium	**Nd**	60	144.24
Neon	**Ne**	10	20.1797
Neptunium*	**Np**	93	237
Nickel	**Ni**	28	58.6934
Niobium	**Nb**	41	92.90638
Nitrogen	**N**	7	14.00674
Nobelium*	**No**	102	259
Osmium	**Os**	76	190.23
Oxygen	**O**	8	15.9994
Palladium	**Pd**	46	106.42
Phosphorus	**P**	15	30.973762
Platinum	**Pt**	78	195.078
Plutonium*	**Pu**	94	244
Polonium*	**Po**	84	209
Potassium	**K**	19	39.0983
Praseodymium	**Pr**	59	140.90765
Promethium*	**Pm**	61	145
Protactinium	**Pa**	91	231.03588
Radium*	**Ra**	88	226
Radon*	**Rn**	86	222
Rhenium	**Re**	75	186.207
Rhodium	**Rh**	45	102.90550
Roentgenium*	**Rg**	111	272
Rubidium	**Rb**	37	85.4678
Ruthenium	**Ru**	44	101.07
Rutherfordium*	**Rf**	104	261
Samarium	**Sm**	62	150.36
Scandium	**Sc**	21	44.955910
Seaborgium*	**Sg**	106	266
Selenium	**Se**	34	78.96
Silicon	**Si**	14	28.0855
Silver	**Ag**	47	107.8682
Sodium	**Na**	11	22.989770
Strontium	**Sr**	38	87.62
Sulfur	**S**	16	32.066
Tantalum	**Ta**	73	180.9479
Technetium*	**Tc**	43	98
Tellurium	**Te**	52	127.60
Terbium	**Tb**	65	158.92534
Thallium	**Tl**	81	204.3833
Thorium	**Th**	90	232.0381
Thulium	**Tm**	69	168.93421
Tin	**Sn**	50	118.710
Titanium	**Ti**	22	47.867
Tungsten	**W**	74	183.94
Uranium	**U**	92	238.0289
Vanadium	**V**	23	50.9415
Xenon	**Xe**	54	131.29
Ytterbium	**Yb**	70	173.04
Yttrium	**Y**	39	88.90585
Zinc	**Zn**	30	65.39
Zirconium	**Zr**	40	91.224

*This element has no stable isotopes. The atomic mass given is that of the isotope with the longest known half-life.